Informatik aktuell

Herausgeber: W. Brauer
im Auftrag der Gesellschaft für Informatik (GI)

Springer
Berlin
Heidelberg
New York
Barcelona
Budapest
Hongkong
London
Mailand
Paris
Singapur
Tokio

Paul Levi Rolf-Jürgen Ahlers
Franz May Michael Schanz (Hrsg.)

Mustererkennung 1998

20. DAGM-Symposium
Stuttgart, 29. September – 1. Oktober 1998

Springer

Herausgeber

P. Levi
M. Schanz
IPVR – Institut für Parallele und Verteilte Höchstleistungsrechner
Universität Stuttgart
Breitwiesenstrasse 20–22, D-70565 Stuttgart

R.-J. Ahlers
AEG Sensorsysteme GmbH
Junkerstrasse 2, D-69469 Weinheim

F. May
DAIMLER BENZ AG
FT3/E
Wilhelm-Runge-Strasse 11, D-89081 Ulm

Die Deutsche Bibliothek - CIP-Einheitsaufnahme

Mustererkennung ... : ... DAGM-Symposium ... / Veranst.: DAGM, Deutsche Arbeitsgemeinschaft für Mustererkennung. - Berlin ; Heidelberg ; New York ; Barcelona ; Budapest ; Hongkong ; London ; Mailand ; Paris ; Singapur ; Tokio : Springer
Beitr. teilw. dt., teilw. engl. - Teilw. im VDE-Verl., Berlin, Offenbach
20. 1998. Stuttgart, 29. September - 01. Oktober 1998. - 1998
(Informatik aktuell)
ISBN-13:978-3-540-64935-9

CR Subject Classification (1998):
A.0, I.2, I.2.7, I.2.10, I.2.11, I.4, I.5, J.3

ISBN-13:978-3-540-64935-9 e-ISBN-13:978-3-642-72282-0
DOI: 10.1007/978-3-642-72282-0

Satz: Reproduktionsfertige Vorlage vom Autor/Herausgeber

SPIN: 10573330 33/3142-543210 – Gedruckt auf säurefreiem Papier

Veranstalter

DAGM, Deutsche Arbeitsgemeinschaft für Mustererkennung

Organisation

Abteilung Bildverstehen, Institut für Parallele und Verteilte Höchstleistungsrechner (IPVR), Universität Stuttgart.

Tagungsleitung

Prof. Dr. P. Levi	IPVR, Universität Stuttgart
Dr. R.-J. Ahlers	A-Tec, Stuttgart
	ASG Luftfahrttechnik und Sensorik GmbH, Weinheim
Dr. F. May	Daimler Benz AG, Ulm

Programmkomitee

R. Albrecht, Innsbruck
B. Burkhardt, Freiburg
D. Fritsch, Stuttgart
H. Giebel, München
W. Förstner, Bonn
S. Fuchs, Dresden
A. Grün, Zürich
H. Haken, Stuttgart
G. Hartmann, Paderborn
B. Höfflinger, Stuttgart
B. Jähne, Heidelberg
W.G. Kropatsch, Wien
C.-E. Liedtke, Hannover
R. Massen, Konstanz
H.-H. Nagel, Karlsruhe
H. Neumann, Ulm
H. Niemann, Erlangen
E. Paulus, Braunschweig
F. Piwonka, Stuttgart
S.-J. Pöppl, Lübeck
P. Pretschner, Braunschweig
B. Radig, München
H. Ritter, Bielefeld
G. Sagerer, Bielefeld
B. Schleifenbaum, Wetzlar
J. Schürmann, Ulm
G. Sommer, Kiel
N. Stein, Wiesbaden
V. Steinbiß, Aachen
H. Tiziani, Stuttgart
F. Wahl, Braunschweig

Tagungsbüro

Dr. M. Schanz	IPVR, Universität Stuttgart
M. Völker	IPVR, Universität Stuttgart

Die Deutsche Arbeitsgemeinschaft für Mustererkennung (DAGM) veranstaltet seit 1978 jährlich ein wissenschaftliches Symposium mit dem Ziel, Aufgabenstellungen, Denkweisen und Forschungsergebnisse aus den Gebieten der Mustererkennung vorzustellen, den Erfahrungs- und Ideenaustausch zwischen Fachleuten anzuregen und den Nachwuchs zu fördern.

Die DAGM wird durch folgende wissenschaftliche Trägergesellschaften gebildet:

DGaO	Deutsche Gesellschaft für angewandte Optik
GMDS	Deutsche Gesellschaft für Medizinische Informatik, Biometrie und Epidemologie
GI	Gesellschaft für Informatik
ITG	Informationstechnische Gesellschaft
DGNM	Deutsche Gesellschaft für Nuklearmedizin
IEEE	The Institut for Electrical and Electronic Engineers, Deutsche Sektion
DGPF	Deutsche Gesellschaft für Photogrammetrie und Fernerkundung
VDMA	Fachabteilung Industrielle Bildverarbeitung/Machine Vision im VDMA
GNNS	German Chapter of the European Neural Network Society

Die DAGM ist Mitglied der International Assosiation for Pattern Recognition (IAPR)

Preisträger des 19. DAGM-Symposiums Mustererkennung 1997
Braunschweig, 15.-17. September 1997

Der mit 5.000 *DM* dotierte **DAGM-Preis** 1997 wurde verliehen an:

J. Haas, J. Hornegger, R. Huber, H. Niemann
(Universität Erlangen-Nürnberg)
für die Arbeit: *Probabilistic Semantic Analysis of Speech*

Die mit 1.000 *DM* dotierten **Anerkennungspreise** wurden verliehen an:

S. Abraham, W. Förstner
(Universität Bonn)
für die Arbeit: *Zur automatischen Modellwahl bei der Kalibrierung von CCD-Kameras*

T. Bülow, G. Sommer
(Christian-Albrechts-Universität zu Kiel)
für die Arbeit: *Das Konzept einer zweidimensionalen Phase unter Verwendung einer algebraisch erweiterten Signalrepräsentation*

H. Haußecker, B. Jähne
(Universität Heidelberg)
für die Arbeit: *A Tensor Approach for Precise Computation of Dense Displacement Vector Fields*

F. Kummert, G.A. Fink, G. Sagerer
(Universität Bielefeld)
für die Arbeit: *Schritthaltende hybride Objektdetektion*

C. Nölker, H. Ritter
(Universität Bielefeld)
für die Arbeit: *Detektion von Fingerspitzen in Videobildern*

Der mit 10.000 *DM* dotierte **Olympus-Preis** der Europa Stiftung *Wissenschaft fürs Leben* wurde verliehen an:

R. Koch
(Katholische Universität Leuven)
für die Arbeit: *Entwurf eines Systems zur automatischen Oberflächenmodellierung dreidimensionaler Körper aus stereoskopischen Rundum-Ansichten*

Guo-Qing Wei
(Deutsches Zentrum für Luft- und Raumfahrt e.V. (DLR), Oberpfaffenhofen)
für die Arbeit: *Shape from Shading-Problem*

Vorwort

Das DAGM-Symposium findet dieses Jahr zum ersten Mal in Stuttgart statt. Das zwanzigjährige Jubiläum nahmen die Vorstandsmitglieder der DAGM zum Anlaß, von der bisherigen Praxis, nur universitäre Veranstalter mit der Organisation und Durchführung der Tagung zu beauftragen, abzurücken und die Tagungsleitung drei Wissenschaftlern, von denen einer an der Universität und zwei in der Industrie tätig sind, zu übertragen. Diese enge Verknüpfung zwischen Wissenschaft und Industrie wird dadurch noch weiter vertieft, daß gleichzeitig zum Symposium die erfolgreiche Industriemesse VISION'98 unter demselben Dach auf dem Gelände der Messe Stuttgart (Killesberg) ausgerichtet wird und beide Veranstaltungen wechselseitig für jede Gruppe offen sind.

Die Verfahren der Mustererkennung haben sich in den letzten Jahren deutlich von ihren anfänglich vorhandenen Unzulänglichkeiten und ihrer geringen Robustheit befreien können und dadurch auf vielen Wegen ihren Eingang in industrielle Anwendungen gefunden. Mittlerweile wird die Bildverarbeitung als eine der Schlüsseltechnologien betrachtet, und die entsprechenden Branchen erwarten zweistellige Zuwachsraten bis ins nächste Jahrtausend hinein.

Im nächsten wissenschaftlichen Entwicklungsschritt geht es primär nicht mehr darum, möglichst effiziente Mustererkennungsalgorithmen zu entwickeln, sondern darum, Systemarchitekturen aufzubauen, die selbstorganisierend im Sinne einer aktiven Exploration sensorspezifische Probleme lösen können. Ein herausragendes Beispiel solcher technischen Kooperationen sind Autonome Mobile Systeme (AMS). Gemeint sind in diesem Rahmen Automobile, die ihren Weg auf der Straße ganz alleine ohne Fahrer finden. Dieser Themenbereich ist nicht nur wissenschaftlich zukunftsweisend, sondern er stellt gleichzeitig ein Forschungsgebiet dar, bei dem sich der Gedanken- und Methodenaustausch zwischen Wissenschaft und Industrie aufgrund der Komplexität der hier auftretenden Probleme über Jahrzehnte hinweg erstrecken muß und nicht nach dem üblichen kurzfristigen Muster der prototypischen Entwicklung, die in wenigen Jahren anwendbar ist, ablaufen kann. Aus diesen Gründen haben die Tagungsleiter diesem Themenfeld einschließlich relevanter Demonstrationen einen ganzen Nachmittag gewidmet.

Zum Symposium wurden insgesamt 97 Beiträge eingereicht, von denen 31 als Vortrag und 30 als Poster vom Programmkomitee angenommen wurden. Dazu kommen sieben eingeladene Vorträge, die von drei Hauptrednern und vier Sprechern im Rahmen des AMS-Tages präsentiert werden.

Die angenommenen Beiträge konzentrieren sich aber nicht nur auf autonome mobile Systeme, sondern überdecken das gesamte Spektrum aktueller Fragestellungen der Mustererkennung, welches sich von den Grundlagen und Algorithmen über die Klassifikation, die aktive Exploration, makro- und mikroskopische Geometrien bis hin zum Sprach- und Bildverstehen erstreckt. Einen breiten Rahmen nehmen ferner, der thematischen Ausrichtung des diesjährigen Symposiums entsprechend, die Anwendungen ein. Neben den vielen bereits „traditionsgemäß" vorhandenen Lösungsansätzen von medizinischen Fragestellungen ist besonders die Zunahme von Beiträgen über industriell eingesetzte optische Prüfverfahren hervorzuheben.

Danken möchten wir vor allem den Autoren für ihre Beiträge und ihr Verständnis für die enge Terminsetzung. Das Programmkomitee wurde wegen des diesjährigen speziellen Veranstaltungscharakters gegenüber den bisherigen Symposien deutlich erweitert. Allen Mitgliedern des Programmkomitees möchten wir für die gute und kooperative Zusammenarbeit unseren Dank aussprechen. In der Planung und Organisation wurden wir von Herrn Dr. Walter Gehring und Herrn Thomas Walter, beide von der Messe Stuttgart, stark unterstützt. Hierfür und für die kostenlose Überlassung von Messeräumen bedanken wir uns sehr. Danken möchten wir ebenfalls auch allen Mitarbeitern und Hilfskräften, insbesondere den Herren M. Völker und V. Avrutin, die uns bei der Organisation des Symposiums tatkräftig unterstützt haben. Last but not least bedanken wir uns herzlich bei allen, die sich mit Exponaten am AMS-Nachmittag beteiligen.

Allen Teilnehmerinnen und Teilnehmern wünschen wir einen erfolgreichen wissenschaftlichen Gedanken- und Erfahrungsaustausch auf dem 20. DAGM-Symposium und einen angenehmen Aufenthalt in Stuttgart.

Stuttgart, im Juli 1998

P. Levi
R.-J. Ahlers
F. May
M. Schanz

Inhaltsverzeichnis

Klassifikation und Neuronale Netze

Aktive Exploration

Anwendungen

3D-Geometrien

Sprach- und Bildverstehen

Grundlagen und Algorithmen (Postersession)

Klassifikation und Neuronale Netze (Postersession)

Aktive Exploration (Postersession)

Anwendungen (Postersession)

3D-Geometrien (Postersession)

Sprach- und Bildverstehen (Postersession)

Eingeladene Vorträge

Maschinelles Sehen zur Steuerung autonomer mobiler Systeme

Volker Graefe und Rainer Bischoff

Universität der Bundeswehr München
Institut für Meßtechnik
Werner-Heisenberg-Weg 39, 85577 Neubiberg
E-Mail: {Graefe | Rainer.Bischoff}@unibw-muenchen.de

Kurzfassung

Die Erforschung und Realisierung „sehender" Maschinen und „intelligenter" Roboter bildet seit 1977 einen Arbeitsschwerpunkt des Instituts für Meßtechnik. Ziel ist zum einen ein grundlegendes Verständnis des Sehens, der Autonomie und der Intelligenz von technischen Systemen, und zum anderen die Realisierung intelligenter sehender Roboter. In diesem Beitrag beschreiben wir drei unserer autonomen mobilen Systeme, die das maschinelle Sehen als Hauptsensormodalität einsetzen, um sich in ihrer jeweiligen Umgebung mit angemessener Geschwindigkeit fortbewegen und selbständig vielfältige Aufgaben erfüllen zu können. Mit ihrer Hilfe haben wir u.a. die Problembereiche der autonomen Mobilität im öffentlichen Straßenverkehr, die Navigation mobiler Roboter in bekannten und unbekannten Umgebungen, das maschinelle Lernen und die Steuerung von Robotern ohne Kenntnis quantitativer Modelle untersucht. Im folgenden geben wir einen Überblick über die Konzeption und Realisierung dieser autonomen Systeme und stellen einige mit ihnen gewonnene Erkenntnisse vor.

1 Einleitung

Ausgehend von der Hypothese, daß sich die Intelligenz von Lebewesen im Laufe der Evolution aus dem Zusammenwirken von Sehen, Bewegungssteuerung und Adaption an die Umwelt entwickelt hat, erforschen wir die Grundlagen eines solchen Zusammenwirkens bei autonomen Robotern. Unser Fernziel ist die Realisierung von Robotern mit einer praktischen Intelligenz von der Art, wie sie auch Lebewesen zum Überleben in ihrer jeweiligen Umwelt befähigt.

Höher entwickelte intelligente Lebewesen sind zumeist mit einem Sehsinn ausgestattet, der es ihnen erlaubt, sich in ihrer Umwelt zu orientieren und sich zweckmäßig und zielorientiert zu bewegen, z.B. zur Nahrungssuche oder zur Vermeidung von Gefahren. Nur in den Bereichen, in denen aufgrund des fehlenden Sonnenlichtes keine visuelle Erfassung der Umgebung möglich ist (z.B. in Höhlen oder im Meer), haben sich andere Sensorsysteme als leistungsfähiger erwiesen, beispielsweise die Verwendung von Ultraschall bei Fledermäusen und Delphinen. Wir sind daher der Auffassung, daß zur Entwicklung und Realisierung intelligenter Roboter Sichtsysteme von entscheidender Bedeutung sind, da sie es den Robotern ermöglichen, komplexe und dynamisch veränderliche Situationen in Echtzeit zu erkennen. Neben der Fähigkeit, mit Sensoren Situationen in einer dynamischen Umwelt zu erkennen, benötigt ein intelligenter Roboter insbesondere auch die Fähigkeit des Lernens, also die Fähigkeit, durch Interaktion mit der Umwelt Kenntnisse zu erwerben bzw. zu erweitern und Verhaltensweisen zu erlernen bzw. zu verbessern. Es ist zu erwarten, daß Roboter, die eine derartige Intelligenz besitzen, in ihrer jeweiligen Umgebung eine ähnliche Adaptionsfähigkeit aufweisen werden wie Lebewesen.

Die Ausrichtung unserer Arbeiten wurde von Anfang an wesentlich durch den Grundsatz bestimmt, daß alle Forschungsergebnisse zum Nachweis ihrer Gültigkeit in praxisnahen Experimenten in der realen Umwelt überprüft und demonstriert werden müssen. Diese Arbeitsweise ist sehr aufwendig, hat aber gegenüber reinen Simulationen den gravierenden Vorteil, zu weitaus solideren und in der Praxis belastbareren Ergebnissen zu führen. Grundvoraussetzung für unsere Arbeiten sind daher Roboter und leistungsfähige Echtzeit-Bildverarbeitungssysteme, die wir zu Beginn selbst entwerfen und bauen mußten, da entsprechende Hardware auf dem Markt nicht verfügbar war. Einige der damit durchgeführten Experimente (u.a. Andockmanöver eines Luftkissenfahrzeugs, Landeanflug eines Flugzeugs und autonome Straßenfahrt) sind in Zusammenarbeit mit dem Institut für Systemdynamik und Flugmechanik der Universität der Bundeswehr realisiert worden [Dickmanns, Graefe 1988]. Im Rahmen des europäischen Verbundprojekts PROMETHEUS, das die Steigerung der Verkehrssicherheit und die Entlastung des Autofahrers zum Ziel hatte, erforschten wir die Grundlagen einer Verkehrssituationserkennung (Abschnitt 2). Danach stand die Navigation mobiler Systeme in bekannten und unbekannten Innenraumumgebungen im Vordergrund, für die wir das Konzept der situationsabhängigen verhaltensbasierten Navigation entwickelten (Abschnitt 3). Das neueste von uns entwickelte System ist ein mobiler Manipulator, mit dem wir die sichtgesteuerte Handhabung verschiedener Objekte an unterschiedlichen Orten einer weitläufigen Einsatzumgebung untersuchen (Abschnitt 4).

2 Autonome Mobilität im Straßenverkehr

2.1 Echtzeit-Bildverarbeitungssysteme

Alle unsere Roboter setzen digitale Bildverarbeitung als leistungsfähige Sensorik ein. Die von Videokameras gelieferten Bilder werden in Echtzeit von Mehrprozessorsystemen ausgewertet. Der von uns gewählte gerätetechnische Ansatz war von Anfang an darauf ausgerichtet, eine hohe Systemleistung durch eine optimal an die Aufgabenstruktur des Rechnersehens angepaßte Systemarchitektur zu erzielen und nicht durch extrem aufwendige, aber unangepaßte Höchstleistungsrechner. Unsere Mehrprozessorsysteme sind aus lose gekoppelten Standard-Mikroprozessoren aufgebaut und deshalb leicht in Hochsprachen zu programmieren. Ein breitbandiger Videobus und ein unabhängiger Systembus (bzw. Link-Verbindungen bei Transputer- und DSP-Systemen) ermöglichen ein effizientes gleichzeitiges Arbeiten aller Prozessoren im System ohne Verzögerungen durch Kommunikationsengpässe.

Diese Hardware-Struktur bildet eine ausgezeichnete Grundlage für die Realisierung besonders leistungsfähiger „objektorientierter" Sichtsysteme [Graefe 1989]. Grundgedanke ist dabei, daß das Gesamtsystem (Hardware und Software) entsprechend den sichtbaren Objekten strukturiert wird, soweit diese für die Aufgabe des sehenden Roboters relevant sind.

Fünf Generationen solcher Echtzeit-Bildverarbeitungssysteme wurden bisher entwickelt. Die erste, 1977 konzipiert und noch mit 8-Bit-Prozessoren ausgestattet, war gleichwohl schon in der Lage, ein instabiles mechanisches System (inverses Pendel) allein durch Rechnersehen zu steuern [Haas 1982]. Die dritte, mit 32-Bit-Prozessoren, weist etwa die tausendfache Leistungsfähigkeit auf und ermöglichte so die gesamte Bildverarbeitung für ein autonom im Autobahnverkehr mitfahrendes Auto [Graefe, Kuhnert 1992]. Während die ersten vier Generationen von Bildverarbeitungssystemen vollständig an unserem Institut entwickelt worden sind, setzen wir nun aus Wirtschaftlichkeitsgründen Industrieprodukte ein. Zur Zeit wird zur Bildverarbeitung und Robotersteuerung ein Mehrprozessorsystem bestehend aus digitalen Signalprozessoren des Typs TMS 320C40 von Texas Instruments eingesetzt, das in einem Standard-PC Platz findet [Bischoff 1997].

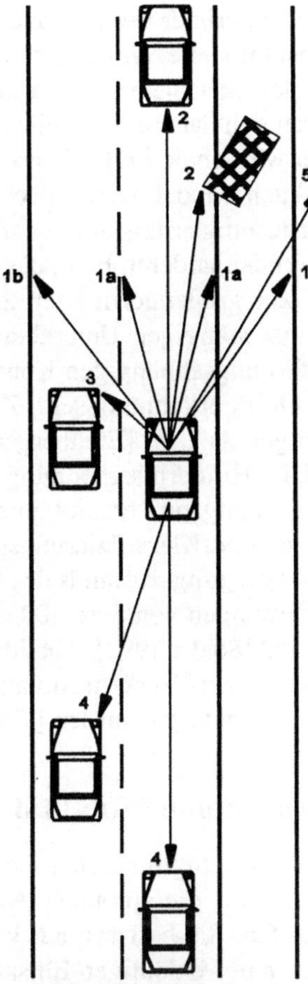

Abbildung 1:
Typische Verkehrssituation mit verschiedenen Objekten in einer komplexen dynamisch veränderlichen Szene, die vom vorgestellten Bildverarbeitungssystem in Echtzeit erkannt wird.

2.2 Objekterkennung in dynamischen Szenen

Beim ersten Schritt der Bildverarbeitung, der Merkmalsextraktion, stehen bei uns Verfahren im Vordergrund, die der Wirkungsweise rezeptiver Felder von biologischen Sichtsystemen ähnlich sind und die auf zeitaufwendige Ganzbildoperationen (z.B. Glättungs- und Segmentierungsverfahren) verzichten. Ein solches Verfahren zur Merkmalsextraktion, die gesteuerte Korrelation, wurde im Rahmen einer systematischen Untersuchung von Methoden zur schnellen, und vor allem robusten, Merkmalsextraktion am Institut für Meßtechnik entwickelt [Kuhnert 1986]. Es läßt sich außerordentlich effizient auf einem Mikrorechner implementieren und ermöglicht die Analyse von Bildfolgen in Echtzeit. Die flexible und robuste, aber normalerweise sehr rechenzeitintensive Korrelation wird – auch ohne Verwendung von Spezialhardware – dadurch echtzeitfähig, daß sie zum einen nur in den jeweils relevanten Teilen des Bildes berechnet wird und daß zum anderen als diskrete Vergleichsfunktion eine dünn besetzte ternäre Maske dient. Eine entscheidende Bewährungsprobe bestand das Verfahren der gesteuerten Korrelation 1987 bei den „Weltrekordfahrten" eines fahrerlosen Straßenfahrzeugs (96 km/h auf Autobahnen und ca. 40 km/h im Universitätsgelände).

Entsprechend dem objektorientierten Grundansatz [Graefe 1989] wird zur Erzielung höchster Robustheit die Merkmalsextraktion von den höheren Systemebenen aus gesteuert und gestützt. Selbst unter erschwerten Bedingungen (Schatten, Reflexionen auf nasser Fahrbahn, texturierter Hintergrund etc.) lassen sich damit gute Ergebnisse erzielen. Primär werden dazu wegen ihrer Vielseitigkeit und Effizienz sogenannte 2D-Objektmodelle eingesetzt, die das Erscheinungsbild des betreffenden physischen Objekts modellieren. Dies können einfache Formmodelle sein; in komplexeren Szenen haben sich daneben Modelle auf der Grundlage von Symmetrien sowie von Statistiken über Merkmale im Bild besonders gut bewährt [Regensburger 1994].

Auch Modellwissen über die Entfernungsabhängigkeit der optischen Abbildung wurde eingesetzt, und zwar bei der Erkennung von hinter dem eigenen Fahrzeug fahrenden anderen Fahrzeugen auf Autobahnen. Hierfür wurde ein einfaches Verfahren zur größennormierten 2D-Objektrepräsentation auf der Grundlage einer entfernungsabhängigen Unterabtastung von Bildern entwickelt. In Verbindung mit einer entfernungsabhängigen Kontrastanpassung bei der Merkmalsextraktion erleichtert es die Objekterkennung sehr. Zugleich ermöglicht es sehr kurze Ausführungszeiten von weniger als 1 ms [Efenberger 1996].

Die Hinderniserkennung beruht auf einem mehrstufigen Verfahren mit einem Entdeckungsprozeß und einem – unter Umständen mehrfach vorhandenen – Verfolgungs- und Klassifizierungsprozeß, denen noch weitere Prozesse, z.B. zur Schätzung des Bewegungszustands des Hindernisses, nachgeschaltet sein können. Entdeckungsentfernungen von über 300 m für kleinere und ca. 700 m für größere Objekte wurden erreicht [Solder 1992]. Abbildung 1 gibt einen Überblick über die von uns bearbeiteten Module zur Verkehrssituationserkennung, die erfolgreich in Autobahnszenen in Echtzeit getestet wurden [Graefe 1992, 1993].

3 Autonome Mobilität in Innenräumen von Gebäuden

Mit den sichtgesteuerten mobilen Robotern *ATHENE I* und *II* wurden die Voraussetzungen für die autonome Navigation in Fabrikgebäuden und Büroumgebungen geschaffen. Sie basieren auf konventionellen Flurförderfahrzeugen (Abbildung 2) und können in weitläufigen Einsatzumgebungen verschiedene Transportaufträge übernehmen [Wershofen 1996]. Hierzu wurde das Konzept der situationsabhängigen verhaltensbasierten Navigation entwickelt [Wershofen, Graefe 1992]. Die Verhaltensauswahl des Roboters erfolgt in jedem Moment auf Grundlage einer ständig mitlaufenden Situationserkennung. Diese Situation ergibt sich im wesentlichen aus den wahrgenommenen Zuständen relevanter Objekte, den aktuellen Handlungsmöglichkeiten und den zu erledigenden Aufgaben. Dies erfordert reichhaltige Sensorinformationen, wie sie primär von einem dynamischen objektorientierten Sichtsystem geliefert werden können [Bischoff et al. 1996].

Abbildung 2:
ATHENE II, ein intelligenter mobiler Roboter, basierend auf einem konventionellen Flurförderfahrzeug und ausgestattet mit einem objektorientiertem Sichtsystem

Zur Navigation verwenden die Roboter topologische Karten, die sie bei einer Erkundungsfahrt selbständig aufbauen und mit Attributen über geometrische Sachverhalte sowie Beschreibungen von sichtbaren Landmarken und Hinweisen auf bestimmte Verhaltensweisen versehen. Ein Bediener überwacht die Roboter dabei, hilft ihnen an kritischen Stellen bei der Verhaltensauswahl und teilt ihnen relevante Stationsnamen mit, so daß ihnen einfache Hol- und Bringdienste unter Verwendung allgemein bekannter Namen wie z.B. „Kopierer" oder „Labor" aufgetragen werden können. Die Kommunikation zwischen Bediener und Roboter wird durch die situationsorientierte verhaltensbasierte Systemarchitektur stark vereinfacht. Insbesondere kann die Kommunikation situationsabhängig und unter Bezugnahme auf Sachverhalte und Objekte der Umgebung erfolgen, wie sie für den Benutzer erkennbar sind, und ohne Bezugnahme auf Roboter-interne Gegebenheiten wie etwa Speicheradressen, Variablennamen, Prozeßbezeichnungen oder Koordinatensysteme [Graefe, Bischoff 1997].

Abbildung 3:
Serviceroboter *HERMES* mit omnidirektionaler Basis, Manipulationssystem mit 13 Freiheitsgraden und Stereosichtsystem auf Schwenk-Neigeplattform (Abmessungen: 170 cm x 70 cm x 70 cm)

4 Mobile Manipulatoren

Serviceroboter rationalisieren und automatisieren Transport-, Handhabungs- und Verarbeitungsaufgaben in den verschiedensten Dienstleistungsbereichen. Mögliche Aufgaben liegen in den Bereichen Reinigung, Überwachung, Baugewerbe, Rehabilitation und Pflege oder Wartung und Reparatur. So vielfältig die Dienstleistungen sind, die an Menschen oder für Einrichtungen erbracht werden können, so unterschiedlich werden die Serviceroboter sein, die an ihre jeweilige Aufgabe und die Umwelt angepaßt sein müssen. Bestimmte Teilsysteme werden jedoch viele Serviceroboter gemein haben, beispielsweise eine mobile Plattform, ein Manipulationssystem, Steuer- und Antriebstechnik, Bedienerschnittstellen und Sensoren. Die dazugehörige Software sollte modular gestaltet sein, damit die einzelnen Teilsysteme je nach Bedarf zu funktionsfähigen Robotern konfiguriert werden können. Es ist eine große technische Herausforderung, solche Serviceroboter zu entwickeln und zu bauen, da Kompetenzen u.a. in vielen Ingenieurwissenschaften und der Informatik gefordert sind.

Gerade die Kombination von Mobilität und die Fähigkeit zur Manipulation stellen wesentliche Grundfunktionen der meisten Serviceroboter dar. So haben wir aufbauend auf unsere Erfahrungen zur situationsabhängigen verhaltensbasierten Navigation mit *ATHENE II* und der unkalibrierten Manipulation mit einem Knickarm-Roboter vom Typ Mitsubishi Movemaster [Graefe 1995] einen neuen vielseitig einsetzbaren Roboter entworfen und gebaut, den wir zur Erforschung und Lösung der noch offenen Probleme im Bereich der Servicerobotik nutzen wollen. Insbesondere bearbeiten wir mit diesem

Roboter zunächst die autonome visuelle Erkundung einer weitläufigen Einsatzumgebung und die unkalibrierte Manipulation von verschiedenen Objekten an unterschiedlichen Orten innerhalb dieser Umgebung. Beim Entwurf und bei der Realisierung dieses mobilen „Manipulators", den wir *HERMES* ("**H**umanoid **E**xperimental **R**obot for Mobile **M**anipulation and **E**xploration **S**ervices") genannt haben (Abbildung 3), richteten wir unser Hauptaugenmerk auf eine modulare und erweiterbare Struktur, sowohl bei der Hardware des Roboters (Antriebsmodule und Informationsverarbeitungssystem) als auch bei der Software [Bischoff 1997].

Die ersten Software-Module, die für *HERMES* geschrieben wurden, ermöglichen einem Bediener, den Roboter in seiner Einsatzumgebung an bestimmte Orte zu fahren und dort einfache Manipulationsaufgaben zu lehren und wiederholt durchführen zu lassen, z.B. das Aufheben eines Balles oder das Öffnen von Schubladen. Erste Implementierungen von sichtgekoppelten Verhaltensmustern erlauben *HERMES*, bekannte Objekte zu fixieren und sich Andockstationen (z.B. Beistelltischen) anzunähern [Graefe, Bischoff 1998].

5 Zusammenfassung und Ausblick

Maschinelles Sehen hat sich als sehr leistungsfähige Sensorik für autonome mobile Systeme erwiesen. Unser Ansatz des objektorientierten Sehens ermöglicht die Interpretation der reichhaltigen Sensorinformation einer Videokamera in Echtzeit und hat u.a. das autonome Mitfahren eines Straßenfahrzeugs im Autobahnverkehr selbst mit aus heutiger Sicht bescheidenen Mitteln und Anforderungen an die Hardware ermöglicht. Es ermöglichte ebenfalls die Analyse von Verkehrssituationen durch gleichzeitiges Entdecken, Klassifizieren und Verfolgen mehrerer Objekte.

Zur Steuerung mobiler Roboter in Fabrikgebäuden und Büroumgebungen wurde das Konzept der situationsabhängigen verhaltensbasierten Navigation auf Grundlage des objektorientierten Sehens entwickelt und mit dem mobilen Roboter *ATHENE II* in unserer Institutsumgebung verifiziert. Es erlaubt einem Benutzer eine Art natürlichsprachlicher Kommunikation mit dem Roboter, solange sich beide auf gemeinsam benannte und bekannte Objekte einer Einsatzumgebung beziehen. Die zukünftige Verbesserung der Wahrnehmungsleistung wird es dem Roboters ermöglichen, seine Einsatzumgebung ohne menschliche Überwachung zu erkunden.

Die Erweiterung der so gewonnenen Mobilität um die Fähigkeit, verschiedene Objekte zu handhaben, ist Ziel unserer gegenwärtigen Forschung. Dazu haben wir den modular aufgebauten Serviceroboter *HERMES* entwickelt, der bereits zahlreiche Eigenschaften besitzt, die von künftigen Servicerobotern erwarten werden, insbesondere was Modularität, Flexibiltät und Erweiterbarkeit betrifft. Wir forschen an robusten Verfahren zur Steuerung mobiler Manipulatoren und haben bereits Verfahren entwickelt, die durch Vermeidung des Einsatzes von Modellwissen und durch den direkten Übergang von Bild-Sensordaten zu Motor-Steuerkommandos so robust sind, daß auf fortlaufende Kalibrierungs- und Wartungsarbeiten verzichtet werden kann. Damit eignet sie sich insbesondere für die Servicerobotik, in der Roboter einerseits ein hohes Maß an Robustheit im Umgang mit unerwarteten Situationen benötigen und andererseits Wartungspersonal nicht zur Verfügung steht.

Literatur

Bischoff, R. (1997). *HERMES* - A Humanoid Mobile Manipulator for Service Tasks. International Conference on Field and Service Robotics. Canberra, pp. 508-515.

Bischoff, R.; Graefe, V.; Wershofen, K. P. (1996). Combining Object-Oriented Vision and Behavior-Based Robot Control. Proc. Int. Conference on Robotics, Vision and Parallel Processing for Industrial Automation. Ipoh, Malaysia, pp 222-227.

Dickmanns, E.D.; Graefe, V. (1988). Applications of Dynamic Monocular Machine Vision. Int. J. Machine Vision and Applications. Vol. 1 (1988), pp 241-261.

Efenberger, W. (1996). Zur Objekterkennung für Fahrzeuge durch Echtzeit-Rechnersehen. Dissertation, Fakultät für Luft- und Raumfahrttechnik der Universität der Bundeswehr München.

Graefe, V. (1989). Dynamic Vision Systems for Autonomous Mobile Robots. Proc. IEEE/RSJ Workshop on Intelligent Robots and Systems, IROS '89. Tsukuba, pp. 12-23.

Graefe, V. (1992). Visual Recognition of Traffic Situations by a Robot Car Driver. Proceedings, 25th ISATA; Conference on Mechatronics. Florence, pp 439-446.

Graefe, V. (1993). Vision for Intelligent Road Vehicles. Proceedings, IEEE Symposium on Intelligent Vehicles. Tokyo, pp 135-140.

Graefe, V. (1995). Object- and Behavior-oriented Stereo Vision for Robust and Adaptive Robot Control. Int. Symp. on Microsystems, Intelligent Materials, and Robots, Sendai, pp. 560-563.

Graefe, V.; Bischoff, R. (1997). A Human Interface for an Intelligent Mobile Robot. 6th IEEE Int. Workshop on Robot and Human Communication. Sendai, Japan, Sept. 1997, pp. 194-197.

Graefe, V.; Bischoff, R. (1998). Vision-Guided Intelligent Robots. Proc. Int. Conf. on Mechatronics and Machine Vision in Practice, Nanking, September 1998 (in Druck).

Graefe, V.; Kuhnert, K.-D. (1992). Vision-based Autonomous Road Vehicles. In I. Masaki (Ed.): Vision-based Vehicle Guidance. Springer-Verlag, pp 1-29.

Haas, G. (1982). Meßwertgewinnung durch Echtzeitauswertung von Bildfolgen. Dissertation, Fak. für Luft- und Raumfahrttechnik der Universität der Bw München.

Kuhnert, K.-D. (1986). A Model-driven Image Analysis System for Vehicle Guidance in Real Time. Proceedings, Second International Electronic Image Week. CESTA, Nice, pp 216-221.

Regensburger, U. (1994). Zur Erkennung von Hindernissen in der Bahn eines autonomen Straßenfahrzeugs durch maschinelles Echtzeitsehen. Dissertation, Fakultät für Luft- und Raumfahrttechnik der Universität der Bundeswehr München.

Solder, U. (1992). Echtzeitfähige Entdeckung von Objekten in der weiten Vorausschau eines Straßenfahrzeugs. Dissertation, Fakultät für Luft- und Raumfahrttechnik der Universität der Bundeswehr München.

Wershofen, K. P. (1996). Zur Navigation sehender mobiler Roboter in Wegenetzen von Gebäuden – Ein objektorientierter verhaltensbasierter Ansatz. Dissertation, Fakultät für Luft- und Raumfahrttechnik der Universität der Bundeswehr München.

Wershofen, K. P., Graefe, V. (1992). An Intelligent Autonomous Vehicle Guided by Behavior-based Navigation. IFToMM-jc International Symposium on Theory of Machines and Mechanisms. Nagoya, pp 244-249.

Image Processing for Driver Assistance

Werner v. Seelen, Uwe Handmann, Thomas Kalinke,
Christos Tzomakas, and Martin Werner

Institut für Neuroinformatik, Ruhr Universität Bochum, 44780 Bochum

Abstract. Systems for automated image analysis are useful for a variety of tasks and their importance is still growing due to technological advances and an increase of social acceptance. Especially in the field of driver assistance systems the progress in science has reached a level of high performance. Fully or partly autonomously guided vehicles, particularly for road-based traffic, pose high demands on the development of reliable algorithms due to the conditions imposed by natural environments. At the *Institut für Neuroinformatik* methods for analyzing driving relevant scenes by computer vision are developed in cooperation with several partners from the automobile industry. We introduce a system which extracts the important information from an image taken by a CCD camera installed at the rear view mirror in a car. The approach consists of a sequential and a parallel sensor and information processing. Three main tasks namely the initial segmentation (object detection), the object tracking and the object classification are realized by integration in the sequential branch and by fusion in the parallel branch. The main gain of this approach is given by the integrative coupling of different algorithms providing partly redundant information.

1 Introduction

Some systems presented in ref. [4, 24, 3] show the principal feasibility of driver assistance systems based on computer vision. Although exclusively vision based systems and algorithms are not yet powerful enough to solve all driving relevant tasks, a large amount of different scenarios can be interpreted sufficiently. Additionally sensors like RADAR and LIDAR extent the contents of sensor information necessary for building a reliable system. The main focus of our system lies in combining various methods for the analysis and interpretation of images and in the fusion of a large spectrum of sensor data to extract most reliable information for the final planning and for predicting of behavior of other vehicles. The great variety of different scenarios as well as the high degree of reliability necessary for the given task require an encompassing and flexible system architecture. The requirements concerning the reliability of the reached solution, the variety of geometric appearances of involved objects and the environmental constraints of both deterministic as well as statistical nature necessitate a multitude of partial solutions based on different representations of the environment. Consequently, complexity and structure of the overall system have to be adaptable to the increasing system complexity in order to allow accommodation of additional modules without degeneration of already accomplished partial solutions. For this reason, even *simple* applications are encumbered by considerations concerning the overall system architecture. Basically, the overall system architecture can be divided into basic, fusion and integration algorithms. Basic methods are those providing specific partial solutions under given constraints. Results and application of the individual algorithms are not independent, resulting in an increase in redundancy making the overall system secure and reliable given a suitable coupling architecture. The necessary methods for fusion and integration ensure a flexible cooperation of the basic building blocks as well as the integrative derivation of results. In a similar vein, a sequential data transmission and system dynamics are necessary in order to build up an overall system and giving solutions to complex tasks.

2 Image Processing for Driver Assistance

The fusion of different sensor information and preprocessing results increases the performance of the system. Basic algorithms for themselves are *specialist* for a specific kind of sensor information. Figure 1 shows the different types of information principles depending on the spatial relationship to the vehicle. With respect to the requirements of various applications optimally adapted algorithms are built. In the area F1 contour based methods are chosen. On the one hand the sparse coding (edges) of the intensity information is sufficient due to the high resolution of the objects in the image and on the other hand it speeds up computation time for real time applications. Here, we mainly use a feature called local orientation coding (LOC) [8]. In the field F2 we use motion detection algorithms to segment overtaking and overtaken vehicles. In contrary to other applications we use a pattern tracking based algorithm which ensures high stability. The long distance field F3 is analyzed by texture based methods. The low spatial resolution makes an edge based processing infeasible. Nevertheless the integrative

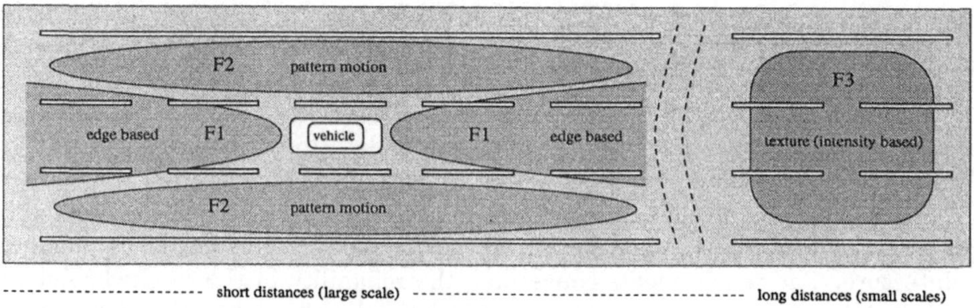

Fig. 1. Separation of the road in fields F1, F2 and F3 in which different algorithms can be applied optimally.

characteristics of texture analysis provides good results by separating the objects from the background by use of their texture. In the area of preprocessing, a multitude of different methods for initial segmentation, object tracking, and object classification has been developed in the context of current research. A few inherent tendencies appear remarkable.

- Previous work often was based on the use of higher features, meaning the generation of a sequence of features beginning at the iconic (image-based) side and continuing to the symbolic side. There are two main reasons to do this. First the historic rooting of image processing in material and surface inspection for quality control has lead to the existence of theoretically well-founded and practically tested algorithms. Second the symbolic features are commonly used for compact coding purposes, so that processed data amounts can be largely reduced for accommodating limited processing resources. The breath-taking evolution of processors has particularly alleviated the impact of this last constraint. In addition, it appears that particularly in the context of limited sensor resolution (i.e., in long distance regions) algorithms can be employed that rely on statistical measures of extensive 'early' (in the chain of processing) feature sets. These algorithms supplement the spectrum of methods explicitly in more traditionally oriented algorithms.
- Often a formulation as an optimization problem can lead to implicitly robust solutions avoiding disadvantages of explicit methods (e.g., the correlation of model with image features, the correspondence problem). In this area the increase in available computational power has contributed to the scientific progress, as well.
- Particularly in natural environments, flexible algorithms possessing a certain learning capability for input data driven adaption are preferably used.

3 The Basic Algorithms

At the *Institut für Neuroinformatik* algorithms providing partial solutions for object detection, tracking and classification have been incorporated in an driver assistance architecture. Namely the following enumeration gives an overview over the applied methods.

- Initial Object Detection: local orientation coding (LOC) [7][8], polygon approximation of contours [5], use of local symmetry [26], pattern motion analysis [26], texture analysis based on local image entropy [18], local variance analysis [25] and local cooccurrence measures [26], shadow analysis [10], color analysis [12], and RADAR mapping [25].
- Object Tracking: Hausdorff distance matching [10], parametric optimization [22] and cross entropy [19].
- Object Classification: local orientation coding [26], Hausdorff distance [26], cooccurrence analysis [9], and parametric optimization [22].

All algorithms can be parted in methods working on differential information (e.g. edges) and integral measurements (e.g. texture). For the application types, initial object detection, tracking and classification, a description of the actually used algorithms is given.

3.1 Initial Object Detection

The main motivation of using multiple simple methods is that the development of designing one basic method solving all conceivable scenarios seems to be impossible. Therefore in order to provide reliable results and to ensure a fast and robust processing a coupling of *specialists* is carried out. Some methods are described shortly in this section.

Local Orientation Coding: The 'raw' gray scale (intensity) images are preprocessed by a method we call local orientation coding (LOC). The image features obtained by this preprocessing are bit strings each representing a binary code for the directional gray-level variation in a pixel neighborhood. In a more formal fashion the operator is defined as

$$b'(n,m) = \sum_{i,j} k(i,j) \cdot u(b(n,m) - b(n+i,m+j) - t(i,j)), (i,j) \in \text{ neighborhood}$$

where $b(n,m)$ denotes the (gray scale) input image, $b'(n,m)$ the output representation, $\mathbf{k}$ a coefficient matrix, $\mathbf{t}$ a threshold matrix and $u(z)$ the unit step function. The output representation consists of labels, where each label corresponds to a specific orientation of the neighborhood. An adaption mechanism for the parameters $\mathbf{t}$ of the coding algorithm yields a high level of flexibility with respect to lighting conditions [8].

Shadow Analysis: The detection of shadows is realized by thresholding the intensity image, some morphological processing and a region clustering stabilized over time. As already introduced in [21] the shadow underneath a vehicle can be used as a sign pattern. For the task of the initial object detection the grey level of the road is analyzed in order to extract a threshold τ for the shadows. Furthermore we select those LOC features that expose an horizontal orientation and correspond to a light-to-dark transition (scanning the image upwards) and group them in clusters. These clusters are imposed in further constraints (i.e. from the camera geometry [17]) and finally build the initial hypotheses (or Regions Of Interest).

Texture Analysis (Entropy, Cooccurrence): Besides operators like intensity derivation (gradients and the LOC) texture analysis as an integrating operator has been used successfully in image processing. The term *texture* is not explicitly defined. Globally texture is a description of image pixels or texture elements (groups of pixels) belonging to a specific texture class due to their spatial arrangement to other elements. Texture depends inherently on scaling. The spatial and intensity relationship between these elements define the kind of texture. Strong variation of intensity in a small area lead to fine textures and

low variations produce coarse textures. Furthermore textures can be parted into properties weak and strong. Weak textures are described mostly by statistical methods. In strong textures the spatial interaction of elements are somewhat regular. Their recognition is usually accompanied by an exact definition of texture primitives (grammars). Actually two different methods for analyzing are commonly used: statistical and syntactic. In our applications we mainly work with statistical texture description. Every kind of texture is represented by a multidimensional feature vector in order to evaluate a statistical pattern recognition for every texture class based on suitable decision rules.
The *local image entropy (LIE)* has been developed at our institute [18]. In this method an estimation of the information contents of a pixel and its neighborhood is given. A saliency map is calculated so that a separation of objects and background can be evaluated. A detection of road-users and the free driving space can be easily done.
One of the fundamental tools in texture analysis, the cooccurrence matrices, were suggested by ref. [13]. In here, the probability of the cooccurrences of pixel pairs under predefined geometrical and intensity constraints are measured. These constraints are determined by the intensity ratio and the spatial relationship (angle and distance) of two image points. A definition of the cooccurrence matrix follows. In an image window **I** of size $M \times N$ and a maximum number of different gray values Q the cooccurrence matrix **P** is calculated under parameters angle α in a given distance d as follows

$$P_{d,\alpha}(i,j) = \frac{\text{number of pairs}_{((x,y),(x',y'))}\text{, verifying }_{(d,\alpha)}\text{ and }_{I(x,y)=i}\text{ and }_{I(x',y')=j}}{\text{number of all pairs in image window}_{((x,y),(x',y'))}}$$

A calculation of texture features is performed in most of the applications under four directions ($\alpha = 0, 45, 90,$ and 135) and different distances $d = 1, 2, \ldots$. A rotation-invariance can be obtained by accumulation of the matrices of the four directions. The amount of scaling variance can be reduced by calculating the matrices over different distances. Julesz showed that the human perception of texture is based on cooccurrence statistics. Haralick, Shunmugan, and Dinstein suggested in [13] 14 different statistical features which can be obtained from the cooccurrence matrices. In our field of research cooccurrence matrices are mainly applied to the initial segmentation. The matrices are calculated in overlapping windows. Features like energy, entropy, contrast, correlation and the highest cooccurrence of [13] are combined for the segmentation process.

3.2 Object Tracking

Algorithms for object tracking are the most important if a stabilization over time or a prediction of e.g. trajectories are demanded. Some methods for object tracking are introduced in this section. As it can be seen in figure 1 the tracking algorithms find their applications depending on the spatial resolution of the images. In the near distance field the Hausdorff distance or order statistics are used as a measurement based on contour codes (LOC). Here we proceed to present the more stable Hausdorff distance tracker that has been tested successfully on a large set of different image sequences (figure 2). For further details of the approach using order statistics see ref. [27]. Supplementary in the long distance field the texture based cross entropy provides optimally results.

Hausdorff Distance: The geometric comparison of shapes is a fundamental tool for model-based object recognition. Most of the methods used in object recognition systems refer to a similarity measure between the model features and the image features [23]. The Hausdorff distance measures the divergence of a set of features with respect to a reference set of features [16]. These sets mostly describe object contours in our application. The comparison of similar object contours yields small distance values, whereby objects with different contours yield larger distances. The directed Hausdorff distance h of one point set A against a point set B is the maximum of the minimum distances of each point of set A to the points of set B. The final Hausdorff distance H is simply the maximum of the two directed distances.

$$h(A,B) = \max_{p \in A} \min_{q \in B} \|p - q\|, \quad H(A,B) = \max(h(A,B), h(B,A))$$

The partial Hausdorff distance performs a ranking of these minimum distances and considers a fraction of them instead of the maximum. Unlike the classical correlation methods the Hausdorff distance uses MinMax operations instead of multiplications, so it is more efficient in time. The partial Hausdorff distance is robust against partially occluded objects and outliers that may arise at the contours due to noise or insufficient feature extraction.

Fig. 2. Objectdetection, objectclassification and objecttracking on german Autobahnen and german Landstraßen

The partial Hausdorff distance can examine object hypotheses in a complex scene. This method was tested successfully with highway-traffic scenes. It was able to recognize vehicles on highways and track them over time. Two degrees of freedom were considered in our schema: translation and scaling of models.

Texture based Object Tracking - Cross Entropy: One of the simple description of textures is obtained by intensity histograms (first order statistics). Especially non-rigid objects like pedestrians and two-wheeled vehicles which consist of a further rotational degree of freedom

Fig. 3. Tracking of pedestrians based on the cross entropy based on intensity distributions and LOC features

compared to other road-users can be tracked using the cross entropy. As described in [20] a matching process can be performed by comparison of two probability distributions. In our application a model distribution at time step $(t-1)$ is compared to several hypotheses at time t. Figure 3 show tracking of pedestrians using intensity and edge probability distributions. As an extension to the proposed method we use instead of statistics given by a histogram correlated statistics given by the cooccurrence matrices. The quality of the estimate of position and scale increases but the calculation time increases as well.

3.3 Neural Classifiers for Vehicles

For the task of classification different methods are used. Feature based and model based solutions have been developed. The LOC-classifier is computational fast method used for a fast estimate of given ROI. It is aimed at separating possible objects from the background. It is independent from the resolution of the objects due to a normalization in size. Additionally two classifiers with higher computational costs perform a reliable classification. The Hausdorff distance classifier processes objects in the near field with high spatial resolution enhancing the ROI image coordinates.

LOC-Classifier: With the given local orientation coding [8], described in section 3.1), a classification of vehicles is realized. The classifier has to cope with partial occlusions, varying illumination conditions, tilt of an object, differently resolved structures depending on the distance of the object under consideration, noise and perturbations induced by the recording and processing equipment, different viewpoints and different kind of cars with different shapes and colors. Additionally, the classifier should be able to generalize from relative few training examples to the necessary features characterizing a car. Therefore, a neural network has been chosen for solving the classification task. It is a feed-forward neural network with one hidden layer trained by the error back-propagation algorithm [14]. These networks are known to be universal approximators for any continuous valued function [15]. Furthermore, it is shown that these structures can, with some small modifications, approximate a-posteriori probabilities in the sense of a Bayesian classifier [6].
The inputs for the classifier are certain subsets of the histograms. The output is the class of the region. The complete system has been implemented and extensively tested on the Mercedes Benz VITA II test vehicle [2]. Different classes of vehicles have been trained. For a further evaluation of the system see ref. [1].

Hausdorff Distance Classifier: Furthermore, the geometric property of the Hausdorff distance leads to the idea of classifying various vehicles into separate classes according to the imposed dissimilarity measure. Because of the need of defining a reference contour for each class we deal here with a model based approach. The design of accurate models (prototypes) is of great importance for our task. At a first step, the Hausdorff distance is used for the classification of cars and trucks. Due to the fact that rear views of cars differ significantly from rear views of trucks, one can expect that the design of generic models for each class can accomplish the separation of the objects of both classes. The classification works according to the following scheme: Each region is compared with two models, i.e. a car model and a truck model. The features of the region and the models have been extracted using the Local Orientation Coding. For more robust results the horizontal features are separated from the verticals, for both the region and the models.

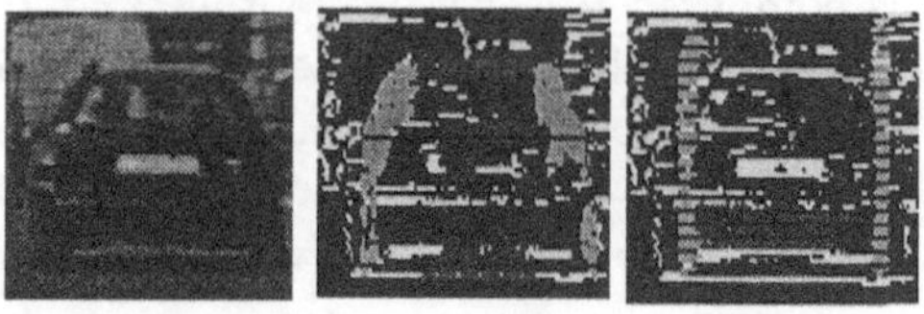

Fig. 4. Hausdorff distance classifier: each region is compared with two models

The Hausdorff distance is computed for each model over all the possible translations inside the region and a certain range of scales. The fractions of the features of the forward and the backward match that verify a given distance threshold constitute the criteria for its classification for each model. These values are learned by a multi-layer perceptron (MLP) network using the back-propagation algorithm.

4 The Concept of Fusion

Data fusion is one of the main goals to be achieved if a large amount of stability and reliability is necessary like in this application of driver assistance systems. On one hand a gain in robustness is reached by creating by creating high redundancy so that poor or missing results of one data stream do not affect the overall result decisively [9]. On the other hand the varying types of objects and background constellations demand a large spectrum of data to be processed to solve the given task. Three different types of neural coupling mechanisms are introduced [11]. The high flexibility, the facility of expansion and the adaptive retraining processes have led to the choice of neural networks.

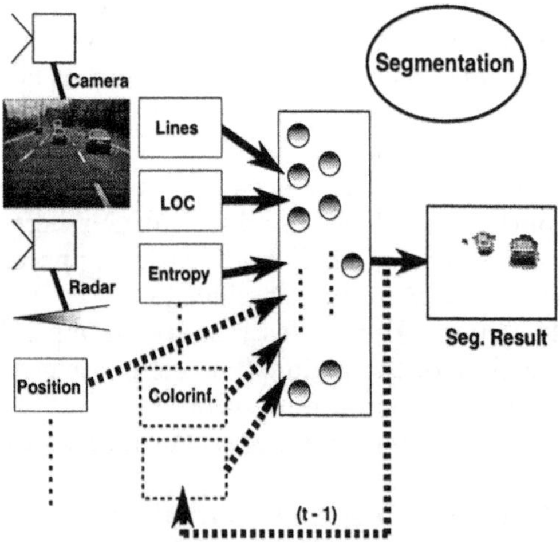

Fig. 5. Coupling model

The aim of fusion in computer vision is to get an improvement of special solutions and single methods with a coupling net (parallel branch). Especially the modular coupling of single processing steps generates redundancy necessary for object recognition.

Within this, greater flexibility and robustness of the image processing modules and high adaptation of the modules regarding to the problems should be achieved. In figure 5 a principle of a fusion process for segmentation is shown [11]. Computer vision modules, generating lines (polygon approximation of the contour) [5], local orientation coding [8], and local image entropy are coupled in a neural network to solve the initial object detection. A feedback over time is realized, additional sensor information could be easily integrated at this level.

5 The Concept of Integration

The overall system is shown is figure 6. The concept of integration (sequential branch) of single steps to a reliable working system is mainly based on feedback of results. As a sensor input the intensity image and radar signals are used. The results of basic preprocessing algorithms are fed into a neural fusion architecture, the initial object detection, that provides hypotheses of possible location of ve-

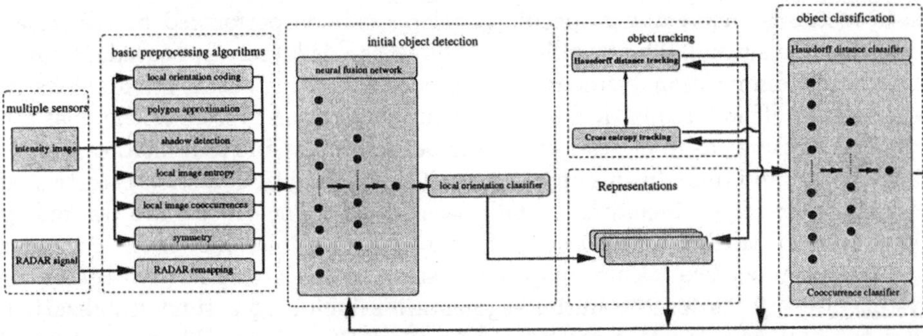

Fig. 6. Overall system

hicles. The very fast calculating LOC-classifier reduces the set of hypothesis. An internal stabilization over time ensures further robustness. In order to confirm the hypothesis an object tracking is performed where the object size and type decides whether the Hausdorff tracking or the cross entropy tracking have to be used. The results scale, position, and confidence are fed into the main stream and to the modular classifier. A neural network determines depending on the results of the object size, the Hausdorff and cooccurrence classifier what type of vehicle has been tracked.

Concerning the calculation rates the object tracking has to be performed for every time step. The initial object detection can work on a slower time rate. Finally the classification provides results on larger time steps due to the fact that a tracked object with high confidence values will not change its class. So the object tracking is the most important task next to the detection process. To ensure a stable tracking over time a Kalman filter is implemented.

The main recoupling stream gathers all the results of the single tasks and is increasing over time. The type of information is changing from an iconic (preprocessing) to a symbolic (classification) description. A global data representation is built. Here the integration of the different processing steps is solved. Task depending pixel oriented maps of attention are implemented. In figure 7 an example of a representation for object tracking of vehicles in front is shown.

Fig. 7. Representation for object tracking

6 Results and Discussion

We presented an overall system integrating the results and experiences of a long period of research in computer vision. Due to the increase of computational power and the development of reliable algorithms a fusion and integration of basic methods each solving specific problems can be performed to realize an overall stable system. The stability and robustness is largely increased. Because the overall computational time is still quite long by using actual standard hardware a spin off was realized. So if a real time operation system is the goal the whole processing has to be restricted to some algorithms due to limited computational power and time. In this application the initial object detection is restricted to a shadow analysis including a LOC-classification. The objects are tracked by the Hausdorff tracker and classified by the Hausdorff classifier in order to use just one preprocessed feature map. On a standard DEC Alpha (500 MHz) the system uses 10 ms for the initial segmentation including a time stabilization, then the LOC-classification needs 2 ms for every ROI, the tracking is performed in about 2 ms per object (we restrict the number of objects to five so that 10 ms for tracking is realistic) and finally the classification takes about 8-12 ms per object. As mentioned before the classification has not to be calculated for every frame. The organization of the global data representation needs 1 ms per frame. This system is quite capable to obey the real time requirements but the processing cannot cope with all different scenarios and we have restricted this application to extra-urban roads and motorways. Nevertheless if the performance of the hardware components will increase the presented overall system is able to cope with most of the scenarios even in more complex situations.

References

1. S. Bohrer, T. Zielke und V. Freiburg. An Integrated Obstacle Detection Framework for Intelligent Cruise Control on Motorways. In *Proceedings of the Intelligent Vehicles Symposium, Detroit*, Seite 276–281, 1995.
2. M.E. Brauckmann, C. Goerick, J. Groß und T. Zielke. Towards all around automatic visual obstacle sensing for cars. In *Proceedings of the Intelligent Vehicles '94 Symposium, Paris, France*, Seite 79–84, 1994.

3. A. Broggi. A Massively Parallel Approach to Real-Time Vision-Based Road Markings Detection. In *Proceedings of the Intelligent Vehicles '95 Symposium, Detroit, USA*, Seite 84–85, 1995.
4. E.D. Dickmanns et al. The Seeing Passenger Car 'VaMoRs-P'. In *Proceedings of the Intelligent Vehicles '94 Symposium, Paris, France*, Seite 68–73, 1994.
5. ELTEC Elektronik GmbH, Mainz. THINEDGE-Processor for Contour Matching. Hardware Manual, Rev. 1A, 1991.
6. M. Finke und K.-R. Müller. Estimating A-Posteriori Probabilities Using Stochastic Network Models. In *Proceedings of the Summer School on Neural Networks, Bolder, Colorado*, Seite 276–281, 1993.
7. C. Goerick. Local orientation coding and adaptive thresholding for real time early vision. Internal Report IRINI 94-05, Institut für Neuroinformatik, Ruhr-Universität Bochum, D-44780 Bochum, Germany, Juni 1994.
8. C. Goerick, D. Noll und M. Werner. Artificial Neural Networks in Real Time Car Detection and Tracking Applications. *Pattern Recognition Letters*, 1996.
9. U. Handmann und T. Kalinke. Fusion of texture and contour based methods for object recognition. In *ITSC'97, IEEE Conference on Intelligent Transportation Systems 1997*, Boston, 1997. IEEE. Session 35: Intelligent Vehicles: Vision(3).
10. U. Handmann, T. Kalinke, C. Tzomakas, M. Werner und W. v. Seelen. Computer vision for driver assistance systems. In *Proceedings of SPIE Vol. 3364*, Orlando, 1998. SPIE. Session Enhanced and Synthetic Vision 1998.
11. U. Handmann, G. Lorenz, T. Schnitger und W. v. Seelen. Fusion of different sensors and algorithms for segmentation. In *IV'98, IEEE International Conference on Intelligent Vehicles 1998*, Stuttgart, 1998. IEEE.
12. U. Handmann, G. Lorenz und W. von Seelen. Fusion von Basisalgorithmen zur Segmentierung von Strassenverkehrsszenen. In *Mustererkennung 1998*, Heidelberg, 1998. Springer-Verlag.
13. Robert M. Haralick, K Shanmugan und Its'hak Dinstein. *Textural Features for Image Classification.* IEEE Transactions on Systems, Man and Cybernetics, 1973.
14. J.A. Hertz, A.S. Krogh und R.G. Palmer. *Introduction to the Theory of Neural Computation.* Addison Wesley, 1991.
15. K. Hornik, M. Stinchcombe und H. White. Multilayer Feedforward Networks are Universal Approximators. *Neural Networks*, 2:359–366, 1989.
16. D.P. Huttenlocher. Comparing Images Using the Hausdorff Distance. *IEEE Transactions on PAMI*, 15(9), September 1993.
17. T. Kalinke und C. Tzomakas. Objekthypothesen in Verkehrsszenen unter Nutzung der Kamerageometrie. Internal Report IRINI 97-07, Institut für Neuroinformatik, Ruhr-Universität Bochum, D-44780 Bochum, Germany, 1997.
18. T. Kalinke und W. von Seelen. Entropie als Maß des lokalen Informationsgehalts in Bildern zur Realisierung einer Aufmersamkeitssteuerung. In *Mustererkennung 1996*, Seite 627–634, Heidelberg, 1996. Springer-Verlag.
19. T. Kalinke und W. von Seelen. Kullbach-Leibler Distanz als Maß zur Erkennung nicht rigider Objekte. In *Mustererkennung 1997*, Seite 501–508, Heidelberg, 1997. Springer-Verlag.
20. T. Kalinke und W. von Seelen. Kullback-Libler Distanz als Maß zur Erkennung nicht rigider Objekte. In *Mustererkennung 1997*, 1997.
21. H. Mori und N. M. Charkari. Shadow and Rhythm as Sign Patterns of Obstacle Detection. In *International Symposium on Industrial Electronics*, Seite 271–277, 1993.

22. D. Noll. *Ein Optimierungsansatz zur Objekterkennung.* Nummer 454 in Fortschrittberichte, Reihe 10. VDI-Verlag, Düsseldorf, 1996. Dissertation, Ruhr-Universität Bochum.
23. D.W. Paglieroni. Distance Transforms: Properties and Machine Vision Applications. *CVGIP*, 54(1):56–74, January 1991.
24. D. Pomerleau. RALPH: Rapidly Adapting Lateral Position Handler. In *Proceedings of the Intelligent Vehicles '95 Symposium, Detroit, USA*, Seite 506–511, 1995.
25. T. Schnitger und U. Handmann. Fusion von Bildanalyseverfahren mittels einer neuronalen Kopplungsstruktur. Internal Report IRINI 98-01, Institut für Neuroinformatik, Ruhr-Universität Bochum, D-44780 Bochum, Germany, April 1998.
26. W. von Seelen et al. Image Processing of Dynamic Scenes. Internal Report IRINI 97-14, Institut für Neuroinformatik, Ruhr-Universität Bochum, D-44780 Bochum, Germany, Juli 1997.
27. M. Werner und W. v. Seelen. Using order statistics for object tracking. In *Proceedings of the Intelligent Vehicles Symposium, Boston*, Seite 323, Digest 1997.

Image Analysis for Micro- and Macrostructure Measurement

H.J. Tiziani

Institut für Technische Optik

Universität Stuttgart

1. INTRODUCTION

Interferometric metrology, confocal microscopy, triangulation and Moiré techniques are well introduced. Furthermore, the detection of defects can be a powerful tool. Computer based analysis is essential as well as new technology and components. The market for shape, defects, deviation or position measurement is continuously growing. High frame rate and high flexibility are needed. Matrix liquid crystal displays (LCD), both electrically or optically addressed are able to display gray values. Digital micromirror devices (DMD) are a basis to lead to a couple of new applications. LCDs enable a instantaneous adjustment of the projection brightness. Both confocal microscopy and white light interferometry are image plane localisating methods. These are of advantage for high resolution microscopic and macroscopic measurements and to determine the shape of rough technical surfaces. Optically addressed crystals enable new possibilities in analog real time image processing. They are able to detect defects in periodic media and microstructures just by using polarization properties. This is of major interest in microelectronic manufacturing today and even more in the near future.

2. FRINGE PROJECTION FOR 3D-SHAPE MEASUREMENT OF TECHNICAL OBJECTS

A number of techniques have been developed for micro- and macrotopography measurements. The choice of an appropriate technique depends on the application, required sensitivity and resolution. Moiré and projected fringe techniques are not as sensitive as interferometric and holographic techniques but are less susceptible to environmental disturbencies.

Projected fringes techniques as an extension of triangulation for out-of-plane topography measurements are well established. Projected fringe patterns can be formed by different methods such as projecting either a grating-like structure or an interference pattern onto the object. Height variations or object deformations lead to a modification of the projected fringes, which in turn are compared with the original or synthetically generated patterns. The image processing for the fringe analysis is well known. Phase shifting techniques are frequently used. Typical contour-line separations can vary between micrometers and millimeters.

For fringe projection, the grating like structure can be generated by means of electrically adressed liquid crystal cells or digital micro mirrors. The fringe spacing can be adjusted in a wide range. In an interferometric setup, the fringe spacing is adjusted by varying the angle of the interfering waves. This is important for extending the unambiguity range as well as for absolute measurements of optically rough surfaces. Gratings with different periods are needed to extend the unambiguity range. A fine grating structure is required for high resolution and a coarse for extending the unambiguity range. Speckles can limit the resolution of the technique.

The technique is established for different technical and biological applications [1, 2, 3]. A new approach for exact object positioning of non-cooperative objects is based on correlation analysis of fringe patterns projected onto 3D objects [4].

A very promising approach for fast examinations of shape and shape deviation is the use of object-adapted fringes together with Moiré techniques.

2.1. Shape adapted Fringe pattern

For fast shape analysis, shape-compensating fringes can be projected. From the shape data of a master object, given by CAD data or from the measurement of a reference object, a distorted fringe pattern ist computed and transfered to a two-dimensional spatial light modulator (SLM) or micromirror system. After the projection of the shape adapted fringes onto the test-object straight aquidistant fringes are obtained on the detector, a CCD chip [3, 5].

In our experiments we used an electrically adressed liquid crystal display (LCD) as well as a digital micromirror device (DMD). The setup includes object illumination, object positioning, object detection, and process control (figure 1).

For a perfect object shape straight and equidistant fringes should be observed and can be compared with a reference grating or the CCD camera chip array in the detector plane. Each shape deviation or defects or a wrong object position results in distorted fringes. The shape deviation can be seen as a moiré fringe pattern leading to a fast shape or defect analysis. The moiré pattern obtained is a direct measure of the shape deviation. No further digital image processing is required. With this scheme, shape control is possible in real time.

Furthermore with a grey shade SLM the local object brightness can be compensated, which is important for shape and position detection.

For shape measurements and comparison used for defect analysis the well-known phase shifting methods can be applied. In our setup, instead of shifting a sinusoidal grating mask, the actual sinus pattern is written into the matrix SLM. No moving parts are required. For absolute shape deviation measurements referring to a synthetic master object the quality of calibration is essential. There are many different methods for calibration of imaging systems (calibration marks, a sequence of vertical and horizontal lines or edges).

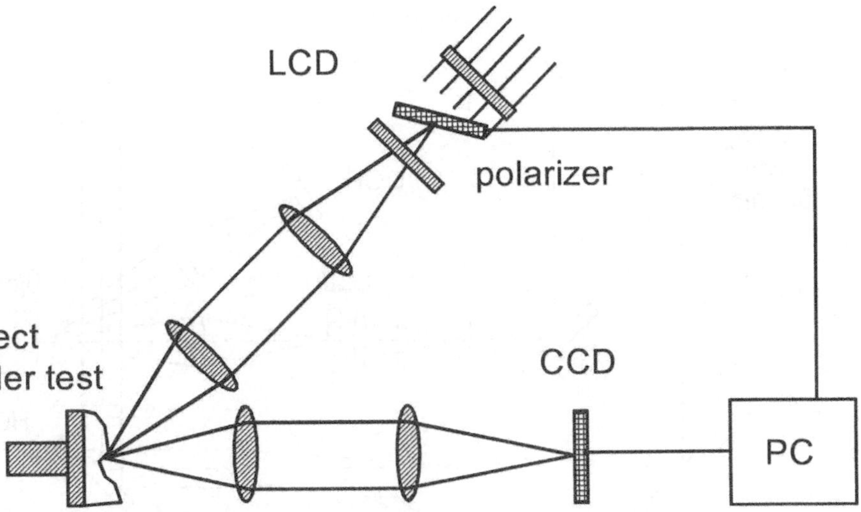

Figure 1: Setup for shape-adapted fringe pattern evaluation

In addition to the visual interpretation of the Moiré fringes, quanitative fringe analysis can be performed. Fourier analysis of the shifted fringe pattern can be used to calculate a shape deviation.

Figure 2 shows an object adapted fringe mask (left) and an image of the mask when projected onto a convex shaped turbine wing. Straight lines are observed.

In addition, holographic projection [5] under grazing incidence leads to an increased sensitivity (figure 3).

The object-adapted fringes are projected coherently onto the master object and a hologram is taken. The orientation of the hologram is almost perpendicular to the test surface. If the hologram is reconstructed, grazing incidence of object-adapted fringes

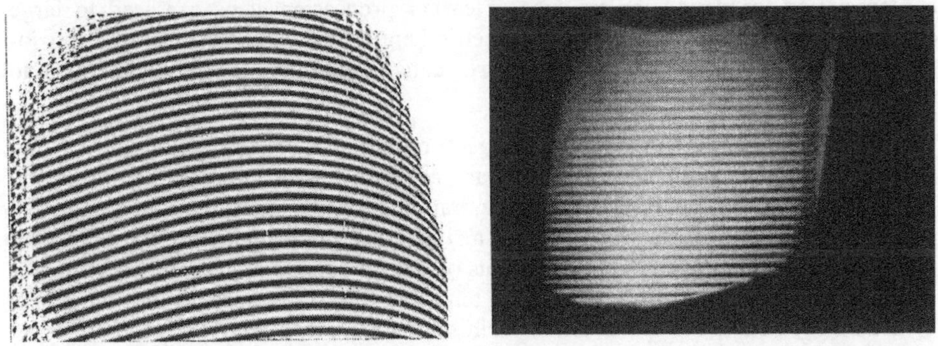

Figure 2: An object adapted fringe mask (left) results in straight lines when projected onto a convex shaped turbine blade

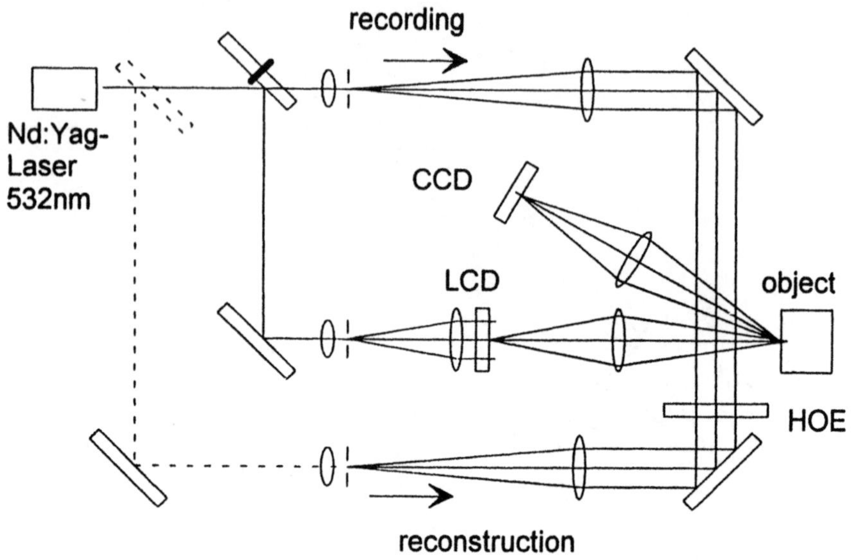

Figure 3: Holographic recording and projection of object-adapted fringes

can be realized. Grazing holographic projection allows for higher triangulation angles and therefore increased sensitivity. Even small shape deviations can be detected and quantitatively evaluated with fast algorithms. The method can be extended to multiple exposed holograms to examine deep objects.

3. SURFACE CHARACTERISATION BY ELLIPSOMETRY

Ellipsometry is appropriate for material analysis. Furthermore. in optical topometry different materials of the surface lead to different phase changes of the reflected waves and to erroneous steps in the topometric heights. Filmlike structures, deposited at will as in the case of micro-lithography or irregular local structures of contamination, oxidation, or residues of earlier processing steps can lead to large errors. To control the errors of the topometric height from fringe projection, Ellipso-height Topometry EHT uses ellipsometric data in addition to the height data to overcome these restrictions.

Figure 4 shows the optical arrangement of the EHT. The principles of our microscopic fringe projection MFP [6] are combined with ellipsometric principle, both acting onto the same imaging configuration given by the microscope objective MO and the CCD- camera which stay in a fixed position in relation to the object for the whole measuring cycle. Thus the points of ellipsometric frames and height frames match exactly. For the height H(x,y) the binary grating **G** is imaged onto the object by the mirror **M** and the spatial filter **SF** in such a way, that the 1st. 0th, and -1st order of the grating spectrum is passed and is positioned exactly in the back focal plane **F** of **MO** but laterally displaced by the distance **d**. In the object space, the projection of the grating is oblique with an angle of incidence ϕ_0 = arcsin(d/f), where f is the focal length of **MO**. The grating image is parallel to the object plane and without distortion for aberration free imaging. This renders calibration easy.

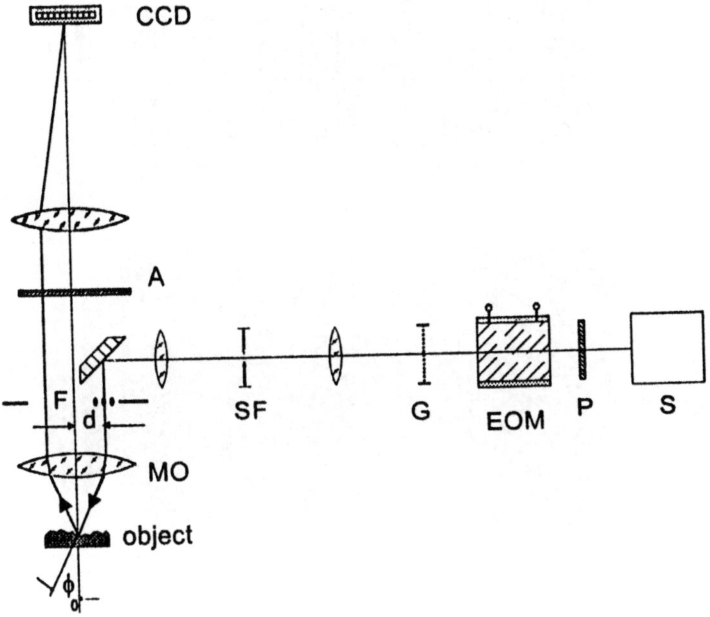

Figure 4: Optical arrangement for Ellispo-Height-Topometry

For ellipsometry the grating is moved out of the beam and a voltage U is applied to the electro optic modulator **EOM**. The light is linearly polarized by the polarizer **P**. The azimuth angles for **P**, fast axis of the **EOM,** and analyser **A** are 45^0, 0^0, and 45^0 respectively. Using Jones calculus the expression for the relative intensity on the camera can be derived [7].

Figure 5 shows typical results for the **EHT** fields: namely the ellipsometric angles $\Psi(x,y)$ and $\Delta(x,y)$. The object field on the microchip is 128 µm x 86 µm, the wavelength of the light was 436 nm. Lateral resolution δ_{lat} < 1 µm. The angle of incidence ϕ_0 for the ellipsometric data is 52° and can be varied in the range of $18° < \phi_0 < 65°$. Assuming aluminum with a thin oxide layer and a rms- microroughness of about 15 nm as material for the conducting strips of the microchip, (see figure 5)., for comparison the theoretical values were calculated from thin film theory and result in $\Psi = 42°$ and $\Delta = 160°$. This can be verified by examination of the topographies with the aid of the greybar scaled for each topography.

The material composition even of complicated structures can be verified and valuable additional information can be gained from the ellipsometry.

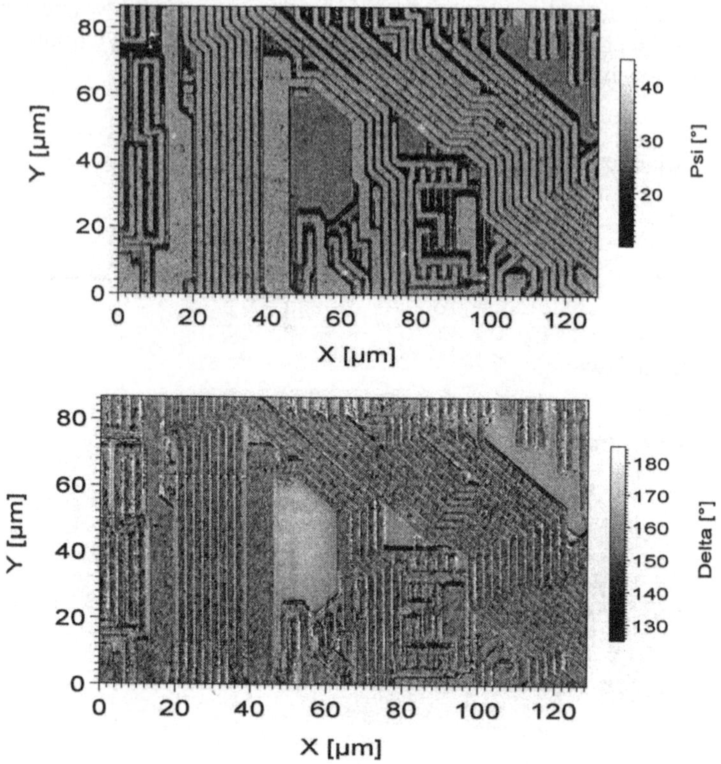

Figure 5: Ψ (top) and Δ of a chip structure

4. IMAGE PLANE LOCALISATION BY CONFOCAL MICROSCOPY AND WHITE LIGHT INTERFEROMETRY

4.1. Confocal microscopy

In recent years the technique of confocal microscopy was established for biological or medicinical applications as well as for the analysis of smooth technical surfaces. It uses highly magnifying microscope objectives with high numerical apertures.

The main difference between confocal microscopy in comparison to classical light microscopy is, that exclusively light beeing reflected in the focal plane is transmitted whereas light from defocused regions is supressed. This so called depth discrimination is obtained by a conjugate arrangement of a point light source, the focused spot on the specimen and a detector pinhole.

Light reflected from the exact focus position is imaged into the detector pinhole and therefore the light flux passing the pinhole to the detector has a maximum. Light reflected from defocused positions is distributed on an area around the pinhole and the light flux reaching the detector is reduced.

The depth response in axial direction is proportional to a $sinc^2$ - function. Significant for the depth resolution is the Full Width at Half Maximum (FWHM) of the signal. The smaller the FWHM is the higher is the axial resolution.

A new method for high resolved surface metrology has been developed at the ITO using microlens arrays in combination with confocal microscopy [8, 9]. One method works in the microscopic range on measurement areas of 100 µm to 1 mm in square. In this case the height resolution is about 10 nm, the lateral resolution 1 µm.

The topographic values are obtained taking an amount of camera frames while the specimen is scanned in axial direction. The height values are calculated by determining the center of gravity of the axial intensity response in each pixel.

Another method is working in the macroscopic range. The extended measurement field is detected by a large array consisting of a few thousends of microlenses. Each microlens acts in connection with the imaging lens as a single confocal system. The method combines the advantage of the high numerical aperture of the single microlens with the possibility of parallel scanning on large areas up to a few centimeters in square. The height resolution in this case is 500 nm.

A wide variety of technical surfaces as well as microchip structures can be measured. Furthermore the 3D-roughness of different metal sheets can be determined. The method can also be adapted to the measurement of curved surfaces (figure 6) [10].

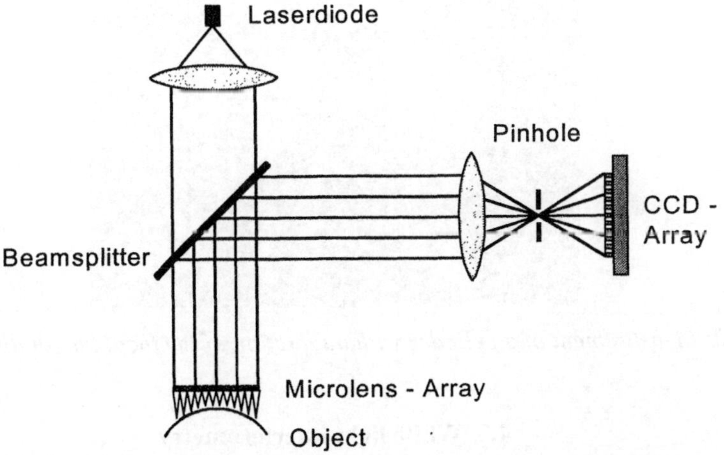

Figure 6: Confocal setup for the measurement of curved surfaces.

Detailed investigations have been made concerning the comparison of the results with tactile stylus instruments. Using several roughness standards a very good coincidence between the confocal measurements and calibrated stylus measurement executed by the PTB (Physikalisch Technische Bundesanstalt, Braunschweig) was obtained. The deviation of calculated roughness parameters was in the order of less than 1%.

Fig. 7 shows an example on a photoresist microstructure. The field size is 226 μm x 231 μm.

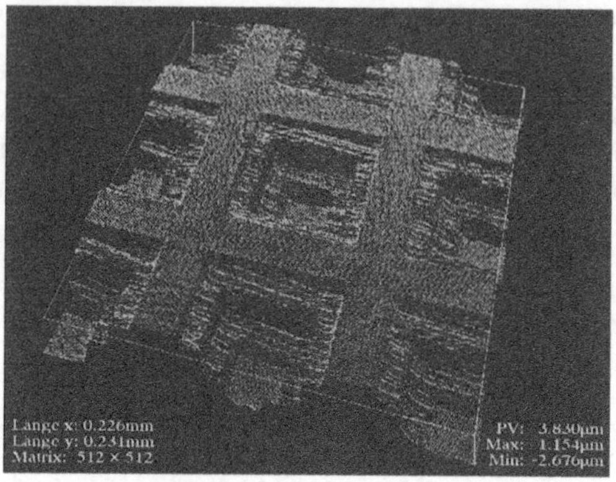

Figure 7: Topography of a photoresist microstructure

A measurement on a curved surface is shown in Fig. 8. The focal length distribution of the microlenses is adapted to the shape of a cylinder. Macro- and microdefects of the cylinder surface can be measured.

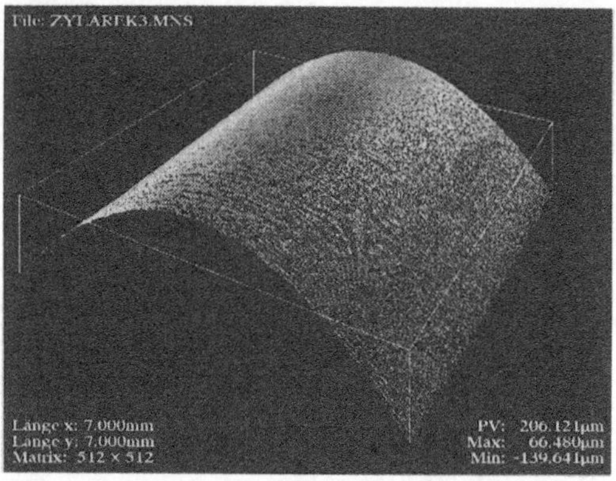

Figure 8: Measurement of a cylinder with adaptation of the focal length distribution

4.2. White light interferometry

A similiar method as the confocal microscopy for image plane localisation is the white light or short coherence interferometry [11]. Classical interferometry can not be used to measure optically rough surfaces. In addition the limited unambiguity range of half a wavelength needs to be taken into account. A light source with a very short coherence length is used. This means that interference only occures if the interferometer is nearly equally balanced. Thus, the occurence of interference is an indicator for the image plane localisation. And the longitudinal measuring range is only limited by the free working distance of the sensor.

The favorite light source for this kind of interferometry are spectral broadband light sources like halogene bulbs, Xe-lamps, superluminescence diodes or LED's with coherence lengths between 1 μm and 30 μm. To achieve a high resolution the coherence length should be as small as possible.

Three dimensional topography measurements are obtained by sequentially changing the optical path difference in a two beam interferometer. A series of frames is acquired for example by a CCD camera and analyzed by an appropriate hard- and software. Each pixel records a signal as is shown in figure 9.

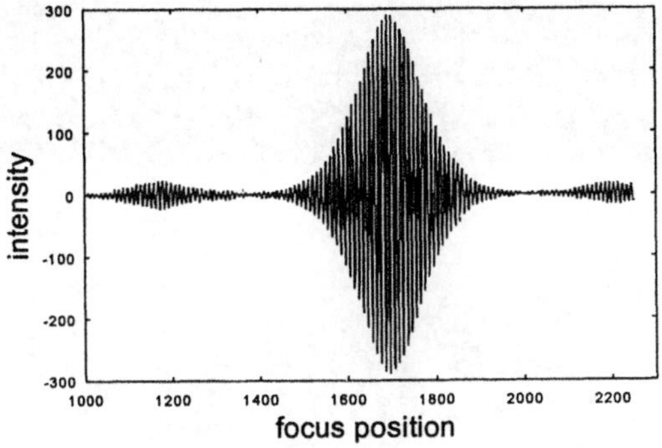

Figure 9: Interference signal for in dependency of the optical path difference.

The aim of the evaluation is the demodulation of the signal to get the maximum of the envelope which indicates a optical path difference of zero. In contrast to the fringe projection technique, the vertical resolution on optical smooth surfaces is independent of the field of view. An extended microscopic set-up was developed covering a field of view ranging from 88 μm × 66 μm up to 14.4 mm × 10.5 mm, which is a zoom range of more than 160×. Thus, microscopic as well as macroscopic measurements can be done in the same setup.

The most serious drawback of the method is the large amount of data that have to be analyzed to get a topography. Typically between 200 MByte and 1 GByte have to be acquired and calculated. Powerful hardware is needed to get fast measurement rates of 0.3 to 4 micrometer per second. In [11, 12] other methods are investigated to achieve high speed measurements up to 100 μm per second.

Not only the fringe contrast can be used for the calculation of the topographic map. Also the phase gives valuable informations about the surface if it is an optical smooth specimen. This means that the vertical resolution can be increased up to 1 nm in these cases.

Figure 10 depicts two measuring examples. One diagram shows a microscopic image of a super fine roughness standard with a mean roughness of 55 nm. The field of view is 370 μm × 275 μm. The other picture shows the topography of a computer generated hologram (CGH). The field of view is 3.7 mm × 2.8 mm.

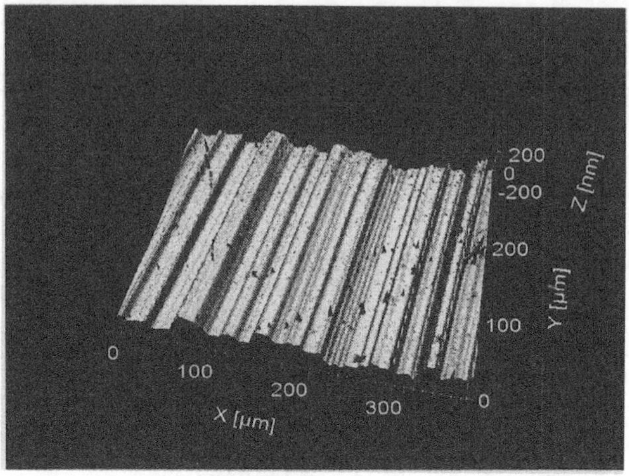

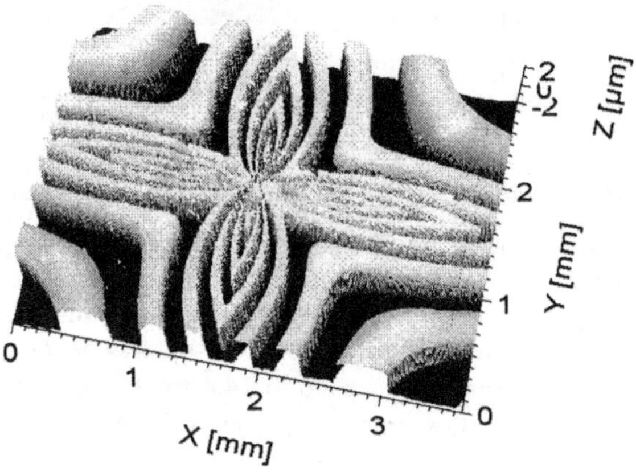

Figure 10: (top) Topography of a super fine roughness standard with a mean roughess of 55 nm.

(bottom) Topography of a computer generated hologram.

Measurements on different engineering surfaces show a good agreement between both image plane localisating methods, the confocal microscopy and the short coherence interferometry. The difference between calculated roughness parameter is usually less than 20 %.

One example of a roughness standard measured with the confocal microscope, the white light interferometer and a tactile stylus sensor is shown in figure 11. Both optical methods measure the shape of the roughness standard in a good conformity.

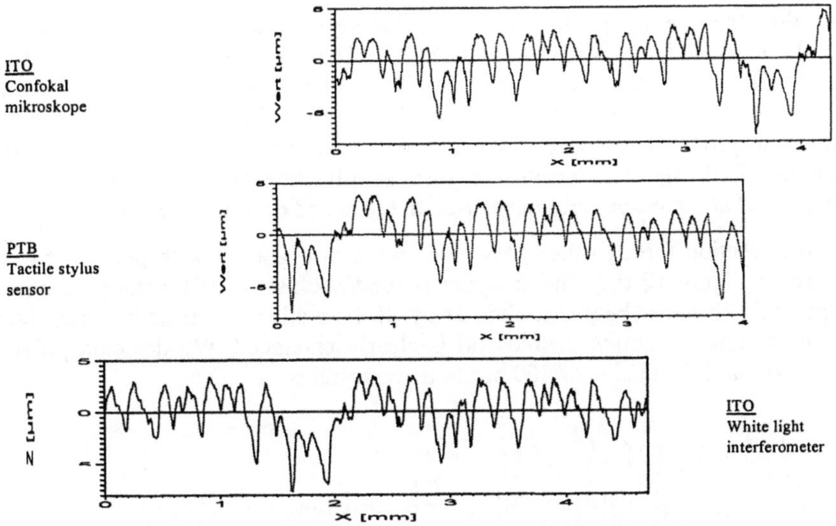

Figure 11: Profile of a roughness standard measures with a confocal microscope, a white light interferometer and a tactile stylus sensor.

5. DEFECT ENHANCEMENT IN PERIODIC MICROSTRUCTURES USING OPTICALLY ADDRESSED CRYSTALS

Defect detection in periodic structures is very important for industrial application in microelectronics today and even more in the near future. If a wafer with diameter 150 mm needs to be inspected with a resolution of 0.5 µm, any quality control has to inspect the enormous number of $2.8*10^{11}$ pixels. Optical inspection methods has advantage compared to digital image processing, because it can use real-time, whole field filter processes.

Self-filtering [13, 14] is able to detect defects by changing the polarisation state of the requested signal, the defect. The filter combines the basic ideas of spatial frequency filtering with the possibilities of modern non-linear optics. Self-filtering means, that the position or orientation of the object, its Fourier transfrom and thus even the object itsself may be unknown for a successful defect enhancement. The filter is always adapted to the object. No part of the setup has to be removed or needs to be adjusted, if a new object is to be inspected.

A typical object to be inspected can be a grating with a single defect, may be dust particles. If this image is Fourier transformed, then the transform consists of intense and discrete peaks due to the decomposition of the periodic structure (grating) and a smeared distribution of low intensity due to the transform of the pointlike structure (defect). In the Fourier plane these two parts are seperated in space, so they can be seperated by a spatial filter. If the intense peaks are blocked and the manipulated spectrum is transformed back into the image space, only defects remain visible.

To filter the intense peaks we apply photorefractive crystals (PRC) as well as optically addressed liquid crystals (OASLM). These elements are positioned in the Fourier plane of an optical processor. Nonlinear birefringence in the crystals leads to a seperation of a periodic structure from defects [13]. This occurs by the fact, that polarization properties between strong peaks and a weak defect signal are different (linear or elliptical). The peaks are then simply attenuated by a linear blocking polarizer and at the output an image with an enhanced defect is generated.

A cross section through the grey-scale image of a grating with period of 6 μm is depicted in Figure 12 (a). The analyzer is rotated close to its blocking position (b). The periodic structure begins to fade away. If the analyser is in its blocking position (c), only a defect remains visible and is clearly enhanced. We detected polystryrol spheres with a defined size of 800 nm in these gratings.

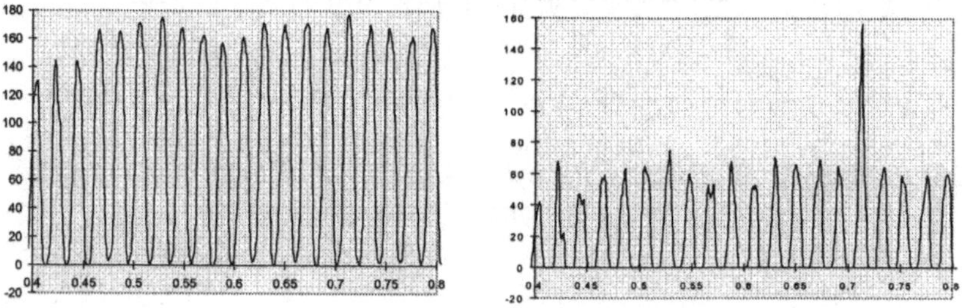

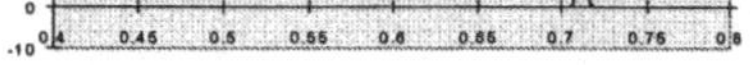

Figure 12 a, b, c (cw): Defect detection in a microscopic grating

Defects in objects with a rotational symmetry (axicons) can be enhanced as well. Non-transparent objects are of special interest. Here information must be extracted from the reflected light. Typical objects to be inspected are CCD-arrays and memory chips [13].

In industrial environment video images are preferred for inspection systems, because many objects can not be examined by laser radiation. Thus to examine video images in real-time, the video image of a periodic object to be inspected is displayed on a liquid crystal display (LCD), which modulates a coherent read-out beam. Thus the LCD acts as a coherence converter for the video image. The modulated coherent beam is Fourier transformed, filtered from all periodic components, and is Fourier transformed again [14]. Figure 13 (a) shows a test object with a periodic structure with and a defect written on the LCD. In figure 13 (b) all periodic elements are surpressed and the defect becomes visible. In both images the defect has been enlarged.

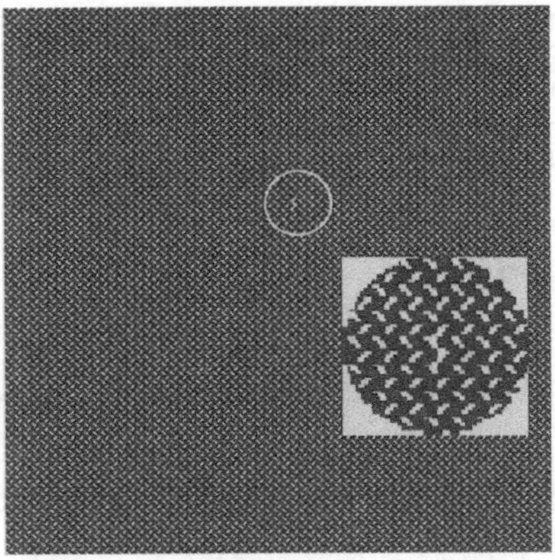

Figure 13a : Structured image with defect (enlarged)

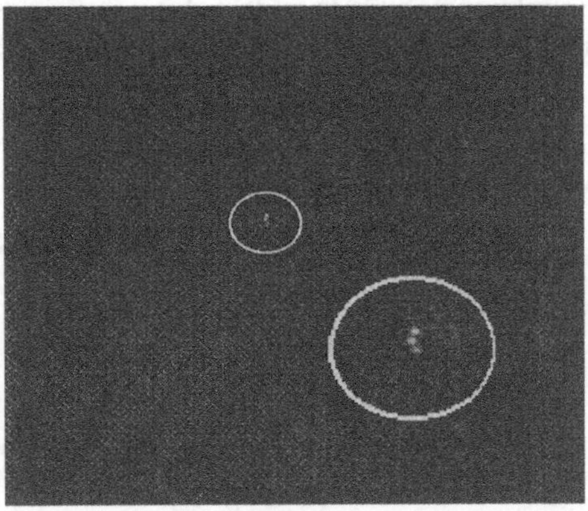

Figure 13 b: Enhanced image with enlarged detail

That the technique is interesting for industrial applications was also demonstrated by a solar cell with a specially designed grid for application in concentrating photovoltaics [14].

Real-time defect detection can be applied for directly illuminated objects as well as on video images based on spatial frequency self-filtering for industrial applications. The approach is interesting for industrial applications due to its very simple self-adjusting setup and high signal-to-noise ratio. Coherent and incoherent light source can be used. OASLM show some advantages against PRC. They will be manufactured with increasing size and enhanced quality in the near future and will be an attractive component for real-time image processing.

6. CONCLUSION

Fringes, adapted to the object, have proven to be of great interest for fast and flexible comparison of objects. If inverse Moiré technique is used, a shape deviation is compared with a regularly grating, which is independent of the object itself. Thus a shape comparison can be performed by simple and fast algorithms. LCDs und DMDs have been applied to determine the object´s topography using structured illumination.

Ellipso-Height-Topometry can be useful to obtain material properties and for correction of the topography of contaminated surfaces.

Confocal microscopy using micro-lens array enables the possibility to measure large areas with best accuracy. It is possible to choose a free working distance if a transfer optic is introduced in the setup. We demonstrated, that micro-lenses may be adapted to the basis geometry of the object in order to reduce measurement time.

White light interferometry enables interferometric measurements in a large spatial as well as longitudinal range. The results on engineering surfaces are very good comparable to the results obtained with confocal microscopy and to reference measurements aquired with tactile stylus sensors.

Defect enhancement using optically addressed crystals is a promising technique of analog image processing. The principle of spatial self-filtering enables the inspection of unknown objects with arbitrary orientation. A single-beam setup and a good signal-to-noise ratio are attractive features for industrial application.

The author acknowledges the financial support by the BMBF and DFG and wants to thank his coworkers K. Leonhardt, M. Schönleber, D. Steudle, E.U. Wagemann, M. Wegner and R. Windecker.

6. LITERATURE

1. Tiziani, H.-J.: "Optical metrology of engineering surfaces: scope and trends," in *Optical Measurement Techniques and Applications*, P.K. Rastogi, ed. (Artech House, Boston, 1997).

2. Windecker, R., Tiziani, H.-J.: „Topometry of technical and biological objects by fringe projection", *Applied Optics*, Vol. 34, No. 19, July 1995, pp.3664

3. Schönleber, M., Tiziani, H.-J.: „Fast and flexible shape control with adaptive LCD fringe masks", *Proc. SPIE 3098*, 35-42 (1997)

4. Haist, T., Tiziani, H.-J.: „Fast positioning of non-cooperative objects", *Proc. SPIE 3098*, 424-430 (1997)

5. Schönleber, M., Wagemann, E.U., Tiziani, H.J.: „Shape deviation measurement with enhanced depth using adapted fringe projection and multiple exposed holograms", *Proc. ISATA*, 2. - 5. June 1998, Düsseldorf

6. Leonhardt, U. Droste and H.J. Tiziani: „ Microshape and rough surface analysis by fringe projection", *Applied Optics,* 33 (1994), 7477-7488.

7. Leonhardt, K., Jordan, H.-J., Tiziani, H.-J.: „Micro-Ellipso-Height-Profilometry", *Optics Communications*, 80 (1991), 205-209.

8. Tiziani, H.-J., Achi, R., Krämer, R.N.: "Chromatic confocal microscopy with microlenses". *Journal of Modern Optics* 43 (1996) 155-163

9. Tiziani, H.-J., Achi, R., Krämer, R.N., Wiegers, L.: "Theoretical analysis of confocal microscopy with microlenses", *Appl. Opt.* 35 (1996) 120-125

10. Hessler T., Rossi M., Pedersen, J., Gale M. T., Wegner, M., Steudle, D., Tiziani H. J.: „Microlens arrays with spatial variation of the optical functions", Pure & *Applied Optics* **6**, pp. 673-681 (1997)

11. Windecker, R., Haible, P., Tiziani, H.J. „Fast coherence scanning interferometry for measuring smooth, rough and spherical surfaces", *Journal of Modern Optics*, 1995, Vol. 42, No. **10**, 2059 - 2069.

12. Windecker, R., Fleischer, M., Franze, B., Tiziani, H.J.: „Two methods for fast coherence tomography and topometry", *Journal of Modern Optics*, Vol. 44, Nr. 5, 1997, 967-977.

13. Wagemann, E.U., Tiziani, H.J.: "Spatial self-filtering using photorefractive and liquid crystals", *Journal of Modern Optics*, in print, 1998

14. Wagemann, E.U., Tiziani, H.J.: "Real time defect detection in video images", *Proc. OPTO 98*, 18. - 20. May 1998, Erfurt

Ein hochsprachenprogrammierbares System zur Vollbildauswertung im Videotakt; Anwendungen zur Interpretation monokularer, semi-strukturierter Bildfolgen bei natürlicher Beleuchtung und schnell bewegter Kamera

[1] Bayerische Motoren Werke Aktiengesellschaft (BMW),
Projektgruppe ELEKTRA, Abt. EW-1, D-80788 München, FRG
[2] Bodenseewerk Gerätetechnik GmbH (BGT),
Bereich FK-E, D-88641 Überlingen, FRG
[3] Fraunhofer Institut für Mikroelektrische Schaltungen und Systeme (IMS),
Finkenstr. 61, D-47057 Duisburg
[4] Fraunhofer Institut für Zuverlässigkeit und Mikrointegration (IZM),
Gustav-Meyer-Allee 25, D-13555 Berlin, FRG
[5] Institut für Neuroinformatik (INI),
Ruhr-Universität Bochum, Geb. ND-04/5868, D-44780 Bochum, FRG
[6] Lehrstuhl für Numerische Mathematik und Analysis (NMA),
Universität Passau, Innstr. 29, D-94032 Passau, FRG

Zusammenfassung Im vorliegenden Beitrag wird ein hochsprachenprogrammierbares System zur schritthaltenden Vollbild-Interpretation natürlich beleuchteter Szenenfolgen im Videotakt vorgestellt. Im einzelnen werden folgende Teilmodule und Subsysteme beschrieben:

- eine hochdynamische, pixellokal autoadaptive CMOS-Kamera mit ca. 120 dB Helligkeitsdynamik ($20 Bits/Pixel$)
- ein hochsprachenprogrammierbarer *Systolic Array* Prozessor (für die pixelbezogenen Verarbeitungsmodule) im PCI-Kartenformat, samt optimierendem Compiler, Simulator und Emulator
- Systemprozeßgerüste unter Linux auf den für die Echtzeit-Anwendungen eingesetzten Hostrechnern (z.B. DEC/Alpha oder Intel/Pentium)
- eine prototypische Anwendung zur bildverarbeitungsbasierten Eigenbewegungsbeobachtung (Translationsrichtung, Rotationsraten)
- eine prototypische, automotive Anwendung zur schritthaltenden Detektion und Kartierung des Straßen- und Spurverlaufs unter partieller monokularer 3D-Rekonstruktion, sowie
- prototypische Anwendungen zur Klassifikation verkehrsrelevanter Hindernisse (Verkehrsteilnehmer)

1 Einleitung und Übersicht

Im Rahmen des BMBF Förderprogramms "Elektronisches Auge" beteiligte sich die BMW AG ab Mai 1994 unter Wahrnehmung der industriellen Leitfunktion an einem F&E-Vorhaben mit der Zielsetzung: "Schritthaltende Interpretation von Fahrszenen mit Anwendungsschwerpunkt Fahrerassistenz" [1]. Der vorliegende Beitrag gibt eine Übersicht über die erzielten Ergebnisse.

Die schritthaltende Bildfolgeninterpretation zur Fahrerassistenz stellt gegenüber anderen Anwendungen aus dem Gebiet der *Computer Vision* eine Reihe erhöhter Anforderungen an die eingesetzten Technologien und die algorithmisch-verfahrenstechnischen Grundlagen, nämlich:

- die Bewältigung szenischer Inhalte extremer bildlokaler Helligkeitsdynamik
- große Bildwiederholraten zur Lösung der Korrespondenzprobleme auch bei hoher Dynamik der Sensor-Eigenbewegung (Geschwindigkeiten bis 250 km/h und bis zu 1 g Querbeschleunigung)
- Tiefenvarianz von annähernd zwei Größenordnungen (5 - 200 m) bei der niedrigen Kameraeinbauhöhe von nur ca. 1 m
- Einhaltung einer Gesamtlatenz der Verarbeitungskette von weniger als 100 ms
- große Variabilität der szenischen Inhalte aufgrund der (durch die Straßen-Infrastruktur) nur teilstrukturierten Umfelder
- erhebliche Zielkonflikte unter den Aspekten Rechenleistung, (Herstellungs-) Kosten, Umweltverträglichkeit (Temperatur, EMV), elektrischer Verlustleistung und Sicherheitskritikalität (Produkthaftpflicht).
- Erfordernis einer vollflächigen Bildauswertung bezüglich der sensornahen und intermediären Verarbeitungsschichten zur system- und kostenbedingten Vermeidung mechanischer Blickrichtungs- und Aufmerksamkeitssteuerung
- hohe Flexibilitätsanforderungen auch an die ikonischen Verarbeitungsprozesse; daher: Voll programmierbare Entwicklungsplattform um Hardware-Entwicklungszyklen von algorithmischen Verfahrensverbesserungen zu entkoppeln und Prozeßwechsel im Fahrbetrieb zu ermöglichen.

Aufgrund der genannten Anforderungen und Zielkonflikte waren die überwiegenden Umfänge des Projektes der Entwicklung geeigneter Technologien und Grundlagenverfahren gewidmet. Die Einzelergebnisse der verschiedenen Hard- und Software-Teilprojekte werden in den folgenden Abschnitten beschrieben.

2 Hochdynamischer, pixellokal autoadaptiver CMOS-Bildsensor (IMS)

In 1997 wurde am FhG/IMS ein CMOS-Bildsensorchip samt Kamera-Hardware realisiert. Die Anforderungen an eine Kamera für den Außenraum bei verschiedensten Beleuchtungsverhältnissen sind extrem hoch. Typischerweise können Einzelbilder gleichzeitig stark unterschiedliche Helligkeitsbereiche mit reflektiertem Sonnenlicht und abgeschatteten Regionen enthalten. Die blendfreie Aufnahme von Nachtszenen bei entgegenkommenden Verkehr stellt eine weitere Anforderung dar, die bewältigt werden mußte. Um sicherzustellen, daß selbst bei solch schwierigen Beleuchtungssituationen der nachgeschalteten Bildauswertung alle wesentlichen in der Szene enthaltenen Signaturen verfügbar gemacht werden können, mußten neue Konzepte für die Bildsensorkomponente und die Kamera-Hardware erarbeitet und implementiert werden.

2.1 Bildsensor-Komponente

Bei der prototypisch realisierten Bildsensorkomponente handelt es sich um einen Standard-CMOS-Chip in einer $1 \mu m$-Technologie. Als lichtempfindliche Bauelemente wurden (wegen ihrer guten Linearität und ihrem geringen Platzbedarf) Photodioden verwendet. Die Bildmatrix besteht für diese Erstversion aus 256 × 256 Pixeln. (Für Folgeversionen ist geplant, die Pixelzahl zunächst zu verdoppeln, dann zu vervierfachen.) Die einzelnen Dioden erzeugen nach einer pixellokal vorgegebenen Belichtungszeit ein helligkeitsproportionales Spannungssignal, welches von einer Ausleseschaltung am Randbereich der Bildmatrix aufbereitet und ausgelesen wird. Die Ausgangssignale aller lichtempfindlichen Elemente einer Zeile werden zeitgleich ausgewertet und können dann sukzessive ausgelesen werden. Um eine möglichst hohe Qualität der auf dem Chip gewonnenen Bildsignale zu erhalten, werden bei deren Aufbereitung im wesentlichen drei besondere Verfahren eingesetzt. Zunächst wird eine Doppelabtastung der Pixelausgangssignale vorgenommen, um Streuungen einzelner Pixel und niederfrequentes Rauschen zu unterdrücken. Weiterhin werden die Pixelausgangssignale je nach Signalwert unterschiedlich verstärkt, so daß ein möglichst großer Signal-Rausch-Abstand (SNR) der Bildsignale auch bei kleineren Signalen erreicht wird. Der erreichte maximale SNR ist größer als 56 dB. Ein limitierender Faktor für den Temperaturbereich ist der Dunkelstrom von Photodioden. Dieser ist jedoch bis zur Ecktemperatur von $+85^oC$ noch so klein, daß er sich in der Bildinformation nur geringfügig bemerkbar macht. Gleichzeitig liefert der Kamerachip ein Signal, welches proportional zur Chiptemperatur ist, so daß der Einfluß der Temperatur extern kompensiert werden kann. Die untere Ecktemperatur von -40^oC ist unkritisch, da CMOS-Schaltungen aufgrund des geringer werdenden Dunkelstroms bei niedrigeren Temperaturen ohnehin zunehmend besser funktionieren. Drittens wird jede Diode innerhalb einer Bildperiode von 20 ms bis zu viermal unterschiedlich belichtet (siehe [3]).

2.2 Kamerahardware

Die Kamera-Hardware steuert den Kamerachip dergestalt, daß jedes Pixel innerhalb einer Bildperiode von $\frac{1}{50}$ Sekunde entsprechend der Mehrfachbelichtung bis zu viermal ausgelesen wird. Werden die im Pixel erzeugten Spannungssignale zur Kamerahardware hin ausgelesen, so muß zunächst der A/D-gewandelte Spannungswert gespeichert werden. Hinzu kommt der von der Aufbereitungsschaltung gewählte Verstärkungsbereich und die von der Ausleselogik gewählte günstigste Belichtungszeit. Die Information stückweise linearer Kennlinien für unterschiedliche Verstärkungen und Belichtungszeiten wird durch die nachgeschaltete Logik in geeigneter Weise genutzt, um den endgültigen Helligkeitswert zusammenzusetzen. Durch die Möglichkeit der unterschiedlichen Belichtung kann man die Übersteuerung der Pixel bei zu großer Helligkeit weitestgehend vermeiden und die erreichbare Helligkeitsdynamik innerhalb eines Bildes auf nahezu 120 dB erweitern. Dies entspricht einer Anzahl von 1.000.000 darstellbaren Helligkeitswerten in einem einzigen Bild (20 $Bits/Pixel$). Im Vergleich dazu besitzen gute bis sehr gute herkömmliche CCD-Kameras eine Helligkeitsdynamik von 48 bis max. 72 dB, was etwa 256 bis 4.000 darstellbaren Helligkeitswerten entspricht (8 bis 12 $Bits/Pixel$).

2.3 Ergebnisse

Als Demonstrationsbeispiel wurde im Labor eine Szene mit großer Helligkeitsdynamik gestellt und mit CCD- bzw. CMOS-Sensor aufgenommen (vgl Abb. 1). Rechts unten im Bild strahlt eine Halogenlampe direkt in das Kameraob-

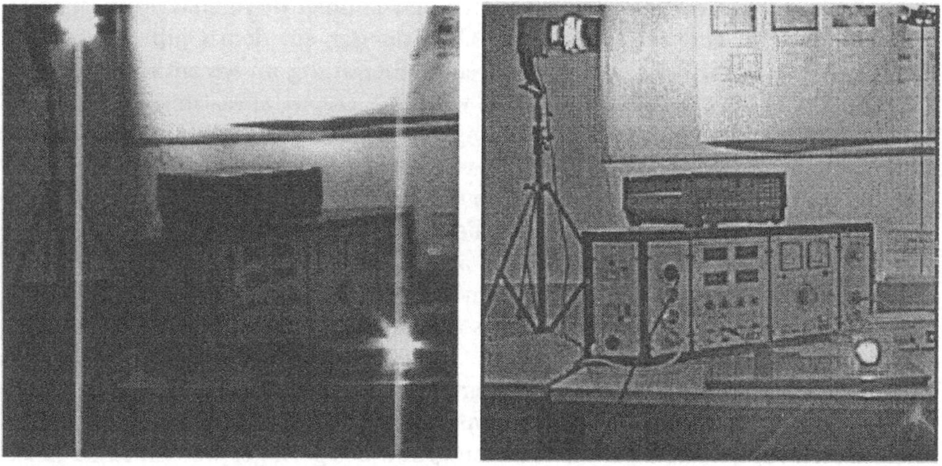

Abbildung 1. Vergleich zwischen herkömmlicher CCD-Kamera (linkes Bild) und dem realisierten hochdynamischen CMOS-Bildsensor (rechtes Bild)

jektiv. Am linken oberen Bildrand bestrahlt ein 1000W Scheinwerfer eine Posterwand. Diese Szene wurde einerseits mit der realisierten CMOS-Kamera und zum Vergleich mit einer handelsüblichen CCD-Kamera mit automatischer globaler Belichtungszeitsteuerung aufgenommen. Das Objektiv der CCD-Kamera muß nahezu geschlossen werden, um bei direkter Einstrahlung Totalblendungen zu vermeiden. Dies verhindert dennoch nicht übersteuerte Spalten (sog. smearing-Effekte), deren Ursache "überlaufende" Ladungsakkumulate im CCD-Kamerachip sind. Aufgrund der großen Helligkeitsdynamik der Bildszene werden viele Bereiche des CCD-Kamerachips übersteuert, während andere Bereiche völlig schwarz bleiben.

Die Blende der CMOS-Kamera ist dagegen vollständig geöffnet. Da der Helligkeitsdynamikbereich der CMOS-Kamera den auf einem Qualitätsmonitor verfügbaren Graustufenbereich von ca. 42 dB übersteigt, müssen die Informationen, die in der Bildszene enthalten sind, durch Hochpaßfilterung dem verfügbaren Graustufenbereich des Bildschirms angepaßt werden. Wie man leicht erkennen kann, enthält das Bild der CMOS-Kamera im Gegensatz zum CCD-Bild nahezu sämtliche Information der Szene.

3 Parallelrechnerhardware für die ikonische Bildverarbeitung (BGT)

Der Einsatz "klassischer" von-Neumann Prozessorarchitekturen ist trotz dramatischer Technologiefortschritte für anspruchsvolle, schritthaltend arbeitende Pixelverarbeitungsalgorithmen suboptimal, was zu einem schlechten Kosten/Nutzen-Verhältnis führt. Dies liegt daran, daß bei den typischen ikonischen Bildverarbeitungsprozessen in erster Linie eine große Anzahl von Operanden mit i.d.R. nur wenigen Bits Wertevorrat verrechnet werden müssen, die überdies für alle Pixel nach einheitlichen Rechenvorschriften zu bearbeiten sind. Derlei Verarbeitungsmerkmale befinden sich im Widerspruch zu der vergleichsweise geringen Operandenverfügbarkeit üblicher Hochleistungsprozessoren in Verbindung mit deren für charakteristische ikonische Verarbeitungsprozesse eigentlich zu breiten Datenpfaden von 32 bzw. 64 bit, sowie ihren Datencachestrategien, die bei typischen Bildmatrixoperationen zu unnötigen Belastungen der Hauptspeicher-Zugriffsbandbreite führen.

Die Verarbeitungsanforderungen an eine adäquate Pixelverarbeitungsmaschine legen eine Rechnerarchitektur nahe, die zum einen massiv parallele Rechenwerke nutzt und zum anderen operandengrößenspezifische Datentransfers und Rechenoperationen erlaubt (n-bit Lese-/Schreibzugriffe, 1×1-bit Logikoperation, $n \times m$-bit Addition, etc.)

Die im Projekt entwickelte Parallelrechnerhardware [2] erfüllt u.a. die oben genannten Anforderungen mit einer weiterentwickelten, hoch effizienten *SIMD*-Architektur (single instruction - multiple data). Die damit für beispielhafte ikonische Bildverarbeitungsprozesse erzielten Rechenzeiten sind je nach Anwendung um Faktoren 100 bis 10000 geringer, als bei herkömmlichen Prozessorarchitekturen (Vergleich 40MHz SIMD- und 50MHz SuperSparc-Prozessor).

Schlüsselkomponente der Rechnerstufe ist der von BGT in das Projekt eingebrachte *Systolic Array* Prozessorchip (SAP). Jeder SAP-Chip beinhaltet 64 parallel arbeitende Prozessorelemente (PE) mit zugeordneten on-chip Speicherstrukturen, die extern beliebig erweitert werden können. Zusätzlich verfügt der SAP-Chip über verschiedene, autonom arbeitende Datenpfade, die simultan zur Datenverarbeitung die Zuführung von Eingangsdaten und Entsorgung der Ausgangsdaten einschließlich Formatwandlung (bit-parallel / pixel-seriell in bit-seriell / pixel-parallel und umgekehrt) bewerkstelligen, sowie Nachbarschaftsoperationen in einer ± 4-Umgebung ohne zusätzliche Zyklen unterstützen.

Vier dieser SAP-Chips sind zusammen mit externer Speichererweiterung auf einem Aufsteckmodul (SAP-Slice) kaskadiert und realisieren eine parallel arbeitende Zeile von 256 Prozessorelementen (PE). Bis zu zwei SAP-Slices können

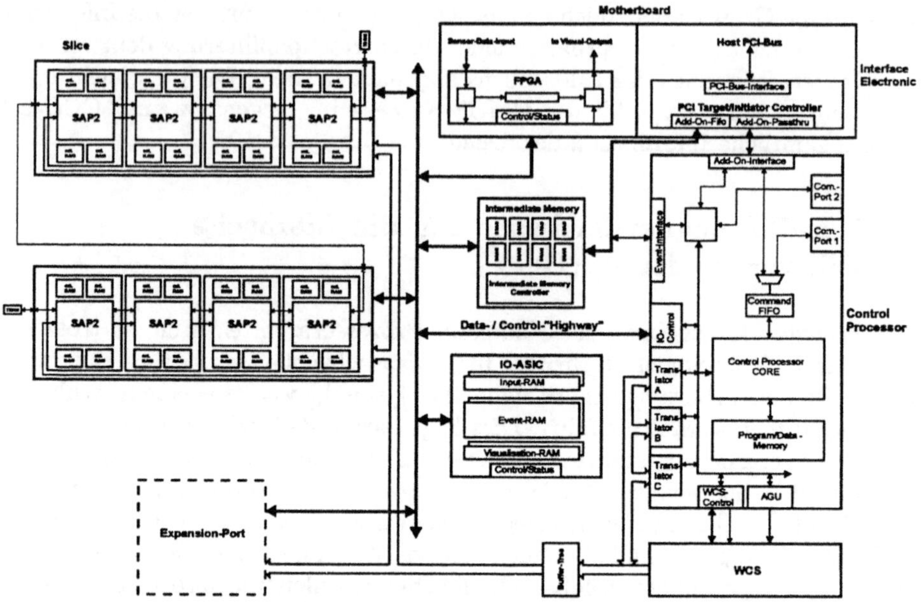

Abbildung 2. Blockschaltbild der Systolic Array Baugruppe

auf der SAP-Trägerplatine (belegt einen PCI-Slot) aufgesteckt werden und erlauben damit Konfigurationen von bis zu 512 PEs. Eine rückseitige Bestückung von zwei weiteren SAP-Slices ist vorgesehen (Konfigurationserweiterung auf 1024 PEs). Die zeilenförmige Aneinanderschaltung von PEs wird durch ein ausgefeiltes Speichermanagement ergänzt, das dem Anwender in Verbindung mit dem mächtigen Compiler die programmierte 2D-Verarbeitung von Bilddaten auf einfachste Art und Weise ermöglicht. Darüberhinaus erlaubt die statische und dynamische Speicherverwaltung bildübergreifende Operationen, sowie das Arbeiten mit Referenzbildern. Diese vielfältigen Funktionen zur Speicherverwaltung sind zusammen mit einer hoch flexiblen Ablaufsteuerung der SAP-Programme und einer dynamischen Parameterverwaltung auf einem einzigen Control-ASIC implementiert. Neben dem Control-ASIC umfaßt die SAP-Trägerplatine weitere Funktionsblöcke, die die Einbindung der Rechnerstufe in handelsübliche Workstations der neuesten Generation unter Berücksichtigung der jeweils projektspezifischen Anwendungsanforderungen erlauben.

Die 14-lagige SAP-Trägerplatine (vgl. Abb. 3) enthält außer dem Control-ASIC folgende weitere Funktionsblöcke:

- eine adaptierbare digitale Sensorschnittstelle
- das I/O-ASIC zur programm- und konfigurationsabhängigen Aufbereitung der SAP-Eingangs- und -Ausgangsdatenströme
- den Instruktionsspeicher für den SAP-Binärcode (WCS)
- 4 *MB* intermediären Speicher zur flexiblen, anwendungsspezifisch formatierten Bereitstellung der SAP-Ausgangsdaten

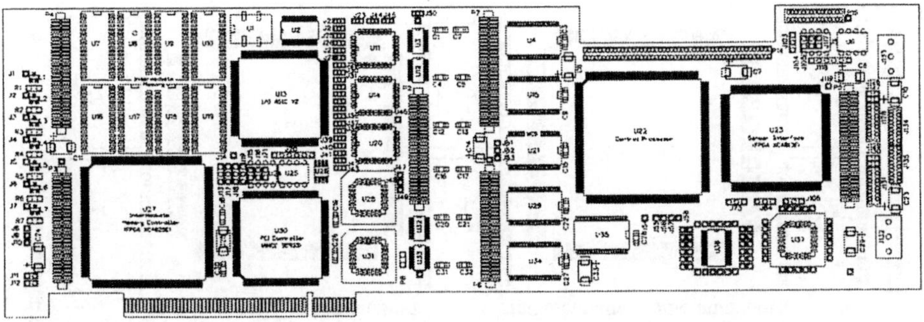

Abbildung 3. CAD-Plot der SAP-Trägerplatine

– das PCI-Bus Interface

Ein Blockschaltbild der SAP-Platine ist in Abb. 2 gezeigt, ein CAD-Plot der Trägerplatine und zweier SAP-Slice-Varianten findet sich in Abb. 3, 4 und 5. Die in Abb. 5 gezeigte Variante des SAP-Slices verwendet anstelle der lediglich auf Komponentenbasis miniaturisierten ASICs und SRAMs, sog. MCMs (vgl. Abschnitt 3.2).

Bauvolumen	$450cm^3$
Verlustleisung	20 W
I/O Rate	160/160 MByte/sec
Maximale Speicherzugriffsbandbreite	10 GByte/sec
Nachbarschaftszugriffe ohne zusätzliche Zyklen	$[-4, +4]$
Rechenleistung (8Bit ADD/SUB)	2,2 GOPS
Rechenleistung (Multiply/Accumulate)	0,4 GOPS

Tabelle1. Leistungsmerkmale der SAP/PCI Platine mit 512 Prozessor-Elementen

3.1 Einsatz der SAP-Rechnerstufe als PCI-Einsteckkarte

Die entwickelte Parallelrechnerstufe wird derzeit in einer Konfiguration mit 512 PEs zur zeilenweise parallelen Verarbeitung von Bilddaten eingesetzt. Die sequentiell angelieferten Pixeldaten werden in einen spaltenparallel anfallenden Datenstrom gewandelt und nach Verrechnung einer geeigneten Anzahl von Zeilen (oder auch Bildern) spaltenparallel ausgegeben, formatgewandelt und im intermediären Speicher ablegt. Der Transfer der Ergebnisdaten in den Hauptspeicher des Wirtsrechners erfolgt im *PCI-Master-Write-Modus* und belastet daher allenfalls die Speicherbandbreite (bei *single-port-memory*), nicht aber die CPU(s)

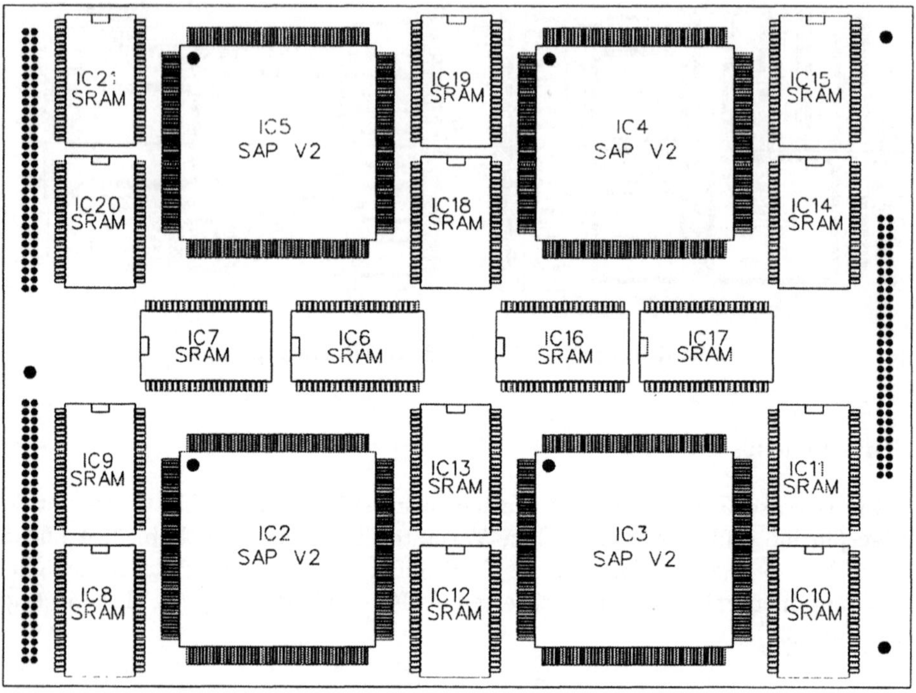

Abbildung 4. SAP-Slice auf Komponentenbasis

des Wirtsrechners. Die mit der entwickelten Baugruppe gemessene PCI-Master-Write-Bandbreite von 128 MB/s realisiert praktisch den theoretischen Spitzenwert (133 MB/s). Tabelle 1 faßt einige der Leistungsmerkmale einer mit 512 Prozessorelementen bestückten SAP-Platine zusammen.

3.2 Modularer Aufbau planarer Multi-Chip-Module (IZM)

Langfristige Zielrichtung dieses Teilprojekts war die über die Entwicklung hochintegrierter ASICs hinausgehende Miniaturisierung mit Methoden der Aufbau- und Verbindungstechnik durch 3-dimensionale Integration (Komponenten-Stapelung). Dem diente ein Konzept zur planaren Hybridintegration in Multi-Chip-Modulen (MCM) [24]. In keramische 4"-Substrate (AIN) werden per YAG-Laser präzise Fensteröffnungen zur Aufnahme ungehäuster IC-Komponenten geschnitten. Eine transparente, laminierte Adhäsionsfolie auf der Substratfrontseite bildet die Referenzfläche für alle einzubettenden Komponenten. Unter der Optik eines Bestückungsaligners werden die Komponenten nacheinander in den Öffnungen mit Hilfe einer Referenzmaske justiert und an der Adhäsionsfolie temporär fixiert. Durch eine Epoxy-Fügung - von der Rückseite in die umlaufenden Fugen eingebracht - werden die Komponenten in ihrer Lage dauerhaft fixiert. Nach Abzug der Adhäsionsfolie liegt wieder eine geschlossene planare

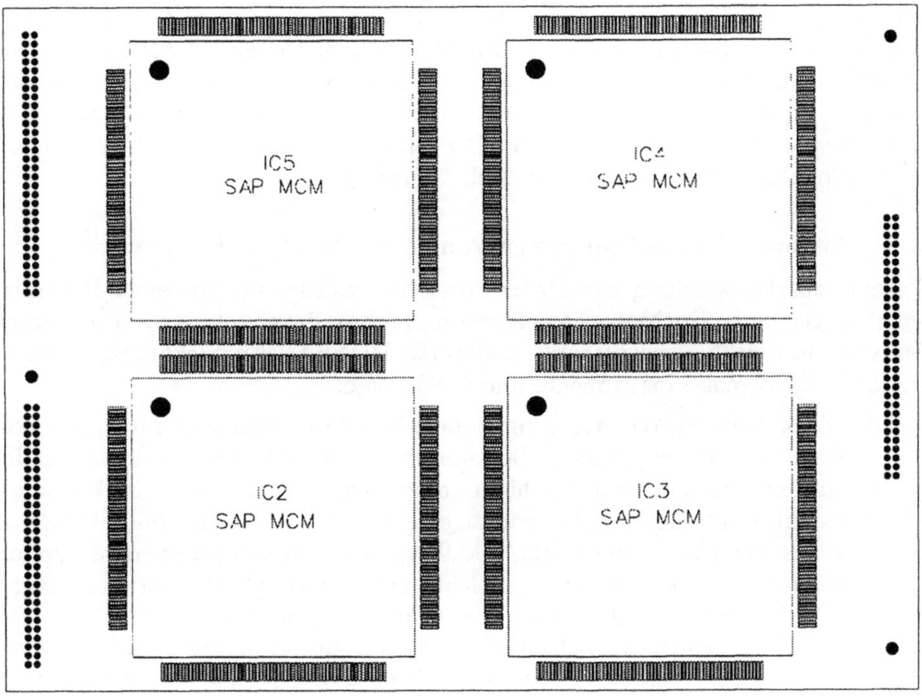

Abbildung 5. SAP-Slice im MCM-Aufbau

Substratoberfläche vor, auf der in starrer Maskentechnik eine Dünnfilmtechnik-Mehrlagenverdratung aufgebaut werden kann, die die Komponenten untereinander verbindet und Bondpads bzw. Lötkugeln in oberster Lage zur Boardkontaktierung des Moduls bietet. Ausschließlich Dunnfilmprozesse werden angewandt, um die IC-Ankontaktierung und die mehrlagige Cu-Verdrahtung im Polyimid-Dielektrikum zu erstellen. Die Ankontaktierung erfordert keine spezielle Vorbehandlung. ICs unterschiedlicher Technologie können miteinander integriert werden. Die Einbettung kapazitiver Bauteile und optischer, elektrischer, thermischer Durchführungen ist ebenso möglich. Planare Einbettmodule realisieren bei dichter Komponentenfügung einen kompakten MCM-Aufbau, dessen Höhe die Substratdicke kaum überschreitet. Diese Einbettmodule, zusätzlich versehen mit Substratdurchkontaktierungen, erlauben durch die Mehrlagenverdrahtung hindurch die Realisierung kurzer, vertikaler Signalwege. Ergänzt durch flächenhaft angeordnete Lötverbindungen auf der Ober- und Unterseite der Module, kann die Stapelung von MCMs mit extrem kurzen, vertikalen Busleitungen erreicht werden.

Im Projektzeitraum konzentrierte man sich zunächst auf den Bau von planaren Einbettmodulen, die zwar noch nicht stapelfähig sind, aber als modulare Einheit (Prozessor, Speicher, Abblock-Kondensatoren) in die vom BGT entwickelte Systemarchitektur eines SIMD-Rechners passen. Dadurch lassen sich die gehäusten Einzelkomponenten auf den SAP-Slices ersetzen. Das SAP MCM der Größe $32 \times 32mm^2$ enthält 4 SRAMs ($64K \times 16$) und 4 SRAMs, die eine SAP Komponente (64 PEs) umgeben. Die erstellten Einbettmodule werden funktionsgeprüft und per Drahtbond in ein Standardgehäuse gebaut. Mit den so gewonnenen SAP MCMs können die SAP-Slices alternativ bestückt werden (ein CAD-Plot eines MCM-Slices ist in Abb. 5 gezeigt).

3.3 Software-Entwicklungswerkzeuge für die SAP-Baugruppe

Parallel zur Entwicklung des SIMD-Prozessors wurde eine zu seiner Programmierung geeignete Hochsprache definiert und die zu ihrer Nutzung erforderlichen Software-Entwicklungswerkzeuge realisiert. Die SAP-Entwicklungsumgebung umfaßt u.a. folgende Instrumente und Werkzeuge:

- die *C*-ähnliche SIMD-Programmiersprache *LISA* (*Language for Image processing on Systolic Arrays*), die zum einen den arithmetisch-logischen Instruktionssatz von *C*, darüber hinaus aber auch Struktur- und Bit-Feld-artige Konstrukte enthält. Den Anwendungen entsprechend werden bildlokale Operationen wie zweidimensionale, bildübergreifenden Operationen wie dreidimensionale Arrays indiziert. Die "Schleifenverwaltung" ist nicht Gegenstand des Quellencodes, sondern wird vom Compiler generiert.
- einen optimierenden *LISA*-Compiler, der nach dem Übersetzungsvorgang die Anzahl der zur Laufzeit erforderlichen Prozessorzyklen, sowie die Größe des bei der Programmausführung benötigten Speicherplatzes ausweist (*Profiler*-Funktion).
- den SAP-Simulator (lauffähig unter Solaris, SunOS, Linux, AIX, HP-UX), welcher die Berechnungsvorgänge der SAP-Hardware nachbildet, die Visualisierung beliebiger Zwischenergebnisse erlaubt und es gestattet, die SAP-Ergebnisdaten in beliebige, auf dem SAP-Hostrechner ablaufende, symbolische Verarbeitungsmodule einzuspeisen.
- den SAP-Emulator, der die SAP-hardwareunterstützte Verarbeitung auf Festplatte gespeicherter, digitaler Bildfolgen ermöglicht, wobei dann u.a. auch ein Bildeinzelschrittmodus zur Verfügung steht, mit dem die Visualisierung von spezifischen ikonischen Zwischenergebnissen möglich ist.

Die Steuerung von SAP-Anwenderprogrammen aus sensorfernen, symbolischen Verarbeitungsprozessen erfolgt über eine *Linux*-Anwenderbibliothek, welche die erforderliche Gerätetreiberfunktionalität zur Verfügung stellt. Eine Übersicht über die verschiedenen SAP-Softwarewerkzeuge und ihr Zusammenwirken ist in Abb. 6 gegeben.

4 Kameramodellierung und Kamerakalibrierung (NMA)

Zur Erhöhung der Vermessungsgenaugigkeit szenischer Inhalte im Fernbereich und zur Nutzung der bei den ikonischen Prozessen berechneten Subpixelgenauigkeiten ist es von entscheidender Bedeutung, möglichst genaue Abbildungsvorschriften zwischen Bildmatrix und Szene zur Verfügung zu haben. Die in diesem Abschnitt beschriebenen Ansätze dienen diesem Ziel.

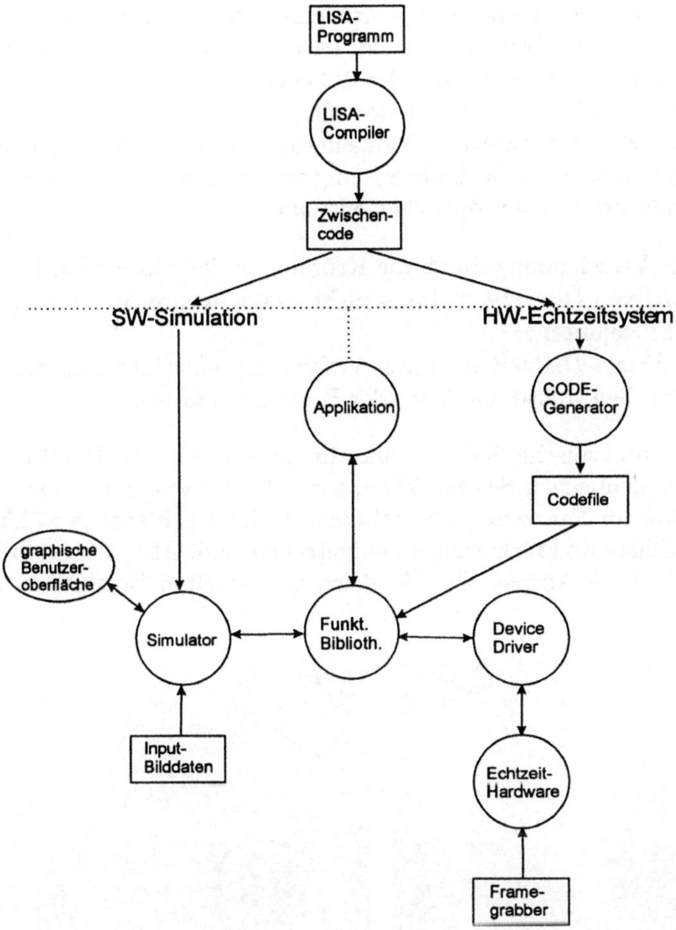

Abbildung 6. SAP-Entwicklungswerkzeuge

Generelle Aufgabe der Kameramodellierung und Kamerakalibrierung ist die approximative Beschreibung und Ermittlung der Abbildungseigenschaften zwischen Punkten eines gegebenen 3D-Koordinatensystems und den Pixelpositionen der in die Sensorebene abgebildeten Objekte. Die Adäquatheit des gewählten Kameramodells, sowie die numerische Genauigkeit der dafür ermittelten Modellparameter ist entscheidend für die Güte der Tiefen- und Bewegungsschätzung aus monokularen oder polyokularen Bildfolgen.

Zur Bildsensorkalibrierung zählen neben der Bestimmung der Kameraparameter aus Bildern von Eichkörpern bekannter Geometrie auch Algorithmen zur subpixel-genauen Detektion projizierter 3D-Marken im Kamerabild und zur automatischen Zuordnung von 2D-Bildpositionen zu den vermessenen 3D-Ortskoordinaten.

Eine zusätzliche Verbesserung der numerischen Stabilität der resultierenden Optimierungsprobleme, sowie der daraus zu berechnenden Kameraparameter, wird durch die Erweiterung der üblichen Einbildkalibrierung auf ein semiautomatisches Mehrbildverfahren erreicht.

Die im Projekt entwickelte Modellierung erlaubt über die perspektivische Projektion hinaus auch die Einbeziehung von Korrekturen für folgende typische Verzeichnungsarten realer optischer Systeme:

- Radiale Verzeichnung durch die Krümmung der Linsenoberflächen;
- *"Decentering Distortion"* durch nicht exakt lineare Anordnung der verwendeten Einzellinsen;
- *"Thin Prism Distortion"* durch Verkippung einzelner Linsenelemente bzgl. der optischen Achse und/oder der Bildsensormatrix.

Für eine prototypische Kalibrieranordnung wurden als 3D-"Eichkörper" ein Würfel mit identischen Seitenflächen, sowie (wahlweise zur Ermittlung der Kameraposition im Fahrzeug) eine fahrzeugstationäre Platte gewählt, auf denen jeweils äquidistante Markierungen aufgetragen sind. Abb. 7 zeigt Bilder der verwendeten Kalibrierkörper. Ein Vergleich der erzielten Ergebnisse der Kalibrie-

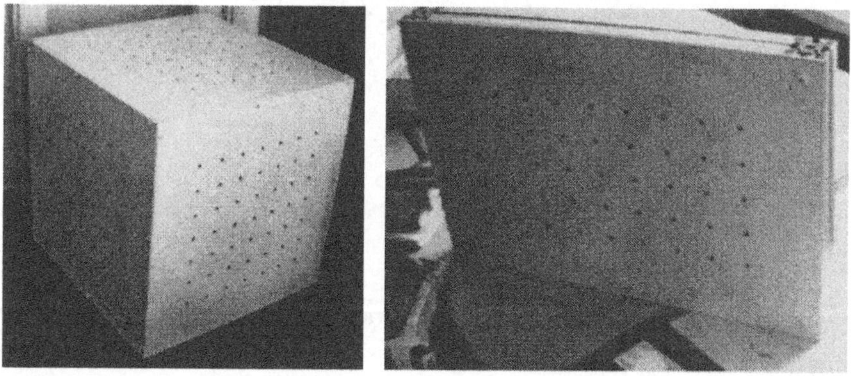

Abbildung 7. Zur Kamerakalibrierung verwendete 3D-"Eichkörper"

rung mit und ohne Berücksichtigung der o.a. Verzeichnungen zeigt, daß letztere bei der Projektion gegebener 3D-Punkte zu Abweichungen der zugehörigen Bildposition in der Größenordnung von mehreren Pixeln (am Bildrand bis zu 15 Pixeln) führen können. Umgekehrt treten bei der Rekonstruktion der 3D-Information aus Bilddaten (beispielsweise bei der Abstandsschätzung von Objekten) allein durch die nicht ausreichend genaue Modellierung der Kameraabbildungseigenschaften relative Fehler von mehr als 10 Prozent auf.

Für die möglichst exakte Eigenbewegungsbestimmung des Fahrzeugs, die 3D-Rekonstruktion des Straßenverlaufs im Fernbereich und eine verbesserte Abstandsschätzung von Hindernissen ist daher die Einbeziehung von optischen Verzeichnungen bei der Kameramodellierung unerläßlich.

5 System-Architektur (BMW)

In Anbetracht des rapiden technologischen Fortschritts im Bereich der Standard-Hochleistungsprozessoren war es von grundsätzlicher Bedeutung, bei der Wahl des Hostrechners auf technisch führende, voll im "Mainstream" des Marktes liegende Systeme zu setzen.

Dabei spielten Verfügbarkeits- und Kostengesichtpunkte ebenso eine Rolle, wie die unbedingte Kontinuität der erzielten Projektergebnisse, d.h. deren Abkopplung von schwer vorhersehbaren Marktentwicklungen im Bereich der Halbleiter- und Rechnerindustrie.
Entsprechend zog dies die Entscheidung nach sich, ein nichtproprietäres Betriebssystem zu verwenden. Nach diversen Evaluierungen vereinbarten die Industriepartner im Projektverbund auf *Linux* und *GNU*-Entwicklungswerkzeuge (inkl. *X11/XFree*) zu setzen.

Als Hardware-Plattformen wurden zunächst DEC/Alpha und Intel/Pentium basierte Systeme gewählt.

Die genannten Festlegungen zogen nach sich:

- die Erfordernis der Implementierung *Linux*-konformer Gerätetreiber für die SAP- und Framegrabber-Baugruppe und die
- die anwendungsgerechte Programmierung der erforderlichen (aufgrund der noch fehlenden POSIX-Erweiterungen nur eingeschränkt echtzeitfähigen) *Linux*-Systemdienste für die schritthaltenden Applikationen.

Insbesondere ging es darum, den strikt synchron anfallenden, die gesamte symbolische Prozeßkette synchronisierenden Ergebnis-Output des *Systolic Array* Prozessors unter Verwendung von applikationsunabhängigen Systemprozessen (Visualisierung und Fahrzeug-Meßdatenerfassung, etc.) von den symbolischen Anwendungsprozessen zu entkoppeln. Die letztgenannte Erfordernis beruht u.a. auf dem Faktum, daß die späten (d.h. sensorfernen) Bildinterpretationsprozesse aufgrund ihrer erhöhten Komplexitätsmaße nur im statistischen Mittel einiger weniger Frames synchron gehalten werden können. Ergebnis entsprechender Untersuchungen ist, daß das zeitliche Antwortverhalten derzeit verfügbarer *Linux*-Distributionen für Bildwiederholraten von 50 bzw. 60 Hz auch unter Verzicht auf die genannten POSIX-Erweiterungen ausreichend ist, und die erwähnte, zeitliche Entkopplung für ganze Klassen von Anwendungen einheitlich realisiert werden kann. Das in Abb. 8 gezeigte Blockdiagramm der hierfür implementierten Systemprozesse dient nicht nur als Grundgerüst für die in Abb. 9 dargestellte Gesamtapplikation zur Bildfolgeninterpretation, sondern bildet auch die Basis für die im Projekt entwickelte Software zur Bild- und Meßdatenerfassung. Im letztgenannten Falle besteht die "Applikation" lediglich aus der Funktion, den anfallenden Bild- und Meßdatenstrom von bis zu 15 *MB/s* durch das *Linux*-Dateisystem schritthaltend auf Festplatte zu speichern. Die dabei durch betriebssystembedingte File-Cache-Strategien auftretenden Fluktuationen der effektiven

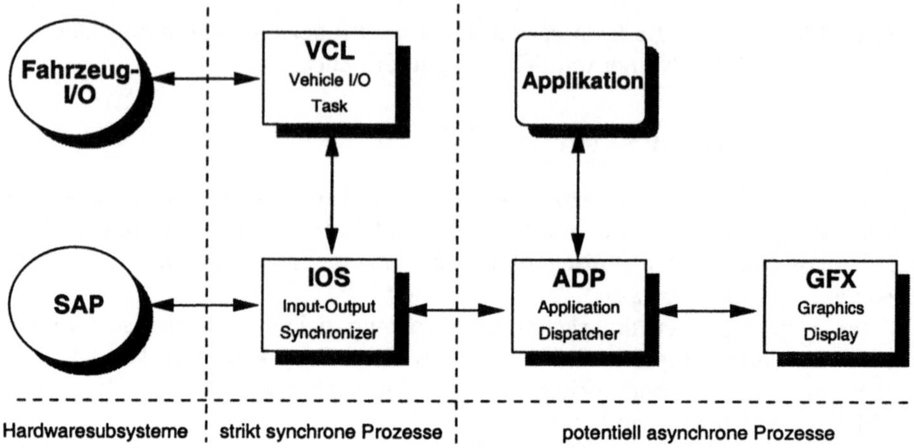

Abbildung 8. Systemprozesse

Schreibdatenrate ziehen transiente Latenzen von bis zu $120ms$ nach sich, die zum einen im statistischen Mittel ohne Datenverluste wieder ausgeglichen werden können und zum anderen aber den vom Frame-Grabber per PCI-Master-Write in den Hauptspeicher kopierten, synchronen Datenstrom in keiner Weise beeinträchtigen.

6 Ikonische Prozesse auf dem Systolic Array Prozessor (BMW/BGT)

Zunächst wird ein Programm zur Berechnung des kantenbasierten optischen Flusses vorgestellt, bei dem die Kantengeschwindigkeits-Normalkomponenten für beliebig auftretende Konturpixel bildflächendeckend im Videotakt bestimmt werden. Dazu erforderlich sind:

- ein Modul zur Berechnung von Binärkanten samt morphologischer Bereinigung, Skelettierung und Attributierung
- ein Modul zur attributorientierten, bildübergreifend eindeutigen Konturpunktzuordnung, inklusive der Bestimmung der aktuellen Positionsverschiebung senkrecht zur Kante, sowie
- eine pro Bild für alle Konturpixel durchgeführte Kalmanfilterung der rohen Verschiebungsdaten zur Gewinnung der dynamisch-optimalen Flußgeschwindigkeitskomponenten.

Zur Charakterisierung der Leistungsfähigkeit des entwickelten Parallelrechners sei erwähnt, daß die beschriebenen Operationen gemäß den Synchronitätseigenschaften des SIMD-Prozessors auf einer einzigen SAP/PCI-Platine in Abhängigkeit von der Bildwiederholrate von 60Hz ca. 8 Millionen mal pro Sekunde durchgeführt werden ($60 \times 512 \times 256 \frac{Pixel}{Bild \times Sekunde}$).

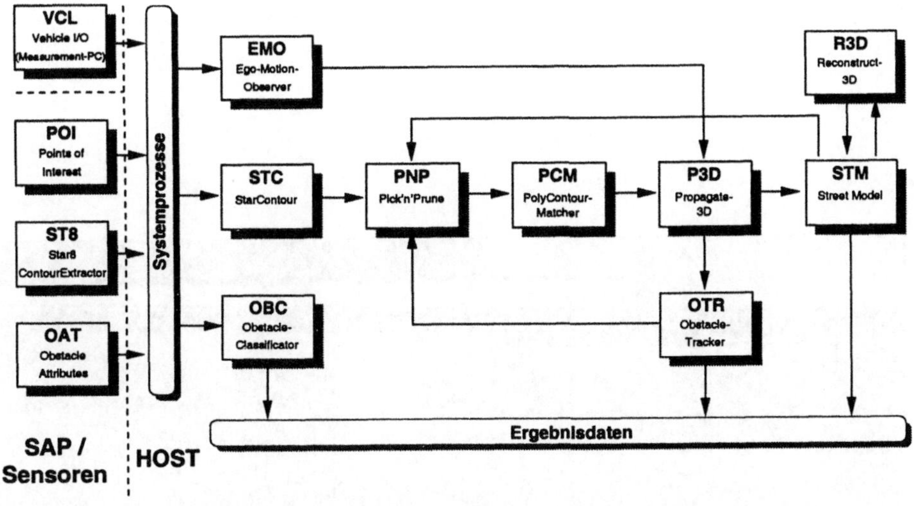

Abbildung 9. Applikationsprozesse "Elektronisches Auge"

In einer anderen in Abb. 9 dargestellten Gesamtapplikation wird die außerordentliche Ausgabebandbreite (128 *bit/Pixel*) des SAP-Prozessors genutzt, um die vorsortierten Ergebnisse verschiedener ikonischer Verarbeitungsprozesse im Bildtakt simultan bereitzustellen. Im einzelnen werden berechnet:

- die zum Aufbau eines Konturnetzwerks erforderliche richtungscodierte Kanten-"Ereignis"-Information (ST8),
- die pixelbasierten Hindernisattribute zur Objektklassifikation (OAT) und
- die "*point-of-interest*"-Bestimmung und Attributierung zur Realisierung eines Bildsignal-basierten Eigenbewegungsbeobachters (POI) (vgl. Abschnitt 7);
- weiterhin wird zu Zwecken der Visualisierung das Orginal-Grauwertbild ausgegeben.

Bei der erwähnten Extraktion der Konturskelett-Information handelt es sich um die Bestimmung von pixellokal ortstreuen Kantenpositionen und Topologien. Die Ortstreue beeinträchtigende Tiefpaßfilterungen werden dabei vermieden. Unter dem Aspekt der Bildkomponentenmarkierung basiert das in α-Version realisierte Verfahren nicht auf einer binären, sondern einer polyvalenten Bildkomponentenzerlegung. Abb. 10 zeigt die mittels SAP-Baugruppe aus dem dargestellten Graubild extrahierten Konturen (Anm.: das zugrunde liegende Bildmaterial wurde noch im suboptimalen Videoformat "*field integration / interlace*" aufgezeichnet).

Abbildung 10. Konturextraktion

Der auf SAP implementierte Algorithmus erfordert ca. 50 9-bit-Operationen und 700 1-bit-Operationen pro Pixel, die sich pro Bildpunkt zu ca. 1150 1-bit-Operationen addieren. Bei einer Bildwiederholrate von 50Hz und 256 Bildzeilen führt dies zu einem erforderlichen Rechendurchsatz von knapp 15 Millionen 1-bit-Instruktionen pro SAP-Prozessorelement, wobei allerdings insgesamt 512 PEs zur Verfügung stehen. Im Gegensatz dazu benötigt eine konventionelle CPU ca. 750 (32- oder 64-bit-) Instruktionen pro Bildpunkt bei 512 $\times$ 256 Pixeln und 50Hz Bildwiederholrate, also ca. 5 Gigainstruktionen. Der "Overhead" für 30-40 Doppelschleifensteuerungen ist hierbei noch nicht berücksichtigt.

Der mittels des ST8-Moduls gegebene Vergleich der auf SIMD-Prozessor bzw. konventioneller CPU erforderlichen Anzahl von Instruktionen zeigt klar die Vorteile der (mit der Bildbreite skalierbaren und der Bildhöhe kaskadierbaren) SAP-Parallelhardware anhand eines konkreten Anwendungsbeispiels.

Das eingangs angeführte Quadruplett von Pixelverarbeitungsmodulen dient der symbolischen Verarbeitungskette als Basis zur simultanen Kartierung des Fahrbahn- und Spurverlaufs, zur Berechnung der Kameraeigenbewegung, zur Hindernisklassifikation und überdies zur Visualisierung von Ergebnisdaten.

7 Bestimmung des fahrzeugdynamischen Zustands; Eigenbewegungsbeobachter (NMA/BMW)

Eine möglichst zeitnahe Bestimmung der Fahrzeug-Rotationen (Gieren, Nicken, Wanken) und der Fahrzeugbewegungsrichtung (Schwimmwinkel) verringert die Komplexität der nachfolgenden symbolischen Algorithmen zur 3D-Rekonstruktion und Fahrbahnkartierung und verkürzt damit deren Laufzeit wesentlich.

Zur Berechnung der Orientierungskinematik hat sich die Verwendung von zeitlich stabilisierten Punktverschiebungen als nützlich erwiesen. Dabei werden mit Hilfe der vom ABS-System gelieferten, skalaren Geschwindigkeit die vektoriell translatorischen und rotatorischen Bewegungsanteile des Fahrzeugs aus Paaren aufeinander folgender Bilder bestimmt.

7.1 Beschreibung der zugehörigen Teilmodule

Extraktion von markanten Punkten aus Bildsequenzen (POI). Für die Detektion markanter Punkte werden Kriterien, basierend auf der expliziten Berechnung von Richtungskrümmungen (partielle zweite Ableitungen des diskreten Grauwertverlaufes) oder aber Eigenschaften des Krümmungstensors verwendet. Für letztere ist das aus der Literatur bekannte Verfahren von Harris und Stephens [4] ein typischer Vertreter.

Attributierung der extrahierten markanten Punkte (POI/EMO). Um eine Zuordnung der pro Halbbild unabhängig voneinander durch die SAP-Parallelhardware extrahierten Punkte über die Zeit zu gewährleisten, werden die extrahierten markanten Punkte mit Attributen versehen, die ihre lokale Bildumgebung beschreiben. Als Merkmale dienen dabei die Koeffizienten einer lokalen Passung des Bildinhalts mit orthogonalen Polynomen niedrigen Grades.

Bildübergreifende Zuordnung mit Subpixelgenauigkeit (EMO). In diesem Teilmodul wird eine Zuordnung von Punkten aufeinander folgender Frames mit Hilfe der bereitgestellten Attribute und unter Verwendung vorher ermittelter Information (d.h. bereits einige Male wiederaufgefundener Punkte) durchgeführt. Eine Verfeinerung der Verschiebungsinformation auf Subpixelgenauigkeit wird erreicht, indem man diejenige Verschiebung berechnet, bei der die Abweichung zwischen den entsprechend translierten Polynomapproximationen minimal wird.

Berechnung der Fahrzeugeigendynamik aus den ermittelten Punktverschiebungen (EMO). Durch eine neuartige, halb-perspektivische Modellierung der bei der Bestimmung der Kameraeigenbewegung zu minimierenden Fehlerfunktion, konnte eine entscheidende Verbesserung der Ergebnisse im Vergleich zu den aus der Literatur bekannten Ansätzen und Verfahren [5–12] erzielt werden. Die verwendete halb-perspektivische Modellierung, die nur Rekonstruktionen von Punkten *vor* der Kamera erlaubt, ist für eine robuste und ausreichend exakte Bestimmung der Kameraeigenbewegung vor allem im Fall von selbstbewegten 3D-Objekten in der betrachteten Szene unabdingbar. Eine ausführliche Darstellung der entwickelten (u.a. semi-perspektivischen) Methoden findet man in [13]; ein weiterer Artikel ist zur Veröffentlichung [14] eingereicht.

7.2 Ergebnisse

Abb. 11 zeigt einen Vergleich der Ergebnisse des oben beschriebenen Verfahrens für 50 Bilder einer Sequenz am Beispiel des Gierwinkels ϕ_x, sowie die zugehörigen Daten des Gyrosensors (Punkte). Die gestrichelten Kurven stellen die berechneten Ergebnisse, bezogen auf alle extrahierten und erfolgreich zugeordneten markanten Punkte dar, basierend auf der Minimierung der hier entwickelten Zielfunktion F_0 (semi-perspektivischer Ansatz) und einem nichtlinearen Standardverfahren (mit zugehöriger Zielfunktion F_W) von Weng, Huang und Ahuja [7]. Die durchgezogenen Kurven repräsentieren die Ergebnisse nach fünf Iterationen unter Berücksichtigung einer Verfeinerung, bei der "Ausreißer" (z.B. Punkte, die auf eigenbewegten Objekten liegen) eliminiert werden.

Bei dem hier entwickelten Ansatz ist eine deutliche Verbesserung der Resultate in Bezug auf Genauigkeit und Stabilität erkennbar. Im Gegensatz dazu werden die mit Hilfe der Zielfunktion F_W berechneten Ergebnisse durch die Elimination von Ausreißern nur unwesentlich verbessert.

Die Genauigkeit der hier durch Bildverarbeitung bestimmten Ergebnisse ist denen handelsüblicher Gyrosensoren durchaus ebenbürtig, in manchen Fällen sogar überlegen, da die entsprechenden Informationen unmittelbar aus dem optischen System gewonnen werden und so Übertragungseffekte (z.B. durch dynamische Filtertechniken) und insbesondere die bei Gyrosensoren auftretenden Temperatureffekte vermieden werden.

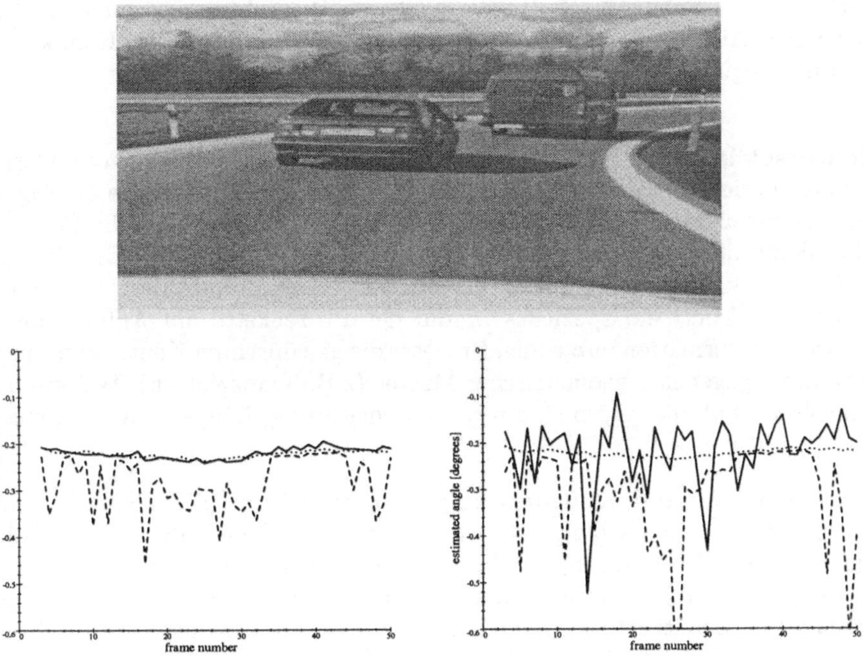

Abbildung 11. Originalbild (oben) und berechnete Ergebnisse der Gierwinkelraten $\Delta\phi_x$ basierend auf den Zielfunktionen F_U (unten links) und F_W (unten rechts); Ergebnisse ohne iterative Verbesserung (gestrichelt), Ergebnisse nach Elimination von Ausreißern (durchgezogen) und entsprechende Daten des Gyrosensors (gepunktet).

8 Straßen- und Hindernismodellierung (BMW/NMA)

Die Straßen- und Hindernismodellierung besteht aus einer Prozeßkette, die sich aus Modulen zur Extraktion von Konturen aus Bilddaten, deren Vorauswahl in interessierenden Bereichen, der Polygonzugpassung zur Datenreduktion, der Zuordnung von Konturteilen zur Gewinnung eines zeitlichen Verlaufs, bis hin zur Rekonstruktion der dreidimensionalen Positionen der Fahrspuren und Hindernisse im ortsfesten 3D-Koordinatensystem zusammensetzt.

Komponentenmarkierung (STC). Das bei BMW entwickelte Modul zur Komponentenmarkierung nutzt die vom SAP-Prozeß ST8 generierten

Kantenskelett-Ereigniscodes, um daraus eine symbolische Repräsentation der Bildregionen samt Attributen und die sie begrenzenden Konturen aufzubauen. Die Konturknoten (End- und Verzweigungspunkte) sind in dieser Darstellung mittels zyklisch und sequentiell vierfach verketteter Listen repräsentiert, die dazwischen liegenden Polygonzüge (8 Freeman-Richtungen) sind lauflängencodiert, um die Datencache-Strategien moderner Workstationprozessoren effizient zu nutzen. Auf Wunsch kann neben der Netzwerktopologie der Randkonturen auch die Regionen-Einschlußhierarchie ausgegeben werden.

Konturselektion (PNP). Bei der Verarbeitung der aus den Vollbildern berechneten reichhaltigen Repräsentation aller Konturskelette stoßen die Algorithmen zur Straßen- und Hindernismodellierung an die ihnen durch die Echtzeitbeschränkung der Gesamtanwendung gesetzte Grenze. Um dieser Einschränkung genügen zu können, ohne auf die Details der Kanteninformation verzichten zu müssen, selektiert ein spezielles Modul der Prozeßkette auf Anforderung der die Kanteninformation nutzenden Prozesse die gewünschten Konturuntermengen anhand vorgegebener geometrischer Muster (z.B. Trapezfolgen). Weitere Spezifika solcher Anforderungen sind u.a. Kantenrichtung, Länge, Gradient, etc.

Approximation der Konturen in jedem einzelnen Bild (PNP). Um den Rechenaufwand für nachfolgende Algorithmen weiter zu reduzieren und damit den Echtzeitanforderungen der Anwendung zu entsprechen, ist es vorteilhaft, für die jeweiligen Teilkonturen eine möglichst kompakte, datenreduzierende Repräsentation bereitzustellen (Quadratmittel-Passung der ausgewählten Kontursegmente mit Polygonzügen und konischen Splines vorgegebenen Fehlermaßes).

Bildübergreifende Zuordnung von Polygonsegmenten (PCM). Für die 3D-Rekonstruktion von Konturen ist die Kenntnis ihrer zeitlichen Korrespondenz erforderlich. Dazu werden die extrahierten und vorgepaßten Kontursegmente in aufeinander folgenden Bildern einander zugeordnet und daraus sogenannte Konturstapel generiert.

Da Kontursegmente durch die Bewegung des Fahrzeugs, die zeitliche Veränderung der Szene, Verdeckungen und Digitalisierungseffekte des öfteren aufbrechen bzw. wieder verschmelzen, muß dafür Sorge getragen werden, daß auch solche Effekte zufriedenstellend gehandhabt werden. Dies impliziert, daß einem momentan betrachteten Polygonsegment im aktuellen Halbbild ein oder mehrere Segmente im vorhergehenden bzw. folgenden Halbbild zugeordnet werden können.

3D-Rekonstruktion von Konturen (P3D/R3D). Die quantitative Nutzung der Orientierungskinematik des Bildsensor-Trägerfahrzeugs, sowie der Kalibrierungsdaten der Kamera, erlaubt das Wiederauffinden beliebiger Punkte einer Kontur über mehrere Frames hinweg.

Unter Nutzung der Tatsache, daß sich ortsfeste Objekte im Bild auf Strahlen vom Fluchtpunkt wegbewegen, kann bei bekannter Fahrzeugkinematik für jeden Punkt einer Kontur die Lage seines Sichtstrahls für das nächste Bild berechnet und darüberhinaus durch den Schitt des projizierten Strahls mit dem korrespondierenden Konturzug ebendieser Punkt wiederaufgefunden werden. Durch wiederholte Anwendung diese Prinzips "künstlicher" Korrespondenzen, können beliebige Punkte eines Polygonzugs über mehrere Bilder hinweg verfolgt und ihnen dabei räumliche Tiefeninformation zugeordnet werden.

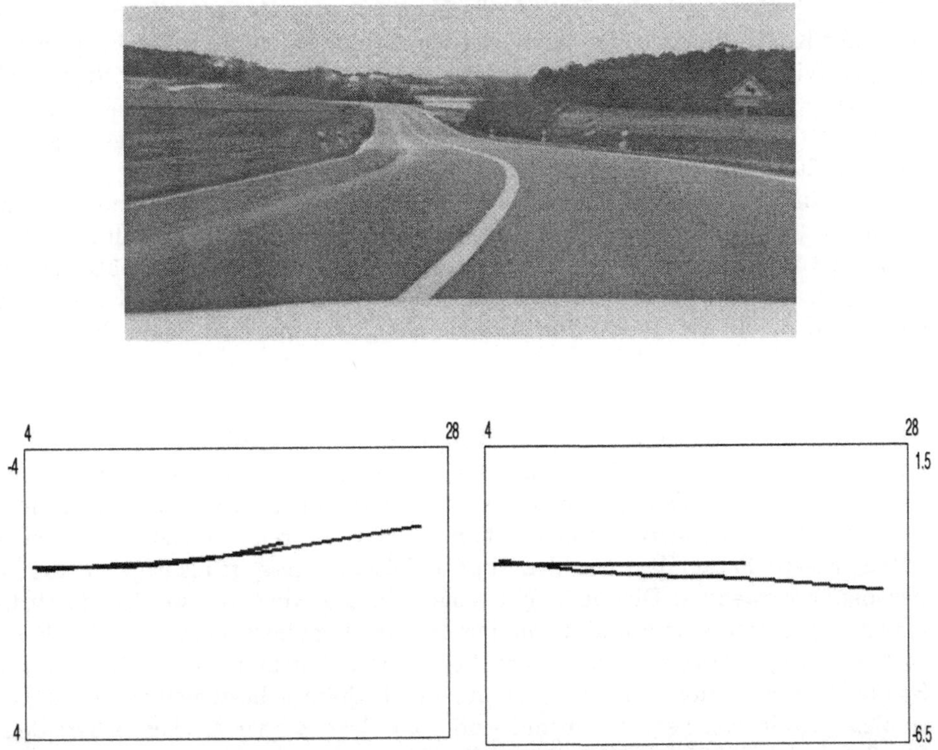

Abbildung 12. Originalbild (oben) und Vergleich der 2D-Reprojektion (kurze Linien) mit der berechneten 3D-Rekonstruktion aus fünf perspektivischen Bildern (lange Linien) am Beispiel der Mittelstreifenkontur. Linkes Bild: Draufsicht; rechtes Bild: Seitenansicht; Angaben in Meter

Abb. 12 zeigt als Beispiel Ergebnisse der Rekonstruktion eines Teils der Mittelstreifenkontur einer Landstraßenszene. Da es sich hierbei um eine im wesentlichen quasistationäre Kurve handelt - man erhält trotz Translationsbewegung der Kamera nahezu identische perspektivische Bilder der Kontur - führen Ansätze ohne Verwendung von Zusatzwissen über die Eigenbewegung der Kamera oder auch solche, die lediglich auf dem optischen Fluß basieren, notwendig zu unbefriedigenden Ergebnissen.

In Abb. 12 dargestellt sind die jeweiligen 3D-Konturen, die man einerseits durch Reprojektion in die Straßenebene (kurze Linie), andererseits mit Hilfe des oben beschriebenen Verfahrens unter Nutzung künstlicher Korrespondenzen bei bekannter Eigendynamik (lange Linie) erhält. Für das gezeigte Beispiel wurden punktweise zugeordnete Konturen aus fünf aufeinanderfolgenden Bildern verwendet. Dies entspricht der zeitlichen Entwicklung der Kontur innerhalb einer $\frac{1}{10}$ Sekunde.

In der Seitenansicht ist deutlich die Fehlschätzung der räumlichen Ausdehnung der Kontur bei der 2D-Reprojektionsmethode erkennbar. Dies liegt an der perspektivischen Verkürzung, die durch die Fahrt über den leichten Hügel auftritt. Die 3D-rekonstruierte Kontur hingegen ist deutlich länger und verläuft bezüglich der augenblicklichen Standebene des Fahrzeugs nach unten, was die Konvexität des Hügels widerspiegelt.

Auch in der Draufsicht ist die Verkürzung der reprojizierten Kontur deutlich erkennbar. Deren Nichtberücksichtigung führte bei der modellgestützten Einordnung von Hindernissen auf der Straße einerseits zu einer Fehlschätzung des Abstands, andererseits aber auch zu einer falschen Interpretation der Krümmung der Straße, die für die Querführung des Fahrzeugs von entscheidender Bedeutung ist.

Modellierung des Straßen- und Spurverlaufs (STM). Die aus den oben beschriebenen Bildverarbeitungsmodulen vorverlesenen Konturen werden mittels Heuristiken den Fahrspuren zugeordnet, welche die Position und Lage, Richtung und die Zusammenhangseigenschaften der Konturen einerseits und geometrische Eigenschaften des aktuellen Spurmodells andererseits als Entscheidungsparameter verwendet. Die einer Spur zugeordneten Konturen werden durch lineare Approximationstechniken an periodische B-Splines gepaßt. Jedes Konturstück trägt dabei nur einen sehr begrenzten Zeitraum (1 - 5 Bilder) zur Spline-Repräsentation bei. Die verwendeten B-Splines bestehen aus einer variablen Anzahl von Segmenten und sind vom Grad 3 oder 4. Dies sichert dem Straßenmodell die Flexibilität, die zur Passung aller in der Realität vorkommenden Straßenkonturen erforderlich ist. Ein beispielhaftes Ergebnis der Straßen- und Spurerkennung ist in Abb. 13 gezeigt. Aus dem errechneten Modell werden die *Regions Of Interests* der erkannten Straßenberandungen, sowie Aufmerksamkeitsbereiche weiterer potentieller Spurberandungen bestimmt. Diese werden als Anforderung an das Modul zur Polygonzugselektion (PNP) rückgekoppelt.

Bildkantenbasierte Detektion und Verfolgung verkehrsrelevanter Hindernisse (OTR). Durch Nachbarschaftsgruppierung horizontaler Kanten der Konturrepräsentation und geeigneten Gruppenstabilisierungen werden verkehrsrelevante Hindernisse detektiert und über die Zeit verfolgt. Dabei findet 3D-Kanteninformation und Fahrzeugbewegungsinformation Verwendung. Das Verfahren zeichnet sich durch kurze Laufzeiten auch bei mehreren verfolgten Ob-

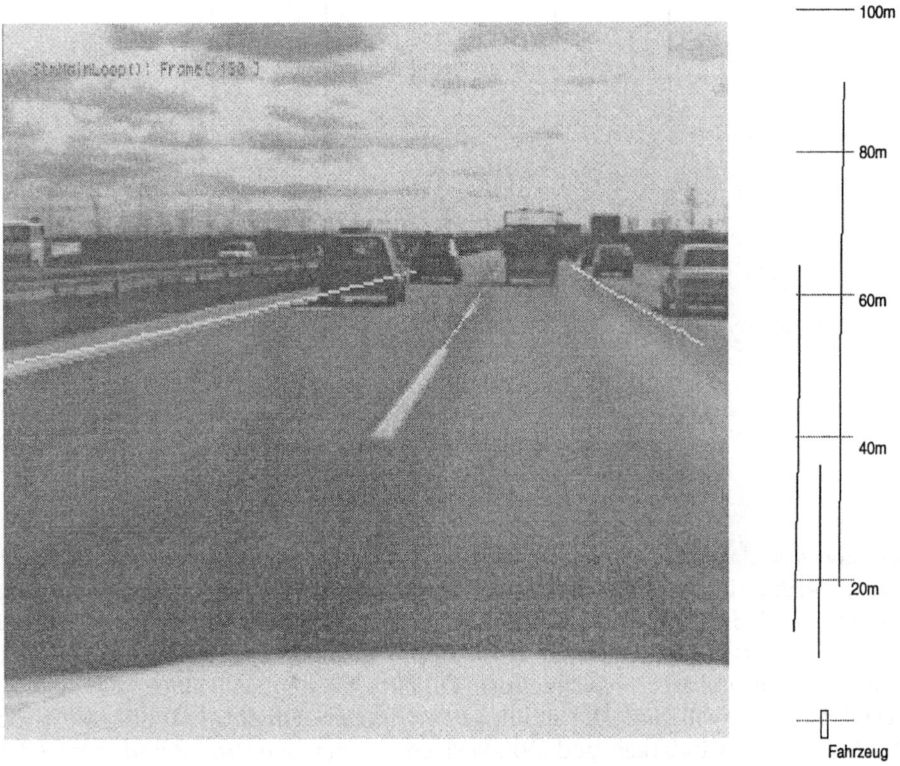

Abbildung 13. Kartierung des Straßen- und Spurverlaufs. Links: Orginalszene mit Spurberandungen; rechts: Spurbegrenzungen in Draufsicht. Man beachte den interpolierenden Effekt der Passung (linke Fahrbahnbegrenzung).

jekten aus. Auf diese Weise kann sehr schnell die Information über die in der Fahrzeugumgebung vorhandenen Hindernisse ermittelt werden (*primal scetch*).

9 Bildpixelbasierte Detektion, Verfolgung und Klassifikation verkehrsrelevanter Hindernisse (INI)

Zielsetzung des in diesem Abschnitt beschriebenen Teilprojekts ist die Detektion und Klassifikation verkehrsrelevanter Hindernisse.

Es wird ein Ansatz vorgestellt, der zunächst Objekthypothesen aus Bildsequenzen generiert, die Hypothesen nutzt, um bildlokal Objektklassifikationen vorzunehmen, die Objekte über die Zeit verfolgt und das vorausgehend erwähnte Klassifikationsergebnis - über mehrere Frames verteilt - verifiziert bzw. falsifiziert. In Abb. 14 sind die Funktionsblöcke des Moduls zur Hinderniserkennung

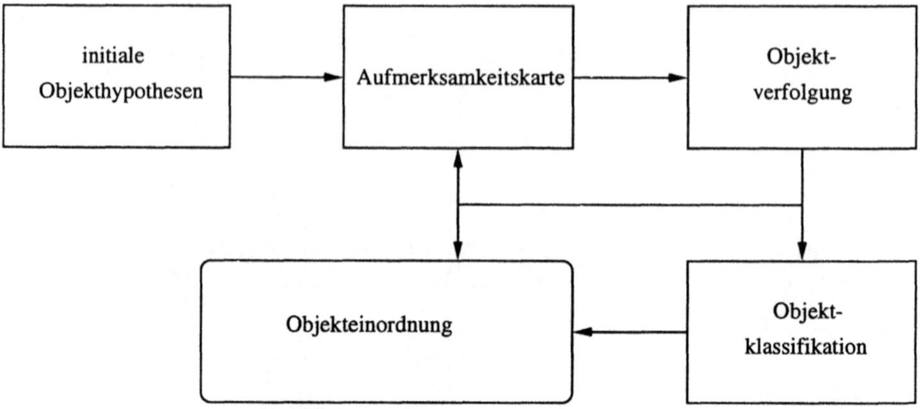

Abbildung 14. Hinderniserkennungssystem

schematisch dargestellt. Detektierte Objekthypothesen werden zur Initialisierung, Verifikation und Stabilisierung in einer Aufmerksamkeitskarte abgelegt. Hierbei wird durch Erzeugung von Redundanz die Robustheit der Ergebnisse der Objektverfolgung durch Rückkopplung signifikant erhöht. Die aus der Aufmerksamkeitskarte resultierenden Objekte werden schließlich über die Zeit verfolgt und klassifiziert. Die simultan gewonnenen Objektattribute, wie Größe, Abstand, Geschwindigkeit und Objektklasse stützen den Prozeß bildübergreifend (siehe Abb. 14). In den folgenden Abschnitten werden die einzelnen Prozeßschritte grob skizziert.

Generierung von Objekthypothesen. Ein erster Schritt in der Verarbeitungskette ist die initiale Detektion potentieller Hindernisse auf der Fahrbahn. Hierfür stehen etablierte Verfahren zur Vefügung; [17, 18]. In der hier beschriebenen Anwendung beschränkt man sich aufgrund der zur Verfügung stehenden, kurzen Rechenzeiten auf die Detektion von Bildbereichen mit ausgeprägtem, negativ-vertikalen Pixelgradienten und beschränkt diese auf den Ausschnitt der Fahrbahn. Da dieser ikonisch orientierte Verarbeitungsschritt am Beginn der Verarbeitungskette steht und mit wenig Daten- bzw. Modellwissen operiert, darf die Reduktion der gesamten Bildinformation auf Aufmerksamkeitsbereiche nur eingeschränkt vollzogen werden. Durch die nachgeschalteten Module ist eine engere Fokusierung auf die interessierenden Objektbereiche zu leisten.

Aufmerksamkeitskarte. Die Bewertung einzelner Bildbereiche in Bezug auf die Aufgabenstellung wird durch eine Aufmerksamkeitskarte realisiert, in welche spezielle Pixelattribute eingetragen werden. In der gewählten Repräsentation werden einkoppelnde Regionen über lokale "Aktivitäten" durch anwendungsspezifisch geeignete Funktionale bewertet und deren Signifikanz verifiziert (vgl. Abb. 15).

Abbildung 15. Aufmerksamkeitskarte am Beispiel der Objektverfolgung

Die Integration verschiedener Bewertungsmodule in das Gesamtsystem wird durchgeführt [19]. Module zur Sensordatenfusion [16] können auf dieser Verarbeitungsstufe in Zukunft ebenfalls integriert werden. Zur zeitlichen Stabilisierung und Verbesserung der Robustheit, führt die Aufmerksamkeitskarte die Zeit als dritte Achse mit und stellt nachfolgenden Verarbeitungsmodulen eine Liste interessierender Bildbereiche zur Verfügung, deren Hindernissignifikanz abfragbar ist. Die Einkopplung in die Objektverfolgung und -klassifikation ist realisiert. Eine Rückkopplung in die Objektdetektion ist prinzipiell möglich und in Zukunft geplant.

Objektverfolgung. Bei der Objektverfolgung ist der geometrische Vergleich von Gestaltsignaturen die zu lösende Hauptaufgabe. In dem gewählten Ansatz [21] wird die Abweichung eines Satzes von Modellmerkmalen zu den Bildmerkmalen mit Hilfe der *Hausdorff-Distanz* gemessen. Hierbei wird zum Zeitpunkt $t = 0$ ein transientes Modell des Objektes akquiriert und über der Zeit unter den Freiheitsgraden der Translation und Skalierung in der Bildsequenz wiedergefunden. Um ein möglichst hohes Maß an Robustheit und Flexibilität zu gewährleisten, wird das Modell anhand der wiedergefundenen Bildmerkmale über die Zeit adaptiert. Hierbei ist ein Verfahren entwickelt worden, welches gestattet das Modell objektselektiv anzupassen, um eine Adaption an Hintergrundbereiche auszuschließen. Der Algorithmus ist auf einer Vielzahl von Bildfolgen unterschiedlicher Geometrie und unterschiedlicher Beleuchtungsverhältnisse erfolgreich getestet worden (Abb. 16).

Objektklassifikation. Bei der Hindernisklassifikation kommen in erster Linie Neuronetz-basierte Methoden zum Einsatz [20, 21]. Grundsätzlich ist zwischen merkmals- und modellbasierten Klassifikatoren zu unterscheiden. Der schnelle merkmalsbasierte neuronale Klassifikator führt mit Hilfe der lokalen Orientierungskodierung [21] eine Initialklassifikation von Fahrzeugen durch. Zeilen-

Abbildung 16. Objektdetektion, -klassifikation und Objektverfolgung auf Autobahnen und Landstraßen

und Spaltenhistogramme werden als Eingangsdaten einem neuronalen MLP-Netz (universeller Approximator kontinuierlicher Funktionen) übergeben. Der Netzausgang beschreibt die Klasse der übergebenen Regionen. Im konkreten Fall wurden drei Klassen (Pkw, Lkw, und Transporter), sowie der Hintergrund gelernt. Der modellbasierte neuronale Klassifikator ordnet Bildregionen mit Hilfe der Hausdorff-Distanz [21] statisch festgelegten Modellen (hier: Limousine, Lieferwagen und Lkw) zu. Ein neuronales Netz bewertet hierbei die Ähnlichkeit zwischen den Modellen und der gegebenen Hypothese. Beide Typen Klassifikatoren wurden unter Verwendung einer Datenbank mit 1500 verschiedenen Objekten, die mit unterschiedlichen Kameratypen und -geometrien aufgenommen wurden, trainiert.

10 Ergebnisse und Diskussion; Ausblick

Zielsetzung des Projektes war die Entwicklung eines prototypischen, hochsprachenprogrammierbaren Systems zur flächendeckenden Interpretation von Bildfolgen für eine schwierige Klasse von Anwendungen. Dabei waren erhebliche Zielkonflikte hinsichtlich der Leistungsfähigkeit, Herstellungskosten, Baugröße, Umweltverträglichkeit und Verlustleistung der zugrundeliegenden Hardware zu bewältigen. Um den entstandenen Versuchsträger einem größeren Kreis (in der Regel wenig hardware-orientiert denkender Software-Entwickler) zuzuführen, wurde das System mit einer leistungsfähigen SW-Entwicklungs- und Produktionsumgebung versehen. Die damit geschaffenen Möglichkeiten ermöglichen nicht nur kürzeste Entwicklungszyklen (Rapid Prototyping) komplexer, vollbildauswertender Systemanwendungen in Software, sondern erlauben erstmals deren unmittelbar folgende schritthaltende, experimentelle Erprobung im Feld.

Dieser Aspekt ist insofern hoch zu bewerten, als die überwiegende Anzahl der in der Literatur veröffentlichten Vollbildverfahren typischerweise auf nur weni-

Abbildung 17. Objektdetektion, -klassifikation und Objektverfolgung an einer Sequenz von 1000 Bildern mit unterschiedlichen Beleuchtungsverhältnissen

gen Bildern oder Bildfolgen erprobt werden und die zugrundeliegenden Verfahren daher erfahrungsgemäß häufig auf die darin enthaltenen Testmuster "trainiert" werden. Da es darüberhinaus keine geschlossene "Systemtheorie des Bildverstehens" gibt, ist man hinsichtlich der Funktions- und Robustheitsuntersuchung prototypischer Gesamtanwendungen notwendig auf umfangreiche experimentelle Untersuchungen angewiesen. Dies gilt zumindest für die hier vorgestellten Applikationen zur Fahrerassistenz und die ihnen eigene immense Szenenvariabilität.

Experimentelle Untersuchungen zu den in diesem Artikel vorgestellten Verfahren erfordern bereits im Open-Loop-Betrieb Echtzeitfähigkeit. In 1994 hätte die beaufsichtigte Verrechnung einer Stunde Bildmaterial auf dann verfügbarer Workstation-Hardware ca. $\frac{1}{2}$ Mannjahr benötigt.

Mit der im Projekt entstandenen Entwicklungs- und Versuchsplattform wurde erstmals die Basis dafür geschaffen, *Computer Vision*-Anwendungen der hier beschriebenen Komplexität, ernsthaft dem Experiment zuzuführen. Die Leistungs- und Systemanforderungen der oben charakterisierten Entwicklungs- und Versuchsplattform wurden u.a. von denen der vorgestellten prototypischen Bildverarbeitungsanwendungen abgeleitet. Die Echtzeitfähigkeit der realisierten Verarbeitungsprozesse zur Eigenbewegungsbeobachtung, zur schritthaltenden Kartierung des Straßen- und Spurenverlaufs und zur Hindernisdetektion ist daher im wesentlichen bereits gesichert. Zeitgleich zur Verfassung dieses Artikels beginnt die Systemintegration der genannten, im Off-line-Betrieb extensiv getesteten Prozesse. Erste Feldversuche sind noch für das 2. Halbjahr 1998 geplant.

Kurz- und mittelfristige Anschlußprojekte beinhalten neben algorithmischen Verfahrensverbesserungen:

- die Kopplung der in Abschnitt 2 beschriebenen CMOS-Kamera mit der SAP-Baugruppe
- die Entwicklung höher auflösender CMOS-Sensorkomponenten (Ziel: bis zu 1024 × 512 Pixel)
- die weitere Miniaturisierung der SAP-Hardware (unter Erhöhung der Taktfrequenz)
- die Entwicklung von Spin-Off Anwendungen deutlich reduzierter Leistungsforderungen, ggf. auch: Integration intermediärer Bildverarbeitungsprozesse

Die in diesem Beitrag beschriebenen Module, Systeme und Anwendungen werden auf der vom 29.09. bis 01.10.1998 zeitgleich zum DAGM Symposium stattfindenden "Vision 98" präsentiert.

11 Danksagung

Die Autoren bedanken sich zunächst beim Bundesministerium für Forschung und Technologie (BMBF) und dem eingeschalteten Projekträger für Informationstechnologie für die Förderung des Vorhabens unter den Kennzeichen IT-ME-01M3010A5, IT-ME-01M3010B und IT-ME-01M3010C0. Das mit der Verbundleitfunkiton betraute BMW Projektteam dankt den zuständigen Entscheidungsträgern der Firmen BMW AG und BGT GmbH für die Bewilligung des Vorhabens und der dafür erforderlichen Ressourcen, die das Projekt und die damit erzielten Ergebnisse im Grundsatz ermöglicht haben. Besonderer Dank gilt den Projektleitern und Entwicklungsteams der Verbundpartner für ihr über die Projektlaufzeit aufrechterhaltenes großes Engagement und ihren unermüdlichen Arbeitseinsatz.

Aufgrund der standort- und institutionsübergreifend engen Zusammenarbeit von zwanzig bis dreißig zumindest zeitweise im Projekt arbeitenden Doktoranden und Entwicklungsingenieuren, haben wir auf die Angabe einer Autorenliste verzichtet. Die Namen der hauptsächlich zum Projekt beitragenden Personen können über die Veröffentlichungen [1–3, 22, 14, 15, 21] zugeordnet werden.

Literatur

1. M. Baur, T. Schumm, R. Wertheimer: Abschlußbericht BMW zu BMBF Förderkennzeichen IT-ME-01M3010A5. Technische Informationsbibliothek (TIB) (1998), Postfach 6080, D-30060 Hannover, FRG
2. Abschlußbericht BGT zu BMBF Förderkennzeichen IT-ME-01M3010B. Technische Informationsbibliothek (TIB) (1998), Postfach 6080, D-30060 Hannover, FRG
3. M. Schanz, T. Eckart, Ch. Nitta, B.J. Hosticka, R. Wertheimer: CMOS Imager with 120 dB Linear Output. Submitted for publication in Proceeding of the ISSCC 99
4. C.G. Harris, M.J. Stephens: A combined corner and edge detector. in: Proceedings, 4th Alvey Vision Conference (1988) 147–152
5. A. Mitiche, P. Bouthemy: Computation and Analysis of Image Motion: A Synopsis of Current Problems and Methods. International Journal Computer Vision, **19** (1996) 29–55

6. Q.T. Luong, O.D. Faugeras: The Fundamental Matrix: Theory, Algorithms and Stability Analysis. International Journal Computer Vision, **17** (1996) 43–75
7. J. Weng, T. S. Huang, N. Ahuja: Motion and Structure from Image Sequences. Springer-Verlag, Berlin, Heidelberg (1993)
8. Z. Zhang, O. Faugeras: 3D Dynamic Scene Analysis. Springer-Verlag, Berlin, Heidelberg (1992)
9. L. S. Shapiro, A. Zisserman, M. Brady: 3D Motion Recovery via Affine Epipolar Geometry. International Journal of Computer Vision, **16** (1995) 147–182
10. Q.-T. Luong, O.D. Faugeras: Self-Calibration of a Moving Camera from Point Correspondences and Fundamental Matrices. International Journal Computer Vision, **22** (1997) 261–289
11. J.M. Orwell, J.F. Boyce, J.F. Haddon: Ego Motion from Near-Degenerate Sequences. in: Proceedings, 13th International Conference on Pattern Recognition, Vienna, Austria (1996) 412–416
12. T. Viéville, O.D. Faugeras, Q.T. Luong: Motion of Points and Lines in the Uncalibrated Case. International Journal Computer Vision, **17** (1996) 7–41
13. R. Wagner, K. Donner, F. Liu: A "Half-Perspective" Approach to Robust Ego-Motion Estimation for Calibrated Cameras. Fakultät für Mathematik und Informatik, Universität Passau, MIP-9718 (1997)
14. R. Wagner, F. Liu, K. Donner: Robust Motion Estimation for Calibrated Cameras from Monocular Image Sequences. erscheint in: Computer Vision and Image Understanding
15. J. Haas: Echtzeit-Korrespondenzprobleme in Bildsequenzen und Subpixelgenauigkeit. Dissertation, Universität Passau, Fakultät für Mathematik und Informatik (1998)
16. U. Handmann and G. Lorenz and W. v. Seelen: Fusion von Basisalgorithmen zur Segmentierung von Strassenverkehrsszenen. in Mustererkennung 1998 (1998), Springer-Verlag, Berlin, Heidelberg
17. T. Kalinke and C. Tzomakas: Objekthypothesen in Verkehrsszenen unter Nutzung der Kamerageometrie. IRINI 97-07 apr (1997), Institut für Neuroinformatik, Ruhr-Universität Bochum
18. U. Handmann and T. Kalinke: Fusion of texture and contour based methods for object recognition. ITSC'97 (1997)
19. U. Handmann and G. Lorenz and T. Schnitger and W. von Seelen: Fusion of different Sensors and Algorithms for Segmentation. ITSC'98
20. T. Kalinke and C. Tzomakas and M. Werner and W. v. Seelen: Texture-Based Object Detection and an Adaptive Model-Based Classification. ITSC'98
21. U. Handmann and T. Kalinke and C. Tzomakas and M. Werner and W. v. Seelen: Computer Vision for Driver Assistance Systems. Proceedings of SPIE Vol. 3364
22. O. Ehrmann, K. Buschick, G. Chmiel, A. Paredes, V. Glaw, H. Reichl: 3D - Multichip Module, Proceedings International Conference on Multichip Modules, Denver, 1995, pp.358-363
23. J. Wolf, F.J. Schmöckle, W. Heinrich, M. Töpper, K. Buschick, A. Owzar, O. Ehrmann, H. Reichl: System Integration for High Frequency Applications, Proceedings International Symposion on Microelectronics, Philadelphia, 1997, pp.29-36
24. M. Töpper, K. Buschick, J. Wolf, V. Glaw, R. Hahn, O. Ehrmann, H. Reichl: Embedding Technology - A Chip-First Approach using BCB, Advancing Microelectronics, Vol.24, No.4 (1997) pp.18-21

Grundlagen und Algorithmen

Junction Characterization Using Polar Pyramid

Weichuan Yu, Kostas Daniilidis, Gerald Sommer

Christian–Albrechts–Universität
Lehrstuhl für Kognitive Systeme
Preußerstraße 1–9, 24105 Kiel
Email: wy@informatik.uni-kiel.de

Abstract. In this paper we present a new approach in characterizing gray-value junctions. Due to the multiple intrinsic orientations present in junctions the response of a filter is needed at every orientation. As a rotation of the filter would considerably increase the computational burden alternative techniques like filter steerability have been proposed. Steerability relies in interpolating the response at an arbitrary orientation from the responses of some basis filters. Unfortunately, current steerability approaches suffer from the consequences of the uncertainty principle: In order to achieve high selectivity in orientation they need a huge number of basis filters increasing, thus, the computational complexity.
The new approach presented here achieves a higher orientational selectivity with a lower complexity. We consider the local polar map of the neighborhood of a junction where the new coordinates are the radius and the angle. Finding the gray-value transitions of a junction can be interpreted as 1D edge detection. Hence, the orientational selectivity problem can be attacked by applying a pyramidal scheme. It is well known that it is always possible to reconstruct a signal using the sampling kernel as an interpolation function. Therefore, our approach can also steer the response of a Gaussian derivative to every orientation. The total algorithmic complexity encompasses the small support 2D-filtering for polar mapping and radial smoothing plus an 1D-differentiation.

1 Introduction

Junctions of gray-value lines or edges are rare events in images carrying important information for many image processing tasks like point matching in object recognition, point tracking in motion analysis, and attentive coding.

In order to utilize junctions for such tasks we must be able to locate their corresponding keypoints (defined as meeting points of lines or edges), to characterize them by means of signatures and to classify them in junction categories. Regarding keypoint detection and localization the reader is referred to Förstner's study [5] and to the comparison of different operators by Rohr [10, 11]. In this paper we address the problem of junction characterization. The resulting signature can be used for further junction classification.

Junctions are local structures with multiple intrinsic scales and orientations [1]. A signature characterizing such a junction can be only obtained by applying a filter at different scales and orientations around the keypoint. This results in an enormous computational complexity. In order to attenuate this burden the concept of steerability has been introduced [6, 9]. Denoting the deformation parameter of a filter with α we define a filter $F(\boldsymbol{x})$ with $\boldsymbol{x} \in \mathbb{R}^n$ as a steerable filter if its deformed versions $F_\alpha(\boldsymbol{x})$ can be expressed as [7]:

$$F_\alpha(\boldsymbol{x}) = \sum_{k=1}^{N} b_k(\alpha) A_k(\boldsymbol{x}) \tag{1}$$

where $A_k(\boldsymbol{x})$ and $b_k(\alpha)$ are called basis filters and interpolation functions, respectively. Applying such a steerable filter on an image $I(\boldsymbol{x})$ yields:

$$\langle F_\alpha(\boldsymbol{x}) | I(\boldsymbol{x}) \rangle = \sum_{k=1}^{N} b_k(\alpha) \langle A_k(\boldsymbol{x}) | I(\boldsymbol{x}) \rangle \tag{2}$$

where $\langle \cdot | \cdot \rangle$ is the usual scalar product. In this way we can reduce the computational cost of applying a whole set of different $F_\alpha(\boldsymbol{x})$ to N scalar products.

For junction characterization many kinds of steerable filters are applied like derivatives of the 2D Gaussian filter [6], elongated Gaussian-derivative kernels [9], the double Hermite function [7] and the steerable wedge filter [12]. All these steerable filters adopt trigonometric functions or complex exponentials of α as interpolation functions $b_k(\alpha)$ (equation 1) such that the basis functions are rotated copies of the original filter or can be solved using Fourier theory.

Steerability provides us with a solid mathematical theory. However, the complexity of its implementation remains high. Due to the uncertainty principle the product of orientational resolution of a steerable filter and its orientational bandwidth has a lower bound. Therefore, in order to achieve a high orientational selectivity we have to apply a huge number of basis filters [7, 9] with large spatial support.

In our previous work [4] we have presented a new approach to characterize junctions. It is based on applying rotated copies of a wedge averaging filter and estimating the derivative with respect to the polar angle. This averaging filter is polar separable. Its radial and angular components are rectangle functions expressed by following and shown in figure 1:

$$p(\rho) = \begin{cases} 1 & R_{min} \le \rho \le R_{max} \\ 0 & \text{otherwise} \end{cases} \tag{3}$$

with R_{min} and R_{max} as its boundaries in radial direction and

$$g_{\theta_i}(\theta) = \begin{cases} 1 & \mathcal{D}(\theta, \theta_i) \le \frac{W}{2} \\ 0 & \text{otherwise} \end{cases} \tag{4}$$

where $\theta, \theta_i \in [0, 2\pi]$, θ_i is the center of the filter and W is the angular width of the mask. Since θ and θ_i are circular angles, we define a $\mathcal{D}(\ )$ to represent the minimal absolute value of their difference.

In this paper we will observe this simple averaging filter from a more interesting point of view and extend it into a pyramidal scheme. The averaging filter is nothing else than the kernel required to conduct a local transformation in polar coordinates (Fig. 1). However, this polar transformation involves a filtering: We smooth and subsample the angular direction and keep just the average of the radial direction. This will allow us to treat oriented structures as structures of a 1D-signal. To treat the orientational selectivity problem we choose a pyramidal representation. The computational burden is much lower than usual Fourier-domain schemes because the filters have smaller supports and part of them act only one-dimensionally. We also show experimentally the superiority of our scheme in many real images.

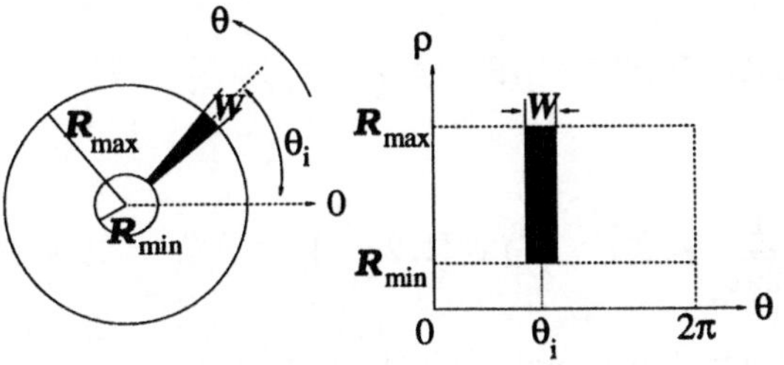

Fig. 1. Mask centered at angle θ_i. **Left:** Original mask with keypoint at the center of the circle. **Right:** Mask with θ and ρ as Cartesian coordinates, R_{max} and R_{min} are radial boundaries of the mask, W is the angular width of the mask.

2 The Polar Pyramid

In the following, we will apply an averaging filter in the radial direction since we do no treat spatial scale in this paper. In the angular direction we replace the averaging filter with a Gaussian function:

$$G_{\theta_i}(\theta) = \frac{1}{\sqrt{2\pi}\sigma} e^{-\frac{(\mathcal{D}(\theta,\theta_i))^2}{2\sigma^2}} \tag{5}$$

where σ is the scale of the Gaussian function, θ_i and $\mathcal{D}(\,)$ are the same as in equation 4.

We will later see that the choice of a Gaussian in establishing the local polar map has many nice effects. Since we want to use the Gaussian function as a FIR-filter we must cut off its support. It is easy to show that in order to keep the energy of the cut-off area below 1% of the total energy the width of the mask must be at least 5σ. According to the sampling theory we can sample the angular space with an interval of σ. Thus, even taking the overlapping of neighboring masks into account we need to cover the whole neighborhood only a few times to obtain the required information.

As we already mentioned in the introduction the main motivation of the proposed approach is the appropriate treatment of the orientational selectivity. The local polar map enables us to look at this problem as a 1D gray value transition description. The Fourier-based steerability approaches approximate a filter with a Fourier series expansion with respect to the angle. The more Fourier-coefficients and basis functions they use the more selective they are in discerning fine gray value transitions in the polar map. However, in order to treat angular transitions of varying scale we introduce a pyramid in the polar domain.

It is well known that one of the most appealing kernels for hierarchical approaches is the Gaussian function [2]. The Gaussian function achieves the lowest bound of the product between spatial support and bandwidth. Burt and Adelson [3] prove that the generating kernel of subsampling can be used as an interpolation function. Moreover, they argue that the interpolation functions can be discrete approximations of Gaussian functions with different scales (Fig. 2). Thus, the continuous orientational information can be reconstructed from all levels of the polar pyramid by interpolating the pyramid elements $P_j(\theta)$ with Gaussian functions of different scales $G_j(\theta)$:

$$\hat{f}_j(\theta) = \sum_n \delta(\theta - n\Delta\theta) P_j(\theta) * G_j(\theta) \qquad j \in [1, 2, \cdots] \tag{6}$$

with

$$G_j(\theta) = \frac{1}{\sqrt{2\pi}\sigma_j} e^{-\frac{\theta^2}{2\sigma_j^2}} \qquad j \in [1, 2, \cdots] \tag{7}$$

where $\hat{f}_j(\theta)$ represents the reconstructed signal from the j-th level.

Because in natural images edges are more important than lines we need to estimate the derivative of $\hat{f}_j(\theta)$ to obtain the required information:

$$D_j(\theta) = \left| \frac{d}{d\theta} \hat{f}_j(\theta) \right| \qquad j \in [1, 2, \cdots] \tag{8}$$

The local maxima in $\hat{f}_j(\theta)$ and $D_j(\theta)$ denote orientations of lines and edges at different levels, respectively.

If we want to build a pure 1D octave Gaussian Pyramid of angles, according to [3] we should have $M2^K + 1$ samples, where K is the number of levels and $M+1$ is the number of samples at the highest level. Taking the periodic property into account we should have $M2^K$ samples as the smoothing outputs. However, usually 360 angular values are used. Observing that $360 = 2^3 \times 3^2 \times 5$ we change the subsampling interval slightly. For instance, for three levels we subsample every two points and for two levels every three.

According to [3] the generating kernel should be normalized, symmetric and unimodal and it should make equal contribution to construct the next higher level. The generating kernels satisfying these constraints for subsampling with factor 2, 3 and 5 can be expressed as following, respectively:

$$S_2 = \frac{1}{16}(1\ 4\ 6\ 4\ 1) \tag{9}$$

$$S_3 = \frac{1}{264}(3\ 22\ 66\ 82\ 66\ 22\ 3) \tag{10}$$

$$S_5 = \frac{(1\ 74\ 299\ 725\ 950\ 1022\ 950\ 725\ 299\ 74\ 1)}{5120} \tag{11}$$

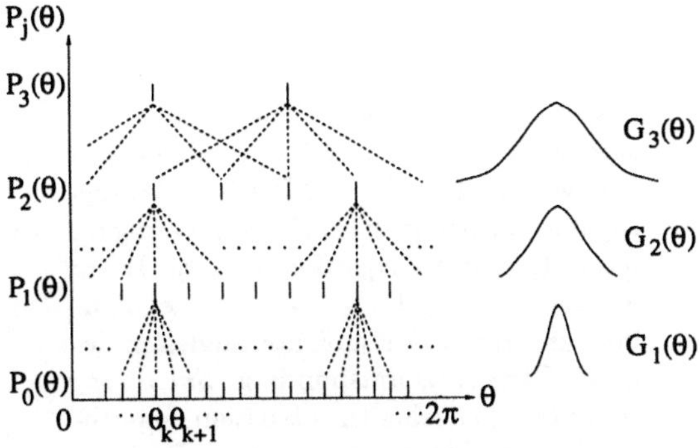

Fig. 2. Left: Polar pyramid structure, $P_0(\theta)$ is the output of smoothing Gaussian filter. $P_j(\theta)(j = 1, 2, \cdots)$ are higher levels after subsampling with generating kernels S_2 or S_3. **Right:** Corresponding interpolation functions at different levels. They are Gaussian functions with different scales.

3 Junction Characterization Examples and Discussions

3.1 Junctions without Scale Variations

For junctions without orientational scale variations the smoothing outputs $P_0(\theta_i)$ and the absolute values of its derivative $D_0(\theta_i)$ $(\theta_i = 0°, \cdots, 359°)$ are adequate for characterizing junctions.

In figure 3 and 4 are examples of synthetic line junctions and edge junctions. The corresponding $P_0(\theta_i)$ and $D_0(\theta_i)$ characterize them correctly. The improvement of replacing the angular averaging filter by a Gaussian smoothing filter is shown in figure 3. The small deviations of orientations in figure 4 are due to the fact that an edge can be only presented by two pixels in the grid, while we can not set the center of a mask between two pixels.

The robustness of our method against keypoint offset is shown in figure 5. A real example is in figure 6. We present results using the steerable wedge filter [12], the averaging method and the polar pyramid scheme. It is clear to see that the polar pyramid reaches the best orientational resolution. In comparison to the averaging method we improve robustness with a moderate increase of complexity which still remains considerably lower than the steerability method.

3.2 Solving Orientational Scale Problem

In figure 7 we show how to solve the orientational scale problem by applying the polar pyramid. Each wide line of the junction is best characterized at the third level ($P_2(\theta)$) with an orientational resolution of $2\sigma_3 = 16°$, while at lower or higher levels the junction is not so discernible. In the future work we further need a criterion to judge the quality of reconstructed information analytically.

The result of a child's left eye corner is presented in figure 8. The corner can be regarded as a combination of irregular wide lines and blurred edges disturbed by noise. Our method provides suitable information for further interpretation.

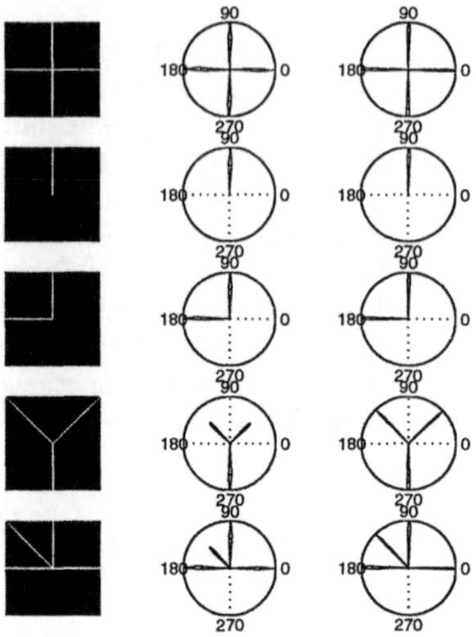

Fig. 3. Left: Synthetic line junction images with the size of 65×65 pixels and line width of 1 pixel. **Middle:** ***Averaging outputs*** using the rectangle averaging filter defined in equation 4 with $W = 6°$, $R_{min} = 3$, $R_{max} = 15$. The numbers around circles indicate angles in degree. **Right:** Corresponding ***Smoothing outputs*** $P_0(\theta)$ using the Gaussian smoothing filter with same parameters. Both filters show good performances. But the normalization of the Gaussian filter is better (row 4 and 5).

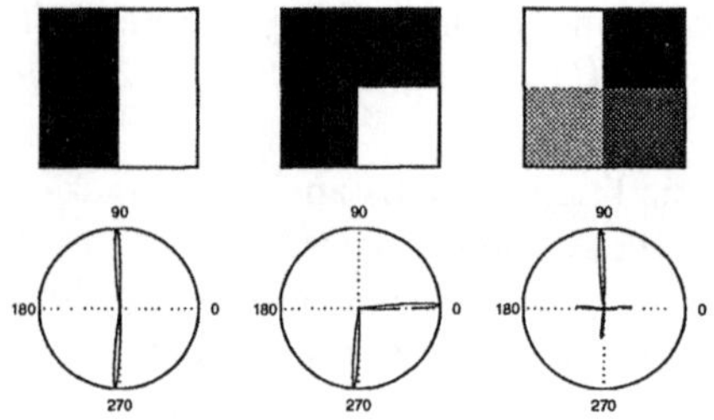

Fig. 4. Top: Synthetic edge junctions. **Bottom:** Polar plots of $D_0(\theta)$. The local maxima show the orientation of edges. $R_{min} = 3$, $R_{max} = 15$.

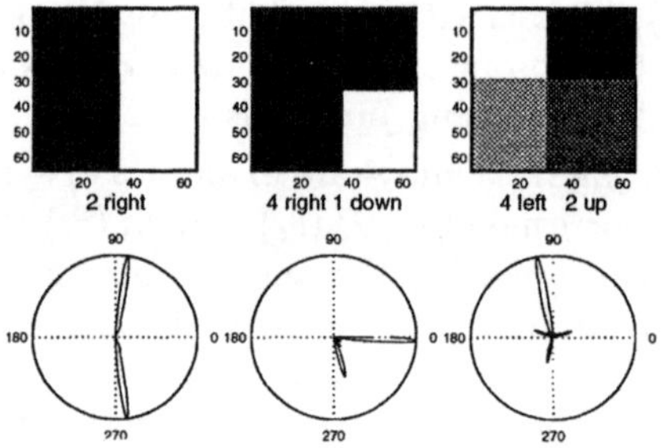

Fig. 5. Top: Different deviation of keypoints from centers of masks. **Bottom:** Corresponding $D_0(\theta)$.

4 Conclusion

The steering of a filter to different orientations enables the characterization of junctions. To avoid rotating a filter at all orientations the theory of steerability allows the interpolation of a filter response at every orientation from a limited set of basis filters. In Fourier-based steerability approaches it turns out that the number of basis filters is proportional to the angular bandwidth, which has an inverse relationship with the orientational selectivity. In practice, to characterize fine structures we need a huge number of basis filters with very large angular support. A second fundamental fact is that intrinsically the Fourier-based methods involve a local polar transformation which maps a filter rotation to a filter translation [8].

We proposed in this paper the explicit implementation of the polar transformation which allows a further 1D-treatment of the orientation problem just in the angular direction. Instead of using a Fourier series expansion we make use of the pyramidal processing theory and introduce the Gaussian as the interpolating function. Simultaneously, the Gaussian function and its derivative are the generating kernels of the pyramid. The hierarchical treatment enables us to characterize both fine and course transitions in gray-value junctions. Experimental results show the superiority of the hierarchical approach over a classic steerability scheme and a similar polar averaging filter in one scale.

Acknowledgment

The financial support of the first author by DAAD and of the second and third authors by DFG grant 320/1-2 is greatly acknowledged. We thank G. Birkelbach and H. Scharr for their helpful discussions. We appreciate Dr. Farid's help in providing the software for the steerable wedge filter.

References

1. D. H. Ballard and L. E. Wixson. Object recognition using steerable filters at multiple scales. In *IEEE Workshop on Qualitative Vision*, pages 2–10, 1993.
2. J. Babaud , A. P. Witkin , M. Baudin and R. O. Duda. Uniqueness of the Gaussian kernel for scale-space filtering. *IEEE Trans. Pattern Analysis and Machine Intelligence*, 8:26–33, 1986.
3. P. J. Burt and E. H. Adelson. The Laplacian pyramid as a compact image code. *IEEE Transactions on Communications*, 31:532–540, 1983.
4. W. Yu, K. Daniilidis and G. Sommer. Rotated wedge averaging method for junction characterization. In *the IEEE Computer Society Conference on Computer Vision and Pattern Recognition (CVPR'98), Santa Barbara, June 23-25*, 1998.
5. W. Foerstner. A framework for low level feature extraction. In *European Conf. on Computer Vision*, volume II, pages 383–394, Stockholm, Sweden, May 2-6, J.O. Eklundh (Ed.), Springer LNCS 801, 1994.
6. W.T. Freeman and E.H. Adelson. The design and use of steerable filters. *IEEE Trans. Pattern Analysis and Machine Intelligence*, 13:891–906, 1991.
7. M. Michaelis and G. Sommer. Junction classification by multiple orientation detection. In *European Conf. on Computer Vision*, volume I, pages 101–108, Stockholm, Sweden, May 2-6, J.O. Eklundh (Ed.), Springer LNCS 800, 1994.
8. M. Michaelis and G. Sommer. A Lie group approach to steerable filters. *Pattern Recognition Letters*, 16:1165–1174, 1995.
9. P. Perona. Deformable kernels for early vision. *IEEE Trans. Pattern Analysis and Machine Intelligence*, 17(5):488–499, 1995.
10. K. Rohr. Recognizing corners by fitting parametric models. *International Journal of Computer Vision*, 9(3):213–230, 1992.
11. K. Rohr. On the precision in estimating the location of edges and corners. *Journal of Mathematical Imaging and Vision*, 7:7–22, 1997.
12. E. P. Simoncelli and H. Farid. Steerable wedge filters for local orientation analysis. *IEEE Trans. Image Processing*, 5(9):1377–1382, 1996.

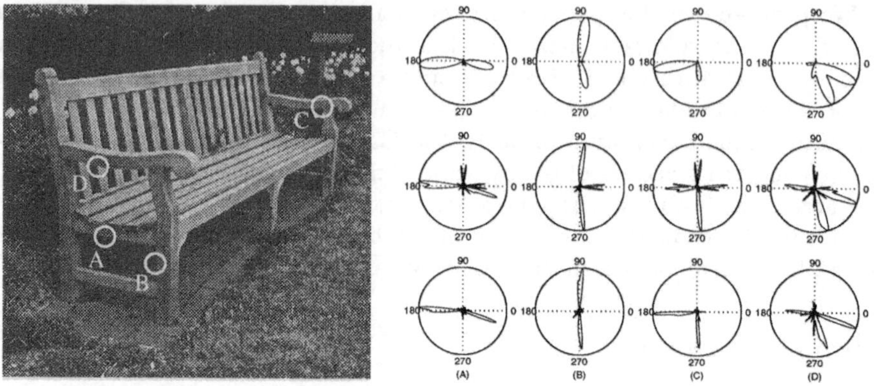

Fig. 6. Left: Parkbench with edge junctions. *A*: horizontal edge; *B*: vertical edge; *C*: corner; *D*: 'T' junction. **Right Row 1:** Steerable wedge filter results using 30 basis filters with 19-tap size. **Right Row 2:** Derivative Outputs using angular averaging filter. The averaging filter is sensitive to high frequency component. **Right Row 3:** Derivative Outputs using Gaussian smoothing filter. The result is better than using averaging filter. $R_{min} = 3, R_{max} = 9$.

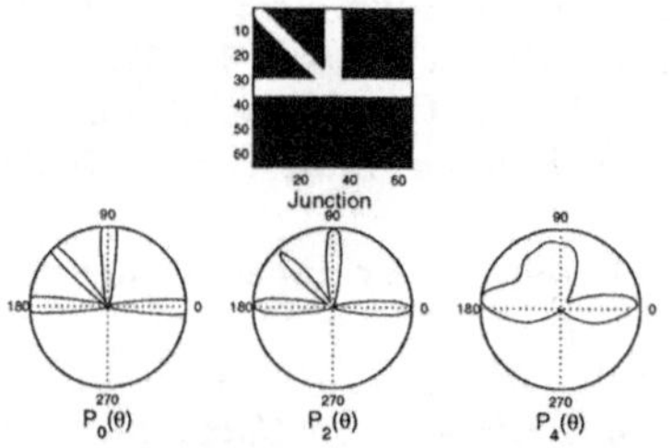

Fig. 7. Top: A junction composed of lines with a width of 7 pixels. **Bottom:** orientational signals reconstructed from the first, third and fifth level of the polar pyramid. Each wide line is best characterized with only one maximum at the third level ($P_2(\theta)$), while at even larger scale ($P_4(\theta)$) the structure is hardly discernible.

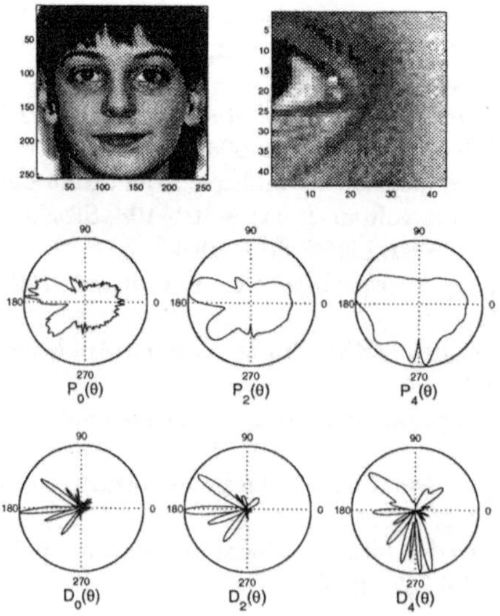

Fig. 8. Top left: Face of a child. **Top right:** Its left eye corner in detail. It can be regarded as a combination of irregular wide lines and blurred edges disturbed by noise. **Middle:** orientational signals reconstructed from the first, third and fifth level of the polar pyramid. The eyelids are clear to see at the third level as two maxima at 135° and 225°. The local maximum near 180° is due the white of the eye. **Bottom:** Corresponding $D_j(\theta), j = 0, 2, 4$.

Probabilistic Hypothesis Verification

Dietrich Büsching

Forschungsgruppe Bildverstehen, TU München,
Orleansstr. 34, 81667 München **

Abstract We describe a method for hypothesis verification for view-based recognition using Bayes rule. For feature positions we assume conditional densities which are based on the observation that deviations in the relative position of feature points increase with the distance of feature points. We advocate the use of curve segments with constant sign of curvature instead of circular or elliptical arc segments. Our approach is validated with real and artificial data of mostly curved objects.

1 Introduction

For most approaches to object recognition based on a geometrical model one can distinguish two major stages: hypothesis generation and hypothesis verification. Object hypotheses can be generated by alignment (e.g. [5]), by searching the space of possible feature correspondences [2] or by grouping of features and indexing in a model database [6]. Each method results in one or more pose hypotheses. The final decision about the correct hypothesis is taken in the verification stage. Some successful practical approaches to verification predict model edges according to a given pose hypothesis and search for corresponding image edges (chamfer matching). In [7] (e.g.) it was pointed out that Bayes rule can be seen as a theoretical justification of these standard methods. The probability of a hypothesis H given some matched image features F_i can be calculated (assuming approximate independence of features):

$$P(H|F_1, .., F_n) \approx \prod_i \frac{P(F_i|H)}{P(F_i)} P(H) \tag{1}$$

Bayes rule offers a sound way to combine the evidence obtained by matching features of different types. More complex features like arc segments are rare in images. According to Bayes rule more weight is given to these features since their prior probability is small. It can be desirable to use several types of features if e.g. polyhedral and curved objects are to be recognized simultaneously. [5] is an important example for such a system. It uses a view-based representation with features of several types. The prior probability densities $p(F_i)$ for position and other feature attributes (orientation, size etc.) are estimated according to typical

** Current address: School of Computer Studies, University of Leeds, Leeds LS2 9JT, UK scenes. Estimation of the conditional densities $p(F_i|H)$ of a model view is based on a comparison of feature attributes in training images that are represented by the model view. This assignment of training images to model views is done according to the minimum description length principle.

Like [5] we use Bayes rule to combine the matching results of several feature types. For the calculation of conditional probabilities we employ a different approach and also use other features. This decision was motivated by some problems with the work described in [5].

- The application of Bayes rule does not give reliable results for the final evaluation of a hypothesis. Instead a variation of chamfer matching is used for verification. Bayes rule is only used to order the matching of features for the incremental computation of pose hypotheses. The failure of the bayesian approach might be caused by the method for evaluating simple curve features (line segments, arc segments or ellipse segments), which ignores the length of curve segments.
- The estimation of conditional probability densities from training images is difficult due to the small number of training exemplars (often less than 10) per model view. Mismatches between features in training images pose an additional problem.
- It seems to be desirable to have a measure of similarity for two views of an object without the need for a prior complicated training procedure.

2 The suggested similiarity measure

We start with a description of the main idea for point features. Like in [5] we assume that a model view represents a part of the possible viewing directions on a 3D-object. Additionally we assume that this part is circular. If the object rotates within this range of directions the 2D projections of feature points on the surface of the object move within ellipses. The orientation, size and excentricity of these ellipses depend on the 3D position of the points. The results of these variations on differences between real and predicted 2D positions also depend on the method used for hypothesis generation. In our approach hypotheses are generated by matching the minimum number of necessary points (two for similarity transformations). We use the two endpoints of a complex feature (see below). If a rotation out of the image plane changes the relative position of these two basis points, this will result in errors of the predicted positions of other feature points. The magnitude of these errors increases in proportion to the distance to the two basis points (see figure 1). Since we do not assume prior knowledge about the 3D structure of the object the directions of these deviations are unknown. Therefore we assume circular (gaussian) probability distributions for these deviations. As explained above the standard deviation of the distribution of a given point is chosen to be proportional to the distance to the basis points.

Simple point features have little discriminatory power. In the implemented recognition system more complex features like line segments or curve segments with constant sign of curvature (called 'convex segments') are used. In other work

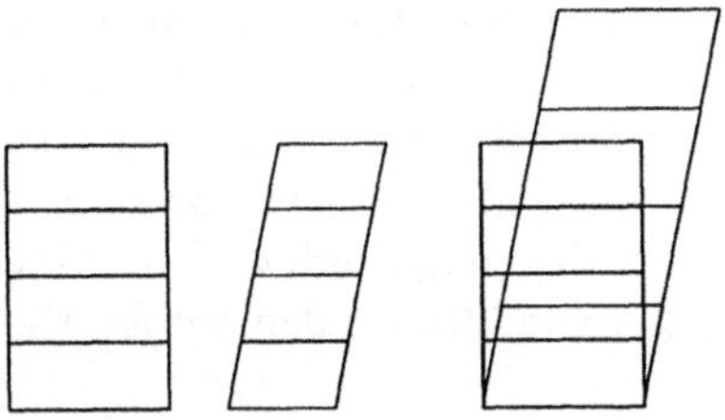

Figure1. Left: model view of an object; middle: rotated object; right: matching of two corresponding lines. The vertical distance between corresponding lines increases with the distance to the matched lines; the difference in the relative position of neighbouring lines remains the same.

(e.g. [5]) curved lines have been approximated by circular or elliptical arcs. The results of section 3 suggest that the representation used is more reliable. Another feature type are pairs of nearly parallel lines enclosing an area of homogeneous colour (homogeneous ribbons). For experiments with presegmented contours bitangent features (see e.g. [6]) were used. These feature types are illustrated in figure 2.

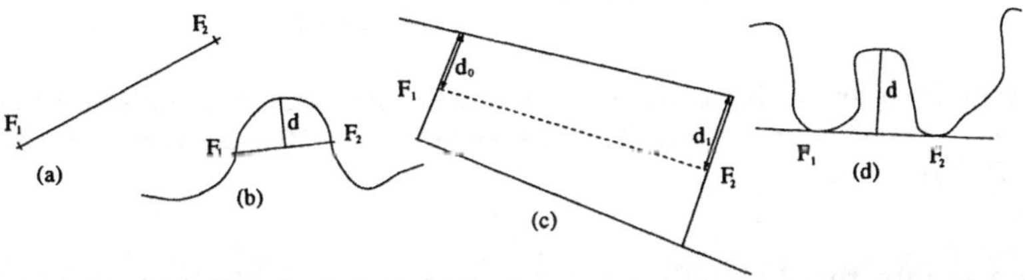

Figure2. Feature types used: (a) line segments, (b) convex segments, (c) homogeneous ribbons, (d) bitangents.

The stability of the detection of features strongly depends on the reliability of the grouping of simple edge points to edge segments. The grouping method used is not very stable. Edges are fragmented due to short gaps or junctions with other edges. In order to decrease the sensitivity of feature detection, input features are analysed in several resolutions. Often small image variations have no influence on features detected in low resolutions. The results of feature detection in different resolutions are combined and multiply detected features are discarded.

The calculation of conditional probabilities for complex features is based on the main idea described for point features. Additionally deviations in the relative position of points within the feature are considered. The magnitude of these deviations is expected to be proportional to the distance of these points (see figure 1). For simplicity we assume independence of all absolute and relative positions considered. Therefore the conditional probability of line segments is computed as the product of the probability for the first endpoint of the segment and the vector from the first to the second endpoint. For both probabilities we use circular gaussian distributions. The standard deviations are proportional to the length of the model vector from B (one of the two basis points) to the first endpoint F_1 respectively the length of the vector from F_1 to the second endpoint F_2. If ΔF_1 and ΔF_2 are the magnitudes of the deviations from the model vectors $\boldsymbol{BF_1}$ and $\boldsymbol{F_1F_2}$, the conditional density for the match of a line segment is calculated as:

$$p(\Delta F_1, \Delta F_2) = \frac{1}{2\pi\delta|\boldsymbol{BF_1}|} e^{-\frac{1}{2}\left(\frac{\Delta F_1}{\delta|\boldsymbol{BF_1}|}\right)^2} \frac{1}{2\pi\delta|\boldsymbol{F_1F_2}|} e^{-\frac{1}{2}\left(\frac{\Delta F_2}{\delta|\boldsymbol{F_1F_2}|}\right)^2} \tag{2}$$

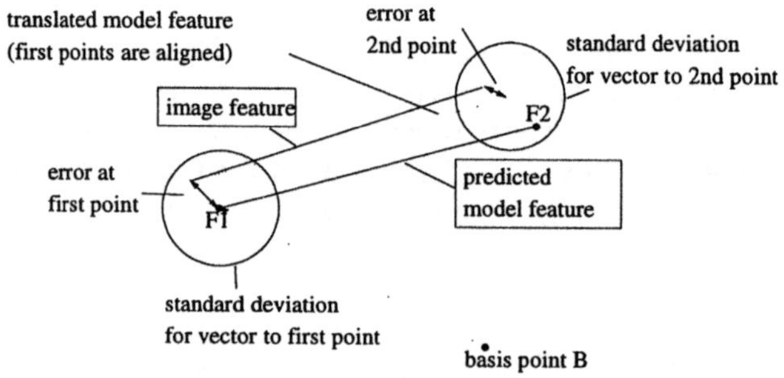

Figure3. Illustration of the distances for a line segment match. Standard deviation for the vector to the first point: $\delta|\boldsymbol{BF_1}|$; standard deviation for the vector to the second point: $\delta|\boldsymbol{F_1F_2}|$

δ is a parameter that influences the tolerance to variations. In the present implementation it is set to 0.15 . For the more complex features the conditional probability consists of a partial probability computed according to the position of endpoints as for line segments with an additional factor representing deviations of the distance attributes. For convex segments and bitangents this is the maximum distance of the curve between the endpoints to the straight line through the two endpoints. Homogeneous ribbons are characterized by their thickness at their two endpoints (see figure 2). Again for these attributes a gaussian distribution with a standard deviation proportional to the expected (model) value is assumed. The probability for a convex segment is:

$$p(d) = \frac{1}{\sqrt{2\pi}\sigma} e^{-\frac{1}{2}\left(\frac{d-d^{mod}}{\sigma}\right)^2}$$

d^{mod} is the attribute value in the reference image used; $\sigma_i = \delta d_i^{mod}$. The corresponding probability for a homogeneous ribbon is:

$$p(d_0, d_1) = \frac{1}{\sqrt{2\pi}\sigma_0} e^{-\frac{1}{2}\left(\frac{d_0-d_0^{mod}}{\sigma_0}\right)^2} \frac{1}{\sqrt{2\pi}\sigma_1} e^{-\frac{1}{2}\left(\frac{d_1-d_1^{mod}}{\sigma_1}\right)^2}$$

The total conditional probability for the more complex features is computed as the product of the probability for the endpoints and the probability for the distance attributes.

As in [5] we use typical scenes to estimate the prior probabilities of features. However a simple naive approach is inappropriate. The positions and attributes of features are described by 4-6 dimensional vectors. A simple estimation of densities in these high dimensional spaces would be quite difficult and inaccurate without large amounts of training data. Therefore as for the conditional densities we compute the probability of features as a product of marginal probabilities. For this calculation feature attributes have to be used that are approximately independent and describe features completely. For line segments we use the position of the first endpoint and the orientation and the length of the line. The probability distributions for the first endpoint and the orientation were chosen as uniform. The density for the length of line segments was estimated according to empirically measured frequencies. Additional approximately independent attributes of the other features used are the relation of feature length and the maximum distance to the feature baseline (for convex segments and homogeneous ribbons) and the relation of the thicknesses at the two endpoints of homogeneous ribbons. Both marginal probabilities are estimated according to empirical frequencies.

Hypotheses are generated by matching the endpoints of two features of the same type in model and object. If these features have attributes, the deviations of the attributes are used to evaluate the hypothesis according to equation 1. Large differences lead to an immediate rejection of the hypothesis. For all other model features the best fitting image feature of the same type is selected. This is the feature where the maximum of all relevant relative deviations ($\frac{\Delta F_1}{|BF_1|}$, $\frac{\Delta F_2}{|F_1 F_2|}$ and $\frac{|d_i - d_i^{mod}|}{d_i^{mod}}$) is smallest. Again the match is evaluated according to equation 1.

3 Experimental results

We did experiments with edge features extracted in colour images and presegmented contour data.

For object recognition in colour images we used a model database of 15 objects. Most objects have simple curved surfaces but some are nearly polyhedral. Some of the objects used are shown in figure 4. For all objects training images were captured with two different camera inclinations from all horizontal directions. The angular differences of horizontal directions and of the two camera inclinations were approximately 15°. Test images were taken with a camera inclination between the two values used for training images and random horizontal directions.

With an automatic procedure [1] model views were generated from training images. The two aims of training are to select stable features and an adequate number of model views (typically about 20) for a given set of training images. A

Figure4. Some of the used objects

model view is generated from a single training image by removing all unstable features.

An important strength of the recognition approach described is the ability to recognize partially occluded objects. Therefore we systematically tested recognition with artificially generated occlusions of test images. In table 1 results are listed for different amounts and directions of occlusion. The differences for the two directions are caused by the fact that most objects have less features on their bottom part than on their top part. Figure 5 offers an example for a successful recognition of one of the objects in the model database in a cluttered scene.

Occlusion	0.0	0.1	0.3	0.5
from left/bottom	0	2.7	6.7	15.3
from right/top	0	2.0	19.3	38.7
mean	0	2.3	13.0	27.0

Table1. Frequency of errors (in %) for different amounts and directions of occlusion

The goal of the experiments with contour data was the estimation of the pose of a human femur bone. For this task good iterative algorithms [3] exist. They require an approximate initial guess of the pose for correct convergence. We wanted to find such an initial estimate by matching a given contour with model contours of the object regularly sampled from all possible directions. Altogether 325 model contours were used. The distance between neighbouring views was 15°. Model and test contours (200 random poses) were generated artificially from a 3D model of the bone. Matching was done with bitangents, convex segments and line segments. In order to evaluate the algorithm we compared the real pose of the test contour with the pose of the most similar model contour (according to the described similarity measure). In table 2 the frequencies of errors in the estimated rotation out of the image plane with different maginitudes are shown. Results for different types of features are compared. For the most successful experiments convex segments were extracted with multiple degrees of smoothing. This multiscale analysis decreases the influences of instabilities of feature detection. The quality of recognition is further improved by the use

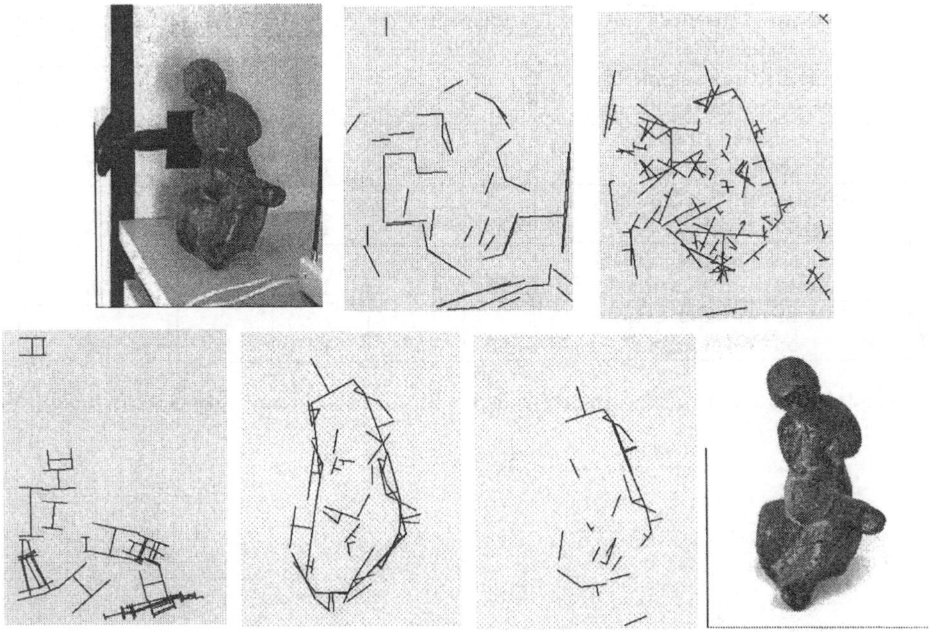

Figure5. Recognition of a small statue with background clutter; top left to bottom right: original image, line segments, convex segments (line to point with maximum distance shown), homogeneous ribbons, best model, matched image features (feature used for hypothesis generation emphasised), training image used for model.

of convex segments instead of the more usual approximation by circular arcs. Presumably this is due to the instability of approximating noncircular curves by circular arcs.

4 Conclusion

The experimental results show that the suggested method for hypothesis verification is feasible. The recognition quality was made possible by using the fact that rotations out of the image plane tend to disturb the relative position of feature points less than their absolute position. While in the experiments discussed relative positions were only used within complex features, in [1] we also describe improvements of recognition due to the use of relative positions of unrelated features. Since many other image deformations preserve relative positions better than absolute positions, we feel that this work could be of interest for tasks beyond the original scope of view-based recognition.

Acknowledgements

This work was supported by a grant of the DFG within the Graduiertenkolleg "Kooperation und Ressourcenmanagement in verteilten Systemen". Thanks to C. Brack for the bone data.

Used features	Deviation from the correct pose (frequency %)				
	< 0,2	0,2 - 0,4	0,4 - 0,6	0,6 - 0,8	> 0,8
1) bitangents, line segments (LS), convex segments (mult. resol.)	27	37	15	7	14
2) as 1) but without bitangents	18	35	19	11	17
3) LS, convex segments, one resolution	22	32	14	9	23
4) LS, circular arcs, multiple resolutions	18	34	15	7	26
5) LS, circular arcs, one resolution	16	27	17	6	34

Table2. Accuracy of pose estimation depending on the feature types used (angles in radians; $0,2 \approx 11,5^o$)

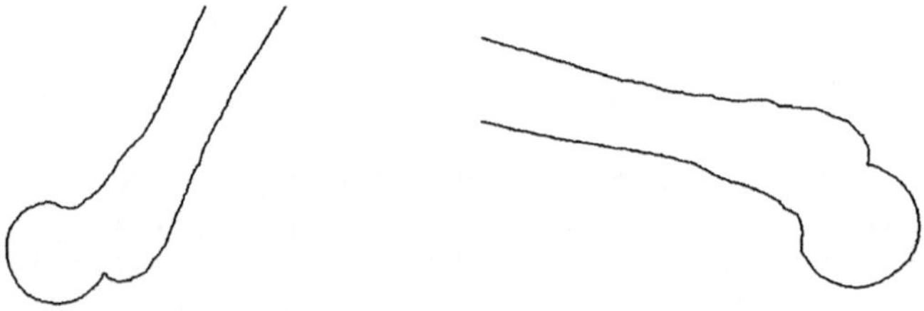

Figure6. Example for a correct pose estimation; left: test contour; right: most similar model; angular deviation: 0,32 (18°)

References

1. Büsching, D.: Verteilte Generierung von Objektmodellen. PhD Thesis (in German, to appear), TU München, 1998.
2. Grimson, W.E.L.: Object Recognition by Computer, MIT Press, 1990.
3. Lavallée, S., Szeliski, R.: Recovering the position and orientation of free-form objects from image contours using 3D distance maps. IEEE Trans. Pattern Analysis and Machine Intelligence, 17, 4, 378–390, 1995.
4. Pope, A.R., Lowe, D.G.: Learning object recognition models from images. In *Proc. ICCV*, 296–301, 1993.
5. Pope, A.R.: Learning to recognize objects in images: acquiring and using probabilistic models of appearance. PhD thesis, University of British Columbia, 1995.
6. Rothwell, C.A., Zisserman, A., Forsysth, D., Mundy, J.: Planar Object Recognition using Projective Shape Representation. International Journal of Computer Vision, 16, 57–99, 1995.
7. Wells, W.: MAP model matching. In *Proc. CVPR*, 486–492, 1991.

Discrete Approach to Curve Evolution

Longin Jan Latecki and Rolf Lakämper

Institut für Angewandte Mathematik, Universität Hamburg
Bundesstr. 55, 20146 Hamburg
{latecki, lakaemper}@math.uni-hamburg.de

Abstract. We propose a simple approach to evolution of digital planar curves that is specially designed to fit discrete nature of curves in digital images. The obtained curve evolution method by digital linearization has many advantages in comparison to curve evolutions in scale-space theories that are usually guided by diffusion equations. We will show that it leads to simplification of shape complexity, analog to evolutions guided by diffusion equations, but with no blurring (i.e., shape rounding) effects and no dislocation of relevant features. Moreover, in our approach the problem to determine the size of discrete steps for numerical implementations does not occur, since our evolution method leads in a natural way to a finite number of discrete evolution steps which are just the iterations of a basic procedure of digital linearization.

1 Scale-space Evolution

In *scale-space theory* a curve or surface Γ is embedded into a continuous family $\{\Gamma_t : t \geq 0\}$ of gradually simplified versions of it. The main idea of *scale-spaces* is that the original curve or surface $\Gamma = \Gamma_0$ should get more and more simplified and noise and small structures should vanish as parameter t increases. Thus, due to different scales (values of t), it is possible to separate small details from relevant shape properties. The ordered sequence $\{\Gamma_t : t \geq 0\}$ is referred to as *evolution* of Γ. Scale-spaces find wide application in computer vision, in particual, due to smoothing ($\Rightarrow$ noise influence is reduced) and elimination of small details ($\Rightarrow$ relevant shape features remain). Some of the main applications are quality enhancement of images, noise removal, and shape description and recognition (see e.g., Sethian [15]).

The scale-space evolution is mostly based on parabolic partial differential equations. The oldest and best-studied are scale-spaces based on a linear diffusion equation (also called geometric heat equation), see Weickert [16]. The solutions of diffusion equations can be obtained by convolution of the original curve (or surface) with a Gaussian function with parameter t (Kimia and Siddiqi [9]). Hence the solutions correspond to Gaussian smoothing of the original curve (or surfaces) with support size t. This leads to a multiscale, curvature-based shape representation. The theory of scale-spaces can be also characterized axiomatically. The first axiomatization was given in Iijima [7], for details see Weickert, et al. [17]. Along with the advantages of evolution based on the linear diffusion equation, there are also some serious problems (Weickert [16], p. 6):

(a) *"Gaussian smoothing does not only reduce noise, but also blurs important features such as edges and, thus, makes them harder to identify. Since Gaussian scale-space is designed to be completely uncommitted, it cannot take into account any a-priori information on structures which are worth being preserved (or even enhanced).*

(b) *Diffusion dislocates features when moving from finer to coarser scales. So features identified at a coarse scale do not give the right location and have to be traced back to the original image [18]. In practice, relating dislocated information obtained at different scales is difficult and bifurcations may give rise to instabilities. These coarse-to-fine tracking difficulties are generally denoted as the correspondence problem."*

To reduce both problems, many anisotropic and nonliear diffusion processes have been proposed for scale-spaces (for an overview see, Weickert [16]). Also reaction-diffusion equations, which lead to reaction-diffusion scale spaces, have been considered (Kimia, et al [8]).

As a solution to both problems, we propose a different approach to scale-space evolution. Our departing point is a discrete nature of digital curves and surfaces in digital images. In opposite to standard approaches in scale-spaces, our evolution is guided neither by differential equations nor Gaussian smoothing, and it is not a discrete version of an evolution by differential equations, as it is the case in Bruckstein, et al. [2]. The main properties of the proposed evolution are (see Figure 1):

- Although it leads to noise elimination, it does not introduce any blurring effects.
- Although irrelevant features vanish during our evolution, there is no dislocation of relevant features.

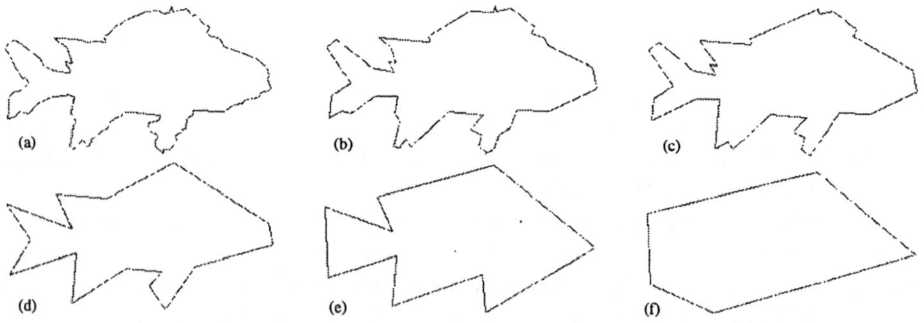

Fig. 1. A few stages of the proposed curve evolution.

There are also some additional problems with scale-spaces:

(c) *"Diffusion filters with a constant steady-state require to specify a stopping time if one wants to get nontrivial results."* (Weickert [16]), p.19)

(d) Problems with stability and computation time of discrete, numeric realizations of diffusion processes.

Clearly, if the stopping time (i.e., stopping parameter t) is too large, it can happen that all relevant features do not any more exist at scale t. This problem is related to a particular case of problem (d): to specify the discrete time steps for t necessary for a stable numeric computation. Since the scale-space theories are continuous theories, i.e., scale (or time) parameter t varies over positive real numbers, the determination of discrete steps is a non-trivial problem; if the steps are too large, it can happen that too many relevant features vanish, and on the other hand, too small discrete steps lead to an inefficient computation. Additionally, a given digital curve (or surface) has some fixed grid resolution that cannot be made infinitely small, and this resolution not always satisfies the requirements for stabile numerical solutions of partial differential equations.

The proposed evolution method leads in a natural way to a finite number of discrete evolution steps which are just the iterations of a basic procedure of digital linearization. Thus, the problem to determine the size of discrete steps does not occur. This also drastically simplifies the problem of stopping time.

2 Curve Evolution by Digital Linearization

In this section, we present our discrete method for curve evolution by digital linearization. We have also developed a similar method for surface evolution by digital planarization, which will be presented in a separate paper, due to a limited number of pages.

Since any digital curve can be regarded as a polygon (with possibly a large number of vertices), it is sufficient to study evolutions of polygonal shapes. The basic idea of the proposed evolution of digital polygons by digital linearization is very simple:

- In every evolution step, a pair of consecutive line segments s_1, s_2 is substituted with a single line segment joining the endpoints of $s_1 \cup s_2$.

The key property of this evolution is the order of the substitution. The substitution is done according to a relevance measure K (defined in Section 4). In every evolution step, two consecutive line segments with the lowest value of the relevance measure K are substituted with a single line segments joining their endpoints. The two consecutive line segments contribute in the less relevant way to the shape of the digital curve.

The fact that every digital curve can be treated as a polygon without loss of information is a well-known result of digital geometry. There exists linear algorithms (with respect to the number of points in the digital curve) for computing decompositions of digital curves into maximal digital line segments. Our computation is based on the algorithm presented in Debled and Reveilles [4]. However, these algorithms are based on characterizations of digital line segments valid for idealized digital images with no blurring and no noise. Gross and Latecki

[6] showed that they also apply to digital line segments obtained by digitizing blurred half-planes. As we will see, the proposed curve evolution method allows us to identify digital lines in presence of noise.

Let $\mathcal{D}_m = s_0, ..., s_{m-1}$ be a decomposition of a digital curve $C = C_m$ into consecutive maximal digital line segments, where $m > 3$. For simplicity, we assume that C is a closed digital curve. (The algorithm and its properties also hold for open digital curves.) The outline of the algorithm that computes a sequence of simplified curves C_k with decompositions $\mathcal{D}_k$ for $k = m, m-1, ..., 3$ (i.e., until $\mathcal{D}_{k-1}$ is a digital triangle) is the following:

Curve Evolution Algorithm($\mathcal{D}_m$)
k = m;
Do
  Find in $\mathcal{D}_k$ a pair s_i, s_{i+1} (mod k) such that $K(s_i, s_{i+1})$ is minimal;
  $\mathcal{D}_{k-1} = \mathcal{D}_k$ with segments s_i, s_{i+1} replaced by line segment s' that joints the endpoints of arc $s_i \cup s_{i+1}$ and no other changes;
  k = k-1;
until $k = 3$.

This algorithm is guaranteed to terminate, since in every evolution step, the number of digital line segments in the curve decomposition decreases by one (one line segment replaces two adjacent segments). It is also obvious that this evolution converges to a convex polygon, since the evolution will reach a state where there are exactly three line segments in the curve decomposition, which clearly form a triangle. Of course, for many curves, a convex polygon with more then three sides can be obtained in an earlier stage of the evolution. Thus, we obtain

Proposition 1. *Curve evolution by digital linearization converges to a convex polygon.* □

This proposition demonstrates mathematical simplicity of the relation between our evolution approach and the geometric properties of the obtained digital curves. Observe that we did not assume that the digital curve is simple (i.e., has no self-intersections). An analog theorem for evolution of continuous planar curves by diffusion equations is a deep and highly non-trivial result of differential geometry (Grayson [5]). It holds only for simple closed smooth curves evolved by the diffusion equation.

Polygonal analogs of the evolution by diffusion equations are presented in Bruckstein, et al. [2]. The experiments in [2] indicate that an arbitrary initial polygon converges to a convex polygon (polygonal circle). However, the proof of this fact in the Euclidean case is an open question. In [2] as well as in evolutions by numerical solutions of differential equations, each vertex of the polygon is mapped to some point in the plane at a single evolution step, whereas in our approach one vertex is removed and the remaining vertices do not change their positions. This is an important difference in favor of our approach, which leads to properties described in the next section.

3 Properties of Curve Evolution by Digital Linearization

We will show in this section that our discrete curve evolution

(P_1) leads to simplification of shape complexity, analog to evolutions guided by diffusion equations, with
(P_2) no blurring (i.e., shape rounding) effects and no dislocation of relevant features,

due to the fact that the remaining vertices do not change their positions. Two more important properties of our curve evolution are based on the relevance measure K defined in Section 4:

(P_3) It is stable with respect to noisy deformations, since noise elimination takes place in early stages of the evolution.
(P_4) It allows to find digital line segments in noisy images, due to the relevance order of the repeated process of digital linearization (e.g., Figure 3).

We begin with some examples to illustrate the properties of our evolution method. A few stages of the proposed curve evolution in Figure 1 illustrate the shape comlexity reduction of the original curve (a). Observe that our curve evolution does not introduce any blurring effects, which result in shape rounding for curves. (for a comparison see the curve evolution on WWW-page http://www.ee.surrey.ac.uk/Research/VSSP/imagedb/demo.html, based on [13]). Observe also that there is no dislocation of the remaining relevant shape features, since the planar position of the remaining points of the digital polygon is unchanged. The stability of relevant shape feature points is demonstrated by marking the corresponding points with the same symbols in Figure 2. Observe also the stability of feature points with respect to noise deformations shown in the second row in Figure 2.

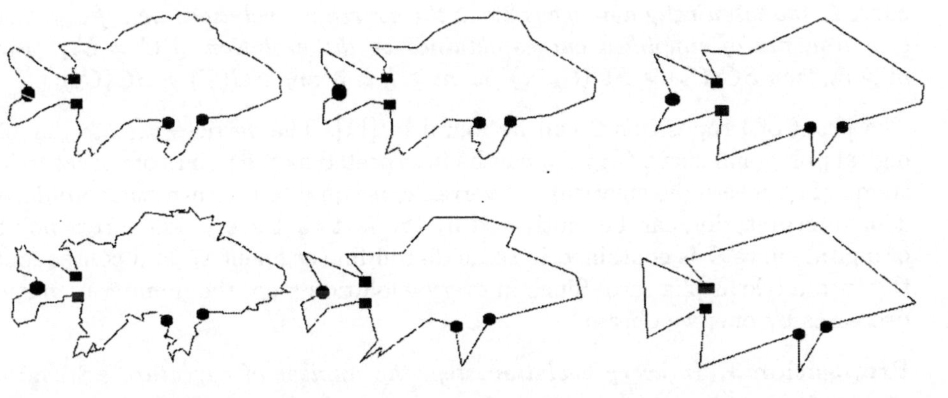

Fig. 2. The same planar position of the points marked with the same symbols demonstrates that there is no displacment of the remaining feature points.

By comparison of the curves (a) and (b) in Figure 3, it can be seen that our evolution method allows us first to eliminate noise influence without changing the shape of objects. If we continue to evolve the curve (b), we will linearize digital arcs that are relevant for the curve shape, which will result in a successive simplification of the curve shape (c) (cf. Brunn et al. [3], Fig. 4).

Fig. 3. (a) → (b): noise elimination. (b) → (c): extraction of relevant line segments.

Now we give a more formal justification of the above properties. The reduction of shape complexity of a digital curve during the evolution process can be justified by Proposition 1. The shape complexity of a polygonal curve can be measured by the sum of the absolute values of the turn angles. For simplicity, we assume that C is a simple closed polygonal curve with vertices $v_0, ..., v_{m-1}$, i.e., the vertices are the endpoints of maximal digital line segments $s_0, ..., s_{m-1}$ in decomposition $\mathcal{D}_m$ of $C = C_m$. Then the **shape complexity** of C is given by $SC(C) = \sum_{i=0}^{m-1} |turn(v_i)|$, where $turn(v_i)$ is the turn angle at vertex v_i. Clearly, the shape complexity of any closed convex curve is 2π and the shape complexity of a closed non-convex curve is greater than 2π.

Proposition 2. *The shape complexity $SC(C)$ of a closed non-convex digital curve C monotonically aproaches 2π in the course of evolution, i.e., if $C_m, ..., C_3$ is a sequence of simplified curves obtained by the evolution of $C = C_m$, where $m > 3$, then $SC(C_k) \geq SC(C_{k-1})$ for $m \geq k > 3$ and $SC(C) > SC(C_3)$.* □

The proof of Proposition 2 can be found in [11]. The vertices $v_0, ..., v_{m-1}$ of a digital polygonal curve $C = C_m$ can be interpreted as the points of curvature extrema of C, where the curvature at vertex v_i is equal to the turn angle $angle(v_i)$. This interpretation can be motivated by the fact that every non-vertex point p of digital curve C is contained in some digital line segment s_i, and consequently the turn angle at p is zero. Since in every evolution step, the number of vertices decreases by one, we obtain

Proposition 3. *In every evolution step, the number of curvature extrema decreases by one.* □

The following proposition is a direct consequence of the evolution procedure:

Proposition 4. *Let $C_m, C_{m-1}, ..., C_3$ be a sequence of simplified curves obtained by evolution of a digital curve $C = C_m$, where $m > 3$. For every vertex v of digital polygonal curve C that also belongs to C_k, the position of v on the plane as vertex of C is the same as the position of v as a vertex of C_k.* □

From Proposition 4, it clearly follows that there is no dislocation of the remaining features during the curve evolution. Thus, in our approach the correspondence problem of coarse-to-fine tracking difficulties does not occur. In contrary, in the course of curve evolution guided by diffusion equations, all points with non-zero curvature change their positions during the evolution. Proposition 4 also explains why our curve evolution does not introduce any blurring (i.e., rounding) effects: In a single evolution step, all vertices remain at their Euclidean positions with exception of a single removed vertex. The two neighbor vertices of the removed vertex are joined by a new line segment, which does not lead to any rounding effects.

4 Relevance Measure

For each two adjacent line segments s_1, s_2 in the decomposition of a digital curve C, we determine the relevance measure $K(s_1, s_2)$ which represents the significance of the contribution of arc $s_1 \cup s_2$ to the shape of C. The value $K(s_1, s_2)$ can be interpreted as the cost required for linearization of arc $s_1 \cup s_2$.

Let $s_1 = AB$ and $s_2 = EF$ be two consecutive line segments in a polygonal decomposition of curve C, such that $B = E$ is their common endpoint and $\beta = \beta(s_1, s_2)$ is the turn angle, i.e., $angle(EF) - angle(AB) = \beta$ (where $angle(s)$ represents the angular direction of line segment s). We determine the circular arclength by which the endpoints B and F must be rotated around points A and E, correspondingly, so that the rotated line segments have the same direction, i.e., until AB' and EF' are parallel, where B' and F' are the images of B and F by the rotations. The circular arclength must be equal for both line segments. The cost function K assigns to pair s_1, s_2 this circular arclength, which is given by the formula

$$K(s_1, s_2) = \frac{\beta(s_1, s_2) l(s_1) l(s_2)}{l(s_1) + l(s_2)}, \tag{1}$$

where l is the length function normalized with respect to the total length of C.

5 Conclusions and Future Work

We presented a discrete approach to scale-space evolution of digital curves. More detalis and evolution examples can be found on our www-side [11] and in [12].

Our evolution method can be also interpreted as hierarchical approximation of the original curve by a polygonal curve whose vertices lie on the original curve. These vertices are determined by our relevance measure. Our approximation is fine-to-coarse and it does not require any error parameters, in opposite to many standard approximations, where starting with some initial coarse approximation to a curve, whereupon line segments that do not satisfy an error criterion are splitted (e.g., [14]). A newer and more sophisticated split-and-merge method for polygon approximation is presented in [1], where multiscale contour approximation is obtained by varying an error parameter t, which defines a scale in a similar manner as it is the case for diffusion scale-spaces. This implies similar problems as for scale-spaces: How to determine the step size for the parameter t? How to find an approperiate t for shape analysis? Additionally, the scale-space property of shape complexity simplification does not result automatically from the approach in [1], but is simply enforced ([1], p. 87): *"New breakpoints, not appearing at finer scales, can occur but are then inserted also at finer levels."*

References

1. A. Bengtsson and J.-O. Eklundh. Shape Representation by Mutliscale Contour Approximation. *IEEE Trans. PAMI* 13, 85–93, 1991.
2. A. M. Bruckstein, G. Shapiro, and C. Shaked. Evolutions of Planer Polygons. *Int. J. of of Pattern Recognition and AI* 9, 991–1014, 1995.
3. A. Brunn, U. Weidner, and W. Förstner. Model-based 2D-Shape Recovery. *Proc. of 17. DAGM Conf. on Pattern Recognition (Mustererkennung)*, Bielefeld, Springer-Verlag, Berlin, 260–268, 1995.
4. I. Debled-Rennesson and J.-P. Reveilles. A Linear Algorithm for Segmentation of Digital Curves. *Intl. J. of Pattern Recognition and AI* 9, 635–662, 1995.
5. M. A. Grayson. The Heat Equation Shrinks Embedded Plane Curves to Round Points. *Journal of Differential Geometry* 26, 285–314, 1987.
6. A. Gross and L. J. Latecki, A realistic digitization model of straight lines, *Computer Vision and Image Understanding* 67, 131–142, 1997.
7. T. Iijima. *Basic theory of pattern observation.* (in Japanese) IECE, 1959.
8. B. B. Kimia, A. R. Tannenbaum, and S. W. Zucker. Shapes, shocks, and deformations, I: The components of shape and the reaction-diffusion space. *Int. J. of Computer Vision* 15, 189-224, 1995.
9. B. B. Kimia and K. Siddiqi. Geometric Heat Equation and Nonlinear Diffusion of Shapes and Images. *Computer Vision and Image Understanding* 64, 305–322, 1996.
10. L. J. Latecki. *Discrete Representation of Spatial Objects in Computer Vision* Kluwer Academic Publisher, Dordrecht, 1998.
11. L. J. Latecki and R. Lakämper. http://www.math.uni-hamburg.de/home/lakaemper/shape
12. L. J. Latecki and R. Lakämper. Hierarchical Shape Decomposition Based on Contour Evolution. *Proc. DAGM Mustererkennung*, Stuttgart 1998.
13. F. Mokhtarian and A. K. Mackworth. A Theory of Multiscale, Curvature-Based Shape Representation for Planar Curves. *IEEE Trans. PAMI* 14, 789–805, 1992.
14. U. Ramer. An Iterative Procedure for the Polygonal Approximation of Plane Curves. *Computer Graphics and Image Processing* 1, 244–256, 1972.
15. J. A. Sethian. *Level Set Methods.* Cambridge University Press, Cambridge, 1996.
16. J. Weickert. A Review of Nonlinear Diffusion Filtering. In B. M. ter Haar Romeny et al. (Eds.) *Scale-Space Theory in Computer Vision*, Springer, Berlin, 3–28, 1997.
17. J. Weickert, S. Ishikawa, and A. Imiya. Scale-Space has been Discovered in Japan. *J. of Mathematical Imaging and Vision*, submitted.
18. A. P. Witkin. Scale-space filtering. *Proc. IJCAI*, Vol. 2, 1019–1022, 1983.

Fast Segmentation Methods Based on Partial Differential Equations and the Watershed Transformation

Joachim Weickert

Department of Computer Science,
University of Copenhagen,
Universitetsparken 1,
DK-2100 Copenhagen, Denmark
E-mail: joachim@diku.dk

Abstract. Segmentation algorithms are presented which combine regularization by nonlinear partial differential equations (PDEs) with a watershed transformation with region merging. We develop efficient algorithms for two well-founded PDE methods. They use an additive operator splitting (AOS) leading to recursive and separable filters. Further speed-up can be obtained by embedding AOS schemes into a pyramid framework. Examples demonstrate that the preprocessing by these PDE techniques eases and stabilizes the segmentation. The typical CPU time for segmenting a 256^2 image on a workstation is less than 2 seconds.

1 Introduction

Segmentation is one of the bottlenecks of many image analysis and computer vision tasks ranging from medical image processing to robot navigation. Ideally it should be efficient to compute and correspond well with the physical objects depicted in the image.

In the last decade much research on PDE-based regularization methods has been carried out. Although the impressive results suggest that they might be attractive as a preprocessing step for many subsequent image analysis methods, not much research has been done which analyses the combination of PDE-based preprocessing methods with other techniques.

In the present paper we address this topic by combining two efficient PDE-based regularization methods with a watershed algorithm, a simple morphological technique for image segmentation.

This leads to segmentation algorithms which can be generalized in a straightforward way to arbitrary dimensional data sets. Their complexity is linear in the pixel number, and they produce identical results if the image is rotated by 90 degrees. A total CPU time of less than 2 seconds for segmenting a 256^2 image on a workstation makes them attractive for many time-critical applications.

The paper is organized as follows. Section 2 sketches the basic ideas behind two well-posed PDE-based regularization methods: a contrast-enhancing nonlinear diffusion filter and a contrast-diminishing diffusion–reaction method with a similarity term. In Section 3 we develop efficient and reliable numerical techniques for these methods. They are based on an additive operator splitting (AOS). For approximating the diffusion-reaction method, these AOS techniques are embedded into a pyramid framework. In Section 4 we discuss the watershed algorithm with region merging, and in Section 5 we illustrate the usefulness of the combined segmentation process by applying it to several examples. The paper is concluded with a summary in Section 6.

2 PDE-Based Regularization

Below two prototypes for well-posed PDE-based regularization techniques are sketched. The first one allows contrast enhancement, while the second one is contrast-diminishing and can be related to energy minimization methods. A more detailed treatment of diffusion filters can be found in [10].

2.1 The Nonlinear Diffusion Filter of Catté *et al.*

In the m-dimensional case the filter of Catté, Lions, Morel and Coll [3] has the following structure:
Let $\Omega := (0, a_1) \times .... \times (0, a_m)$ be our image domain and consider a (scalar) image $f(x)$ as a bounded mapping from Ω into the real numbers $\mathbb{R}$. Then a filtered image $u(x,t)$ of $f(x)$ is calculated by solving a diffusion equation with the original image as initial state, and reflecting boundary conditions:

$$\partial_t u = \sum_{l=1}^{m} \partial_{x_l} \Big(g(|\nabla u_\sigma|^2)\, \partial_{x_l} u \Big) \tag{1}$$

$$u(x,0) = f(x), \tag{2}$$

$$\partial_n u|_{\partial\Omega} = 0, \tag{3}$$

where n denotes the normal to the image boundary $\partial\Omega$.
The "time" t is a scale parameter: larger values lead to simpler image representations. In order to reduce smoothing at edges, the diffusivity g is chosen as a decreasing function of the edge detector $|\nabla u_\sigma|$, where ∇u_σ is the gradient of a Gaussian-smoothed version of u. We use the diffusivity

$$g(s^2) := \begin{cases} 1 & (s^2 = 0) \\ 1 - \exp\left(\frac{-3.315}{(s/\lambda)^8}\right) & (s^2 > 0). \end{cases} \tag{4}$$

For such rapidly decreasing diffusivities, smoothing on both sides of an edge is much stronger than smoothing across it. This selective smoothing process prefers intraregional smoothing to interregional blurring. The factor 3.315 ensures that the flux $\Phi(s) := sg(s^2)$ is increasing for $|s| \le \lambda$ and decreasing for $|s| > \lambda$. Thus, λ is a contrast parameter separating low-contrast regions with (smoothing) forward diffusion from high-contrast locations where backward diffusion may enhance edges [8]. After some time this filter creates segmentation-like results which are piecewise almost constant. For $t \to \infty$, however, the image reaches a flat steady-state [10]. Well-posedness results for this filter can be found in [3, 10] and a scale-space interpretation in terms of an extremum principle as well as decreasing variance, decreasing energy, and increasing entropy is given in [10].

2.2 A Diffusion-Reaction Filter

Sometimes nonlinear diffusion filtering is modified by supplement it with a reaction term which keeps the steady-state solution for $t \to \infty$ close to the original image f:

$$\partial_t u = \operatorname{div}\left(g(|\nabla u|^2)\,\nabla u\right) + \beta(f-u) \qquad (\beta > 0). \tag{5}$$

This equation can be regarded as the descent method of the energy functional

$$E_f(u) := \int_\Omega \left(\tfrac{\beta}{2}\cdot(u-f)^2 + \Psi(|\nabla u|)\right) dx, \tag{6}$$

where $\Psi(|\nabla u|)$ is a potential with $\nabla(\Psi(|\nabla u|)) = g(|\nabla u|^2)\nabla u$. The first term in the integrand of the energy functional is a similarity term which causes the regularized image to be close to the original one, while the second term tends to increase its smoothness.

From a theoretical viewpoint, it is advantageous to choose a convex potential function Ψ, since this guarantees well-posedness and stable algorithms [9]. For nonconvex potentials as in [6], several well-posedness questions are open. An example for a diffusivity leading to a convex potential is

$$g(|\nabla u|^2) := \frac{1}{\sqrt{1+|\nabla u|^2/\lambda^2}} \qquad (\lambda > 0). \tag{7}$$

This diffusivity is used in our experiments. It should be noted that diffusivities $g(s^2)$ with a corresponding convex potential $\Psi(s)$ yield a monotonously increasing flux function $\Phi(s)$. Thus, contrast enhancement is not possible.

3 Efficient Algorithms for PDE-Based Regularization

3.1 An AOS Scheme for the Pure Diffusion Filter

Let us first consider finite difference approximations to the m-dimensional diffusion filter of Catté *et al.*. A discrete m-dimensional image can be regarded as a vector $f \in \mathbb{R}^N$, whose components f_i, $i \in \{1, ..., N\}$ display the grey values at the pixels. Pixel i represents the location x_i. Let h_l denote the grid size in the l direction. We consider discrete times $t_k := k\tau$, where $k \in \mathbb{N}_0$ and τ is the time step size. By u_i^k and g_i^k we denote approximations to $u(x_i, t_k)$ and $g(|\nabla u_\sigma(x_i, t_k)|^2)$, respectively, where the gradient is replaced by central differences. The simplest discretization of the diffusion equation (1) with reflecting boundary conditions is given by

$$u_i^{k+1} = u_i^k + \tau \sum_{l=1}^{m} \sum_{j \in \mathcal{N}_l(i)} \frac{g_j^k + g_i^k}{2h_l^2} (u_j^k - u_i^k) \tag{8}$$

where $\mathcal{N}_l(i)$ consists of the two neighbours of pixel i along the l direction (boundary pixels may have only one neighbour). In vector–matrix notation (8) becomes

$$u^{k+1} = \Big(I + \tau \sum_{l=1}^{m} A_l(u^k)\Big)\, u^k, \tag{9}$$

where A_l describes the diffusive interaction in l direction. Unfortunately, the step size of this scheme has to be very small for stability reasons [11]: it must satisfy $\tau \leq 1/\sum_{l=1}^{m} \frac{2}{h_l^2}$. For most practical applications, this restriction requires many iterations, such that the explicit scheme is rather slow.
In order to address the preceding problem we consider instead of (9) the *additive operator splitting (AOS) scheme*

$$u^{k+1} = \frac{1}{m} \sum_{l=1}^{m} \Big(I - m\tau A_l(u^k)\Big)^{-1} u^k. \tag{10}$$

The operators $I - m\tau A_l(u^k)$ describe one-dimensional diffusion processes along the x_l axes. Under a consecutive pixel numbering along the direction l they come down to strictly diagonally dominant tridiagonal matrices. They can be inverted in linear complexity by a Gaussian algorithm for tridiagonal systems. Moreover, (10) has the same first-order Taylor expansion in τ as (8): both methods are $O(\tau + h_1^2 + ... + h_m^2)$ approximations to the continuous equation. Since (10) is an *additive* splitting, all coordinate axes are treated in exactly the same manner. This is in contrast to conventional splitting techniques which are *multiplicative*. They may produce different results if the image is rotated by 90 degrees.
Recently a general framework for discrete nonlinear diffusion scale-spaces has been discovered, which guarantees that the discretization reveals the same scale-space properties as its continuous counterpart [10]. One can verify [11] that the AOS scheme creates such a discrete nonlinear diffusion scale-space for every (!) step size τ. As a consequence, it preserves the average grey level μ, satisfies a causality property in terms of a maximum–minimum principle, and converges to a constant steady state. Moreover, the process is a simplifying, information-reducing transform with respect to many aspects: The p-norms $\|u^k\|_p := (\sum_{i=1}^{N} |u_i^k|^p)^{1/p}$ and all even central moments $M_{2n}[u^k] := \frac{1}{N}\sum_{j=1}^{N}(u_j^k - \mu)^{2n}$ are decreasing in k, and the entropy $S[u^k] := -\sum_{j=1}^{N} u_j^k \ln u_j^k$, a measure of uncertainty and missing information, is increasing in k (if all f_j are positive).
Many nonlinear diffusion problems require only the elimination of noise and some small-scale details. Usually this can be accomplished with less than 10 iterations in sufficient precision. On an HP 9000/889 workstation, this does not take more than 1 second for a 256^2 image. For further algorithmic details on AOS schemes, the reader is referred to [11].

3.2 A Pyramid AOS Scheme for the Diffusion–Reaction Method

It is possible to extend the preceding AOS framework also to the diffusion–reaction filter (5). It is not hard to see that this can be achieved by using

$$u^{k+1} = \frac{1}{m}\sum_{l=1}^{m}\left(I - \frac{m\tau}{1+\tau\beta}A_l(u^k)\right)^{-1}\frac{u^k+\tau\beta f}{1+\tau\beta}. \tag{11}$$

In contrast to the pure diffusion filter, however, we are now interested in approximating the steady-state solution for $t \to \infty$. Even with large time step sizes, the diffusion process will mainly act within a fairly small vicinity around each pixel. Thus, many iterations are required if the image size is large. In order to speed up the process, we may embed the AOS schemes into a pyramid framework. The idea is as follows:

- downsample the image by creating a Gaussian pyramid [2] with the smoothing mask $(\frac{1}{4}, \frac{1}{2}, \frac{1}{4})$
- adapt the filter parameters to the downsampled image; experiments have shown that λ should be divided by 1.5 and β by 4 if one reduces the image size by a factor 2.
- start with a coarse level (e.g. a 4×4 image), and apply AOS iterations until convergence is obtained.
- expand this solution to the next finer level by linear interpolation, and use it as initial value for AOS iterations at this level.
- proceed in the same way until convergence at the finest level is reached.

Typically, only a few iterations (around 10) are necessary in order to have convergence in terms of a sufficiently small residue at each level. The total complexity remains linear, since the Gaussian pyramid decomposition can be performed in linear complexity. For regularizing a 256^2 image on an HP 9000/889 workstation, a CPU time of 1–1.5 seconds is sufficient.

4 Watershed Segmentation with Region Merging

The previously discussed PDE-based regularization techniques lead to piecewise almost constant image with less noise and less fine-scale details. In order to create a true segmentation, we have to process the regularized image by a technique which gives an edge map without dangling edges. This edge map should lead to a partitioning of the entire image into a finite number of regions, it should handle the semantically important corners and junctions gracefully, and – last but not least – it should be fast.

We found a watershed technique [1] based on the gradient magnitude very useful for these purposes. Such a technique regards the gradient magnitude image as a landscape where the intensity values correspond to the elevation. Areas where a rain drop would drain to the same minimum are denoted as catchment basins, and the lines separating adjacent catchment basins are called watersheds.

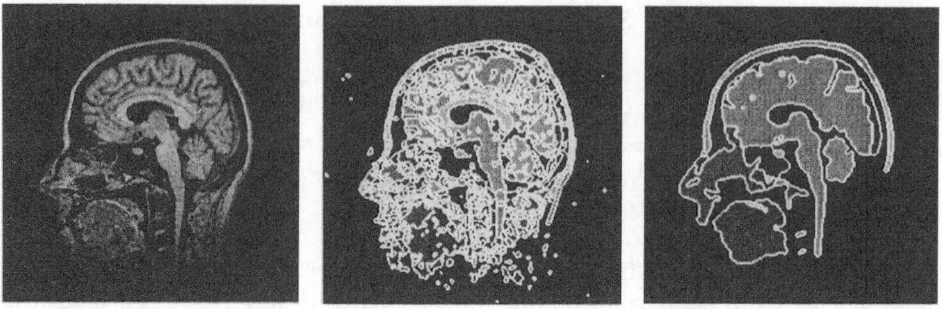

Fig. 1. LEFT: Test image. MIDDLE: Segmented. RIGHT: Diffused and segmented.

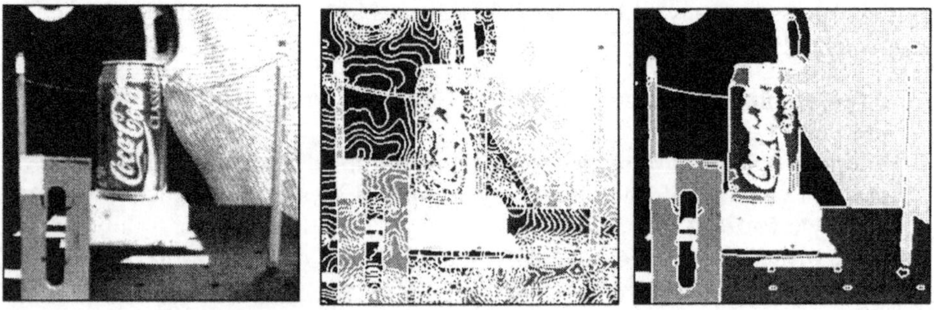

Fig. 2. LEFT: Test image. MIDDLE: Diffused, segmented without merging. RIGHT: Diffused, segmented with merging.

We used an implementation as proposed by Fairfield [4]. The basic idea of this algorithm is sliding downhill on the gradient squared surface until one arrives at a local minimum. Then one replaces all pixels along this path by the image intensity at its corresponding extremum. This algorithm has linear complexity.

Watershed algorithms often create too many segments. Although this oversegmentation is less dominant in the PDE-regularized image than in the original one, it may still lead to problems. They can be circumvented by supplementing the watershed segmentation with a region merging step. In such a step, adjacent regions are merged if their contrast difference is below a specified threshold. This contrast parameter can be related to the contrast parameter λ of the PDE-based regularization, thus it does not constitute an additional parameter. First our algorithm identifies all candidates for merging, and performs the final region merging afterwards. Thus, the whole procedure is independent of the order in which the algorithm runs through the pixels, and it produces a segmentation which is invariant under image rotations by 90 degrees. Since the merging step is also of linear complexity, the total complexity remains linear.

A watershed segmentation of a 256^2 image with subsequent region merging takes about 0.5 CPU seconds on an HP 9000/889. Thus, the total segmentation time including the PDE-based regularization is less than 2 seconds.

Fig. 3. Hallway scene. LEFT: Original image. MIDDLE: Diffusion–reaction, segmented. RIGHT: Diffusion, segmented.

5 Experiments

Figure 1 illustrates how preprocessing by nonlinear diffusion filtering greatly reduces the number of segments in a medical MR image. We also observe that under nonlinear diffusion the segment boundaries remain well located and need not be traced back in order to improve their localization as it is common with watershed segmentation based on linear diffusion [5, 7]. The segments correspond well with the real physical objects and segmentation of elongated objects does not create problems.

In Figure 2 it is shown that the merging step can be essential for avoiding the oversegmentation problems in the watershed algorithm. Nonlinear diffusion may create almost piecewise constant areas, but small fluctuations within such an area introduce many semantically irrelevant catchment basins. Such fluctuations can also be caused by quantization errors, e.g. by storing grey values in a bytewise manner. Merging adjacent regions with similar grey values constitutes a simple remedy for these problems.

Finally, Figure 3 gives a comparison between the two PDE-based regularization techniques. We observe a typical behaviour that can also be experienced with many other test images: the contrast-reducing diffusion-reaction method gives smoother results and is better capable of eliminating small-scale structures, while the edge-enhancing pure nonlinear diffusion process appears to be advantageous for recovering elongated structures. It leads to a very realistic segmentation of the man walking along the hallway.

6 Summary

We have presented efficient algorithms for two PDE-based regularizations. They greatly simplify subsequent segmentation tasks, such that already a simple watershed algorithm with region merging gives good results. Thanks to the AOS and pyramid AOS schemes these segmentation techniques are very fast. This makes them attractive for many time-critical applications. All axes are treated equally, since the result is independent of the pixel ordering, and the algorithm has linear complexity in any dimension.

Acknowledgements. This work was done within the EU-TMR network VIRGO. The hallway image was kindly provided by Ole Fogh Olsen and Mads Nielsen, and the watershed algorithm is a modification of an implementation by Hans Oltmans.

References

1. S. Beucher, *Segmentation tools in mathematical morphology*, C.H. Chen, L.F. Pau, P.S.P. Wang (Eds.), Handbook of pattern recognition and computer vision, World Scientific, Singapore, 443–456, 1992.
2. P.J. Burt, E.H. Adelson, *The Laplacian pyramid as a compact image code*, IEEE Trans. Comm., Vol. 31, 532–540, 1983.
3. F. Catté, P.-L. Lions, J.-M. Morel, T. Coll, *Image selective smoothing and edge detection by nonlinear diffusion*, SIAM J. Numer. Anal., Vol. 29, 182–193, 1992.
4. J. Fairfield, *Toboggan contrast enhancement for contrast segmentation*, Proc. ICPR 10 (Atlantic City, June 16–21, 1990), Vol. 1, 712–716, 1990.
5. L.D. Griffin, A.C.F. Colchester, G.P. Robinson, *Scale and segmentation of grey-level images using maximum gradient paths*, Image Vision Comput., Vol. 10, 389–402, 1992.
6. N. Nordström, *Biased anisotropic diffusion - a unified regularization and diffusion approach to edge detection*, Image Vision Comput., Vol. 8, 318–327, 1990.
7. O.F. Olsen, *Multiscale watershed segmentation*, J. Sporring, M. Nielsen, L. Florack, P. Johansen (Eds.), Gaussian scale-space theory, Kluwer, Dordrecht, 191–200, 1997.
8. P. Perona, J. Malik, *Scale space and edge detection using anisotropic diffusion*, IEEE Trans. Pattern Anal. Mach. Intell., Vol. 12, 629–639, 1990.
9. C. Schnörr, *Unique reconstruction of piecewise smooth images by minimizing strictly convex non-quadratic functionals*, J. Math. Imag. Vision, Vol. 4, 189–198, 1994.
10. J. Weickert, *Anisotropic diffusion in image processing*, Teubner-Verlag, Stuttgart, 1998.
11. J. Weickert, B.M. ter Haar Romeny, M.A. Viergever, *Efficient and reliable schemes for nonlinear diffusion filtering*, IEEE Trans. Image Proc., Vol. 7, 398–410, 1998.

Fusion von Basisalgorithmen zur Segmentierung von Straßenverkehrsszenen

Uwe Handmann, Gesa Lorenz und Werner von Seelen

Institut für Neuroinformatik, Ruhr Universität Bochum, 44780 Bochum

1 Einleitung

Derzeitige Projekte am Institut für Neuroinformatik in Bochum beschäftigen sich mit der Analyse von Straßenverkehrsszenen mittels Computer Vision [12]. Dies impliziert, wegen der durch die natürliche Umwelt aufgestellten Randbedingungen, hohe Anforderungen an die zu entwickelnden Algorithmen. Im speziellen wird versucht, Verkehrsteilnehmer aus Videobildern zu extrahieren und die so gewonnenen Objekthypothesen weiter zu attributieren (z.B. Objektklasse, Abstand, Geschwindigkeit, Gefahrenpotential hinsichtlich der beabsichtigten Eigentrajektorie etc.), um im Hinblick auf den Einsatz in Assistenzsystemen in Fahrzeugen eine möglichst genaue Beschreibung der Umwelt zu erreichen. Nicht nur die große Vielfalt der unterschiedlichen Umweltszenarien, sondern auch das hohe Maß an Sicherheit, das die gestellte Aufgabe erfordert, bedingen ein breitbandiges und flexibles Gesamtsystem [6]. Ein Lösungsvorschlag wird im folgenden behandelt.

2 Fusionsansatz

Im wesentlichen läßt sich das Gesamtsystem in Basisalgorithmen und Algorithmen zur Fusion unterteilen. Als Basis dienen Verfahren, die spezielle Teillösungen bei gegebenen Randbedingungen liefern. Die Ergebnisse der einzelnen Algorithmen sind nicht unabhängig, so daß eine Fusion der Ergebnisse eine Erhöhung der Redundanz zur Folge hat, die ein Gesamtsystem sicher und zuverlässig macht. Die hierbei einzusetzenden Methoden der Fusion sorgen sowohl für ein flexibles Zusammenspiel als auch für eine integrative Ergebnisbildung der Basiskomponenten.

Ein Vorschlag zum Aufbau eines Gesamtsystems unter Verwendung eines Fusionsansatzes ist in Abbildung 1 dargestellt. Hierbei werden auf unterschiedlichen Kopplungsstufen die Ergebnisse der Basisalgorithmen, wie z.B. die Entropie mit den Linien zur Segmentierung oder die Hausdorff-Distanz mit der Kreuzentropie zum Tracking (Objektverfolgung), fusioniert. Im letzten Verarbeitungsschritt wird mit Hilfe symbolischer Informationen das Analyseergebnis bewertet und ein Verhaltensvorschlag erzeugt. Als Kopplungsstruktur werden neuronale Netze eingesetzt, da hier die Kopplung lernbar und flexibel ist [5]. Ein online Training ist ebenso möglich wie eine Rückkopplung über der Zeit.

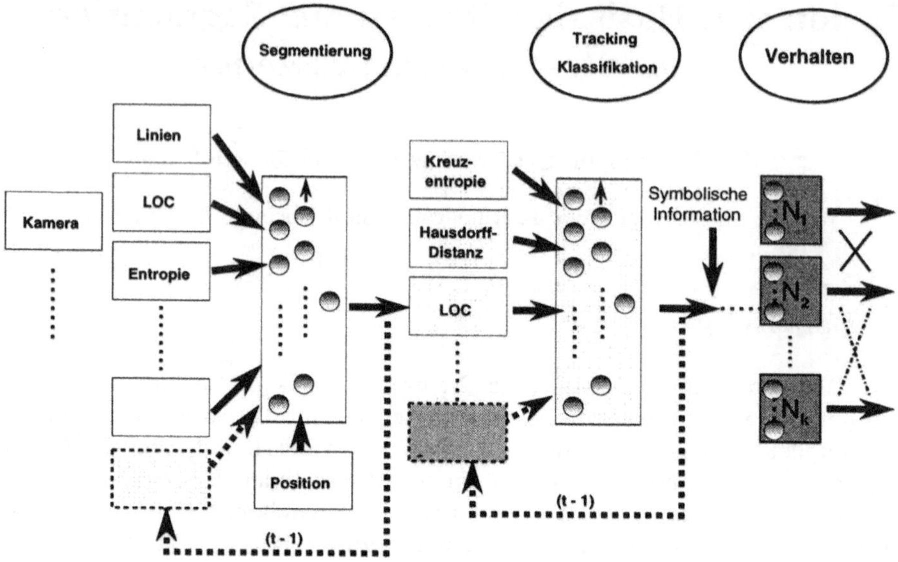

Abbildung1. Kopplungsmodell

Einen wesentlichen Teilaspekt der automatischen Szenenanalyse stellt die Bildsegmentierung dar. Dabei wird unter Segmentierung eine Aufteilung des Bildes in Objekthypothesen und Hintergrund verstanden. Die generierten Objekthypothesen werden innerhalb des Gesamtsystems in weiteren Verarbeitungsschritten klassifiziert (verifiziert oder verworfen) und gegebenenfalls über der Zeit verfolgt. Die Fusion von Basisalgorithmen wird im folgenden exemplarisch am Beispiel der Bildsegmentierungsaufgabe durchgeführt.
Ein Fusionsprozeß kann auf verschiedenen hierarchischen Ebenen erfolgen ([1], Seite 6ff). Allgemein wird zwischen einer Fusion auf Daten-, Merkmals- oder Entscheidungsebene unterschieden.
Im vorliegenden Fall wird ein Fusionsprozeß auf der Merkmalsebene gewählt (Abbildung 2), um den Vorteil der Datenreduktion im Vergleich zu einer Fusion auf der Datenebene zu nutzen. Dazu werden relevante Ergebnismerkmale der Basisalgorithmen zu einem Merkmalsvektor zusammengesetzt und mit Hilfe eines neuronalen Netzes fusioniert. Eine Schwellwertbildung am Ausgang (Entscheidung) ordnet die einzelnen Pixel des Bildmaterials dem Hintergrund bzw. einem relevanten Segment zu.

3 Segmentierung

Am Beispiel der Segmentierung werden im folgenden Abschnitt zunächst die verwendete Kopplungsstruktur sowie die Basisalgorithmen vorgestellt und motivierende Segmentierungsergebnisse präsentiert. In einem weiteren Abschnitt

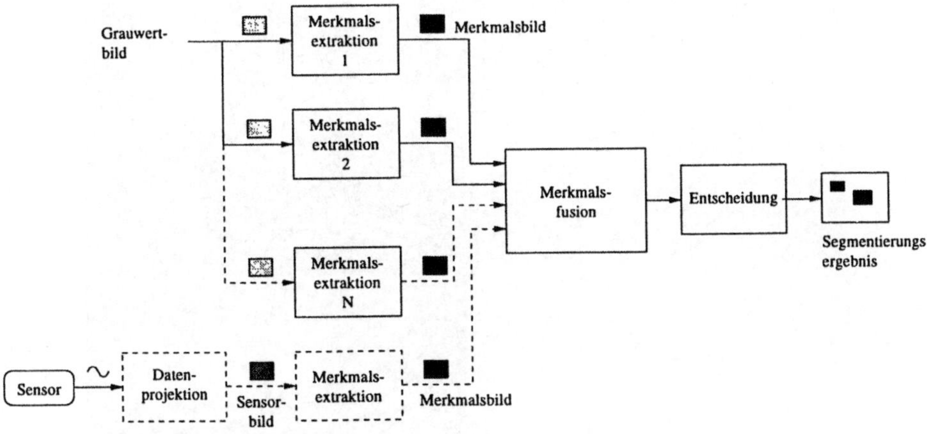

Abbildung2. Prinzip einer Multisensor-Fusion auf der Merkmalsebene

wird am Beispiel der Einkopplung von Farbinformation gezeigt, daß die Kopplungsstruktur flexibel erweiterbar ist und die integrative Ergebnisbildung die Qualität der Segmentierung erhöht.

3.1 Segmentierung basierend auf Grauwertbildsequenzen

Im diesem Ansatz koppeln differenzierende und integrierende Basisalgorithmen (Abbildung 3) in ein neuronales Netz ein, um die Segmentierungsaufgabe zu lösen. Das Linienbild (Polygonzugapproximation der Kontur) und die Lokale Orientierungskodierung (LOC [3] [4]) finden als differenzierende Merkmale Verwendung. Die Entropie [8] wird als integrierendes Merkmal eingesetzt. Für jedes Pixel wird ein zwölfdimensionaler Eingangsvektor

$$\mathbf{u}(x,y) = (\mathbf{u_1}(x,y)^T, x, y)^T$$

für das Kopplungsnetz erzeugt. Der 10-dimensionale Vektor $\mathbf{u_1}(x,y)$ beinhaltet die Kontur- und Texturinformationen, die Variablen x und y repräsentieren die Pixelkoordinaten. Der Vektor $\mathbf{u_1}(x,y)$ ist definiert durch

$$\mathbf{u_1}(x,y) = \sum_{(i,j)\in R} \mathbf{v}(i,j),$$

wobei R die lokale Nachbarschaft (9×9) des Bildpunktes (x,y) und $\mathbf{v}(i,j)$ einen binären Vektor beschreibt. Die Vektorelemente $v_1(x,y),\ldots,v_4(x,y)$ kodieren reduzierte LOC-Merkmale, $v_5(x,y),\ldots,v_9(x,y)$ kodieren die Entropiewerte und $v_{10}(x,y)$ wird gesetzt, wenn der Bildpunkt (x,y) Teil eines Linienelements ist.

Als Kopplungsstruktur wird ein Multilayer Perzeptron (MLP, [7], Seite 138ff) eingesetzt. Es findet eine 12-5-1-Struktur der Neuronen Verwendung. Die zwölfdimensionalen Merkmalsvektoren werden über zwölf Eingangsneuronen und fünf

Abbildung3. Grauwert, LOC, Linien, Entropie

versteckte Neuronen zum Ausgangsneuron vorwärts durch das Netz propagiert. Die Aktivität des Ausgangsneurons beschreibt die Zugehörigkeit der Pixel zu den initialen Segmenten. Durch eine Schwellwertbildung wird die Entscheidung getroffen.
Das Training des Kopplungsnetzes findet überwacht mit Hilfe des Back Propagation Algorithmus statt ([7], Seite 142ff). Mit von Hand segmentierten Daten wird die Fehlerfunktion zwischen Ausgangsaktivität und dem Sollzustand am Ausgang ermittelt und rückwärts durch das Netz propagiert.
Diese Kopplungsstruktur ist flexibel erweiterbar und auf neue Probleme leicht adaptierbar. Abbildung 4 verdeutlicht das Segmentierungsergebnis von Fahrzeugen an einer Sequenz von 200 Bildern einer Straßenverkehrsszene, wobei jedes fünfzigste Bild dargestellt ist. Es werden alle relevanten Objekte stabil segmentiert. Partielle Fehlsegmentierungen bei Objekten mit geringem Kontrast oder fehlender Struktur sind jedoch möglich (Abbildung 4, Spalte 3).
Sollen komplexere Szenarien analysiert werden, so reicht es deshalb nicht aus, eine Analyse nur mit Standardkameras wie den herkömmlichen CCD[1]-Kameras durchzuführen. Einerseits ist der Dynamikbereich dieser Kameras gering und andererseits wird nur die Lichtintensität der Szenarien ausgewertet. Abhilfe kann der Einsatz höherwertiger Kameras (z.B. HDRC[2], Farbkamera) und die Hinzunahme weiterer Sensoren schaffen.

[1] Charged Couple-Device
[2] High Dynamic Range Camera

Abbildung4. Segmentierung mit Hilfe eines MLP-Kopplungsnetzes

3.2 Integration von Farbsättigung zur Segmentierung

Ein wichtiges Merkmal, welches der Mensch zur Szenenanalyse und Steuerung seiner Aufmerksamkeit verwendet, ist die Farbinformation. Sowohl in der Natur als auch in der vom Menschen geschaffenen Umwelt wird die Signalwirkung der Farbe gezielt zur Erregung von Aufmerksamkeit eingesetzt. So werden wichtige Objekte im Straßenverkehr, wie Verkehrsschilder, Bremslichter, Rettungsfahrzeuge, etc. mit Signalfarben versehen. Zur Detektion dieser für die Verkehrsteilnehmer wichtigen Objekte wird der Farbwert erfolgreich zur Segmentierung eingesetzt [10].

Bei allgemeineren Segmentierungsaufgaben, hier der Segmentierung von Fahrzeugen, wurde Farbe als Merkmal bisher selten genutzt, da Farbe im Unterschied zur Form i.a. nicht objektspezifisch ist. Fahrzeuge zeichnen sich vor allem durch eine besondere Geometrie, wie der Anordnung von horizontalen und vertikalen Linien aus, die in Linien und LOC-Merkmalen kodiert sind. Zusätzlich besitzen sie eine im Vergleich zur umgebenden Straße große Strukturiertheit, die durch die lokale Bildentropie ausgedrückt wird. Ein Teil der Fahrzeuge zeichnet sich zudem durch einen markanten Farbwert aus. Wird die Farbinformation in Verbindung mit den anderen Merkmalen verwendet, so stellt sie ein wichtiges ergänzendes Merkmal dar. Sie ist in der Lage große Objekte wie LKWs, die im allgemeinen eine geringe Struktur aufweisen, zu segmentieren.
Das hier vorgeschlagene Fusionskonzept bietet die Möglichkeit, die Farbinformation neben den bereits vorhandenen differentiellen und integrativen Merkmalen in das Kopplungsnetz zu integrieren.

Farbe ist eine dreidimensionale Größe. Zu ihrer eindeutigen Beschreibung muß daher ein 3-D Koordinatensystem definiert werden. Häufig handelt es sich dabei aus Rechenzeitgründen um das hardwareorientierte RGB Koordinatensystem. Ein an der menschlichen Farbvorstellung orientierter Farbraum ist der HSV-Raum, der eine Farbe durch die Attribute Farbton, Sättigung und Helligkeit beschreibt ([2], Seite 590ff). Für die Extraktion farbiger Objekte, kann die dreidimensionale Farbbeschreibung auf die Verwendung der Sättigungskoordinate

$$S = \frac{\max(R, G, B) - \min(R, G, B)}{\max(R, G, B)}$$

reduziert werden.

Abbildung5. Links: Originalbild, Rechts: Binarisiertes Sättigungsbild

Das Sättigungsbild **S** wird durch eine Schwellwertoperation (Schwelle s_{min}) binarisiert. Das binarisierte Bild **W** wird durch

$$w(x, y) = \begin{cases} 1 \text{ falls } s(x, y) > s_{min} \\ 0 \text{ sonst} \end{cases},$$

beschrieben und als zusätzliches Merkmal

$$u_2(x, y) = \sum_{(i,j) \in R} w(i, j),$$

in das Kopplungsnetzwerk eingespeist, wobei die lokale Nachbarschaft R eines Bildpunktes ebenfalls Berücksichtigung findet. Es ergibt sich für jedes Pixel ein 13-dimensionaler Eingangsvektor

$$\mathbf{u}(x, y) = (\mathbf{u_1}(x, y)^T, u_2(x, y), x, y)^T$$

für das Kopplungsnetz.

Die Wahl des Schwellwertes erweist sich als unkritisch für die anschließende Weiterverarbeitung und kann auch in unterschiedlichen Bildsequenzen konstant gehalten werden. Abbildung 5 zeigt das Segmentierungsergebnis auf der Basis der Farbsättigung.
Es werden vor allem diejenigen Objektbereiche detektiert, welche durch textur- und kantenbasierte Verfahren aufgrund mangelnder Struktur häufig dem Hintergrund zugeordnet werden, wie z.B. die LKW-Rückfront. Es werden allerdings auch Hintergrundbereiche segmentiert, wie die Vegetation am Randstreifen. Dieses Problem kann jedoch bei bekanntem Fahrbahnverlauf, geeigneten

Abbildung6. Segmentierung mit Hilfe eines MLP-Kopplungsnetzes mit/ohne Bewertung der Farbsättigung

zusätzlichen Basisalgorithmen oder durch Einkopplung weiterer Sensoren (z.B. RADAR [11], LIDAR- oder Infrarotsensoren [9]) gelöst werden.
Abbildung 6 zeigt eine Gegenüberstellung der Ergebnisse am Beispiel der Segmentierung von PKWs und LKWs einer Sequenz von 200 Bildern einer Straßenverkehrsszene, wobei jedes fünfzigste Bild dargestellt ist. Während bei beiden Ansätzen die Segmentierung der PKWs bei guten Kontrastverhältnissen gelingt, zeigt die Integration von Farbsättigung zur Segmentierung eine deutliche Steigerung der Ergebnisse bei schlecht ausgeleuchteten Bereichen (Abbildung 6, Spalte 3), sowie der Segmentierung von LKWs mit stark homogenen Flächen (Abbildung 6, Zeile 3).

4 Diskussion

Es wird ein System zur Fusion von Basisalgorithmen zur Segmentierung vorgestellt und die Funktionsfähigkeit anhand zweier Beispiele demonstriert. Durch Nutzung eines breiten Merkmalspektrums (Kanten, Textur, Farbe) wird eine verbesserte Segmentierung in natürlichen Umwelten erreicht. Die Wahl eines neuronalen Netzes erlaubt eine Adaption auf weiterführende Aufgabenstellungen (z.B. Innenstadtverkehr). Seine Flexibilität erlaubt die einfache Integrationsmöglichkeit von Farbinformation, sowie anderer Verfahren, um eine weitere Verbesserung der Segmentierung zu erreichen. Der Austausch der Netzgewichte, sowie die Möglichkeit des online Lernens ist möglich. Weitergehend ist eine Fusion verschiedener Sensordaten denkbar (z.B. RADAR-, LIDAR- oder Infrarotsensoren), um auftretende Fehlsegmentierungen im Randbereich eliminieren zu können.

Das vorgestellte Fusionsnetz ist Teil eines Gesamtsystems. Um eine zuverlässige Analyse von Straßenverkehrsszenen zu erreichen, müssen segmentierte Bereiche klassifiziert (Hintergrund, Hindernis, Fahrzeug) und über der Zeit verfolgt werden. Diese Analyseergebnisse können einfach in das Fusionsnetz integriert werden und so das Segmentierungsergebnis weiter stabilisieren.

Literatur

1. B. V. Dasarathy. *Decision Fusion.* IEEE Computer Society Press, Los Alamitos, 1994.
2. J.D. Foley, A. van Dam, S.K. Feiner und J.F. Hughes. *Computer Graphics: Principles and Practice.* Addison-Wesley Publishing Company, Inc., 1990.
3. C. Goerick. Local orientation coding and adaptive thresholding for real time early vision. Internal Report IR–INI 94–05, Institut für Neuroinformatik, Ruhr–Universität Bochum, D–44780 Bochum, Germany, Juni 1994.
4. C. Goerick, D. Noll und M. Werner. Artificial Neural Networks in Real Time Car Detection and Tracking Applications. *Pattern Recognition Letters*, 1996.
5. U. Handmann und T. Kalinke. Fusion of texture and contour based methods for object recognition. In *ITSC'97*, Boston, 1997. IEEE. Session 35: Intelligent Vehicles: Vision(3).
6. U. Handmann, T. Kalinke, C. Tzomakas, M. Werner und W. v. Seelen. Computer vision for driver assistance systems. In *Proceedings of SPIE Vol. 3364*, Orlando, 1998. SPIE. Session Enhanced and Synthetic Vision 1998.
7. S. Haykin. *Neural Networks.* Macmillan College Publishing Company, Inc., 1994.
8. T. Kalinke und W. von Seelen. Entropie als Maß des lokalen Informationsgehalts in Bildern zur Realisierung einer Aufmersamkeitssteuerung. In *Mustererkennung 1996*, Seite 627–634, Berlin, Heidelberg, 1996. Springer-Verlag.
9. L. C. Ramac, M. K. Uner, P. K. Varshney, M. Alford und D. Ferris. Morphological filters and wavelet based image fusion for concealed weapons detection. In *Proceedings of SPIE Vol. 3376*, Orlando, 1998. SPIE. Session Sensor Fusion 1998.
10. W. Ritter, F. Stein und R. Janssen. Traffic Sign Recognition Using Colour Information. *Mathl. Comput. Modelling*, 22(4–7):149–161, 1995.
11. T. Schnitger und U. Handmann. Fusion von Bildanalyseverfahren mittels einer neuronalen Kopplungsstruktur. Internal Report IRINI 98-01, Institut für Neuroinformatik, Ruhr-Universität Bochum, D-44780 Bochum, Germany, April 1998.
12. W. von Seelen, W. Gillner, C. Goerick, U. Handmann, T. Kalinke, D. Kastrup, I. Leefken, G. Lorenz, D. Noll, C. Tzomakas, B. Völpel, M. Werner und C. Winkel. Image Processing of Dynamic Scenes. Internal Report IRINI 97-14, Institut für Neuroinformatik, Ruhr-Universität Bochum, D-44780 Bochum, Germany, Juli 1997.

Parallele und global konvergente iterative Minimierung nichtlinearer Variationsansätze zur adaptiven Glättung und Segmentation von Bildern

J. Heers, C. Schnörr, and H.S. Stiehl

Universität Hamburg, FB Informatik, AB Kognitive Systeme
Vogt-Kölln-Str. 30, D-22527 Hamburg, Germany
email: {heers, schnoerr, stiehl}@informatik.uni-hamburg.de

Zusammenfassung Wir betrachten nicht-quadratische, konvexe Variationsansätze zur Bildsegmentation und untersuchen numerische Verfahren, die eine effiziente Berechnung globaler Minima auf parallelen Architekturen erlauben. Unser Augenmerk gilt dabei Verfahren mit globaler Konvergenz, d.h. Konvergenz gegen die eindeutige Lösung bei beliebigen Startwerten. Im Kontext der (semi-)automatischen Bildverarbeitung ist eine solche Eigenschaft sehr wichtig.
Wir charakterisieren verschiedene Verfahren aus der aktuellen Literatur als spezielle Fälle eines allgemeinen Iterationsschemas. Darunter fallen der von Geman und Reynolds eingeführte *auxiliary variable*–Ansatz und die Linearisierungstechnik von Geman und Young. Als Ergebnis können wir die globale Konvergenz dieser Verfahren unter abgeschwächten Voraussetzungen zeigen.
Zur Beschleunigung der Laufzeit unserer numerischen Verfahren wurden effiziente Krylov-Verfahren für die linearen Teilprobleme auf parallelen Architekturen implementiert. Experimentelle Ergebnisse bezüglich Konvergenzraten und *speed–up* werden vorgelegt. Aufgrund der Verwandschaft des Segmentationsansatzes mit Restaurationsverfahren, die auf dem Maß der Totalen Variation basieren, sind unsere Ergebnisse auch für diese breite Klasse von Bildverarbeitungverfahren relevant.

1 Einleitung

Bildsegmentierung und Merkmalsextraktion sind Schlüsselprobleme in der Bildverarbeitung. Variationsansätze [1–4] [5] bieten eine mathematisch klare Problemformulierung, die andere ad–hoc–Formulierungen subsumiert. Ein gemeinsames Problem dieser Variationsansätze ist der relativ hohe Berechnungsaufwand gängiger Minimierungsverfahren. Stochastische Optimierung [1] ist für typische Bilddimensionen von 512×512 Pixel praktisch nicht durchführbar und deterministische *annealing*–Verfahren [2, 4] können nicht garantieren, daß ein "gutes" lokales Minimum gefunden wird. Deshalb betrachten wir nicht–quadratische, aber konvexe Funktionale zur Bildsegmentation und Merkmalsextraktion [6–9].

Wir untersuchen Verfahren zur Berechnung der eindeutigen Funktion u, welche das folgende konvexe Funktional minimiert [7, 8]:

$$J(v) = \frac{1}{2} \int_{\Omega} (v - g)^2 + \lambda(|\nabla v|) \, dx \; , \tag{1}$$

Dabei bezeichnet g ein gegebenes Bild, und und die Funktion $\lambda(\cdot)$ wird definiert durch

$$\lambda(t) = \begin{cases} \lambda_h^2 t^2 & , 0 \le t \le c_\rho \, , \\ \lambda_l^2 t^2 + c_\psi(2t - c_\rho) & , 0 < c_\rho \le t \, , \end{cases} \quad (c_\psi = (\lambda_h^2 - \lambda_l^2) c_\rho) \, . \tag{2}$$

Dieser Minimierungsansatz beinhaltet im wesentlichen zwei Parameter: λ_h bestimmt den Grad der Glättung, wohingegen c_ρ die Sensitivität des Prozesses gegenüber Signalübergängen kontrolliert. An signifikanten Signalübergängen wird der Glättungsprozess anisotrop und stoppt graduell in Richtung des Bildgradienten. Dadurch werden wesentliche Signalstrukturen erhalten. λ_l ist eine kleine positive Konstante (z.B. $\lambda_l = 0.1$), deren Bedeutung für die hier präsentierten Untersuchungen nicht weiter relevant ist.

Eine Analyse [10, 8] des Funktionals in (1) zeigt, daß es eine geeignete Approximation des prototypischen, aber bzgl. der Berechenbarkeit sehr komplexen Segmentationsmodells von Mumford und Shah [3] bietet. Entsprechend hat sich der Minimierungsansatz (1) in zahlreichen Anwendungen als nützlich im Hinblick auf die adaptive Glättung und Segmentierung mehrdimensionaler verrauschter Signale erwiesen (siehe z.B. Abb. 1). Dies motivierte unsere Untersuchungen von effizienten numerischen Verfahren zur Berechnung globaler Minima von (1).

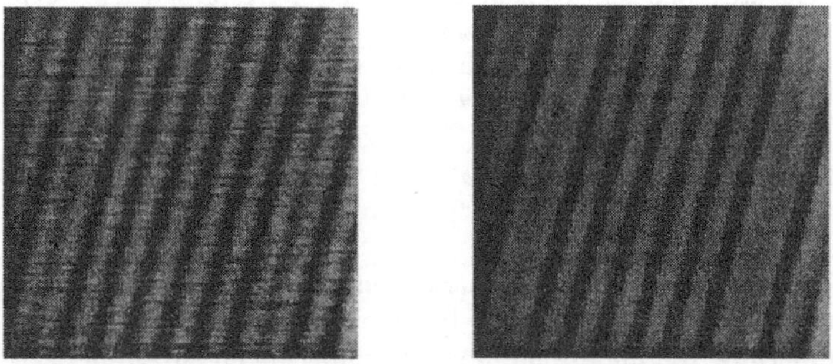

Abbildung 1. Adaptive kantenerhaltende Bildglättung. Quelle: Zentrum f. Mikrostrukturforschung an der Universität Hamburg, Prof. Dr. R. Wiesendanger

2 Eine Klasse global konvergenter Iterationen

2.1 Problemformulierung

Das globale Minimum u von (1) ist gleichzeitig die eindeutige Lösung der folgenden Variationsgleichung [7,8]:

$$L(u,v) = \int_{\Omega} (u-g)v + \rho(|\nabla u|)\nabla u \nabla v \, dx = 0 \ , \ \forall v \, . \tag{3}$$

Die Funktion $\rho(t)$ wird durch $\rho(t) = \lambda'(t)/(2t)$ definiert. Aus der Diskretisierung von (3) mit stückweise–linearen Ansatzfunktionen (Finite Elemente) erhalten wir ein System von N nichtlinearen Gleichungen,

$$\mathbf{L_h}(\mathbf{u}) = \mathbf{0} \ , \ \mathbf{u} \in \mathbb{R}^N \ , \tag{4}$$

wobei N die Anzahl der Pixel bezeichnet.

Wir betrachten nun die Aufgabe, Verfahren zur effizienten Berechnung der eindeutigen Lösung $\mathbf{u}$ von (4) zu entwerfen. Wir verweisen hier noch einmal darauf, daß wir nur an Berechnungsverfahren mit globaler Konvergenz interessiert sind, d.h. Konvergenz gegen die Lösung $\mathbf{u}$ unabhängig von einem beliebigen Startpunkt. Im allgemeinen schließt dies Verfahren zweiter Ordnung, wie z.B. das Newtonverfahren aus.

2.2 Ein allgemeines Iterationsschema

Ein allgemeines Iterationsschema zur Lösung nichtlinearer Variationsgleichungen ist in [11] beschrieben:

$$B(u^k, u^{k+1}, v) = B(u^k, u^k, v) - \omega_k L(u^k, v) \ , \tag{5}$$

für $k = 1, 2, \ldots$. Konkrete Definitionen von $B(\cdot)$ im Kontext des Problems (3) werden in den Abschnitten 3.1 und 3.2 gegeben. Unter bestimmten mathematischen Bedingungen an $B(\cdot)$ [12] läßt sich eine lineare Konvergenzrate zu $u^k \to u$ zeigen [11]. Die diskretisierte Version von (5) notieren wir als:

$$\mathbf{B^k u^{k+1}} = \mathbf{B^k u^k} - \omega_\mathbf{k} \mathbf{L_h}(\mathbf{u^k}). \tag{6}$$

Dieses Verfahren entspricht gerade dem Newtonverfahren, wenn man für $\mathbf{B^k}$ die Jacobimatrix der nichtlinearen Abbildung $\mathbf{L_h}(\mathbf{u^k})$ an der Stelle $\mathbf{u^k}$ einsetzt und $\omega_k = 1$ wählt. In diesem Sinne können konkrete Versionen $\mathbf{B^k}$ auch als Approximationen an die Jacobimatrix unter der Nebenbedingung „globale Konvergenz“ verstanden werden. Als einen weiteren Spezialfall erhält man das einfache nichtlineare Jacobiverfahren (Abstiegverfahren) für $\mathbf{B^k} = \mathbf{I_N}$.

3 Theoretische Ergebnisse

Für ein *konvexes* Funktional wie z.B. (1) umfaßt das allgemeine Iterationsschema (5) des vorangegangen Kapitels eine Reihe verschiedener Verfahren, die in der jüngeren Literatur vorgestellt wurden. Als Konsequenz dieser Beschreibung kann der Nachweis globaler Konvergenz solcher Verfahren unter abgeschwächten Bedingungen gezeigt werden [12]. Im diesen Abschnitt beschreiben wir zwei spezielle iterative Verfahren, die in diesen Rahmen fallen.

3.1 Adaptive Linearisierung

Die Wahl von $B(u, v, w) := \int_\Omega vw + \rho(|\nabla u|)\nabla v \nabla w \, dx$ und $\omega_k = 1$ führt zu einer linearen Konvergenzrate. Die Iteration (5) läßt sich dann wie folgt umformulieren:

$$\int_\Omega u^{k+1}v + \rho(|\nabla u^k|)\nabla u^{k+1}\nabla v \, dx = \int_\Omega gv \, dx \tag{7}$$

Dieses Verfahren ist in der mathematischen Elastizitätstheorie als Kačanov–Verfahren bekannt [13]. „Adaptiv“ bedeutet hier die Abhängigkeit des linearen Operators $B(u, v, w)$ von der aktuellen Approximation u^k.

3.2 Nicht–adaptive Linearisierung

Durch $B(v, w) := \int_\Omega vw + \alpha \nabla v \nabla w \, dx$ und $\omega_k = 1$ erhält man eine Vereinfachung des oberen Verfahren. Die Iteration (5) lautet nun wie folgt:

$$\int_\Omega u^{k+1}v + \alpha \nabla u^{k+1}\nabla v \, dx = \int_\Omega \sigma(|\nabla u^k|)\nabla u^k \nabla v + gv \, dx \tag{8}$$

mit $\sigma(t) = \alpha - \rho(t)$. Damit die Konvergenz dieses Iterationsverfahrens gezeigt werden kann, muß der Parameter $\alpha > \lambda_h^2$ gewählt werden. Experimentelle Ergebnisse haben einen Zuwachs der Konvergenzgeschwindigkeit für $\alpha \to \lambda_h^2$ gezeigt. Ein Vorteil dieses Verfahrens ist, daß sich der lineare Operator $B(v, w)$ nicht verändert und dadurch in der Diskretisierung aufwendige Berechnungen für Matrixkompilierungen und Vorkonditionierern nicht wiederholt berechnet werden müssen. Auf der anderen Seite ist die Konvergenzrate i.a. deutlich kleiner (vgl. Abschnitt 5).

4 Interpretation als Minimierung mit Hilfsvariablen

In diesem Abschnitt diskutieren wir zwei wichtige Konzepte, die jüngst zur Minimierung nichtquadratischer Funktionale mittels sog. Hilfsvariable (***auxiliary variables***) vorgeschlagen worden sind. Im konvexen Fall (1) können wir über die Eliminierung der Hilfsvariablen zeigen, daß diese beiden Konzepte auf die Verfahren (7) und (8) führen. Als Konsequenz der Ergebnisse aus Abschnitt 3 können wir somit die globale, lineare Konvergenz dieser Ansätze im konvexen Fall zeigen.

4.1 Multiplikative Hilfsvariable

Geman und Reynolds [14] schlugen einen auf adaptiven Hilfsvariablen basierenden "kanten–erhaltenden" Ansatz für Restaurationsprobleme vor:

$$J(v,w) = \frac{1}{2}\int_\Omega \Big\{(v-g)^2 + w|\nabla v|^2 + \psi_{GR}(w)\Big\}dx \,.$$

Hierbei bezeichnet w die Hilfsfunktion und ψ_{GR} wird so gewählt, daß die zweistufige Iteration $v_{k+1} = min_v\, J(v, w^k)$, $w^{k+1} = min_w\, J(v_{k+1}, w)$ der Minimierung des originalen Funktionals $J(v)$ entspricht. Untersuchungen von Charbonnier et al. [15] führten zu einen Beweis der globalen Konvergenz dieses Ansatzes für bestimmte konvexe Funktionale. Dabei können die Hilfsvariablen explizit durch $w^k = \rho(|\nabla u^k)$ bestimmt werden. Die Anwendung dieser Ergebnisse auf Restaurationsformulierungen wurde von Dobson in [16] veröffentlicht, worin ein iteratives Verfahren abgeleitet ist, welches mit (7) korrespondiert.

Die Anwendung der Ergebnisse aus Abschnitt 3 auf den Hilfsvariablenansatz von Geman und Reynolds erlaubt uns die Bedingungen von Charbonnier et al. für globale Konvergenz abzuschwächen, so daß u.a. das uns interessierende konvexe Funktional (1) berücksichtigt wird. Mit anderen Worten: Im konvexen Fall können wir die globale Konvergenz für (nicht notwendigerweise strikt) monoton fallende und fast überall stetige Funktion $\rho(\cdot)$ in (3) zeigen.

4.2 Entkoppelte Hilfsvariablen

Eine andere Verwendung von Hilfsvariablen wurde von Geman und Yang unter der Bezeichnung *half quadratic regularization* vorgestellt [17]:

$$J(v,w) = \frac{1}{2}\int_\Omega \Big\{(v-g)^2 + \alpha|w - \nabla v|^2 + \psi_{GY}(w)\Big\}dx \,.$$

Cohen erweiterte diesen Ansatz auf eine allgemeine Klasse zweistufiger Optimierungsverfahren für Probleme der Bildverarbeitung [18]. Im konvexen Fall (1) können wir die Äquivalenz dieses Ansatzes zu (8) nachweisen. Damit ist auch gemäß Abschnitt 3 die globale, lineare Konvergenz der *half–quadratic regularization*–Formulierung bzgl. (1) gezeigt.

5 Experimentelle Ergebnisse

In diesem Abschnitt skizzieren wir zunächst einige Details unserer Implementierungen. Dann fassen wir verschiedene experimentelle Ergebnisse unter zwei Aspekten zusammen: (i) Effizienz verschiedener iterativer Verfahren in Bezug auf die Berechnungszeiten und (ii) *speed–up* bei Multiprozessorarchitekturen.

5.1 Details der Implementation, Vorkonditionierung und Lösungsverfahren für die linearen Teilschritte

Die oben beschriebenen Verfahren wurden auf einer "SGI PowerChallenge" implementiert. Zur Berechnung der linearen Gleichungssysteme (CG-Verfahren mit Block–Jacobi Vorkonditionierung) in (6) verwendeten wir das Softwarepaket "Portable and Extensible Toolkit for Scientific Computation (PETSC)", welches auf dem "Message Passing Interface (MPI)" aufsetzt.

Als Testdaten benutzten wir ein CCD–Kamera Bild einer Laborszene mit 288×360 Bildpunkte. Um die Nichtlinearität des Gleichungssystem zu betonen, haben wir einen durchschnittlichen Glättungsaufwand ($\lambda_h = 4.0$) und eine hohe Sensitivität des Ansatzes gegenüber Signalstrukturen ($c_\rho = 1.0$) für unsere Tests gewählt.

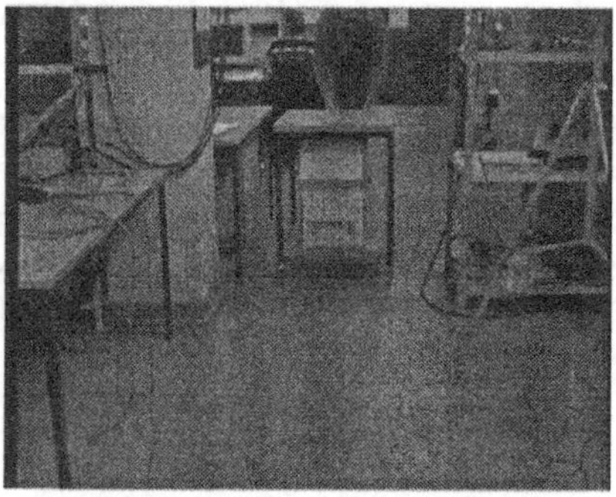

Abbildung 2. *CCD-Kamera-Bild einer Laborszene (288x360 Pixel)*

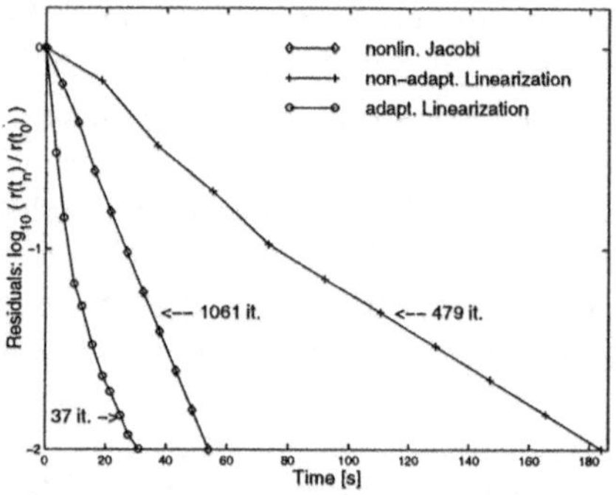

Abbildung 3. *Reduzierung der Residuen als Funktion der Berechnungszeit für drei iterative Verfahren. Es wurden jeweils 10 Prozessoren benutzt.*

5.2 Effizienz der numerischen Verfahren

Tabelle 1 veranschaulicht die Auswirkungen verschiedener (bisher implementierter) Vorkonditionierer auf den Berechnungsaufwand pro Iterationsschritt. Dieses Ergebnis zeigt, daß mit der Block–Jacobi–Vorkonditionierung, die im Gegensatz zur ILU–Vorkonditionierung gut parallelisiert werden kann, der Aufwand um die Hälfte reduziert wird.

Abbildung 3 zeigt wie die verschiedenen numerischen Verfahren das Anfangsresiduum $\mathbf{r_0} = ||\mathbf{L_h}(\mathbf{u^0})||_\infty$ in Abhängigkeit der Berechnungszeit reduzieren. In diesem Experiment wurde auf 10 Prozessoren parallel gerechnet. Es ist zu sehen, daß die adaptive Linearisierung die besten Ergebnisse erzielt, gefolgt von dem nichtlinearen Jacobiverfahren. Die nicht–adaptive Linearisierung reduziert zwar (geg. dem Jacobiverfahren) deutlich die Anzahl der Iterationen, allerdings kann der zusätzliche Aufwand zur Lösung der linearen Systeme dabei nicht ausgeglichen werden.

5.3 Parallelisierung und *speed–up*

In Abbildung 4 ist zu sehen, daß eine signifikante Verbesserung der Berechnungszeiten durch Parallelisierung der Verfahren erreicht wird. Diese Ergebnis eines nahezu linearen *speed–up* belegt die nahezu optimale Implementierung der benutzten linearen Gleichungslöser.

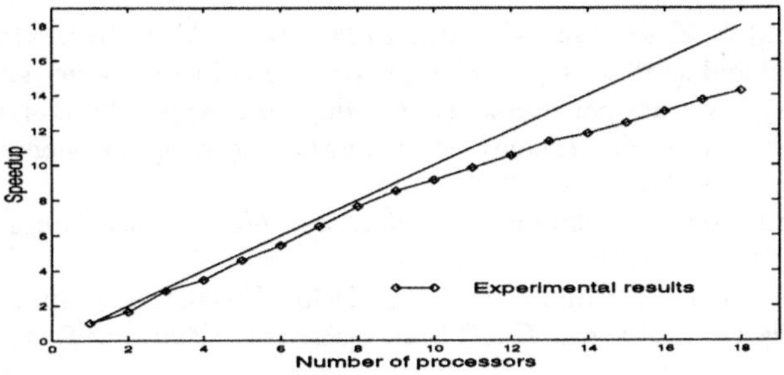

Abbildung 4. *speed–up–Faktor als Funktion der Anzahl der Prozessoren (z.Z. maximal 18 Prozessoren).*

Vorkond.	None	Jacobi	BJacobi	ILU
Zeit/Schritt	14.5s	13.1s	7.9s	7.6s

Tabelle 1. Berechnungszeiten pro Iterationsschritt für verschiedene Vorkonditionierer (1 Prozessor).

Bildgröße	144x180	288x360
$\lambda_h = 4$ $c_\rho = 4$	8.2s	30.5s
$\lambda_h = 2$ $c_\rho = 4$	4.5s	16.2s

Tabelle 2. Berechnungszeiten für veschiedene Bildgrößen und zwei Parametersätze (λ_h, c_ρ) (10 Prozessoren).

6 Zusammenfassung und Ausblick

In der vorliegenden Arbeit wurde eine generisches Schema zur iterativen Minimierung nichtlinearer Variationsansätze vorgestellt. Es wurde gezeigt, daß mit einer parallelen Implementierung (10 Prozessoren) einer adaptiven Linearisierungstechnik die Verarbeitung von Bilder moderater Größe ($\approx$160$\times$160 Pixel) in 5–10 Sekunden bewerkstelligt werden kann (vgl. Tabelle 2). Eine vorsichtige Extrapolation dieser Ergebnisse erlaubt infolge des guten *speed–up* (siehe Abb. 4) die Annahme, daß mit dem gegenwärtigen Verfahren und einem gut ausgestatteten Mehrprozessorsystem Bilder der Grösse 128$\times$128 Bildpunkte in Quasi–Echtzeit ($<$ 1 Sek.) verarbeitet werden können.

In der nächsten Zeit werden verschiedene Optionen zur Optimierung der Verfahren (7) und (8) untersucht werden. Dabei werden wir uns auf Mehrgitterverfahren und adaptive Gitterweitensteuerung konzentrieren.

Danksagung J. Heers und C. Schnörr bedanken sich für die Unterstützung durch die Deutsche Forschungsgemeinschaft ((DFG) Sti 147/1–2 und Schn 457/1–4).

Literatur

1. S. Geman and D. Geman. Stochastic relaxation, Gibbs distributions and the Bayesain restoration of images. *IEEE Trans. Patt. Anal. Mach. Intell.*, 6(6):721–741, 1984.
2. A. Blake and A. Zisserman. *Visual Reconstruction.* MIT Press, 1987.
3. D. Mumford and J. Shah. Optimal approximations by piecewiese smooth functions and associated variational problems. *Comm. Pure Appl. Math.*, 42:577–685, 1989.
4. D. Geiger and A. Yuille. A common framework for image segmentation. *Int. J. of Comp. Vision*, 6(3):227–243, 1991.
5. J.-M. Morel and S. Solimini. *Variational Methods in Image Segmentation.* Birkhäuser,Boston, 1995.
6. R.L. Stevenson, B.E. Schmitz, and E.J. Delp. Discontinuity preserving regularization of inverse problems. *IEEE Trans. Systems, Man and Cyb.*, 24(3):455–469, 1994.
7. C. Schnörr. Unique reconstruction of piecewise-smooth images by minimizing strictly convex noquadratic functionals. *J. of Math. Imag. Vision*, 4:189–198, 1994.
8. C. Schnörr. Repräsentation von bilddaten mit einem konvexen variationsansatz. In *Mustererkennung*, pages 21–28, Berlin–Heidelberg, 1996. Springer.
9. S.Z. Li, Y.H. Huang, and J. Fu. Convex energy functionals in the DA model. In *Proc. IEEE Int. Conf. Image Processing*, 1995.
10. C. Schnörr. A study of a convex variational approach for image segmentation and feature extraction. *J. of Math. Imag. Vision*, 8(3):271–292, 1998.
11. V.V. Shaidurov. *Multigrid Methods for Finite Elements.* Kluwer Academic Publisher, 1995.
12. J. Heers, C. Schnörr, and H.S. Stiehl. Investigating a class of iterative schemes and their parallel implementation for nonlinear variational image smoothing and segmentation. Technical report, Universität Hamburg,Comp. Sci. Dept.,AB KOGS, 1998. in press.
13. E. Zeidler. *Nonlinear Functional Analysis and its Applications.* Springer Verlag, 1995.
14. D. Geman and G.Reynolds. Constrained restoration and the recovery of discontinuites. *IEEE Trans. Patt. Anal. Mach. Intell.*, 14:367–383, 1992.
15. P. Charbonnier, L. Blanc-Feraud, G. Aubert, and M. Barlaud. Determistic edge–preserving regularization in computed imaging. In *Proc. 12th Intern. Conference of Pattern Recognition*, pages C–188–191, 1994.
16. D.C. Dobson and C.R. Vogel. Convergence of an iteratve method for total variation denoising. *SIAM J. of. Numerical Analysis*, 34:1779–1791, 1997.
17. D. Geman and C. Yang. Nonlinear image recovery with half–quadratic regularization. *IEEE Trans. Image Proc.*, 4:932–945, 1995.
18. L.D. Cohen. Auxiliary variables and two–step iterative algorithms in computer vision problems. *J. of Math. Imag. Vision*, 6(1):59–83, 1996.

Calibration Errors in Structure from Motion

Steffen Abraham, Wolfgang Förstner

Institut für Photogrammetrie, Universität Bonn
Nussalle 15, 53115 Bonn
e-mail: Steffen.Abraham@ipb.uni-bonn.de,wf@ipb.uni-bonn.de

Abstract. In this paper we investigate the relation between camera calibration and structure from motion. A method is presented to analyze the effect of systematic errors and uncertainty in camera calibration on 3D-reconstruction and motion parameters. In two simple examples from stereo with lateral and forward motion the approach is demonstrated. The approach can easily be extended to more complex situations and used for planning and online diagnostics in calibration and structure from motion.

1 Introduction

Geometric reconstruction is a central task in Computer Vision motivating the large amount of papers on geometric analysis of stereo and motion in the last decade. The calibration of camera systems plays an important role in getting reliable metric information from 3D–Objects. This paper investigates the relation between camera calibration and structure from motion (SfM) as a representative example for performing sensitivity analysis. Specifically, it is assumed that the calibration is performed off–line leading to calibration parameters which then are used in image sequence analysis. The uncertainty of the calibration parameters can be described by the first two moments of their distribution, namely the *bias*, i. e. an error in the mean value, and the *covariance matrix*, i. e. the effect of random errors of the observations during calibration. The uncertainty of the calibration propagates to the estimated motion parameters and the 3D-structure, for which the uncertainty are described again by bias and covariance matrices.

Whereas calibration is an old and not the central issue of the paper, the necessity for analyzing the sensitivity with respect to calibration errors has been stressed only recently, for example directly in [ACDR94] or indirectly in [Har94]. The effect of calibration errors on the result of motion estimates (cf.[KH94]) and SfM ([FA96], [SS97,Stu97]) has been addressed in previous work. However, in all cases simplifications have been made, e. g. by restricting to only a few parameters, especially to errors in the principle point or the principle distance, or more general, to the calibration parameters distinguishing projective and Euclidean reconstruction.

Sensitivity analysis may be achieved in several ways:

– *Algebraic methods* (cf. e. g. [Stu97]) give closed form expressions for bias and covariance matrices. They therefore have the advantage of giving insight into the internal structure of the system and the domain of validity of the

analysis. But algebraic methods require idealized situations and therefore cannot be used for online self diagnosis.
- *Monte Carlo simulations* for determining bias and covariance matrices are able to handle all types of models, errors and configurations and therefore are not limited to specific application[1] but have the disadvantage that a generalization of their results is limited.
- *Numerical determination* of bias and covariance matrices share advantages and disadvantages of algebraic and simulation techniques. They assume comparably small errors and smooth functions and require the specification of a certain representative geometric setup. However, their results have more predictive power than simulations, they are applicable for online diagnosis and also for quite general nonlinear problems. The lack of representativity and predictive power may be compensated by linking their results to algebraically achieved ones.

We want to demonstrate this combination of algebraic and numerical analysis by choosing the analysis of the interrelation of calibration and SfM as an example.

2 Theoretical Background

2.1 Statistical Tools

This section describes the necessary tools from statistics. Stochastical variables are underscored, $\underline{x}$, if this seems necessary for clarification; the covariance matrix of a vector $\underline{x}$ is given by $\mathsf{D}(\underline{x}) = \boldsymbol{\Sigma}_{xx}$; a general distribution of $\underline{x}$ with first and second moments $\boldsymbol{\mu}_x$ and $\boldsymbol{\Sigma}_{xx}$ is designated with $\underline{x} \sim M(\boldsymbol{\mu}_x, \boldsymbol{\Sigma}_{xx})$.

Propagation of True and Random Errors. Given a nonlinear function $\underline{y} = \boldsymbol{f}(\underline{x})$ of a stochastic vector $\underline{x} = \boldsymbol{\mu}_x + \underline{e_x}$ with distribution $\underline{x} \sim \overline{M(\boldsymbol{\mu}_x, \boldsymbol{\Sigma}_{xx})}$ the following holds for small deterministic or stochastic errors:

$$\boldsymbol{\Delta y} = \boldsymbol{J}_{yx}\boldsymbol{\Delta x} \qquad \underline{y} \sim M(\boldsymbol{f}(\boldsymbol{\mu}_x), \boldsymbol{J}_{yx}\boldsymbol{\Sigma}_{xx}\boldsymbol{J}_{yx}^{\mathsf{T}}) \tag{1}$$

with the Jacobian $\boldsymbol{J}_{yx} = (\partial y_i / \partial x_j)$. Observe that this error propagation of means and variances is independent on the actual distribution, thus does not require the stochastical variables to be normally distributed. Possible mutual dependencies between the input variables are taken into account by the covariances on the off diagonal elements of the covariance matrix.

Parameter Estimation. Parameter estimation aims at inferring the most likely values of unknown parameters, collected in a U-vector $\boldsymbol{\beta} = (\beta_j)$ from observations, collected in a N-vector $\boldsymbol{y} = (y_i)$. It is most easily set up if an explicit expression for the observations as a function of the unknown parameters is available, specifying the first moments $\mathsf{E}(\underline{y}) = \boldsymbol{f}(\boldsymbol{\beta})$. Together with the covariance matrix $\mathsf{D}(\boldsymbol{y}) = \boldsymbol{\Sigma}_{yy}$ it establishes the nonlinear model, which, when making errors $\boldsymbol{e}$ in the observations explicit by $\underline{y} = \mathsf{E}(\underline{y}) + \underline{e}$, reads as

$$\underline{y} = \mathsf{E}(\underline{y}) + \underline{e}; \qquad \underline{y} =\sim M(\boldsymbol{f}(\boldsymbol{\beta}), \boldsymbol{\Sigma}_{yy}) \tag{2}$$

[1] even are able to handle all types of gross errors

We obtain a linear model using approximate values $\beta^{(0)}$: $\beta = \beta^{(0)} + \Delta\beta$ and the Jacobian $A = (\partial y_i / \partial \beta_j)$:

$$\underline{\Delta y} = \mathsf{E}(\underline{\Delta y}) + \underline{e} = \underline{y} - f(\beta^{(0)}) \sim M(A\Delta\beta, \Sigma_{yy}) \qquad (3)$$

It is assumed to hold also for the estimated quantities, thus we have for the first moments:

$$\Delta y = A\widehat{\Delta\beta} + \widehat{e} \qquad (4)$$

The best linear unbiased estimate (BLUE, cf. e. g. [MA76]) for the unknown parameters is given by $\widehat{\beta} = \beta^{(0)} + \widehat{\Delta\beta}$ with

$$\widehat{\Delta\beta} = (A^{\mathsf{T}}\Sigma_{yy}^{-1}A)^{-1}A^{\mathsf{T}}\Sigma_{yy}^{-1}\Delta y \quad \text{with} \quad \Sigma_{\widehat{\beta}\widehat{\beta}} = N^{-1} \doteq (A^{\mathsf{T}}\Sigma_{yy}^{-1}A)^{-1} \qquad (5)$$

Observe that the covariance matrix of the estimated parameters is equal to the inverse of the normal equation matrix N which is always available when estimating parameters using – possibly weighted – least squares. It can be derived by error propagation from eq. (5) using eq. (1).

Self Calibration and SfM. The general setup of *self calibration* from several images, containing SfM as a special case, leads to a natural partitioning of the unknowns $\beta^{\mathsf{T}} = (t^{\mathsf{T}}, k^{\mathsf{T}}, s^{\mathsf{T}})$ into transformation parameters t representing the orientation, motion or external parameters of all camera stations, parameters k for describing the object structure, e. g. by coordinates of 3D-points and additional *calibration parameters* s for describing the the internal orientation (cf. sect. 3.1) of the camera. Instead of $\Delta y = A\widehat{\Delta\beta} + \widehat{e}$ (cf. eq. (4)) the linear model now may be written as

$$\Delta y = B\widehat{\Delta t} + C\widehat{\Delta k} + H\widehat{\Delta s} + \widehat{e} \qquad (6)$$

This setup allows to flexibly model all types of deviations from the pin hole model, e. g. the case where two cameras of a stereo image sequence have different characteristics which keeps constant over a certain time interval.

Obviously eqs. (2) to (6) can be used for both, self calibration and SfM. The geometry in SfM is usually weak, this is why the parameters $\widehat{\Delta s}$ are often calibrated off-line.

2.2 Effect of Calibration Errors

The model (6) is also used to investigate the effect of errors and inaccuracies in the calibration on the 3D–coordinates $\widehat{\Delta k}$ and/or motion parameters $\widehat{\Delta t}$.

The Effect of Bias in Calibration Parameters. Assume the model (6) holds in SfM for estimating the unknown parameters β. If the calibration of the camera is incorrect, the observations y are biased by $H\Delta s$. With eq. (5) the influence on the unknown parameters, actually only containing t and k is given by

$$\Delta\widehat{\beta} = (A^{\mathsf{T}}\Sigma_{yy}^{-1}A)^{-1}A^{\mathsf{T}}\Sigma_{yy}^{-1}H\Delta s \qquad (7)$$

This expression can be used to evaluate the influence of each parameter s_k on each unknown parameter β_j. For example we can evaluate the effect of changes in the focal distance in the camera to the 3D–reconstruction.

The effect of Uncertainty in Calibration Parameters. The uncertainty in the calibration parameters can be expressed by the second order moments $\underline{s} \sim M(s, \Sigma_{ss}(s))$. The effect on the coordinates can be evaluated the same way as the random observational errors under (1) by using the *actual covariance matrix of the observations*, which takes the random observational and the random systematic errors into account: From $\Delta \underline{y} = \mathsf{E}(\Delta y) + H \underline{\Delta s} + \underline{e}$ (eq. (6)) one obtains $\Sigma_{yy}^{\text{true}} = \Sigma_{yy} + H \Sigma_{ss} H^{\mathsf{T}}$ by error propagation. This leads to the actual covariance matrix of the unknown parameters

$$\Sigma_{\widehat{\beta\beta}}^{\text{true}} = \begin{pmatrix} \Sigma_{\hat{t}\hat{t}} & \Sigma_{\hat{t}\hat{k}} \\ \Sigma_{\hat{k}\hat{t}} & \Sigma_{\hat{k}\hat{k}} \end{pmatrix}^{\text{true}} = N^{-1} A^{\mathsf{T}} \Sigma_{yy}^{-1} \cdot \Sigma_{yy}^{\text{true}} \cdot \Sigma_{yy}^{-1} A N^{-1} \tag{8}$$

$$= N^{-1} + N^{-1} A^{\mathsf{T}} \Sigma_{yy}^{-1} H \Sigma_{ss} H^{\mathsf{T}} \Sigma_{yy}^{-1} A N^{-1} \doteq \Sigma_{\widehat{\beta\beta}}(e) + \Sigma_{\widehat{\beta\beta}}(s) \tag{9}$$

which only in case the calibration parameters are estimated with infinite accuracy ($\Sigma_{\widehat{\beta\beta}}(s) = 0$) equals N^{-1}. We can recognize the two sources of uncertainty in 3D–reconstruction, one from the feature measurement in image processing and one from the calibration step.

3 The Examples

In this section we demonstrate our approach in two simple examples. The results are compared with an algebraic solution.

3.1 The setup

We investigate lateral and forward motion with a single camera. The structure and the motion are estimated simultaneously from two images. We consider a 'cube' of data–points in a distance of 10 meters and a camera motion about 1 meter. The coordinate system is fixed between the two projection centers. In the adjustment process there is a rank deficiency of 7. We fix the two projection centers and the rotation around the basis (figure 1).

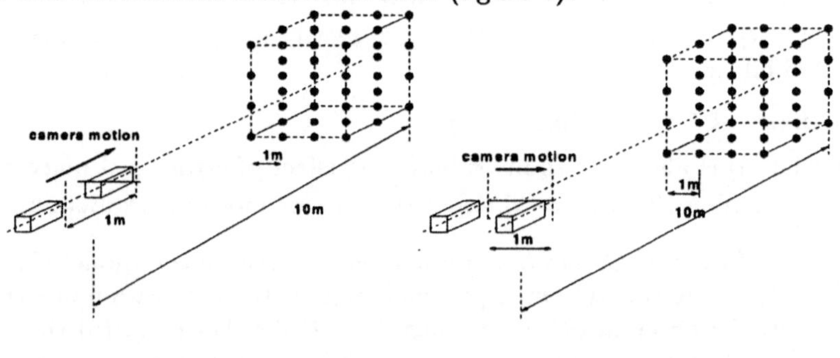

Fig. 1. Experimental setup for forward motion and lateral motion

A pinhole camera–model is used in the investigation:

$$x_{ij} = c \frac{r_1(\Omega_j)(p_i - p_{0j})}{r_3(\Omega_j)(p_i - p_{0j})} + \Delta x_{ij} \qquad y_{ij} = c \frac{r_2(\Omega_j)(p_i - p_{0j})}{r_3(\Omega_j)(p_i - p_{0j})} + \Delta y_{ij} \tag{10}$$

where x_{ij} and y_{ij} are the coordinates of the image of the i-th point in camera in position j. $\boldsymbol{R}(\boldsymbol{\Omega}_j)$ is the rotation matrix for image j depending on the rotation parameters $\boldsymbol{\Omega}_j$, $\boldsymbol{p}_i$ is the 3-vector of the coordinates of the object point i, $\boldsymbol{p}_{0j}$ is the 3-vector of the coordinates of the projection center and $\Delta x_{ij}(\boldsymbol{s})$ and $\Delta y_{ij}(\boldsymbol{s})$ are correction terms. In our example we consider a physical motivated model. It includes 4 Parameters to describe the interior orientation of the camera:

$$\Delta x_{ij} = x_H + x_{ij} A_1 (r^2 - r_0^2) \qquad \Delta y_{ij} = y_H + y_{ij} A_1 (r^2 - r_0^2) \tag{11}$$

(c–principal distance, x_H, y_H–principal point, A_1–radial distortion, $r^2 = x_{ij}^2 + y_{ij}^2$, r_0-constant, depending on the size of the image [WE80]). We took the numerical values from calibration of a real camera–system ([AF97], $c = 1144pel$, $x_H = 3.5pel$, $y_H = 10.8pel$, $A_1 = -6.1e-8$).

3.2 Algebraic Prediction

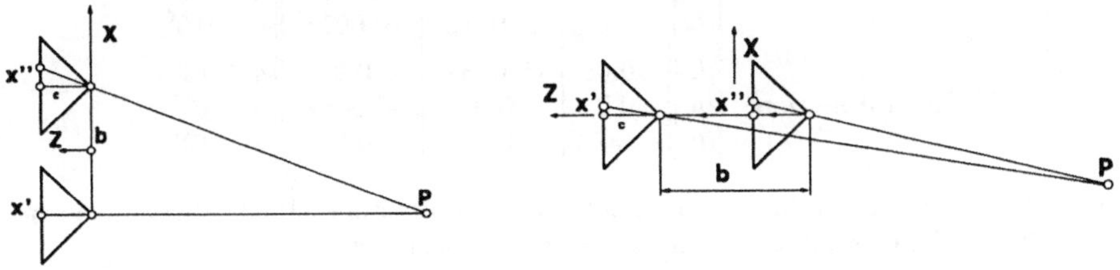

Fig. 2. The geometry for lateral (left) and for forward motion (right)

First we are making some predictions about the effect of errors in the principal distance c on the 3D–reconstruction. By fixation of the motion parameters the 3D–coordinate of one point can easily be calculated by triangulation (cf. fig. 2): for forward motion:

$$X = \frac{b}{c} \frac{x' x''}{(x' - x'')}; \qquad Z = \frac{b}{2} \frac{x' + x''}{(x' - x'')} \tag{12}$$

and for lateral motion:

$$X = \frac{b}{2} \frac{(x'' + x')}{(x'' - x')}; \qquad Z = \frac{bc}{x'' - x'}. \tag{13}$$

Observe that the Z coordinate in forward motion and the X and Y coordinate in lateral motion are independent from the principal distance c. Changes and errors in the calibration of this parameter do not effect the estimation in these coordinates. On the other hand, c directly effects the X, Y coordinates (forward motion)

$$X + \Delta X = \frac{b}{c + \Delta c} \frac{x' x''}{(x' - x'')}; \qquad \Delta X = \frac{\Delta c}{c + \Delta c} X \tag{14}$$

and the Z–coordinates in (lateral motion):

$$Z + \Delta Z = \frac{b(c + \Delta c)}{x'' - x'}; \qquad \Delta Z = \frac{\Delta c}{c} Z \tag{15}$$

The size of the error depends also on the position of the points in space. Points with a greater distance from the motion trajectory are more effected. We can do this analysis for all interior orientation parameters. But an analytic solution without fixation of the motion parameters is more complicated, so at this point we explore the numerical method.

3.3 Effect of calibration errors on 3D–Coordinates

Bias. The elements $(\partial\beta_j/\partial s_k)$ in $\boldsymbol{N}^{-1}\boldsymbol{A}^{\mathsf{T}}\boldsymbol{\Sigma}_{yy}\boldsymbol{H}\boldsymbol{\Delta s}$ (eq. (7)) directly give the influence on individual estimates. Table 1 collects the **maximum** influence numbers for the interior orientation parameters (according (11)) onto the 3D–coordinates. The coefficient A_1 needs a comment for interpretation: A value of $2.2\,10^{-8}$ leads to a maximum distortion of 1 [pel] in the image plane. From all

		∂c	∂x_H	∂y_H	∂A_1
	∂x	0.0013	-6.5 10^{-9}	-2.6 10^{-8}	-2.9 10^{5}
Forward motion	∂y	0.0013	-1.3 10^{-8}	-3.2 10^{-8}	-2.7 10^{5}
	∂z	0.0	-0.0012	-0.0013	-1.4 10^{6}
	∂x	0.0	-0.0082	0.0	4.6 10^{6}
Lateral motion	∂y	0.0	0.0	-0.0087	4.6 10^{6}
	∂z	-0.0087	0.0	0.0	3.1 10^{7}

Table 1. Maximum influence of calibration parameters $c[pel]$, $x_H[pel]$, $y_H[pel]$, $A_1[1/pel^2]$ on the coordinates of the 3D–reconstruction $[m]$

influence numbers (not only the maximum numbers) we get some interesting results:

- Errors in the principle distance c do not influence the Z–coordinate in forward motion and the X and Y coordinate in lateral motion. The size of the influence number depends on the X and Y point coordinate in forward motion and the Z coordinate in lateral motion. The result is analogical to the algebraic solution
- Errors in the principal point coordinates causes errors in the Z–coordinate in forward and X and Y–coordinates in lateral motion. Points far from the motion trajectory are more effected.
- Calibration errors in A_1 have an appr. 10 times larger influence on the coordinates in the case of lateral motion than for forward motion.

The maximal effect of calibration errors could be observed on the 'corner' of our 'point–cube'. This can be intuitively expected.

Uncertainty. According to eq. (8) we can propagate the uncertainty from the calibration to the 3D–reconstruction. We consider a example point $(1.5, 1.5, 10)[m]$ on the corner of the cube and assume a precision in point measurement of about $0.5pel$ in the image plane, which is a reasonable assumption for non–signalized points. The camera is calibrated with a medium precision (see standard deviations in table 2). The first column reports the estimated precision of the point coordinates without taking the calibration errors into account (table 2). The

		(*)	(**)				(***)
			c $\sigma c = 55pel(5\%c)$	x_H $\sigma_{xH} = 25pel$	y_H $\sigma_{yH} = 25pel$	A_1 $\sigma_{A1} = 1e-7$	Σ
Forward	σ_x	0.117	0.072	0.0	0.0	0.023	0.193
	σ_y	0.119	0.072	0.0	0.0	0.025	0.195
	σ_z	0.313	0.0	0.031	0.033	0.108	0.431
Lateral	σ_x	0.105	0.0	0.206	0.0	0.463	0.507
	σ_y	0.088	0.0	0.0	0.218	0.455	0.505
	σ_z	0.572	0.481	0.0	0.0	3.084	3.122

Table 2. Uncertainty in camera calibration and effect on 3D–reconstruction (point $(1.5, 1.5, 10)[m]$), precision of 3D–reconstruction (standard–deviation $[m]$), (*)–error free calibration, (**)–effect of uncertainty in one single parameter, (***)–all error–sources integrated

influence of the uncertainty from the single calibration parameters are shown in the next columns. The last column describes the estimated precision including both influences, point measurement and calibration.

Observe that the critical parameters in our example are the calibration of the principal distance c and the radial distortion coefficient A_1. For lateral motion the effect of the camera–calibration is more critical than for forward motion.

3.4 Errors by linearization ?

One could argue, that the simple linearization step would falsify our error approximation. This can faced by adjusting the structure and the motion parameters with modified interior orientation within a adjustment by nonlinear optimization. Then the reconstruction error can be compared with the linear approximation. We did this adjustment with larger modifications of the interior orientation parameters. The effect of the modifications on our example point for forward motion are reported in table 3. The results for lateral motion are similar. The approximation from the linear algorithm are also shown. Comparing the results it can be observed, that the differences between linear approximation and the nonlinear adjustment can be neglected in practise.

	$\Delta c = -100pel$	$\Delta x_H = 25pel$	$\Delta y_H = 25pel$	$\Delta A_1 = 1e-7$
$\Delta x_{nonlinear}$	-0.144	0.010	0.006	-0.024
$\Delta x_{Approx.}$	-0.130	0.000	0.000	-0.024
$\Delta y_{nonlinear}$	-0.144	0.005	0.011	-0.025
$\Delta y_{Approx.}$	-0.130	0.000	0.000	-0.025
$\Delta z_{nonlinear}$	0.000	-0.123	-0.130	-0.109
$\Delta z_{Approx.}$	0.000	-0.120	-0.130	-0.110

Table 3. Nonlinear adjustment in comparison to linear approximation, Modifications in 3D–reconstruction caused by changes of the interior orientation, forward motion, example point $(1.5, 1.5, 10)[m]$

4 Conclusions

The paper presented methods for sensitivity analysis and demonstrated its applicability for evaluating the effect of systematic and stochastic calibration errors in structure from motion estimation. The proposed methodology is useful during planning, for online self diagnosis and for final evaluation and is transferable to all types of parameter estimation problems in Computer Vision. To find analytic solutions is often hard. The numerical analysis is very flexible and different functional models can be investigated in a short time.

We also got some interesting results for structure from motion. The effect of errors and uncertainties in the camera parameters are very different. Errors in the principle distance modify directly the reconstruction. The sensitivity for calibration errors for forward and lateral motion is different. For example for forward motion the effect of errors in the principal point coordinates are very small. Without calibrating this parameters (assumption image center) a 3D–reconstruction would be reasonable depending on the requirements in the application. This could be checked in a planning stage by the proposed approach.

The next step in our work is to apply these tools in real applications. There are interesting questions to answer, for example: which precision of camera calibration is needed to fulfill the requirements in an application or can the desired measurement precision be reached by the camera setup.

References

[ACDR94] Y. Aloimonos, R. Chellappa, L. S. Davis, and A. Rosenfeld. Maryland progress in image understanding. In *ARPA Image Understanding Workshop*, pages 9–20, 1994.

[AF97] S. Abraham and W. Förstner. Zur automatischen Modellwahl bei der Kalibrierung von CCD–Kameras. In *DAGM Symposium Mustererkennung*. Springer Verlag Berlin, 1997.

[FA96] C. Fermüller and Y. Aloimonos. Algorithms-independent stability analysis of structure from motion. Technical Report CS-TR-391, Center fro Automation Research, U. of Maryland, 1996.

[Har94] R. Hartley. An algorithm for self calibration from several views. In *Proc. CVPR*, pages 908–912, 1994.

[KH94] R. Kumar and A. R. Hanson. Robust methods for estimating pose and a sensitivity analysis. *CVGIP-IU*, 60(3):313–342, 1994.

[MA76] E. M. Mikhail and F. Ackermann. *Observations and Least Squares*. University Press of America, 1976.

[SS97] T. Svoboda and P. Sturm. Badly calibrated camera in ego-motion estimation – propagation of uncertainty. In *Proc. CAIP*. Springer, 1997.

[Stu97] P. Sturm. Critical motion sequences for monocular self-calibration and uncalibrated euclidean reconstruction. In *Proc. CVPR*, pages 1100–1105, 1997.

[WE80] W. Wester-Ebbinghaus. Photographisch–numerische Bestimmung der geometrischen Abbildungseigenschaften eines optischen Systems. *Optik*, 55(3):253–259, 1980.

Fast Approximation of Support Vector Kernel Expansions, and an Interpretation of Clustering as Approximation in Feature Spaces

Bernhard Schölkopf[1], Phil Knirsch[2], Alex Smola[1], and Chris Burges[2]

[1] GMD FIRST, Rudower Chaussee 5, 12489 Berlin, http://svm.first.gmd.de
[2] Bell Labs, Lucent Technologies, Crawfords Corner Rd., Holmdel NJ 07733, USA

Abstract. Kernel-based learning methods provide their solutions as expansions in terms of a kernel. We consider the problem of reducing the computational complexity of evaluating these expansions by approximating them using fewer terms. As a by-product, we point out a connection between clustering and approximation in reproducing kernel Hilbert spaces generated by a particular class of kernels.

1 Introduction

Recent years have seen a surge in the interest in learning methods based on Mercer kernels, i.e. functions $k(\mathbf{x}, \mathbf{y})$ which for all data sets $\{\mathbf{x}_1, \ldots, \mathbf{x}_\ell\} \subset \mathbf{R}^N$ give rise to positive matrices $K_{ij} := k(\mathbf{x}_i, \mathbf{x}_j)$ [10]. Using k instead of a dot product in $\mathbf{R}^N$ corresponds to mapping the data into a possibly high-dimensional space F by a (usually nonlinear) map $\Phi : \mathbf{R}^N \to F$, and taking the dot product there, i.e. [2]

$$k(\mathbf{x}, \mathbf{y}) = (\Phi(\mathbf{x}) \cdot \Phi(\mathbf{y})). \tag{1}$$

Any linear algorithm which can be carried out in terms of dot products can be made nonlinear by substituting an a priori chosen kernel. Examples thereof are the Gaussian kernel $k(\mathbf{x}, \mathbf{y}) = \exp(-\|\mathbf{x} - \mathbf{y}\|^2/(2\sigma^2))$ and the polynomial kernel $k(\mathbf{x}, \mathbf{y}) = (\mathbf{x} \cdot \mathbf{y})^d$ (which corresponds to a dot product in a feature space spanned by *all* products of order d in the original input features [2, 11]). We can think of Φ as a map $\Phi : \mathbf{x} \mapsto k(\mathbf{x}, .)$ into a Hilbert space F of functions $\sum_i \alpha_i k(\mathbf{x}_i, .)$ with a dot product satisfying $\langle k(\mathbf{x}, .), k(\mathbf{y}, .)\rangle_F = k(\mathbf{x}, \mathbf{y})$ [11]. By virtue of this property, F is called a *reproducing kernel Hilbert space* (e.g. [10]).

Let us mention two examples of algorithms using Mercer kernels:

Support Vector (SV) classifiers [14] construct a maximum margin hyperplane in F. In input space, this corresponds to a nonlinear decision boundary of the form

$$f(\mathbf{x}) = \operatorname{sgn}\left(\sum_{i=1}^{\ell} \gamma_i k(\mathbf{x}, \mathbf{x}_i) + b\right), \tag{2}$$

where the $\mathbf{x}_i$ are the training examples. Those with $\gamma_i \neq 0$ are called *Support*

Vectors; in many applications, most of the γ_i, which are found by solving a quadratic program, turn out to be 0. Excellent classification accuracies in both OCR and object recognition have been obtained using SV machines [11]. A generalization to the case of regression estimation, leading to similar function expansion, exists [14].

Kernel Principal Component Analysis [12] carries out a linear PCA in the feature space F. The extracted features take the nonlinear form

$$f_k(\mathbf{x}) = \sum_{i=1}^{\ell} \gamma_i^k k(\mathbf{x}_i, \mathbf{x}), \tag{3}$$

where, up to a normalization, the γ_i^k are the components of the k-th Eigenvector of the matrix $(k(\mathbf{x}_i, \mathbf{x}_j))_{ij}$.

In both techniques, there is a price that we pay for the nonlinearity: since vectors in F are only given implicitly by expansions in terms of images of training examples under Φ, we need to evaluate the kernel function many times. In the NIST benchmark of 60000 handwritten digits, SV machines are more accurate than any other single classifier [11], however, they are inferior to neural nets in run-time classification speed [5]. In applications where the latter is an issue, it is thus desirable to come up with acceleration methods [4, 5, 9].

The present paper gives an analysis for the case of the Gaussian kernel, which has proven to perform very well in applications [13], and proposes and tests an iteration procedure for computing fast approximations of kernel expansions.

2 The Reduced Set (RS) Method

Given a vector $\Psi \in F$, expanded in images of input patterns $\mathbf{y}_i \in \mathbf{R}^N$,

$$\Psi = \sum_{i=1}^{N_y} \gamma_i \Phi(\mathbf{y}_i), \tag{4}$$

with $\gamma_i \in \mathbf{R}, \mathbf{y}_i \in \mathbf{R}^N$, one can try to approximate it by [4]

$$\Psi' = \sum_{i=1}^{N_z} \beta_i \Phi(\mathbf{z}_i), \tag{5}$$

with $N_z < N_y$, $\beta_i \in \mathbf{R}, \mathbf{z}_i \in \mathbf{R}^N$. To this end, one can minimize [4]

$$||\Psi - \Psi'||^2 = \sum_{i,j=1}^{N_y} \gamma_i \gamma_j k(\mathbf{y}_i, \mathbf{y}_j) + \sum_{i,j=1}^{N_z} \beta_i \beta_j k(\mathbf{z}_i, \mathbf{z}_j) - 2 \sum_{i=1}^{N_y} \sum_{j=1}^{N_z} \gamma_i \beta_j k(\mathbf{y}_i, \mathbf{z}_j). \tag{6}$$

The crucial point is that even if Φ is not given explicitely, (6) can be computed (and minimized) in terms of the kernel. Consider first the case $N_z = 1$, i.e.

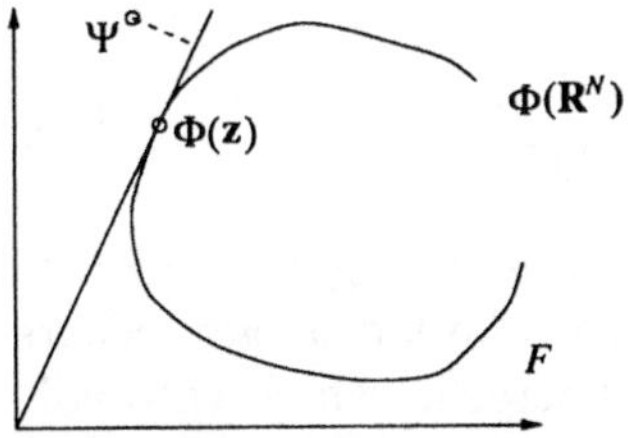

Fig. 1. Given a vector $\Psi \in F$, we try to approximate it by a multiple of a vector $\Phi(\mathbf{z})$ in the image of input space ($\mathbf{R}^N$) under the nonlinear map Φ by finding $\mathbf{z}$ such that the projection distance of Φ onto Span($\Phi(\mathbf{z})$) is minimized.

$\Psi' = \beta\Phi(\mathbf{z})$. We observe that rather than minimizing (6), we can minimize the distance between Ψ and the orthogonal projection of Ψ onto Span($\Phi(\mathbf{z})$) (Fig. 1),

$$\left\| \frac{(\Psi \cdot \Phi(\mathbf{z}))}{(\Phi(\mathbf{z}) \cdot \Phi(\mathbf{z}))} \Phi(\mathbf{z}) - \Psi \right\|^2 = \|\Psi\|^2 - \frac{(\Psi \cdot \Phi(\mathbf{z}))^2}{(\Phi(\mathbf{z}) \cdot \Phi(\mathbf{z}))}. \tag{7}$$

To this end, we maximize

$$\frac{(\Psi \cdot \Phi(\mathbf{z}))^2}{(\Phi(\mathbf{z}) \cdot \Phi(\mathbf{z}))}, \tag{8}$$

which can be expressed in terms of the kernel. The maximization of (8) over $\mathbf{z}$ is preferable to the one of (6) over $\mathbf{z}$ and β, since it comprises a lower-dimensional problem, and since $\mathbf{z}$ and β have different scaling behaviour. Once the maximum of (8) is found, it is extended to the minimum of (6) by setting (cf. (7)) $\beta = (\Psi \cdot \Phi(\mathbf{z}))/(\Phi(\mathbf{z}) \cdot \Phi(\mathbf{z}))$. The function (8) can either be minimized using standard techniques (as in [4]), or, for particular choices of kernels, using fixed-point iteration methods, as shown in the next section.

3 Clustering as Approximation in Feature Space

For kernels which satify $k(\mathbf{z}, \mathbf{z}) = 1$ for all $\mathbf{z} \in \mathbf{R}^N$ (e.g. Gaussian kernels), (8) reduces to

$$(\Psi \cdot \Phi(\mathbf{z}))^2. \tag{9}$$

For the extremum, we have $0 = \nabla_{\mathbf{z}}(\Psi \cdot \Phi(\mathbf{z}))^2 = 2(\Psi \cdot \Phi(\mathbf{z}))\nabla_{\mathbf{z}}(\Psi \cdot \Phi(\mathbf{z}))$. To evaluate the gradient in terms of k, we substitute (4) to get the sufficient condition $0 = \sum_{i=1}^{N_y} \gamma_i \nabla_{\mathbf{z}} k(\mathbf{y}_i, \mathbf{z})$. For $k(\mathbf{x}_i, \mathbf{z}) = k(\|\mathbf{x}_i - \mathbf{z}\|^2)$ (e.g. Gaussians, or $(\|\mathbf{x}_i - \mathbf{z}\|^2 + 1)^c$ for $c = -1, -1/2$), we obtain $0 = \sum_{i=1}^{N_y} \gamma_i k'(\|\mathbf{y}_i - \mathbf{z}\|^2)(\mathbf{y}_i - \mathbf{z})$, leading to

$$\mathbf{z} = \frac{\sum_{i=1}^{N_y} \gamma_i k'(\|\mathbf{y}_i - \mathbf{z}\|^2)\mathbf{y}_i}{\sum_{i=1}^{N_y} \gamma_i k'(\|\mathbf{y}_i - \mathbf{z}\|^2)}. \tag{10}$$

For the Gaussian kernel $k(\mathbf{x}_i, \mathbf{z}) = \exp(-\|\mathbf{x}_i - \mathbf{z}\|^2/(2\sigma^2))$ we thus arrive at

$$\mathbf{z} = \frac{\sum_{i=1}^{N_y} \gamma_i \exp(-\|\mathbf{y}_i - \mathbf{z}\|^2/(2\sigma^2))\mathbf{y}_i}{\sum_{i=1}^{N_y} \gamma_i \exp(-\|\mathbf{y}_i - \mathbf{z}\|^2/(2\sigma^2))}, \tag{11}$$

and devise an iteration

$$\mathbf{z}_{n+1} = \frac{\sum_{i=1}^{N_y} \gamma_i \exp(-\|\mathbf{y}_i - \mathbf{z}_n\|^2/(2\sigma^2))\mathbf{y}_i}{\sum_{i=1}^{N_y} \gamma_i \exp(-\|\mathbf{y}_i - \mathbf{z}_n\|^2/(2\sigma^2))}. \tag{12}$$

The denominator equals $(\Psi \cdot \Phi(\mathbf{z}_n))$ and thus is nonzero in a neighbourhood of the extremum of (9), unless the extremum itself is zero. The latter only occurs if the projection of Ψ on the linear span of $\Phi(\mathbf{R}^N)$ is zero, in which case it is pointless to try to approximate Ψ. Numerical instabilities related to $(\Psi \cdot \Phi(\mathbf{z}))$ being small can thus be approached by restarting the iteration with different starting values.

Interestingly, (12) can be interpreted in the context of clustering (e.g. [3]). It resembles an EM iteration for the determination of the center of a single Gaussian cluster trying to capture as many of the $\mathbf{y}_i$ with positive γ_i as possible, and simultaneously avoids those $\mathbf{y}_i$ with negative γ_i. For SV classifiers, the sign of the γ_i equals the label of the pattern $\mathbf{y}_i$. It is this sign which distinguishes (12) from plain clustering or parametric density estimation. The occurence of negative signs is related to the fact that we are not trying to estimate a parametric density but the *difference* between two densities (modulo normalization constants).

To see this, we define the sets $pos = \{i : \gamma_i > 0\}$ and $neg = \{i : \gamma_i < 0\}$, and the shorthands $p_{pos}(\mathbf{z}) = \sum_{pos} \gamma_i \exp(-\|\mathbf{y}_i - \mathbf{z}\|^2/(2\sigma^2))$ and $p_{neg}(\mathbf{z}) = \sum_{neg} |\gamma_i| \exp(-\|\mathbf{y}_i - \mathbf{z}\|^2/(2\sigma^2))$. The target (9) then reads $(p_{pos}(\mathbf{z}) - p_{neg}(\mathbf{z}))^2$, i.e. we are trying to find a point $\mathbf{z}$ where the difference between the (unnormalized) "probabilities" for the two classes is maximal, and estimate the approximation to (4) by a Gaussian centered at $\mathbf{z}$. Moreover, note that we can rewrite (11) as

$$\mathbf{z} = \frac{p_{pos}(\mathbf{z})}{p_{pos}(\mathbf{z}) - p_{neg}(\mathbf{z})}\mathbf{y}_{pos} + \frac{p_{neg}(\mathbf{z})}{p_{neg}(\mathbf{z}) - p_{pos}(\mathbf{z})}\mathbf{y}_{neg}, \tag{13}$$

where

$$\mathbf{y}_{pos/neg} = \frac{\sum_{pos/neg} \gamma_i \exp(-\|\mathbf{y}_i - \mathbf{z}\|^2/(2\sigma^2))\mathbf{y}_i}{\sum_{pos/neg} \gamma_i \exp(-\|\mathbf{y}_i - \mathbf{z}\|^2/(2\sigma^2))}. \tag{14}$$

4 Multi-Term RBF Approximations

So far, we have dealt with approximations $\Psi' = \beta\Phi(\mathbf{z})$. Now suppose we want to approximate a vector

$$\Psi_1 = \sum_{i=1}^{\ell} \alpha_i \Phi(\mathbf{x}_i) \tag{15}$$

by an expansion of the type (5) with $N_z > 1$. To this end, we iterate [4] by $\Psi_{m+1} := \Psi_m - \beta_m \Phi(\mathbf{z}_m)$. Here, $\mathbf{z}_m$ denotes the $\mathbf{z}$ found for Ψ_m, obtained by iterating (12). To apply (12), Ψ_m needs to be utilized in its representation in terms of mapped input images,

$$\Psi_m = \sum_{i=1}^{\ell} \alpha_i \Phi(\mathbf{x}_i) - \sum_{i=1}^{m-1} \beta_i \Phi(\mathbf{z}_i), \tag{16}$$

i.e. we need to set $N_y = \ell+m-1$, $(\gamma_1, \ldots, \gamma_{N_y}) = (\alpha_1, \ldots, \alpha_\ell, -\beta_1, \ldots, -\beta_{m-1})$, and $(\mathbf{y}_1, \ldots, \mathbf{y}_{N_y}) = (\mathbf{x}_1, \ldots, \mathbf{x}_\ell, \mathbf{z}_1, \ldots, \mathbf{z}_{m-1})$.

The coefficient $\beta_m = \beta(\mathbf{z}_m)$ could be computed as $(\Psi \cdot \Phi(\mathbf{z}))/(\Phi(\mathbf{z}) \cdot \Phi(\mathbf{z}))$. However, if the $\Phi(\mathbf{z}_1), \ldots, \Phi(\mathbf{z}_m)$ are not orthogonal in F, then the best approximation of Ψ_1 in their Span is *not* obtained by computing orthogonal projections onto each direction. Instead, we need to compute the optimal coefficients $\boldsymbol{\beta} = (\beta_1, \ldots, \beta_m)$ anew in each step. To this end, we evaluate the derivative of the distance in F, $\frac{\partial}{\partial \beta_k} \|\Psi_1 - \sum_{i=1}^m \beta_i \Phi(\mathbf{z}_i)\|^2 = -2\Phi(\mathbf{z}_k)(\Psi_1 - \sum_{i=1}^m \beta_i \Phi(\mathbf{z}_i))$, and set it to 0. Substituting (15) (with $\boldsymbol{\alpha} = (\alpha_1, \ldots, \alpha_\ell)$) and using matrix notation, $K^z_{ij} := (\Phi(\mathbf{z}_i) \cdot \Phi(\mathbf{z}_j))$ and $K^{zx}_{ij} := (\Phi(\mathbf{z}_i) \cdot \Phi(\mathbf{x}_j))$, this leads to $K^{zx}\boldsymbol{\alpha} = K^z\boldsymbol{\beta}$, hence[1]

$$\boldsymbol{\beta} = (K^z)^{-1} K^{zx} \boldsymbol{\alpha}. \tag{17}$$

The iteration is stopped after N_z steps, either specified in advance, or by monitoring when $\|\Psi_{m+1}\|$ (i.e. $\|\Psi_1 - \sum_{i=1}^m \beta_i \Phi(\mathbf{z}_i)\|$) falls below a specified threshold. The solution vector takes the form (5).

5 Approximating Several Vectors Simultaneously

In many applications, we actually would like to approximate more than just one vector in feature space, and we are interested in keeping the total number of approximation vectors small. For instance, in SV digit recognition, one typically trains 10 binary classifiers, and combines the responses. Previous studies have dealt with this by estimating 10 approximations separately [4, 5], which need not be optimal. Even more pronounced is the case of Kernel PCA [12]. There, a network extracting n features computes n projections on vectors in feature space. Approximating each one separately would most likely be wasteful. Approximating the vectors $\Psi^1, \ldots, \Psi^n$ simultaneously results in a network where the computationally expensive first layer is shared by the different sub-networks.

First note that in most cases, already the above algorithm can be used in this respect. Often, the terms in the approximation of one vector will not be orthogonal to the other vectors, and can thus be used, essentially for free, to decrease the discrepancy also for the other vectors, by computing corresponding values of the β_i as in (17).

More principled is the following approch: as a straightforward generalization of (7), we minimize the sum of the projection distances,

$$\sum_{k=1}^{n} \left\| \frac{(\Psi^k \cdot \Phi(\mathbf{z}))}{(\Phi(\mathbf{z}) \cdot \Phi(\mathbf{z}))} \Phi(\mathbf{z}) - \Psi^k \right\|^2 = \sum_{k=1}^{n} \left(\|\Psi^k\|^2 - \frac{(\Psi^k \cdot \Phi(\mathbf{z}))^2}{(\Phi(\mathbf{z}) \cdot \Phi(\mathbf{z}))} \right), \tag{18}$$

[1] If the discrepancy Ψ_{m+1} has not yet reached zero, then K^z will be invertible. If K^z does not have full rank, one can use the Pseudoinverse, or select the solution which has the largest number of zero components.

which amounts to maximizing $\sum_{k=1}^{n}(\Psi^k \cdot \Phi(\mathbf{z}))^2/(\Phi(\mathbf{z})\cdot\Phi(\mathbf{z}))$. For the Gaussian RBF (radial basis function) kernel, the gradient takes the form $0 = \nabla_{\mathbf{z}}(\Psi^k \cdot \Phi(\mathbf{z}))^2 = 2(\Psi^k \cdot \Phi(\mathbf{z}))\nabla_{\mathbf{z}}(\Psi^k \cdot \Phi(\mathbf{z}))$. For $\Psi^k = \sum_{i=1}^{N_y^k} \gamma_i^k \Phi(\mathbf{y}_i^k)$, we thus obtain

$$\mathbf{z} = \frac{\sum_{k=1}^{n}\sum_{i,j=1}^{N_y^k} \gamma_i^k \exp(-\|\mathbf{y}_i^k - \mathbf{z}\|^2/(2\sigma^2))\exp(-\|\mathbf{y}_j^k - \mathbf{z}\|^2/(2\sigma^2))\mathbf{y}_j^k}{\sum_{k=1}^{n}\sum_{i,j=1}^{N_y^k} \gamma_i^k \exp(-\|\mathbf{y}_i^k - \mathbf{z}\|^2/(2\sigma^2))\exp(-\|\mathbf{y}_j^k - \mathbf{z}\|^2/(2\sigma^2))}. \quad (19)$$

6 Experiments

In our experiments on the USPS handwritten digit database (e.g. [11]) of 7291 training patterns and 2007 test patterns (size 16×16), we approximated the SV expansions (2) of ten binary classifiers, each trained to separate one digit from the rest. We used the Gaussian kernel $k(\mathbf{x}, \mathbf{y}) = \exp(-\|\mathbf{x} - \mathbf{y}\|^2/128)$ (cf. [13]), and the approximation technique described in Sec. 4.

Table 1 shows classification error results for approximations using varying numbers of RS vectors. For large numbers, the accuracy of the original system can be approached closely. The original SV system had 254 SVs per classifier. To get a speedup by a factor of 10, we have to use the system with 25 RS vectors (RS-25), in which case the classification accuracy drops moderately from 4.4% to 5.1%, which is still competitive with convolutional neural networks on

Table 1. Top: numbers of SVs for the original SV RBF system. Bottom: numbers of test errors for each binary recognizer, and test error rates for 10-class classification. First row: original SV system, with 254 SVs on average; following rows: systems with varying numbers of RS (approximation) vectors (RS-n stands for n vectors) per binary recognizer (optimal threshold re-computed on the training set), computed by iterating one-term approximations, separately for each recognizer. Last two rows: with a subsequent global gradient descent, the results can be further improved (see text).

digit	0	1	2	3	4	5	6	7	8	9	average
#SVs	219	92	316	309	288	340	213	206	304	250	254

	0	1	2	3	4	5	6	7	8	9	10-class
SV-254	16	13	30	17	32	22	11	12	26	17	4.4%
RS-10	26	13	45	49	35	54	22	24	39	24	7.1%
RS-15	17	16	43	50	49	37	14	18	45	35	6.4%
RS-20	27	11	38	30	35	43	12	16	30	25	5.6%
RS-25	21	12	38	32	31	22	12	18	33	28	5.1%
RS-50	18	10	33	28	32	23	12	15	35	27	5.0%
RS-100	14	13	26	22	30	26	11	14	28	23	4.8%
RS-150	13	14	28	32	27	24	12	14	29	26	4.7%
RS-250	12	13	26	26	32	25	11	14	26	24	4.6%
RS2-25	14	14	31	22	30	23	11	14	26	17	4.7%
RS2^{old}-25	16	13	32	19	31	26	11	15	25	18	5.0%

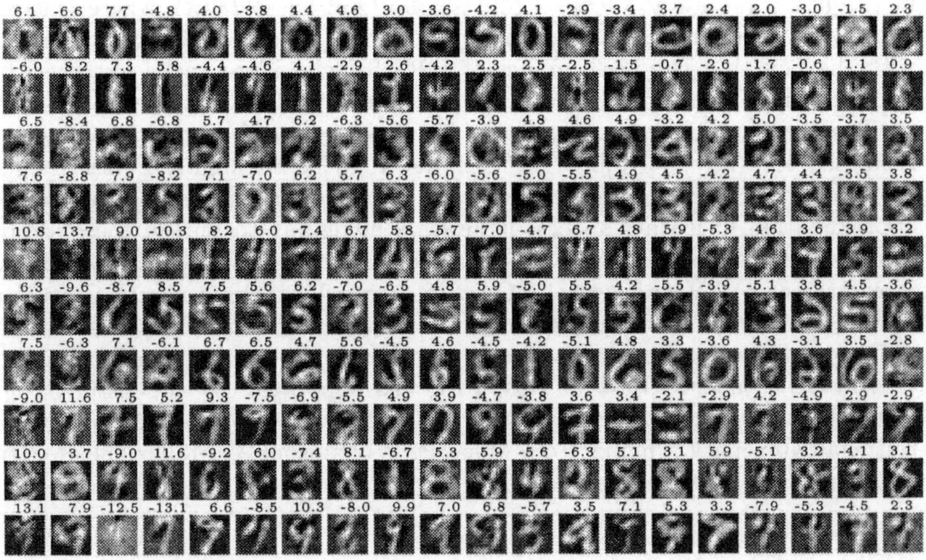

Fig. 2. Complete display of RS-20, with coefficients (Top: recognizer of digit 0,. . . , bottom: digit 9). Note that positive coefficients (roughly) correspond to positive examples.

that data base [7]. Moreover, we can further improve this result by adding the second phase of the traditional RS algorithm, where a global gradient descent is performed in the space of all $(\mathbf{z}_i, \beta_i)$ [4, 5] (computationally more expensive than the first phase by about two orders of magnitude): this led to an error rate of 4.7%. For the considered kernel, this is slightly better than the traditional RS method, which yielded 5.0% (for polynomial kernels, the latter method led to 4.3% at the same speedup [4]). Figure 2 shows the RS-20 vectors of the 10 binary classifiers. As an aside, note that unlike the approach of [4], our algorithm produces images which do look meaningful (i.e. digit-like).

7 Discussion

We have proposed an approach for approximating kernel expansions, and shown experimentally that it did speed up a Gaussian RBF SV machine. Note that in the Gaussian RBF case, the approximation can never be as good as the original: no approximation with $N_z < N_y$ will lead to $\Psi_m = 0$, since the kernel matrix $K_{ij} = (k(\mathbf{y}_i, \mathbf{y}_j))$ has full rank [8]. Otherwise, *exact* reductions can be obtained, as noted in [11], and, independently, [6]: if $K\boldsymbol{\alpha} = 0$, then we can express one column of K, and hence one vector $\Phi(\mathbf{y}_i)$, as a linear combination of the other ones; e.g. by picking the i^{max} with largest $|\alpha_i|$: $\Phi(\mathbf{y}_{i^{max}}) = -\sum_{i \neq i^{max}} \alpha_i \Phi(\mathbf{y}_i)/\alpha_{i^{max}}$.

As in [5], good results are obtained even though the objective function does not decrease to zero (in our experiments, it was reduced by a factor of 2 to 20 in the first phase, depending on how many RS vectors were computed; the global gradient descent yielded another factor 2 – 3). We conjecture that this is due to the following: in classification, we are not interested in $\|\Psi - \Psi'\|$, but in $\int |\text{sgn}(\sum_{i=1}^{N_y} \gamma_i k(\mathbf{x}, \mathbf{y}_i) + b) - \text{sgn}(\sum_{j=1}^{N_z} \beta_i k(\mathbf{x}, \mathbf{z}_i) + \tilde{b})| dP(\mathbf{x})$, where P is the underlying probability distribution of the patterns (cf. [1]). This is consistent with the fact that the performance of a RS SV classifier can be improved by re-computing an optimal threshold $\tilde{b}$ (which we have done in our experiments).

The previous RS method [4, 5] can be used for any SV kernel; the new one is limited to $k(\mathbf{x}, \mathbf{y}) = k(||\mathbf{x} - \mathbf{y}||^2)$. However, it led to slightly better results, interpretable RS images, and an interesting connection between clustering and approximation in feature spaces. It appears intriguing to pursue the question whether this connection could be exploited to form more general types of approximations of SV and kernel PCA expansions by making use of Gaussians of variable widths. Moreover, it should be worthwhile to carry out experiments testing simultaneous approximation approaches as the one proposed in Sec. 5.

Acknowledgement. Part of this work was supported by the DFG (# Ja 379/71), the NeuroColt2 program, the ARC, and carried out at The Australian National University. Thanks to S. Mika, B. Williamson, K. Müller and R. Herbrich for discussions and comments.

References

1. K. Blackmore, R. Williamson, and I. Mareels. Decision region approximation by polynomials or neural networks. *IEEE Trans. Inf. Theory*, 43:903 - 907, 1997.
2. B. Boser, I. Guyon, and V. Vapnik. A training algorithm for optimal margin classifiers. In D. Haussler, editor, *Proceedings of the 5th Annual ACM Workshop on COLT*, pages 144–152, Pittsburgh, PA, July 1992. ACM Press.
3. J. M. Buhmann. Data clustering and learning. In M. A. Arbib, editor, *The Handbook of Brain Theory and Neural Networks*, pages 278–281. MIT Press, 1995.
4. C. J. C. Burges. Simplified support vector decision rules. In L. Saitta, editor, *Proceedings, 13th ICML*, pages 71–77, San Mateo, CA, 1996. Morgan Kaufmann.
5. C. J. C. Burges and B. Schölkopf. Improving the accuracy and speed of support vector learning machines. In M. Mozer, M. Jordan, and T. Petsche, editors, *Advances in NIPS 9*, pages 375–381, Cambridge, MA, 1997. MIT Press.
6. T. Frieß. Personal communication. 1998.
7. Y. LeCun, B. Boser, J. Denker, D. Henderson, R. Howard, W. Hubbard, and L. Jackel. Backpropagation applied to handwritten zip code recognition. *Neural Computation*, 1:541 - 551, 1989.
8. C. A. Micchelli. Interpolation of scattered data: distance matrices and conditionally positive definite functions. *Constructive Approximation*, 2:11–22, 1986.
9. E. Osuna and F. Girosi. Reducing run-time complexity in support vector machines. In B. Schölkopf, C. Burges, and A. Smola, editors, *Advances in Kernel Methods — Support Vector Learning*. MIT Press, Cambridge, MA, 1998. to appear.
10. S. Saitoh. *Theory of Reproducing Kernels and its Applications*. Longman Scientific & Technical, Harlow, England, 1988.
11. B. Schölkopf. *Support Vector Learning*. R. Oldenbourg Verlag, Munich, 1997.
12. B. Schölkopf, A. Smola, and K.-R. Müller. Nonlinear component analysis as a kernel eigenvalue problem. *Neural Computation*, 10:1299 - 1319, 1998.
13. B. Schölkopf, K. Sung, C. Burges, F. Girosi, P. Niyogi, T. Poggio, and V. Vapnik. Comparing support vector machines with gaussian kernels to radial basis function classifiers. *IEEE Trans. Sign. Processing*, 45:2758 - 2765, 1997.
14. V. Vapnik. *The Nature of Statistical Learning Theory*. Springer, New York, 1995.

Klassifikation und Neuronale Netze

Discrete Mixture Models for Unsupervised Image Segmentation*

Jan Puzicha[+], Thomas Hofmann[*], and Joachim M. Buhmann[+]

[+] Institut für Informatik III
University of Bonn, Germany
{jan,jb}@cs.uni-bonn.de

[*] Artificial Intelligence Laboratory
Massachusetts Institute of Technology
hofmann@ai.mit.edu

Abstract. This paper introduces a novel statistical mixture model for probabilistic clustering of histogram data and, more generally, for the analysis of discrete co–occurrence data. Adopting the maximum likelihood framework, an alternating maximization algorithm is derived which is combined with annealing techniques to overcome the inherent locality of alternating optimization schemes. We demonstrate an application of this method to the unsupervised segmentation of textured images based on local empirical distributions of Gabor coefficients. In order to accelerate the optimization process an efficient multiscale formulation is utilized. We present benchmark results on a representative set of Brodatz mondrians and real–world images.

1 Introduction

Grouping of homogeneous image regions is an important task in low–level computer vision that is widely pursued to solve the problem of image segmentation, in particular in the context of textured images. Two steps have to be considered in order to address this problem:

- Most fundamentally, a mathematical notion of homogeneity or similarity between image regions is required in order to formalize the segmentation problem. Especially for textured images the similarity measure has to capture the significant variability within a texture, without loosing the ability to discriminate between different textures.
- In a second step, after a similarity measure is defined, an efficient algorithm for partitioning or clustering has to be derived to solve the computational problem. The selection of a suitable clustering method is tightly coupled to the chosen similarity measure.

A successful approach has to rely on a similarity measure which is powerful enough to discriminate a wide range of textures, while preserving the computational tractability of the overall segmentation algorithm.

* It is a pleasure to thank Hans du Buf for providing the aerial image mixtures in Fig. 2. This work has been supported by the German Research Foundation (DFG) under grant #BU 914/3–1 and by a M.I.T. Faculty Sponser's Discretionary Fund.

Numerous approaches to unsupervised texture segmentation have been proposed over the past decades. In the classical approaches, locally extracted features are spatially smoothed and interpreted as *vectors in a metric space* [5, 6], thereby characterizing each texture by a specific average feature vector or *centroid*. The most commonly used distortion measure is the (weighted) squared Euclidean norm which effectively models the data by a Gaussian mixture model, where each Gaussian represents exactly one texture. The method of choice for clustering vectorial data is the K–means algorithm and its variants, which have been exploited for texture segmentation in [5, 6].

Since the Gaussian mixture assumption turns out to be inadequate in many cases, several alternative approaches have utilized *pairwise proximity data*, usually obtained by applying statistical tests to the local feature distribution at two image sites [1, 4, 7]. As a major advantage, these methods do not require the specification of a suitable vector-space metric. Instead, similarity is defined by the *similarity of the respective feature distributions*. For pairwise similarity data agglomerative clustering [7] and, more rigorously, optimization approaches to graph partitioning [1, 4, 12] have been proposed in the texture segmentation context, which we refer to as *pairwise dissimilarity clustering* (PDC). Although these methods are directly applicable to proximity data, they are only tractable in image segmentation problems if they avoid the computation of dissimilarities for all possible pairs of sites [9].

The major contribution of this paper is a general approach to the problem of grouping *feature distributions*, extending a technique known as distributional clustering in statistical language modeling [8]. In contrast to methods based on feature vectors and pairwise dissimilarities this approach is directly applicable to histogram data and empirical distributions. In comparison to K–means clustering, distributional clustering naturally includes component distributions with multiple modes rather than fitting segments with an univariate Gaussian mode. As a major advantage compared to PDC it requires no external similarity measure, but exclusively relies on the *feature occurrence statistics*. Another important consideration for a clustering approach to image segmentation are real-time constraints. Given the respective data (vectors, histograms or proximities) all algorithms require only a few seconds for optimization [9]. While vector–based methods suffer from inferior quality, it is the data extraction process of PDC which is prohibitive for real–time applications like autonomous robotics. Using the histogram data directly avoids the necessity for pairwise comparisons altogether while achieving segmentations of similar quality compared to PDC. In addition, the proposed mixture model provides a generative statistical model for the observed features by defining a *texture specific distribution*. This can be utilized in subsequent processing steps such as boundary localization [11].

2 Mixture Models for Histogram Data

To stress the generality of the proposed model we temporarily detach the presentation from the specific problem of image segmentation. Consider therefore the following more abstract setting: $\mathcal{X} = \{x_1, \ldots, x_N\}$ denotes a finite set of abstract objects with arbitrary labeling and $\mathcal{Y} = \{y_1, \ldots, y_M\}$ describes the domain of nominal scale feature(s). Each $x_i \in \mathcal{X}$ is characterized by a number of observations (x_i, y_j) summarized in the sufficient statistics of counts n_{ij}. Effectively, this defines for each x_i an *empirical distribution* or *histogram* over $\mathcal{Y}$ defined by $n_{j|i} \equiv n_{ij}/n_i$ where $n_i \equiv \sum_j n_{ij}$.

The proposed mixture model, which is referred to as Asymmetric Clustering Model (ACM)[1], explains the observed data by a finite number of component probability distributions on the feature space. The generative model is defined as follows:

1. select an object $x_i \in \mathcal{X}$ with probability p_i,
2. choose the cluster C_α according to the cluster membership of x_i,
3. select $y_j \in \mathcal{Y}$ from a cluster-specific conditional distribution $q_{j|\alpha}$.

In addition to the parameters $\mathbf{p} = (p_i)$ and $\mathbf{q} = (q_{j|\alpha})$ let us introduce indicator variables $M_{i\alpha} \in \{0, 1\}$ for the class membership of x_i ($\sum_\alpha M_{i\alpha} = 1$). The probability of a feature occurrence (x_i, y_j) is then given by

$$P(x_i, y_j | \mathbf{M}, \mathbf{p}, \mathbf{q}) = p_i \sum_{\alpha=1}^{K} M_{i\alpha}\, q_{j|\alpha} \quad . \tag{1}$$

The ACM has $K(M-1)$ continuous parameters for the component densities $q_{j|\alpha}$, $N-1$ parameters for the probabilities p_i, and N sets of indicator functions encoding an one–out–of–K choice each. An essential assumption behind the ACM is that observations for x_i are conditionally independent given the continuous parameters and the cluster assignment $(M_{i\alpha})$ of x_i.

Returning to the texture segmentation problem, we see that we can identify the set of objects $\mathcal{X}$ with the set of image locations or sites and the set $\mathcal{Y}$ with possible values of discrete or discretized texture features computed from the image data. The distributions $n_{j|i}$ then represent a histogram of features occurring in an *image neighborhood* or *window* around some location x_i [1, 4, 7]. The framework is, however, general enough to cover distinctive application domains like information retrieval [3] and natural language modeling [8].

In the context of texture segmentation, each class C_α corresponds to a different texture which is characterized by a specific distribution $q_{j|\alpha}$ of features y_j. Since these component distributions of the mixture model are not constrained, they can virtually model any distribution of features. In particular, no further parametric restrictions on $q_{j|\alpha}$ are imposed. There is also no need to specify an additional noise model or, equivalently, a metric in feature space.

---

[1] It is called asymmetric because clustering structure is inferred solely in the $\mathcal{X}$–space. The ACM is the most suitable model for image segmentation out of a family of novel mixture models developed for general co–occurrence data [3].

3 Maximum Likelihood Estimation for the ACM

To fit the model specified by (1) we apply the maximum likelihood principle and determine the parameter values with the highest probability to generate the observed data. We start with the log–likelihood function which is given by

$$\mathcal{L} = \sum_{i=1}^{N}\sum_{j=1}^{M} n_{ij} \sum_{\alpha=1}^{K} M_{i\alpha} \log q_{j|\alpha} + \sum_{i=1}^{N} n_i \log p_i \ . \tag{2}$$

Maximum likelihood equations are derived from (2) by differentiation, using Lagrange parameters to ensure a proper normalization of the continuous model parameters $\mathbf{p}$ and $\mathbf{q}$. The resulting stationary equations are given by

$$\hat{p}_i = \frac{n_i}{\sum_k n_k}, \tag{3}$$

$$\hat{q}_{j|\alpha} = \frac{\sum_{i=1}^{N} \hat{M}_{i\alpha} n_{ij}}{\sum_{i=1}^{N} \hat{M}_{i\alpha} n_i} = \sum_{i=1}^{N} \frac{\hat{M}_{i\alpha} n_i}{\sum_{k=1}^{N} \hat{M}_{k\alpha} n_k} n_{j|i} , \tag{4}$$

$$\hat{M}_{i\alpha} = \begin{cases} 1 & \text{if } \alpha = \arg\min_\nu \left\{ -\sum_{j=1}^{M} n_{j|i} \log \hat{q}_{j|\nu} \right\} \\ 0 & \text{else} \end{cases} \tag{5}$$

From (3) we see that the probabilities p_i are estimated independently of all other parameters. The maximum likelihood estimates of the class-conditional distributions $\hat{q}_{j|\alpha}$ are linear superpositions of all empirical distributions for objects x_i belonging to cluster $\mathcal{C}_\alpha$. Eq. (4) thus generalizes the *centroid condition* from K–means clustering. Notice however, that the components of $\hat{q}_{j|\alpha}$ define probabilities for feature values and do not correspond to dimensions in the original feature space. Eq. (4) averages over feature *distributions*, not over feature *values*. The formal similarity to K–means clustering is extended by (5), which is the analogon to the nearest neighbor rule.

The ACM is similar to the distributional clustering model formulated in [8] as the minimization of the cost function

$$\mathcal{H} = \sum_{i=1}^{N}\sum_{\alpha=1}^{K} M_{i\alpha} D\left[n_{j|i} | q_{j|\alpha}\right] \ . \tag{6}$$

Here D denotes the cross entropy or Kullback–Leibler (KL) divergence. In distributional clustering, the KL–divergence as a distortion measure for distributions has been motivated by the fact that the centroid equation (4) is satisfied at stationary points[2]. Yet, after dropping the p_i parameters in (2) and a (data dependent) constant we derive the formula

$$\mathcal{L} = -\sum_{i=1}^{N} n_i \sum_{\alpha=1}^{K} M_{i\alpha} D\left[n_{j|i} | q_{j|\alpha}\right] \ . \tag{7}$$

[2] Note that this is not a unique property of the KL–divergence as it is also satisfied for the Euclidean distance.

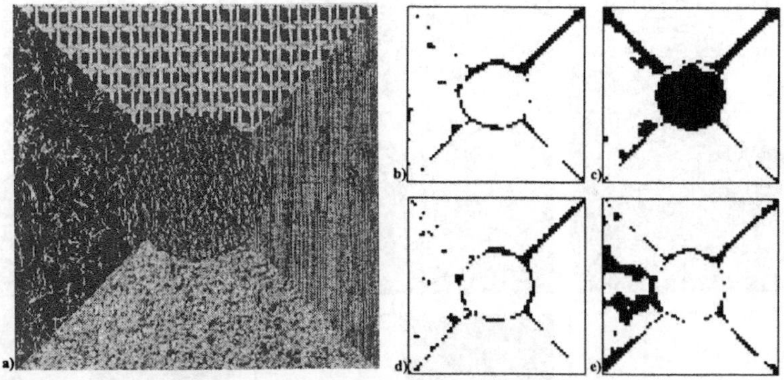

Fig. 1. Typical segmentation results with $K = 5$ for the algorithms under examination: (a) original image, (b) annealed ACM, (c) AM for ACM, (d) PDC and (e) K–means. Misclassified blocks w.r.t. ground truth are depicted in black.

This proves that the choice of the KL-divergence as a distortion measure simply follows from the likelihood principle. The analogy between the stationary conditions for the ACM and for K–means clustering also holds for the model fitting algorithm. The likelihood can be maximized by an *alternating maximization* (AM) update scheme which calculates assignments for given centroids according to the nearest neighbor rule (5) and recalculates the centroid distributions (4) in alternation. Both algorithmic steps increase the likelihood and convergence to a (local) maximum of (7) is thus ensured.

A technique which allows us to improve the presented AM procedure by avoiding unfavorable local minima is known as *deterministic annealing* (DA). The key idea is to introduce a temperature parameter T and to replace the minimization of a combinatorial objective function by a substitute known as the *generalized free energy.* Details on this topic in the context of data clustering can be found in [10, 8, 4]. Minimization of the free energy corresponding to (7) yields the following equations for probabilistic assignments:

$$\mathbf{P}\left(M_{i\alpha} = 1 | \, \mathbf{p}, \mathbf{q}\right) = \frac{\exp\left(-n_i D\left[n_{j|i} | \hat{q}_{j|\alpha}\right] / T\right)}{\sum_{\nu=1}^{K} \exp\left(-n_i D\left[n_{j|i} | \hat{q}_{j|\nu}\right] / T\right)} \tag{8}$$

This partition of unity is a very intuitive generalization of the nearest neighbor rule in (5). For $T \to 0$ the arg–min operation performed in the nearest neighbor rule is recovered. Since in DA solutions are tracked from high to low temperatures, we finally maximize the log–likelihood at $T = 0$. Notice that the DA procedure also generalizes the Expectation Maximization (EM) algorithm which is obtained for $T = 1$. In this case (8) corresponds to the computation of posterior probabilities for hidden variables $M_{i\alpha}$ in the E–step.

4 Unsupervised Segmentation of Textured Images

We applied the ACM model to the unsupervised segmentation of textured images on the basis of histograms. Now the set of objects $\mathcal{X}$ corresponds to the

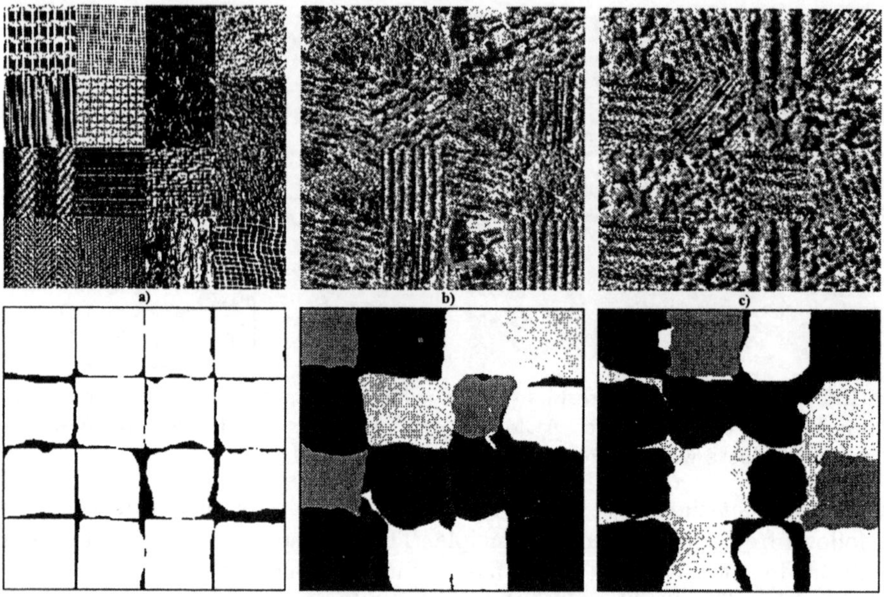

Fig. 2. Typical segmentation results: (a) on a mondrian of 16 different Brodatz textures (misclassified blocks w.r.t. ground truth are depicted in black), (b) and (c) mondrians of 7 different textures taken from aerial images (no ground truth available).

set of image locations. Typically, an identical number of features is observed for all sites, simplifying the equations to $p_i = n_i = 1/N$. Algorithms based on distributions $n_{j|i}$ of features have been successfully used in texture analysis [1, 7, 4]. In the experiments, we have adopted the framework of [5, 4, 9] and utilized an image representation based on the modulus of complex Gabor filters. For each site the empirical distribution of coefficients in a surrounding (filter–specific) window is determined. All reported segmentations are based on a filter bank of twelve Gabor filters with four orientations and three scales. Each filter output was discretized into 20 equally sized bins. Assuming independent filter channels results in a feature space $\mathcal{Y}$ of size $M = 240$. The benchmark results are obtained on images which were generated from a representative set of 86 micro–patterns taken from the Brodatz texture album. A database of random mixtures (512×512 pixels each) containing 100 entities of five textures each (as depicted in Fig. 1) was constructed. For the K–means algorithm a spatial smoothing step was applied before clustering, see [5]. For all cost functions the multiscale annealing optimization scheme was implemented, where coarse grids up to a resolution of 8x8 grid points have been used. It is a natural assumption that adjacent image sites contain identical texture with high probability. This fact can be exploited to significantly accelerate the optimization of the likelihood by maximizing over a suitable nested sequence of subspaces in a coarse–to–fine manner, where each of these subspaces is spanned by a greatly reduced number of indicator variables. This strategy is formalized by the concept of *multiscale optimization* [2] and it essentially leads to cost functions redefined on a coarsened

	Median	20% quantile
AM for ACM	8.9%	18%
annealed ACM	6.7%	6%
annealed PDC	6.0%	6%
annealed K–means	11.7%	28%

Table 1. Errors by comparison with ground truth over 100 randomly generated images with $K = 5$ textures, 512x512 pixels and 64x64 assignments.

version of the original image. In contrast to most multiresolution optimization schemes the *original* log–likelihood is optimized at all grids, only the variable configuration space is reduced. For ACM–based segmentation, cost–functions of identical algebraic structure are obtained at all levels. Deterministic annealing and multiscale optimization are combined in the concept of multiscale annealing. The resulting algorithm provides an acceleration factor of 5 – 500 compared to single scale optimization. For details we refer to [9, 3].

The question examined in detail is concerned with the benefits of the ACM in comparison to other clustering schemes. A typical example with $K = 5$ clusters is given in Fig. 1. It can be seen that the segmentations achieved by ACM and PDC are highly similar. Most errors occur at texture boundaries where texture information is mixed due to the spatial support of the Gabor filters and the extent of the neighborhood used for computing the local feature statistics. The K–means clustering cost function exhibits substantial deficits to correctly model the segmentation task. These observations are confirmed by the benchmark results in Tab. 1. We report the median, since the distributions of the empirical errors are highly asymmetric. In addition, the percentage of segmentations with an error rate larger than 20% is reported, which we define as the percentage of segmentations where the essential structure is not detected. ACM and PDC yield similar results with a statistically insignificant difference. For the ACM a median error of 6.7% was achieved compared to 6.0% for PDC. The percentage of segmentations, where the essential structure has been detected, is in both cases as high as 94%. The K–means model yields significantly worse results with a median error of 11.7%. Moreover, in 28% of the cases the essential structure was not detected. The quality of the ACM model is confirmed by the results on more difficult segmentation tasks in Fig. 2. The mixture of $K = 16$ different Brodatz textures has been partitioned accurately with an error rate of 7.9%. The errors basically correspond to boundary sites. The results obtained for the mondrians of aerial images are satisfactory. Disconnected texture regions of the same type have been identified correctly, while problems again occur at texture boundaries.

We conclude, that (annealed) ACM combines the expressive power of pairwise similarity clustering with the efficiency of the conventional K–means clustering and provides a fast, accurate and reliable algorithm for unsupervised texture segmentation. Compared to PDC the time–consuming algorithmic step of computing pairwise dissimilarities between objects has been avoided. Moreover, statistical group information is provided for subsequent processing steps making

	K–means	ACM	PDC
Underlying data type	vector	histogram	proximity
Computational complexity of data extraction	lowest	medium	highest
Computational complexity of optimization	lowest	medium	highest
Segmentation quality	low	high	high
Generative statistical model provided	yes	yes	no
Implementation effort	low	low	high

Table 2. Advantages and disadvantages of the clustering algorithms.

the ACM an interesting alternative to PDC in the segmentation context. The advantages and disadvantages of all three clustering methods are summarized in Tab. 2. It has been confirmed, that global optimization algorithms like multiscale annealing are essential for reliable computation of high quality segmentations. As a general clustering scheme this model can be extended to color and motion segmentation, region grouping and even integrated sensor segmentation simply by choosing appropriate features.

References

1. D. Geman, S. Geman, C. Graffigne, and P. Dong. Boundary detection by constrained optimization. *IEEE Trans. Pattern Analysis and Machine Intelligence*, 12(7):609–628, 1990.
2. F. Heitz, P. Perez, and P. Bouthemy. Multiscale minimization of global energy functions in some visual recovery problems. *Computer Vision and Image Understanding*, 59(1):125–134, 1994.
3. T. Hofmann and J. Puzicha. Statistical models for co-occurrence data. AI–Memo 1625, MIT, 1998.
4. T. Hofmann, J. Puzicha, and J. Buhmann. Deterministic annealing for unsupervised texture segmentation. In *Proc. EMMCVPR'97, LNCS 1223*, pages 213–228, 1997.
5. A. Jain and F. Farrokhnia. Unsupervised texture segmentation using Gabor filters. *Pattern Recognition*, 24(12):1167–1186, 1991.
6. J. Mao and A. Jain. Texture classification and segmentation using multiresolution simultaneous autoregressive models. *Pattern Recognition*, 25:173–188, 1992.
7. T. Ojala and M. Pietikäinen. Unsupervised texture segmentation using feature distributions. Tech. Rep. CAR-TR-837, Center for Robotics and Automation, University Maryland, 1996.
8. F. Pereira, N. Tishby, and L. Lee. Distributional clustering of English words. In *Proc. Association for Computational Linguistics*, pages 181–190, 1993.
9. J. Puzicha and J. Buhmann. Multiscale annealing for real–time unsupervised texture segmentation. Technical Report IAI–97–4, Institut für Informatik III (a short version appeared in: Proc. ICCV'98, pp. 267–273), 1997.
10. K. Rose, E. Gurewitz, and G. Fox. A deterministic annealing approach to clustering. *Pattern Recognition Letters*, 11:589–594, 1990.
11. P. Schroeter and J. Bigun. Hierarchical image segmentation by multi-dimensional clustering and orientation-adaptive boundary refinement. *Pattern Recognition*, 28(5):695–709, 1995.
12. J. Shi and J. Malik. Normalized cuts and image segmentation. In *Proc. IEEE Conf. Computer Vision and Pattern Recognition (CVPR'97)*, pages 731–737, 1997.

Integration of Multiple Feature Detection by a Bayesian Net for 3d Object Recognition

B. Krebs, M. Burkhardt and F. M. Wahl
e-mail:{B.Krebs, M. Burkhardt, F.Wahl}@tu-bs.de

Institute for Robotics and Computer Control,
Technical University Braunschweig,
Hamburger Str. 267, D-38114 Braunschweig, F.R.G.

Abstract. This paper proposes a general framework to build a 3d object recognition system from a set of CAD object definitions. Various, reliable features from object corners, edges and 3d rim curves are introduced which provide sufficient information to allow identification and pose estimation of CAD designed industrial parts. As features relying on differential surface properties tend to be very vulnerable with respect to noise, we model the statistical behavior of the data by means of Bayesian nets, representing the relations between objects and observable features. This allows to identify objects by a combination of several features considering the significance of each single feature with respect to the object model base. On this basis robust and powerful 3d CAD based object recognition systems can be build.

1 Introduction

The task to perform recognition for a 3d scene is a highly complex process which involves various types of sensor processing and interpretation algorithms. Sensor data is corrupted by noise leading to ambiguous recognition results. A reliable 3d object recognition system has to exploit several sources of information or has to collect new sensor data until enough evidence for a correct interpretation of the scene is acquired. A major problem is how to integrate and validate uncertain information from different sources. Especially, the recognition of free-form objects has become a challenge because features relying on differential surface properties tend to be very vulnerable with respect to noise(e. g. [4, 5]).

To cope with these problems, the statistical behavior of the features has to be modeled within the recognition process. Using *Bayesian nets*, not only the statistical behavior of the feature values but also the dependencies between features and objects can be modeled. In most object recognition systems using Bayesian nets, the net nodes represent subparts of an object [1, 11, 14]. However, observable features need not to coincide with object subparts, i.e. several features of one subpart are not statistical independent, as it is assumed when constructing objects from subparts. Furthermore, general free-form objects can not be constructed from simple sets of primitives.

In this paper we propose a new method to automatically construct a Bayesian net from a set of CAD object definitions by representing elementary features in terms of general relations. Dependencies between tuples of several relations allow to build a hierarchical structure which can be mapped onto a graph representation defining the Bayesian net. We define various features of free-form objects which easily can be computed from CAD and sensor data. Thus, the recognition can rely on different sources of information while their integration is performed by the Bayesian net.

Only relational properties of objects and features are modeled in a Bayesian net to separate position dependent from position independent information. The nodes represent subsets of object hypotheses and the links point from more discriminative to more general subsets of hypotheses. This allows to use efficient well-known propagation algorithms [13]. In this sense, recognition can be viewed as the task to minimize the uncertainty to select only the most probable set of object hypotheses.

2 Bayesian Nets from CAD Models

The Bayesian net, encoding the dependencies between observable features and objects, is build in an off-line preprocessing and is optimized with respect to a given CAD

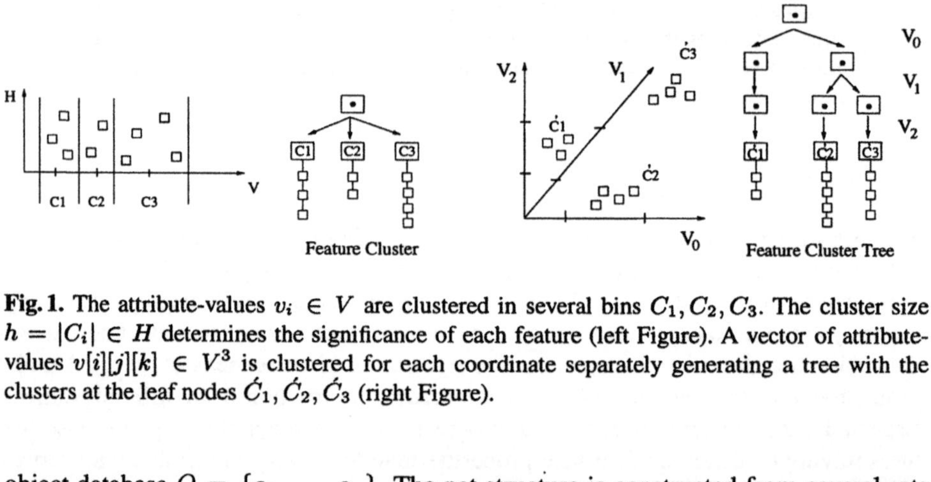

Fig. 1. The attribute-values $v_i \in V$ are clustered in several bins C_1, C_2, C_3. The cluster size $h = |C_i| \in H$ determines the significance of each feature (left Figure). A vector of attribute-values $v[i][j][k] \in V^3$ is clustered for each coordinate separately generating a tree with the clusters at the leaf nodes $\acute{C}_1, \acute{C}_2, \acute{C}_3$ (right Figure).

object database $O = \{o_0, \ldots, o_n\}$. The net structure is constructed from several sets of relations $R = \{R_0, \ldots, R_m\}$ on the surface primitives $S = \{s_0, \ldots, s_k\}$ given by CAD object definitions. An attributed, named relation

$$R_i = (n, \{t_{ij}\}, f) \subseteq S^k \tag{1}$$

on the set of surfaces S consists of a name n, the set of tuples $\{t_{ij}\}$ representing object hypotheses and a function f to retrieve the attribute-values. An attribute for a tuple $A = f(t_{ij})$ is either a simple value, a vector or a set of values. A tuple t_{ij} contains the object elements with relation R_i, a transformation T describing a local position with respect to some object coordinate system, the attribute-values A and a *likelihood*

$L \in [0,1]$ determining the degree of membership of

$$t_{ij} = (\{s_0, \ldots, s_{k_t}\}, T, A, L) \in \{t_{ij}\}. \tag{2}$$

The relations are computed from the topological structure of the surface elements of each object as e.g. described by the *attributed adjacent graph* (AAG) [12]. Each surface is represented by a node in the AAG and an edge connects two nodes, if the two corresponding surfaces are neighbors. The tuples for the relations are extracted from neighbor surfaces.

The computed attribute-values have to be mapped onto reliable features because the measurements from sensor data tend to be noisy. A subset of attribute-values A_s is mapped onto an *elementary feature* F with a single representative value $\bar{a}$, i.e. the mean of A_s. Thus, only the observation of an elementary feature $F_i = [\bar{a}, v_a, A_s] \subset F^+$ in the sensor data has to be performed in the recognition process.

The expected sensor error ε_s of a computed attribute-value has to be smaller than the variance v_a of the attribute-values belonging to an elementary feature. Furthermore, the number of attribute-values $|A_s|$ of F determine the *significance* of the feature. The significance determines the degree of reduction of a set of object hypotheses obtained by observing the feature. The elementary features are constructed from the attribute-values by an *adaptive clustering*, based on the *separation index* D_1 proposed by Dunn [2]:

$$D_1 = \min_{1 \leq i \leq c-1} \left(\min_{1 \leq j \leq c-1} \left(\frac{dis(F_i, F_j)}{\max_{1 \leq k \leq c} diag(F_k)} \right) \right), \tag{3}$$

$$diag(F_k) = \max_{x,y \in F_k} d(x,y), \tag{4}$$

$$dis(F_i, F_j) = \min_{x \in F_i, y \in F_j} d(x,y), \tag{5}$$

with $d(x,y) = |x - y|$ being the distance norm. D_1 is the quotient of the minimal distance between two attribute-values of different classes and the maximal distance between attribute-values of the same class. The partition is considered to be "good" only if compact and well separated classes exist, i.e. $1.0 \leq D_1$, and the separation distance $\varepsilon_s \leq \min_{i \neq j} dis(F_i, F_j)$ is greater than the sensor error. The clustering provides a set of features where each feature represents a cluster of attribute-values (see Figure 1 left).

An attribute with a vector of values is clustered for each coordinate separately (see Figure 1 right). Multidimensional hash-tables for feature indexing can be constructed from the resulting cluster-tree (like in [4]). Nevertheless, the representation of uncertain feature selection can be modeled by mapping the feature clusters onto the nodes of a Bayesian network. Thus, the uncertainty with respect to the feature significance can be modeled within the recognition. Furthermore, dependencies between features can be expressed within the graph in an explicit way. Tuples having other tuples as elements define a hierarchical structure modeling the dependencies between different feature clusters. The hierarchical structure encodes different node layers within the Bayesian net.

3 Rim and Edge Features

If two free-form patches s_0 and s_1 meet at a discontinuity they define a *rim curve*. The rim can be represented by a 3d cubic B-spline curve which efficiently can be computed from CAD data [3]. All rim curves of the CAD models are subdivided into *significant subcurves* in order to handle the difficult case of overlapping objects. Relations between curves are determined if different curves share subcurves with same shape, i.e. curve with same curvature and torsion.

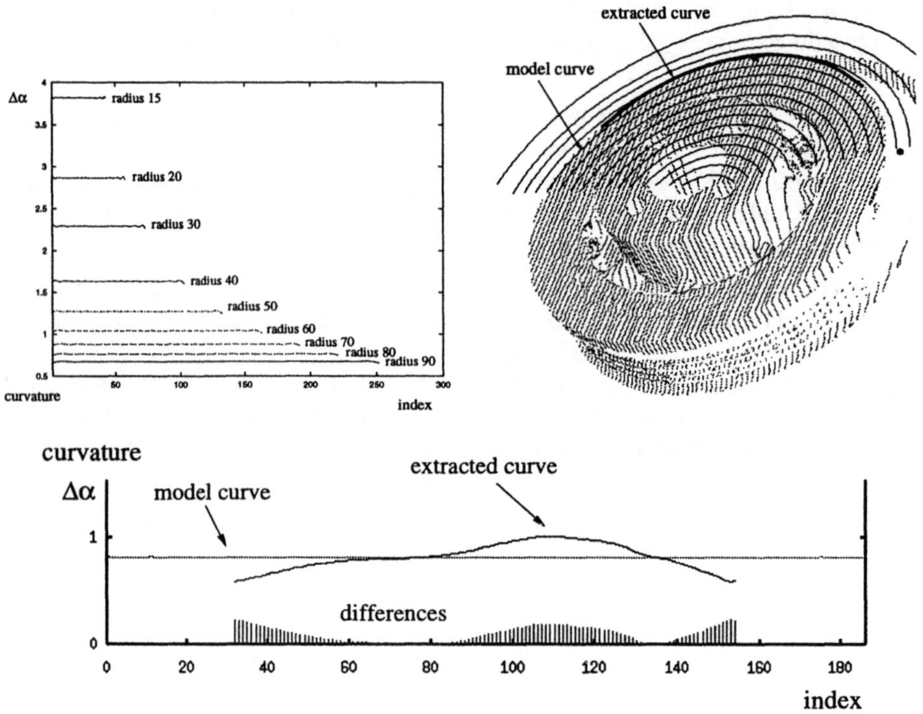

Fig. 2. An example database contains circles of various radii with distinctive deviation angles $\Delta\alpha_i$ at each index (Figure top left). The matching selects the correct corresponding model curve and provides the transformation onto the extracted curve (Figure top right). The differences of the deviation angles between model and extracted curve are small (bottom Figure).

Each subcurve is sampled equidistantly, i.e. $ds \approx \Delta s = ||p_{i+1} - p_i|| = const$. At each sample point $p_i \in [p_o, \ldots, p_{n_s}]$ the *Frenet-Frame* F_i is computed [3]:

$$F_i = \left[t_i = \frac{\dot{p}_i}{||\dot{p}_i||}\,,\; b_i = \frac{\dot{p}_i \times \ddot{p}_i}{||\dot{p}_i \times \ddot{p}_i||}\,,\; m_i = t_i \times b_i \right]. \tag{6}$$

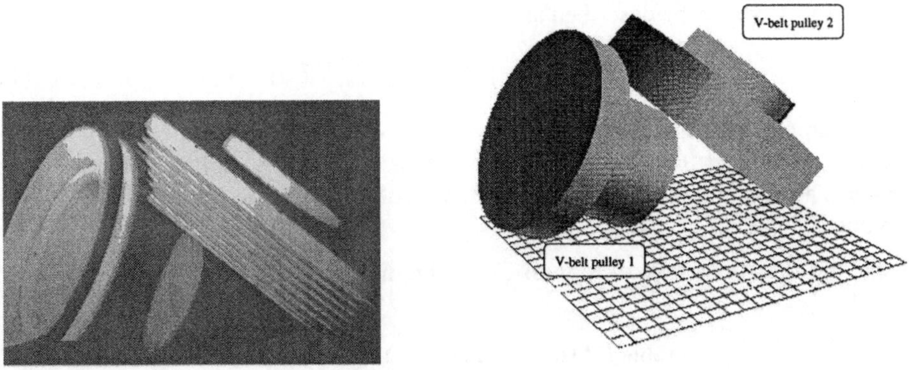

Fig. 3. An example scene containing two V-belt pulleys (left) and the recognition results (right).

The curvature $\kappa = \frac{d\alpha}{ds}$ and the torsion $\tau = \frac{d\beta}{ds}$ at each sample point p_i are the angular velocities of $\boldsymbol{t}$ and $\boldsymbol{b}$ with

$$d\alpha_i = arccos\left(\frac{< \boldsymbol{t}_i, \boldsymbol{t}_{i+1} >}{|\boldsymbol{t}_i||\boldsymbol{t}_{i+1}|}\right) \text{ and } d\beta_i = arccos\left(\frac{< \boldsymbol{b}_i, \boldsymbol{b}_{i+1} >}{|\boldsymbol{b}_i||\boldsymbol{b}_{i+1}|}\right). \tag{7}$$

Each subcurve is uniquely represented by two sequences of deviation angles:

$$f_{s_\kappa} = [\Delta\alpha_0, \ldots, \Delta\alpha_{n_s}] \text{ and } f_{s_\tau} = [\Delta\beta_0, \ldots, \Delta\beta_{n_s}] \tag{8}$$

with $\Delta\phi \approx d\phi$ since $ds \approx \Delta s = \text{const}$. E.g. a circular rim curve has constant curvature and zero torsion describing a cylindrical object part. A database for recognizing those objet parts contains circular rims with different radii. Nevertheless, the significance to distinguish between different rim curves decreases when the radius r increases since $\kappa \sim \frac{1}{r}$ (Figure 2 top left).

The *similarity* s_f between two feature sequences f_s and f_c is defined by

$$s_f(i) = \sum_j s_f(i,j) \tag{9}$$

$$s_f(i,j) = \begin{cases} 1 \; (f_s[i] - f_c[i+j])^2 \leq \varepsilon_{max} \\ 0 \qquad\qquad else \end{cases}. \tag{10}$$

A subcurve can be identified by a correlation of its feature vectors. The correlation simultaneously provides a matching evidence e_m [8]. Thus, the correct matching curve can be selected by picking the highest matching evidence (Table 1 contains the matching results for the example in Figure 2). The differences in the deviation angles of two corresponding curves are small < 0.2 and are caused by sensor errors (Figure 2 bottom).

The matching evidence is used to specify the likelihood of the observation of a subcurve S given the curve C of object O:

$$L_S(C|S) = L_S(O|S) = \begin{cases} 1.0 & S \text{ derived from } C \\ e_m \in [0,1] & S \text{ similar to } C \end{cases} \tag{11}$$

Thus, a subcurve can be described as a tuple

$$t_S = (\{s_0, s_1\}, F_m, [f_{s_\kappa}, f_{s_\tau}], L_S). \tag{12}$$

The hierarchical structure of objects, curves and subcurves conveniently can be mapped onto a Bayesian net [6]. The recognition process can identify arbitrary objects by rim curves. The two V-belt pulleys can be identified correctly and their positions can be computed (Figure 3).

If the first derivative is zero too, no Frenet-Frame can be computed and a rim curve

Table 1. Matching results for the V-belt pulley

Model curve	0	1	2	3	4	5	6	7	8	9	10	11	12	13	14	15
Matching evidence	0.5	0.5	0.5	0.5	0.5	0.5	0.5	0.6	0.7	0.8	0.8	1	0.8	0.8	0.8	0.7

degenerates to an *edge* E. The edge is defined by the intersection of the two adjacent tangential planes with normals n_0 and n_1 and their deviation angle α. For each edge two tuples are generated at the end points p_s and p_e:

$$t_{E_{s|e}} = (\{s_0, s_1\}, [n_0, n_1, n_0 \times n_1, p_{s|e}], \alpha, L_E) \tag{13}$$

The likelihood L_E is computed from all tuples $\{t_i\}$ after the clustering by

$$L_E(O|E_j) = 1.0 - \prod_i (1.0 - l_i) \quad \forall t_{E_j} \in O, \qquad l_i = 1.0 - \frac{|\alpha - \bar{\alpha}|}{\varepsilon_s}. \tag{14}$$

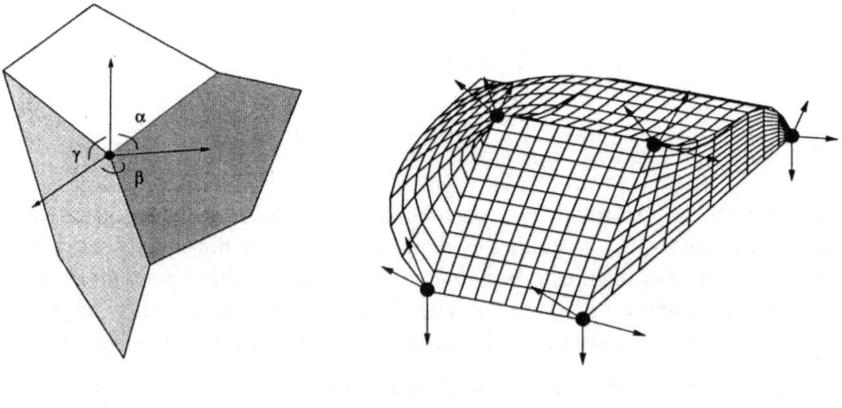

Fig. 4. A feature at an object corner is defined by three intersecting tangential surfaces.

4 Corner Features

A corner of a free-form object is defined, if three surface patches meet at a discontinuity (Figure 4). The intersection ot the three tangential planes define a local coordinate system consisting of the intersection point p_s and their three normals. A corner-tuple is

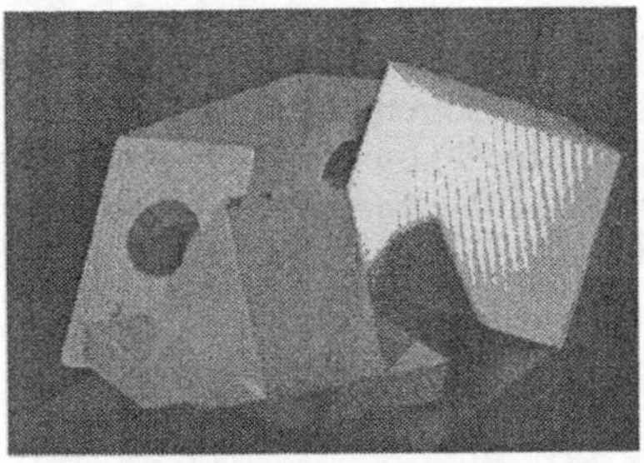

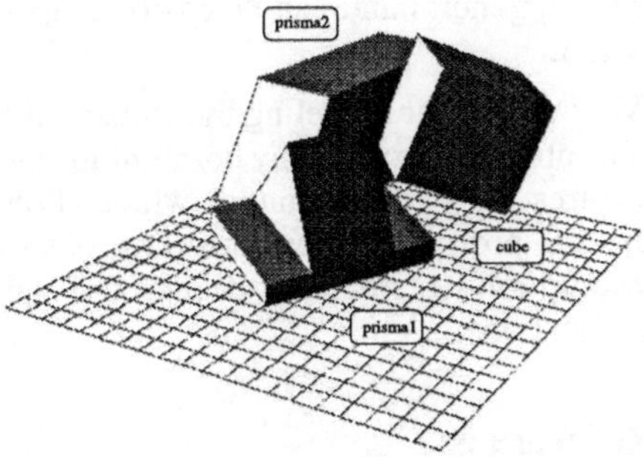

Fig. 5. An example scene with three polyhedral objects (left) and their computed locations (right).

defined as

$$t_I = (\{s_0, s_1, s_2\}, [n_0, n_1, n_2, p_s], A_f, L_I). \quad (15)$$

The set of angles between the normals $A = \{\alpha_0, \alpha_1, \alpha_2\}$ defines the corner attribute. A clustering of all angle values $A^+ = \bigcup_i A_i$ provides the set of possible angle features A_f. The feature for each corner is a multi-set (bag) of A_f, represented by a vector of dimension $|A_f|$. Each coordinate represents the frequency of $a_f \in A_f$. The likelihood L_I is computed after the clustering from all tuples $\{t_i\}$ by

$$L_I(O|I_j) = 1.0 - \prod_i (1.0 - l_i) \quad \forall t_{I_j} \in O, \qquad l_i = \min_{a_i \in A} \left(1.0 - \frac{|\alpha_i - \bar{\alpha}|}{\varepsilon_s}\right). \quad (16)$$

These features provide a robust method to identify objects by corners. The integration of different corner matches is done via the Bayesian net allowing a robust recognition; an example recognition of three polyhedral objects is depicted in Figure 5.

5 3d Recognition System

The recognition system based on the proposed Bayesian net uses a task oriented decision system which collects new sensor data incrementally until enough evidence for recognizing objects is collected [7]. The recognition is started with an initial view to extract elementary features. Rim curves describing surface discontinuities are extracted by a scan-plane segmentation with a subsequent 3d-spline approximation [8]. Corner and edge features are computed on the basis of a planar patch segmentation of the range image [10]. The net allows to incorporate relational properties as well as dependencies between different models within the recognition process itself. Thus, in the recognition process the set of possible hypotheses is reduced to a set of likely hypotheses only.

Subsequently, the correct hypotheses from the set of likeliest hypotheses are determined by a *Fuzzy ICP* algorithm in combination with a *measure of reference-evidence* [9]. The Bayesian net has been implemented using the HUGIN software system [13].

The proposed methods provide a very efficient approach for CAD model based recognition in range data. Nevertheless, the concept of handling uncertainty with Bayesian nets is quite general and can be easily adapted to other fields of recognition and computer vision.

We showed that modeling the statistical behavior of features with a Bayesian net leads to a robust recognition by combining several features of different nature. All computed features carry 3d information which allow to compute object poses in addition very easily. The proposed algorithms have been integrated in a robust and powerful 3d object recognition system; it has been tested elaborately to validate its robustness in real world applications.

References

1. D. M. Chelberg. Uncertainty in interpretation of range imagery. In *Proc. International Conference on Computer Vison, Osaka, Japan*, pages 634–657, 1990.
2. J. C. Dunn. Well separated clusters and optimal fuzzy partitions. *Int. J. of Cypernetics*, 4(1):95–104, 1974.
3. G. E. Farin. *Curves and Surfaces for Computer Aided Geometric Design, a Practical Guide 3rd. ed.* Academic Press, New York, 1993.
4. L. Grewe and A. Kak. Interactive learning of a multiple-attributed hash table classifier for fast object recognition. *Int. J. of Computer Vision and Image Understanding*, 61(3):387–416, 1995.
5. A. Gueziec and N. Ayache. Smoothing and matching of 3-d space curves. *Int. J. of Computer Vision*, 12(1):79–104, 1994.
6. B. Krebs, M. Burkhardt, and F. M. Wahl. A bayesian network for 3d object recognition in range data. In *Proc.International Conference on Computer Analysis of Images and Patterns, Kiel, Germany*, pages 361–368, 1997.
7. B. Krebs, B. Korn, and M. Burkhardt. A task driven 3d object recognition system using bayesian networks. In *Proc. International Conference on Computer Vison, Bombay, India*, pages 527–532, 1998.
8. B. Krebs, B. Korn, and F.M. Wahl. 3d b-spline curve matching for model based object recognition. In *Proc. International Conference on Image Processing, Santa Barbara, USA*, pages 716–719, 1997.
9. B. Krebs, P. Sieverding, and B. Korn. A fuzzy icp algorithm for 3d free form object recognition. In *Proc. International Conference on Pattern Recognition, Vienna, Austria*, pages 539–543, 1996.
10. B. Krebs and F. M. Wahl. *Advances in Computer Vision*, chapter CAD Based 3d Object Recognition on Range Images, pages 221–230. Advances in Computer Science. Springer, 1997.
11. W. B. Mann and T. O. Binford. An example of 3d interpretation of images using bayesian networks. In *DARPA Image Understanding Workshop*, pages 793–801, 1992.
12. M. M. Marafat and R. L. Kashyap. Geometric reasoning for recognition of three-dimensional object features. *IEEE Transactions on Pattern Analysis and Machine Intelligence*, 12(10):949–965, 1990.
13. K. G. Olesen, S. L. Lauritzen, and F. V. Jensen. Hugin: A system creating adaptive causual probabilistic networks. In *Proc. International Conference on Uncertainty in Artificial Intelligence*, pages 223–229, 1992.
14. R. D. Rimey and C. M. Brown. Control of selective perception using bayes nets and decision theory. *Int. J. of Computer Vision*, 12(2/3):173–208, 1994.

Automatische Bewertung von Bildverarbeitungsroutinen zur konturorientierten Segmentierung mit FuzzySOLUTION*

H. Münkel, C.-E. Liedtke und U. Rost

Institut für Theoretische Nachrichtentechnik und Informationsverarbeitung, Universität Hannover, Appelstr. 9a, 30167 Hannover, Germany

Zusammenfassung Derzeit herrscht ein allgemeines großes Interesse an der Bewertung von Bildverarbeitungsverfahren. In den letzten 20 Jahren wurde weltweit eine Vielzahl von Verfahren erarbeitet und es kommen immer neue dazu. Bei der Lösung eines konkreten Bildverarbeitungsproblems stellen sich daher oft die Fragen, welches Verfahren für eine vorgegebene Aufgabenstellung am besten geeignet ist und wie man Bildverarbeitungsverfahren beurteilen könnte.
Für den Spezialfall der konturorientierten Segmentierungsverfahren wird ein Lösungsansatz für dieses Problem vorgestellt. Das System FuzzySOLUTION ist in der Lage Bildverarbeitungsverfahren aufgrund einer vorgegebenen Aufgabenstellung zu konfigurieren, deren freie Parameter zu adaptieren und anschließend die einzelnen Verfahren hochsprachlich zu bewerten.

1 Einleitung

Derzeit herrscht ein allgemeines großes Interesse an der Bewertung von Bildverarbeitungsverfahren [1], [2], [3], [4], [5]. In den letzten 20 Jahren wurde weltweit eine Vielzahl von Verfahren erarbeitet und es kommen immer neue dazu (siehe z.B. [6]). Dadurch stellt sich für die Entwickler von Bildverarbeitungssystemen häufig die Frage, ob die Verfahren tatsächlich immer neu sind und welches das beste für die eigene Aufgabenstellung ist. Dieses Problem führt zur allgemeineren Frage: "Wie könnte man Bildverarbeitungsverfahren beurteilen?"

Zur objektiven Beurteilung braucht man ein Gütemaß. In der optischen Meßtechnik (z.B. Tiefenmessung durch Stereo, Bewegungsmessung, Höhen, Breitenmessung, usw.) ist die Güte durch Messung mit alternativen Verfahren direkt am physikalischen Objekt oft leicht überprüfbar. Wie sieht es aber aus, wenn nur ein Teil der Verarbeitungskette (z.B. die Segmentierung) betrachtet wird oder wenn die Ergebnisse nicht-meßtechnische oder sogar subjektive Qualitäten aufweisen. Wie muß man dann Verfahren beurteilen?

* Die Entwicklung von SOLUTION wurde unterstützt durch die Deutsche Forschungsgemeinschaft. Besonderer Dank geht an das National Research Council of Canada und insbesondere an R.A. Orchard für die Entwicklung und Zurverfügungstellung von FuzzyCLIPS, in dem ein wesentlicher Teil von FuzzySOLUTION realisiert ist.

Eine objektive Beurteilung des Auswahlprozesses ist zwar nicht möglich, wir beobachten aber, daß ein Bildverarbeitungsexperte bei der Entwicklung eines Systems die nach seiner Meinung günstigsten Verfahren aussucht und auch auf der Basis der Beurteilung von Zwischenresultaten offensichtlich in der Lage ist, die Richtigkeit seiner Auswahl zu verifizieren. Deshalb wollen wir uns für die Fälle, für die es keine anderen besseren und objektiven Kriterien gibt, an der Vorgehensweise des Menschen bei der Bewertung von Bildverarbeitungsverfahren orientieren.

Für ein automatisches System zur Beurteilung von Bildverarbeitungsverfahren bedeutet dies, daß es ähnliche Beurteilungskriterien verwenden sollte, wie sie auch der menschliche Experte verwendet. Auf jeden Fall sollte der Experte in der Lage sein die von diesem System getroffenen Entscheidungen nachzuvollziehen. Dies bedeutet, daß eine explizite Wissensrepräsentation angestrebt werden muß. Da außerdem ein Mensch eher in hochsprachlichen Begriffen wie *Gut* und *Schlecht* oder *Stark* und *Schwach* anstatt in abstrakten Zahlen denkt, sollte auch das automatische System diese Begriffe verwenden.

Die Schlußfolgerungen eines menschlichen Experten beruhen nicht nur auf absolut sicherem Wissen. Daher soll das System weiterhin in der Lage sein, Unsicherheiten z.B. in Form von Fuzzy Sets und Certainty Values zu modellieren.

Wir gehen davon aus, daß eine Bewertung von Bildverarbeitungsverfahren nur im Kontext einer spezifischen Aufgabenstellung sinnvoll ist. Das heißt, das Bewertungssystem muß an die jeweilige Aufgabe vom Anwender angepaßt werden können. Dies kann z.B. bedeuten, daß die einzelnen Beurteilungskriterien unterschiedliche Prioritäten (Gewichte) erhalten oder aber das unterschiedliche Werte (Zielvorgaben) angestrebt werden sollen.

Als Ziel soll hier zunächst einmal die Gewinnung von Konturbildern aus Grauwertbildern betrachtet werden. Für die unterschiedlichsten Anwendungen wurden in der Vergangenheit spezielle Konturfindungsverfahren hergeleitet, die dann für die spezielle Anwendung sehr gut, für andere Anwendungen weniger gut funktionierten. Wir gehen davon aus, daß es genügt einen Satz von Standardverfahren (hier unterteilt in Vorverarbeitungs-, Konturverstärkungs-, Konturfindungs- und Nachbearbeitungsverfahren) festzulegen und dann für die jeweilige Aufgabenstellung zu bewerten, welche Kombination von Verfahren mit welchen Parametereinstellungen die beste ist. Eine Beurteilung der Qualität kann dabei nur für den gesamten Pfad und nicht für einzelne Operatoren erfolgen, da die Einzelverfahren in der Regel nicht unabhängig voneinander sind.

Hierfür muß im ersten Schritt eine Konfiguration, d.h. die Auswahl aller geeigneten Operatoren und deren Verknüpfung zu sinnvollen Operatorpfaden durchgeführt werden. Die Konfiguration kann anhand von Eigenschaften der zu verarbeitenden Grauwertbilder, Randbedingungen wie z.B. Rechenzeitanforderungen und A-Priori-Wissen über die Operatoren erfolgen. Durch die Konfiguration wird lediglich die Menge der insgesamt möglichen Verfahren aufgrund von Vorwissen reduziert.

Im zweiten Schritt müssen die freien Parameter an die aktuelle Aufgabenstellung adaptiert werden, da andernfalls auch gut geeignete Verfahren schlechte Ergebnisse liefern könnten.

Nachdem alle in Frage kommenden Verfahren konfiguriert und adaptiert worden sind, kann eine abschließende Bewertung und ein Vergleich der Verfahren durchgeführt werden.

Im folgenden wird das System FuzzySOLUTION vorgestellt, das die oben beschriebenen Anforderungen an ein System zur automatischen Bewertung von Bildverarbeitungsverfahren erfüllt. Es ist in der Lage eine vergleichende Bewertung von konturorientierten Segmentierungsverfahren automatisch durchzuführen und für eine vorgegebene Aufgabenstellung das am besten geeignete Verfahren auszuwählen. Im Abschnitt 2 wird die Architektur von FuzzySOLUTION vorgestellt, in 3 auf die eigentliche Bewertung näher eingegangen und im Abschnitt 4 einige Ergebnisse vorgestellt. Abschließend wird im Abschnitt 5 eine kurze Zusammenfassung gegeben.

2 Architektur von FuzzySOLUTION

Am Institut für Theoretische Nachrichtentechnik und Informationsverarbeitung wurden in den letzten drei Jahren die Systeme SOLUTION [7] und FuzzySOLUTION entwickelt. Beide Systeme sind in der Lage Low- Level- Bildverarbeitungspfade zur konturorientierten Segmentierung für vom Anwender vorgegebene Aufgabenstellungen automatisch aus einzelnen Bildverarbeitungsoperatoren zu konfigurieren und deren Parameter zu adaptieren. Sie basieren auf den Erfahrungen, die mit dem Vorgängersystem Conny [8] gesammelt wurden.

Ein Großteil des entwickelten Programmcodes wird von beiden Systemen gemeinsam benutzt. Die wesentlichen Unterschiede liegen dabei in der Verwendung von Fuzzy Sets und einer Fuzzy Inferenz Maschine für die bessere Modellierung unsicheren Wissens, verbunden mit anderen Adaptionsstrategien in der Adaptionseinheit von FuzzySOLUTION. Außerdem verfügt FuzzySOLUTION im Gegensatz zu SOLUTION bisher nicht über automatische Einheiten zum Lernen von Adaptionswissen.

Im folgenden wird nur das Konfigurations- und Adaptionssystem FuzzySOLUTION näher beschrieben.

2.1 Gesamtsystem

Abb. 1 zeigt die Struktur von FuzzySOLUTION. Über das Benutzerinterface spezifiziert der Anwender die zu lösende Aufgabe. Die Aufgabenstellung besteht aus der Festlegung des Eingangsbildes, Eigenschaften des Eingangsbildes (z.B. Rauschen), Randbedingungen (z.B. Rechenzeit) und Anforderungen an die Eigenschaften des Ergebnisbildes (z.B. die gewünschte Breite der Konturen).

Mit diesen Informationen stößt die Steuereinheit als erstes die Konfigurationseinheit an, die aus der Menge aller möglichen Operatorpfade diejenigen auswählt, die als am besten geeignet erscheinen um die Aufgabenstellung zu erfüllen. Danach wird von der Steuereinheit die Adaptionseinheit angestoßen, die für jeden konfigurierten Pfad die Parameter seiner Operatoren an die vorgegebene Aufgabenstellung anpaßt.

Zum Schluß werden alle adaptierten Pfade durch die Bewertungseinheit beurteilt, d.h. es wird festgestellt, wie gut sie die vorgegebene Aufgabenstellung erfüllen. Der Pfad, der das Konturbild mit der, im Sinne der Aufgabenstellung, besten Qualität erzeugt hat, wird dem Anwender zusammen mit dem zugehörigen Ergebnisbild über das Benutzerinterface präsentiert.

Das ganze System arbeitet wissensbasiert. Die Wissenbasis besteht dabei aus explizitem Wissen über die Konfiguration und Adaption und aus Wissen über die eigentliche Bildverarbeitung. Während das Wissen über die Bildverarbeitung in Form von Klassenhierachien und Prozeduren abgelegt ist, ist das Adaptions- und Konfigurationswissen in Form von Regeln kodiert.

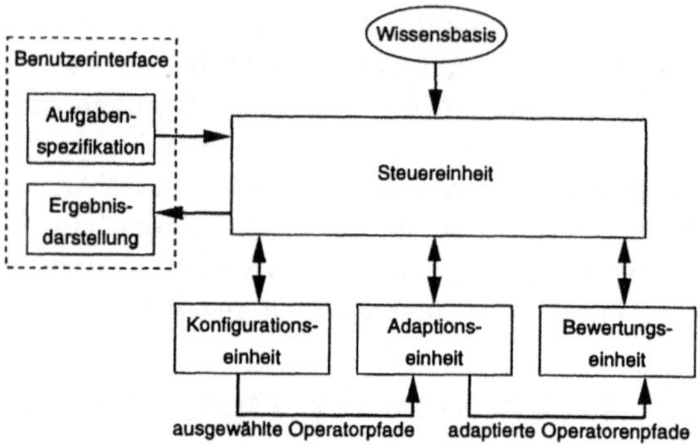

Abbildung1. Struktur von FuzzySOLUTION

2.2 Konfiguration

Für die Konfiguration werden der Aufgabenstellung Informationen über die Eigenschaften des Eingangsbildes entnommen. Diese Informationen liegen in hochsprachlicher Form vor, z.B. "das Rauschen ist stark", "die Kantenanstiegsbreiten sind gering" und "die Kantenanstiegshöhen sind groß". Weitere Informationen aus der Aufgabenstellung betreffen Forderungen nach der Rechenzeit oder der Lagegenauigkeit von Objektecken und Kanten. Auch hier werden hochsprachliche Angaben verarbeitet, z.B. "die Rechenzeit ist irrelevant", "es ist wichtig, daß die Lagegenauigkeit der Ecken hoch ist" usw.

Aufgrund dieser Informationen werden im ersten Schritt ungeeignete Operatoren ausgeschlossen. Im zweiten Schritt werden aus den verbliebenen Operatoren alle gültigen Operatorenpfade erzeugt und im letzten Schritt die freien Parameter initialisiert.

Da die Konfigurationseinheiten von SOLUTION und FuzzySOLUTION identisch sind, wird hier für eine detailliertere Beschreibung der Konfiguration auf [7] verwiesen.

2.3 Adaption

In Abb. 2 ist der Ablauf der Adaption in FuzzySOLUTION dargestellt. Im ersten Schritt wird aus dem Grauwertbild durch Ausführung des Bildverarbeitungspfades ein Konturbild erzeugt. Dabei wird der initiale Parametersatz P_0 verwendet.

Mit Hilfe von speziellen Sensoren (Meßroutinen, s.a. Abschnitt 3) werden Eigenschaften des Konturbildes (z.B. die Breite der Konturen) gemessen. Die gemessenen Werte werden auf einen Wertebereich von 0 bis 4.0 skaliert und in Fuzzy Sets (Crisp-Werte mit einem Zugehörigkeitswert von 1.0) umgewandelt.

Aus der vom Anwender vorgegebenen Aufgabenstellung (in der z.B. steht: *Es ist wichtig, daß die Breite der Konturen den Wert Nicht hat.*) wird das Globale

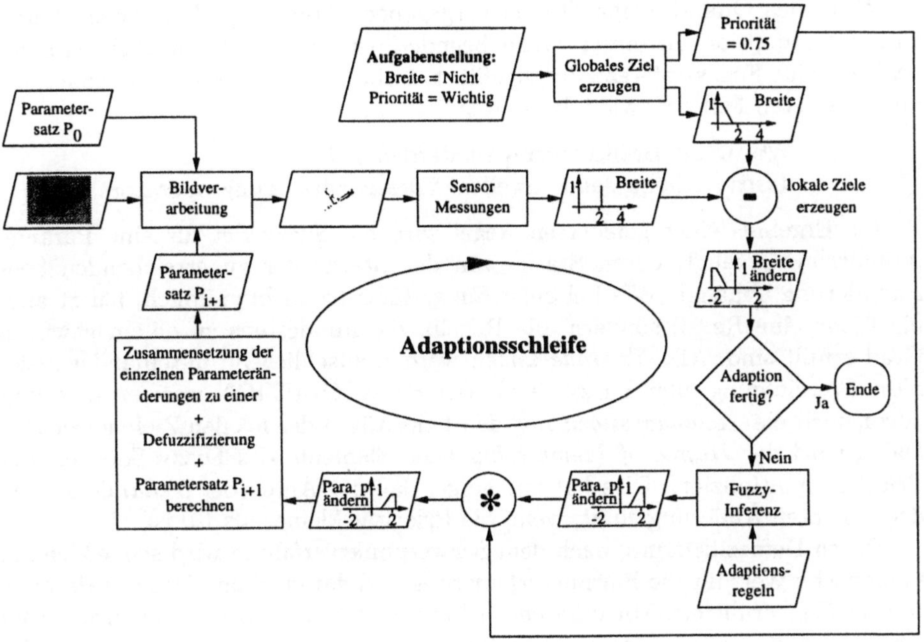

Abbildung2. Adaption in FuzzySOLUTION

Ziel für die Adaption erzeugt. Dabei werden die hochsprachlichen Werte der Konturbildbeschreibungen (z.B.: *Breite = Nicht*) in Fuzzy Sets umgewandelt. Die hochsprachlich vorgegebenen Prioritäten werden in reelle Zahlen von 0.0 bis 1.0 umgewandelt. So wird z.B. aus dem hochsprachlichen Wert *Wichtig* der numerische Wert 0.75.

Die Fuzzy Sets der Zielvorgabe werden mit den gemessenen Fuzzy Sets verglichen. Dies geschieht, indem durch eine Subtraktion die Fuzzy Sets der *lokalen Ziele* (notwendige Änderungen der Meßwerte, um das globale Ziel zu erreichen) ermittelt werden. Für die Breite ergibt sich so z.B. ein Fuzzy Set, das die Stärke der notwendigen Änderung der Breite angibt, um ausgehend von dem momentanen Zustand des Konturbildes den gewünschten Zielzustand zu erreichen. Die Fuzzy Sets der lokalen Ziele liegen in einem Wertebereich (Universe of Discourse) von -4 bis 4. Dabei bedeuten negative Werte, daß der zugehörige Sensorwert verringert und positive Werte, daß er erhöht werden soll.

Aus allen lokalen Zielen wird im nächsten Schritt ein Qualitätsmaß Q nach Gl. 1 berechnet. Dabei ist N die Anzahl aller lokalen Ziele, l_i der numerische Wert des defuzzifizierten lokalen Zieles Z_i und p_i seine Priorität aus der Aufgabenstellung. Es stellt im Prinzip ein mit den Prioritäten gewichtetes Abstandsmaß zwischen den Sollvorgaben und den Ist-Werten da und kann Werte zwischen 0.0 (schlecht) und 1.0 (gut) annehmen.

$$Q = 1 - \frac{\sqrt{\sum_{i=1}^{N}(l_i \cdot p_i)^2}}{4 \cdot \sqrt{\sum_{i=1}^{N} p_i^2}} \tag{1}$$

Wenn das Qualitätsmaß Q eine vorgegebene Mindestqualität erreicht hat, wird die Adaption an dieser Stelle beendet. Ist dies nicht der Fall, wird als nächstes eine Fuzzy-Inferenz durchgeführt. Dabei werden Regeln verwendet, die im einfachsten Fall z.B. wie folgt lauten:

Wenn die Breite verringert werden soll,
dann erhöhe den Schwellwert des Binärisierungsoperators.

Als Ergebnis einer gefeuerten Regel wird ein Fuzzy Set für eine Parameteränderung erzeugt, deren Stärke von der Stärke der zu erreichenden Sensoränderung abhängt. Wie bei einer Fuzzy-Inferenzmachine üblich, feuert aber nicht nur eine Regel, sondern alle Regeln, die mindestens zu einem gewissen Grad erfüllt sind. Alle *Teiländerungen* werden anschließend miteinander fusioniert (Vereinigung aller Fuzzy Sets). Bei FuzzySOLUTION werden allerdings zuvor noch die *Teiländerungen* mit den Prioritäten der lokalen Ziele gewichtet. Dabei wird der *Degree of Membership* aller Elemente des Fuzzy Sets mit der Priorität multipliziert. Dies hat zur Folge, daß der Anteil der *Teiländerung* an der gesamten Änderung sinkt, wenn die Priorität kleiner als 1.0 ist.

Durch Defuzzifizierung nach dem Schwerpunktverfahren wird schließlich der numerische Wert für die Parameteränderung und daraus dann der neue Parametersatz P_{i+1} ermittelt. Mit dem neuen Parametersatz beginnt das ganze wieder von vorn und wird solange durchgeführt, bis die oben genannte Abbruchbedingung erfüllt ist.

3 Das hochsprachliche Bewertungssystem

Das hochsprachliche Bewertungssystem von FuzzySOLUTION wird sowohl zur iterativen Optimierung der Parameterwerte während der Adaption als auch für die Bewertung und den Vergleich der verschiedenen Bildverarbeitungsverfahren (Operatorpfade) nach der Konfiguration und Adaption verwendet. Es besteht aus zur Zeit 19 verschiedenen Sensoren (Meßroutinen), die Eigenschaften im Konturbild (Ergebnisbild) messen.

Die Sensoren lassen sich in zwei Gruppen aufteilen. Zur ersten Gruppe gehören Sensoren, die hauptsächlich eingesetzt werden, um schnell und zuverlässig extreme Bildzustände erkennen zu können. In diesen Zuständen besitzt das Konturbild entweder sehr viele Konturpunkte, so daß die Berechnung der übrigen Sensoren extrem lange dauern würde, oder aber nur sehr wenige Konturpunkte, so daß die übrigen Sensoren entweder gar keine oder nur unzuverlässige Werte liefern. Diese Sensoren werden daher hauptsächlich während der Adaption benötigt, um das System aus diesen Extremzuständen zu bringen. Zur Qualitätsbeurteilung spielen sie höchstens eine untergeordnete Rolle. Hierfür sind hauptsächlich die Sensoren der zweiten Gruppe gedacht. Sie werden aber auch für die Adaption benötigt, wenn sich das System nicht in einem Extremzustand befindet.

Die Sensoren liefern z.B. Aussagen über die Breite, den Verflechtungsgrad, die Ausgefranstheit, die Geschlossenheit, die Zerstückelung und die Vollständigkeit von Konturen, aber auch Aussagen über den Anteil an verschiedenen Störkonturen, den Anteil an Konturpunkten pro Bildfläche oder den Anteil von Konturpunkten auf langen Geraden und großen Ellipsen.

Die Meßwerte der Sensoren können in die vier hochsprachlichen Werte *Nicht*, *Schwach*, *Mittel* und *Stark* abgebildet werden. Die eigentliche Bewertung erfolgt durch einen Vergleich der gemessen Werte mit den hochsprachlichen Vorgaben (Sollwerten) des Anwenders.

Wie bereits oben erläutert, kann der Wert Q aus Gl. 1 als Maß für die insgesamt erzielte Qualität verwendet werden. Bei einer automatischen Auswahl des besten Pfades wird Q als Entscheidungskriterium verwendet. Für den Menschen sind allerdings die hochsprachlichen Werte der Sensoren aussagekräftiger.

4 Ergebnisse

In Abb. 3 sind von links nach rechts ein stark verrauschtes Eingangsgrauwertbild, das Konturbild eines adaptierten Pfades, bestehend aus einem Gauß-Filter, einem Sobel-Operator und einem Rangfolgefilter sowie das Konturbild eines adaptierten Pfades, bestehend aus einem Gauß-Filter und einem Nichtlinearen-Laplace-Operator, dargestellt. Man erkennt deutliche Unterschiede in der Qualität der Ergebnisbilder. Während durch den Nichtlinearen-Laplace alle Konturen geschlossen wurden, sind die Konturen im Ergebnisbild des Pfades mit dem Sobel-Operator stark zerstückelt.

Abbildung3. Grauwertbild und zwei Konturbilder von verschiedenen Pfaden

Sensorname	Wert mit Sobel	Wert mit NLL	Zielvorgabe
Geschlossenheit	Nicht	Stark	Stark
Vollständigkeit	Mittel	Stark	Stark
Körnung	Schwach	Nicht	Nicht
Breite	Mittel	Nicht	Nicht
Ausfransung	Nicht	Mittel	Nicht
Qualitätsmaß Q	0.92	0.954	

Tabelle1. Bewertung der Ergebnisbilder aus Abb. 3

Die Unterschiede spiegeln sich auch in den Werten der Sensoren wieder. Einige davon werden in Tab. 1 zusammen mit den Zielvorgaben des Benutzers dargestellt. Mit Ausnahme der Ausfransung wurden alle Zielvorgaben vom Pfad mit dem Nichtlinearen-Laplace-Operator (NLL) erfüllt. Der Pfad mit dem Sobel-Operator kann die Zielvorgaben dagegen nur für die Ausfransung erfüllen. Die Zielvorgaben für die übrigen, hier nicht aufgelisteten Sensoren wurden in diesem Beispiel von beiden Pfaden erfüllt.

Auch das Qualitätsmaß Q (berechnet nach Gl. 1 aus vierzehn Sensoren mit der Priorität 1.0) zeigt, daß der Pfad mit dem NLL die Aufgabenstellung besser erfüllt. Der relative geringe Unterschied liegt darin begründet, daß hier lediglich 5 von 14 gleichgewichteten Sensoren vom Zielwert abweichen.

5 Schlußbetrachtungen

In dieser Veröffentlichung wurden ein Ansatz zur Bewertung von Bildverarbeitungsverfahren und das System FuzzySOLUTION vorgestellt, das nach diesem Ansatz arbeitet. Anhand eines Beispiels wurde in Abschnitt 4 gezeigt, daß es möglich ist mit FuzzySOLUTION für eine vorgegebene Aufgabenstellung verschiedene Bildverarbeitungsverfahren zu bewerten und miteinander zu vergleichen. Die hierfür notwendige Konfiguration und Adaption der Verfahren kann dabei ebenfalls mit FuzzySOLUTION erfolgen.

Da neben der Berechnung eines abstrakten Qualitätsmaßes auch hochsprachliche Aussagen über die Qualität gemacht werden, ist die Bewertung für den Menschen nachvollziehbar.

Der vorgestellte Ansatz macht Bewertungen aufgrund von theoretischen [9] oder statistischen Untersuchungen [10] nicht überflüssig. Im Gegenteil, die Ergebnisse derartiger Untersuchungen werden in FuzzySOLUTION für die Vorauswahl und Kombination verschiedener Verfahren (Konfiguration) verwendet. FuzzySOLUTION realisiert einen pragmatischen Ansatz, der Systementwicklern ein Werkzeug in die Hand gibt, um für seine spezifische Aufgabenstellung eine geeignete Lösung zu finden.

Literatur

1. Wolfgang Förstner: "10 Pros and Cons Against Performance Characterization of Vision Algorithms", Workshop on Performance Characteristics of Vision Algorithms, Cambridge, 1996.
2. V. Ramesh and R.M. Haralick: "Performance Evaluation of Edge Operators", Proceedings of the DARPA IUW 93, Washington DC, April 93, pp. 1071-79.
3. R. Haralick: "Performance Characterization Protocol in Computer Vision", Proceedings of the DARPA IUW 94.
4. M. Salotti, F. Bellet, C. Garbay: "Evaluation of Edge Detectors: Critics and Proposal", Workshop on Performance Characteristics of Vision Algorithms, Cambridge, 1996.
5. P. Courtney, N. Thacker, A. Clark: "Algorithmic Modeling for Performance Evaluation", Machine Vision and Applications 9, pp. 219-228, 1997.
6. I. Pitas: "Digital Image Processing Algorithms", Prentice Hall, 1993.
7. U. Rost, H. Münkel: "Knowledge Based Configuration of Image Processing Algorithms", Proceedings of the ICCIMA 1998.
8. C.-E. Liedtke, A. Blömer: "Architecture of the Knowledge Based Configuration System for Image Analysis CONNY", Proceedings of the 11th ICPR, International Conference on Pattern Recognition, The Hague, Vol I, pp. 375–378, Sept. 1992.
9. V. Ramesh and R.M. Haralick: "A Methodology for Automatic Selection of IU Algorithm Tuning Parameters", Proceedings of the DARPA IUW 94.
10. X. Liu, T. Kanungo, R.M. Haralick: "Statistical Validation of Computer Vision Software", Proceedings of the DARPA Image Understanding Workshop, Palm Springs, Vol. II, pp. 1533-1540, 1996.

Detektion von Personengruppen mit parameterfreien statistischen Klassifikatoren und Waveletmerkmalen

D. Faulhaber, P.Weierich, H. Niemann

Bayerisches Forschungszentrum für Wissensbasierte Systeme
Forschungsgruppe Wissensverarbeitung
Am Weichselgarten 7, D-91058 Erlangen
e-mail:faulhaber@forwiss.uni-erlangen.de

Zusammenfassung In diesem Beitrag untersuchen wir, in wie weit Waveletmerkmale in Verbindung mit parameterfreien statistischen Klassifikatoren geeignet sind, Personengruppen in Bildern einer Videosequenz vom Szenenhintergrund zu unterscheiden. Anhand einer manuell segmentierten Stichprobe, die 2850 Bilder umfaßt, werden parameterfreie statistische Klassifikatoren überwacht trainiert. Die Klassifikatoren basieren auf Bin-Histogrammen und Markovketten und zeichnen sich durch eine sehr geringe algorithmische Komplexität und hohe Geschwindigkeit aus. Es wurden mittlere Erkennungsraten von über 80% im Test erreicht.

1 Einleitung

Ziel des Projektes "Intelligenter Bahnsteig"[1] ist die videobasierte Überwachung von Ereignissen auf einem U-Bahnsteig. Betriebsrelevante Ereignisse sollen mit digitaler Bildverarbeitung automatisch erkannt und in bezug auf die Sicherheit der Fahrgäste beurteilt werden. Betriebsrelevante Ereignisse sind das Eindringen von Personen in den Gleisbereich, die Ein- und Ausfahrt des Zuges und der Fahrgastwechsel vom und zum Zug während des Zugaufenthaltes im U-Bahnhof.

Die zugrundeliegende Vorstellung unseres Ansatzes ist, daß Personengruppen eine vom Hintergrund unterscheidbare Textur haben. Die prinzipielle Eignung von Waveletmerkmalen zur Unterscheidung von Texturen wurde bereits mehrfach nachgewiesen (z.B. [4, 2]).Wir untersuchen im folgenden in wie weit Waveletmerkmale und statistische Klassifikatoren geeignet sind, diese Aufgabe zu lösen. In Abschnitt 2 wird die Wavelettransformation mengentheoretisch charakterisiert, und die zu klassifizierenden Merkmale werden definiert. Mit Hilfe einer manuell segmentierten Stichprobe trainieren wir parameterfreie statistische Klassifikatoren, die in Abschnitt 4 beschrieben sind. Zuvor geben wir in Abschnitt 3 eine Beschreibung der Bilddaten und erläutern, wie die manuelle Segmentierung der Stichprobe vorgenommen wurde. Experimentelle Ergebnisse werden in Abschnitt 5 vorgestellt.

[1] Verbundprojekt gemeinsam mit der VAG Nürnberg, plettac electronics Fürth und FORWISS Erlangen, gefördert durch den Freistaat Bayern

Ein texturbasierter Ansatz zur Personengruppendetektion ist uns nicht bekannt. Ein weiterführender Ansatz zur Personengruppenbeobachtung findet sich in [5], wo davon ausgegangen wird, daß bewegte Bildbereiche Personengruppen entsprechen. Dort wird die Anzahl der beobachteten Personen geschätzt. Wir gehen bei unserem Ansatz davon aus, daß auch Nichtpersonenobjekte Bewegungsinformation liefern werden. Daher wollen wir mit unserem Ansatz zunächst Personen unter den bewegten Objekten finden.

2 Waveletmerkmale

In diesem Abschnitt definieren wir die verwendeten Waveletmerkmale nach Mallat ([3]). Dabei gehen wir davon aus, daß der Leser mit der Waveletterminologie bereits vertraut ist. Eine Einführung in die Wavelettransformation findet sich z.B. in [1].

Sei $\mathbf{R}$ die Menge der reellen Zahlen und $\mathbf{Z}$ die Menge der ganzen Zahlen. Sei $L^2(\mathbf{R}^2)$ die Menge der zweidimensionalen, meßbaren, quadratisch integrierbaren Funktionen. Sei V_{2^j} eine Folge von ineinandergeschachtelten Unterräumen, für die gilt: $\forall j : V_{2^{j+1}} \subset V_{2^j} \wedge V_{2^j} \in L^2(\mathbf{R}^2)$. Sei O_{2^j} das orthogonale Komplement von $V_{2^{j+1}}$ in $V_{2^j} : \forall j : V_{2^{j+1}} \oplus O_{2^{j+1}} = V_{2^j}$. Sei $\Phi(x)$ eine eindimensionale Skalierungsfunktion und sei $\Phi_{2^j}(x) = 2^j\Phi(2^jx)$. Sei ferner $\Psi(x)$ ein eindimensionales Wavelet und sei $\Psi_{2^j}(x) = 2^j\Psi(2^jx)$. Dann lassen sich zweidimensionale Skalierungsfunktionen und Wavelets mittels $\Phi(x,y) = \Phi(x)\Phi(y), {}_1\Psi(x,y) = \Phi(x)\Psi(y), {}_2\Psi(x,y) = \Psi(x)\Phi(y)$ und ${}_3\Psi(x,y) = \Psi(x)\Psi(y)$ definieren. Sei $\langle g(u,v), f(u,v)\rangle = \int\int g(u,v)f(u,v)dudv$ das innere Produkt zweier Funktionen. Die Approximation einer Funktion $f(x,y) \in L^2(\mathbf{R}^2)$ geschieht mittels des Operators

$$V_{2^j} \ni A_{2^j}f = (\langle f(u,v), \Phi(u-2^{-j}n, v-2^{-j}m)\rangle)_{n,m\in\mathbf{Z}}. \tag{1}$$

Das dabei verlorengehende Detailsignal $(A_{2^j}f(x,y) - f(x,y)) \in O_{2^j}$ ist durch die drei Operatoren ${}_iD_{2^j}$ gegeben:

$${}_iD_{2^j}f = (\langle f(u,v), {}_i\Psi(u-2^{-j}n, v-2^{-j}m)\rangle)_{n,m\in\mathbf{Z}}, i = 1,2,3. \tag{2}$$

Die Berechnung der Operatoren A_{2^j} und ${}_iD_{2^j}$ mittels eines QMF Filterpaars ist in [3] beschrieben.

Mit Hilfe von V_{2^j} und $O_{2^j}, j \in \mathbf{Z}$ kann eine redundante Waveletdekomposition von $L^2(\mathbf{R})^2$ defniert werden:

$$L^2(\mathbf{R}^2) = \bigcup_j V_{2^j} \oplus \bigcup_j O_{2^j}. \tag{3}$$

Die Dekomposition ist redundant, weil jede Dekomposition $V_{2^{j+1}} \oplus O_{2^{j+1}} = V_{2^j}$ auf beiden Seiten der Gleichung die gleiche Information enthält. Wir nehmen an, die Redundanz sei für die Klassifikation hilfreich.

Die Waveletdekomposition eines Bildes $s_0(x,y)$ ist in Bild 1 dargestellt. In diesem Fall definieren wir Merkmalvektoren $\mathbf{c}$ durch Zusammenstellung von Waveletkoeffizienten aus Signalapproximat. und Projektionen $s_i, i = 1,2,3,\ldots,o$ an einem Ort (x,y) des Eingabebildes durch:

$$
\begin{aligned}
\mathbf{c}(x,y) = (&s_0(x,y),\\
&s_1(\lfloor\frac{x}{2}\rfloor,\lfloor\frac{y}{2}\rfloor), s_2(\lfloor\frac{x}{2}\rfloor,\lfloor\frac{y}{2}\rfloor), s_3(\lfloor\frac{x}{2}\rfloor,\lfloor\frac{y}{2}\rfloor), s_4(\lfloor\frac{x}{2}\rfloor,\lfloor\frac{y}{2}\rfloor),\\
&s_5(\lfloor\frac{x}{2^2}\rfloor,\lfloor\frac{y}{2^2}\rfloor),\ldots, s_8(\lfloor\frac{x}{2^2}\rfloor,\lfloor\frac{y}{2^2}\rfloor)\\
&\ldots,\\
&s_o(\lfloor\frac{x}{2^{t(o)}}\rfloor,\lfloor\frac{y}{2^{t(o)}}\rfloor))\\
= (&c_k)_{k=0,\ldots,n-1}
\end{aligned} \tag{4}
$$

wobei $t(.)$ angibt, welcher Operator $A_{2^{t(.)}}, D_{2^{t(.)}}$ auf $s_0(x,y)$ angewendet wurde. t bezeichnet dabei die "Baumtiefe". Wenn wir verdeutlichen wollen, daß $\mathbf{c}(x,y)$ zu einem Eingabebild $s_0(x,y)$ gehört, schreiben wir "$\mathbf{c}(s_0,x,y)$".

3 Stichprobe

Zur Validierung der entwickelten Verfahren wurde eine Kamera auf einem Nürnberger U-Bahnsteig installiert. Dabei wurde die Installation so vorgenommen, daß sie etwa die Hälfte des $90m$ langen U-Bahnsteigs aufnimmt. Während des Weihnachtsverkehrs 1997 wurden mehrere Stunden des Betriebs auf ein SVHS Videoband aufgenommen und mit $25Hz$ zu einem Format von 352×288 Grauwertpixeln digitalisiert.

Um die Klassifikationsverfahren zu bewerten, wurden Personengruppen manuell durch umschließende Polygone markiert. Die Markierung wurde in einer Sequenz von 14250 Einzelaufnahmen mit einer Abtastrate von $5Hz$ vorgenommen. Somit standen insgesamt 2850 manuell segmentierte Aufnahmen zur Verfügung.

Eine Beispielaufnahme ist in Abbildung 2 zu sehen, wobei die zu beobachtenden Personengruppen durch Polygone umschlossen sind. Mit Hilfe der Polygone können binäre Masken generiert werden, die für jeden Merkmalvektor $\mathbf{c}(x,y)$ angeben zu welcher Klasse er gehört. Durch die Markierung entsteht ein Zweiklassenproblem.

4 Parameterfreie statistische Klassifikatoren

Wir nehmen an, das resultierende Zweiklassenproblem ließe sich anhand der Verteilung $p(\mathbf{c}|\Omega_\kappa), \kappa = 0,1$ der Merkmalvektoren $\mathbf{c}$ in den Klassen Ω_0 (= Szenenhintergrund) und Ω_1 (= Personengruppen) entscheiden. In der Erkennungsphase wird für einen gegebenen Merkmalvektor $\mathbf{c}(x,y)$ der Wavelettransformation des Eingabebildes $s_0(x,y)$ die Zugehörigkeit zu den Klassen $\Omega_\kappa, \kappa = 0,1$ durch einen trainierten parameterfreien Bayesklassifikator entschieden, d.h. für einen gegebenen Merkmalvektor $\mathbf{c}(x,y)$ ist die Größe $p(\mathbf{c}|\Omega_\kappa)p_\kappa, \kappa = 0,1$ zu maximieren. p_κ ist die a priori Wahrscheinlichkeit der Klasse und $p(\mathbf{c}|\Omega_\kappa)$ die klassenbedingte Verteilungsdichte. Im folgenden setzen wir p_κ für jedes Bild der Videosequenz als bekannt voraus und untersuchen, wie $p(\mathbf{c}|\Omega_\kappa)$ in einer Trainingsphase erlernt werden kann.

4.1 Bin-Histogramme

Die Modellierung von klassenbedingten Verteilungsdichten $p(\mathbf{c}|\Omega_\kappa)$ mit eindimensionalen Bin-Histogrammen geht von statistischer Unabhängigkeit der Komponenten c_k des Merkmalvektors $\mathbf{c}$ aus:

$$p(\mathbf{c}|\Omega_\kappa) = \Pi_{k=0}^{n-1} p(c_k|\Omega_\kappa). \tag{5}$$

Dadurch wird die Schätzung der n-dimensionalen Dichte $p(\mathbf{c}|\Omega_\kappa)$ auf die Schätzung von n eindimensionalen Dichten $p(c_k|\Omega_\kappa)$ zurückgeführt. Eine solche eindimensionale Dichte kann durch ein Bin-Histogramm mit b Bins approximiert werden.

Es stellt sich die Frage, ob die Annahme statistischer Unabhängigkeit der Komponenten des Merkmalvektors zulässig ist. Wornell [6] zeigt, daß die Koeffizienten einer Wavelettransformation nahezu dekorrelliert sind, falls das gegebene Signal zur Klasse der $1/f$-Prozesse gehört. Damit ist eine nichtlineare statistische Abhängigkeit allerdings nicht ausgeschlossen. Deswegen untersuchen wir als zu den Bin-Histogrammen komplementären Ansatz eine Modellierung von $p(\mathbf{c}|\Omega_\kappa)$ durch Markovketten, die prinzipiell in der Lage sind, auch nichtlineare Abhängigkeiten zu erlernen.

4.2 Markovketten

Die Modellierung von klassenbedingten Verteilungsdichten $p(\mathbf{c}|\Omega_\kappa)$ mit Markovketten geht von einer statistischen Abhängigkeit aufeinanderfolgender Merkmalvektorkomponenten $c_k, c_{k+1}, k = 0, \ldots, n-2$ aus. Durch Quantisierung jeder Merkmalvektorkomponente in b diskrete Werte kann dies durch eine Markovkette mit b Zuständen und $n-1$ stochastischen Matrizen $P(k)$ modelliert werden:

Sei $Q = \{q_k(c_k) : \mathbf{R} \mapsto \{0, \ldots, b-1\}; k = 0, \ldots, n-1\}$ eine Menge von Quantisierungsfunktionen, wobei jedes q_k eine Komponente c_k des Merkmalvektors $\mathbf{c}$ quantisiere. Sei $U = \{u_0, \ldots, u_{b-1}\}$ eine endliche Menge von Zuständen und $T = \{P(k) = (p_{ij}(k))_{0 \leq i,j < b}; k = 0, \ldots, n-2\}$ eine Menge von $n-1$ stoch. Transitionsmatrizen. $p_{ij}(k)$ ist die Wahrscheinlichkeit einer Transition vom Zustand u_i in den Zustand u_j im k-ten Schritt. Dann ist $M = (U, T, Q)$ eine Markovkette oder ein stochastischer Automat.

Für einen gegebenen Merkmalvektor $\mathbf{c}$ ist die Akzeptanzwahrscheinlichkeit durch den Automaten M gegeben durch

$$p(\mathbf{c}|M) = \Pi_{k=0}^{n-1} p_{q_k(c_k), q_{k+1}(c_{k+1})}(k) \tag{6}$$

Dies bedeutet, daß aufeinanderfolgende Merkmalvektorkomponenten c_k, c_{k+1} durch ihre zugehörige Quantisierungsfunktion q_k, q_{k+1} quantisiert werden. Dadurch ergeben sich Markovkettenzustände $u_k = q_k(c_k)$ und $u_{k+1} = q_{k+1}(c_{k+1})$. Dann ist $p_{u_k, u_{k+1}}(k)$ die Wahrscheinlichkeit einer Transition vom Zustand u_k in den Zustand u_{K+1} im k-ten Schritt. Die Multiplikation über alle Zustandspaare $(k, k+1)$, gegeben durch den Merkmalvektor $\mathbf{c}$, gibt die Wahrscheinlichkeit des Pfades $(q_k(c_k))_{k=0,\ldots,n-1}$ im vollvernetzten Graph zwischen Zuständen U des Automaten M an.

Die Parameter des Automaten M werden wie folgt aus einer endlichen Stichprobe $\Omega = \{\mathbf{c}_0, \ldots, \mathbf{c}_r\}$ geschätzt:

$$\begin{aligned}
\mathbf{c} &= (c_0, \ldots, c_{n-1}) \in \Omega \\
{}_{max}c_k &= max\{c_k; \mathbf{c} \in \Omega\} \\
{}_{min}c_k &= min\{c_k; \mathbf{c} \in \Omega\} \\
q_k(c_k) &= \left\lfloor \frac{c_k - {}_{min}\, c_k}{{}_{max}c_k - {}_{min}\, c_k} \cdot b \right\rfloor \\
a_{ij}(k) &= |\{\mathbf{c} = (c_1, \ldots, c_n) \in \Omega_\kappa; \exists 0 \leq k < b : q_k(c_k) = u_i \wedge q_{k+1}(c_{k+1}) = u_j\}| \\
o_i(k) &= \sum_{j=0}^{b-1} a_{ij}(k) \\
p_{ij}(k) &= \frac{a_{ij}(k)}{o_i(k)}
\end{aligned}$$

Im Falle unseres Zweiklassenproblems werden zwei Markovketten M_1 und M_2 für die Klassen Ω_0 und Ω_1 als Modelle für $p(\mathbf{c}|\Omega_1)$ und $p(\mathbf{c}|\Omega_1)$ geschätzt. Dies geht nur falls klassifizierte Stichproben vorliegen.

5 Ergebnisse

Für die Berechnung der Erkennungsrate in einer bearbeiteten Bildsequenz S werden pro Einzelbild $s \in S$ zwei binäre Bilder verglichen, zum einen das aus der manuellen Segmentierung der Personengruppen resultierende Bild $b(x, y)$, zum anderen das durch Maximierung von $p(\mathbf{c}|\Omega_\kappa), \kappa = 0, 1$ resultierende binäre Klassifikationsergebnisbild $k(x, y)$. Eine Systemübersicht ist in Abbildung 2 dargestellt. Die Erkennungsraten pro Bild wurden wie folgt berechnet:

$$\begin{aligned}
k(x,y) &= \begin{cases} 1 \text{ falls } (x,y) \in \text{Personengruppe} \\ 0 \text{ sonst} \end{cases} \\
b(x,y) &= \begin{cases} 1 \text{ falls } p_1 p(\mathbf{c}(s,x,y)|\Omega_1) > p_0 p(\mathbf{c}(s,x,y)|\Omega_0) \\ 0 \text{ sonst} \end{cases} \\
\theta_0(s) &= |\{k(x,y); k(x,y) = b(x,y) \wedge b(x,y) = 0\}| / |\{b(x,y); b(x,y) = 0\}| \\
\theta_1(s) &= |\{k(x,y); k(x,y) = b(x,y) \wedge b(x,y) = 1\}| / |\{b(x,y); b(x,y) = 1\}| \\
\theta(s) &= p_0\theta_0(s) + p_1\theta_1(s)
\end{aligned}$$

$p_0(s)$ und $p_1(s)$ geben die a-priori Wahrscheinlichkeiten der Klassen Ω_0 und Ω_1 pro Bild s an. Die in den Tabellen angegebenen Erkennungsraten ergeben sich durch eine Mittelung über die Bildsequenz S: $\theta_0 = \frac{1}{|S|}\sum_{s\in S}\theta_0(s)$, $\theta_1 = \frac{1}{|S|}\sum_{s\in S}\theta_1(s)$, $\theta = \frac{1}{|S|}\sum_{s\in S}\theta(s)$.

Die in Abschnitt 3 beschriebene Bildsequenz wurde in disjunkte Trainings- und Teststichproben aufgeteilt. Dabei wurden die ersten 1425 manuell markierten Bilder als Trainingssequenz und die letzten 1425 manuell markierten Bilder als Testsequenz verwendet. Während der Trainingsphase wird ein Zugaufenthalt beobachtet, in der Testphase sind zwei Zugaufenthalte aufgenommen. Während der gesamten Aufnahmezeit sind sowohl viele als auch wenige Personen auf dem Bahnsteig zu sehen.

Es wurden 16 in der Literatur bekannte Wavelets zur Detektion von Personengruppen verglichen. Da die Bilder eine Größe von $352 \times 288 = 2^5 \cdot 11 \times 2^5 \cdot 9$ Pixeln haben, können die Wavelettransformationsbäume bis zu einer Tiefe $t(o) = 5$ entwickelt werden. Die Verarbeitungszeit eines Bildes in der Klassifikationsphase beträgt weniger als $1s$ auf einer *Sun Ultra Sparc III* mit 296 Mhz Taktfrequenz.

Die Ergebnisse sind in den Tabellen 1 und 2 zusammengefaßt, wobei sowohl für den Bin-Histogramm-Ansatz als auch für den Markovketten-Ansatz $b = 50$ angenommen wurde. Weitere Experimente zeigten, daß der Parameter b in weiten Grenzen variiert werden kann, ohne daß sich die Erkennungsraten signifikant verändern.

Die Tabellen zeigen, daß der Markovketten-Ansatz dem Bin-Histogramm-Ansatz in bezug auf die Erkennungsraten θ_1 überlegen ist. So erhält man z.B. für θ_1 im Test für das HAAR-Wavelet fast 10% höhere Erkennungsraten für den Markovketten-Ansatz. θ_0 dagegen ist für den Markovketten-Ansatz und den Bin-Histogramm-Ansatz fast gleich. D.h. es gibt zwischen den Komponenten des aus der Wavelettransformation resultierenden Merkmalvektors nichtlineare Abhängigkeiten. Die in Abbildung 1 gezeigte Linearisierung des Waveletbaumes ist nur eine Möglichkeit. Es stellt sich die Frage, ob es für die Erkennung günstigere Linearisierungen gibt. Dies muß in weiteren Arbeiten untersucht werden. Die Frage nach einem am besten geeigneten Wavelet ist nicht eindeutig beantwortbar, denn mehrere Wavelets liefern vergleichbare Erkennungsraten. In unserem Fall erscheint das HAAR-Wavelet als das Geeignetste, denn es besitzt den kürzesten Faltungskern, und ist damit am schnellsten berechenbar.

References

1. C.S. Burrus, R.A. Gopinath, and H. Guo. *Introduction to Wavelets and Wavelet Transforms.* Prentice Hall Publishers, New Jersey, 1998.
2. N. Fatemi-Ghomi. *Performance measures for Wavelet-based Segmentation Algorithms.* PhD thesis, Centre for Vision, Speech and Signal processing, School of Electronic Engineering, Information technology and Mathematics, University of Surrey, 1997.
3. Stéphane G. Mallat. A theory for multiresolution signal decomposition: The wavelet representation. *IEEE Transactions on Pattern Analysis and Machine Intelligence,* 11(7):674–693, 1989.
4. T. Randen. *Filter and Filter Bank Design For Image Texture Recognition.* PhD thesis, Norwegian University of Science and Technology, Stavanger College, 1997.
5. C.S. Regazzoni and A. Tesei. Distributed data fusion for real-time crowding estimation. *Signal Processing,* (53):47–63, 1996.
6. G. Wornell. *Signal processing with fractals, A Wavelet based Approach.* Prentice Hall Signal Processing series, New Jersey, 1996.

	Training			Test		
Wavelet	θ_1 in %	θ_0 in %	θ in %	θ_1 in %	θ_0 in %	θ in %
Haar	62.8	87.5	87.4	70.7	72.1	76.0
Daub,4-Tap	76.0	90.2	90.2	76.8	77.7	80.7
Daub,6-Tap	**80.1**	**91.3**	91.2	76.3	80.5	82.5
Daub,8-Tap	77.6	92.2	**92.2**	**80.6**	**83.1**	**84.8**
Coif,6-Tap	72.8	89.0	89.0	75.1	74.9	78.5
John,8-Tap	58.6	87.0	86.9	71.4	71.5	75.7
John,12-Tap	58.6	86.7	86.6	70.3	70.8	75.1
Andrew,10-Tap	58.8	86.6	86.5	70.6	70.9	75.3
Andrew,8-Tap	72.5	89.0	88.9	73.3	74.8	78.1
Smithb,8-Tap	64.6	87.8	87.5	70.2	72.3	76.1
Gop,8-Tap	65.3	**91.3**	90.8	72.7	79.0	81.2
Zhu,4-Tap	52.7	87.6	86.6	63.3	73.9	76.1
Simmonic,9-Tap	63.2	87.7	87.4	70.5	72.4	76.4
Daub,9-Tap	60.0	87.4	86.9	68.7	71.6	75.5
Villasenor,9-Tap	64.7	87.3	87.1	71.5	71.7	76.0
Zhu,5-Tap	55.9	87.8	87.2	67.2	72.3	75.9

Tabelle1. Waveletbasierte Personengruppendetektion mit Bin-Histogrammen, 21 dimensionale Merkmalvektoren, $t(o) = 5$

	Training			Test		
Wavelet	θ_1 in %	θ_0 in %	θ in %	θ_1 in %	θ_0 in %	θ in %
Haar,2-Tap	81.9	90.0	90.8	80.2	71.7	77.0
Daub,4-Tap	87.3	91.5	92.2	**84.4**	75.9	80.7
Daub,6-Tap	87.8	91.9	92.6	83.8	77.1	81.5
Daub,8-Tap	85.1	**92.5**	**93.0**	83.2	**79.7**	**83.0**
Coif,6-Tap	86.2	91.2	91.8	83.8	74.8	79.7
John,8-Tap	74.1	90.4	90.8	79.3	72.1	77.4
John,12-Tap	74.2	90.4	90.7	77.8	71.7	76.9
Andrew,10-Tap	71.8	90.1	90.5	78.4	71.9	77.0
Andrew,8-Tap	**86.3**	91.1	91.7	82.3	74.3	79.2
Smithb,8-Tap	84.3	89.8	90.6	81.4	70.5	76.6
Gop,8-Tap	85.7	90.8	91.3	80.9	77.3	80.5
Zhu,4-Tap	71.5	90.4	90.3	73.6	72.5	76.8
Simmonic,9-Tap	82.9	89.5	90.0	78.6	70.3	76.1
Daub,9-Tap	78.9	89.2	89.7	77.9	69.8	75.6
Villasenor,9-Tap	82.9	89.4	90.0	80.0	69.5	75.7
Zhu,5-Tap	80.8	89.3	89.8	78.3	69.8	75.6

Tabelle2. Waveletbasierte Personengruppendetektion mit Markovketten, 21 dimensionale Merkmalvektoren, $t(o) = 5$

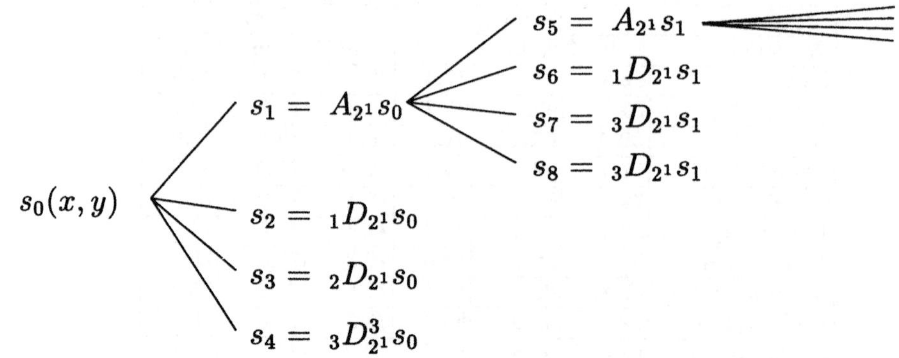

Abbildung1. Redundante Waveletdekomposition des Signals $s_0(x, y)$

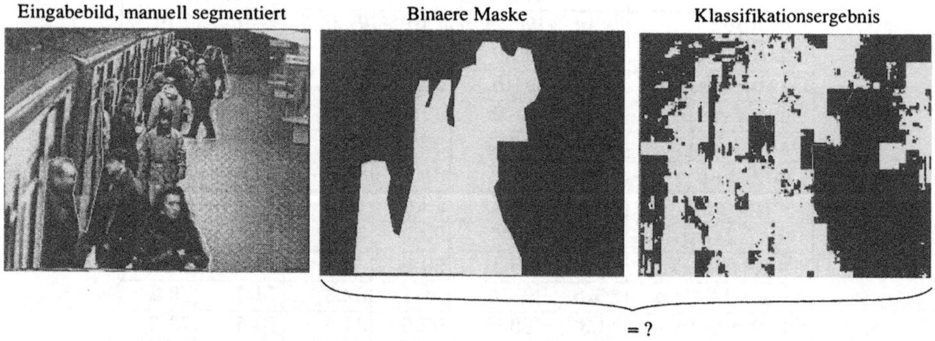

Abbildung2. Systemüberblick

Objektklassifikation mit Mischverteilungen

J. Dahmen, K. Beulen, H. Ney

Lehrstuhl für Informatik VI der RWTH Aachen,
Ahornstraße 55, D-52056 Aachen,
e-mail: {dahmen, beulen, ney}@informatik.rwth-aachen.de

Zusammenfassung. Ziel dieser Arbeit ist die Untersuchung der Fragestellung, inwieweit 'konventionelle' Klassifikatoren aus der statistischen Mustererkennung für die Objektklassifikation in der Bildverarbeitung einsetzbar sind. Dazu wurde ein auf Mischverteilungen basierender Klassifikator implementiert. Zur Merkmalsanalyse wurden die hochdimensionalen Bilddaten geeignet in einen niederdimensionalen Merkmalsraum projiziert. Die Klassifikation der so dimensionsreduzierten Merkmalsvektoren erfolgt über die Bayes'sche Entscheidungsregel. Auf der 'Chair-Image-Database' des Max-Planck-Instituts [1] erzielen wir eine Fehlerrate von 0.64%.

1 Einführung

In der vorliegenden Arbeit wird die Anwendbarkeit 'konventioneller' statistischer Klassifikatoren auf das Problem der Objektklassifikation in der Bildverarbeitung überprüft. Zur Evaluierung des von uns implementierten Klassifikators verwenden wir die 'Chair-Image-Database' des Max-Planck-Instituts, da für dieses Corpus bereits Resultate des Max-Planck-Instituts [1] und der Daimler-Benz-AG [2] vorliegen. Diese Gruppen setzen Verfahren wie neuronale Netze [1], Support-Vektoren [1] oder polynomiale Klassifikatoren [2] ein und erzielen so Fehlerraten zwischen 0.8% und 9% (siehe Kapitel 5). In dem von uns implementierten Klassifikator werden die Merkmalsvektoren der zu klassifizierenden Objekte (im folgenden auch Beobachtungen genannt) vereinfachend als gaußverteilt angesehen und die Parameter dieser Verteilung in einer Trainingsphase mittels des Expectation-Maximization-Algorithmus geschätzt. Die so gewonnenen Modelle werden dann in der Erkennung zur Realisierung der Bayes'schen Entscheidungsregel eingesetzt. Da die Merkmalsvektoren im Falle des von uns gewählten Corpus sehr hochdimensional sind, entwickeln wir zur Merkmalsreduktion die Beobachtungsvektoren nach einer geeignet zu wählenden Orthonormalbasis (siehe Kapitel 4). Unter Verwendung von Mischverteilungen und klassenspezifischen Varianzen erreichen wir eine Fehlerrate von 0.64%.

2 Datensammlung und Aufgabenstellung

Die von uns verwendete 'Chair-Image-Database' des Max-Planck-Instituts besteht aus computergenerierten Bildern von Bürostühlen aus 25 Klassen (`ftp://`

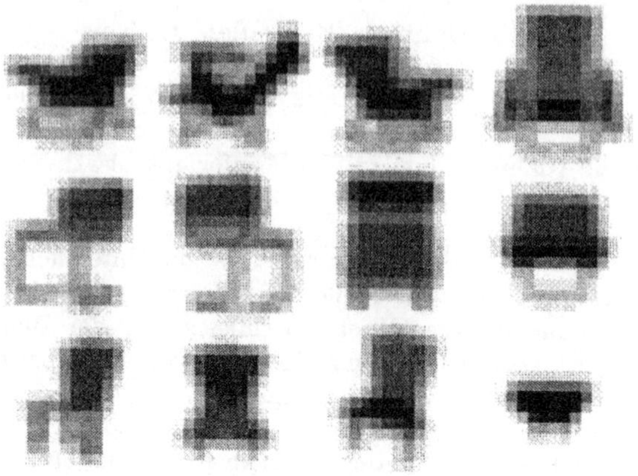

Abb. 1. Beispielbilder verschiedener Stuhlklassen (16 · 16 Pixel, Bilder einer Zeile gehören zur gleichen Klasse)

`ftp.mpik-tueb.mpg.de/pub/chair_dataset`). Jede Klasse enthält Aufnahmen eines unter verschiedenen Blickwinkeln betrachteten Stuhls (Abbildung 1). Alle Bilder sind auf ein Format von 16 · 16 Pixel und auf 256 Grauwerte normiert. Zusätzlich zum eigentlichen Grauwertbild werden noch vier richtungsabhängige Kantenbilder [1] verwendet, so daß die resultierenden Merkmalsvektoren 1280-dimensional sind. Die 'Chair-Image-Database' enthält verschiedene, unterschiedlich große Trainingscorpora und ein Testcorpus. Das Testcorpus besteht aus 2500 Bildern, zeigt also jeden Stuhl aus 100 verschiedenen Blickrichtungen, während die Anzahl der Blickrichtungen für die Trainingscorpora zwischen 25 und 400 liegt. In unseren Experimenten setzen wir sowohl das Corpus mit 89 Blickwinkeln (Train89) als auch das mit 400 Blickwinkeln (Train400) pro Klasse ein. Die Aufgabenstellung besteht darin, mit Hilfe der Trainingscorpora einen Klassifikator zu entwerfen, der dann mit dem Testcorpus evaluiert werden kann.

3 Bayes'sche Entscheidungsregel und Mischverteilungen

Für die Klassifikation der Daten verwenden wir die Bayes'sche Entscheidungsregel [3, S.10-43]. Sind sowohl die *a-priori* Wahrscheinlichkeit $p(k)$ als auch die klassenbedingte Wahrscheinlichkeit $p(x|k)$ bekannt, so läßt sich ein Beobachtungsvektor x mit folgender Regel klassifizieren:

$$\hat{k} = \underset{k}{\operatorname{argmax}} \{p(k)p(x|k)\} \tag{1}$$

Da weder die a-priori Klassenwahrscheinlichkeit $p(k)$ noch die klassenbedingte Wahrscheinlichkeit $p(x|k)$ bekannt sind, müssen diese aus den vorhandenen Trainingsdaten gelernt werden. Die a-priori Wahrscheinlichkeit wird über relative Häufigkeiten geschätzt, während die klassenbedingten Wahrscheinlichkeiten mit Gaußschen Mischverteilungen modelliert werden.

3.1 Gaußsche Mischverteilungen

Eine Mischverteilung (2) ist die gewichtete Summe von Einzelverteilungen mit Parametermenge ϑ_{ki}, wobei für die Gewichte die Bedingungen $\sum_i c_{ki} = 1$ sowie $c_{ki} > 0$ erfüllt sein müssen:

$$p(x|k) = \sum_{i=1}^{I_k} c_{ki} \cdot p(x|k, i, \vartheta_{ki}) \tag{2}$$

Wir verwenden für die Einzelverteilungen $p(x|k, i, \vartheta_{ki})$ die Gaußverteilung (3) und bezeichnen diese als Mischverteilungskomponenten oder Dichten:

$$p(x|k, i, \vartheta_{ki}) = \frac{1}{\sqrt{(2\pi)^D \det(\Sigma_{ki})}} \exp\left[-\frac{1}{2}(x - \mu_{ki})^T \Sigma_{ki}^{-1} (x - \mu_{ki})\right] \tag{3}$$

Die Parametermenge einer Mischverteilung k umfaßt also neben den Parametern μ_{ki} und Σ_{ki} auch Gewichte c_{ki}, so daß im folgenden $\vartheta_{ki} = (c_{ki}, \mu_{ki}, \Sigma_{ki})$ gelte. Im Gegensatz zu Einzelverteilungen können Mischverteilungen auch multimodale Verteilungen beschreiben, so daß ihre Verwendung in der Regel zu besseren Resultaten führt. Das Schätzen von empirischen Kovarianzmatrizen erweist sich jedoch oftmals als schwierig. Zur Behebung dieses Problems gibt es drei Möglichkeiten:

a) Schätze nur eine Diagonalmatrix statt der vollen Kovarianzmatrix, betrachte also nur einen Varianzvektor.
b) *klassenspezifisches Pooling*: Bestimme für jede Klasse k nur ein Σ_k, also: $\Sigma_{ki} = \Sigma_k \; \forall \; i \in I_k$.
c) *globales Pooling*: Bestimme nur eine einzige Kovarianzmatrix Σ, also: $\Sigma_{ki} = \Sigma \; \forall \; k$ und $\forall \; i \in I_k$.

Wir werden im folgenden stets Möglichkeit a) in Kombination mit b) oder c) wählen. Da eine diagonale Kovarianzmatrix nur für unkorrelierte Merkmale geeignet ist, werden die Daten vor der Parameterschätzung einer Whitening-Transformation [4, S.26-29] unterworfen. Die mittlere klassenbedingte Kovarianzmatrix

$$\hat{\Sigma} = \frac{1}{N} \sum_{n=1}^{N} (x_n - \mu_{k_n})(x_n - \mu_{k_n})^t \tag{4}$$

der so transformierten Daten ist dann die Einheitsmatrix. Dabei sei N die Anzahl der Trainingsdaten, x_n das n. Trainingsdatum, k_n der zu x_n gehörige Klassenindex und μ_{k_n} der empirische Mittelwertsvektor der Klasse k_n.

4 Der Klassifikator

4.1 Systemarchitektur

Der zu entwerfende Klassifikator hat die in Abbildung 2 dargestellte Struktur. Die Transformationsmatrix U entspricht dabei der aus den Trainingsdaten gewonnenen Whitening-Transformationsmatrix (Kapitel 3). Da wir Eigenvektoren zum Eigenwert Null nicht berücksichtigen, erfolgt in diesem Schritt bereits eine Dimensionsreduktion auf 1231 Dimensionen. Zur Bestimmung der Transformationsmatrix T betrachten wir K Vektoren der Form $\mu_k - \mu$, wobei μ_k der empirische Mittelwertsvektor der Klasse k und μ der empirische Mittelwertsvektor aller Beobachtungen sei. Für den von diesen Vektoren aufgespannten Unterraum wird eine Orthonormalbasis bestimmt. Um die durch Rundungsfehler bedingten numerischen Probleme der üblichen *Gram-Schmidt-Orthogonalisierung* zu vermeiden, verwenden wir hierzu eine *Singular-Value-Decomposition* (SVD) [5, S.59-67]. Man erhält maximal $K-1$ Basisvektoren [4, S.451]. Die lineare Transformation T ist nun die Projektion der Merkmalsvektoren in diesen Unterraum, der im Fall der von uns verwendeten Daten genau $K-1$ dimensional ist.

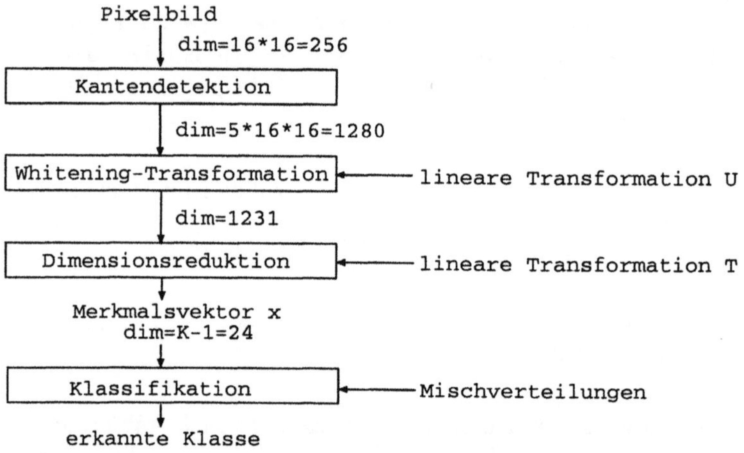

Abb. 2. Grundstruktur des Gesamtsystems

4.2 Training der Mischverteilungen

Die noch zu bestimmenden Verteilungsparameter werden mit dem EM-Algorithmus, einem Maximum-Likelihood-Schätzverfahren für Daten mit verborgenen Variablen, geschätzt. Wir geben Schätzformeln für die zu bestimmenden Parameter an, die sich aus der Anwendung des EM-Algorithmus auf unser Problem ergeben. Weiterhin beschreiben wir eine vereinfachte Form dieses Algorithmus, die auf einer Maximum-Approximation beruht.

Der Expectation-Maximization-Algorithmus: Der EM-Algorithmus ist ein iteratives Schätzverfahren zur Bestimmung der Parameter einer Wahrscheinlichkeitsverteilung mit verborgenen Variablen. Seine Anwendung auf Mischverteilungen wird in [6] beschrieben, wobei der Index der zu einer Beobachtung gehörigen Dichte als verborgene Variable interpretiert wird. Diese Zuordnung wird als Wahrscheinlichkeit $p(i|x,k,\vartheta_{ki})$ ausgedrückt, wobei k der Klassenindex, i der Index der Mischverteilungskomponente und ϑ_{ki} deren Parameter sind. Die Schätzformeln für die Parameter der Verteilung $p(x|k,i,\vartheta_{ki})$ ergeben sich für den EM-Algorithmus zu:

$$p(i|x,k,\vartheta_{ki}) = \frac{c_{ki} \cdot p(x|k,i,\vartheta_{ki})}{\sum_{i'} c_{ki'} \cdot p(x|k,i',\vartheta_{ki'})} \tag{5}$$

$$\bar{\mu}_{ki} = \sum_{n=1}^{N_k} \frac{p(i|x_n,k,\vartheta_{ki})}{\sum_{n'} p(i|x_{n'},k,\vartheta_{ki})} x_n \tag{6}$$

$$\bar{\Sigma}_{ki} = \sum_{n=1}^{N_k} \frac{p(i|x_n,k,\vartheta_{ki})}{\sum_{n'} p(i|x_{n'},k,\vartheta_{ki})} [x_n - \bar{\mu}_{ki}][x_n - \bar{\mu}_{ki}]^T \tag{7}$$

$$\bar{c}_{ki} = \frac{1}{N_k} \sum_{n=1}^{N_k} p(i|x_n,k,\vartheta_{ki}) \tag{8}$$

wobei N_k die Anzahl der Beobachtungen der Klasse k und x_n der n. Merkmalsvektor dieser Klasse sei. Mit einer initialen Schätzung der Parameter (c_{ki}, μ_{ki}, Σ_{ki}) wird zunächst ein $p(i|x,k,\vartheta_{ki})$ bestimmt. Mit $p(i|x,k,\vartheta_{ki})$ können die Parameter ϑ_{ki} neu geschätzt werden, die wiederum eine weitere Schätzung von $p(i|x,k,\vartheta_{ki})$ erlauben. Dieser Vorgang wird nun bis zur Konvergenz wiederholt. Sowohl die Anzahl der Mischverteilungskomponenten als auch die initialen Werte ihrer Parameter werden über das sukzessive Spalten von Mischverteilungskomponenten bestimmt. Begonnen wird dabei mit dem Schätzen einer Einzelverteilung. Diese wird sukzessive gespalten, d.h. eine Einzelverteilung ki wird durch das Stören des Mittelwertsvektors μ_{ki} über einen Störvektor ϵ durch zwei Einzelverteilungen ersetzt. Dieses Verfahren hat sich für die Modellierung von Emissionswahrscheinlichkeiten in der Spracherkennung bewährt [7]. Man erhält zwei Mittelwertsvektoren $\mu_{ki}^+ = \mu_{ki} + \epsilon$ sowie $\mu_{ki}^- = \mu_{ki} - \epsilon$, also eine Mischverteilung mit den beiden Dichten $p(x|\mu_{ki}^+, \Sigma_{ki})$ und $p(x|\mu_{ki}^-, \Sigma_{ki})$. Nun werden die Parameter dieser Dichten mittels (5)-(8) neu geschätzt. Dieser Vorgang wird iteriert, bis die gewünschte Anzahl Dichten erreicht ist. Dabei werden alle Einzelverteilungen i der Klasse k gespalten, für die gilt:

$$\sum_{n=1}^{N_k} p(i|x_n,k,\vartheta_{ki}) \geq N_{Split} \tag{9}$$

Maximum-Approximation: Bei der Maximum-Approximation wird jeder Trainingsvektor nur einer Mischverteilungskomponente zugeordnet, d.h. $p(i|x,k,\vartheta_{ki})$ wird modelliert als:

$$p(i|x,k,\vartheta_{ki}) = \begin{cases} 1 & : \quad i = \text{'beste' Einzelverteilung} \\ 0 & : \quad \text{sonst} \end{cases} \tag{10}$$

Die Rechtfertigung dieser Vorgehensweise ergibt sich aus dem exponentiellen Abfall der Gaußverteilung, der zur Folge hat, daß in der Regel ein dominierendes $p(i|x,k,\vartheta_{ki})$ existiert. Vorteil der Maximum-Approximation ist, daß die relativ aufwendige Bestimmung der $p(i|x,k,\vartheta_{ki})$ ersetzt wird durch eine einfache Bestimmung des maximalen $p(x|k,i,\vartheta_{ki})$. Diese reduziert sich bei Logarithmierung von Gleichung (3) im wesentlichen auf die Bestimmung des quadrierten euklidischen Abstands zwischen Mittelwerts- und Beobachtungsvektor.

5 Ergebnisse

Den Beginn unserer Experimente bilden Gaußsche Einzelverteilungen, die mit den Originalmerkmalsvektoren, also ohne vorhergehende Dimensionsreduktion, trainiert werden. Als Problem erweisen sich dabei die synthetisch erstellten Bilddaten, in denen Merkmale mit einer empirischen Varianz gleich Null existieren. Dies sind in der Regel Pixel, die am Bildrand gelegen sind und in allen Bildklassen zum Bildhintergrund gehören. Um dieses Problem zu umgehen, setzen wir zwei Verfahren ein: Im ersten Fall werden Nullvarianzen durch extrem kleine Varianzen approximiert, alternativ werden solche Merkmale 'ausmaskiert', d.h. nicht zur Klassifikation verwendet. Wie die in Tabelle 1 dargestellten Ergebnisse zeigen, erweist sich das Ausmaskieren als die deutlich bessere Wahl. Die Tatsache, daß auf dem Train89-Corpus bessere Ergebnisse erzielt werden als auf dem Train400-Corpus ist dadurch zu erklären, daß die Blickwinkel für das kleine Trainingscorpus und das Testcorpus äquidistant, die für das Train400-Corpus jedoch zufällig gewählt sind. Eine vorhergehende Dimensionsreduktion auf 24 Merkmale mittels der in Kapitel 4 vorgestellten linearen Transformation T reduziert die Fehlerrate auf dem Train400-Corpus auf 3.36% (Einzelverteilung, klassenspezifisches Pooling). Abbildung 3 zeigt die Fehlerrate für die Verwendung von Mischverteilungen auf den dimensionsreduzierten Daten. Dabei werden

Tabelle 1. Fehlerraten mit verschiedenen Trainingscorpora und Varianzberechnungen (Einzelverteilungen, keine Dimensionsreduktion)

Trainingscorpus	Varianz-maskierung	Varianz-modellierung	Fehlerrate [%]
Train400	nein	klassenspezifisch	54.72
	ja	global	24.28
		klassenspezifisch	15.08
Train89	nein	klassenspezifisch	55.08
	ja	global	22.60
		klassenspezifisch	14.24

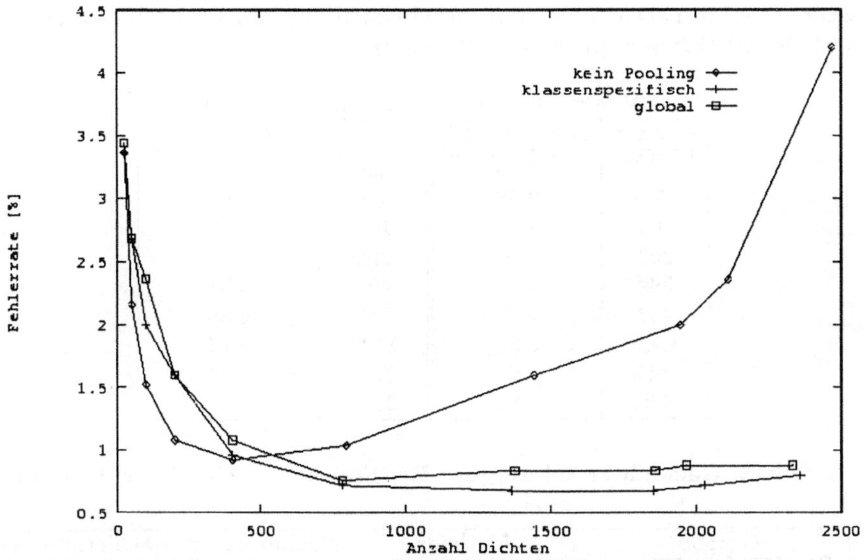

Abb. 3. Fehlerraten mit Train400 in Abhängigkeit von der Zahl der Dichten für verschiedene Arten des Varianzpoolings (EM-Algorithmus)

sowohl die Anzahl der Dichten als auch die Art des Varianzpoolings variiert. Das optimale Ergebnis von 0.64% (Tabelle 2) wird mit insgesamt 2292 Dichten und klassenspezifischer Varianz erreicht. Auf dem Train89-Corpus erzielen wir eine Fehlerrate von 1.72% (456 Dichten). Wie Tabelle 2 zeigt, gibt es auf den von uns verwendeten Daten keine nennenswerten Unterschiede zwischen dem Einsatz von EM-Algorithmus und Maximum-Approximation. Tabelle 3 zeigt den Vergleich unserer Ergebnisse mit Resultaten anderer Arbeitsgruppen, die auf denselben Daten erzielt wurden.

6 Zusammenfassung und Ausblick

Wir haben in dieser Arbeit einen auf Mischverteilungen basierenden Ansatz zur Objektklassifikation vorgestellt. Obwohl dieser auf Maximum-Likelihood-Training beruht, im Gegensatz zu den Verfahren der anderen Arbeitsgruppen also nicht diskriminativ ist, sind unsere Ergebnisse mit diesen vergleichbar bzw. überlegen. Die beste Fehlerrate von 0.64% erzielen wir mit Gaußschen Mischverteilungen und klassenspezifischen Varianzen. Neben einem Einsatz des Verfahrens auf realen Daten sind für die Zukunft weitere Verbesserungen des Algorithmus geplant, beispielsweise die Realisierung eines diskriminativen Trainings oder eine weitere Vorverarbeitung der Merkmalsvektoren (Invarianz gegenüber Rotation, Translation und Skalierung, Normierung der Daten, etc.).

Tabelle 2. Fehlerraten für Maximum-Approximation und EM-Algorithmus mit Train400 (klassenspezifische Varianz)

EM-Algorithmus		Maximum-Approximation	
Dichten	Fehlerrate [%]	Dichten	Fehlerrate [%]
25	3.36	25	3.36
50	2.68	50	2.68
100	2.00	100	1.92
200	1.60	200	1.40
399	0.96	400	0.88
782	0.72	780	0.84
1365	0.68	1406	0.76
1853	0.68	2292	0.64
2358	0.80	2373	0.80

Tabelle 3. Gegenüberstellung der Ergebnisse verschiedener Gruppen

Gruppe	Methode	Trainingscorpus	Fehlerrate [%]
Max-Planck-Institut [1]	Support-Vektoren	Train89	1.0
	Neuronales Netz	Train89	9.0
Daimler-Benz AG [2]	polynom. Klassifikator	Train89	1.68
		Train400	0.80
diese Arbeit	Mischverteilungen	Train89	1.72
		Train400	0.64

Literatur

1. V. Blanz, B. Schölkopf, H. Bülthoff, C. Burges, V. Vapnik, T. Vetter, "Comparison of View–Based Object Recognition Algorithms using Realistic 3D Models," C. von der Malsburg, W. von Seelen, J.C. Vorbrüggen, B. Sendhoff (eds.): *Artificial Neural Networks - ICANN'96* , Springer Lecture Notes in Computer Science Vol. 1112, S.251-256, Berlin, 1996.
2. U. Kressel, persönliche Mitteilung, Daimler-Benz-AG Research and Technology, Abteilung F3M/T, Ulm, März 1998.
3. R.O. Duda, P.E. Hart, *Pattern Classification and Scene Analysis*, John Wiley & Sons, 1973.
4. K. Fukunaga, *Introduction to Statistical Pattern Recognition*, Academic Press, San Diego, 1990.
5. W.H. Press, S.A. Teukolsky, W.T. Vetterling, B.P. Flannery, *Numerical Recipes in C*, University Press, Cambridge, 1992.
6. A.P. Dempster, N.M. Laird, D.B. Rubin, "Maximum Likelihood from Incomplete Data via the EM Algorithm," *Journal of the Royal Statistical Society*, 39(B), 1-38, 1977.
7. H. Ney, "Acoustic Modelling of Phoneme Units for Continuous Speech Recognition," L. Torres, E. Masgrau, M.A. Lagunas (eds.): *Signal Processing V: Theories and Applications*, Elsevier Science Publishers B.V., 1990.

Optimierung der Konvergenzgeschwindigkeit von Backpropagation

R. Linder, S. J. Pöppl
Institut für Medizinische Informatik der Medizinischen Universität zu Lübeck

Zusammenfassung

Für Backpropagation-Netzwerke wird eine neue Methode zur neuronenspezifischen Anpassung der Lernrate vorgestellt, welche eine deutliche Beschleunigung der Konvergenz insbesondere für große und vielschichtige Netzwerke erlaubt. Gezeigt wird dies am Zwei-Spiralen-Benchmark sowie an zwei Real-world-Problemen aus Medizintechnik und Physik.

Einleitung

1986 publizierte Rumelhart [24] den bereits zuvor von Robbins, Werbos und Parker beschriebenen Backpropagation-Algorithmus für die Gewichtsanpassung mehrschichtiger Feedforward-Netzwerke. Die meisten künstlichen neuronalen Netze (KNN) basieren heute auf Backpropagation (BPN) und seinen vielfältigen Varianten. Einer der wichtigsten Nachteile dieses Algorithmus liegt in der vergleichsweise geringen Konvergenzgeschwindigkeit, die v.a. bei Einsatz mehrerer verdeckter Schichten das Antrainieren eines KNN sehr erschwert. Seit der Einführung von BPN werden in der Literatur entsprechende Lösungsvorschläge diskutiert (siehe Reviews [1, 2, 16]). Neben einer optimierten Initialisierung der Gewichte, modifizierten Fehlerfunktionen und der Reduzierung von *k*-Klassenproblemen auf *k* Zwei-Klassenprobleme wurde eine Vielzahl weiterer Optimierungsmöglichkeiten publiziert und teilweise in Kombination eingesetzt. Es wurden direkte Methoden zur Festlegung der Gewichte entwickelt und mit iterativen Verfahren kombiniert. Intensiv wurde über eine Anpassung der Lernrate nachgedacht [17, 33], auch evolutionäre Algorithmen wurden dafür eingesetzt [11]. Entweder wird die Lernrate generell angepaßt [30] oder gewichtspezifisch [9]. In der einfachsten Variante wird die Lernrate durch den Input des Zielneurons einer gewichteten Verbindung dividiert [27]. Andere Ansätze berücksichtigen die Anzahl der Verknüpfungen von der vorhergehenden und zur nachfolgenden Schicht [7, 29] oder ersetzen den Verhulst-Term $a*(1-a)$ durch eine Konstante [3]. Insgesamt erwiesen sich alle vorgenannten Optimierungsversuche mit Ausnahme von Quickprop [5] und RPROP [23] nicht generell als effektiv [25].

In der vorliegenden Arbeit wird ein neuer, sehr einfacher Algorithmus vorgestellt, welcher über eine neuronenspezifische Anpassung der Lernrate (NSAL) die Konvergenzgeschwindigkeit von BPN - auch unter Verwendung vieler verdeckter Schichten - deutlich erhöht. Im folgenden soll unter Konvergenzgeschwindigkeit dabei nicht die erforderliche Anzahl an Rechenschritten, sondern an Lernepochen verstanden werden. Dazu berechtigt die rasche Entwicklung im Bereich der Hardware, die bei einer maximal parallelisierten Abarbeitung der Rechenschritte [19] die Lernzeit maßgeblich mit der Anzahl der erforderlichen Epochen korrelieren läßt. Am Beispiel eines Benchmark-Tests und zweier Real-world-Probleme soll gezeigt werden, daß BPN mit NSAL gegenüber Standard-BPN nur vergleichsweise wenige Lernepochen benötigt.

Methodik

Backpropagation (BPN)

Im folgenden wurde für BPN neben einer Konstanten c_1 für die Lernrate l eine Konstante c_2 als Momentum-Term verwendet. Entsprechend den Empfehlungen aus der Literatur galt $c_2 = 0{,}9$, wohingegen c_1 für die jeweilige Fragestellung empirisch optimiert wurde. Die Gewichte wurden stets im Intervall [-0,01; +0,01] nach einer von IBM beschriebenen Methode [8] initialisiert. Als Transferfunktion wurde ausschließlich die Sigmoid-Funktion verwendet. Die Bias-Neuronen hatten einen konstanten Output von 1,0.

Die Neuanpassung der Gewichte erfolgte jeweils am Ende einer Lernepoche, d.h. kummuliert nach der Präsentation aller Lernmuster. Diese Vorgehensweise wird auch als Fullpropagation, offline oder batch-mode Technik bezeichnet [2, 16]. Sie ist eine notwendige Voraussetzung für den Einsatz von NSAL. Um NSAL besser erläutern zu können, sollen im folgenden kurz die Grundzüge von BPN vorgestellt und erforderliche Variablenbezeichnungen eingeführt werden.

Für ein dreischichtiges Feedforward-Netzwerk mit einer Inputschicht i, einer verdeckten Schicht v und einer Outputschicht o sei die Anzahl der Neuronen n pro Schicht s gleich f_s, die Anzahl aller Muster sei g. BPN berechnet die Fehler δ der einzelnen Neurone von der Output- hin zur Inputschicht. Für jedes Muster m ergibt sich der Fehler der Outputneurone als Differenz aus dem vorgegebenen Zieloutput z_o und der tatsächlichen neuronalen Aktivität a_o der Outputneurone.

$$\delta_{o,n,m} = z_{o,n,m} - a_{o,n,m} \tag{1}$$

Bei Verwendung der Sigmoid-Funktion ist der Fehler für die verdeckten Elemente gegeben durch

$$\delta_{v,n,m} = a_{v,n,m} \times \left(1 - a_{v,n,m}\right) \times \sum_{j=1}^{f_o} (\delta_{o,j,m} \times w_{v,i,o,j,e}) \tag{2}$$

wobei $w_{v,i,o,j,e}$ das Gewicht zwischen dem i-ten Neuron der verdeckten Schicht und dem j-ten Neuron der Outputschicht in der Epoche e sei. Die Anpassung der Gewichte erfolgt über

$$w_{v,i,o,j,e} = w_{v,i,o,j,e} + \Delta w_{v,i,o,j,e} \tag{3}$$

mit

$$\Delta w_{v,i,o,j,e} = l_{v,i,o,j} \times \sum_{m=1}^{g} (\delta_{o,n,m} \times a_{v,n,m}) + c_2 \times \Delta w_{v,i,o,j,e-1} \quad \Big| \; l_{v,i,o,j} = c_1 \tag{4}$$

Entsprechendes gilt für die Anpassung der Gewichte zwischen Input- und verdeckter Schicht sowie Input- und Outputschicht. Sind die Neuronen jeder Schicht mit allen

Neuronen der benachbarten Schichten verbunden, spricht man von einer *vollständigen Verknüpfung*. Bei den nachfolgenden KNNs bestehen darüber hinaus alle schichtübergreifenden Verbindungen, d.h. eine *totale Verknüpfung*.

Neuronenspezifische Anpassung der Lernrate (NSAL)
So, wie in einem Sozialgefüge der einzelne wie auch die Gesamtheit davon profitieren, wenn sich die einzelnen Individuen bevorzugt an erfolgreichen Individuen ausrichten, besteht der Grundgedanke von NSAL darin, die Gewichte derjenigen Verbindungen bevorzugt zu adaptieren, welche zu "erfolgreichen" Neuronen führen. Als Maß für den Erfolg (engl. success) S soll gelten:

$$S_{s,n} = \frac{1}{\sqrt{\sum_{m=1}^{g} \delta_{s,n,m}^2}} \quad \Big| S_{\max} = 10^{10} \tag{5}$$

Wie sich in Vorversuchen zeigte, haben Outputneurone im Mittel fünffach größere Fehlerbeträge als verdeckte Neuronen. Die Konvergenz läßt sich beschleunigen, wenn $S_{o,n}$ für Outputneurone durch eine Konstante $c_3 = 5$ dividiert wird. Dadurch kann c_1 größer gewählt werden. Die NSAL für Gewichte von Verknüpfungen, die zu Neuronen in der Outputschicht führen, berechnet sich damit wie folgt:

$$l_{v,i,o,j} = c_1 \times \frac{S_{o,j}}{c_3} \tag{6}$$

Mit zunehmendem Lernerfolg verringern sich die Beträge der Fehler δ, so daß der Erfolg S entsprechend Gleichung (5) schnell zunimmt. Um die Einstellung der Gewichte nicht zu sehr zu beschleunigen (und damit die Konvergenz zu gefährden), wurden die Lernraten zusätzlich mit dem im Lernverlauf abnehmenden Fehler E der Ausgabeneuronen multipliziert.

$$E = \sum_{i=1}^{m} \sum_{j=1}^{n_o} \left(z_{o,n,m} - a_{o,n,m}\right)^2 \tag{7}$$

Außerdem wurden die Lernraten mit einem für jede Verknüpfung konstanten Zufallswert im Intervall [0; 1] multipliziert, um die Robustheit des Algorithmus zu verbessern.
Für die folgenden Fragestellungen wurden Prototypen mit Microsoft® Visual C++ geschrieben, die Ausführung erfolgte auf einem Pentium II - Prozessor.

Das Zwei-Spiralen-Problem
Dieser Benchmark-Test ist ein Zwei-Klassenproblem von Alexis Wieland, bei dem je 97 Punkte (x, y) auf zwei ineinander verschränkten Spiralen nach ihrer Spiralenzugehörigkeit klassifiziert werden müssen [13], siehe Abb. 2 links. Aufgrund seiner extremen Nicht-Linearität zählt er zu den schwierigsten Benchmarks für KNNs.
Verwendet wurde ein 12schichtiges BPN-Netzwerk mit zwei Inputneuronen, pro

verdeckter Schicht 20 Neuronen und einem Ausgabeneuron mit dem Zieloutput 1 für Punkte der ersten Spirale und 0 für Punkte der zweiten Spirale. Mit einem Output ≥ 0,5 wurde ein Punkt der ersten Spirale, andernfalls der zweiten zugeordnet. Die Anlernphase mit NSAL wurde gestoppt, wenn die Differenz zwischen Ziel- und Ist-Output für alle 194 Punkte weniger als 0,4 betrug. c_l = 100/194. Es wurden 100 KNNs antrainiert mit jeweils unterschiedlichen Gewichtsinitialisierungen.

Klassifikation gasförmiger Substanzgemische
Ausgangspunkt waren olfaktorische Messungen mittels PEN (portable electronic nose) der WMA Airsense Analysentechnik GmbH. Gemessen wurde das Gasgemisch in dem sich über einer Probe befindlichen Dampfraum. Die Messung erfolgte mit 10 unterschiedlichen Halbleiter-Gassensoren, welche auf das Vorhandensein gasförmiger Verbindungen mit einer Änderung ihres elektrischen Widerstandes reagieren (reelle Zahlen). Untersucht wurden 13 verschiedene Bakterien- und Pilzstämme (13 Klassen). Jede Klasse wurde durch 7-8 Proben repräsentiert, insgesamt wurden 100 Proben untersucht.
Ein total verknüpftes 10-8-8-13 BPN mit NSAL wurde ebenso wie ein Standard-BPN derselben Topologie mit den o.g. Widerstandswerten nach der Leaving-one-out-Methode trainiert und validiert. Für die Klassifikation des Outputs galt das Prinzip „Winner takes it all". c_l = 10/100 (BPN mit NSAL) bzw. c_l = 20/100 (Standard-BPN). Trainiert wurde über 100 Epochen.

Interpretation von Mößbauerspektren
Gegenstand der Untersuchung waren diamagnetische Mößbauerspektren, welche durch die folgenden drei Parameter gekennzeichnet sind: die Quadrupolaufspaltung ΔE_Q, die Feldstärke des angelegten Magnetfeldes B (beide bekannt) und der mittels KNN zu bestimmende Asymmetrieparameter η. Mit einem Kern-Hamilton-Operator-Formalismus wurden 14.000 realitätsnah verrauschte Spektren simuliert: 10.000 zum Anlernen des KNN (Lerndaten), 2.000 zur Überprüfung der Generalisierungsleistung im Rahmen des stopped training [26] (Generalisierungsdaten) und weitere 2.000 zum Testen der Klassifikationsleistung (Testdaten). Für die Simulation der Trainingsdaten wurden die Werte für ΔE_Q, B und η in gleichmäßigen Abständen aus vorgegebenen Intervallen genommen, für die Generalisierungs- und Testdaten wurden die Parameter innerhalb der Intervallgrenzen mit einem Zufallszahlengenerator gewählt. Jedes Spektrum wurde durch 6 Lorentzlinien vorgegebener Linienbreite angepaßt. Dazu wurden mit einem Davidson-Fletcher-Powell-Minimierungsverfahren nach Wagner [31] jeweils Position und Intensität der Linien bestimmt. Die Eingabevektoren für das KNN bestanden somit aus 6 * 2 = 12 Parametern, ΔE_Q und B. Soll-Output war η, normalisiert im Intervall [0,4; 0,6], so daß sich die Funktionswerte des Outputs auf den weitgehend linearen Abschnitt der Sigmoid-Funktion beschränkten. Outputs außerhalb dieses Intervalls erhielten den Wert der entsprechenden Intervallgrenze.
Als KNNs wurden ein total verknüpftes 14-20-1 BPN mit NSAL ebenso wie ein Standard-BPN derselben Topologie eingesetzt. c_l = 1,0/10.000 (BPN mit NSAL) bzw. c_l = 2,0/10.000 (Standard-BPN). Trainiert wurde über 10.000 Epochen.

Ergebnisse

Sowohl bei den BPN-Netzwerken mit als auch ohne NSAL kam es in allen Fällen zur Konvergenz. Für das Zwei-Spiralen-Problem konnte sie mit 97 Epochen am schnellsten erreicht werden (maximal 372 Epochen), der Mittelwert betrug 141,92 Epochen, die Standardabweichung 36,97 Epochen. Abb. 1 zeigt das Lernverhalten im Vergleich zu zwei Trainingsphasen eines total verknüpften Standard-2-5-5-5-1-BPN-Netzwerkes von Lang und Witbrock [13].

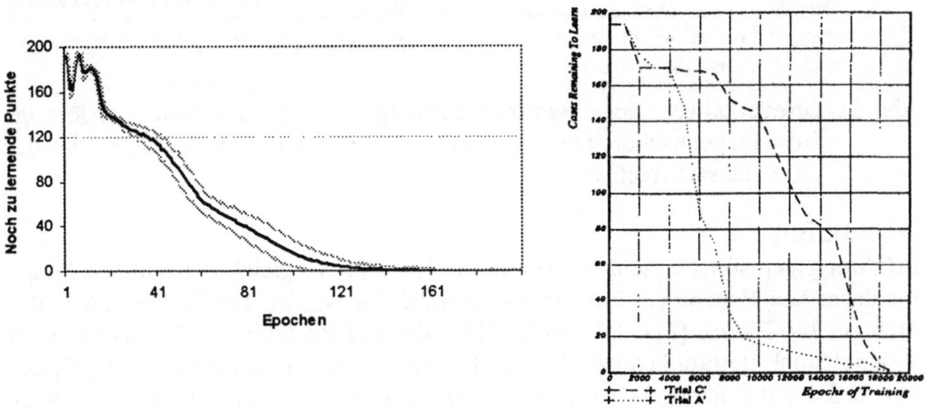

Abb. 1: Lernverhalten mit Angabe der Standardabweichungen (links) im Vergleich zum Standard-BPN von Lang und Witbrock [13] (rechts)

Zur Schätzung der Generalisierungsleistung wurden unterschiedliche Punkte als Input vorgegeben und die jeweilige Aktivität des Outputneurons weiß (Aktivität > 0,5) oder andernfalls schwarz dargestellt, siehe Abb. 2.

Abb. 2: Links: 2 Spiralen (Klassen) mit je 97 Punkten (x, y).
Mitte: Aktivität des Outputneurons für Inputs von $-7 \leq x \leq 7$ und $-7 \leq y \leq 7$ nach 1.900 Epochen im Vergleich zum Standard-BPN von Lang und Witbrock [13] nach 19.000 Epochen (rechts)

Das Lernverhalten der beiden Real-world-Probleme ist in Abb. 3 dargestellt.

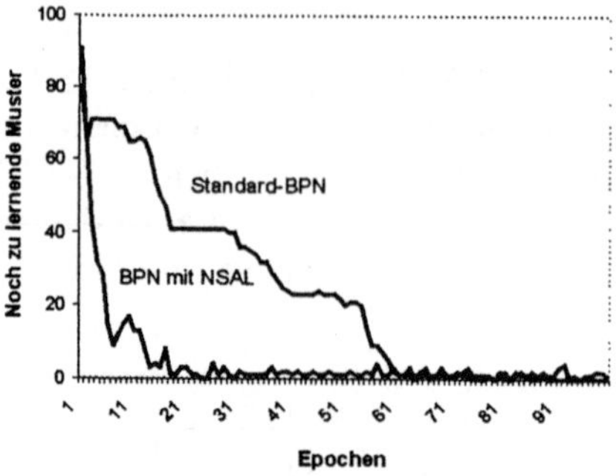

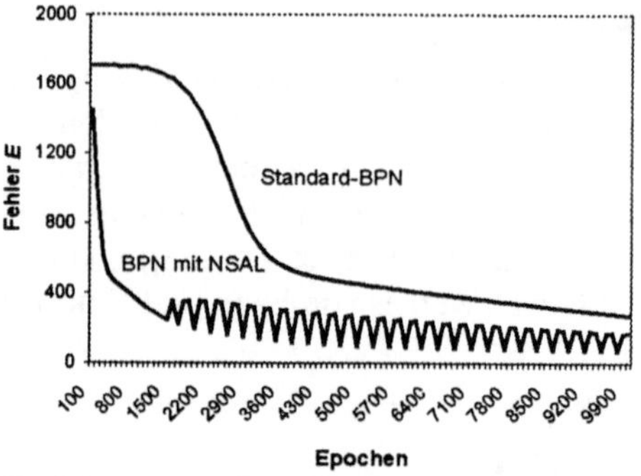

Abb. 3: Lernverhalten der Real-world-Probleme: Klassifikation gasförmiger Substanzgemische (links) und Fehler E der Mößbauerspektren aus dem Testdatensatz (rechts)

Diskussion

Erfahrungsgemäß ist es sehr schwer, in der Literatur vergleichbare Untersuchungen für denselben Benchmark-Test zu finden [20]. So wurden für das Zwei-Spiralen-Problem 146 Punkte [21], 150 Punkte [10] oder 192 Punkte [12, 15] verwendet, ein anderes Stopkriterium benutzt [18], die Konvergenzgeschwindigkeit nicht [22] oder nur in Sekunden angegeben [4]. Verglichen werden kann der von BPN mit NSAL erzielte Mittelwert von 142 Epochen mit den gemittelten Ergebnissen von Lang und Witbrock [13] für Standard-BPN (20.067 Epochen), BPN mit Cross Entropy (10.800 Epochen), Quickprop (7.867 Epochen), dem Ergebnis von Fahlmann und Lebiere [6] für Counterpropagation (1.700 Epochen) sowie nicht publizierten Ergebnissen für Standard-BPN von Frostrom (13.900 Epochen) und Leighton (4.010 Epochen) [25]. Mit 18.783 Gewichten bei 194 Inputmustern ist die hier vorgestellte Netzwerktopologie deutlich überdimensioniert. Einige Autoren befürworten den Einsatz überdimensionierter Netzwerke bei Verwendung von stopped training [14, 26, 32]. Möglicherweise werden - vergleichbar mit einem Pruning-Algorithmus - durch NSAL viele Verbindungen, die zu Neuronen mit sehr hohen Delta-Fehlern führen, funktionell ausgeschaltet. Dies gilt es noch eingehender zu untersuchen. Am Beispiel der beiden Real-world-Probleme konnte gezeigt werden, daß BPN mit NSAL gegenüber Standard-BPN hinsichtlich der Konvergenzgeschwindigkeit und im Fall der Mößbauerspektren in bezug auf die Generalisierungsleistung bevorteilt ist. Da Standard-BPN mit vielen verdeckten Schichten große Konvergenzprobleme hat, wurden die Vergleiche zwischen BPN mit und ohne NSAL für die Real-world-Probleme mit nur ein bzw. zwei verdeckten Schichten durchgeführt. Bei Einsatz von NSAL sollte man sich jedoch die Vorteile weiterer verdeckter Schichten [28] nutzbar machen. Wie ein weiterer hier nicht näher beschriebener Ansatz zeigte, konvergierte ein 20schichtiges, nicht schichtübergreifend verknüpftes BPN-Netzwerk mit NSAL für das XOR-Problem. Diese Ergebnisse berechtigen zu der Hoffnung, mit NSAL ein Werkzeug gefunden zu haben für große und v.a. vielschichtige Netzwerktopologien, das einhergehend mit den Neuentwicklungen im Bereich der Multiprozessor-Architekturen im Sinne von VLSI-Implementationen oder optoelektronischen Modulen ein performantes Antrainieren mächtiger KNNs erlaubt.

Danksagung

Für die freundliche Unterstützung und Bereitstellung des Datenmaterials danken wir besonders Herrn Dr.-Ing. Andreas Walte von der WMA Airsense Analysentechnik GmbH aus Schwerin sowie Herrn Dr. rer. nat. Hauke Paulsen vom Institut für Physik der Medizinischen Universität zu Lübeck.

Literatur

[1] Alpsan D, Towsey M, Ozdamar O, Tsoi AC, Ghista DN: Efficacy of Modified Backpropagation and Optimisation Methods on a Real-world Problem. Neural Networks 8(6):945-962, 1995.

[2] Barnard E, Holm JEW: A comparative study of optimization techniques for backpropagation. Neurocomputing 6:19-30, 1994.

[3] Chen JR, Mars P: Stepsize variation methods for accelerating the back propagation algorithm. Proceedings of the International Joint Conference on Neural Networks, Washington, DC, Vol. 1, 601-604, 1990.

[4] Chen K, Yang L, Yu X, Chi H: A self-generating modular neural network architecture for supervised learning. Neurocomputing 16:33-48, 1997.

[5] Fahlmann SE: Faster-Learning Variations on Back-Propagation: An Empirical Study. In: Touretzky D, Hinton GE, Sejnowski T (Hrsg.): Proc. Of the 1988 Connectionist Models Summer School, Morgan Kaufmann, 38-51, 1989.

[6] Fahlman SE, Lebiere Ch: The cascade-correlation learning architecture. In Touretzky DS (Hrsg.): Advances in Neural Information Processing Systems 2, Morgan Kaufmann, 524-532, 1990.

[7] Higashino J, de Greef BL, Persoon EH: Numerical analysis and adaptation method for learning rate of back propagation. Proceedings of the International Joint Conference on Neural Networks, Washington, DC, Vol. 1, 627-630, 1990.

[8] IBM: Reference manual: Random number generation and testing. C20-8011, 1959.

[9] Jacobs R: Increased Rates of Convergence Through Learning Rate Adaption. Neural Networks 1, 295-307, 1988.

[10] Karayiannis NB: Accelerating the Training of Feedforward Neural Networks Using Generalized Hebbian Rules for Initializing the Internal Representations. IEEE Trans. Neural Networks 7(2):419-426, 1996.

[11] Kim HB, Jung SH, Kim TG, Park KH: Fast learning method for back-propagation neural network by evolutionary adaptation of learning rates. Neurocomputing 11:101-106, 1996.

[12] Kuncheva LI: Initializing of an RBF network by a genetic algorithm. Neurocomputing 14:273-288, 1997.

[13] Lang KJ, Witbrock MJ: Learning to Tell Two Spirals Apart. In: Touretzky D, Hinton GE, Sejnowski T (Hrsg.): Proc. Of the 1988 Connectionist Models Summer School, Morgan Kaufmann, 52-59, 1989.

[14] Lawrence S, giles CL, Tsoi AC: What size neural network gives optimal generalization? Convergence properties of backpropagation. Technical Report UMIACS-TR-96-22 and CS-TR-3617, Institute for Advanced Computer Studies, University of Maryland, College Park, MD 20742

[15] Lengellé R, Denceux T: Training MLPs Layer by Layer Using an Objective Function for Internal Representations. Neural Networks 9(1):83-97, 1996.
[16] Looney CG: Stabilization and speedup of convergence in training feedforward neural networks. Neurocomputing 10:7-31, 1996.
[17] Magoulas GD, Vrahatis MN, Androulakis GS: Effective Backpropagation Training with Variable Stepsize. Neural Networks 10(1):69-82, 1997.
[18] Osowski S, Bojarczak P, Stodolski M: Fast Second Order Learning Algorithm for Feedforward Multilayer Neural Networks and ist Applications. Neural Networks 9(9):1583-1596, 1996.
[19] Petrowski A, Dreyfus G, Girault C: Performance Analysis of a Pipelined Backpropagation Parallel Algorithm. IEEE Trans. Neural Networks 4(6):970-981, 1993.
[20] Prechelt L: Some notes on neural learning algorithm benchmarking. Neurocomputing 9:343-347, 1995.
[21] Rathbun TF, Rogers SK, DeSimio MP, Oxley ME: MLP iterative construction algorithm. Neurocomputing 17:195-216, 1997.
[22] Ridella S, Rovetta S, Zunino R: Circular Backpropagation Networks for Classification. IEEE Trans. Neural Networks 8(1):84-97, 1997.
[23] Riedmiller M, Braun H: A direct adaptive method for faster backpropagation learning: the RPROP algorithm, Proc. 1993 IEEE Int. Conf. Neural Networks, vol 1, San Franzisco 586-591, 1993.
[24] Rumelhart DE, Hinton GE, Williams RJ: Learning Representations by Back-Propagating Errors. Nature 323:533-536, 1986.
[25] Sarle, W.S., ed. (1997), Neural Network FAQ, part 2 of 7: Learning, periodic posting to the Usenet newsgroup comp.ai.neural-nets, URL: ftp://ftp.sas.com/pub/neural/FAQ.html
[26] Sarle WS: Stopped Training and Other Remedies for Overfitting. Proceedings of the 27th Symposium on the Interface of Computer Science and Statistics, 352-360 ,1995.
[27] Sietsma J, Dow RJF: Creating artificial neural networks that generalize. Neural Networks 2: 67-79, 1991.
[28] Tamura S, Tateishi M: Capabilities of a Four-Layered Feedforward Neural Network: Four Layers Versus Three. IEEE Trans. Neural Networks 8(2):251-255, 1997.
[29] Tesauro G, Janssens B: Scaling relationships in back propagation learning. Complex Systems 2:39-44, 1988.
[30] Vogl TP, Mangis JK, Zigler AK, Zink WT, Alkon DL: Accelerating the convergence of the back-propagation method, Biol. Cybernet. 59(4/5):257-264, 1988.
[31] Wagner F, Inst. f. Theoretische Physik, Christian-Albrechts-Universität zu Kiel
[32] Weigend A: On overfitting and the effective number of hidden units. Proceedings of the 1993 Connectionist Models Summer School, 335-342, 1994.
[33] Yu X-H, Chen G-A, Cheng S-X: Dynamic Learning Rate Optimization of the Backpropagation Algorithm. IEEE Trans. Neural Networks 6(3):669-677, 1995.

Aktive Exploration

Wissensbasierte aktive Szenenanalyse

D. Paulus, U. Ahlrichs, B. Heigl, H. Niemann

Lehrstuhl für Mustererkennung (LME, Informatik 5)
Martensstr. 3, Universität Erlangen–Nürnberg, 91058 Erlangen
Tel.: +49 (9131) 85–7894 — Fax: +49 (9131) 303811
paulus@informatik.uni-erlangen.de
http://www5.informatik.uni-erlangen.de

Übersicht In diesem Beitrag wird ein Gesamtsystem zur wissensbasierten Szenenexploration vorgestellt, das aus drei Hauptkomponenten besteht. Positionen von Objekten werden datengetrieben aufgrund ihrer Farbe hypothetisiert; Farbe wird auch zur Bildsegmentierung eingesetzt. Eine zweite Komponente, die dem aktiven Sehen zuzuordnen ist, steuert eine Schwenk/Neige–Kamera, die auf einem Linearschlitten montiert ist; hiermit wird zum einen eine Schätzung der Entfernung der Objekte von der Kamera errechnet, zum anderen werden anschließend die Objekthypothesen formatfüllend aufgenommen. Ein drittes Modul zur wissenbasierten Bildanalyse verwendet eine Objektrepräsentation in Form von semantischen Netzen um abschließend mittels einer explizit repräsentierten Explorationsstrategie die einzelnen Objekte zu erkennen. Experimente in einer Büorumgebung belegen, daß diese Architektur dazu verwendet werden kann, flexibel Objekte zu erkennen.

1 Einleitung

Das hier vorgestellte Projekt zählt in seiner Ausrichtung zur wissensbasierten Bildanalyse und zum aktiven Sehen und kombiniert zahlreiche Module. Ziel ist es, eine Szene aktiv zu erkunden und dabei datengetriebene und wissensbasierte Techniken miteinander so zu verbinden, daß eine gezielte Exploration aufgrund von explizit repräsentiertem Wissen über Objekte und Strategie möglich ist. Als mögliches Anwendungsgebiet ist hierbei an mobile Systeme gedacht, die aufgrund visueller Information navigieren und agieren. Hierzu ist es nützlich, Aktionen wie beispielsweise die Blickrichtungsänderung oder Geschwindigkeitsregelung explizit in der Wissensbasis zu repräsentieren. Die Strategie, mit der die Aktionen ausgewählt werden, sollte ebenfalls wählbar sein.

In der vorgestellten Anwendung wird mit einer rechnergesteuerten Kamera, die in der Ecke eines Raumes auf einem Linearschlitten montiert ist, Tiefe berechnet. Die Position einiger Objekte, die dem System zuvor isoliert präsentiert wurden, wird zunächst aufgrund ihrer Farbe hypothetisiert. Unter Verwendung der Tiefeninformation werden diese Bereiche in der Szene vergrößert aufgenommen. Die so ermittelten Detailaufnahmen werden mit Modellen für Objekte verglichen, um falsche Hypothesen auszuschalten. Das System gestattet eine einheitliche Repräsentation von Wissen über die Szene

[1] Diese Arbeit wurde unterstützt durch die *Deutsche Forschungsgemeinschaft* im Rahmen des Sonderforschungsbereichs 182.

und über diejenigen Aktionen, die zur Exploration benötigt werden. Hierzu zählen die Auswahl der Bildverarbeitungsmodule, die Einstellung der externen Kameraparameter und die Strategie, mit der Objekte gesucht werden.

Eine umfangreiche Literatursammlung zum Thema „wissensbasierte Bildanalyse" in [2] enthält wenige Zitate europäischer Autoren, obwohl zahlreiche Systeme hier veröffentlicht sind, z.B. in [8, 12, 13]. In [7] wird beispielsweise ein Gesamtsystem vorgestellt, das durch seine Verwendung von semantischen Netzen zur Wissensrepräsentation dem hier vorgestellten ähnlich ist; Unterschiede ergeben sich für die einzelnen Komponenten ebenso wie in der gewählten Beispielanwendung.

Im folgenden ist das Anwendungsszenario eine Büroumgebung; typische Objekte sind in Bild 1 dargestellt, wobei *drei* rote Objekte gewählt wurden, um eine Identifikation allein aufgrund der Farbe (Abs. 2) unmöglich zu machen. Errechnete Tiefenwerte und segmentierte Detailaufnahmen (Abs. 3) dienen als Eingabe zur wissensbasierte Überprüfung der Hypothesen (Abs. 4). Ergebnisse und Experimente werden in Abs. 5 vorgestellt.

Bild 1. Objekte der Büroumgebung (siehe Hinweis in der Fußnote 0).

2 Generierung von Objekthypothesen

In [14] wird das Verfahren der Histogramm–Rückprojektion eingeführt, mit dem Hypothesen für die Position von Objekten in 2D Bildern gebildet werden. Das Verfahren bedient sich zweier Farb–Histogramme und sucht in einem Farbbild $[\boldsymbol{f}_{ij}]_{1\leq i\leq M, 1\leq j\leq N}$ nach möglichen Positionen eines Objekts, das durch sein Farbhistogramm $\boldsymbol{T} = [T_k]_{k=1\ldots K}$ charakterisiert ist. Zusätzlich wird noch eine Annahme für die Größe des Objekts im Bild verwendet, die durch eine Maske $\boldsymbol{D}_r$ angegeben wird. Zur Schätzung der Größe von $\boldsymbol{D}_r$ kann die in Abs. 3 ermittelte Tiefeninformation verwendet werden. Die Funktion h bildet einen Farbvektor $\boldsymbol{f}$ auf den Index $k = 1 \ldots K$ im Histogramm ab und erlaubt somit eine beliebige Quantisierung des Histogramms. Das Prinzip des Verfahrens ist in Bild 2 als Struktogramm dargestellt. Für das Anwendungsbeispiel findet sich unten in Bild 7 eine Rückprojektion und Faltung mit der Maske $\boldsymbol{D}_r$ für das erste Objekt aus Bild 1.

---

[0] Die Farbbilder sind im WWW unter der Adresse `http://www5.informatik.uni-erlangen.de/Images/` zu finden.

Gegeben: Bild $[\boldsymbol{f}_{ij}]$, Modellhistogramm $\boldsymbol{T}$ eines Objekts,
Gesucht Position des Objekts (i_t, j_t)

Berechne Farbhistogramm $\boldsymbol{H} = [\boldsymbol{H}_k]_{k=1\ldots K}$ vom aktuellen Bild

FOR Jedes Histogramm "Bin" $k \in \{1, \ldots, K\}$

  $R_k = \min\{\frac{T_k}{H_k}, 1\}$ (Bilde Verhältnishistogramm $\boldsymbol{R} = [R_k]_{k=1\ldots K}$)

FOR Alle Positionen (i, j) im Bild

  $A_{i,j} := R_{h(\boldsymbol{f}_{i,j})}$, wobei $\boldsymbol{f}_{i,j}$ den Farbvektor an der Stelle (i, j) bezeichnet

$\boldsymbol{B} := \boldsymbol{D}_r \star \boldsymbol{A}$, wobei $\star$ die Faltung bezeichnet

$(i_t, j_t) := \text{argmax}_{i,j}(B_{i,j})$

Bild 2. Objektlokalisierung mit Histogramm–Rückprojektion nach [14].

3 Einzelobjekte

Das folgende Ziel ist es nun, die Hypothesen mit der Schwenk/Neige/Zoom–Kamera so anzusteuern, daß jede Objekthypothese formatfüllend und zentriert dargestellt wird. Hierzu soll der Bildbereich einer vorgegebenen Breite in der Szene entsprechen, wozu die Ermittlung von Tiefe notwendig wird.

Die Positionen der zu untersuchenden Hypothesen seien definiert durch die Bildkoordinaten der Mittelpunkte der durch die Histogrammrückprojektion gewonnenen Regionen. In einem ersten Schritt wird für jede Hypothese die Kamera so bewegt, daß das Expansionszentrum des Zoomobjektivs mit diesem ObjektMittelpunkt übereinstimmt. Die hierfür erforderliche Kalibrierung der internen Kameraparameter erfolgte bei minimaler Zoom–Einstellung mittels des Verfahrens von Tsai [15].

Zur Tiefengewinnung werden gut verfolgbare farbige Punkte der Szene bei minimaler Brennweiteneinstellung verfolgt [4], während die Kamera auf dem Linearschlitten bewegt wird. Bild 8 zeigt die ermittelten 3D–Punkte. Für eine Objekthypothese wird der in der Projektion nächstliegende 3D–Punkt gesucht und dessen Entfernung als Schätzung für die Entfernung der Hypothese verwendet. Die Zoomeinstellung der Kamera wird so gewählt, daß das Objekt im Bild in einer vorgegebenen Größe erscheint.

Zur Ermittlung der Brennweite in Abhängigkeit von der Zoomeinstellung existiert ein Verfahren Willson [16], was jedoch zu praktischen Problemen führt, da für alle Zoom– und Fokuseinstellungen ein Kalibriermuster so im Bild plaziert werden muß, daß gleichzeitig eine genügend genaue Segmentierung möglich ist. Anstelle dieses Verfahrens wird hier der Vergrößerungsfaktor zwischen zwei sich nur gering unterscheidenden Zoomeinstellungen für beliebige Linien eines Bildes experimentell bestimmt, welche sich durch Verbindung von je zwei gut verfolgbaren Punkten ergeben. Die lokalen Vergrößerungsfaktoren werden logarithmiert, anschließend durch eine lineare Funktion approximiert und letztlich über diese Funktion integriert, was zu einer Reduktion und Elimination lokaler Fehler führt, die sich durch Ausmultiplizieren über mehrere Schritte akkumulieren würden. Nach Anwendung der Exponentialfunktion erhält man daraus den Vergrößerungsfaktor gegenüber der minimalen Zoomeinstellung (Bild 3). Die jeweilige Brennweite verhält sich zur Vergrößerung proportional mit der minimialen Brennweite als Proportionalitätsfaktor.

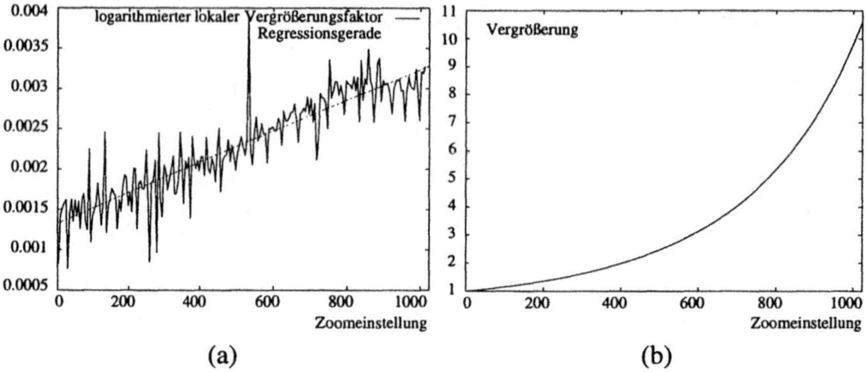

Bild 3. (a): Approximation der logarithmierten lokalen Vergrößerungsfaktoren, (b): geschätzte Vergrößerungsfunktion

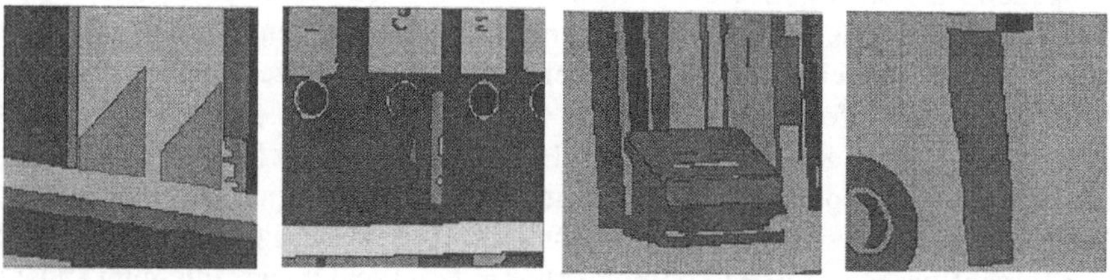

Bild 4. Segmentierung der Hypothesen in der Büroumgebung mit einem auf Farbe angepaßten Split–and–Merge Algorithmus [3]. Die dazugehörigen Ausgangsbilder sind in Bild 6 zu sehen.

4 Verifikation der Objekte

Ziel des Moduls zur Verifikation der Objekthypothesen ist es, die optimale Zuordnung zwischen Regionen, die von einer Segmentierung der Nahaufnahmen geliefert werden (Bild 4), und den grau unterlegten Ovalen in Bild 5 zu bestimmen. Der dargestellte Graph ist ein semantischen Netzes nach der Definition in [8]. Die Bewertung der Zuordnungen erfolgt über die Bewertung des Konzepts „Büroszene", das zur Zeit das Zielkonzept der Analyse ist (s.a. Abs. 6). Für die Berechnung dieser Bewertung wird auf Attribute, Relationen und deren Bewertungen in den Bestandteilen und Konkretisierungen zurückgegriffen. In dem Konzept „roteRegion" wird der euklidische Abstand des mittleren Farbwertes einer Region von einem als rot definierten Farbwert beurteilt. Als Farbraum wird hier der mit der Intensität normierte RGB–Farbraum verwendet.

Die Ähnlichkeit der Regionen mit den durch die Konzepte „Locher", „Abroller" und „Klebestift" repräsentierten Objekte wird über die Höhe und Breite der Objekte bewertet und ist somit in der aktuellen Implementierung abhängig von der Ansicht. Bei dieser Berechnung fließen die in Abschnitt 3 bestimmten Tiefenwerte in die Attributberechnung mit ein.

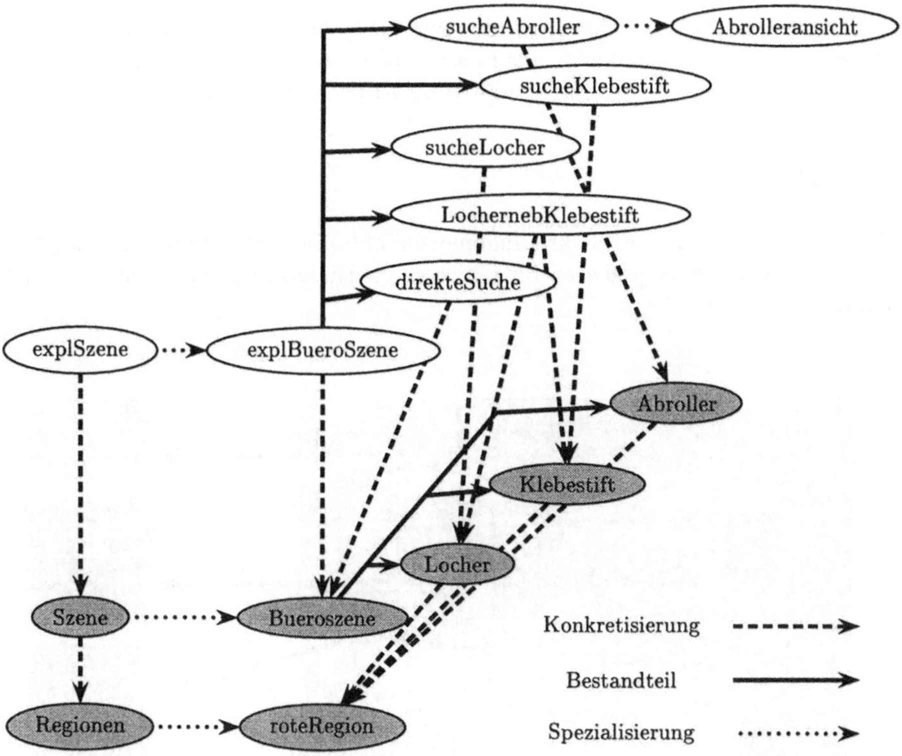

Bild 5. Kombination von Aktionen– und Szenenrepräsentation in einer Wissensbasis; gestrichelte Linien bezeichnen Konkretisierungskanten, gepunktete Spezialisierungskanten und durchgezogene Bestandteilskanten

In den meisten anderen wissensbasierten Systemen wurde bisher lediglich das Wissen über *Objekte* in der Szene genutzt. Durch den Übergang von passiven zu aktiven Systemen besteht nun der Wunsch, auch die Aktionen als aktive Komponenten in die Wissensbasis einzubeziehen. Ziel ist es deshalb, auf Basis des repräsentierten Wissens über die Szene eine *einheitliche* Repräsentation von Aktionen und Objekten zu entwickeln. In Bild 5 sind die Konzepte „sucheAbroller“ zur Lokalisation des Abrollers und „Abrolleransicht“ zur Generierung von neuen Ansichten auf ein Objekt Beispiele dieser Form der Aktionsrepräsentation. Aktionen und Szene werden dabei auf unterschiedlichen Abstraktionsebenen des semantischen Netzes angesiedelt. Die Verbindung zwischen diesen Abstraktionsebenen erfolgt über Konkretisierungskanten, zum Beispiel zwischen „sucheKlebestift“ und „Klebestift“.

Zur Suche der Objekte in dem von den Farbregionen aufgespannten Suchraum wird eine A*–Kontrolle verwendet, die auf der Berechnung des Instantiierungspfades des Netzes basiert. Die Konzepte werden anschließend in der durch den Instantiierungspfad vorgegebenen Reihenfolge bottom–up instantiiert. Dies bedeutet insbesondere, daß gegenwärtig keinerlei Restriktionen während der Analyse von den Konzepten propagiert wird. Die Bewertung der Suchbaumknoten entspricht der des Zielkonzepts, wobei solange kein Zielkonzept vorhanden ist, alle Bewertungen der noch nicht ermittelten Attribute auf Eins gesetzt werden, um eine optimistische Restkostenschätzung zu gewährleisten.

5 Experimente

Bild 6 zeigt links eine Überblicksaufnahme, die mit minimaler Brennweite aufgenommen wurde und vier Hypothesen für Objekte. Im Hintergrund rechts ist ein Teil des Kalibriermusters zu sehen.

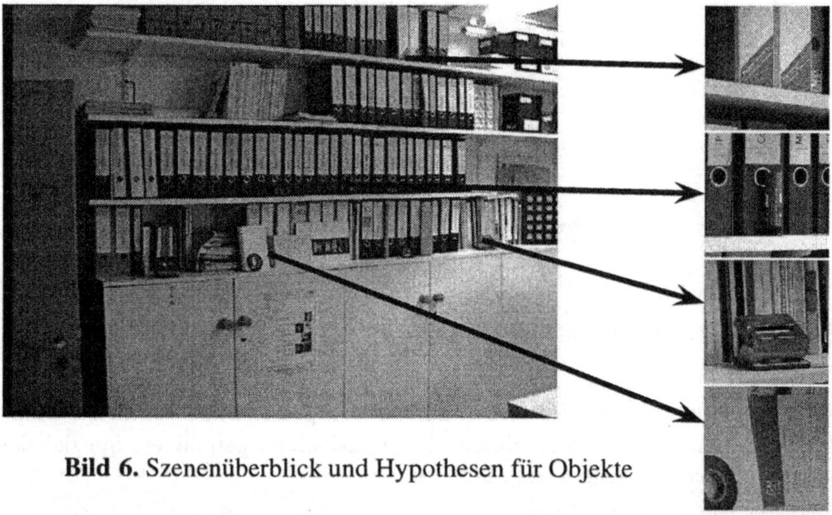

Bild 6. Szenenüberblick und Hypothesen für Objekte

Bild 7. Rückprojektion

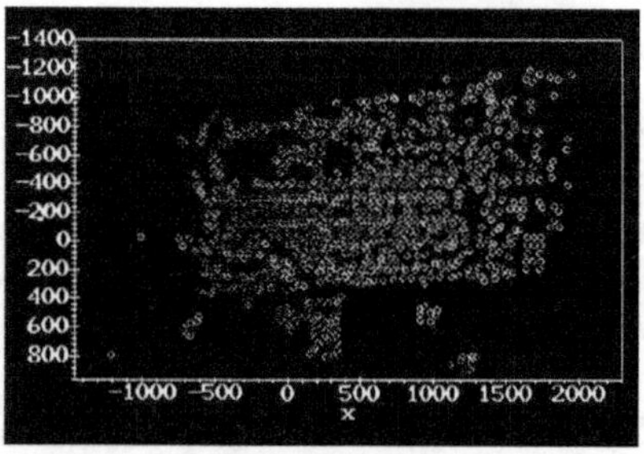

Bild 8. 3D–Punkte

In 17 zu diesem Zweck durchgeführten Versuchen wurden die drei in Bild 1 dargestellten Objekte und der Abroller für das Klebeband in zwei Größen verwendet. Die Objekte waren teilweise in mehrfacher Ausfertigung vorhanden. Durch die datengetriebene Hypothesengenerierung wurden von 90 Objekten 19 nicht erkannt. 17 Objekte kamen in mehr als einer Hypothese gleichzeitig vor.

Um die Größe des Suchbaums auf ungefähr 300–600 Knoten zu reduzieren, wurden in den Experimenten zwei weitere Heuristiken eingeführt, um die roten von den anderen Regionen trennen. Die Farbsegmentierung für die Bestimmung der Regionen lieferte für die verschiedenen Versuche zwischen 54 und 152 Regionen, die an die primitiven Knoten des semantischen Netzes gebunden wurden. Durch die Bewertung der Farbe ließ sich die Zahl jedoch drastisch reduzieren, zum Beispiel von 54 auf 8 Regionen. Bei der Verifikation der Hypothesen wurden bei 7 Versuchen alle Objekte richtig zugeordnet. Insgesamt gesehen wird in 68 % der Fälle die richtige Region einem Objekt zugeordnet. Die Ursache liegt in der hohe Verwechslungsrate für den Locher, der in 7 Fällen für den größeren der beiden Abroller gehalten wurde.

Die Gesamtdauer für einen Verarbeitungszyklus liegt bei ca. 2 min. auf einer SGI O2 (R10000, 195 MHz). Große Teile dieser Zeit werden zur Bewegung der Motoren, mit Warten auf das Erreichen der Zielposition, zur Medianfilterung der Rückprojektion und zur Segmentierung der Farbregionen verwendet.

Bei der Integration der zahlreichen Module zu einem Gesamtsystem war die durchgängige Software–Architektur in objekt–orientierter Programmierung, wie sie in [10] beschrieben wird, von großem Nutzen. Auch der Modul zur wissensbasierten Verarbeitung in semantischen Netzen bedient sich einer Klassenhierarchie in C++.

6 Zusammenfassung und Ausblick

Es wurde ein System vorgestellt, das in drei wesentlichen Komponenten nämlich der Bildung von Objekthypothesen aufgrund von Farbe (Abs. 2), der Fovealisierung auf einzelne Objekte unter Verwendung von 3D–Information (Abs. 3) und der Verifikation mit explizit repräsentierter Suchstrategie und Objektmodellen (Abs. 4) den gesamten Bereich von der Bildaufnahme über die Kamerasteuerung, Tiefenermittlung und Segmentierung bis hin zur aktiven wissensbasierten Bildanalyse abdeckt. Farbbildverarbeitung wurde zur Segmentierung von Objekten und zur Tiefenermittlung eingesetzt. Ideen des aktiven Sehens dienen zur Steuerung der Kamera, wodurch zum einen Tiefe ermittelt wird und zum anderen Objekte formatfüllend ins Blickfeld gerückt werden. Die Wissensbasis wurde durch ein semantisches Netz realisiert, wobei sowohl Objekte als auch Explorationsstrategie und Ziel repräsentiert wurden. Als Beispiel dienten Objekte, die in einem Büro in einer unstrukturierten Umgebung gesucht wurden. Gegenwärtig wird die Wissensbasis zur Verifikation von Objekthypothesen eingesetzt. Damit wurde gezeigt, daß diese Vorgehensweise im Prinzip dazu geeignet ist, Bildanalyse und Szenenexploration miteinander zu verbinden. Die Integration von Aktionen wird zukünftig dadurch erreicht, daß der Knoten "explSzene" in Bild 5 als Zielkonzept der Analyse gewählt wird.

Der Einsatz eines Farbnormierungsverfahrens [9] und seine Auswirkungen auf die Ergebnisse der Rückprojektion und der Farbsegmentierung ist Gegenstand aktueller Untersuchungen. Weiterhin soll das System dahingehend erweitert werden, daß durch den Einsatz einer holistischen Objekterkennung [1, 7] oder mit statistischen Objektmodellen [1, 5, 11] eine Invarianz bezüglich der Beobachtungsrichtung erzielt wird.

Dank

Wir danken den Kollegen im SFB 360 in Bielefeld für die fruchtbaren Diskussionen über die Einsatzmöglichkeiten semantischer Netze zur Repräsentation von Szenenwissen und von Aktionen zur aktiven Bildanalyse.

References

1. U. Büker, J. Dunker, G. Hartmann, E. Seidenberg: *Aktives Sehen für die 3-D Objekterkennung in hybrider Architektur*, in B. Mertsching (Hrsg.): *Aktives Sehen in technischen und biologischen Systemen*, Proceedings in Artificial Intelligence, Infix, Sankt Augustin, Dezember 1996, S. 174–181.
2. D. Crevier, R. Lepage: *Knowledge-Based Image Understanding Systems: A Survey*, *CVIU*, Bd. 67, Nr. 2, August 1997, S. 161–185.
3. J. Denzler, B. Heigl, D. Paulus: *Farbsegmentierung für aktives Sehen*, in V. Rehrmann (Hrsg.): *Erster Workshop Farbbildverarbeitung*, Bd. 15 von *Fachberichte Informatik*, Universität Koblenz–Landau, 1995, S. 9–12.
4. B. Heigl, D. Paulus: *Punktverfolgung in Farbbildsequenzen*, in D. Paulus, T. Wagner (Hrsg.): *Dritter Workshop Farbbildverarbeitung*, IRB-Verlag, Stuttgart, 1997, S. 87–92 & 105.
5. J. Hornegger: *Statistische Modellierung, Klassifikation und Lokalisation von Objekten*, Shaker Verlag, Aachen, 1996.
6. *Proceedings of the International Conference on Image Processing (ICIP)*, IEEE Computer Society Press, Lausanne, Schweiz, September 1996.
7. F. Kummert, G. Fink, G. Sagerer: *Schritthaltende hybride Objektdetektion*, in E. Paulus, F. Wahl (Hrsg.): *Mustererkennung 1997*, Braunschweig, 1997, S. 137–144.
8. H. Niemann: *Pattern Analysis and Understanding*, Bd. 4 von *Springer Series in Information Sciences*, Springer, Heidelberg, 1990.
9. D. Paulus, L. Csink: *On Color Normalization*, in S. Tanás, B. Jósef (Hrsg.): *Magyar Képfeldolgozók és Alakfelismerõk Orszßgos Konferencißja, Konferenciakiadvßny*, Budapest, 1997, S. 222–229.
10. D. Paulus, J. Hornegger: *Pattern Recognition of Images and Speech in C++*, Advanced Studies in Computer Science, Vieweg, Braunschweig, 1997.
11. J. Pösl, H. Niemann: *Statistical 3–D Object Localization Without Segmentation Using Wavelet Analysis*, in *Computer Analysis of Images and Patterns (CAIP)*, Springer, Kiel, September 1997, S. 440–447.
12. C. Schroeder, B. Neumann: *On the Logics of Image Interpretation: Model Construction in a Formal Knowledge-Representation Framework*, in ICIP 96 [6], S. 18A2.
13. G. Socher, G. Sagerer, F. Kummert, T. Fuhr: *Talking About 3D Scenes: Integration of Image and Speech Understanding in a Hybrid Distributed System*, in ICIP 96 [6], S. 18A2.
14. M. J. Swain, D. H. Ballard: *Color Indexing*, *International Journal of Computer Vision*, Bd. 7, Nr. 1, November 1991, S. 11–32.
15. R. Y. Tsai: *A Versatile Camera Calibration Technique for High-Accuracy 3D Machine Vision Metrology using Off-the-Shelf TV Cameras and Lenses*, *IEEE Journal of Robotics and Automation*, Bd. Ra-3, Nr. 3, August 1987, S. 323–344.
16. R. Willson: *Modeling and Calibration of Automated Zoom Lenses*, PhD thesis, Carnegie Mellon University, Pittsburgh, 1994.

Ortsvariante Karte als effektives Datenformat zur Integration visueller Navigationsaufgaben

Christian Toepfer*, Gregory Baratoff*, Moritz Wende, Heiko Neumann*

Abteilung Neuroinformatik, Universität Ulm, D-89069 Ulm
{ct, baratoff, mwende, hneumann}@neuro.informatik.uni-ulm.de

Zusammenfassung Die Eigenbewegung eines Beobachters durch eine strukturierte Umgebung verursacht ein globales Bewegungsmuster auf dessen Retina, den *optischen Fluss*. Die Auswertung dieses Flusses liefert wichtige Informationen zur Steuerung verhaltensbezogener Aufgaben, wie beispielsweise Kollisionsvermeidung und Navigation in offenen Umgebungen. Eine Fahrtrichtungsregelung, die den optischen Fluss in der linken und rechten Peripherie ausgleicht, erreicht eine sichere Navigation zwischen seitlichen Hindernissen hindurch. Der zentrale optische Fluss erlaubt dagegen die Vermeidung von Hindernissen in Fahrtrichtung. Eine Analyse der allgemeinen Struktur des Flussfeldes in dieser Situation führt uns dazu, eine spezielle Form der komplex-logarithmischen Transformation des Bildes als effizientes, einheitliches Datenformat für diese Aufgaben zu untersuchen. Unsere Roboter-Simulation zeigt die Geschwindigkeit und Zuverlässigkeit der vorgeschlagenen Architektur.

1 Motivation und Einordnung der Arbeit

Wichtige Aspekte der visuellen Informationsverarbeitung in technischen und biologischen Systemen treten hervor, wenn man sie nach Gibson [9] als Teil eines Zyklus aus Wahrnehmung und Handlung betrachtet. In unserem Szenario hat ein simulierter autonomer Agent die Aufgabe, durch die Auswertung eines monokularen Grauwertbildes kollisionsfrei in verschiedenen, dem Agenten unbekannten Umgebungen zu navigieren. Aloimonos hat seiner Idee einer zweckgebundenen Darstellung (*purposive representation*) [1] folgend betont, dass es dabei nicht nötig ist, die 3D-Umgebung zu rekonstruieren [2]. Eine effiziente interne Repräsentation folgt dagegen aus einer Analyse der Anforderungen, die sich dem Agenten beim Lösen der Navigationsaufgaben stellen.

- Echtzeitfähigkeit: Eine schnelle Reaktion des Agenten ist gerade bei der Kollisionsvermeidung überlebenswichtig. Wegen der beschränkten Informationsverarbeitungsressourcen kann aber auf einem hochauflösendes Bild keine umfassende Flusserkennung in Echtzeit durchgeführt werden. Eine Kompression weniger relevanter Bildteile ist daher nötig.

---

* Ch. Toepfer, G. Baratoff und H. Neumann arbeiten im Sonderforschungsbereich *Integration symbolischer und subsymbolischer Informationsverarbeitung in adaptiven sensomotorischen Systemen* (SFB 527), der von der Deutschen Forschungsgemeinschaft finanziert wird.

- Flussstruktur: Der optische Fluss, der durch die Eigenbewegung entsteht, hat eine charakteristische, radiale Struktur. Dieses Vorwissen sollte im Sinne eines *matched filter* bei der Flussdetektion ausgenutzt werden.
- Relevanz: Die Richtung einer selbstgewählten Eigenbewegung ist dem Agenten bekannt und genau in dieser Richtung liegt das maximale Gefahrenpotential für eine Kollision. Gleichzeitig ist aber auch ein großes Sichtfeld nötig, um seitlichen Hindernissen rechtzeitig ausweichen zu können.

In diesem Artikel werden wir zeigen, dass eine modifizierte Form einer komplex-logarithmischen Transformation des Bildes (*Logmap*) diesen Anforderungen gerecht wird. Diese Bildtransformation beruht auf einer Modellierung der Projektion des retinalen Bildes in das kortikale Areal V1 in Primaten [14].

Zwei Kontrollverfahren zur Steuerung des Roboters bieten sich besonders an: (a) Eine elementare Strategie zum Navigieren zwischen seitlichen Begrenzungen besteht darin, die Bewegungsrichtung so zu steuern, dass der optische Fluss in der linken und rechten Peripherie gleich groß gehalten wird [7][8]. (b) Zur Detektion von drohenden Kollisionen in Fahrtrichtung lässt sich allein aus der Expansionsrate im Bildzentrum die *time to contact* (TTC) bestimmen. Dieses Verfahren wurde erfolgreich auf Robotern erprobt [5], auch in Kombination mit dem Balancieren der peripheren Flüsse [6]. Experimente mit Insekten [15] belegen, dass auch in biologischen Systemen ähnliche Verfahren erfolgreich sind. Jain et al. [11] haben eine komplex-logarithmisch transformierte Karte zur Bestimmung der TTC bei reiner Translation eingeführt. Aufgrund der radialen Struktur des optischen Flusses bieten die Abbildungseigenschaften [12] dieser Logmap besondere Vorteile bei der TTC-Berechnung [16].

Unsere Arbeit zeigt, wie man eine modifizierte Logmap als einheitliches Datenformat dazu einsetzen kann, sowohl den peripheren als auch den zentralen optischen Fluss zu detektieren, um die beiden eben beschriebenen Navigationsleistungen effizient zu ermöglichen. Die Analyse des durch Eigenbewegung verursachten Flussfeldes im nächsten Abschnitt begründet die Einführung dieses Datenformats in Abschnitt 3. Die Implementation in einem Simulator (Abschnitt 4) zeigt die Robustheit und Geschwindigkeit des Verfahrens (Abschnitt 5).

2 Flussfeld: Vereinfachungen und Eigenschaften

Bewegt sich ein Beobachter mit Translationsgeschwindigkeit $\boldsymbol{V} = (V_x, V_y, V_z)^T$ und Rotationsgeschwindigkeit $\boldsymbol{\Omega} = (\Omega_x, \Omega_y, \Omega_z)^T$ durch eine stationäre Umwelt gemäß $\dot{\boldsymbol{R}} = -\boldsymbol{V} - \Omega \times \boldsymbol{R}$, so ergibt sich durch Projektion $\boldsymbol{r} = \boldsymbol{R}/Z$ auf die Bildebene der optische Fluss [10]. Für die von uns behandelten Agenten nehmen wir die folgenden Vereinfachungen an. (a) Die Bewegung findet in der X-Z-Ebene statt (*ground plane motion*):, d.h. $V_y = 0$, und es sind nur Rotationen um die vertikale Achse gestattet, d.h. $\Omega_x = \Omega_z = 0$. (b) Die Bewegung erfolgt in Blickrichtung, d.h. $V_x = V_y = 0$. Dadurch vereinfacht sich das Flussfeld zu

$$\dot{x} = xV_z/Z - (1 + x^2)\ \Omega_y\ , \tag{1}$$

$$\dot{y} = yV_z/Z - xy\,\Omega_y\ . \tag{2}$$

Bei einer nur translatorischen Bewegung ist der Fluss auch bei beliebiger Tiefenstruktur rein radial, und für ein Hindernis mit nahezu konstanter Tiefe Z ergibt sich das bekannte *looming* mit der Proportionalitätskonstanten $\tau = Z/V_z$ (*time to contact*) zwischen Flussvektor und Abstand vom Zentrum.

Bei zusätzlicher Rotation lässt sich auch ohne eine explizite Bestimmung beider Parameter τ und Ω_y die *time to contact* durch Mittelung bestimmen:

$$1/\overline{\tau} = < \dot{x}/x > \tag{3}$$

Diese einfache Bestimmung der TTC ist also rotationsinvariant, weil sich an den zwei zum vertikalen Meridian symmetrischen Geraden bei x und $-x$ die Rotationsbeiträge $(1 + x^2)\, \Omega_y/x$ kompensieren, die in der Mittelung auftreten.

Falls sich das zentrale Hindernis durch eine Ebene beschreiben lässt mit dem Normalenvektor $\boldsymbol{n} = (n_x, n_y, n_z)^T$ und dem Abstand d zum Beobachter, lautet der translatorische Anteil des Flusses [10]

$$\dot{x} = (n_x x + n_y y + n_z)\, x V_z/d \; , \tag{4}$$

$$\dot{y} = (n_x x + n_y y + n_z)\, y V_z/d \; . \tag{5}$$

Die Mittelung ergibt

$$< \dot{x}/x > = (n_x < x > + n_y < y > + n_z)\; V_z/d = 1/\tau \; ,$$

denn $< x > = < y > = 0$ wie oben, und $d/n_z = d/\cos\theta = Z_0$ ist die Tiefe des Wandpunktes bei $(x, y) = (0, 0)$, wo man bei reiner Translation in der Zeit Z_0/V_z mit der Wand kollidieren würde.

Man kann nicht hoffen, an zwei symmetrischen Punkten immer gleichzeitig diese sich kompensierenden Vektoren zu detektieren. Jedoch zeigen diese Überlegungen, dass die Mittelung der Störterme einen nur kleinen Beitrag liefert, wenn man annehmen kann, dass die Verfälschungen durch Aperturproblem und fehlende Textur zufällig erfolgen; in unserer texturreichen Simulation ist diese Voraussetzung erfüllt (vgl. Abb. 4).

3 Logmap als Datenstruktur

Die Echtzeitanforderung macht eine Kompression des Eingangsbildes nötig, die aufgrund der diskutierten Bedingungen jedoch nicht über das Bild gleichförmig sein sollte. Bei translatorischer Bewegung gegen eine Wand sind die Flussvektoren in der Bildmitte nach Gleichung (4) und (5) so klein, dass sie nicht weiter reduziert werden sollten. Andererseits nimmt die Größe der Flussvektoren mit dem Abstand $r = \sqrt{x^2 + y^2}$ vom Zentrum des Bildes zu (für eine senkrechte Wand genau proportional zu r, andernfalls abhängig von der Tiefenstruktur $Z(x, y)$.). Eine Transformation sollte also die Größe der Vektoren mit dem Faktor $1/r$ komprimieren. Auch die geringere Relevanz der peripheren Vektoren spricht für eine Kompression. Die radiale Struktur des translatorischen Flusses legt außerdem eine Transformation nahe, die Winkel- und Radius-Richtung separiert.

Die konforme Transformation, die genau diese Separation und Kompression leistet, schreibt sich am einfachsten als komplexe Abbildung [16]

$$w = \log(z) \ , \tag{6}$$

die einem (retinalen) Punkt $z = x + \mathrm{i}y$ des Bildes den (kortikalen) Punkt $w = u + \mathrm{i}v$ zuordnet. Die Ableitung $\mathrm{d}w/\mathrm{d}z = 1/z$ zeigt, dass die Flussvektoren genau mit dem gewünschten Faktor $1/r$ komprimiert werden.

Diese Darstellung hat den Nachteil, dass der Logarithmus bei $z = 0$ eine Singularität besitzt, so dass der Bereich um das Bildzentrum (Fovea) ausgespart bleiben muss. Eine alternative, immer noch konforme Transformation, erreicht man durch Addition einer reellen Konstanten a zum Argument des Logarithmus [14]

$$\zeta = \xi + \mathrm{i}\eta = \log(z + \sigma a) - \log(a) \ . \tag{7}$$

Die Behebung der Singularität im Ursprung erkauft man sich mit dem Zerfall des Bildes in zwei Hälften: $\sigma = 1$ für die rechten Hemisphäre ($\mathrm{Re}\, z > 0$) und $\sigma = -1$ für die linke. Die Subtraktion der Konstanten $\log(a)$ sorgt für die Abbildung des Bildnullpunkts in den Nullpunkt der Logmap.

Diese Darstellung verhält sich für kleine $|z| \ll a$ wie eine einfache Skalierung des kartesischen Bildes: $\zeta \propto z$. Man verliert im zentralen Bildbereich also den Vorteil der Radius-Winkel-Separation, gewinnt aber eine verzerrungsarme Abbildung der gesamten Fovea. In der Peripherie ($|z| \gg a$) dagegen nähert sich die Darstellung dem einfachen Logarithmus $\zeta \approx \log(z)$, so dass die Koordinatenachsen dieser Logmap wieder nahezu axial und radial in Bezug auf den Bildmittelpunkt sind.

Die Transformation des translatorischen Anteils des optischen Flusses nach Gleichung (1) und (2) in die Logmap ergibt

$$\dot{\zeta} = \frac{V_z}{Z}(1 - \sigma \exp(-\zeta)) \ .$$

In der Peripherie ($|\zeta| \gg 1$) fällt der Exponentialterm weg: Der Fluss wird rein reell, was die Tatsache widerspiegelt, dass das translatorische Flussfeld rein radial ist. Dies erlaubt es, sich bei der Flussdetektion in der Logmap auf die ξ-Richtung zu beschränken, d.h. die Suche wird eindimensional.

4 Architektur unserer Implementation

Die Logmap-Transformation lässt sich durch eine einfache Tabelle implementieren, die jedem retinalen Pixel ein kortikales zuordnet. Über das retinale Urbild eines kortikalen Pixels wird mit Hilfe einer weiteren LUT gemittelt. Die Konstante a lässt sich durch zwei Forderungen festlegen: (1) Um auch die diskretisierte Logmap konform zu halten, wählt man die Abtastintervalle beider Koordinaten $\Delta\xi$ und $\Delta\eta$ gleich groß. Für typische Bildgrößen bedeutet dies, dass die Anzahl N_η der Logmap-Pixel in Winkelrichtung größer als in radialer Richtung ist. Unsere Tests zeigen, dass wir mit einer Logmap-Größe von 65 Pixeln radial und

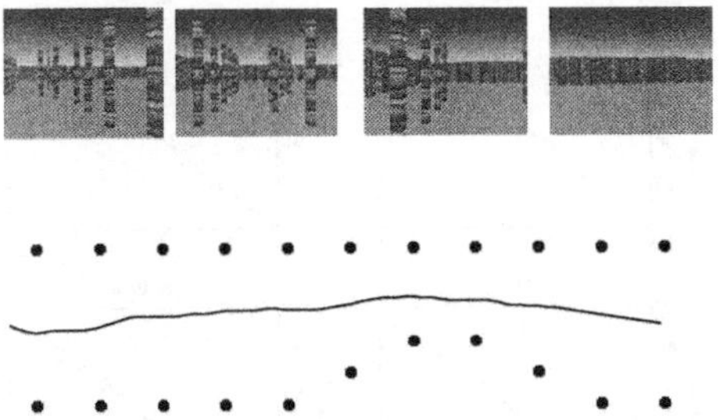

Abb. 1. Richtungsregelung: Die vier Bilder oben zeigen das simulierte Kamerabild während der Fahrt durch eine Säulenreihe. Darunter ist der Pfad des Roboters in Draufsicht gezeigt. Trotz der großen Abstände zwischen den Säulen sorgt des Steuerverfahren für eine sichere Navigation des Roboters durch die Kollonade.

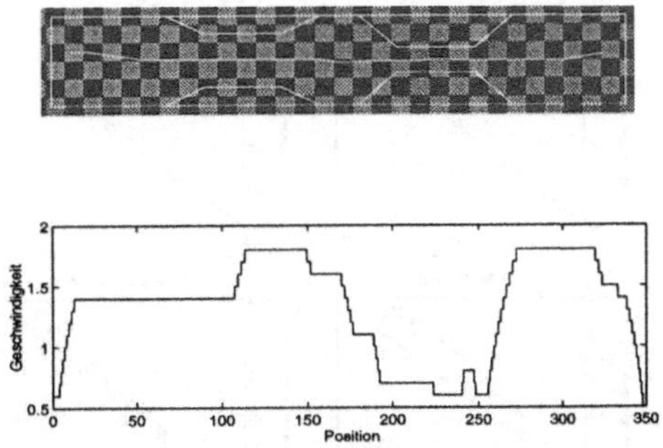

Abb. 2. Geschwindigkeitsregelung: Die Robotergeschwindigkeit wird gesteuert, indem die Summe der beiden peripheren Flussrepräsentanten konstant gehalten wird. In den Engstellen reduziert sich daher die Geschwindigkeit. Durch den vorwärtsgerichteten Blick bremst und beschleunigt der Roboter bereits im Voraus.

150 Pixeln axial je Hemisphäre das gesamte Verfahren in Echtzeit durchführen können. (2) Die Forderung nach einer maximalen fovealen Auflösung, d.h. einer Jakobideterminante der Transformation $\det J = 1$, führt dann zur Festlegung $a = N_\eta/(2\pi)$.

Zur Bestimmung des optischen Flusses in der Logmap haben wir Little's biologisch inspiriertes Korrelationsverfahren implementiert [4]. Dadurch erhalten wir neben dem Flussvektor selbst noch einen Konfidenzwert, den wir in der statistischen Auswertung des Flussfeldes nutzen. Der optische Fluss ist im fovealen Bereich so klein, dass man Subpixelgenauigkeit erreichen sollte. Wir speichern dazu die letzten fünf fovealen Bereiche des logarithmisch transformierten Bildes [5]. Bei geringer Rotation reichte eine räumliche Suchweite von einem Pixel aus, wir suchen daher in einem Bereich der Gesamtgröße $3 \cdot 3 \cdot 5$ nach der besten Korrelation. Trotz der logarithmischen Transformation bleiben die Flussvektoren in der Peripherie so groß, dass wir nur mit dem letzten Bild vergleichen, aber einen größeren räumlichen Bereich absuchen, z.B. 1×10 Logmap-Pixel.

Aus dem Flussfeld extrahieren wir die TTC nach Gleichung (3) und je einen Repräsentanten für den linken und rechten peripheren Fluss Φ_L und Φ_R. Diese drei Parameter können als die vereinfachte, zweckorientierte Darstellung des Agenten für eine Welt aus einer frontalen und zwei seitlichen Begrenzungen interpretiert werden. Der in C implementierte Algorithmus erreicht Echtzeit auf einer Sun Ultra 1 mit 167 MHz. Die Logmap-Transformation benötigt im Mittel 22 ms, die TTC-Berechnung im Zentrum 52 ms und die Extraktion eines linken und rechten peripheren Flussrepräsentanten 11 ms.

Die Kontrollarchitektur zur Steuerung des Roboters baut auf dieser internen Darstellung der Welt auf: Überschreitet die Differenz $|\Phi_\mathrm{R} - \Phi_\mathrm{L}|$ einen Schwellwert, wird die Fahrtrichtung um einen zur Zeit festen Winkel von 9° geändert. Die Abweichung Summe $\Phi_\mathrm{R} + \Phi_\mathrm{L}$ von einem Sollwert kann über einen Proportio-

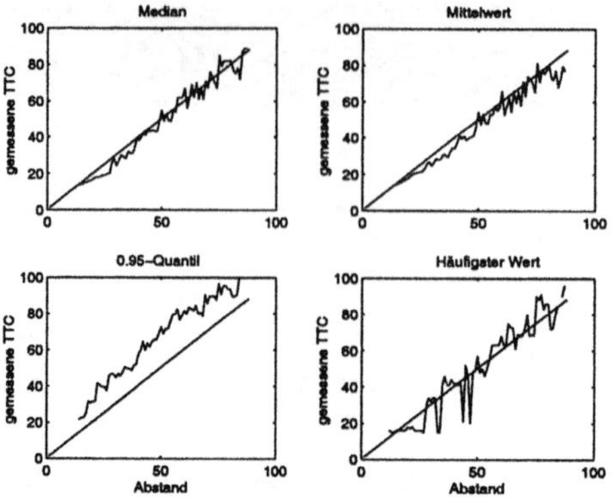

Abb. 3. Time to contact: Vergleich der verschiedenen statistischen Verfahren zur Gewinnung der TTC. Der Roboter fährt senkrecht auf die Wand zu.

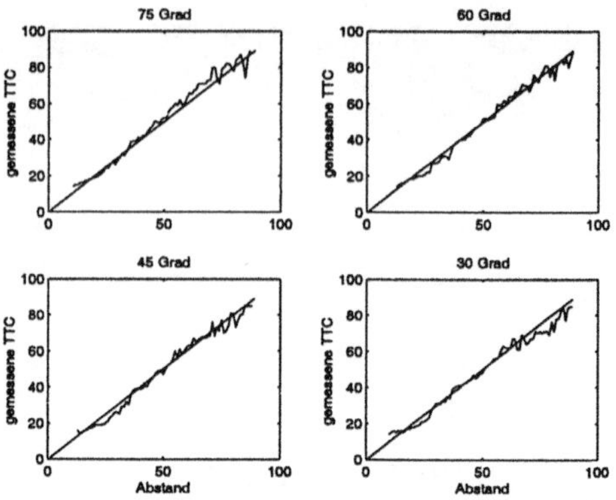

Abb. 4. Unabhängig vom Winkel zwischen Fahrtrichtung und Wand zeigt der Median eine gute Schätzung der wahren TTC.

nalregler die Geschwindigkeit steuern. Diese wird nur innerhalb eines gewissen Bereichs nachgeregelt, um bei zu nahen Hindernissen nicht stehenzubleiben und um bei großer Entfernung zu allen Hindernissen nicht übermäßig schnell zu fahren. Man erreicht eine langsamere, d.h. vorsichtigere Fahrt durch Engstellen (Abb. 2). Die TTC-Messung wird zeitlich tiefpassgefiltert, um sie robuster zu machen, und dann mit einem Schwellwert verglichen. Falls dieser unterschritten wird, dreht sich der Roboter von dem Hindernis fort. Im Augenblick haben wir eine 135°-Drehung implementiert. Der Drehsinn wird mit Hilfe eines Speichers der letzten 10 peripher induzierten Drehungen entschieden.

5 Test in einem Roboter-Simulator

Zum Test unserer Architektur haben wir eine Simulationsumgebung in Matlab entwickelt, die durch Texturmapping simulierte Kamerabilder bei der Fahrt durch ein Labyrinth von Ziegelwänden generiert. Die Festlegung der drei Kontrollparameter Φ_{L}, Φ_{R} und τ aus einem dichten Flussfeld ist ein überbestimmtes Problem, für das wir verschiedene statistische Verfahren getestet haben. Zur Kollisionsvermeidung sollte eine zweckgebundene Darstellung den (skalierten) Abstand zu den nächsten Hindernissen repräsentieren. Dies würde die Wahl des Maximums der peripheren Flüsse und des Minimums der TTC nahelegen.

Um gegenüber Rauschen robust zu sein, ist es im Fall der peripheren Flüsse jedoch sinnvoller, ein 95-%-Quantil zu wählen, d.h. die 5% größten Flüsse werden ignoriert. Dadurch werden Hindernisse sichtbar, die mindestens diesen Anteil der Peripherie ausfüllen. Dies ermöglicht z.B. eine sichere Fahrt durch eine Säulenreihe wie in Abb. 1, während sonarbasierte Ansätze oder Bildinterpolationsverfahren [13] Schwierigkeiten haben mit Hindernissen, die nur 5% des Sichtkegels ausmachen. Für große Tiefenvariationen wie im Beispiel machen Mittelung oder gar Median der peripheren Flussvektoren keinen Sinn, da der Roboter dann für die Säulen im Vordergrund praktisch blind würde.

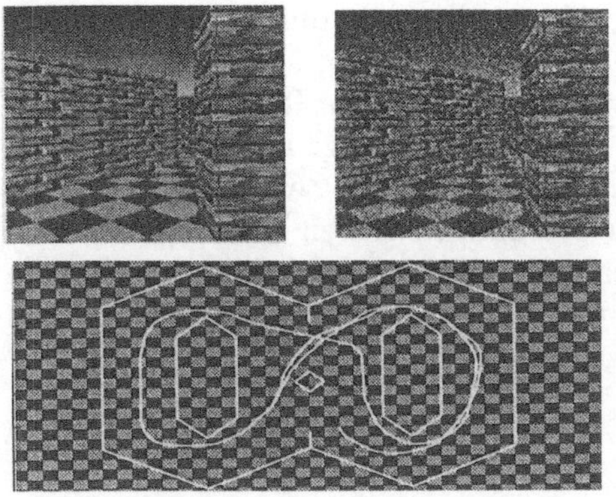

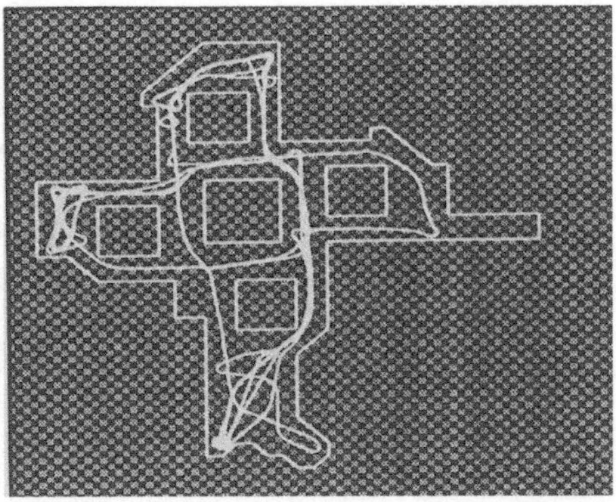

Abb. 5. Robustheit: Das simulierte Kamerabild (links oben) wurde mit einem normalverteilten Rauschen mit $\sigma = 10\%$ der maximalen Intensität überlagert (Mitte oben). Trotz Schwankungen im Pfad navigiert der Roboter kollisionsfrei (unten).

Abb. 6. Langzeitverhalten: Simulierte Roboterfahrt durch das Erdgeschoß des Informatikgebäudes in Ulm. Es wurde eine Fahrstrecke von 400 m mit 12.5 Bildern pro Sekunde bei einer mittleren Geschwindigkeit von 0.5 m/s simuliert.

Im Falle der TTC müssen auch die Überlegungen zur Elimination der durch Rotation und Wandneigung verursachten Störterme in Abschnitt 2 einbezogen werden. Für ein 95%-Quantil oder das Maximum im Histogramm der TTC-Werte ist die Mittelung, die zum Wegfallen dieser Störterme führt, nicht gewährleistet. Wie Abbildungen 3 und 4 zeigen, liefert tatsächlich der Median aller TTC-Messungen die besten Ergebnisse in der Simulation. Der Mittelwert ist fast ebenso gut, aber weniger robust gegenüber Ausreißern, die sich aus fehlerhaften Flussmessungen des Korrelationsverfahrens ergeben. Ein 95%-Quantil überschätzt natürlich die wahre TTC und ist, wie die Abbildung zeigt, ebenso wie der am häufigsten gemessene TTC-Wert weniger zuverlässig.

6 Ergebnis und Ausblick

Im Rahmen der von uns untersuchten Verhaltensleistungen der Navigation und Kollisionsvermeidung führte die Analyse der Anforderungen der Echtzeitfähigkeit, der Flussstruktur und der Relevanz der Bildbereiche auf eine ortsvariante Karte nach dem Vorbild der retino-kortikalen Projektion in Primaten. Auf diesem *einheitlichen* Datenformat lassen sich sowohl der zentrale als auch der periphere optische Fluss in einer den Anforderungen entsprechenden Auflösung detektieren. Inspiriert durch die Verhaltensleistungen von Insekten [15] haben wir eine zweckorientierte Darstellung der Umwelt in Form von drei Kontrollgrößen gewählt. Der Test in der simulierten Realität zeigt, dass diese minimale Repräsentation ein flexibles, robustes und schnelles Verhalten erlaubt.

Wir arbeiten derzeit an der Verbesserung unserer Architektur in folgenden Bereichen: (a) Wir portieren das Verfahren auf einen Roboter, um so die tatsächliche Realwelt- und Echtzeitfähigkeit zu zeigen. (b) Unsere jetzige schwellwertbasierte Auswertung der Kontrollgrößen wollen wir durch neuronale Repräsentation auf der Basis von Amari-Feldern ersetzen [3]. Neben der dadurch erreichten gleichmäßigeren Robotersteuerung erlaubt diese Repräsentation insbesondere die Integration mit anderen sensorischen Eingängen und externen Steuerkommandos oder Zielvorgaben. (c) Wir testen zur Zeit Verfahren, auch die Eigenrotation Ω_y aus dem Fluss zu bestimmen. Neben einer verbesserten Schätzung der Kontrollgrößen stellt dies einen Beitrag für höhere als die von uns untersuchten Verhaltensleistungen dar, wie etwa eine Wegintegration mit darauf folgender Rückkehr zum Ausgangspunkt.

Literatur

1. J. Aloimonos. Purposive and qualitative active vision. In *Proc. 10th Int. Conf. Patt. Recog.*, pages 345–360, June 1990.
2. Y. Aloimonos. Is visual reconstruction necessary?: Obstacle avoidance without passive ranging. *J. of Robotics Systems*, 9(6):843–858, 1992.
3. S. Amari. Dynamics of pattern formation in lateral-inhibition type neural fields. *Biol. Cybernetics*, 27:77–87, 1977.
4. H. Bülthoff, J. Little, and T. Poggio. A parallel algorithm for real-time computation of optical flow. *Nature*, 337(6207):549–553, 1989.
5. T. A. Camus. *Real-Time Optical Flow.* PhD thesis, Department of Computer Science, Brown University, 1994.
6. D. Coombs, M. Herman, T. Hong, and M. Nashman. Real-time obstacle avoidance using central flow divergence and peripheral flow. In *Fifth Int. Conf. on Computer Vision*, pages 276–283, Los Alamitos, CA, 1995.
7. D. Coombs and K. Roberts. Centering behavior using peripheral vision. In *Proc. of CVPR 1993*, New York, 1993. June 15–17.
8. A. P. Duchon. Maze navigation using optical flow. In *Proc. 4th Intl. Conf. on Simulation of Adaptive Behavior*, Cambridge, MA, 1996. MIT Press/Bradford Book.
9. J. J. Gibson. *The Perception of the Visual World.* Houghton Mifflin, Boston, Massachusetts, 1950.
10. B. Horn. *Robot Vision.* MIT Press, Cambridge, Massachusetts, 1986.
11. R. Jain, S. L. Bartlett, and N. O'Brian. Motion stereo using ego-motion complex logarithmic mapping. *IEEE Trans. Pattern Analysis and Machine Intelligence*, 9:356–369, 1987.
12. H. A. Mallot, W. v. Seelen, and F. Giannakopoulos. Neural mapping and space-variant image processing. *Neural Networks*, 3:245–263, 1990.
13. M. G. Nagle and M.V. Srinivasan. Structure from motion: determining the range and orientation of surfaces by image interpolation. *Journal of the Opt. Soc. of America*, 13:25–34, 1996.
14. E. Schwartz. Visual cortex from a computational perspective. In A. Peters and K. Rockland, editors, *Cerebral Cortex*, volume 10. Plenum, New York, 1994.
15. M. V. Srinivasan. How bees exploit optical flow: Behavioral experiments and neural models. *Phil. Trans. of the Royal Society of London B*, 337:253–259, 1992.
16. M. Tistarelli and G. Sandini. On the advantages of polar and log-polar mapping for direct estimation fo time-to-impact from optical flow. In *IEEE Trans. Pattern Analysis and Machine Intelligence*, volume 14, pages 402–410, 1993.

Vision-Based Detection of Navigation Codes for Mobile Robots

T. Cord, P. Levi
Forschungszentrum Informatik (FZI)
Dept. Mobility Management and Robotics
Haid-und-Neu-Str. 10-14
D-76131 Karlsruhe, Germany

D. E. Lazic
Universität Karlsruhe
Institut für Algorithmen und Kognitive Systeme
Am Fasanengarten 5
D-76128 Karlsruhe, Germany

1 Introduction

A new breed of robots whose characteristics, objectives and operational modes drastically differ from more conventional industrial robots is getting to be more and more of interest. Conventional robotics, which has been successful in improving productivity and relieving human workers from tedious routine tasks in manufacturing, now faces a new challenge of automating areas of labor other than those directly linked to production.
Service robots are defined as reprogrammable, sensor-based mechatronic devices that perform useful services to human activities in an everyday environment. Human-machine symbiosis and the relatively unstructured environment in which service robots must operate are among the major differences between service robots and industrial robots. Safety, reliability, cost, appearance and user interfaces are key issues in the design of service robots. They are individually designed to perform tasks in a specific environment and should be able to perform services fully or semiautomatically. Service robots are for example those used for inspection, maintenance, housekeeping, office automation and aiding of senior citizens or physically challenged individuals.
Navigation, in particular the method of determining the position and orientation, is one of the vital concepts when designing mobile robot systems. In most mobile robot applications two basic position estimations are employed in combination: absolute and relative positioning. Relative positioning is a simple method which measures features inherent to the robot, such as distances, velocities and accelerations to determine the momentary position of the vehicle relative to a known starting position. This computation is called odometry. However, due to slippage between the robot's wheels and the ground, the translations and rotations measured by the encoders may not accurately reflect the robot's actual motions. Over time, the cumulative effects of small errors can result in a large error in the estimate of the robot's position. Because of this, odometry is considered to be too unreliable to use for real-world navigation tasks.
However, even if odometry is insufficiently precise for reliable navigation, it can provide a coarse location estimate. Some method for relocalization is needed to compen-

sate for the errors introduced. Therefore, the method of absolute positioning uses sensors detecting features in the robot's environment. These may either exist naturally (unstructured environment) or they may have been inserted for navigation purposes. In a guided robot system the environment is always structured to define all possible trajectories. In a path-independent system, where the greater flexibility of the system necessitates sophisticated navigation, both types of environment are possible. A structured environment can contain inserted elements or marks, both active and passive. The former actively emit signals which are received by the robot, the latter are passive and thus easily detectable by the robot's sensors.

A frequently used sensor is a 2D camera, since it provides environment data that is very rich in information. A common approach is to observe the robot's natural environment with a camera and to detect its features using image processing methods. This is not always easy, because noise and difficulties in interpreting the sensory information make the navigation unreliable. Another possibility is the use of ultrasonic range readings or scanning laser sensors. The individual readings can be done quickly, need little preprocessing and behave predictably, though with limited accuracy.

The use of the Global Positioning System (GPS) could be promising. The best accuracy one might anticipate with an available Differential GPS is insufficient for the requirements of a reliable navigation system. Furthermore, for indoor applications GPS is not feasible, since its signals are not readily detectable.

In order to increase the reliability of navigation, it therefore seems sensible to adapt the robot's environment to the robot and thus facilitate the task of navigation where possible. For example, the use of beacons to identify doors or individual rooms adds very little cost to the overall system and greatly simplifies the task of navigation. There are many other similar environmental modifications or support technologies that simplify the robot's tasks while at the same time decreasing the total cost of the system. Although a general sensor-based and autonomous navigation was a desirable aim, the additional complexity and expense would have to be justified with the usefulness of the system.

A convenient way to correct the drift of a mobile robot is to use marked floor tiles, as was done in the FROG (Free Ranging on Grid) navigation system. The marked floor tiles (for instance a visible chess board grid) represent information about the environment in a way that is compliant with the pattern recognition capabilities of the vehicle. Alternatives to this visible grid are the magnetic grid and the transponder grid, also found in these kinds of systems. Nonetheless, the momentary position of the vehicle is determined only incrementally which inevitably leads to the accumulation of errors and thus to low accuracy in the determination of the position. In addition, no capabilities for error control are provided.

In [1] a navigation system of an AGV with the ability to recover its position and orientation anywhere in the environment is described. The floor is permanently encoded with the terms of a pseudorandom binary array (PRBA) which are represented by binary visible symbols on the floor. However, it is not always possible to recognize at least one complete pseudorandom window when large objects are covering the symbols on the floor. The PRBA-based encoding technique offers only a limited reliability since no error control capabilities are available.

The concept introduced in this article makes use of a color CCD camera which observes the floor of the robot's environment. This localization system uses floor tiles with random patterns which are interpreted as random codes. The detection of the tile patterns is done using image analysis and adequate decoding methods. The described concept provides a highly reliable system with exact and rapid position determination through its powerful error control capability. The system is low in cost as it only requires one camera with image processor per robot. Due to its high reliability and exactness difficult situations such as docking and precise maneuvers in service robot applications can be handled effortlessly.

2 Visible Random Navigation Code

In the described localization system the floor of the robot's environment is covered with square floor tiles. Each tile contains a unique visible pattern which can be produced by some random process in the production. Such floorcoverings made of rubber are available on the market in various versions designed according to aesthetic criteria. This solution seems to be suitable for service robots which operate in environments where humans have to live. The specific tile pattern is used to encode the absolute position (X_i and Y_j coordinates) of the tile's center point T_{ij} (see Fig. 1).

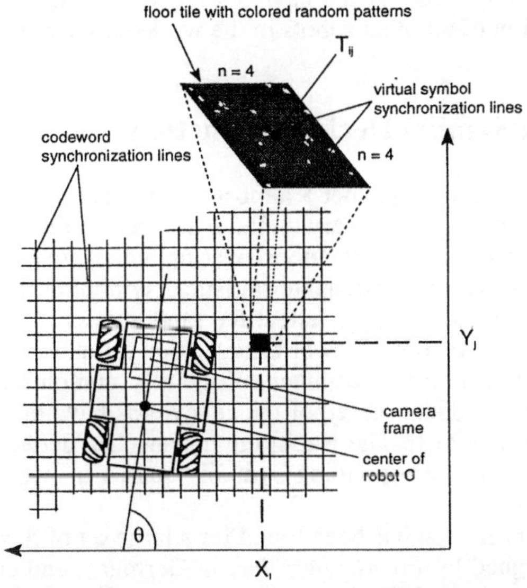

Fig. 1: Visible Random Navigation Code (VRNC) with an extracted tile

The applied random codes should have error detecting and error correcting capabilities in order to control errors if the tile patterns are not completely recognizable (e.g. when dirt or shadows cover the pattern). By means of such a *visible random navigation code (VRNC)* the robot receives an image of the entire tile closest to it and of parts of neighboring tiles with a CCD-camera. An image processor extracts first the boundaries of the entire tile (codeword synchronization lines) and afterwards the virtual symbol boundaries (symbol synchronization lines). When the $N = n \, x \, n$ symbol squares of the entire tile are localized, the symbol detection can be provided by an adequate decision function f_d determining which of the q possible symbols correspond to each of the N symbol squares. In this way the codeword inscribed in the closest tile is received. It may be noisy, i.e. some of the symbols could be erroneously detected with the symbol detection error probability p. With the knowledge of all possible codewords of the whole VRNC and by comparison of the extracted (noisy) codeword error detection and correction can be provided. This decoding process results in a decoded codeword which differs from the original one with some block decoding error probability P_e. The decoding algorithm is based on lookup-tables. The main goal in the design of a VRNC is the selection of the symbol decision function f_d and the decoding method in order to reduce P_e to a value as small as possible.

After the successful decoding process the robot recovers data about the absolute coordinates of the (virtual) center point of the closest tile. With the knowledge of the camera position on the robot the orientation θ and the exact position of the center C of the robot can be estimated by simple geometric reasoning. The accuracy of these calculations is greater than the distance between a few pixels in the image. The described VRNC allows the robot to estimate its absolute position and its orientation with respect to the environment to a highly reliable and accurate extent.

Finally, the decoded position information can be broadcast in the same form to the central station and other robots using wireless data communication. The error protection capabilities of the VRNC are usable again on the noisy radio channel. In this way, each robot obtains (as a by-product of the navigation code) reliable and accurate information about the position of all other robots in the workshop in real-time.

3 Selection of a Symbol Decision Function

A symbol from the symbol Alphabet S almost never corresponds to the same random pattern. Because of that, it is necessary to find q different statistical features of the considered random symbol pattern of the floorcovering which are easily detectable (distinguishable) in order to define an adequate statistical symbol decision function f_d. If the symbols, which are extracted by f_d, reflect the randomness of the production process then the corresponding random codewords have good error control capabilities according to *Shannon's random coding argument* of the information theory [2]. Such a code is good if the Hamming distances d_H of the codewords have the binomial distribution with the mean value $N/2$ [3]. The Hamming distance d_H between two random codewords is the number of corresponding symbol squares in which symbols from two codewords are different.

An optimal decision function has been found for a large set of tiles of the floorcovering *noraplan viva,* designed by *Freudenberg Inc.* in Germany, and consisting of blue and yellow spots on dark ground. Each random tile is subdivided in $N = n\,x\,n$ squares as shown in Fig. 2. The symbol synchronization lines shown in the figure are not inscribed in the floorcovering. The feature extraction determines the related feature vector for each square. The feature vector consists of two elements. The first component of the vector represents the number of yellow spots on the square and the second one is defined by the number of blue spots. A spot which covers more than one square is considered in all the corresponding feature vectors.

The decision function f_d determines for each feature vector of the feature space Ω a symbol s_i of the codeword.

$$S = \{s_i\}_{i=1}^{q} \tag{1}$$

The decision function f_d is defined by:

$$\begin{aligned} f_d &: \Omega \rightarrow S \\ f_d(t) &\rightarrow s \end{aligned} \tag{2}$$

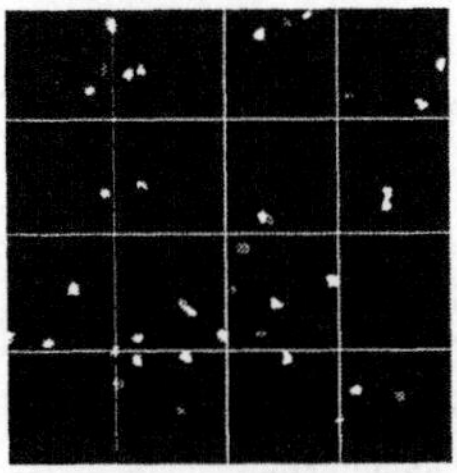

(2, 1)	(2, 0)	(2, 1)	(2, 1)
(1, 0)	(1, 0)	(1, 1)	(2, 0)
(4, 1)	(4, 0)	(2, 3)	(0, 0)
(1, 1)	(2, 2)	(2, 0)	(2, 1)

Fig. 2: Random floor tile with corresponding feature vectors

An example of the decision function f_d is shown in (3). This function uses only the number of yellow spots.

$$f_d : \mathrm{N} \otimes \mathrm{N} \rightarrow \{s_0, s_1, s_2\}$$

$$f_d((k_1, k_2)) = \begin{cases} s_0, & k_1 < 2 \\ s_1, & k_1 = 2 \\ s_2, & k_1 > 2 \end{cases} \tag{3}$$

The symbols are sorted line by line to define the codeword. Fig. 3 shows the resulting codeword for the tile in Fig. 2.

(2, 1)	(2, 0)	(2, 1)	(2, 1)
(1, 0)	(1, 0)	(1, 1)	(2, 0)
(4, 1)	(4, 0)	(2, 3)	(0, 0)
(1, 1)	(2, 2)	(2, 0)	(2, 1)

f_d →

s_1	s_1	s_1	s_1
s_0	s_0	s_0	s_1
s_2	s_2	s_1	s_0
s_0	s_1	s_1	s_1

Fig. 3: Calculation of the symbols of the codeword from the feature vectors

A random code consisting of M codewords of the length N is defined as follows:

$$B(R, N)_q = \{\vec{x}_m\}_{m=1}^{M} \tag{4}$$

$$\vec{x}_m = (x_{m1}, x_{m2}, \ldots, x_{mN}), \; x_{mk} \in S \tag{5}$$

The codewords of the random code should differ in as many symbols as possible in order to have optimal error control capabilities. The set of Hamming distances of all codewords of the random code is given by

$$D_H = \{d_H(\vec{x}, \vec{y}) : \vec{x}, \vec{y} \in B_f(R, N)_q\} \tag{6}$$

Fig. 4 shows the absolute number $H(d)$ of the Hamming distances d when the different codewords of the random codes B_{f1} (B_{f2} respectively) are compared in pairs. These random codes have been produced with a set of *456* tiles. Each tile has been subdivided in $N = 6 \times 6$ squares and for each square two binary symbols are calculated. The length of a codeword is *72* Bit. The symbol decision functions f_1 and f_2 differ in the thresholds used.

The error control capability of a code $B_f(R, N)_q$ is influenced by the smallest Hamming distance in D_H [4]. The error control capability of B_{f1} is better than the error control capability of B_{f2} since the minimal Hamming distance in B_{f2} is much smaller than the

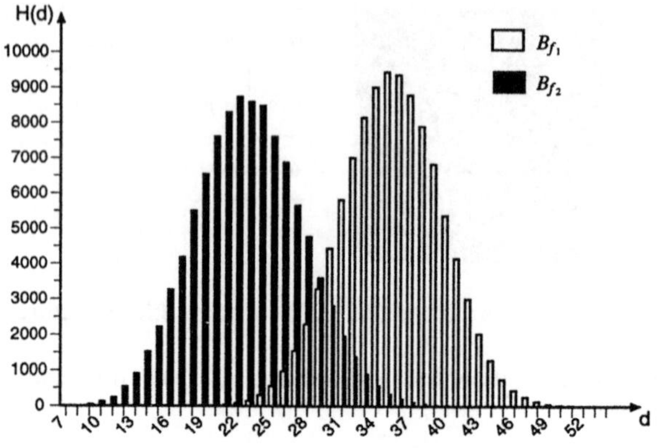

Fig. 4: Comparison of the two random codes B_{f1} and B_{f2}

minimal Hamming distance in B_{f1}. The tiles which are used in the robot's environment can be sorted before they are placed on the floor in order to eliminate codewords which are similar.

The choice of an optimal decision function f_d is of great importance for the error control capability of the random code. Another important factor is the size of the N symbol squares. If the squares are too small or too large then for a lot of squares the same feature vector is extracted and the resulting codewords have a small Hamming distance. It has been shown that a decision function with two thresholds and a subdivision of a tile in $N = 6 \, x \, 6$ squares produces optimal random codes for the floorcovering *noraplan viva* [5]. The length of a codeword is *72* Bit. If the tiles are sorted before they are put down on the floor in the robot's environment, a minimal Hamming distance of *20* can be guaranteed. During the sorting process about *2%* of the tiles are eliminated.

4 Detection of Colored Floor Tiles

The task of the color detector is to identify the colors of the floor tiles in order to estimate the inscribed coded position information. The objective of the color detection system is to segment the image and to classify each segmented region into one of the classes representing the colored spots of the random tiles.

Edge-based methods can hardly satisfy the needs, since the colored squares are very close to another and their boundaries may be ill defined because of dirt and shadows on the floor. The regions in the image are sometimes not uniform in the sense that fragments with large color variations exist in these regions because of different lighting conditions. Therefore an effective method that utilizes chromatic information has been developed to segment the images and classify these segmented regions into the different classes. The underlying idea of the detection of the colored spots on the floor is to classify each pixel value. The gained result of classification is not directly suitable for an abstract evaluation in terms of localization information. There are more processing steps necessary in order to obtain this. A series of morphological operations is used to delete any disturbances. After this, the classification result is interpreted in order to estimate the symbols of the codeword on the floor tile (Fig. 6).

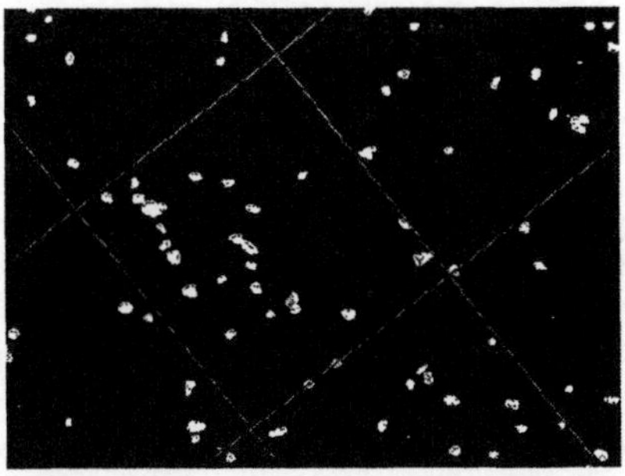

Fig. 5: Detection of a tile on the floor and its joints

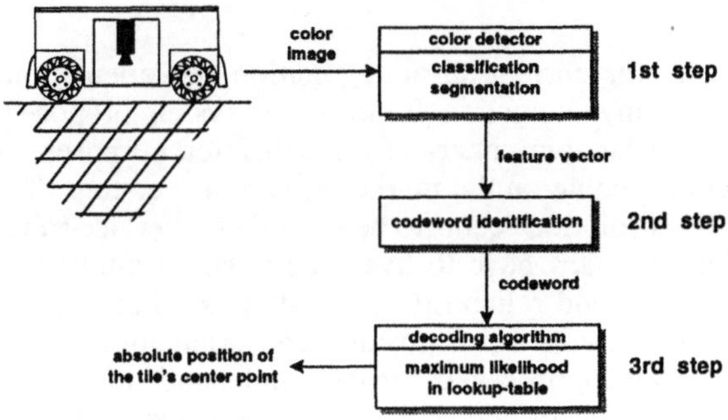

Fig. 6: Flow chart of the localization process

The color-based *Bayes-Classification* is based on the assumption that the decision problem is posed in probabilistic terms, and that all of the relevant probability values are known, as are the a-priori probability of each class and the probability density of the color features. From these values it is possible to calculate an a-posteriori probability of the feature vector for each class. The feature vector will be assigned to that class, whose a-posteriori probability is greatest. The a-priori probabilities of the q classes are known and the class-conditional probabilities were assumed to be normally distributed, i.e. the Bayes-Classifier is exactly defined if for each class there exists its mean μ and its covariance matrix K.

In order to adapt the Bayes-Classifier to a given kind of tiles (colors) the parameters are learned by the system. To obtain the parameters n samples are given for each class and the parameters are estimated automatically by the following forms:

$$\mu = \frac{1}{n}\sum_{i=1}^{n} v_i , \quad K = \frac{1}{n}\sum_{i=1}^{n}(v_i - \mu)(v_i - \mu)^T \qquad (7)$$

If the different features in the feature vectors are linearly dependent, the covariance matrix is singular and thus non invertable.

The classical color model decomposes colors into R (red), G (green), and B (blue) tristimuli values. In general, several features can be calculated from the RGB model for different purposes. For example, intensity, hue, and saturation are convenient for representing human color perception. In [6] a set of effective color features is proposed that is called $I_1I_2I_3$ *color features,* by using the Karhunen-Loève transformation. This color model is efficient for color image segmentation and is adopted in the method for the detection of the colored squares. The original *RGB* values are transformed into the $I_1I_2I_3$ color space by means of the following linear equations:

$$I_1 = \frac{R+G+B}{3},\ I_2 = \frac{R-B}{2},\ I_3 = \frac{2G-R-B}{4} \tag{8}$$

The described color image processing system showed its effectiveness in the segmentation of images and the classification of these segmented regions into one of the q classes representing the symbols on the floor. The pixel classification based on feature space distribution does not take into consideration the spatial relationship among points on the image and results in a fragmented image. By means of morphological post-processing and a spot analysis this problem is solved adequately.

5 Conclusion

The determination of the current absolute position and orientation of a mobile robot provides the basis for any navigation purpose. The system proposed in this article uses floor tiles that contain random patterns for localization purposes. Such floorcoverings made of rubber are available on the market in various versions designed according to aesthetic criteria. This solution seems to be suitable for service robots which operate in environments where humans have to live. The position information can be detected with sufficient accuracy and reliability in real-time so that an autonomous vehicle is able to determine robustly its exact position and orientation in a given environment, such as factories, hospitals, offices, or airports, independently of any guidance system. Therefore, a good base is provided for navigation and path planning because the knowledge of the current position at any time allows an easy and accurate surveillance of a given route.

The conclusion we drew from our experiments is that the environmental modification has to be considered an integral part of the design of a service robot system. The environment can directly support the robot to increase the robustness of the robot's navigation system. Although a general sensor-based autonomous navigation capability is a desirable aim it is hard to justify the extra complexity and expense required for it.

6 References

[1] Petriu, E. M., McMath, W. S., Yeung, S. K., Trif, N., Bieseman, T.: Two-Dimensional Position Recovery for a Free-Ranging Automated Guided Vehicle, *IEEE Transactions on Instrumentation and Measurement,* Vol. 42, pp. 701-706, 1993

[2] Wosencraft, J. M., Jacobs, I. M.: *Principles of Communication Engineering,* Wiley, New York, 1965

[3] Lazic, D. E., Senk, V.: A Direct Geometrical Method for Bounding the Error Exponent for any Specific Family of Channel Codes - Part I: Cutoff Rate Lower Bound for Block Codes, *IEEE Transactions on Information Theory,* Vol. 38, No. 5, 1992

[4] MacWilliams, F. J., Sloane, N. J. A.: *The Theory of Error-Correcting Codes,* North-Holland, Amsterdam, 1988

[5] Cord, T.: *Absolute Lokalisation mobiler Roboter durch Codierung der Einsatzumgebung,* Infix, Sankt Augustin, Germany, 1997

[6] Ohta, Y., Kanade, T., Sakai, T.: Color information for region segmentation, *Comp. Vision Graphical Image Processing*, Vol. 13, 1980

Anwendungen

Untersuchung der Detektionsleistung von 3D-Operatoren zur Ermittlung von Punktlandmarken in MR- und CT-Bildern

Thomas Hartkens, Karl Rohr, H. Siegfried Stiehl

Arbeitsbereich Kognitive Systeme, Fachbereich Informatik, Universität Hamburg
Vogt-Kölln-Str. 30, D-22527 Hamburg
{hartkens,rohr,stiehl}@informatik.uni-hamburg.de

Zusammenfassung Während die Leistungsfähigkeit von 2D-Operatoren zur Detektion von Punktlandmarken in tomographischen Bildern bereits untersucht wurde (z.B. [10]), sind Untersuchungen bzgl. der Detektionsleistung von 3D-Operatoren kaum bekannt. In dieser Arbeit werden neun verschiedene 3D-Differentialoperatoren dahingehend untersucht, ob und inwieweit sie geeignet sind, Landmarken in 3D-Bildern zu detektieren. Dazu verwenden wir Maße, die unterschiedliche Aspekte der Detektionsleistung der Operatoren widerspiegeln.

1 Motivation und Einleitung

Für die punktbasierte Registrierung von medizinischen Bildern ist die Detektion korrespondierender Punkte eine notwendige Voraussetzung. Untersuchungsgegenstand in dieser Arbeit ist die *Detektion* anatomischer Landmarken in 3D MR- und CT-Bildern des menschlichen Gehirns. Im Unterschied hierzu wurde in [9] die verfeinerte Lokalisierung von Landmarken untersucht. Als Landmarken dienen ausgezeichnete Punkte, an denen die Oberflächen von anatomischen Strukturen stark gekrümmt sind (wie z.B. Spitzen der Seitenventrikel). Um die Landmarke zu detektieren, werden 3D-Operatoren auf die Bilder angewendet. Diese berechnen für jeden Punkt aus den partiellen Ableitungen der Bildfunktion ein Maß für die Krümmung. Prinzipiell gehen wir von einer semi-automatischen Prozedur aus, bei der zuvor eine interessierende Region (ROI), in der eine Landmarke zu finden ist, von dem Benutzer vorgegeben wird.

Wir untersuchen in dieser Arbeit 3D-Operatoren dahingehend, (a) ob sie grundsätzlich geeignet sind, korrespondierende Punkte in MR- und CT-Bilder zu ermitteln und (b) wie gut sie die vorgegebenen Landmarken detektieren. Dazu wenden wir die in Tabelle 1 aufgeführten neun Differentialoperatoren sowohl auf synthetische als auch MR- und CT-Bilder an und berechnen aus den Detektionen anhand der in Abschnitt 2 vorgeschlagenenen Maße die Detektionsleistung der Operatoren. Insgesamt werden alle Operatoren auf jeweils 308 synthetische 3D-Bilder (Tetrader, Ellipsoide und hyperbolische Paraboloide mit unterschiedlich starken Deformationen und verschiedenen Rauschstufen) und auf 301 tomographische 3D-Teilbilder (MR- und CT-Bilder mit ebenfalls unterschiedlich starken Deformationen und Rauschstufen) angewendet.

Ansatz „mittlere Krümmung“ [11],[4],[7],[14],[13]		
• $k_{mittlere\,Kruemmung} =$	$\frac{1}{2\lvert\nabla g\rvert^3}\left[g_x^2(g_{yy}+g_{zz})+g_y^2(g_{xx}+g_{zz})+g_z^2(g_{xx}+g_{yy}) -2(g_xg_yg_{xy}+g_xg_zg_{xz}+g_yg_zg_{yz}\right]$	
• $k_{KitchenRosenfeld3D} =$	$k_{mittlere\,Kruemmung}\cdot 2\lvert\nabla g\rvert$	
• $k_{Blom3D} =$	$k_{mittlere\,Kruemmung}\cdot 2\lvert\nabla g\rvert^3$	mit $\nabla g = (g_x, g_y, g_z)^T$
Ansatz „Gaußkrümmung“ [7],[14],[3]		
• $k_{Gausskruemmung} =$	$\frac{1}{\lvert\nabla g\rvert^4}\left[g_x^2\cdot(g_{yy}g_{zz}-g_{yz}^2)+2g_yg_z\cdot(g_{xz}g_{xy}-g_{xx}g_{yz}) +g_y^2\cdot(g_{xx}g_{zz}-g_{xz}^2)+2g_xg_z\cdot(g_{yz}g_{xy}-g_{yy}g_{xz}) +g_z^2\cdot(g_{xx}g_{yy}-g_{xy}^2)+2g_xg_y\cdot(g_{xz}g_{yz}-g_{zz}g_{xy})\right]$	
• $k_{Gausskreummung*Grad^4} =$	$k_{Gausskruemmung}\cdot\lvert\nabla g\rvert^4$	
Ansatz „Förstner/Rohr“ [8],[12],[13]		
• $k_{V1} =$	$\frac{det(\underline{C})}{trace(\underline{C})}.$	
• $k_{Rohr3D} =$	$det(\underline{C})$	
• $k_{Foerstner3D} =$	$\frac{1}{trace(\underline{C}^{-1})} = \frac{det(\underline{C})}{trace(\underline{C}^{adj.})}$	mit $\underline{C} = \overline{\nabla g(\nabla g)^T}$
Ansatz „Beaudet“ [2]		
• $k_{Beaudet3D} =$	$det(H_g)$	H_g Hesse-Matrix

Tabelle1. 3D-Differentialoperatoren *Die Operatoren berechnen aus den ersten bzw. ersten und zweiten partiellen Ableitungen der Bildfunktion für jeden Punkt des Bildes ein Maß für die Krümmung der Strukturen im Bild. Die Operatoren lassen sich in vier Gruppen einteilen, entsprechend vier unterschiedlichen Ansätzen für die Berechnung des Maßes. Innerhalb der Gruppen unterscheiden sich die Operatoren z.B. durch eine unterschiedliche Gewichtung mit dem Betrag des Gradienten der Grauwertfunktion. Einige Operatoren sind Erweiterungen von sogenannten Eckenoperatoren für 2D-Bilder. Wir bezeichnen diese Operatoren mit dem Namen der ursprünglichen Autoren und dem Suffix „3D“.*

2 Maße für die Detektionsleistung

Um die Operatoren miteinander vergleichen zu können, legen wir geeignete Maße für die Detektionsleistung fest. Dabei betrachten wir zwei unterschiedliche Aspekte der Detektionsleistung:

Anzahl korrespondierender Punkte
Nach [8] sollen die detektierten Punkte u.a. invariant gegenüber Verzerrungen und stabil bzgl. Rauschen sein, d.h. auch nach einer Deformation sowie in verrauschten Bildern sollten dieselben Punkte detektiert werden. Wir untersuchen diese Eigenschaften, indem wir die 3D-Bilder algorithmisch durch eine elastische Transformation verformen bzw. mit Gauß-Rauschen überlagern und automatisch die Anzahl korrespondierender Punkte in dem Orginalbild und dem veränderten Bild ermitteln (siehe auch [10] für den Fall von 2D Bildern).

Klassische Detektionsleistung
Die in der Literatur vorgeschlagenen Maße zur Beurteilung der Detektionsleistung von Operatoren (siehe z.B. [1], [6], [15], [16]) beziehen sich zumeist auf die Detektion von Grauwertkanten. In dieser Arbeit übertragen wir diese Maße

auf die Detektion von Punktlandmarken und fassen die Maße unter dem Begriff „klassische Detektionsleistung" zusammen.

Alle Landmarken werden jeweils innerhalb eines Detektionsgebietes betrachtet (z.B. 7 × 7 × 7 Volumenelemente). Liegt eine Detektion innerhalb dieses Gebietes, so gilt die Landmarke als detektiert und wir bezeichnen diese Detektion als *korrekte Detektion.* Befindet sich keine Detektion innerhalb des Detektionsgebietes, so ist dies eine *Fehl-Detektion* dieser Landmarke. Einen detektierten Punkt außerhalb der Detektionsgebiete bezeichnen wir als *Falsch-Detektion.* Die Verwendung von Detektionsgebieten hat den Vorteil, daß relativ geringe Lokalisierungsfehler das Detektionsergebnis nicht verfälschen.

Nach Anwendung der Operatoren und Auswertung der Ergebnisse erhalten wir folgende Größen: Anzahl aller Detektionen (n_d), Anzahl der Detektionen, die die Landmarken korrekt detektieren, d.h. innerhalb des Detektionsgebiets liegen ($n_{d,in}$), und Anzahl der mindestens einmal detektierten Landmarken ($n_{l,detect}$). Die Anzahl der Landmarken bezeichnen wir mit n_l. Drei Ausdrücke beschreiben die Detektionsleistung der Operatoren: der Anteil der Detektionen, die die Landmarke korrekt detektieren (P_{in}), der Anteil der detektierten Landmarken (P_{detect}) und die durchschnittliche Anzahl der Mehrfach-Detektionen pro Landmarke ($P_{multiple}$):

$$P_{in} = \frac{n_{d,in}}{n_d}, \qquad P_{detect} = \frac{n_{l,detect}}{n_l}, \qquad P_{multiple} = \frac{n_{d,in}}{n_l} \tag{1}$$

3 Untersuchung der Operatoren

Zur Berechnung der partiellen Ableitungen der Bildfunktion werden in dieser Untersuchung Beaudet-Operatoren mit der Maskengröße 5 × 5 × 5 verwendet (siehe [2]). Die Operatoren des Ansatzes „Förstner/Rohr" [8], [12], [13] mitteln die partiellen Ableitungen über eine bestimmte Umgebung. Die Größe dieser Umgebung haben wir mit 3 × 3 × 3 Voxel festgelegt. Um die Extrema der Operatorwerte zu ermitteln, wird für jeden Punkt überprüft, ob dieser der größte Wert in seiner nächsten Umgebung ist (Größe 5 × 5 × 5). Operatorwerte werden bei der Extrema-Suche nur dann mit einbezogen, wenn sie größer als ein bestimmter Schwellenwert sind. Dieser Schwellenwert ergibt sich prozentual aus allen Operatorantworten, in unserem Fall 1% des größten Operatorbetrags. Es werden 22 synthetische und 43 tomographische Teil-Bilder verwendet. Die 3D synthetischen Bilder bestehen aus Tetraedern (Öffnungswinkel α =30, 40, 50, 60, 70, 80 und 90), Ellipsoiden (Halbachsenlänge a =8, 9, 10, 11, 12, 13, 14, 15 und 16, $b = 8$, $c = 40$) und hyperbolischen Paraboloiden (Parameter $(a, b) = (1, 1)$, $(2, 2)$, $(3, 3)$, $(1, 2)$, $(1, 3)$ und $(2, 3)$), die jeweils mit einer Gaußfunktion der Standardabweichung $\sigma = 0.7$ geglättet werden. Tetraeder und Ellipsoide repräsentieren anatomische Spitzen, während hyperbolische Paraboloide Sattelpunkte modellieren. In vier 3D MR-Bildern und einem 3D CT-Bild schneiden wir an 10 verschiedenen Landmarken jeweils einen Ausschnitt der Größe 40 × 40 × 40 aus und betrachten diesen Bildausschnitt als ein einzelnes Bild, in dem genau eine Landmarke zu detektieren ist.

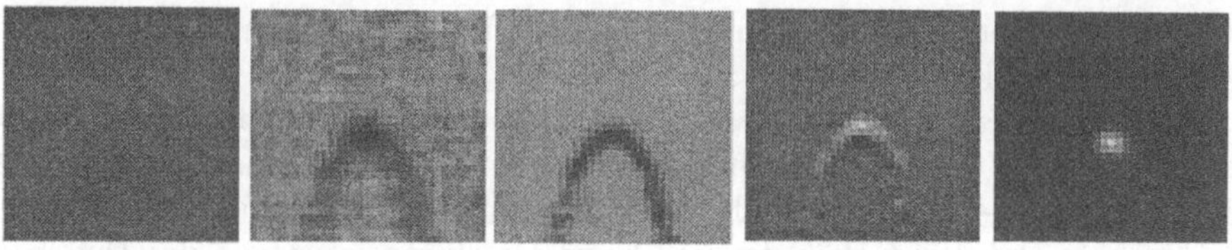

Abbildung1. Operatorantworten an einem Ellipsoid *Das 3D-Bild wurde mit Gaußrauschen der Varianz $\sigma_n^2 = 100$ überlagert. In den Abbildungen ist der Schnitt der Operatorbilder an der Spitze des Ellipsoids mit den Halbachsenlängen $a = 10$, $b = 8$ und $c = 40$ dargestellt.*

3.1 Operatorantworten

Die Operatoranworten in Abb. 1 zeigen das prinzipielle Verhalten einiger Operatoren aus Tabelle 1 bei Rauschen. Beim Operator *mittlere Krümmung* ist für die Spitze eines Ellipsoids (in diesem Fall die Landmarke) unter Rauschen kaum ein Signal in den Operatorantworten zu erkennen. Erst die Multiplikation mit dem Grauwertgradienten (Operator *Kitchen&Rosenfeld3D*) verbessert die Ergebnisse. Jedoch zeigt auch dort das Rauschen signifikante Auswirkungen auf die Operatorantworten. Die stärkere Bewertung des Grauwertgradienten beim Operator *Blom3D* unterdrückt das Rauschen an den Stellen, an denen der Grauwertgradient annähernd Null ist. Beim Operator *Beaudet3D* wird (wie auch bei den Operatoren *Kitchen&Rosenfeld3D* und *Blom3D*) nicht nur die Spitze des Ellipsoids, sondern auch die Oberfläche detektiert. Jedoch liefert dieser Operator an der Spitze des Ellipsoids zwei Extrema: ein Minimum und ein Maximum. Die Operatoren des Ansatzes „Förstner/Rohr" (siehe Operator *V1* in Abb. 1) zeigen die besten Ergebnisse: sie liefern fast ausschließlich an der Spitze des Ellipsoids signifikante Werte und reagieren kaum auf das Rauschen im Bild.

3.2 Anzahl korrespondierender Punkte

Wir untersuchen die Stabilität der Operatoren bzgl. Deformationen, indem wir die Bilder dreimal algorithmisch mittels einer elastischen Transformation (siehe [5]) deformieren und die Anzahl der korrespondierenden Punkte in allen Bildern automatisch ermitteln. Eine automatische Auswertung der Ergebnisse ist möglich, da die elastischen Transformationen in analytischer Form angebbar sind und somit die Orte der transformierten Landmarken exakt berechnet werden können. In der Untersuchung der Stabilität bzgl. Rauschen gehen wir entsprechend vor und überlagern die Bilder zusätzlich dreimal mit Gauß-Rauschen unterschiedlicher Varianz. Da die Operatoren unterschiedlich viele Detektionen liefern, teilen wir die Anzahl der korrespondierenden Punkte durch die Gesamtzahl der Detektionen und erhalten so den Anteil korrespondierender Punkte. Wir berechnen jeweils getrennt für die Tetraeder, die Ellipsoide, die hyperbolischen Paraboloide, die MR- sowie die CT-Bilder den Mittelwert der Ergebnisse und tragen die Mittelwerte in 10 Diagramme ein (als Beispiel für die hyperbolischen Paraboloide

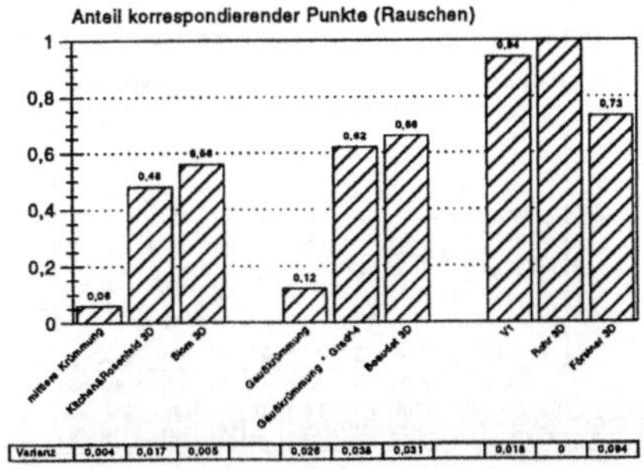

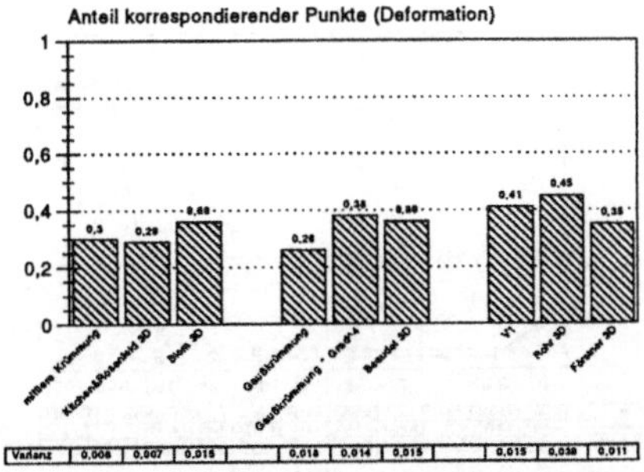

Abbildung2. Anteil korrespondierender Punkte in den verrauschten Bildern mit einem hyperbolischen Paraboloiden (links) und in den deformierten MR-Bildern (rechts) *In den Diagrammen sind die Mittelwerte und die Varianzen über alle untersuchten Bilder eingetragen.*

und die MR-Bilder siehe Abb. 2). Aufgrund der unterschiedlichen Rauschstufen und Deformationen erhalten wir insgesamt 154 synthetische und 301 tomographische Bilder in diesem Teil der Untersuchung.

Die Untersuchungsergebnisse für die synthetischen Bildern zeigen, daß die Operatoren *mittlere Krümmung* und *Gaußkrümmung* nicht geeignet sind, korrespondierende Punkte in verrauschten Bildern zu liefern (siehe z.B. Abb.2 links). Die Multiplikation mit dem Betrag des Grauwertgradienten (Operator *Kitchen&Rosenfeld3D*, *Blom3D*, *Gaußkrümmung*Grad^4)* verbessert deutlich die Ergebnisse. Die Operatoren des Ansatzes „Förstner/Rohr“ (Operator *V1*, *Rohr3D*, *Förstner3D*) liefern insbesondere in den verrauschten Bildern den größten Anteil korrespondierender Punkte. Dagegen unterscheiden sich die Ergebnisse der Operatoren in den MR-Bildern – sowohl bei der Untersuchung der Stabilität bzgl. Deformation als auch bzgl. Rauschen – nicht entsprechend deutlich (siehe z.B. Abb.2 rechts für die deformierten MR-Bilder). Die Operatoren *V1* und *Rohr3D* detektieren nur einen knapp höheren Anteil korrespondierender Punkte als die übrigen Operatoren. Die Operatoren *mittlere Krümmung* und *Gaußkrümmung* liefern allerdings in den MR- und CT-Bildern sehr viele Detektionen, wodurch dann Korrespondenzen rein zufällig entstehen können. Dies begründet z.T. den im Vergleich zu den synthetischen Bildern hohen Anteil korrespondierender Punkte in den MR- und CT-Bildern.

3.3 Klassische Detektionsleistung

Zur Untersuchung der klassischen Detektionsleistung ermitteln wir aus den Detektionen der Operatoren die Werte für P_{in}, P_{detect} und $P_{multiple}$ (siehe Gl. 1) und mitteln diese Werte jeweils getrennt für die Tetraeder, die Ellipsoide, die hyperbolischen Paraboloide, die MR- und CT-Bilder (für die MR-Bilder siehe Abb. 4). Die Landmarke gilt dabei als detektiert, wenn mindestens eine Detektion innerhalb des Detektionsgebiets (Größe $7 \times 7 \times 7$) der Landmarke liegt. Die synthetischen Bilder werden zusätzlich mit 10 unterschiedlichen Rauschstufen

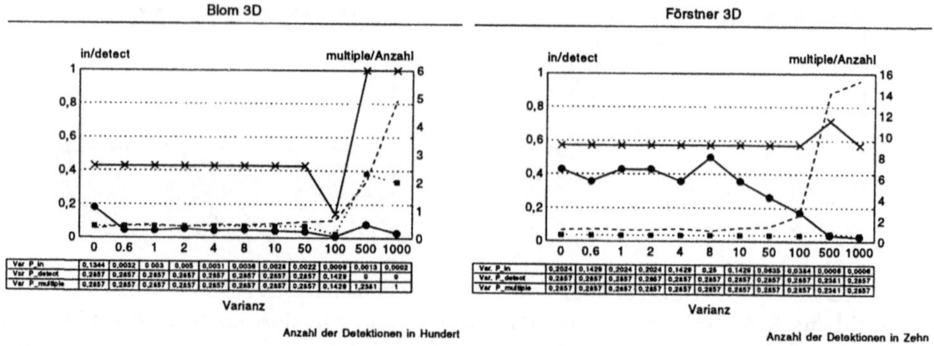

Blom 3D:

Varianz	0	0.6	1	2	4	8	10	50	100	500	1000
Var P_in	0,1344	0,0032	0,003	0,005	0,0031	0,0038	0,0028	0,0022	0,0008	0,0013	0,0002
Var P_detect	0,2857	0,2857	0,2857	0,2857	0,2857	0,2857	0,2857	0,2857	0,1429	0	0
Var P_multiple	0,2857	0,2857	0,2857	0,2857	0,2857	0,2857	0,2857	0,2857	0,1429	1,2381	1

Förstner 3D:

Varianz	0	0.6	1	2	4	8	10	50	100	500	1000
Var. P_in	0,2024	0,1429	0,2024	0,2024	0,1429	0,25	0,1429	0,0635	0,0384	0,0008	0,0006
Var. P_detect	0,2857	0,2857	0,2857	0,2857	0,2857	0,2857	0,2857	0,2857	0,2857	0,2341	0,2857
Var P_multiple	0,2857	0,2857	0,2857	0,2857	0,2857	0,2857	0,2857	0,2857	0,2857	0,2341	0,2857

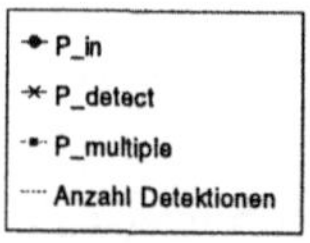

Abbildung3. Klassische Detektionsleistung der Operatoren Blom3D und Förstner3D in den Bildern mit einem Tetraeder *Die Mittelwerte der drei Maße sind in Abhängigkeit vom Rauschen aufgetragen. Außerdem ist die Anzahl der Detektionen in den Bildern, über die gemittelt wurde, mitangegeben. Die Werte für P_{in} und P_{detect} beziehen sich auf die linke Achse und $P_{multiple}$ und die Anzahl der Detektionen auf die rechte Achse des Diagramms. Die Anzahl der Detektionen ist in Hundert bzw. Zehn angegeben.*

(σ_n^2= 0.6, 1, 2, 4, 8, ,10, 50, 100, 500, 1000) überlagert und die Mittelwerte in Abhängigkeit vom Rauschen in Diagrammen aufgetragen (siehe Abb. 3). Damit werden in diesem Teil der Untersuchung insgesamt 242 synthetische und 43 tomographische Bilder verwendet (29 Diagramme).

Da die Operatoren des Ansatzes „Förstner/Rohr" primär die Landmarke detektieren (siehe z.B. Abb. 1) und kaum auf Rauschen reagieren, liefern diese Operatoren auch i.a. den größten Anteil korrekter Detektionen P_{in} (vgl. z.B. Operator *Blom3D* mit Operator *Förstner3D in* Abb. 3). Erst bei starkem Rauschen detektiert auch der Operator *Förstner3D* häufiger Punkte außerhalb des Detektionsgebiets, wodurch der Anteil korrekter Detektionen sinkt. Die Operatoren

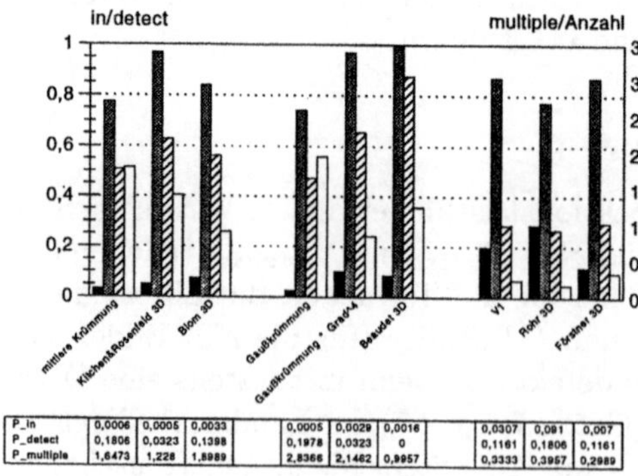

	mittlere Krümmung	Kitchen&Rosenfeld 3D	Blom 3D		Gaußkrümmung	Gaußkrümmung * Grad^4	Beaudet 3D		V1	Rohr 3D	Förstner 3D
P_in	0,0006	0,0005	0,0033		0,0005	0,0029	0,0016		0,0307	0,091	0,007
P_detect	0,1806	0,0323	0,1398		0,1978	0,0323	0		0,1161	0,1806	0,1161
P_multiple	1,6473	1,228	1,8989		2,8366	2,1462	0,9957		0,3333	0,3957	0,2989

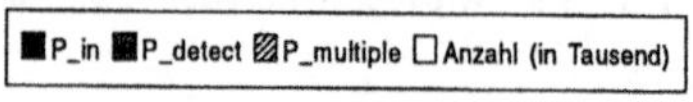

Abbildung4. Klassische Detektionsleistung in den MR-Bildern *Neben den Mittelwerten der Maße aus Abschnitt 2 sind in der Abbildung außerdem die Anzahl aller Detektionen in den Bildern, über die gemittelt wurde, angegeben. Unter dem Diagramm sind die Varianzen der Mittelwerte aufgeführt.*

der Ansätze „mittlere Krümmung“ und „Gaußkrümmung“ reagieren jedoch empfindlicher auf Rauschen und liefern teilweise soviele Detektionen, daß in den Detektionsgebieten zufällig ein Punkt detektiert wird. Dies drückt sich in den Diagrammen dadurch aus, daß bei stärkerem Rauschen sprunghaft alle Landmarken detektiert werden (siehe Operator *Blom3D* in Abb.3: $P_{detect} = 1$ für $\sigma_n^2 \geq 500$). Daß diese Operatoren viele Falsch-Detektionen liefern und dadurch auch zufällig Landmarken detektiert werden, zeigen auch die Ergebnisse in den MR-Bildern (siehe Abb. 4): Zwar detektieren die Operatoren *Kitchen&Rosenfeld3D, Gaußkrümmung*Grad^4* und *Beaudet3D* im Vergleich zu den Operatoren des Ansatzes „Förstner/Rohr“ eine größere Anzahl an Landmarken, jedoch ist der Anteil der korrekten Detektionen deutlich geringer. Zum Beispiel liegen nur etwa 5% der Detektionen des Operators *Kitchen&Rosenfeld3D* innerhalb eines Detektionsgebiets ($P_{in} = 0,05$). Im Vergleich dazu sind es beim Operator *Rohr3D* etwa 30% der Detektionen. Außerdem treten bei den Operatoren der Ansätze „mittlere Krümmung“ und „Gaußkümmung“ häufiger Mehrfach-Detektionen der Landmarke auf als bei den Operatoren des Ansatzes „Förstner/Rohr“ (z.B. $P_{multiple} \approx 3$ beim Operator *Beaudet3D*).

4 Zusammenfassung

Aus den insgesamt 39 ausgewerteten Diagrammen lassen sich zusammenfassend folgende Aussagen ableiten:

1. Die Operatoren des Ansatzes „Förstner/Rohr“ reagieren am stabilsten auf Rauschen und liefern daher signifikant weniger Detektionen als die Operatoren der Ansätze „mittlere Krümmung“ und „Gaußkrümmung“. Die Operatoren des Ansatzes „Förstner/Rohr“ detektieren dabei den größten Teil der Landmarken (in MR-Bildern über 70%), wodurch sie i.a. einen höheren Anteil korrekter Detektionen liefern.
2. Vergleicht man die Operatoren des Ansatzes „Förstner/Rohr“ miteinander, so reagiert i.a. der Operator *Förstner3D* am empfindlichsten auf starkes Rauschen im Bild und liefert daher im Vergleich zu den Operatoren *V1* und *Rohr3D* insbesondere in den tomographischen Bildern einen geringeren Anteil korrespondierender Punkte.
3. Die Multiplikation mit dem Betrag des Grauwertgradienten beim Ansatz „mittlere Kümmung“ und „Gaußkrümmung“ verbessert die Ergebnisse (vgl. [10]). Die Operatoren *Blom3D* und *Gaußkrümmung*Grad^4* ermitteln sowohl in den deformierten als auch in den verrauschten Bildern einen größeren Anteil korrespondierender Punkte als die Operatoren *mittlere Krümmung* und *Gaußkrümmung*. Der Operator *Blom3D* liefert bessere Ergebnisse als der Operator *Kitchen&Rosenfeld3D*.
4. Der Operator *Beaudet3D* führt in der gesamten Untersuchung zu schlechteren Ergebnissen als die Operatoren des Ansatzes „Förstner/Rohr“.

Die Untersuchung hat gezeigt, daß insbesondere die Operatoren des Ansatzes „Förstner/Rohr“ geeignet sind, korrespondierende Punkte zu liefern und die vorgegebenen Landmarken am besten detektieren.

Danksagung
Diese Arbeit enstand im Rahmen des von den Philips Forschungslaboratorien Hamburg finanzierten Projektes IMAGINE.

Literatur

1. I. E. Abdou and W. K. Pratt. Qualitative design and evaluation of enhancement/thresholding edge detector. *Proc. IEEE*, 67(5):753–763, May 1979.
2. P. Beaudet. Rotationally invariant image operators. In *Proc. 4th Int. Joint Conf. Patt. Recog.*, pages 579–583, November 1978. Kyoto, Japan.
3. W. Beil, K. Rohr, and H.S. Stiehl. Investigation of approaches for the localization of anatomical landmarks in 3D medical images. In *Proc. Computer Assisted Radiology and Surgery (CAR'97)*, pages Seite 265–270, Juni 1997.
4. J. Blom. *Topological and Geometrical Aspects of Image Structure.* PhD thesis, Utrecht University, 1992.
5. F. L. Bookstein. Principal warps: Thin-plate splines and the decomposition of deformations. *IEEE Transactions on Pattern Analysis and Machine Intelligence , June 1989*, PAMI-11(6):567–585, 1989.
6. E. S. Deutsch and J. R. Fram. A quantitative study of the orientational bias of some edge detector schemes. *IEEE Transactions on Computers*, 27(3):205–213, March 1978.
7. L. M. J. Florack, B. M. ter Haar Romeny, J. J. Koenderink, and M. A. Viergever. General intensity transformations and differential invariants. *Journal of Mathematical Imaging and Vision*, 4(2):171–187, 1994.
8. W. Förstner. A feature based correspondence algorithm for image matching. In *Int. Arch. Photogramm. Remote Sensing*, volume 26, pages 150–166, 1986.
9. S. Frantz, K. Rohr, and H. S. Stiehl. On the localization of 3D anatomical point-landmarks in medical imagery using multi-step differential approaches. In *Proc. 19. DAGM-Symposium Mustererkennung, Braunschweig/Germany*, pages 340–347. Springer-Verlag Berlin Heidelberg, Sept. 1997.
10. T. Hartkens, K. Rohr, and H. S. Stiehl. Evaluierung von Differentialoperatoren zur Detektion charakteristischer Punkte in tomographischen Bildern. In *Proc. 18. DAGM-Symposium Mustererkennung, Heidelberg/Germany*, pages 637–644. Springer-Verlag Berlin Heidelberg, Sept. 1996.
11. L. Kitchen and A. Rosenfeld. Grey level corner detection. *Pattern Recognition Letters*, 1:95–102, December 1982.
12. K. Rohr. Untersuchung von grauwertabhängigen Transformationen zur Ermittlung des optischen Flusses in Bildfolgen. *Diplomarbeit, Institut für Nachrichtensysteme, Universität Karlsruhe, Germany*, 1987.
13. K. Rohr. On 3D differential operators for detecting point landmarks. *Image Vision Computing*, 15(3):219–233, March 1997.
14. J. P. Thirion and A. Gourdon. Computing the differential characteristics of isointensity surfaces. *Computer Vision and Image Understanding*, 61(2):190–202, March 1995.
15. S. Venkatesh and L. J. Kitchen. Edge evaluation using necessary components. *Computer Vision, Graphics, and Image Processing; Graphical Models and Image Processing*, 54(1):23–30, January 1992.
16. O. A. Zuniga and R. M. Haralick. Corner detection using the facet model. In *Proc. IEEE Conf. on Computer Vision and Pattern Recognition*, pages 30–37, 1983.

Erkennung deformierbarer Konturen und Oberflächen in verrauschten Bilddaten durch Einsatz modellgekoppelter anisotroper Merkmalsextraktoren

M. Schreckenberg, G. von Dziembowski, D. Meyer-Ebrecht

Lehrstuhl für Meßtechnik
Rheinisch-Westfälische Technische Hochschule (RWTH), 52056 Aachen
Email: schreckenberg@lfm.rwth-aachen.de

Zusammenfassung. Die zuverlässige Detektion von Oberflächen oder Konturen in verrauschten Bildern erfordert den Einsatz von globalem Kontextwissen. Während bestehende Ansätze dieses Wissen für die Objektidentifikation in Merkmalsräumen einsetzen, die in unabhängigen Vorverarbeitungsschritten berechnet wurden, stellen wir einen Ansatz vor, der das Kontextwissen bereits bei der Extraktion lokaler Bildmerkmale verwenden kann. Zur Beschreibung des globalen Kontextes kommt dabei ein aktives Oberflächenmodell zum Einsatz, das mittels FEM realisiert wurde. Es wird exemplarisch die Anwendung auf Ultraschalldaten vorgestellt.

Schlüsselwörter: Anisotroper Merkmalsextraktor, aktive Oberflächen, Automatische Konturfindung

1 Einleitung

Bei verrauschten Bilddaten ist die lokale Bildinformation zu unzuverlässig um damit alleine eine Kontur- oder Oberflächenerkennung durchführen zu können. Vielmehr ist es notwendig, a priori Wissen sowohl über das Abbildungsverfahren als auch über das abgebildete Objekt hinzu zu nehmen. Während Wissen über den Abbildungsprozess im wesentlichen die Auswertung der lokalen Bildmerkmale betrifft, kann Information über das abgebildete Objekt den Erkennungsprozeß durch globales Kontextwissen stabilisieren. Speziell bei stückweise glattberandeten Objekten läßt sich dieses Kontextwissen sehr elegant über adäquat deformierbare Kontur- bzw. Oberflächenmodelle einbringen.

Da die Erkennung von Objektoberflächen im 3D-Raum eine Verallgemeinerung der Konturerkennung im 2D-Raum darstellt, beschränken wir uns bei unserer Beschreibung im weiteren auf die Erkennung von Objektoberflächen.

In der Literatur beschriebene Ansätze gehen dabei i.a. in zwei Stufen vor. Zunächst wird aus den Bilddaten unter Verwendung lokaler Extraktoren ein Merkmalsbild berechnet. Erst danach kommt das deformierbare Modell zum Einsatz, das so verformt wird, daß es bestmöglich zu dem ermittelten Merkmalsbild paßt [1, 2, 3, 4]. Dabei ist es prinzipiell gleichgültig, ob die optimale Anpassung in Analogie zu physikalischen Modellen über die Minimierung einer Energiebilanz bestimmt wird oder allgemeiner eine die Abweichung quantifizierende Kostenfunktion verwendet wird. Wesentlich ist, daß die Analyse der lokalen Bildmerkmale ohne die dem Objektmodell inhärenten Kontextinformationen auskommen muß.

Bei dem hier vorgestellten Ansatz wird ein lokaler Merkmalsextraktor eingesetzt, der an das deformierbare Objektmodell angekoppelt ist. Dadurch stehen ihm bei der Exploration der Bilddaten stetig genauer werdende Schätzwerte für lokale Position und Orientierung der gesuchten Objektoberfläche zur Verfügung. Insbesondere die lokale Orientierung erlaubt es anisotrope Extraktoren zu verwenden, die eine hohe örtliche Auflösung bei gleichzeitig hoher Signifikanz ermöglichen.

Das verwendete Verfahren wird hier exemplarisch am Beispiel der Endokarderkennung in transösophagealen Echokardiographiesequenzen erläutert.

2 Bildmaterial

Die Echokardiographie liefert 2D-Bildsequenzen des schlagenden Herzens (Abb. 1). Von besonderem Interesse ist dabei die linke Herzkammer (LV), da diese am häufigsten von pathologischen Veränderungen betroffen ist.

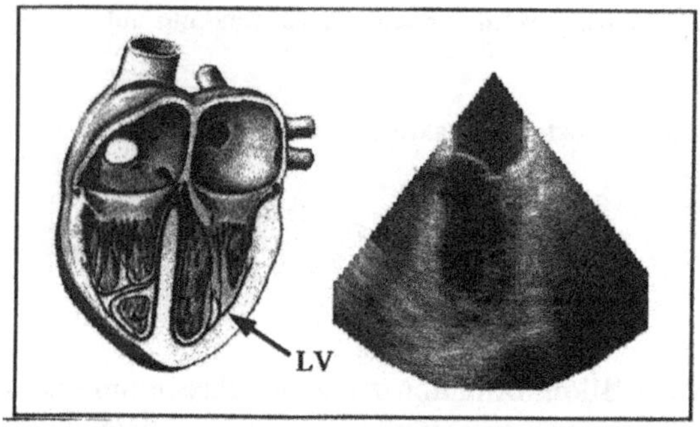

Abb. 1: Herz und zugehöriges Ultraschallbild

Die Bildgebung bei Ultraschall beruht auf Änderungen der lokalen akustischen Impedanz und hat damit differenzierenden Charakter, was den stark verrauschten Bildeindruck erklärt. Dazu kommen Abschattungseffekte, Interferenzmuster (Speckles), Mehrfachechos und mit dem Anschallwinkel variierende Reflexionseigenschaften des Gewebes.

Um auf diesem Bildmaterial eine stabile Konturdetektion zu realisieren, haben wir nicht nur den räumlichen Kontext innerhalb einer 2D-Schicht modelliert, sondern zusätzlich den zeitlichen Kontext. Im Gegensatz zu Ansätzen aus der Literatur [3] wird dazu aber kein kinematisches Konturmodell verwendet, sondern ein statisches Oberflächenmodell. Die Einzelbilder der Zeitsequenz werden dazu hintereinander zu einem Volumen angeordnet, wobei Anfang und Ende der Sequenz zum Abklingen der Randeffekte periodisch verlängert werden (Abb. 2).

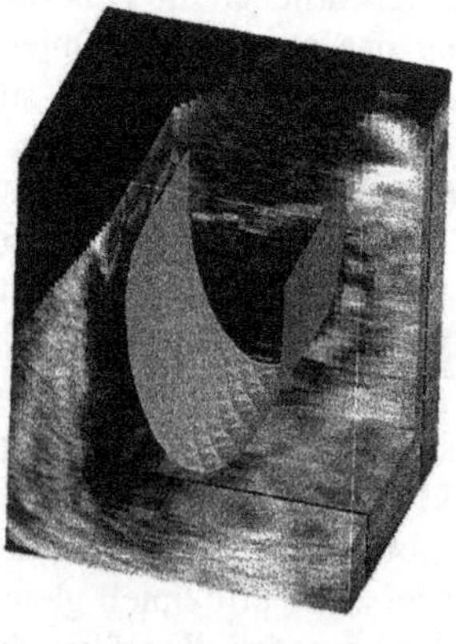

Abb. 2: Zeitsequenz als Volumen (aufgeschnitten)

Gesucht wird hier die Grenze zwischen dem blutgefüllten Innenraum (Cavum) und dem Herzmuskel (Myokard). Die dort befindliche auskleidende Haut wird Endokard genannt.

3 Der Merkmalsextraktor

Der hier eingesetzte anisotrope Merkmalsextraktor vergleicht die lokalen Bildmerkmale in zwei benachbarten Teilmengen A und B. Im hier vorgestellten Beispiel sind diese Gruppen quaderförmig (Abb. 3). Dabei wird davon ausgegangen, daß die "Materialstärke" des Objektes mindestens der Höhe einer quaderförmigen Teilmenge entspricht.

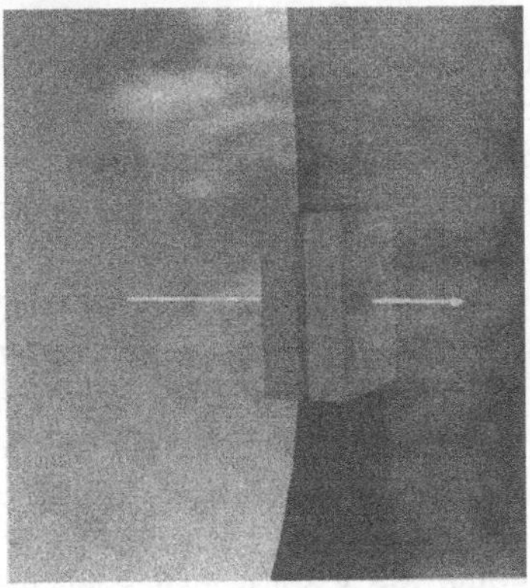

Abb. 3: anisotroper Merkmalsextraktor

Der Extraktor befindet sich genau dann auf einer Objektgrenzfläche, wenn Menge A komplett außerhalb und Menge B komplett innerhalb des Objektes liegt. Die maximale Sensitivität wird daher erreicht, wenn die Trennfläche zwischen den Mengen A und B genau parallel zu der gesuchten Objektoberfläche orientiert ist. Steht sie senkrecht dazu, wird keine Oberfläche detektiert. Zudem kann bei nicht massiven, d.h. schalenförmigen Objekten unterschieden werden, ob es sich um innere oder äußere Grenzfläche handelt. Diese selektiven Eigenschaften sind weitere wichtige Vorteile des anisotropen Extraktors.

Dem Extraktor kann nun durch Angabe eines Normalenvektors $\vec{n}$ ein Schätzwert für die Orientierung der gesuchten Grenzfläche vorgegeben werden. Entsprechend orientierte Objektoberflächen können so mit maximaler Sensitivität gesucht werden. Störstrukturen, die senkrecht dazu liegen, werden dabei ignoriert.

Der Extraktor liefert also zu Ort $\vec{x}$ und Orientierung $\vec{n}$ ein Maß Q für die Wahrscheinlichkeit, daß dort eine entsprechend orientierte Grenzfläche vorliegt. Dabei bezeichnet P_i die Wahrscheinlichkeit, daß sich Teilmenge i noch komplett außerhalb des Objektes befindet.

$$Q(\vec{x},\vec{n}) = P_A(\vec{x},\vec{n}) - P_B(\vec{x},\vec{n}) \tag{1}$$

Für Grenzflächen, die statt zum gesuchten Übergang Hintergrund/Objekt zum Übergang Objekt/Hintergrund gehören, wird Q negativ.

3.1 Bestimmung der Wahrscheinlichkeitsfunktion

Für die Bestimmung der Wahrscheinlichkeit P_i muß a priori Wissen über das Abbildungssystem und die Objekteigenschaften verwendet werden.

Bei medizinischen Ultraschallaufnahmen wird hier vereinfachend angenommen, daß der Betrag der Rückstreuamplitude - und damit auch die Grauwertamplituden im Bild - Rayleigh-verteilt sind. Die freien Parameter Mittelwert und Varianz weisen dabei einen linearen Zusammenhang auf.

Diese Annahme basiert darauf, daß Rückstreuung im wesentlichen an mikroskopisch kleinen Streuzentren erfolgt, deren Abmessung unterhalb der Auflösungsgrenze der verwendeten Wellenlänge liegt. Die Anzahl dieser Streuzentren muß dabei gegen unendlich streben. Das rückgestreute Signal ergibt sich als Interferenzmuster (Speckles) der Einzelstreuungen. Unter diesen Voraussetzungen können Objekte dann unterschieden werden, wenn sie sich hinsichtlich der Dichte der Rückstreuzentren unterscheiden.

Dieser Ansatz berücksichtigt in keiner Weise strukturelle Informationen wie z.B. Textur oder geometrische Information wie die relative Lage zum Transducer. Es soll hier aber gezeigt werden wie auch mit vergleichsweise primitiven Bildmerkmalen, bei gleichzeitiger Ausnutzung der Kontextinformation, stabile Ergebnisse erzielt werden können.

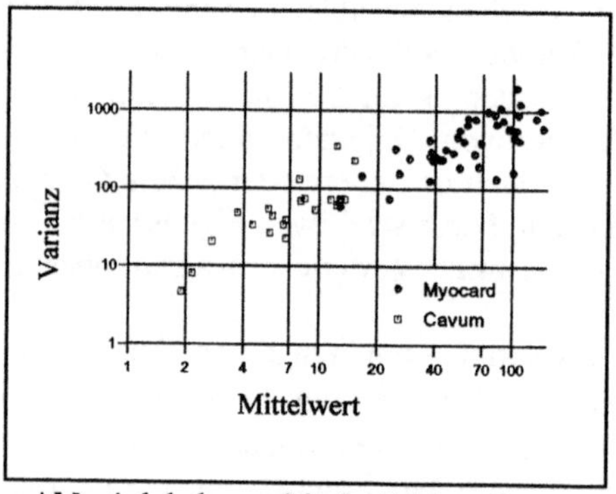

Abb. 4: lokale empirische Stichproben

Abb. 4 zeigt die Ergebnisse von lokalen empirischen Stichproben im Bildmaterial. Die Stichprobengröße entspricht hierbei der Größe einer der Teilmengen des beschriebenen Merkmalsextraktors. Je kleiner diese Teilmenge ist, desto instabiler werden die empirischen Schätzungen von lokalem Mittelwert und lokaler Varianz. Der lineare Zusammenhang zwischen Mittelwert und Varianz ist im doppelt logorithmischen Maßstab aber noch gut zu erkennen. Gleichzeitig zeigt die Graphik, daß erwartungsgemäß die Schätzung des 2. zentralen Momentes unzuverlässiger ist, als die des 1. Momentes.

Für die Ermittlung der Wahrscheinlichkeit P_i wird hier daher der lokale Mittelwert verwendet. Insbesondere bei kleinen Stichprobengrößen liefert dabei der Median eine zuverlässigere Schätzung des lokalen Mittelwertes als die direkte Berechnung .

Über Grauwerthistogramme lassen sich nun die Wahrscheinlichkeiten ermitteln, daß ein Voxel mit dem lokalen mittleren Grauwert G zum Objekt (hier Myocard) bzw. zum Hintergrund (Cavum) gehört (Abb. 5).

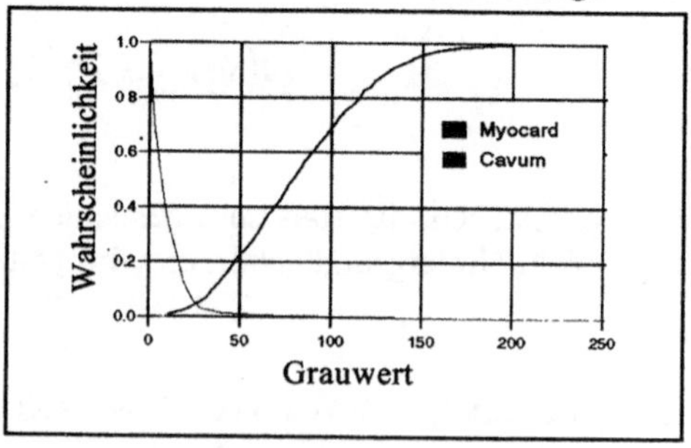

Abb. 5: Wahrscheinlichkeitskurven

Die Wahrscheinlichkeitskurve verläuft für Hintergrundvoxel deutlich steiler. Sie ist daher als Basis für die Ermittlung der Wahrscheinlichkeit P_i geeigneter. Alternativ kann auch die bedingte Wahrscheinlichkeit, daß das betrachtete Voxel zum Hintergrund und nicht zum Objekt gehört, verwendet werden.

4 Das deformierbare Modell

Wir machen folgende Annahmen:

- Die Objektoberfläche ist glatt und zusammenhängend (mind. C_1-stetig).
- Die lokale Bildinformation ist durch Specklerauschen und Aufnahmefehler (lokale Signalausfälle, Abschattungen, Mehrfachechos) beeinträchtigt.

Daraus ergeben sich notwendige Eigenschaften des Oberflächenmodells, das Bereiche mit lokalen Störungen überbrücken und statistische Fluktuationen der Objektgrenze approximierend unterdrücken muß.

Das hier realisierte aktive Oberflächenmodell bildet das physikalische Verhalten einer sowohl elastischen als auch biegesteifen Schale nach, die unter Einwirkung der Bildinformation verformt wird [1, 2].

Die Lage des Oberflächenmodells ist über die Abbildung X vollständig definiert.

$$X:\Omega=[0,1]\times[0,1]\rightarrow\Re \quad X(u,v)=(x(u,v),y(u,v),z(u,v)) \tag{2}$$

Physikalisch betrachtet nimmt das Oberflächenmodell den Verformungszustand an, bei dem äußere und innere Energie im Gleichgewicht stehen. Die innere Energie ergibt sich dabei aus den Materialeigenschaften und den vorherrschenden Dehnungen, Biegungen und Scherungen, die äußere Energie resultiert aus den angreifenden Kräften, die diese Verformungen verursachen.

$$E_{ext}(X) = \int\int \vec{x}\vec{f}\,(u,v)dudv \tag{3}$$

$$E_{int}(X) = \iint \alpha_{01}\left|\frac{\partial X}{\partial u}\right|^2 + \alpha_{10}\left|\frac{\partial X}{\partial v}\right|^2 + \beta_{02}\left|\frac{\partial^2 X}{\partial u^2}\right|^2 + \beta_{11}\left|\frac{\partial^2 X}{\partial u \partial v}\right|^2 + \beta_{20}\left|\frac{\partial^2 X}{\partial v^2}\right|^2$$

Das Schalenmodell wurde mit der Methode der Finiten Elemente approximiert. Dabei wurden Hermite C_1 Dreieckselemente mit 3 Knoten und 6 Freiheitsgraden pro Knoten und Raumrichtung verwendet [1, 6].

5 Ankopplung des Extraktors an das Modell

Zu jedem Punkt auf der Modelloberfläche können Position und Normalenrichtung berechnet werden. Der Extraktor wird in Normalenrichtung innerhalb eines vorgege-

benen Suchradius bewegt. Als Orientierung wird ebenfalls die Normalenrichtung verwendet. Der Extraktor liefert sowohl die Position zurück, bei der sich mit größter Wahrscheinlichkeit eine Objektgrenzfläche befindet, als auch die Größe der Wahrscheinlichkeit selbst als Qualitätsmaß. Eine negative Güte Q nach (1) wird hier ignoriert.
Aus diesen Angaben wird nun die angreifende Kraft berechnet. Der Abstand der gefundenen Position von der Modelloberfläche wird dabei über ein Abstandsgesetz gewichtet. Damit wird berücksichtigt, daß auch die lokale Modellposition eine stetig genauer werdende Schätzung für die Position der tatsächlichen Objektoberfläche darstellt. Näher liegende Objektoberflächen also wahrscheinlicher sind als entfernte. Das Abstandsgesetz wurde hier dem Gravitationsgesetz nachempfunden, wobei die punktförmige Probemasse das Zentrum der das Gravitationsfeld verursachenden räumlichen Masseverteilung tatsächlichen erreichen kann.
Die Richtung der Kraftvektoren wird auf die Normalenrichtung der Modelloberfläche gesetzt (Blendenproblem). Die berechneten Kräfte greifen nun wieder am Oberflächenmodell an und verursachen eine Verformung.

6 Resultate

Abbildungen 6 und 7 zeigen die Anwendung auf synthetische Daten. Die Grenzflächen innerhalb eines binären Datenkubus sind dunkelgrau dargestellt, die Modelloberfläche hellgrau. In Abbildung 6 sind drei Schritte des iterativen Verformungsprozesses zu sehen, wobei die Modelloberfläche zur besseren Darstellung nach oben verschoben wurde. Abbildung 7 zeigt die endgültige Position.

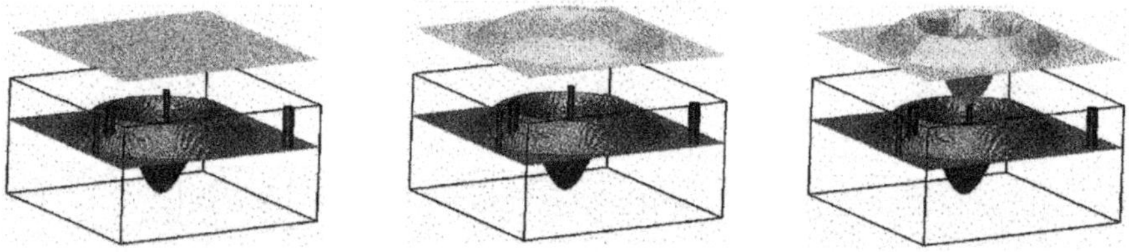

Abb. 6: Drei Iterationsschritte des Verformungsprozesses. Störstrukturen, die weitgehend senkrecht zur gesuchten Oberfläche stehen, werden ignoriert.

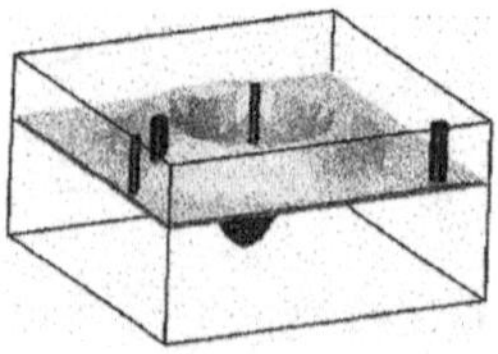

Abb. 7: Angepaßtes Modell

Es ist deutlich zu erkennen, daß bei der Oberflächendetektion die säulenförmigen Störstrukturen ignoriert werden, da ihre Grenzflächen weitgehend senkrecht zur gesuchten Oberfläche stehen.

Abbildung 8 zeigt 3 Schritte der iterativen Verformung bei der Ultraschallsequenz. Durch die Selektivität des Extraktors spielt es keine Rolle, ob das Modell innerhalb oder außerhalb des Cavum initialisiert wird.

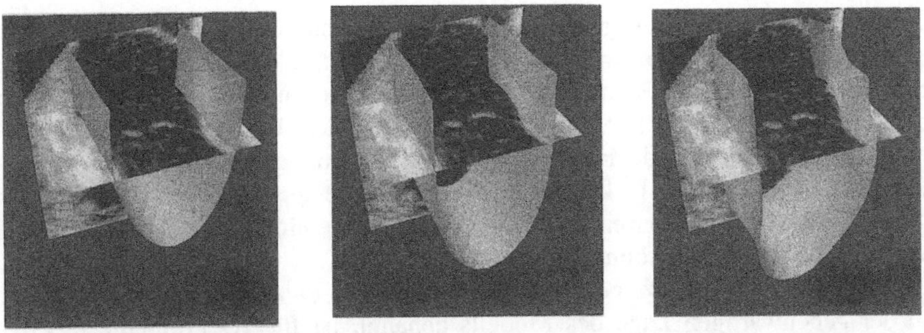

Abb. 8: Iterative Verformung des Oberflächenmodells in der Ultraschallsequenz

In den Abbildungen 9 und 10 ist die Stabilität des Verfahrens gegenüber lokalen Störungen gezeigt. In Abbildung 9 läßt sich ein sogenannter Papillarmuskel ausmachen (Pfeil). Er stellt keine Bildstörung im eigentlichen Sinne dar, denn die zugehörige anatomische Struktur ist tatsächlich vorhanden. Er gehört allerdings nicht mehr zum Myocard hinzu. Aufgrund des globalen Objektmodells und der Selektivität des Merkmalsextraktors wird hier der Papillarmuskel korrekterweise ignoriert. Er widerspricht der Forderung nach lokaler Glattheit und der überwiegende Teil seiner Grenzflächen verläuft nicht genau parallel zur Endokardfläche. Abbildung 10 zeigt das Verhalten bei einem lokalen Signalausfall durch Abschattung (Pfeil). Aufgrund des räumlichen und zeitlichen Kontextes wird diese Stelle überbrückt.

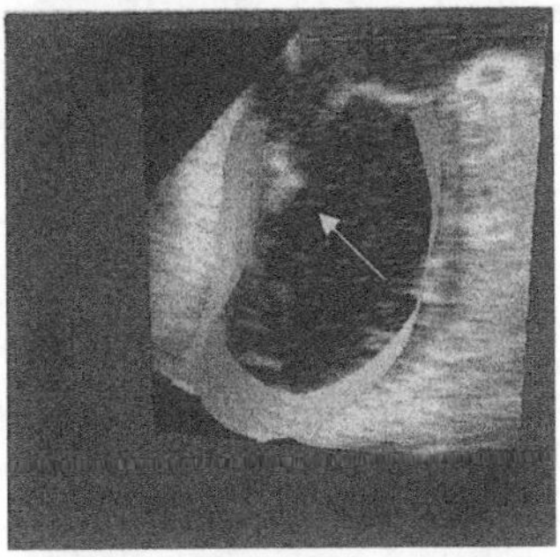

Abb. 9: Störung: Papillarmuskel

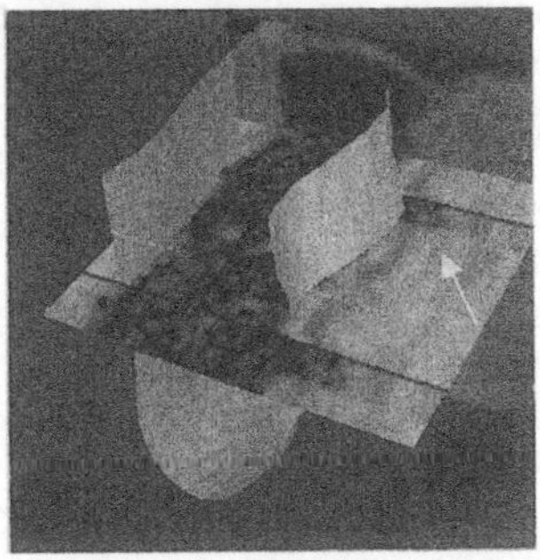

Abb. 10: Störung: Signalausfall

In Abbildung 11 sind die gefundenen Konturen in die entsprechenden Einzelbilder der Echokardiographiesequenz eingeblendet.

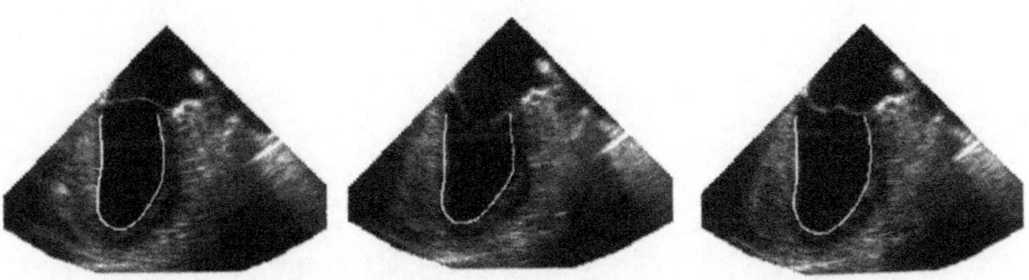

Abb. 11: Echokardiographische Einzelbilder mit eingeblendeten Konturen

Zusammenfassend kann gesagt werden, daß der vorgestellte Extraktor durch Verwenden von globaler Kontextinformation bereits bei der Extraktion lokaler Bildmerkmale eine hohe Signifikanz erreicht. Besonders bei stark verrauschten Bilddaten und entsprechend hoher statistischer Unsicherheit der lokalen Bildmerkmale kann dieses Verhalten gewinnbringend eingesetzt werden. Die Anwendung ist dabei nicht auf FEM Modelle beschränkt. Es können ebensogut andere approximierende deformierbare Oberflächenmodelle verwendet werden, sofern diese ein Gütemaß in ihrer Kostenfunktion berücksichtigen können.

Ein verwandter Ansatz, bei dem die Berechnung lokaler Bildenergien ebenfalls von der gegenwärtigen Lage des Modells abhängt, ist für 2D Probleme in [5] beschrieben. Der dort verwendete Algorithmus besitzt allerdings keine global stabilisierenden Eigenschaften, eignet sich also nicht zum Kompensieren lokaler Störungen.

7 Literatur

1. McInerney T, Terzopoulos D: A dynamic finite element surface model for segmentation and tracking in multidimensional medical images with application to cardiac 4D image analysis, Computerized Medical Imaging and Graphics, Vol. 19, No. 1, pp.69-83,1995
2. Cohen L, Cohen I: Finite-Elemente Methodes for Active Contour Models and Balloons for 2-D and 3-D Images, IEEE Trans. PAMI, Vol. 15, No. 11, 1993
3. Leymarie L, Levine MD: Tracking Deformable Objects in the Plane Using an Active Contour Model, IEEE Trans. PAMI, Vol. 15, No. 6, 1993
4. Chalana V et al.: A Multiple Active Contour Model for Cardiac Boundary Detection on Echocardiographic Sequences, IEEE Trans. Med. Img., Vol. 15, No. 3, 1996
5. Ronfard R: Region-Based Strategies for Active Contour Models, International Journal of Computer Vision,13:2, 229-251 (1994)
6. Dhatt G:Finite Elemente Methode Displayed, Wiley & Sons 1985, ISBN 047190110 5

Real-Time-Tracking von Augenbewegungen: Automatische Mikroperimetrie mit dem Scanning-Laser-Ophthalmoskop

A. Dölemeyer, H. Liebau, F. Toonen*, S. Wolf*, D. Meyer-Ebrecht.

Lehrstuhl für Meßtechnik
*Augenklinik der Medizinischen Fakultät

Rheinisch-Westfälische Technische Hochschule (RWTH), 52056 Aachen
Email: doelemeyer@rwth-aachen.de

Zusammenfassung: In diesem Beitrag wird ein effizienter Algorithmus zur Detektion und Kompensation von Augenbewegungen in On-Line-Bildsequenzen der Retina vorgestellt. Dabei wurde die Ermittlung des Verschiebungsvektors in die Bestimmung orthogonaler Komponenten aufgeteilt, deren sukzessive Berechnung anhand der Lage von Gefäßkanten durchgeführt wird. Bei der Implementierung in ein Mikroperimetrie-Untersuchungssystem auf der Rechnerplattform WindowsNT konnte ein Real-Time-Verhalten erreicht werden. Nicht auswertbare Positionen hervorgerufen durch Lidschlag oder saccadische Augenbewegungen werden automatisch erkannt und zurückgewiesen. Falsche Positionsbestimmungen konnten über ein Zuverlässigkeitsmaß vollständig ausgeschlossen werden.

Einleitung

Die Untersuchung der Funktion der Wahrnehmung durch das Visuelle System des Menschen bildet einen wichtigen Abschnitt bei der Diagnose verschiedener Augenerkrankungen [1]. Diese Perimetrie genannte Untersuchung ist ein psychophysisches Meßverfahren, bei dem der Patient das Erkennen eines Lichtstimulus vor einer Grundhelligkeit bestätigen muß [2].

Die Verwendung des Scanning-Laser-Ophthalmoskop (SLO) zur Bestimmung der Lichtunterschiedsempfindlichkeit der Retina eröffnet durch die hohe Ortsauflösung neue Möglichkeiten bei der Untersuchung auch kleiner Gesichtsfeldausfälle im Bereich der Makula [3,4]. Dies ist besonders bei der Diagnose der (senilen) Makula-Degeneration und der diabetischen Retinopathie von großer Bedeutung. Die unvermeidlichen Augenbewegungen des Patienten schränken jedoch die Validität der erzielbaren Ergebnisse stark ein.

Ein großer Vorteil bei der Nutzung des SLOs für die (Mikro-)Perimetrie liegt in der Verfügbarkeit eines Videobildes der Retina während des gesamten Untersuchungszeitraums. Damit kann eine direkte Funduskontrolle auf Augenbewegungen erfolgen.

Bei der Mikroperimetrie definiert der Untersucher einen Bereich der Retina, der in einem zu wählenden Stimulationsraster getestet werden soll. Innerhalb des Rasters werden dem Patienten/Probanden Lichtpunkte präsentiert, deren Wahrnehmung er mit Hilfe eines Drucktasters bestätigt. An jedem Rasterpunkt werden die Intensitäten nun solange variiert, bis ein Schwellwert für die Wahrnehmung festgestellt werden kann. So erhält man eine Darstellung des Gesichtsfelds des Patienten. Abbildung 1 zeigt ein Fundusbild mit überlagertem Testergebnis. Neben der Wahrnehmungsschwelle an den Rasterpunkten sind auch die Fixationsorte für die einzelnen Stimuli abgebildet.

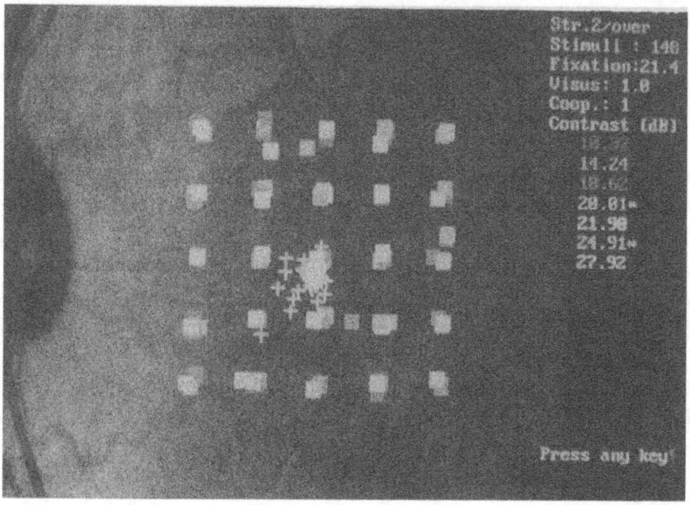

Abbildung 2: Ergebnis einer mikroperimetrischen Untersuchung. Die Wahrnehmungsschwelle ist farblich kodiert. Die Fixationsorte sind als Kreuze markiert.

Augenbewegungen führen zu Artefakten und beeinflussen das Untersuchungsergebnis deutlich, bis hin zur Unbrauchbarkeit ganzer Untersuchungen. Der individuelle Stimulus, der sich bei einer Dauer von 200 ms über mehrere Videobilder erstreckt, wird bei Auftreten einer Bewegung verwischt, so daß seine Intensität geringer wird und der Testfleck sich vergrößert. Augenbewegungen zwischen einzelnen Stimuli führen dazu, daß das vorgewählte Raster nicht mehr eingehalten wird, wovon auch die Bestimmung der einzelnen Schwellwerte betroffen ist (Abb. 2).

Abbildung 2: Artefaktbildung durch Augenbewegungen: Verwischung eines Stimulus über mehrere Bilder (links), Verschiebung der Rasterpunkte (rechts)

Eine manuelle Nachführung, wie sie in experimenteller Software bereits zum Einsatz kommt [4], bringt erhebliche Belastungen für den Untersucher mit sich. Er muß für jeden Stimulus entweder nach der Präsentation oder direkt davor einen Referenzpunkt im Bild, beispielsweise eine Gefäßgabelung, markieren. Das verlangt während der üblichen Untersuchungszeit von 20 min und der Präsentation von bis zu 200 Stimuli die volle Aufmerksamkeit des Untersuchers. Die Untersuchungszeit verlängert sich deutlich, was sich über die zeitliche Beschränkung der Aufmerksamkeit des Patienten auch negativ auf die Validität der Ergebnisse auswirkt.

Ansätze für die automatische Erkennung von Augenbewegungen sind in der Ophthalmologie bereits verfolgt worden [7, 8]. Ziel war es hierbei, die Laserkoagulation – den therapeutischen Einsatz eines höherenergetischen Lasers zur Verödung von Gefäßen - zu unterstützen. Die dabei realisierten Auswertungsraten von maximal 5 Bildern/sec sind für die Mikroperimetrie jedoch nicht ausreichend.

Ansatz für die Bestimmung des Verschiebungsvektors

Um das Ziel einer schritthaltenden Erkennung der Funduspositon im Videodatenstrom zu erreichen, muß die Bestimmung des Verschiebungsvektors während eines Halbbildes erfolgen. Für die Berechnung stehen somit nicht mehr als 20 ms zur Verfügung. Das verwendete Verfahren muß also soweit als möglich – auch numerisch – vereinfacht werden.

Ein Ansatz für ein schnelles Verfahren läßt sich bei Verwendung von a-priori-Wissen über den Bildinhalt von Fundusbildern entwickeln. Die Fundusmorphologie mit der Lage des Sehnerven und der Makula und dem Verlauf der retinalen Gefäße zeigt Strukturen, deren Verschiebung in einem Bild effizient bestimmt werden kann. Beispielsweise kann die Verschiebung einer L-förmigen Struktur durch die Verschiebung der beiden Arme des L entlang von orthogonalen Richtungen direkt bestimmt werden (Abb. 3).

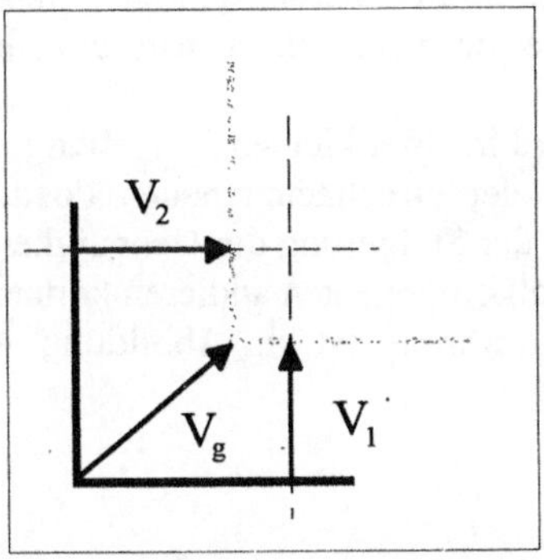

Abbildung 3: Die Verschiebung V_g des L-förmigen Objekts kann aus den orthogonalen Komponenten V_1 und V_2 bestimmt werden.

Um die Verschiebung der Arme zu bestimmen, braucht nur die Lage der Kanten entlang einer zur Kantenrichtung senkrechten Suchlinie ermittelt zu werden. Die

Verschiebungskomponenten können dann zur Gesamtverschiebung zusammengesetzt werden.

Die oben beschriebene L-förmige Struktur läßt sich in der Regel in den Fundusbildern in Form von Gefäßabschnitten wiederfinden. Da bei der Mikroperimetrie häufig in der Nähe der Makula untersucht wird, sind in den Fundusbildern der Sehnerv und die von ihm ausgehenden großen retinalen Gefäße, die sich in die vier Bildquadranten verzweigen, abgebildet. Diese Gefäße verlaufen über eine gewisse Länge annähernd gerade und es lassen sich meist - verteilt auf zwei Quadranten - Abschnitte finden, die zueinander senkrecht verlaufen.

Die Gefäßkanten haben einen ausreichend guten Kontrast, um ihre Position entlang einer Linie mit einem einfachen, kreuzkorrelationsbasierten Template-Matching-Verfahren zu finden.

Algorithmus

Die Bestimmung des Verschiebungsvektors läuft in mehreren Stufen ab. Die Komponenten des Verschiebungsvektors werden einzeln berechnet. Dabei ergeben sich ein oder mehrere Kandidaten für die neue Gesamtposition.

Für das Aufsetzen des Trackings werden zwei Positionen auf Gefäßkanten ausgewählt, die den oben beschriebenen Gefäßverlauf zeigen. An diesen Stellen wird automatisch die Richtung der Gefäße bestimmt. Senkrecht zu den Gefäßrichtungen wird nun ein eindimensionales Template erzeugt, das die Kante vollständig einschließt. Entlang der Linie wird mit dem Template die neue Position der Gefäßkante bestimmt, wobei der normierte Kreuzkorrelationskoeffizient als Zuverlässigkeitsmaß benutzt wird. Das Maß mit der höheren Qualität wird als erste Komponente der Verschiebung akzeptiert.

In der zweiten Stufe wird senkrecht zur ersten ermittelten Verschiebung nach der Position der zweiten Gefäßkante gesucht. Dabei wird die schon errechnete Verschiebungskomponente mit berücksichtigt. Insgesamt ergeben sich so Kandidatenpositionen, die im nächsten Schritt auf ihre Gültigkeit hin überprüft werden.

In der dritten Stufe wird in einer kleinen Umgebung mit einem zweidimensionalen Template die Richtigkeit der errechneten neuen Position überprüft. Dieser Schritt dient darüber hinaus auch der Steigerung der Genauigkeit, da sich Ungenauigkeiten in der Bestimmung der Einzelkomponenten addieren können.

Der Ablauf des Algorithmus ist in Abbildung 4 noch einmal schematisch dargestellt.

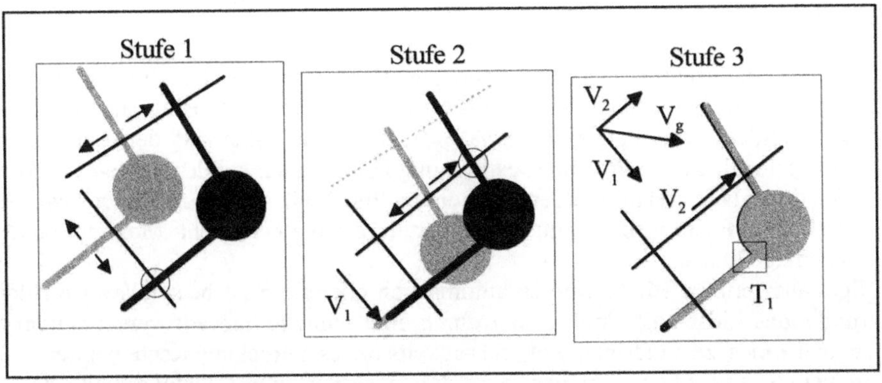

Abbildung 4: Die 3-stufige Suche zur Bestimmung der resultierenden Verschiebung V_g aus den orthogonalen Komponenten V_1 und V_2. Das Template T_1 dient zur Validierung der ermittelten Verschiebung. Zu beachten ist, daß in Stufe 1 die Richtungen nur näherungsweise orthogonal sein müssen.

Implementierung in das Untersuchungssystem

Am Lehrstuhl für Meßtechnik wurde in Zusammenarbeit mit der Augenklinik der RWTH Aachen eine Steuersoftware zur Durchführung der Mikroperimetrie mit dem SLO entwickelt [5]. Sie sorgt für die Erzeugung und Auswahl des Testrasters und der für die Untersuchung notwendigen Einstellungen und wertet die Reaktionen des Patienten über einen angeschlossenen Drucktaster aus. In dieses System wurde das automatische Tracking integriert [6]. Nachdem der Untersucher den zu testenden Bereich und das Raster ausgewählt hat, setzt er das Tracking auf. Dazu wählt er die Gefäßkanten aus und selektiert wenn möglich eine signifikante Gefäßverzweigung, die sich eindeutig wiederfinden läßt. Ist dies erfolgt, übernimmt der Trackingprozeß die weitere Steuerung der Untersuchung. Die Stimuli werden mit unterschiedlicher Intensität präsentiert, wobei über den Stimuluszeitraum von 200ms alle Bilder aufgezeichnet werden. Diese werden dann auf Augenbewegungen überprüft und das normale Tracking von Bild zu Bild fortgesetzt.

Das Perimetriesystem wie auch der Trackingalgorithmus wurden unter dem Betriebsystem WindowsNT implementiert, um einerseits die Vorteile der einfachen Erstellung eines Userinterface nutzen zu können und andererseits von preiswerten Standardkomponenten für die Bildaufnahme des Videosignals Gebrauch machen zu können.

Problematisch ist die mangelnde Echtzeitfähigkeit von WindowsNT. Eine Real-Time-Verarbeitung kann nicht garantiert werden. Bei geringer Systemauslastung und der Ausnutzung von speziellen Mechanismen können jedoch die Zeitbedingungen eingehalten werden. Durch die Anwendung von Multithreading mit schnellen Kontextwechseln und dem Umgehen von Windows-Kommunikationswegen durch die Verwendung von Statusvariablen konnte eine schritthaltende Verarbeitung erzielt werden.

Ergebnisse

Mit dem vorgestellten Algorithmus können Verschiebungen bis zu etwa 20% der horizontalen Bilddimension der CCIR-Auflösung von 768 Bildpunkten sicher bestimmt werden. Dies entspricht etwa 6 Grad Raumwinkel auf der Retina. Die geforderte schritthaltende Positionsbestimmung konnte auch bei der Implementierung unter Windows NT erreicht werden. Auf einem 200 MHz –Pentium-System wird für die vollständige Positionsbestimmung im Mittel 18,5 ms gebraucht. Maximal beträgt die Rechenzeit 20 ms.

Nicht auswertbare Bilder werden automatisch erkannt. Sind beispielsweise Bilder aufgrund eines Lidschlags dunkel oder durch eine schnelle Saccade verzerrt, können sie anhand eines zu niedrigen Zuverlässigkeitsmaßes zurückgewiesen werden. Ist dadurch ein Stimulus betroffen, wird auch dieser verworfen und später erneut getestet.

Der Algorithmus wurde an Videomitschnitten verschiedener mikroperimetrischer Untersuchungen getestet. Darüber hinaus wurde die Rauschempfindlichkeit mit künstlichen Testsequenzen überprüft.

Der Einfluß von Rauschen auf das Ergebnis ist gering. Erst ab einer Standardabweichung, die um einen Faktor 3 über der beobachtbaren liegt, tauchen rauschbedingte Fehler in der Positionsbestimmung auf.

In Abbildung 5 sind die Ergebnisse der Positionsbestimmung für zwei Sequenzen dargestellt, die auch die Extremfälle beinhalten.

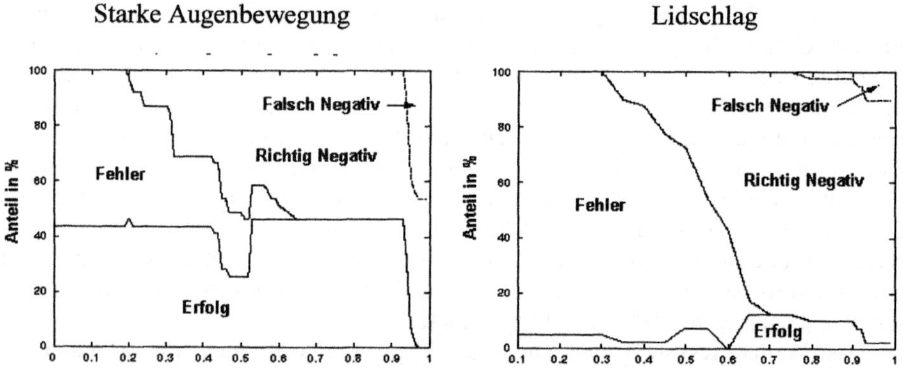

Abbildung 5: Ergebnisse der Positionsbestimmung für zwei Sequenzen. Aufgetragen sind die prozentualen Anteile der verschiedenen Erkennungsklassen über dem Zuverlässigkeitsmaß.

In der ersten Sequenz treten große Augenbewegungen auf, bei denen die gewählten Gefäße den Bildbereich verlassen, so daß keine übereinstimmende Position mehr gefunden werden kann. In der zweiten Sequenz tauchen durch häufigen Lidschlag viele schwarze Bilder auf, die sich in nur noch wenigen richtig erkennbaren und erkannten Positionen widerspiegeln.

Zusammenfassung

In diesem Beitrag wurde ein Verfahren vorgestellt, mit dem das Real-Time-Tracking von Augenbewegungen in Videosequenzen der Retina realisiert werden konnte. Dieser Algorithmus wurde in eine Untersuchungssoftware zur Durchführung der Mikroperimetrie mit dem SLO integriert.

Durch das Real-Time-Tracking kann nun das Potential des SLO für die Perimetrie voll ausgenutzt werden. Die jetzt erzielbare Einhaltung des Stimulationsrasters erlaubt eine hohe Ortsauflösung bei der Eingrenzung auch kleiner Skotome. Die durch Läsionen oder andere morphologische Änderungen herbeigeführte Verschlechterung der Sehfähigkeit kann objektiv quantifiziert werden.

Der Untersucher wird bei seiner Aufgabe deutlich entlastet. Im wesentlichen ist er zu Beginn der Untersuchung bei der Auswahl des zu untersuchenden Areals und dem Aufsetzen des Trackings aktiv. Das ermöglicht es auch geschultem Hilfspersonal, diese Untersuchung durchzuführen.

Die Zeit, die bisher für die manuelle Nachführung aufgewendet werden mußte, kann somit für die Präsentation zusätzlicher Stimuli verwendet werden. Das erlaubt die Ausnutzung feinerer Raster sowie feinere Abstufung der Stimulushelligkeiten, um die Wahrnehmungsschwelle genauer zu bestimmen.

Weitere Arbeiten sind notwendig, um das System für den klinischen Einsatz tauglich zu machen. Für das Wiederaufsetzen nach dem Verlust des Trackings müssen Strategien weiterentwickelt werden, die Statistiken über Augenbewegungen bei Vorgabe einer Fixationshilfe für die Schätzung der momentanen Position nutzen.

Referenzen

1. Reim M: Augenheilkunde, Enke, Stuttgart, 1993
2. Dannheim F: Computer-Perimeter. Ophthalmologisch-optische Instrumente, Rassow B (Hg.). Enke, Stuttgart, 1987
3. van de Velde FJ, Jalkh AE, Elsner AE: Microperimetry with the scanning laser ophthalmoskop. Perimetry Update, Proc Ixth Intenational Perimetric Society Meeting. Kugler Pub, pp. 93-101, 1990
4. Rohrschneider K, Becker M, Fendrich T, Völcker HE: Kinetische funduskontrollierte Perimetrie mit dem Scanning-Laser-Ophthalmoskop. Klinische Monatsblätter für Augenheilkunde, 2 (207):102-110. Okt 1995
5. Buchholtz E: Konzeption und Aufbau einer Steuereinheit zur durchführung der Mikroperimetrie mit dem Scanning-Laser-Ophthalmoskop. Diplomarbeit am Lehrstuhl für Meßtechnik, RWTH Aachen, 1996
6. Liebau H: Integration eines Verfahrens zur automatischen Verfolgung von Augenbewegungen in eine Steuereinheit für die Mikroperimetrie, Diplomarbeit am Lehrstuhl für Meßtechnik, RWTH Aachen, 1997
7. Markow MS, Rylander HG, Welch AJ: Real-Time Algorithm for Retinal Tracking. IEEE Trans. Biomed. Eng. 40(12), 1993
8. Barrett SF, Jerath MR, Rylander HG, Welch AJ: Digital tracking and control of retinal images. Optical Engineering 33(1), 150-159, 1994

Arbeitsraumüberwachung mit Ultraschall-Sensorarrays

Hauke Schmidt[1], Roland Klinnert[1], Wolfgang M. Grimm[1], Bernhard Wirnitzer[2]

[1]Robert Bosch GmbH, FV/FLI, Postfach 10 60 50, D-70049 Stuttgart

[2] FH-Mannheim, Institut für Digitale Signalverarbeitung, Windeckstr. 110, 68163 Mannheim

email: hschmidt@fli.sh.bosch.de

Zusammenfassung

Mit der Zunahme der Automatisierung mechanischer Vorgänge durch die Robotik im direkten Interaktionsbereich des Menschen gewinnen auch Verfahren zum sicherheitstechnischen Schutz des Menschen und zur Erkennung möglicher Gefahrensituationen an Bedeutung. Dieser Beitrag beschreibt ein Verfahren zur Arbeitsraumüberwachung mit Hilfe einer Anordnung mehrerer Ultraschallsensoren, die in einem sehr weiten Winkelbereich detektieren können. Die im Pulsecho-Betrieb arbeitenden Sensoren liefern, jeder für sich, zwar sehr zuverlässige Informationen über die Entfernung zwischen detektiertem Objekt und Sensor, die Winkellage der Objekte kann jedoch innerhalb des Detektionsbereiches nicht bestimmt werden. Eine gleichzeitige Auswertung der von verschiedenen Sensoren des Arrays gesendeten Pulse durch Kreuzechoauswertung ermöglicht eine hinreichend genaue Ermittlung der räumlichen Lage von Objekten. Die Detektion wird mit MAP-Schätzverfahren für die Objektlage und -erkennungswahrscheinlichkeit durchgeführt. Die Funktion der diskutierten Verfahren wird am Beispiel einer Arbeitsraumüberwachung eines autonomen Betankungsroboters nachgewiesen*.

Einleitung

Maschinen und Roboter im industriellen Bereich arbeiten bis heute noch vorwiegend isoliert und entkoppelt vom Menschen hinter Schutzzäunen oder in abgesperrten Arbeitsräumen. Im Zuge eines verstärkten Einsatzes von Robotern im wachsenden Service- und Dienstleistungsbereich sind hingegen nicht nur neue Paradigmen zur Aufteilung von Arbeit zwischen Mensch und Maschine notwendig, sondern auch neuartige „intelligente" Sicherheitsvorrichtungen für die Mensch-Maschine Interaktion. Ziel ist es hierbei, den sicherheitstechnischen Schutz des Menschen bei einer interaktiven Zusammenarbeit mit Maschinen und Robotern „Hand in Hand" durch die Einführung „virtueller Schutzräume" zu gewährleisten. Die Schlüsselkomponenten hierfür sind Multisensorsysteme, deren Daten mit speziellen Algorithmen fusioniert und ausgewertet werden.

Während im industriellen Umfeld oft nur Kontaktschalter und Lichtschranken zur Auslösung von Not-Aus Zuständen zum Einsatz kommen, um ausgebildetes Personal vor Unfällen zu schützen, reicht dies bei Servicesystemen nicht mehr aus, da sie im alltäglichen Lebensumfeld zum Einsatz kommen sollen. Zur Erfassung von Gefahrenzustän-

---

* Die Arbeiten werden im Rahmen des Verbundprojektes INSERVUM (Intelligente Serviceumgebungen) vom BMBF gefördert (Förderkennzeichen 01 IN 601 B 5).

den werden daher vorwiegend Laserscanner, Videokameras und Infrarotsensoren verwendet. Alternativ dazu werden im folgenden Ultraschallsensoren betrachtet, die einfache, schnelle und kostengünstige Realisierungen ermöglichen. Im Vergleich zu einer Videobildauswertung können Ultraschallsensoren Objekte schneller detektieren. Ferner sind sie aufgrund ihres Einsatzes im Kfz-Bereich mechanisch sehr robust und kostengünstig im Vergleich zu Laserscannern. Sie sind prädestiniert für redundante Überwachungskonzepte, die nicht nur sicherheitstechnischen Schutz bieten, sondern auch „sicher" sind im Sinne einer systemimmanenten Störunanfälligkeit und möglicherweise auch sicherer bei Ausfällen von Teilkomponenten infolge von Sabotage oder anderweitiger Manipulation.

Echoauswertung der Eigenpulse

Das Ziel der Echoauswertung ist die Detektion und Lokalisierung von Objekten. Im einfachsten Anwendungsmodus eines Puls-Echo-Sensors sind der Sende- und Empfangssensor identisch. Der Sensor sendet einen kurzen Ultraschallpuls und mißt die Zeit Δt bis zum Empfang eines oder mehrerer Echos. Bei der naiven Interpretation eines Echos wird angenommen, daß sich ein Objekt auf einem Kreisbogen im Abstand

$$d = \frac{1}{2} c \Delta t \qquad (1)$$

vom Sensor befindet (c ist die Schallgeschwindigkeit). Der Bereich auf dem Kreisbogen, auf dem Objekte detektiert werden können, ist dabei nur innerhalb der Sende- und Empfangskeule des Sensors möglich. Komplexer wird die Deutung, wenn Sekundärechos, beispielsweise durch die Reflexion an einer Wand, empfangen werden. Die obige Interpretation würde dann nur ein "Spiegelbild" des tatsächlichen Objekts beschreiben. Die Erkennung von Mehrfachreflexionen wird erst durch eine detaillierte Kenntnis der Umgebung möglich. Im Rahmen der hier dargestellten Betrachtungen gehen wir daher von der naiven Deutung (1) aus.

Die empfangenen Echos sind zum einen mit einer Meßungenauigkeit behaftet, die zu einer Unsicherheit der Lageschätzung führt. Andererseits gibt es durch Fremdsender und Mehrfachreflexionen verursachte falsche Echos sowie Echoauslöschungen, aufgrund derer die tatsächliche Existenz eines Objektes nur mit einer gewissen Wahrscheinlichkeit angenommen werden kann. Die Objektbeschreibung beinhaltet daher neben der ermittelten Lage auch Schätzwerte für die Lageunsicherheit und Objektwahrscheinlichkeit, die im folgenden hergeleitet werden. Einen ähnlichen Ansatz verfolgen Leonard et al. [1], in dem anstelle einer Wahrscheinlichkeit eine "Kredibilität" nach einem heuristischen Verfahren berechnet wird.

Die Wahrscheinlichkeit für das Vorhandensein eines Objekts O, wenn ein Echo E empfangen wurde (mit Komplementärereignis $\overline{O}$, daß kein Objekt vorliegt) ist

$$P(O \mid E) = \frac{P(E \mid O) P(O)}{P(E \mid O) P(O) + P(E \mid \overline{O}) P(\overline{O})} = \frac{1}{1 + \frac{1 - P(O)}{P(O)} \frac{P(E \mid \overline{O})}{P(E \mid O)}}, \qquad (2)$$

wobei P(O) die a priori Wahrscheinlichkeit für das Vorhandensein des Objekts ist. Ebenso wird die Wahrscheinlichkeit dafür betrachtet, daß ein Objekt vorhanden ist, wenn - aufgrund einer Auslöschung oder mangelnder Reflexionseigenschaften des Objekts - kein Echo empfangen wurde (Ereignis $\bar{E}$):

$$P(O \mid \bar{E}) = \frac{P(\bar{E} \mid O)P(O)}{P(\bar{E} \mid O)P(O) + P(\bar{E} \mid \bar{O})P(\bar{O})} = \frac{1}{1 + \frac{1-P(O)}{P(O)} \frac{1-P(E \mid \bar{O})}{1-P(E \mid O)}} \tag{3}$$

Um beide Größen bestimmen zu können, betrachten wir das nach einem gesendeten idealen Puls empfangene Echosignal R:

$$R = r(1-a) + n(1 - r(1-a)) \tag{4}$$

Darin sind r, a und n binäre, stochastische und voneinander unabhängige Größen; r gibt das Empfangssignal bei einer idealen ungestörten Übertragung an. Falls r = 1, kann das Empfangssignal durch ein Störsignal a ausgelöscht werden. Andererseits kann für r = 0 ein Störpuls n, verursacht z.B. von einem fremden Sender oder einer Sekundärreflexion, zum Empfang eines falschen Echos R = 1 führen. Es seien

$$r, n, a \in (0,1), \; P(r=1) = P_0, \quad P(a=1) = P_1, \quad P(n=1) = P_2 \tag{5}$$

Die Größen P_1 (Auslöschwahrscheinlichkeit) und P_2 (Störpulswahrscheinlichkeit) sind Konstanten, die abhängig von den jeweils vorliegenden Umgebungsbedingungen gewählt werden. Wie zuvor muß man die beiden interessierenden Fälle unterscheiden :

1. Es liegt ein Objekt vor; dann ist r = 1 und es folgt:

$$E(R) = P(E \mid O) = E(1 - a + na) = 1 - E(a)(1 - E(n)) = 1 - P_1(1 - P_2) \tag{6}$$

2. Es liegt kein Objekt vor und r = 0. Daraus folgt:

$$E(R) = P(E \mid \bar{O}) = E(n) = P_2 \tag{7}$$

Eingesetzt in (2) und (3) erhält man:

$$P(O \mid E) = \frac{1}{1 + \frac{1-P(O)}{P(O)} \frac{P_2}{1 - P_1(1-P_2)}} \tag{8}$$

$$P(O \mid \bar{E}) = \frac{1}{1 + \frac{1-P(O)}{P(O)} \frac{1}{P_1}} \tag{9}$$

Für die Schätzung der Objektlage bietet sich der Kalmanfilteransatz an [4]. Ausgehend von einer Schätzung des Abstands d eines - als statisch angenommenen - Objekts vom Sensor und der Schätzunsicherheit σ erhält man aufgrund eines Echos Δt und der Meßunsicherheit σ_t die neuen Schätzungen

$$\hat{d} = d + \frac{2\hat{\sigma}^2}{c\sigma_t^2}\left(\Delta t - \frac{2d}{c}\right); \qquad \hat{\sigma}^2 = \frac{c^2\sigma^2\sigma_t^2}{c^2\sigma_t^2 + 4\sigma^2} \tag{10}$$

Bei mehr als einem Echo und/oder mehreren Objekten muß eine Zuordnung von Objekt und Echo getroffen werden. Wir ordnen jedem Objekt O_j das ihm nächste Echo (mit Laufzeit Δt_i) zu, wenn die Differenz Δ_{ij} zwischen Objektabstand und äquivalentem Echoabstand im Rahmen der Meß- und Schätzunsicherheit liegt:

$$\Delta_{ij} = d_j - \frac{c\,\Delta t_i}{2};\quad \Delta t_i \to O_j \Leftrightarrow i = \arg\min \Delta_{ij}^2 \wedge \Delta_{ij}^2 \le k\left(\sigma_j^2 + \left(\frac{c}{2}\sigma_{ti}\right)^2\right) \tag{11}$$

Ein typischer Wert für die Toleranzbreite k ist 2. Ein Objekt O wird somit durch das Tripel (d, σ, P) beschrieben, das durch folgende Fallunterscheidung festgelegt wird:

1. Einem Objekt wird ein Echo zugeordnet. Die Aktualisierung der Objektparameter (d, σ, P) erfolgt nach den Gleichungen (8) und (10).
2. Einem Objekt wird kein Echo zugeordnet. Die Größen d und σ werden nicht verändert. P wird durch Gleichung (9) aktualisiert. Wenn P unter eine Schwelle P_u fällt, wird das Objekt aus der Liste der Objekte gelöscht.
3. Ein Echo wird keinem Objekt zugewiesen. Für das Echo wird ein neues Objekt generiert. Die Initialisierung der Objektgrößen erfolgt durch $d = c\Delta t/2$, $\sigma = c\sigma_t/2$ und $P = P_v$, wobei die a priori Wahrscheinlichkeit P_v situationsabhängig gewählt werden muß.

Objektklassifikation durch Kreuzechoauswertung

Bei Verwendung von mehreren Sensoren erreichen jeden Empfänger neben den Echos der Eigenpulse zusätzlich auch Kreuzechos der Pulse anderer Sender. Man erhält so zusätzliche Information über die Objektlage. Wir bezeichnen Δt_{ij} als Echozeit eines Pulses von Sender i in Empfänger j, wobei i, j = 1...n. Das mit Δt_{ij} gemessene Echo stammt von einem Objekt, das sich auf einem Ellipsenbogen befindet, in deren Brennpunkten die beiden Sensoren i und j im Abstand D_{ij} liegen. Für i = j liegt ein Eigenpulsecho vor, und die Ellipse wird zu einem Kreis.

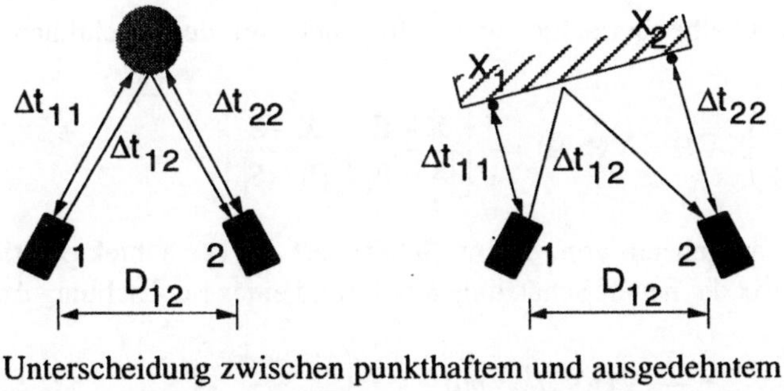

Abb.1: Unterscheidung zwischen punkthaftem und ausgedehntem Objekt

Bei Verwendung von zwei Sensoren erhält man pro Objekt bis zu vier Meßwerte Δt_{11}, Δt_{22} und $\Delta t_{12} = \Delta t_{21}$, mit denen sich bereits einfache Schlüsse über die grobe Objektform im Bereich der auftreffenden Pulse ziehen lassen [5]:

$$\Delta t_{12} - \Delta t_{11} - \Delta t_{22} \begin{cases} = 0 \Rightarrow \textit{Objekt punktf rmig}\ (\textit{z.B. Pfahl}) \\ > 0 \Rightarrow \textit{Objekt ausgedehnt} \end{cases} \tag{12}$$

$$\Delta t_{12}^2 - \frac{D^2}{c^2} - 4\Delta t_{11}\Delta t_{22} \begin{cases} > 0 \Rightarrow \textit{Objekt konkav} \\ = 0 \Rightarrow \textit{Objekt gerade}(\textit{z.B. Wand}) \\ < 0 \Rightarrow \textit{Objekt konvex} \end{cases} \tag{13}$$

Zwei der in Gleichung (13) und (14) beschriebenen Fälle sind in Abb. 1 dargestellt. Die in Abb.1 rechts dargestellte Wand kann durch eine Gerade beschrieben werden. Wählt man ein kartesisches Koordinatensystem so, daß Sensor 1 an der Position (0,0) und Sensor 2 an (D, 0) liegen, kann die Gerade gemäß der Hesseschen Normalform

$$(m,\ n)\begin{pmatrix} x \\ y \end{pmatrix} = \left(\frac{\Delta t_{11} - \Delta t_{22}}{D\Delta t_{11}},\ \frac{\sqrt{(D/c)^2 - (\Delta t_{11} - \Delta t_{22})^2}}{D\Delta t_{11}} \right)\begin{pmatrix} x \\ y \end{pmatrix} = 1 \tag{14}$$

angegeben werden, die zwischen den Auftreffpunkten

$$X_1 = \frac{1}{m^2+n^2}\begin{pmatrix} m \\ n \end{pmatrix} \quad \textit{und} \quad X_2 = \frac{1}{m^2+n^2}\begin{pmatrix} n^2 D + m \\ n - mnD \end{pmatrix} \tag{15}$$

der Pulse mit den Laufzeiten Δt_{11} und Δt_{22} gültig ist.

Falls die Anwendung dieser Gleichungen z.B. aufgrund fehlender Messungen nicht möglich ist, läßt sich die Objektlage in der Regel wieder mit Hilfe der Kalman-Filterung bestimmen. Die implizite Annahme hierbei ist, daß das (punktförmige) Objekt die gemessenen Echos am gleichen - zu schätzenden - Ort reflektiert hat. Aufgrund dieser Überlegung befinden sich mögliche Objekte überall dort, wo Schnittpunkte der durch die gemessenen Echos definierten Kreis- und Ellipsenbögen liegen.

Zunächst wird die Beobachtungsgleichung H(**X**) bestimmt, die bei gegebener Lage **S** des Senders und **E** des Empfängers die Messung Δt aus der Objektlage **X** angibt:

$$H(\mathbf{X}) = \frac{|\mathbf{X}-\mathbf{S}| + |\mathbf{X}-\mathbf{E}|}{c} \tag{16}$$

Der Einfachheit halber betrachten wir im folgenden nur den zweidimensionalen Fall in x und y. Für

$$\mathbf{X} = \begin{pmatrix} x \\ y \end{pmatrix},\ \nabla\mathbf{H} = \nabla\mathbf{H}(\mathbf{X}) = \frac{1}{c}\left[\frac{\mathbf{X}-\mathbf{E}}{|\mathbf{X}-\mathbf{E}|} + \frac{\mathbf{X}-\mathbf{S}}{|\mathbf{X}-\mathbf{S}|}\right] \tag{17}$$

ergeben sich ausgehend von einem Schätzwert **X** der Objektposition und **V** der Kovarianzmatrix die neuen Schätzungen aufgrund einer Beobachtung Δt zu

$$\hat{\mathbf{V}}^{-1} = \mathbf{V}^{-1} + \frac{1}{\sigma_T^2}\nabla\mathbf{H}\,\nabla\mathbf{H}^T \quad \textit{mit} \quad \mathbf{V} = \begin{pmatrix} \sigma_{xx} & \sigma_{xy} \\ \sigma_{xy} & \sigma_{yy} \end{pmatrix} \tag{18}$$

$$\hat{\mathbf{X}} = \mathbf{X} + \hat{\mathbf{V}}\,\nabla\mathbf{H}(\Delta t - H(\mathbf{X})) \tag{19}$$

Für eine sinnvolle Schätzung müssen mindestens zwei unabhängige Messungen Δt_i eines Objektes j vorliegen. Die Schätzwerte werden dann für jede Einzelmessung nach Gleichungen (18) und (19) neu berechnet.

Wie bei dem Eigenechoempfang ist eine Zuordnung von Echos und Objekten notwendig. Analog zu Gleichung (11) verwenden wir die Bedingung

$$\Delta_{ij} = \left|\mathbf{X}_j - \mathbf{E}_i\right| + \left|\mathbf{X}_j - \mathbf{S}_i\right| - \frac{c\,\Delta t_i}{2};$$
$$\Delta t_i \to O_j \Leftrightarrow i = \arg\min \Delta_{ij}^2 \;\wedge\; \Delta_{ij}^2 \le k\left(\sigma_{xxj} + \sigma_{yyj} + \left(\frac{c}{2}\sigma_{ti}\right)^2\right) \qquad (20)$$

Die Aktualisierung erfolgt wie oben nach einer Fallunterscheidung:

1. Objekte mit zugeordneten Echos werden nach den Gleichungen (18) - (20) aktualisiert.
2. Bei Objekten ohne Echozuordnung werden die Größen **X** und **V** nicht verändert. P wird durch Gleichung (9) aktualisiert. Wenn P unter eine Schwelle P_u fällt, wird das Objekt aus der Liste der Objekte gelöscht.
3. Für nicht zugeordnete Echos werden neue Objekte an Kreuzungspunkten der entsprechenden Echoellipsen mit Wahrscheinlichkeit $P = P_v$ generiert. Die Kovarianzmatrix **V** wird mit $\sigma_{xx} = \sigma_{yy} = (c\sigma_t/2)^2$ und $\sigma_{xy} = 0$ initialisiert.

Ultraschallsensorik

Die oben beschriebenen Verfahren sollen im Rahmen einer Arbeitsraumüberwachung mit Ultraschallsensorik zum Einsatz kommen. Die hierzu verwendeten Sensoren zeichnen sich durch große Öffnungswinkel von 120° in der Horizontalen und 60° in der Vertikalen aus, so daß bereits jeder einzelne Sensor einen großen Raumbereich abdecken kann [1]. Jeder einzelne Sensor kann Pulse von etwa 300 µs Dauer, die durch ein mit Wechselspannung erregtes piezokeramisches Plättchen erzeugt werden, sowohl aussenden als auch deren Echos empfangen. Beim sogenannten Puls-Echo Betrieb müssen die Schwingungen des Resonators nach dem Senden möglichst schnell abklingen, damit der Sensor wieder empfangsbereit ist. Die dafür erforderliche Zeitspanne bestimmt die untere Grenze einer Abstandsmessung, die für die verwendeten Resonatoren bei etwa 20 cm liegen. Bei kleineren Abständen können Gegenstände nur bei Verzicht auf eine absolute Abstandsangabe detektiert werden. Die realisierte Arbeitsfrequenz der Sensoren von etwa 45 kHz stellt einen Kompromiß dar, der der abnehmenden Meßauflösung und voluminöseren Bauweise bei niedrigeren Frequenzen einerseits und dem schlechteren Wirkungsgrad bedingt durch zunehmende Luftdämpfung bei höheren Frequenzen andererseits Rechnung trägt. Die Sensoren werden über eine bidirektionale Sensor-Echoleitung angesteuert. Die empfangenen Echosignale werden im Sensor durch einen integrierten Schaltkreis u.a. durch eine Bandpaßfilterung und eine Schwellwertlogik vorverarbeitet [2]. Der Sensor liefert ein zeitkontinuierliches wertediskretes (binäres) Signal, das die Zeitpunkte des Eintreffens von Echosignalen wiedergibt.

Anwendung in einem Tankroboterszenario

Mit Hilfe der oben beschriebenen Sensoren und Auswerteverfahren wird eine Überwachung des Arbeitsraumes eines Betankungsroboters realisiert. Der Aufbau der Anlage ist in Abbildung 2 skizziert. Man erkennt die Lage und den Erfassungsbereich der

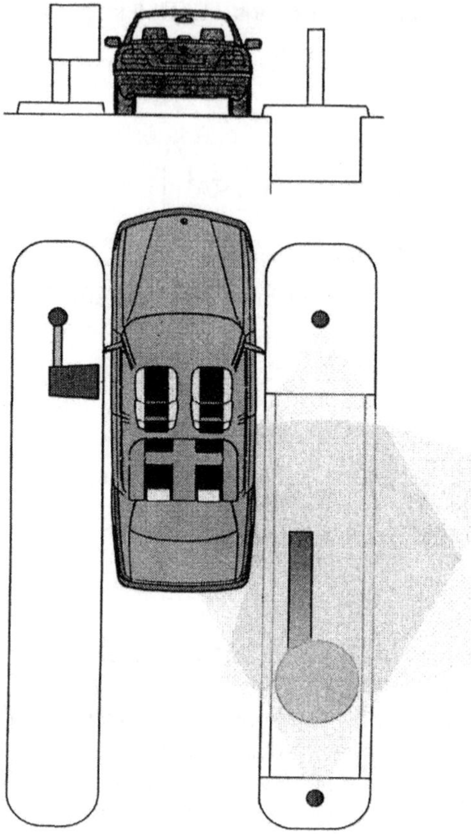

Abb.2: Skizze der Anlage mit Sensoranordnung und -erfassungsbereich

Ultraschallsensoren, sowie den Roboterarm. Da dessen Lage, wie auch die des Fahrzeugs und anderer fester Objekte, in allen Phasen der Betankung bekannt ist, können Reflexionen von Personen, die sich im Gefahrenbereich befinden, eindeutig von denen der bekannten Objekte unterschieden werden. Erste Untersuchungen zeigen, daß ein in den Erfassungsbereich eindringendes Objekt bereits nach weniger als 25 ms sicher erkannt wird. Nimmt man eine maximale Geschwindigkeit von 10 m/s an, so bewegt sich das Objekt in dieser Zeit um 25 cm. Der durch den Erfassungsbereich aufgespannte virtuelle Schutzraum erstreckt sich mit mindestens 60 cm Abstand von jedem Punkt des Roboterarms somit hinreichend weit, um die Sicherheit der Anlage zu gewährleisten.

Ausblick und weitere Arbeiten

Bei dem Betrieb der Ultraschallsensoren können Störungen durch Mehrfachreflexionen oder durch gleichzeitig sendende Sensoren (sogenannte Interferenzen) auftreten. Das Problem der Mehrfachreflexion wurde oben bereits erwähnt. Zur Unterdrückung von Interferenzen können Codiertechniken eingesetzt werden. Der Sender kodiert dabei das ausgesandte Signal und der Empfänger kann durch die Decodierung erkennen, von welchem Sender das Signal stammt. Zwei prinzipielle Möglichkeiten sind die Codierung der Trägerfrequenz oder die Verwendung von Pulsfolgen [11]. Abb.3 zeigt das Ergebnis der bei der Robert Bosch GmbH entwickelten stochastischen Codierung [12]. Zwei bewegte Störsender (weiße Kreise) führen zu Pseudoobjekten im Empfänger (schwarzer Kreis). Die drei Ultraschalleinheiten senden dabei mit der gleichen Pulsfolgefrequenz. Gegenüber dem Normalbetrieb (Abb.3b) hebt die Störunterdrückung (Abb. 3c) die Echos der beiden eingezeichneten Laufwege hervor, während die Störpulse als stochastisches Muster erscheinen.

(a)

(b)

(c)

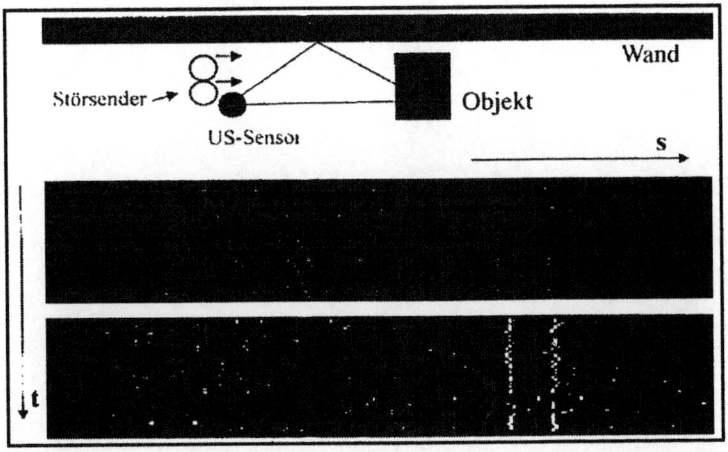

Abb. 3: Interferenzfreier Betrieb von Bosch Ultraschallsensoren. (a): Anordnung der Sensoren. (b): Gleichzeitiger Betrieb von drei Puls-Echo-Sensoren. (c): Interferenzfreier Betrieb von drei Sensoren durch stochastische Codierung.

Literatur

[1] J.Leonard, H.Durrant-Whyte: Dynamic Map Building for an Autonomous Mobile Robot, Intern. Journal of Robotics Research, Vol. 11, No. 4, 1992

[2] E.Tschiskale: Kurzbeschreibung des US-Sensors 2.1, Robert Bosch GmbH. Interner Bericht für das BMBF-Projekt INSERVUM, März 1996.

[3] M.Noll: Abstände hörbar gemacht - Ultraschall hilft beim Einparken, Bosch-Zünder, 10/1997, S. 8.

[4] Y.Bar-Shalom,T.Fortmann: Tracking and Data Association, Academic Press, Orlando, 1988

[5] D.Meyer,et al.:Offenlegungsschrift DE4023538 A1, Deutsches Patentamt, 1992

[6] G.Mauris, E.Benoit, L.Foulloy: An Intelligent Ultrasonic Range Finding Sensor for Robotics, IFAC 13th Triennial World Congress, San Francisco, 1996

[7] H.Beom, K.Koh, H.Cho: The Improvement of sonar based Mobile Robot Localization Method by multiple Beacons, IFAC 13th Trien. World Congress, San Francisco, 1996

[8] M.Mergel, C.Koburg, U.Vogel: Automatisches Einparken, AMS '94, 10. Fachgespräch über Autonome Mobile Systeme, Stuttgart, Okt. 1994

[9] G.Boer, G.Albada, et.al: The MARIE Autonomous Mobile Robot, Proc. of the Intern.Conf. on Intelligent Autonomous Systems, Pittsburgh, 1993

[10] B.Kröse, K.Compagner, F.Groen: Accurate Estimation of Environment Parameters from Ultrasonic Data, Proc. of the Intern.Conf. on Intelligent Autonomous Systems, Pittsburgh, 1993

[11] B.Menz, J.Rappold: Application of Simultaneous Multiplex Techniques to Ultrasonic Reflection Tomography and Model Based Object Recognition; Proceedings of World Congress of Ultrasonics, August 1997, Yokohama, Japan;

[12] B.Wirnitzer, H.Schmidt, R.Klinnert, W.Grimm: Interference Cancelation in Ultrasonic Sensor Arrays by Stochastic COding and Adaptive Filtering, IEEE International Conf. on Intelligent Vehicles, Stuttgart 1998

3D-Geometrien

Circular Coded Target and Its Application to Optical 3D-Measurement Techniques

SUNG JOON AHN and WOLFGANG RAUH

Fraunhofer Institute for Manufacturing Engineering and
Automation IPA, Nobelstr. 12, 70569 Stuttgart, FRG
Tel.: +49-711-970 1859, Fax: +49-711-970 1004
E-mail: {sja; wor}@ipa.fhg.de

One of the primary but tedious tasks for the user and developer of an optical 3D-measurement system is to find the homologous image points in multiple images - a task that is frequently referred to as the correspondence problem. With the solution, the error-free correspondence and accurate measurement of image points are of great importance, on which the qualitative results of the succeeding 3D-measurement are immediately dependent. In fact, the automation of measurement processes is getting more important with developments in production and hence of increasing topical interest. In this paper, we present a circular coded target for automatic image point measurement and identification, its data processing and application to some optical 3D-measurement methods.

1 Introduction

At the moment, basic software and hardware components that are capable of constructing an optical 3D-measurement system are available on the market, and provide numerous combinations of measurement principles and methods, thus making possible the construction of a special or general purpose measurement system. Beside the software and hardware components any point target for the 3D-measurement is also considered as a part of the measurement system, since the overall measuring performance and the resulting accuracy levels are directly linked to the quality of center point determination. In addition, the measurement procedure of image points has blocked the automation and hence general acceptance of the optical 3D-measurement techniques in industrial applications.

In recent years, the demand for a *coded target* guaranteeing automatic error-free correspondence and accurate image point measurement has been dramatically increased [1,3-9]. Until now, some coded targets are accepted more or less for the industrial application of optical 3D-measurement techniques [2-9] (fig. 1). To realize fully automatic 3D-measurement procedures and an accurate image point measurement, we have developed a new circular coded target [10] (fig. 2), and have found an absolute acceptance from industry even in the toughest application of the coded target with superimposed random pattern [11]. In this paper, we intend to describe the geometric construction of the target, its image processing and application to some 3D-measurement methods.

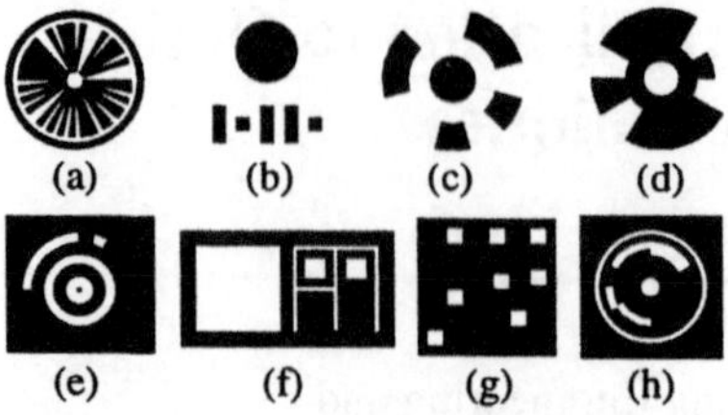

Fig. 1 Various types of coded target:
(a) Trenkler, 1981; (b) Wong, 1988;
(c) Schneider, 1991; (d) Knobloch, 1992;
(e) van den Heuvel, 1992; (f) Homainejad, 1995; (g) Fraser, 1997; (h) Niederöst, 1997.

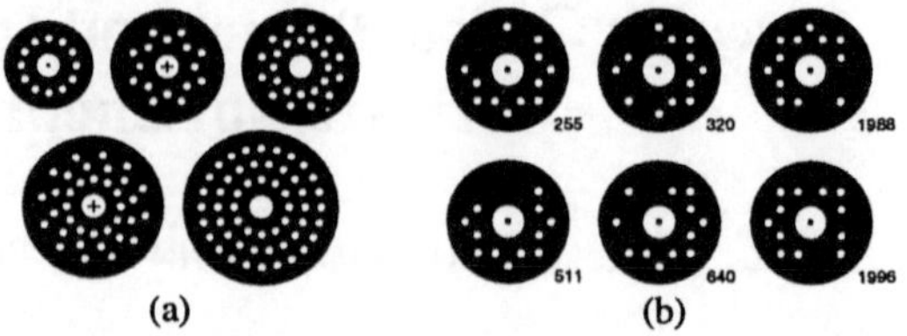

Fig. 2 The proposed circular coded target (Ahn, 1998): (a) Examples with 12-, 16-, 24-, 36- and 54-bit code length; (b) Coding examples of the 16-bit circular coded target with 12 turned on coding points and 2 coding layers. The bit position 0 (upper center) is the parity bit, and 2, 6, 10 and 12 (clockwise) are turned on always.

2 Geometric construction

Out of consideration for a robust and reliable image processing, the target is geometrically constructed only of circular elements, i.e. circular point mark, circular coding points and circular background plate (fig. 2). The superiority of the circular object target over other geometric features for the purpose of image point measurement is also reported in the literature [12]. The identification number of the target is coded down into the circular arrangement of the coding points around the point mark. Additionally, the size and arrangement of the coding point around the point mark are also geometrically prescribed (fig. 3b-3e). We have fully utilized all these geometric constraints on the target construction with the image processing, and have achieved a very robust and reliable image point measurement and identification.

3 Image processing

To minimize the necessary computing costs, binary image processing is preferably applied here, except the subpixel accurate edge detection of the contour points of the point mark. First, we know from the given imaging conditions the approximate size of the image ellipse of the used point mark (fig. 3a), assuming that an image ellipse is the projection of an object circle onto the image plane. In addition, such an image ellipse is a relatively compact image feature for a natural scene, where the compactness can be represented by the *form factor* as in equation (1).

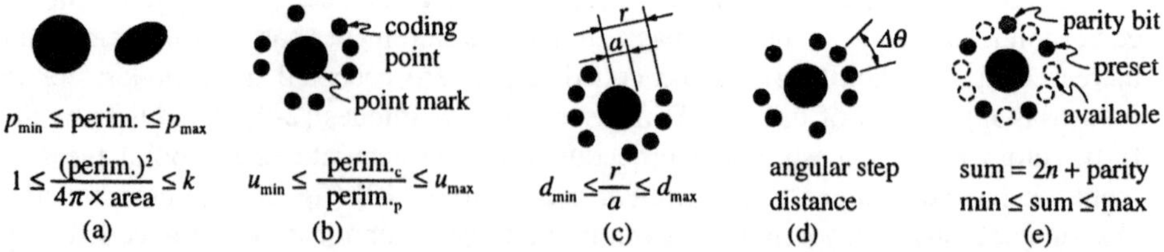

Fig. 3 Geometric features of the circular coded target: (a) Size and form factor of the point mark; (b) Normalized size of the coding points against the point mark; (c) Normalized distance of the coding points from the center of the point mark; (d) Angular step distance of the coding points; (e) Total sum and preset pattern of the bit positions including the parity bit.

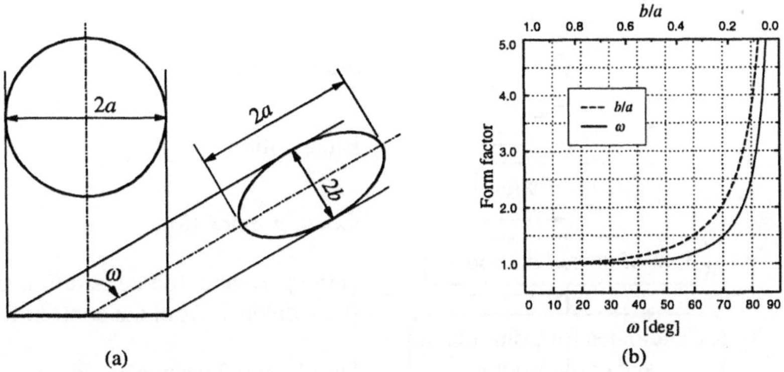

Fig. 4 Form factor of an image ellipse: (a) Ellipse as the image of an object circle assuming a parallel projection; (b) Dependence of the form factor on the projection angle.

$$\text{Form factor} = \frac{(\text{perimeter})^2}{4\pi \times \text{area}} \tag{1}$$

The form factor of an ellipse, the image of an object circle, increases with the imaging angle monotonously and moderately up to 1.5 at the angle of 70° (fig. 4). Thus, if we put bounds to the size and form factor of image objects (fig. 3a), we can detect all point mark candidates in the image plane, whose validity will be individually qualified in the subsequent processing steps. Next, around a point mark candidate we search for smaller image objects as the coding points of the point mark satisfying the membership criteria (fig. 3b-3c). After a proper coordinate transformation, an inverse affine transformation by using the 5 ellipse parameters of the point mark, we may obtain the reconstructed coded target (fig. 5). Through a certification and interpretation of the arrangement of the coding points (fig. 3d-3e), we will get the identification number of the target. A simplified work flow of the image processing for the circular coded target is given in fig. 6.

For an accurate image point measurement of the point mark and for a reliable rearrangement of the coding points (see fig. 5), an accurate estimation of the 5 ellipse parameters, e.g. ellipse fitting, of the image point mark is inevitable. Here we have applied an *orthogonal distances ellipse fitting* [13], which has an objective function of the orthogonal error distances and results in an unbiased optimal parameter estimation.

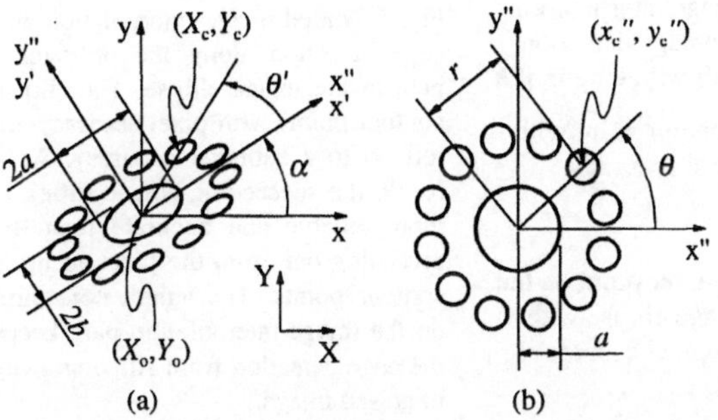

Fig. 5 Reconstruction of the circular coded target: (a) Definition of the coordinate systems by using the estimated 5 ellipse parameters (X_O, Y_O, a, b and α) of the point mark; (b) Rearranged coding points around the point mark. A valid coding point must find itself in a ring shaped zone around the point mark (see also fig. 3c), and reversibly, a valid point mark must have a valid arrangement of the coding points around it (see also fig. 3d-3e).

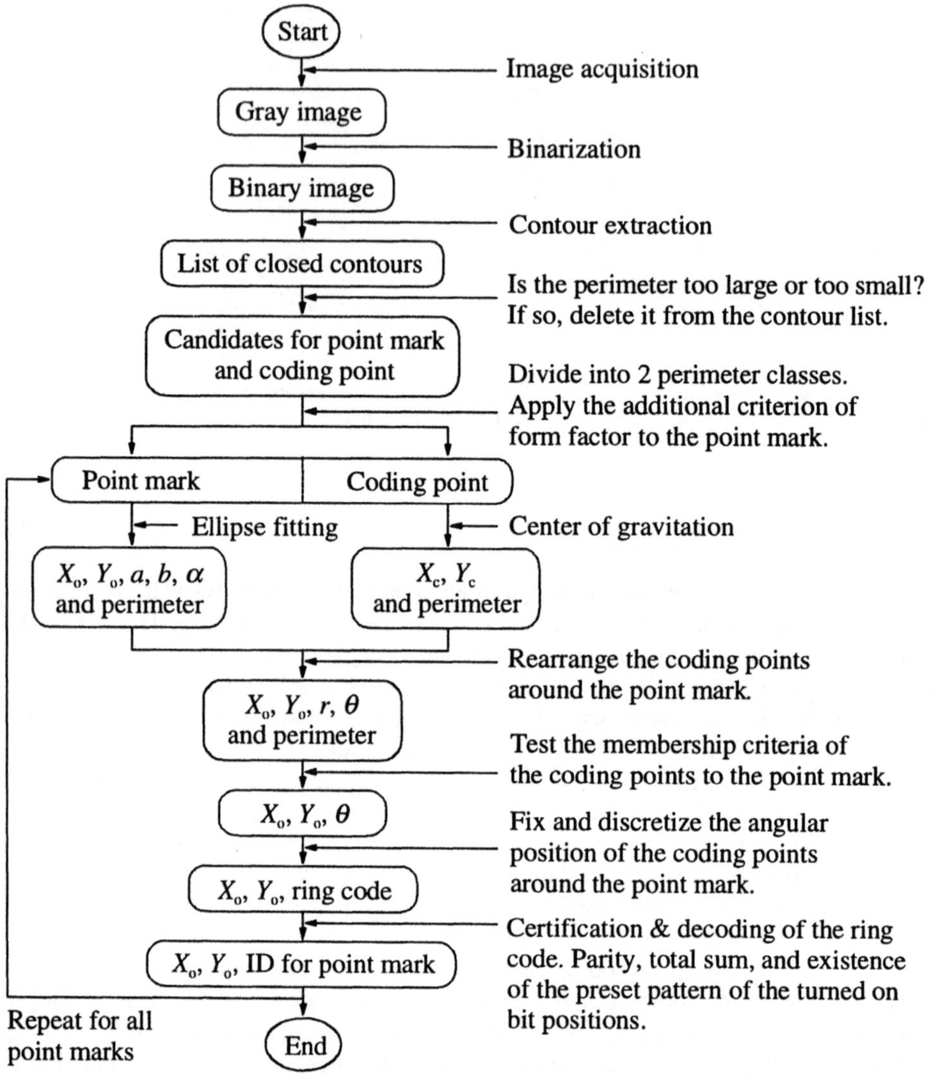

Fig. 6 Work flow of the image processing for the circular coded target.

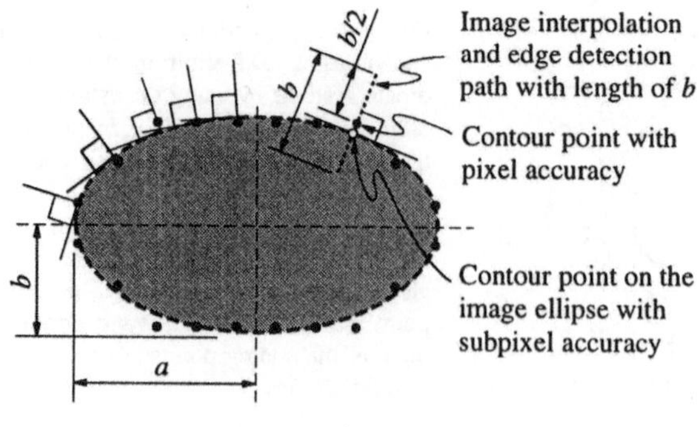

Fig. 7 Forced image interpolation and edge detection along the orthogonal path to the image ellipse. The ellipse contour points with pixel accuracy are refined to a subpixel accuracy. As a result, the succeeding ellipse fitting is more reliable and accurate than the preceding one from the pixel accurate contour points. The length constraint on the image interpolation path keeps the edge detection from straying away in noised image.

Additionally, the *forced orthogonal edge detection* is applied to the contour of the point mark by using the ellipse parameters from a preceding ellipse fitting with pixel accurate contour points (fig. 7). As a result, a robust and meaningful edge detection is also made possible in a very noised image, e.g. image of coded targets with superimposed random pattern. The succeeding ellipse fitting is more reliable and accurate than the preceding one from the pixel accurate contour points.

4 Applications

In general, in order to estimate the parameters of a system, some observations must be made by all system components, whose parameters are essentially involved in the functionality of the system. Given the system model, the model parameters can be estimated through an analysis of the observations.

In the case of stereo or multi-imaging vision systems, several widely accepted system models (camera model) and system parameter estimation (camera calibration) algorithms are already available. What we have yet to do is determine the homologous image point coordinates for some given object points, and this task can be taken over by the circular coded target and its image processing (see fig. 8a, a typical object points field for camera calibration, and fig. 8b, result of the image points measurement and identification).

As a slightly unusual application of the circular coded target, the calibration problem of the programmable LCD stripe projector [14] is considered. With various stripe projection methods, e.g. Coded Light Approach (CLA), phase-shift method and Moiré method, the projector needs to be precalibrated, since the projector parameters are directly involved in the object point determination. If we like to make use of the camera model and calibration algorithm also for the projector, the projector must image the object points field, despite it having no sensing ability. Consequently, we need a special operational method, the *virtual imaging* [15]. A regular grid pattern will be projected onto the object surface, and each projected quadrangle on the object surface can be regarded as the projection of a single virtual pixel of the projector image. Thus, if we locate an object point within the projected grid pattern, it may be said that the projector has imaged the object point (virtual imaging). To make the overall virtual imaging procedures easier, we use the circular coded target as the object point, and vertically/horizontally programmable LCD stripe projector and CLA for the registration of the virtual pixel of the projector (fig. 8).

A more difficult but very successful industrial application of the circular coded target is camera calibration for the *random pattern method.* The hardware setup of the measurement system is comprised of two high-resolution CCD cameras and a random pattern projector. The projector distinctly marks the object surface, and the two cameras simultaneously capture the patterned object surface, hence providing a stereoscopic view of the scene [11]. For the purpose of surface reconstruction through image correlation, the camera parameters at the moment of the stereoscopic imaging should be estimated. This means both the object targets for camera calibration and the patterned object surface for image correlation are imaged at the same time in each image (fig. 9a). To locate the target areas in the textured image, we have searched for the circular background plate (fig. 9b-9c). In the image processing, the constraints on

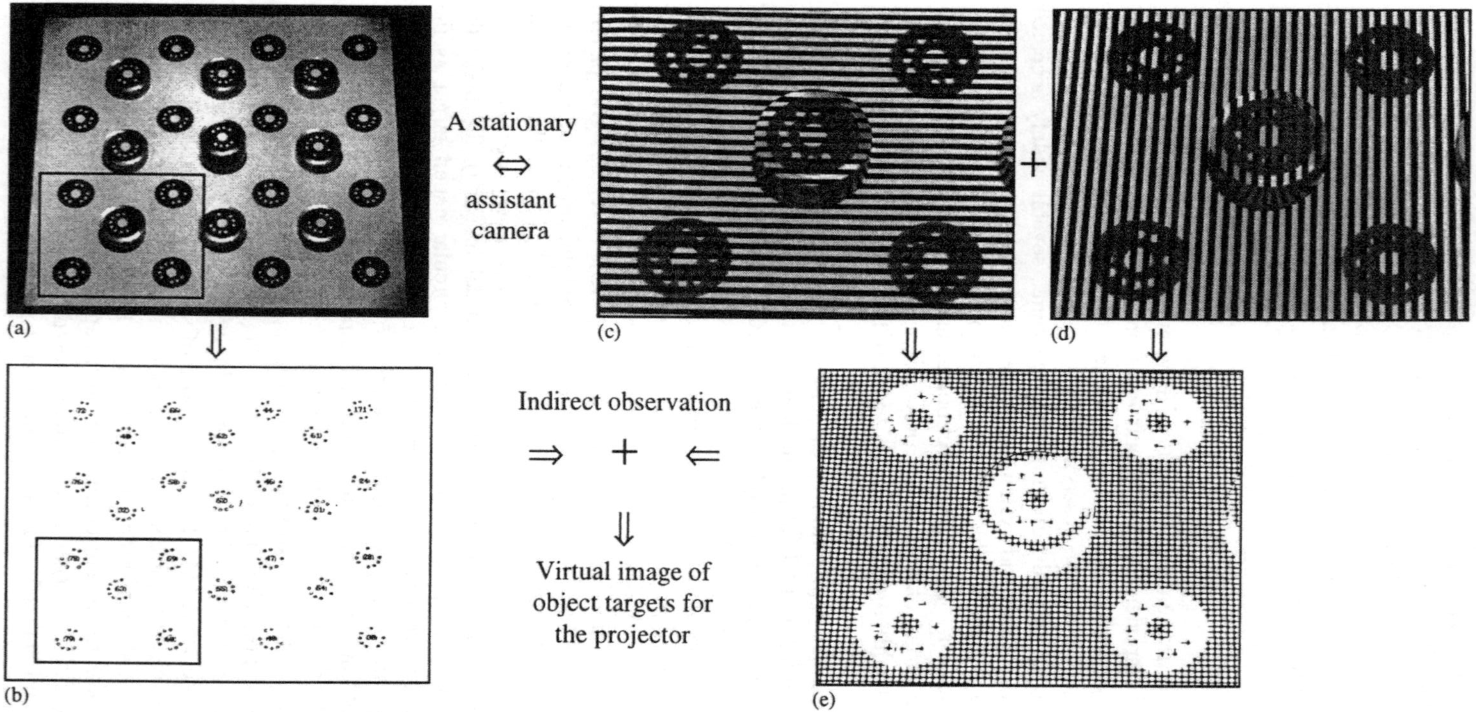

Fig. 8 Virtual imaging for vertically/horizontally programmable stripe projector: (a) Camera image of the object points field with circular coded targets; (b) Positioned and identified targets in the camera image plane; (c) Horizontal and; (d) Vertical projection of stripe patterns, where each stripe is registered through the Coded Light Approach (CLA); (e) As a curved grid pattern in the camera image plane mapped coordinate system of the projector virtual image. By means of the virtual imaging, a projector can indirectly observe the object points in collaboration with a stationary assistant camera. In the camera image plane, the object target centers and the coordinate system of the projector virtual image are mapped and superimposed together, where each quadrangle of the grid pattern in fig. 8e is the projection of a single virtual pixel of the projector and uniquely registered by CLA. Thus, if we can locate the object target centers in the grid pattern, the projector has indirectly imaged the object target centers. The subpixel accurate determination of the target center in the projector virtual image plane can be achieved through an interpolation between the stripe edge lines or through the combination of the CLA with the phase-shift method. The stationary camera needs not necessarily to have been or to be calibrated.

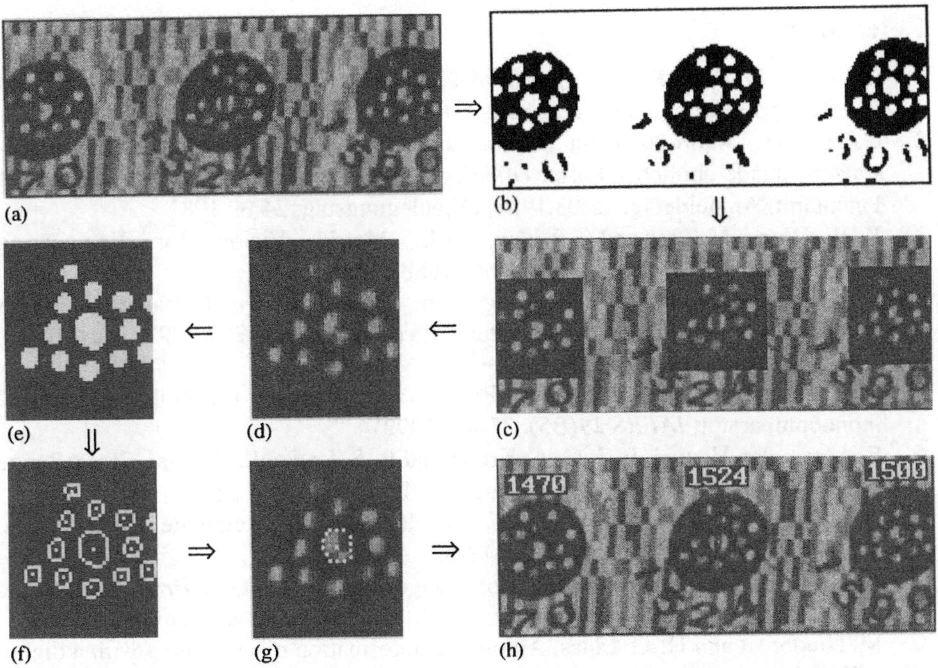

Fig. 9 Image processing of the circular coded target with superimposed random pattern: (a) Gray image (225×90 pixel) of coded targets with superimposed random pattern; (b) Optimal thresholding concerning the size and form factor of the target background plate; (c) Masked image areas for target; (d) Local gray image block; (e) Local optimal thresholding concerning the size and form factor of the point mark and the coding points, and their separation; (f) Contours of the point mark and the coding points; (g) Result of the forced orthogonal edge detection, and the estimated center of the point mark; (h) Identified and positioned target center.

target geometry as described in section 2 and 3 play an important role and guarantee a reliable object detection and identification (fig. 9d-9h). The forced edge detection enables the point mark contours to be recovered with subpixel accuracy even in the textured image (fig. 9g).

5 Discussion and conclusion

Alongside the accuracy and stability of measurement results, automation of measurement procedures is indispensable for the future industrial application of optical 3D-measurement techniques. In this paper, we have presented a new circular coded target, its image processing and application to some optical 3D-measurement methods. The special geometric construction of the target and the proper image processing make possible an accurate and reliable image point measurement and identification. The programmable arrangement of the coding points provides the adaptability of the target for all possible measuring tasks occurring in practice. The circular coded target is successfully integrated in a commercial digital photogrammetric system for surface reconstruction [16], and furthermore, we expect a general acceptance of the target with industrial applications with regard to accuracy, reliability and the automation of the measuring task.

References

1. C. Fraser and D. C. Brown, Industrial Photogrammetry: New Developments and Recent Applications, *Photogrammetric Record* **12**(68), 197-217 (1986).
2. G. Trenkler, Anordnung zur automatischen Sortierung von Stückgütern, insbesondere quaderähnlicher Form, *Offenlegungsschrift* DE 30 10 112 A1, Deutsches Patentamt, Anmeldetag: 15.03.1980, Offenlegungstag: 24.09.1981.
3. K. W. Wong, M. Lew and G. Wiley, 3-D Metric Vision for Engineering Construction, *IAPRS* **27**(B5), 647-656 (1988).
4. C.-T. Schneider, 3-D Vermessung von Oberflächen und Bauteilen durch Photogrammetrie und Bildverarbeitung, *Proc. IDENT/VISION '91*, May 14-17, 1991, Stuttgart, Germany, 90-93.
5. M. Knobloch and T. Rosenthal, MIROS: A new software for Rollei RS1 digital monocomparator, *IAPRS* **29**(B5), 35-42 (1992).
6. F. A. van den Heuvel, R. J. G. A. Kroon and R. S. Le Poole, Digital Close Range Photogrammetry Using Artificial Targets, *IAPRS* **29**(B5), 222-229 (1992).
7. A. S. Homainejad and M. R. Shortis, Development of a template for automatic stereo matching, *IAPRS* **30**(5W1), 318-322 (1995).
8. C. Fraser, Automation in digital close-range photogrammetry, *Proc. First Trans Tasman Surveyors Conference*, Apr. 12-18, 1997, Newcastle, Australia.
9. M. Niederöst and H.-G. Maas, Automatic deformation measurement with a digital still video camera, *Optical 3-D Measurement Techniques IV*, A. Gruen and H. Kahmen (Eds.), Herbert Wichmann Verlag, Heidelberg, 266-271 (1997).
10. S. J. Ahn and B. Oberdorfer, *Optoelektronisch erfaßbares Identifizierungs- oder Zielelement sowie Verfahren zu seiner Erfassung*, Deutsches Patentamt, Patentschrift DE 196 32 058 C1, Anmeldetag: 09.08.1996, Veröffentlichungstag der Patenterteilung: 05.03.1998.
11. S. J. Ahn and M. Schultes, A new circular coded target for the automation of photogrammetric 3D-surface measurements, *Optical 3-D Measurement Techniques IV*, A. Gruen and H. Kahmen (Eds.), Herbert Wichmann Verlag, Heidelberg, 225-234 (1997).
12. C. B. Bose and I. Amir, Design of Fiducials for Accurate Registration Using Machine Vision, *IEEE Trans. PAMI* **12**(12), 1196-1200 (1990).
13. S. J. Ahn, W. Rauh and B. Oberdorfer, Least Squares Fitting of Circle and Ellipse, *Proc. 20. DAGM Symp. Mustererkennung*, P. Levi et. al. (Eds.), Sep. 29-Oct. 1, 1998, Stuttgart, Germany.
14. N. N., *Bedienungsanleitung: Linienprojektoren LCD-320, LCD-640, LCD-1280*, ABW Automatisierung Bildverarbeitung Dr. Wolf GmbH, Frickenhausen, Germany (1997).
15. H.-P. Duwe, Vorstellung eines neuen 3-D Sensors zur Kombination photogrammetrischer Verfahren mit dem "kodierten Licht Ansatz", *Proc. 2. ABW Workshop Optische 3D-Formerfassung großer Objekte*, H. Wolf (Ed.), Technische Akademie Esslingen, 25.-26. Jan. 1996, ABW Automatisierung Bildverarbeitung Dr. Wolf GmbH, Frickenhausen, Germany.
16. N. N., *MATCH-I: Digitization of Industrial Surfaces. Innovation in industrial photogrammetry*, Inpho GmbH, Stuttgart, Germany (1997).

Dreidimensionale Lagebestimmung zu demontierender Objekte mit einem aktiven Stereokamerasystem

R. Stemmer, R. Trapp, S. Drüe, G. Hartmann
Universität-GH Paderborn, Fachbereich Elektrotechnik
Pohlweg 47-49, 33098 Paderborn
Email: stemmer@get.uni-paderborn.de

Zusammenfassung: *In diesem Beitrag wird ein aktives Stereokamerasystem zur visuellen Steuerung eines Demontageroboters beschrieben. Durch ein neuentwickeltes Stereoverfahren wird im Sinne einer Grob-Fein-Strategie eine erste grobe Tiefenschätzung durchgeführt und mögliche Hindernisse im Arbeitsbereich des Roboters lokalisiert. Parallel dazu werden die zu demontierenden Objekte erkannt und nach der Erkennung durch das aktive Sehsystem fixiert. Durch Erhöhung des Zoomfaktors und Filterung der Bilder in hoher Auflösung werden die dreidimensionale Position und Lage der Demontageobjekte im Raum genauer ermittelt. Die durch das Stereoverfahren errechneten Weltkoordinaten werden in Roboterkoordinaten transformiert, um eine Demontage des Objektes durch den Roboter zu ermöglichen. Die Positions- und Lagebestimmung der Objekte ist für eine Demontage von Rädern hinreichend genau.*[1]

1 Einleitung

Bei vielen Robotikanwendungen (z.B. Autoproduktion) werden vom Manipulator immer die gleichen Arbeitsschritte abgearbeitet. In diesen konventionellen Systemen kann der Roboter jedoch nicht ohne weitere Sensorik auf Veränderungen seiner Umgebung reagieren und muß notwendigerweise an die neue Situation adaptiert werden. Da bei der Autodemontage hingegen die Arbeitsumgebung für jeden Fahrzeugtyp variiert, muß die Roboterkoordination flexibel auf die veränderte Demontageszene reagieren. Neben einer toleranten Erkennungsstrategie muß die dreidimensionale Struktur der Szene erfaßt werden, um Greifvorgänge ohne Kollisionen realisieren zu können.
Zur Rekonstruktion der Tiefeninformation werden zwei unterschiedliche Ansätze verfolgt. Bei aktiven Verfahren werden Sensoren eingesetzt, die selbst Energie emittieren z.B. Ultrasschall- oder Lasersensoren. Diese Sensoren nutzen Laufzeitmessungen oder Triangulationsverfahren zur Tiefenbestimmung [MDW94]. Diese Verfahren weisen jedoch bei der Raddemontage den Nachteil auf, daß sie die Entfernung zu den durch Verschmutzung und Oxydation unebenen Felgenoberflächen nur ungenau bestimmen können. Darüber hinaus ist in einem solchen Ansatz eine aufwendige Fusion von Distanz- und Bildinformation notwendig. Durch den ausschließlichen Einsatz des Stereo-

1. Diese Arbeit wurde im Rahmen des BMBF Projektes 01 IN 506 B 2 DEMON gefördert.

kamerasystems zur Szenenanalyse und somit Verzicht auf weitere Distanzsensoren an der Roboterhand werden zusätzliche Fahrwege des Roboters und somit Zeit beim Demontageprozeß gespart.
Passive Verfahren nutzen nur das von der Umgebung reflektierte Licht. Die verschiedenen Ansätze („shape from X") rekonstruieren aus mindestens einem Bild der Szene die dreidimensionale Information, welche durch die Projektion auf die zweidimensionale Sensorfläche verloren geht [JH97].
In der beschriebenen Demontageanwendung wird die dreidimensionale Information unter Anwendung eines binokularen „shape from stereo"-Ansatzes mit einem aktiven Stereokamerakopf rekonstruiert. Nach einer Filterung der Grauwertbilder mit Gabor-Filtern werden die Filterantworten lokal korreliert. Mit diesem Ergebnis wird ein Selbstorganisationsprozeß initialisiert. Ein Ansatz von Reimann und Haken [RH94] wurde erweitert, um Fehlzuweisungen durch Okklusionen zu minimieren. Anschließend werden die Disparitäten durch Ausnutzung des linearen Phasenverlaufs des Korrelationsergebnisses subpixelgenau ermittelt. Weitere Verfahren zur Tiefenbestimmung durch Stereoskopie basierend auf Gabor- bzw. quadraturgefilterten Bildern werden in [Fle94, FJJ91, HS96, JJ94, San88, TM93] beschrieben.
Zur Schätzung der räumlichen Orientierung der Objekte wird ein „geometrisches Filter" auf die detektierte Punktmenge angewendet und eine Ebene im Sinne der kleinsten Fehlerquadrate durch die Punkte gelegt. Die ermittelten Weltkoordinaten bezogen auf den Stereokopf und die Lage des Objektes werden abschließend durch Rotationen und translatorische Verschiebungen in Roboterkoordinaten transformiert.

2 Systemüberblick

Das Gesamtsystem besteht aus einem Roboter mit sechs Freiheitsgraden und einem aktiven Stereokamerakopf. Das Sehsystem ist schräg hinter dem Roboter angebracht, so daß es am Roboter vorbei auf die Demontageszene blicken kann (siehe Abb. 1). Da es außerhalb des Arbeitsbereiches des Roboters befestigt ist, kann die Gefahr einer Kollision mit dem Roboter ausgeschlossen werden.
Das Stereokamerasystem besitzt vier mechanische Freiheitsgrade. Die Kameras können unabhängig voneinander um jeweils eine vertikale Vergenzachse geschwenkt werden. Somit können Objekte mit asymmetrischen Vergenzwinkeln fixiert werden. Eine gemeinsame Neigeachse erlaubt es, die Kameras vertikal auszurichten. Eine gemeinsame Schwenkachse ermöglicht die Fokussierung von Objekten mit symmetrischen Vergenzwinkeln. Die Kameras besitzen Mehrlinsensysteme mit jeweils drei optischen Freiheitsgraden (Blende, Fokus, Zoom). Eine detailliertere Beschreibung des Aufbaus wurde in [TD96] vorgestellt.
Neben der dort beschriebenen Kalibrierung der extrinsischen und intrinsischen Kameraparameter wurde zusätzlich die Auswirkung des Kamerazooms auf die Parameter des Lochkameramodells untersucht. Die Motorpositionen und die aktuellen Werte für Zoom, Fokus und Blende werden vor der Berechnung der Tiefenkarte aus Encodern ausgelesen und mit diesen Werten die aktuellen Parameter des Lochkameramodells aus Look-up-tables bestimmt.

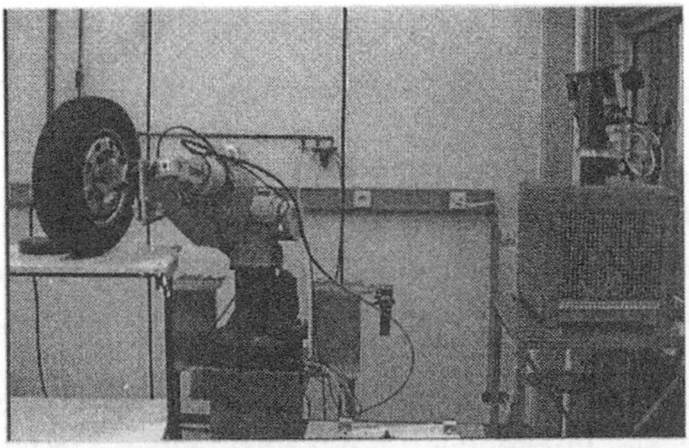

Abb. 1: Demontageszene mit Roboter und Stereokamerakopf

Zur weiteren Fehlerreduktion bei der Objektvermessung werden die Effekte der radialen Linsendistorsion aus den Bildern eliminiert. Die tangentiale Komponente der Verzeichnung wird vernachlässigt. Mittels eines bilinearen Interpolationsverfahren werden die Bilder in einem Schritt gleichzeitig entzerrt und rektifiziert.

3 Positions- und Lagebestimmung

Sowohl bei der Rekonstruktion der Tiefeninformation wie auch bei der Erkennung [Göt98] der zu demontierenden Objekte wird eine Grob-Fein-Strategie verfolgt. In einem ersten Blick wird eine grobe Schätzung der Tiefe vorgenommen. Nach Fixation der zu demontierenden Objekte und Detektion möglicher Hindernisse werden die Tiefenkarten mit hoherer Auflösung aber kleinerem Bildausschnitt berechnet, so daß die Position des Objektes in Weltkoordinaten mit hoher Genauigkeit gemessen werden kann. Zudem werden die Objekte durch den aktiven Stereokamerakopf fixiert. Eine Erhöhung des Zoomfaktors erlaubt die Vermessung von Details.

3.1 Vorverarbeitung des Stereobildpaares

Um die Tiefeninformationen aus einem Stereobildpaar zu rekonstruieren, ist es notwendig, die Verschiebung korrespondierender Bildpunkte in den Bildern zu ermitteln. Im ungünstigsten Fall sind die Verschiebungen in horizontaler und vertikaler Richtung ungleich Null. Zur Reduktion der Suche auf eine Dimension und somit zur Verringerung des Rechenaufwandes werden die Bilder vor der Stereobildverarbeitung rektifiziert. Dazu werden die Bilder, die bei nicht parallel ausgerichteten Vergenzachsen aufgenommen wurden, durch die Rektifikation in Bilder eines Sehsystems mit parallelen Achsen transformiert [AH88].
Nach der Entzerrung und Rektifikation werden die Stereobilder mit einem Satz orientierter Gabor-Filter [Gab46] in unterschiedlichen Auflösungen gefiltert. Diese Filter bieten die Möglichkeit, Bilder frequenzselektiv bei optimaler Auflösung im Orts- und Ortsfrequenzbereich zu filtern. Im Gegensatz zu gebiets- und konturbasierten Verfahren erlauben Phasendifferenzmethoden darüber hinaus Disparitätsmessungen, die genauer als das zugrunde liegende Abtastraster sind.

3.2 Rekonstruktion der Tiefeninformation durch Stereoskopie

Im Unterschied zu klassischen Phasendifferenzmethoden werden bei diesem Verfahren in einem ersten Schritt die Amplituden der Filterantworten in einem lokalen Fenster korreliert. Eine Tiefenschätzung anhand der Phaseninformation beruht auf dem Verschiebungssatz der Fouriertransformation. Aufgrund der lokalen Impulsantwort der Filter wird der Verschiebungssatz jedoch nicht exakt erfüllt. Darüber hinaus wird die maximal detektierbare Disparität bei ausschließlicher Auswertung der Phaseninformation durch die Modulationsfrequenz beschränkt.
Die Faltung des Produktes von linker Filterantwort r_l und räumlich verschobener konjugiert-komplexer rechter Filterantwort r_r mit einer reellen Fensterfunktion w liefert ein komplexes Ähnlichkeitsmaß ρ_{lr} zwischen den gefilterten Bildern. Dieses Ähnlichkeitsmaß wird durch die lokale Signalenergie der Filterantworten normiert:

$$\rho_{lr}(\boldsymbol{x}, \boldsymbol{d}) = \frac{w(\boldsymbol{x}) * (r_l(\boldsymbol{x}) r_r^*(\boldsymbol{x} + \boldsymbol{d}))}{\sqrt{w(\boldsymbol{x}) * |r_l(\boldsymbol{x})|^2} \sqrt{w(\boldsymbol{x}) * |r_r(\boldsymbol{x} + \boldsymbol{d})|^2}} \tag{1}$$

Der Vektor $\boldsymbol{x} = [x, y]^{\mathrm{T}}$ repräsentiert die Bildkoordinaten und die Disparität $\boldsymbol{d} = [d_x, d_y]^{\mathrm{T}}$ gibt den zweidimensionalen Verschiebungsvektor der rechten Filterantwort an. Für den Fall, daß die Bilder vor der Filterung und Korrelation rektifiziert wurden, ist die vertikale Komponente des Vektors Null und die Faltung reduziert sich auf eine eindimensionale Berechnung. Das reelle Fenster w wird in diesem Fall durch eine eindimensionale Gauß-Funktion beschrieben, die der Einhüllenden der Impulsantwort der Filter in horizontaler Richtung entspricht.
Sind die Filterantworten in einem lokalen Bereich ähnlich, so wird eine geringe Phasendifferenz erwartet. Dies bedeutet, daß der Realteil des Ähnlichkeitsmaßes ein Maximum und der Imaginärteil einen Nulldurchgang aufgrund des annähernd linearen Phasenverlaufs der Filterantworten aufweist.
Durch Rauschen, Verzerrungen und periodischen Strukturen in den Bildern können Fehlzuweisungen von korrespondierenden Punkten auftreten. Durch Berücksichtigung von Bedingungen der physikalischen Umwelt können diese Mehrdeutigkeiten aufgelöst werden. Die zwei wichtigsten Bedingungen für die Stereoskopie sind die Eindeutigkeitsbedingung und die Kontinuitätsbedingung [MP79]. Verschiedene weitere Ansätze verfolgen in ähnlicher Weise die Strategie, das globale Minimum von Kostenfunktionen zu bestimmen [GG84, Ter86]. Zur Lösung des Mehrdeutigkeitsproblems wird für jede Bildkoordinate $\boldsymbol{x}$ und jede mögliche Disparität $\boldsymbol{d}$ eine reelle, zeitabhängige Variable $\xi(\boldsymbol{x}, \boldsymbol{d}, t)$ definiert. Zum Zeitpunkt $t = 0$ werden die Variablen mit dem Realteil des Korrelationsergebnisses initialisiert:

$$\xi(\boldsymbol{x}, \boldsymbol{d}, t = 0) = f\left(\sum_i c_i(\mathrm{Re}\{\rho_{lr_i}(\boldsymbol{x}, \boldsymbol{d})\})\right) \tag{2}$$

Es wird die Summe über die Ergebnisse der Korrelation bezüglich der Filterorientierungen -30°, 0° und +30° gebildet. Die verschiedenen Orientierungen können über den Parameter c_i individuell gewichtet werden. Die Funktion f dient dazu, die Ähnlichkeitswerte in einen positiven Zahlenbereich abzubilden. In diesem Verfahren werden dazu die negativen Werte zu Null gesetzt. Die zeitliche Dynamik der Variablen wird durch folgendes gekoppelte, nichtlineare Differentialgleichungssystem beschrieben:

$$\begin{aligned}\dot{\xi}(x, d, t) = \Bigg\{ & A - C\xi^2(x, d, t) \\ & - \frac{B}{3K}\sum_{d' \neq d} \xi^2(x, d', t) \\ & - \frac{B}{3K}\sum_{d' \neq d} \xi^2(x + d - d', d', t) + \xi^2\left(x + \frac{1}{2}(d - d'), d', t\right) \\ & + \frac{D}{L}\sum_{x' \in U} \xi(x', d, t) \Bigg\}\xi(x, d, t)\end{aligned} \tag{3}$$

Die Parameter A, B, C, D sind positive Konstanten und K und L geben die Anzahl der in den jeweiligen Prozeß einbezogenen Variablen an. Der erste Term ist verantwortlich für ein exponentielles Wachstum, der zweite hingegen beschränkt die Amplitude von $\xi(x, d, t)$. Der Term in der zweiten Zeile berücksichtigt die Eindeutigkeitsbedingung. Dieser Term verursacht einen Wettbewerb zwischen allen Variablen der selben Bildkoordinate. Die Terme der dritten Zeile verhindern in diesem Wettbewerb einen Gewinn von Variablen in Bildbereichen mit Okklusionen [TDH98]. Der Term in der letzten Zeile berücksichtigt schließlich die Kontinuitätsbedingung, indem die Variablen gleicher Disparität in einem kleinen Umfeld positiv miteinander gekoppelt werden. Dieser Selbstorganisationsprozeß entspricht mit Ausnahme der Unterdrückung von Okklusionskorrespondenzen durch den Term in der dritten Zeile im wesentlichen dem Ansatz von Reimann und Haken [RH94]. Die Ruhelagen der Differentialgleichung $\xi_r(x, d)$ erfüllen die Bedingung $\dot{\xi}(x, d, t) = 0$. Von dieser Differentialgleichung existieren drei verschiedene Ruhelagen:

$$\begin{aligned}\xi_{r_1}(x, d) &= 0 \\ \xi_{r_{2/3}}(x, d) &= \pm\sqrt{\frac{\Gamma(x, d)}{C}}\end{aligned} \tag{4}$$

mit

$$\begin{aligned}\Gamma(x, d) = {} & A + \frac{D}{L}\sum_{x' \in U} \xi_r(x', d) - \frac{B}{3K}\sum_{d' \neq d} \xi_r^2(x, d') \\ & - \frac{B}{3K}\sum_{d' \neq d} \xi_r^2(x + d - d', d') + \xi_r^2\left(x + \frac{d}{2} - \frac{d'}{2}, d'\right)\end{aligned} \tag{5}$$

Da der Prozeß ausschließlich mit positiven Werten initialisiert wird, kann gezeigt werden, daß die negative Ruhelage niemals erreicht wird [RH94]. Es werden demnach nur noch die Ruhelage $\xi_{r1}(x, d)$ und die positive Ruhelage $\xi_{r2}(x, d)$ berücksichtigt. Die

Parameter in Beziehung (3) werden so gewählt, daß zwar ein Wettbewerb zwischen den Variablen ermöglicht, die Stabilität der Ruhelagen aber zudem gewährleistet ist. In diesem Wettbewerb ist es wahrscheinlich, daß eine Variable mit starkem kooperativem Umfeld und hohem Initialisierungswert gewinnt. Der Gewinner, dessen Disparität übernommen wird, konvergiert in die Ruhelage $\xi_{r2}(\boldsymbol{x}, \boldsymbol{d})$. Alle weiteren Variablen in diesem Wettbewerb erreichen im allgemeinen die Ruhelage $\xi_{r1}(\boldsymbol{x}, \boldsymbol{d})$, so daß der Selbstorganisationsprozeß das Auftreten von Fehlzuweisungen minimiert.
Nach Erreichen des stationären Zustandes im Selbstorganisationsprozeß wird der Imaginärteil des Ähnlichkeitsmaßes an Positionen von Variablen mit Werten ungleich Null interpoliert. Diese Positionen korrespondieren mit hoher Wahrscheinlichkeit mit Maxima im Realteil. Der Nulldurchgang des Imaginärteils gibt die neue subpixelgenaue Disparität an. Der mittlere quadratische Fehler der Quantisierungsfehler kann bei Einsatz des subpixelgenauen Verfahrens um bis zu 60% reduziert werden [TDH98].

3.3 Bestimmung der dreidimensionalen Lage

Um einen Greifer kollisionsfrei an ein zu demontierendes Objekt heranzufahren, muß nicht nur dessen Position parallel zur Bildebene, sondern auch die Verkippung in die Tiefe ermittelt werden. Bei der Raddemontage ist es zum Beispiel notwendig, die Radmuttern senkrecht zur Felgenoberflächen zu entschrauben. Um die Daten der Erkennung im zweidimensionalen Bild mit den dreidimensionalen Tiefeninformationen zu fusionieren, werden einfache geometrische Strukturen in den Filterantworten gesucht. Die dreidimensionalen Koordinaten an den Positionen der Strukturen bilden eine Punktmenge, die als eine verrauschte Abtastung der Objektkonturen im dreidimensionalen Raum interpretiert werden kann.
Für die Raddemontage wird im Grauwertbild ein Kreis bzw. Ellipse an der Falz zwischen Reifen und Felge detektiert. Diese Falz wird als eben angenommen. Zur Bestimmung der Lage der Falz und somit des Rades wird im Sinne des kleinsten Fehlerquadrates eine Ebene durch die detektierte Punktmenge gelegt. Durch eine „geometrische Filterung" basierend auf der Annahme eines kontinuierlichen Konturverlaufs werden fehlerhafte Punkte innerhalb der Punktmenge eliminiert. Die Punkte P_k werden in einer Reihenfolge sortiert, die dem Verlauf der Kontur entspricht. Die Abweichung zwischen den Verbindungsvektoren zu den Nachbarpunkten berechnet sich über den Betrag des Vektorprodukts der Vektoren:

$$D_k(\lambda) = \sum_{\delta=1}^{\lambda} \left|(P_k - P_{k-\delta}) \times (P_{k+\delta} - P_k)\right| \tag{6}$$

Der Parameter λ beschreibt eine symmetrische Nachbarschaft der Größe $2\lambda + 1$ um den Punkt P_k. Die Punkte, deren Differenzen D_k größer als ein bestimmtes Vielfaches des Medians aller Differenzen sind, werden aufgrund der Annahme eines kontinuierlichen Konturverlaufs aussortiert.
Abschließend müssen die ermittelten Punkte und Oberflächennormale, die sich auf ein Koordinatensystem mit dem Ursprung zwischen den beiden optischen Zentren beziehen, in Roboterkoordinaten transformiert werden. Die Parameter der affinen Transformation werden mit einem ähnlichen Verfahren wie bei der Kamerakalibrierung bestimmt und mit Hilfe des Gauß-Newton-Minimierungsverfahrens optimiert [TD96].

4 Ergebnisse

In diesem Beitrag wurde ein Verfahren zur genauen Bestimmung der dreidimensionalen Position und Lage von Objekten mit Hilfe eines neuen Stereoverfahrens beschrieben. Durch Einsatz dieses Verfahrens werden Fehlzuweisungen korrespondierender Bildpunkte minimiert. Die Wahrscheinlichkeit für falsche Tiefenschätzungen wird dadurch reduziert. Eine weitere Distanzmessung mit Sensoren an der Roboterhand wird nicht benötigt, so daß die Arbeitszeit durch Einsparungen von Roboterbewegungen reduziert wird. Besonders bei der Demontage von Autorädern, die eine dreidimensionale Struktur aufweisen, werden mit dem neuen Stereoansatz und einer Grob-Fein-Strategie zuverlässigere Distanzinformationen gewonnen, da die aufgrund der dreidimensionalen Oberfläche der Radfelgen vermehrt auftretenden Okklusionen sicher vom Zuordnungsprozeß ausgeschlossen werden.
In den bisherigen Untersuchungen konnte gezeigt werden, daß die Genauigkeit des Stereoverfahrens und der Transformation der Koordinaten für eine Demontage von Autorädern ausreichend ist. Die Streuung des Fehlers in der Bildebene liegt unter 1mm und in der Tiefe bei ca. 2mm bei einem Abstand von ca. 1,5 Meter zwischen Felge und Stereokamerakopf. Erste Untersuchungen an verkippten Autorädern haben gezeigt, daß der Kippwinkel zur Bildebene (Lenkradeinschlag, Sturz) in einem Bereich von ±20° bei gleicher Entfernung auf unter ein Grad genau bestimmt werden kann. Die durch Untersuchungen ermittelten Fehler sind in den folgenden Graphiken dargestellt:

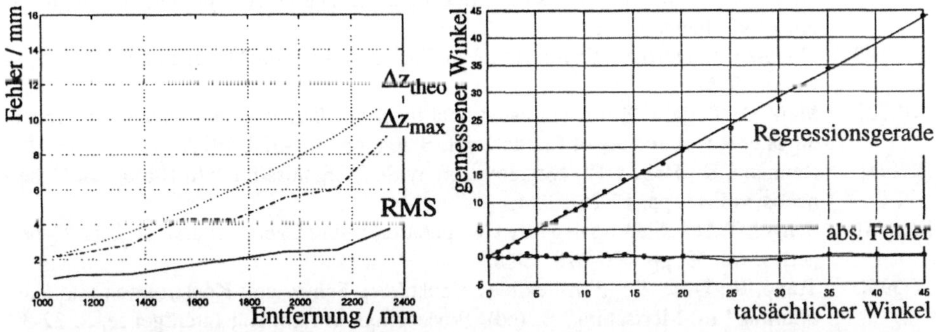

Abb. 2: Fehlermaße zur Tiefenschätzung und Bestimmung des Kippwinkels

In der linken Abbildung ist der Verlauf des theoretischen und des gemessenen Maximalfehlers und die Streuung für den gemessenen Fehler für die Tiefenmessungen aufgetragen. Die rechte Abbildung zeigt den ermittelten und den eingestellten Kippwinkel bei der Raddemontage in einer mittleren Auflösung.
In Testreihen zur Raddemontage konnten die Radmuttern bei dem genannten Abstand durch Anwendung der beschriebenen Grob-Fein-Strategie sicher gegriffen werden. Der Einsatz des Verfahrens zur Steuerung eines autonomen Demontagemanipulators ist mit diesen Ergebnissen möglich.

5 Literatur

[AH88] Ayache, N.; Hansen, C.: „Rectification of Images for Binocular and Trinocular Stereovision.“ In Proceedings of the 9th Intern. Conf. on Pattern Recognition, pp. 11-16, 1988.

[FJJ91] Fleet, D. J.; Jepson, A. D.; Jenkin, M. R. M.: "Phase-Based Disparity Measurement." In CVGP: Image Understanding, vol. 53, no. 2, pp. 198-210, March 1991.

[Fle94] Fleet, D. J.: „Disparity from Local Weighted Phase Correlation.“ In IEEE International Conference on Systems, Man and Cybernetics, pp. 48-56, 1994

[Gab46] Gabor, D.: „Theory of communication.“ In J. IEE, vol. 93, pp. 429-457, London, 1946.

[GG84] Geman, S.; Geman, D.: „Stochastic Relaxation, Gibbs Distributions, and the Bayesian Restoration of Images.“ In IEEE Transactions on Patttern Analysis and Machine Intelligence, vol. PAMI-6, no. 6, pp. 721-741, November 1984.

[Göt98] Götze, N.; Drüe, S.; Hartmann, G.: „Erkennung und Lagebestimmung realler Objekte durch tolerante Repräsentation und Modellbasiertes Gruppieren.“ Angenommen zur DAGM 1998.

[HS96] Hansen, M.; Sommer, G.: „Active depth estimation with gaze and vergence control using Gabor filters.“ In 13th Int. Conf. on Pattern Recognition, Vol. A, Vienna, pp. 287-291, 1996.

[JH97] Jähne, B.; Herrmann, H.: „3D-Exploration mit einer sich bewegenden Kamera.“ Symposium Bildverarbeitung ‘97 - Forschen, Entwickeln, Anwenden.

[JJ94] Jenkin, M. R. M.; Jepson, A. D.: „Recovering Local Surface Structure through Local Phase Difference Measurements.“ In CVGIP: Image Understanding, vol. 59, no. 1, pp. 72-93, January, 1994.

[MDW94] Manyika, J.; Durrant-Whyte, H.: „Data Fusion and Sensor Management.“ Ellis Horwood, 1994.

[MP79] Marr, D.; Poggio, T.: „A computational theory of human stereo vision.“ In Proceedings of Royal Society of London, vol. B 204, pp. 301-328, 1979.

[RH94] Reimann, D.; Haken, H.: „Stereovision by Self-Organisation.“ In Biological Cybernetics, vol. 71, pp. 17-26, 1994.

[San88] Sanger, T. D.: „Stereo Disparity Computation Using Gabor Filters.“ In Biol. Cybernetics, vol. 59, pp. 405-418, 1988.

[TD96] Trapp, R.; Drüe, S.: „Ein flexibles binokulares Sehsystem: Konstruktion und Kalibrierung“ In: Mertsching, B. (ed): Proceedings in Artificial Intelligence, S. 32-39, Sankt Augustin (Infix), Dezember 1996.

[TDH98] Trapp, R.; Drüe, S.; Hartmann, G.: „Stereo Matching with Implicit Detection of Occlussions“. In: Burkhardt, H.; Neumann, B. (Hrsg.): Computer Vision - ECCV ‘98. Proceedings, Volume II, S. 17-33, Springer Verlag, Berlin, Juni 1998.

[Ter86] Terzopoulos, D.: „Regularization of Inverse Visual Problems Involving Discontinuities.“ In IEEE Trans. Pat. Anal. and Mach. Int., vol. PAMI-8, no. 4, pp. 413-424, 1986.

[TM93] Theimer, W. M.; Mallot, H. P.: „Vergence Guided Depth Reconstruction Using a Phase Method.“ In Neural Networks and their Industrial and Cognitive Applications: Neuro Nimes ‘93, pp. 299-308, 1993.

Sprach- und Bildverstehen

Erkennung handgeschriebener Wortsequenzen

Urs-Viktor Marti und Horst Bunke

Institut für Informatik und angewandte Mathematik
Universität Bern, Neubrückstrasse 10, CH-3012 Bern
marti,bunke@iam.unibe.ch

Zusammenfassung Die vorliegende Arbeit stellt ein System vor, welches in der Lage ist, mit einfachen Satzmodellen eine Sequenz von handgeschriebenen Wörtern zu erkennen. Hierzu werden Satzmodelle, welche ursprünglich in der Spracherkennung entwickelt wurden, verwendet. Durch die Verwendung dieser statistischen Modelle kann eine Reduzierung der Fehlerrate des Worterkenners von 23% auf 18% bzw. 15% erreicht werden.

1 Einleitung

Immer noch stellt die Erkennung von Handschriften eine grosse Herausforderung in der Mustererkennung dar. Die ersten Systeme, die sich mit Handschriftenerkennung beschäftigten, konnten einzelne Buchstaben und aus einzelnen Buchstaben zusammengesetzte Wörter erkennen[LB93]. In der Zwischenzeit ist es auch möglich, Wörter, die in Kursivschrift geschrieben sind, zu lesen[BS89]. Trotz der Erfolge, die man in den letzten Jahren mit Systemen erzielte, die von Hand geschriebene Scheckbeträge[KBH96] oder Teile von Postadressen[SGS93] lesen können, ist es noch ein weiter Weg bis zum Lesen allgemeiner Texte. Ein Grund besteht darin, dass die bereits existierenden Systeme nur in einem sehr engen Kontext arbeiten oder aber sehr spezielles Wissen verwenden. Der nächste anstehende Schritt in der Entwicklung von Lesesystemen geht in die Richtung, ganze Sätze erkennen zu können[GSS95, SB93]. Um diesem Ziel näher zu kommen, setzen wir in der hier vorgestellten Arbeit Techniken ein, die aus Spracherkennung stammen[CDDEB+87, Jel90, RJ93, Hul92].
Im vorliegenden Paper stellen wir ein System vor, welches mit Hilfe eines Vokabulars und einfacher Satzmodelle in der Lage ist, ganze Sätze zu lesen. Das System besteht aus drei Komponenten: einer Vorverarbeitung, die die Bilddaten für die Trainings- und Testphase vorbereitet, einem Worterkenner basierend auf Hidden-Markov-Modellen und einer Satzmodelierung, die auf statistischen Methoden beruht.
In den nächsten zwei Kapiteln beschreiben wir die Vorverarbeitung der Daten und die Erkennung von einzelnen Wörtern mit Hidden-Markov-Modellen. Das Kapitel 4 erläutert die verwendeten Satzmodelle. Die Experimente und Resultate, die zu den beschriebenen Satzmodellen durchgeführt wurden, werden im Kapitel 5 erläutert. Zum Schluss beschreiben wir im Kapitel 6 die Schlussfolgerungen die aus dieser Arbeit gezogen werden können, und geben einen Ausblick auf zukünftige Arbeiten.

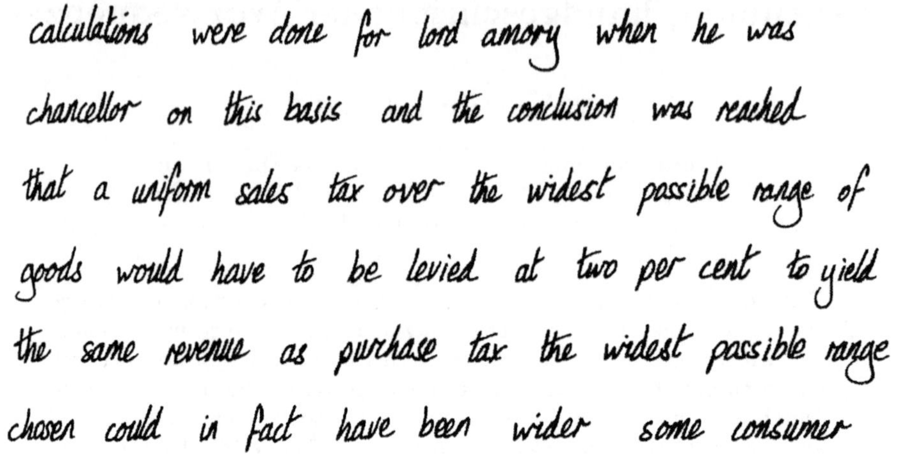

Abbildung1. Teil aus einem digitalisierten Textbild

2 Vorverarbeitung

Da für die im Rahmen dieser Arbeit durchgeführten Experimente eine Datenbank verwendet wird, in der die Wörter nur von einem Schreiber stammen und sehr schön geschrieben sind (siehe Abb. 1), kann auf eine komplizierte Vorverarbeitung verzichtet werden. Es ist jedoch für weitere Entwicklungen nötig, die Vorvearbeitung zu verbessern, wenn das System schreiberunabhängig sein soll.

Im vorliegenden System besteht die Vorverarbeitung aus vier Teilen: der Binärisierung, der Zeilenextraktion, der Positionierung und Skalierung der Schriftbilder sowie einer Datenkompression.

Als erster Schritt wird eine Binärisierung der Bilder vorgenommen. Diese wird mittels Otsus Algorithmus realisiert[Ots79]. Die Extraktion der Textzeilen und die Schätzung der Basislinien erfolgt mithilfe eines Histogramms der gesetzten Bildpunkte in jeder Bildzeile. Da die Schrift sehr regelmässig ist, werden die Schriftbilder auf eine konstante Bildhöhe skaliert.

Mit einem Fenster konstanter Breite, das über das ganze Textbild einer Zeile von links nach rechts geschoben wird, wird der zu jeder Zeit für das System sichtbare Bereich festgelegt. Diese Technik ist allgemein unter dem Namen "sliding window technique" bekannt. In unserer Arbeit verwenden wir eine Fensterbreite von einem Pixel. Um in den nachfolgenden Verarbeitungsschritten weniger Daten verarbeiten zu müssen, werden die einzelnen Fensterinhalte komprimiert. Zur Komprimierung verwenden wir eine Karhunen-Loève-Transformation, welche mit einem Neuronalen Netz realisiert wird. Für Details siehe [Sch96], Kapitel 9.

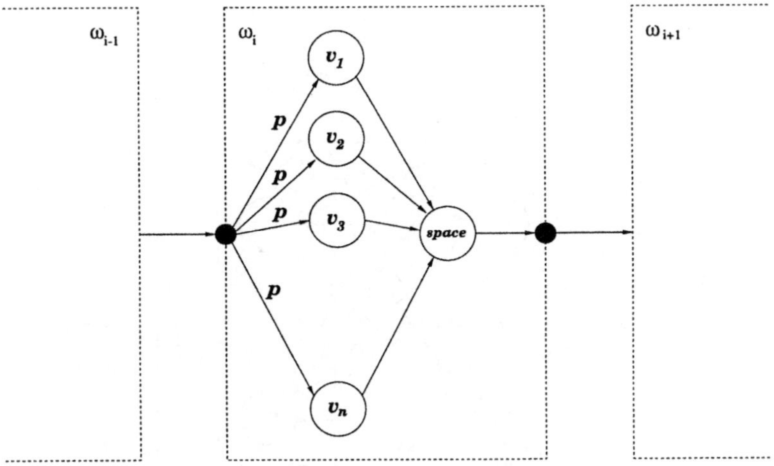

Abbildung2. Einfaches Satzmodell ("space" bedeutet Leerraum)

3 Hidden-Markov-Modelle

Hidden-Markov-Modelle haben sich mittlerweile zu einem Standardwerkzeug bei er Erkennung von Kursivschrift entwickelt.
In Handschrifterkennern, welche auf einem kleinen Vokabular arbeiten, ist es möglich, für jedes Wort ein eigenes Hidden-Markov-Modell zu erstellen. Existieren für jedes Wort genügend viele Instanzen, ist es möglich, die Modelle zuverlässig zu trainieren. In Systemen, welche mit grösserem Vokabular arbeiten, ist es nicht mehr möglich, jedes Wort als eigenständiges Modell zu realisieren. Hierzu fehlen in den meisten Fällen Daten, um die Modellparameter zuverlässig abschätzen zu können. Aus diesem Grund teilt man die Wörter in kleinere Einheiten und reduziert hiermit die Anzahl der Parameter. In der Spracherkennung werden die gesprochenen Wörter in Silben aufgeteilt, während sich in der Handschrifterkennung eine Aufteilung in Buchstaben anbietet. Für jeden Buchstaben wird so ein lineares Hidden-Markov-Modell mit 8 Zuständen erstellt. Diese werden dann zu den einzelnen Wörtern zusammengesetzt. Für das Trainieren verwenden wir den Baum-Welch-Algorithmus, für das Erkennen den Viterbi-Algorithmus[RJ93].

4 Statistische Satzmodelle

Um die Erkennung von Sätzen zu ermöglichen, ist es nötig, die als Hidden-Markov-Modell dargestellten Wörter in ein Satzmodell aufzunehmen. Hierzu werden die Wörter in ein Netzwerk integriert, mit dem die Eingabe in der Erkennungsphase verglichen wird.
Als erstes wurde ein sehr einfaches Satzmodell, welches als Referenzmodell dienen soll, realisiert. Bei diesem Modell gehen wir von der Voraussetzung aus, dass

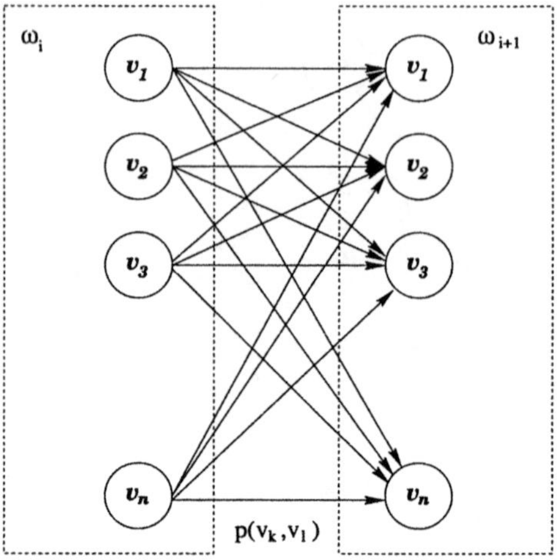

Abbildung3. Bigramm Satzmodell ("space" ist in jedem Wort enthalten)

jedes Wort v_i mit gleicher Wahrscheinlichkeit p auftritt und dass jedes Wort an jeder Stelle des Satzes auftreten kann. Weiter nehmen wir an, dass zwischen zwei Wörtern eine Leerstelle auftreten muss. Diese Leerstelle ist ein eigenständiges Wort mit einem zugehörigen Hidden-Markov-Modell. Sei $V = \{ v_1, \ldots, v_n\}$ die Menge aller möglichen Wörter (Vokabular). Weiter sei $\omega = \omega_1 \ldots \omega_m$ eine Sequenz zu erkennender Wörter, wobei $\omega_i \in V$ für $i = 1, \ldots, m$. Eine grafische Darstellung dieses einfachen Modelles ist in Abb. 2 wiedergegeben.
Die Wahrscheinlichkeit für das Auftreten einer gegebenen Wortsequenz $\omega = \omega_1\omega_2 \ldots \omega_m$ kann in diesem Modell wie folgt berechnet werden:

$$P(\omega) = \prod_{i=1}^{m} P(\omega_i) = \frac{1}{n^m} \tag{1}$$

mit $P(\omega_i) = 1/n = p$, wobei n die Anzahl der in V enthaltenen Wörter ist. Im weiteren Verlauf dieser Arbeit wird sei dieses Modell *einfaches Satzmodell* genannt.

Das zweite in dieser Arbeit verwendete Satzmodell ist das sogenannte Back-off-Bigrammmodell (siehe [Jel90], Appendix E). Dieses basiert auf dem einfachen Bigrammmodell, benutzt jedoch noch zusätzliche Parameter, um dem Fehlen von Trainingsdaten entgegenzuwirken. Abb. 3 zeigt eine grafische Darstellung dieses Bigrammmodelles.
Um dieses Satzmodell zu erstellen, wird in einem Trainingstext das Auftreten von zwei unmittelbar nacheinander stehenden Wörtern gezählt. Hieraus lässt sich die Wahrscheinlichkeit $P(\omega_i|\omega_{i-1})$, dass ein Wort ω_i unmittelbar auf ein Wort ω_{i-1} folgt, berechnen. So kann die Wahrscheinlichkeit eines ganzen Satzes ω wie folgt berechnet werden:

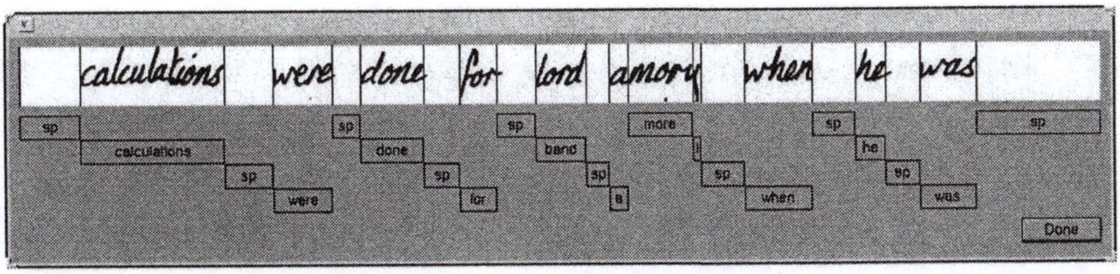

Abbildung4. Mit dem einfachen Satzmodell erhaltenes Ergebnis ("sp" entspricht dem Leerraum)

$$P(\omega) = P(\omega_1) \prod_{i=2}^{m} P(\omega_i|\omega_{i-1}) \tag{2}$$

Es ist klar, dass mit dieser Methode nur Paare von Wörtern erkannt werden können, die auch im Trainingstext aufgetreten sind. Ebenso muss der für das Training verwendete Text sehr gross sein, um ein verlässliches Satzmodell zu erhalten. Oft ist es jedoch nicht möglich, genügend grosse Trainingstexte zu erhalten, so dass zum einen zu erkennende Wortpaare im Trainingstext nicht aufgetreten sind und zum anderen die Wahrscheinlichkeiten nicht verlässlich berechnet werden können. In diesem Fall müssen andere Trainingsmethoden gefunden werden. Eine Möglichkeit besteht darin, dass eine untere Grenze t definiert wird, welche angibt, wie viele Instanzen eines Wortpaares im Trainingstext mindestens auftreten müssen. Für alle Paare, welche unter dieser Grenze liegen, wird die Wahrscheinlichkeit $P(v_i|v_j)$ auf die Wahrscheinlichkeit $P(v_i)$ zurückgeführt, welche mit einem Faktor $b(v_j)$ gewichtet wird. Dieser Faktor $b(v_j)$ wird als *Back-off-Gewicht* bezeichnet und berechnet sich wie folgt:

$$b(v_j) = \frac{1 - \sum_{v_i \in B} P(v_i|v_j)}{1 - \sum_{v_i \in B} P(v_i)}, \qquad B = \{v_i|N(v_i, v_j) > t\} \tag{3}$$

$N(.)$ steht hierbei für die Funktion, die Wörter zählt, und $N(v_i, v_j)$ gibt an, wie viele Male das Wort v_i dem Wort v_j folgt.
Die Wahrscheinlichkeit, dass ein Wort v_i auftritt unter der Annahme, dass das Wort v_j vorausgegangen ist, berechnet sich somit wie folgt:

$$P(v_i|v_j) = \frac{N(v_i, v_j)}{N(v_j)} \qquad \text{falls } N(v_i, v_j) > t \tag{4}$$

$$P(v_i|v_j) = b(v_j)P(v_i) \qquad \text{sonst} \tag{5}$$

5 Experimente und Resultate

Für unsere Experimente benutzen wir eine Datenbank, welche von A.W. Senior erstellt wurde[Sen94]. Sie besteht aus 22 handgeschriebenen Textseiten; 12 davon

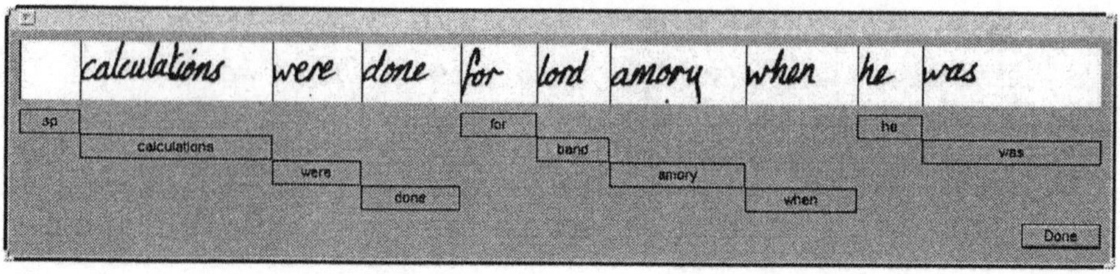

Abbildung5. Mit dem Back-off-Bigramm-Satzmodell erhaltenes Ergebnis ("sp" ist inden Wörtern enthalten)

enthalten Text, der nur in Kleinbuchstaben geschrieben ist, die anderen Seiten enthalten sowohl Gross- als auch Kleinbuchstaben. Alle 22 Textseiten sind von einer einzigen Person geschrieben. Für unsere Experimente benutzen wir den Teil der Datenbank, der nur in Kleinbuchstaben geschrieben ist. Auf diesen Seiten stehen 711 verschiedene Wörter mit total 2029 Wortinstanzen. Wir haben 27 Hidden-Markov-Modelle erstellt, für jeden Buchstaben eines und ein zusätzliches, um die Leerstellen zwischen den Wörtern zu modellieren. Für das Training dieser 27 Hidden-Markov-Modelle benutzen wir 8 Textseiten. Mit den verbleibenden 4 Seiten wird das System getestet. Da es sehr schwer ist, Satzzeichen in fliessendem Text zu erkennen, nehmen wir an, dass eine Textzeile einem Satz entspricht.

Als erstes haben wir die Leistungsfähigkeit des Worterkenners getestet. Dazu wurden die Wörter manuell aus den Seiten extrahiert und dem System übergeben. Die erzielte Erkennungsrate liegt etwa bei 77% der Wörter.

Danach haben wir das einfache Satzmodell eingesetzt. Mit diesem Modell wurden nur etwa 74% der Wörter richtig erkannt. Es ist zu beachten, dass hier nicht mehr segmentierte Wörter, sondern Textzeilen dem System übergeben werden. Deshalb ist die hier zu lösende Aufgabe schwieriger, da sowohl jede Zeile in Wörter zu segmentieren ist, als auch die Wörter zu erkennen sind. Es hat sich gezeigt, dass unter dem einfachen Satzmodell kurze Wörter vom Erkenner bevorzugt werden (siehe Abb. 4). Dies geschieht aus zwei Gründen. Erstens kann ein kurzes Wort Teil eines langen Wortes sein, und zweitens treten Situationen auf, in denen schmale Zwischenräume in Wörtern auftreten, die als Leerräume erkannt werden. In diesen Fällen teilt der Erkenner ein langes Wort in mehrere kürzere.

Im dritten Experiment haben wir das Back-off-Bigramm-Satzmodell wie in Abschnitt 4 beschrieben verwendet. Die Leerräume zwischen den Wörtern wurden durch Integrieren des Hidden-Markov-Modells für Leerräume in die Wortmodelle realisiert. Hierbei konnte das Problem, dass kurze Wörter bevorzugt werden, weitgehend eliminiert werden (siehe Abb. 5). Wörter, welche im einfachen Satzmodell in mehrere kleinere Fragmente zerteilt werden, können nun als ganze Wörter richtig erkannt werden. Die Worterkennungsrate fiel mit 82% auch entsprechend besser aus. Die Fehlerrate konnte im Vergleich zum ersten Experiment um etwa 22% reduziert werden.

HMM auf segmentierten Wörtern ohne Satzmodell	77%
HMM mit einfachem Satzmodell	74%
HMM mit generellem Bigramm Satzmodell	82%
HMM mit aufgabenspezifischem Bigramm Satzmodell	85%

Tabelle1. Worterkennungsraten der verschiedenen Experimente

Neben der Bilddatenbank, welche die Bilddaten der von Hand geschriebenen Texte enthält, wird für die hier beschriebenen Experimente noch zusätzlich eine Textsammlung benötigt, um die Satzmodelle zu erstellen. In obigen Experimenten haben wir hierzu den ganzen Lancaster-Oslo/Bergen Corpus [JLG78] benutzt. Diese Textsammlung enthält etwa 1 Million Wortinstanzen in etwa 51'000 Sätzen. Alle Sätze der Bilddatenbank sind im Corpus enthalten.
Die Texte des Lancaster-Oslo/Bergen Corpus gehören verschiedenen Sparten von Texten an. Deshalb können diese für das Satzmodell verwendeten Trainingsdaten als relativ allgemein angesehen werden - bezüglich des Texts der benutzten Bilddatenbank. Es ist bekannt, dass mit spezifischerem Wissen zum Satzbau bessere Erkennungsresultate erwartet werden dürfen [Sto97]. Deshalb haben wir in einem weiteren Experiment den am meisten spezifischen Kontext gewählt, um die Bigramm-Satzmodelle zu trainieren, nämlich den Text der Datenbank selbst. Wir benutzten sowohl Trainings- wie auch Testdaten, da eine Menge alleine nicht ausgereicht, um das Back-off-Bigrammmodell zu erstellen. In diesem Experiment haben wir eine Erkennungsrate von 85% erreicht. Tabelle 1 fasst alle oben beschriebenen Experimente noch einmal zusammen.

6 Schlussfolgerungen

In der vorliegenden Arbeit haben wir statistische Satzmodelle bei der Handschrifterkennung eingesetzt. Diese Modelle sind sehr einfach aus Texten zu generieren und benötigen deshalb nicht viel Rechenzeit beim Training. Durch den Einsatz dieser statistischen Modelle konnte die Fehlerrate eines Systems zur Erkennung handschriftlicher Texte von 23% auf 18% reduziert werden. Diese Verbesserung wurde mit einem sehr allgemeinen Satzmodell erhalten. Mit einem spezifischeren Kontextwissen gelang es uns, die Fehlerrate auf 15% zu reduzieren.
Trotz dieser Erfolge steht für zukünftige Arbeiten noch eine ganze Reihe von Problemen an. Insbesondere sollen mögliche Weiterentwicklungen des Satzmodells untersucht werden. Auch sollen zukünftige Arbeiten auf einer neuen Datenbank erfolgen, die von verschiedenen Personen geschriebene Sätze enthält.

References

[BS89] R. Bozinovic and S.N. Srihari. Off-line cursive script word recognition. *IEEE Trans. on Pattern Analysis and Machine Intelligence*, 11(1):68–83, January 1989.

[CDDEB$^+$87] H. Cerf-Danon, A.-M. Derouault, M. El-Beze, B. Merialdo, and S. Soudoplatoff. Speech recognition experiment with 10'000 words dictionary. In P.A. Devijver and J. Kittler, editors, *Pattern Recognition Theory and Applications*, volume F30 of *NATO ASI Series.* Springer Verlag, 1987.

[GSS95] V. Govindaraju, R.K. Srihari, and S.N. Srihari. Handwritten text recognition. In A. L. Spitz and A. Dengel, editors, *Document Analysis Systems*, pages 288–304. World Scientific, 1995.

[Hul92] J.J. Hull. Incorporation of a markov model of language syntax in a text recognition algorithm. In *Proc. Symp. Document Analysis and Information Retrieval*, pages 174–185, 1992.

[Jel90] F. Jelinek. Self-organized language modeling for speech recognition. In A. Waibel and K.-F. Lee, editors, *Readings in Speech Recognition*, pages 450–506. Morgan Kaufmann Publishers, Inc., 1990.

[JLG78] S. Johansson, G.N. Leech, and H. Goodluck. *Manual of Information to accompany the Lancaster-Oslo/Bergen Corpus of British English, for use with digital Computers.* Department of English, University of Oslo, Oslo, 1978.

[KBH96] G. Kaufmann, H. Bunke, and T.M. Ha. Recognition of cursively handwritten words using a combined normalization/perturabation approach. In *Proc. of the IWFHR-5*, pages 17–22, 1996.

[LB93] E. Lecolinet and Olivier Baret. Cursive word recognition: Methods and strategies. In Sebastiano Impedovo, editor, *Fundamentals in Handwriting Recognition*, pages 235–263. NATO ASI Series, 1993.

[Ots79] N. Otsu. A threshold selection method from gray-level histograms. *IEEE Transaction on Systems, Man and Cybernetics*, SMC-9:pp. 62–66, 1979.

[RJ93] L. Rabiner and B.-H. Juang. *Fundamentals of Speech Recognition.* Prentice Hall, 1993.

[SB93] R.K. Srihari and C. M. Baltus. Incorporating syntactic constraints in recognizing handwritten sentences. In *Proc. of IJCAN-93, Chambery*, volume 2, pages 1262–1267, 1993.

[Sch96] J. Schuermann. *Pattern Classification.* John Wiley and Sons, Inc., 1996.

[Sen94] A.W. Senior. *Off-line Cursive Handwriting Recognition using Recurrent Neural Networks.* PhD thesis, Trinity Hall, Cambridge, England, September 1994.

[SGS93] S.N. Srihari, V. Govindaraju, and A. Shekhawat. Interpretation of handwritten addresses in us mailstream. In *ICDAR*, pages 291–304, 1993.

[Sto97] A. Stolcke. Linguistic knowledge and empirical methods in speech recognition. *AI MAGAZINE*, WINTER 1997:25–31, 1997.

Invariante Erkennung handskizzierter Piktogramme mit Anwendungsmöglichkeiten in der inhaltsorientierten Bilddatenbankabfrage

Stefan Müller, Gerhard Rigoll, Denis Mazurenok und Daniel Willett

Fachgebiet Technische Informatik, Fachbereich Elektrotechnik
Gerhard-Mercator-Universität - GH Duisburg
Bismarckstr. 90, D-47057 Duisburg
email: {stm,rigoll,dmazur,willett}@fb9-ti.uni-duisburg.de

Zusammenfassung. In diesem Bericht präsentieren wir ein neuartiges Merkmalsextraktionsverfahren sowie einen Hidden-Markov-Modell (HMM) basierten Klassifizierer für die rotations-, translations-, und skalierungsunabhängige Erkennung von handskizzierten Piktogrammen. Die Merkmalsvektoren werden dabei in einem festen Winkelabstand entlang der Verbindungslinie zwischen dem Schwerpunkt und einem das Piktogramm einschließenden Kreis entnommen. Für die Erkennung werden HMM-Techniken verwendet, die es ermöglichen, neben dem Klassifikationsergebnis auch einen Schätzwert für den Rotationswinkel anzugeben. Die Evaluierung der vorgestellten Methoden erfolgt mittels zweier großer Piktogramm-Datenbasen, die jeweils aus 20 Klassen bestehen und deren Klassenelemente durch die handschriftliche Eingabe große Formvariationen aufweisen. Die erzielten Erkennungsraten liegen über 99% und sind denen, die mit Momentenmethoden in Kombination mit neuronalen Netzen erzielt wurden, überlegen. Als Beispiel für eine mögliche Anwendung der vorgestellten Techniken wird ein experimentelles Bilddatenbanksystem vorgestellt, das intuitiv über Skizzen des Benutzers angefragt werden kann.

1 Einführung

Die invariante Erkennung von Mustern wird als komplexe Aufgabe angesehen, die viele Anwendungen, wie z.B. maschinelle Zeichenerkennung (OCR), Zielidentifikation und industrielle Produktinspektion ermöglicht. Die zur Lösung dieses Problems vorgeschlagenen Methoden reichen von (geometrischen-) Momenten und Integraltransformationen bis hin zu invarianten Klassifizierern (siehe auch [1]). Die Aufgabe, handskizzierte Piktogramme invariant zu erkennen stellt aufgrund der Einbeziehung großer Variationen innerhalb der einzelnen Klassen durch die handschriftliche Eingabe zusätzliche, hohe Anforderungen. Die im folgenden kurz beschriebenen Veröffentlichungen befassen sich ebenfalls mit dieser Thematik und verwenden auch einen Teil der Klassen unserer Datenbasen.

1.1 Relevante Arbeiten anderer Autoren

He und Kundu beschreiben in [2] eine Methode, die Konturen mittels HMM und autoregressiven (AR)-Modellen klassifiziert. Sie verwenden lediglich eine

8-Klassen-Datenbasis, die einen Teil unserer in Abb. 1 vorgestellten 20-Klassen-Datenbasis bildet. Da die Merkmale in [2] aus Radien zwischen dem Schwerpunkt des Piktogramms und dessen Randkurve bestehen, können nur Symbole klassifiziert werden, die aus einer geschlossenen Randkurve bestehen und keine Piktogramme wie die in Abb. 1 dargestellten Klassen *Haus*, *Drachen* oder *Fisch*. Die angestrebte Rotationsinvarianz wird durch einen Vorverarbeitungsprozeß realisiert, der sukzessive in drei Schritten mittels geometrischer Betrachtungen sowie daraus abgeleiteten Entscheidungen die Piktogramme relativ zu einer festgelegten Achse ausrichtet. Lee und Lovell führen in [3] Experimente auf einer sehr ähnlichen Piktogramm-Datenbasis durch und verwenden ebenfalls dieselben acht Klassen. Wie auch in [2] werden als Merkmale die Radien zwischen dem Schwerpunkt und der Randkurve verwendet und die Rotationsinvarianz wird mit denselben Methoden ermöglicht. Der größte Unterschied zu [2] liegt in der Verwendung eines Vektorquantisierers zur Klassifikation anstelle der HMMs. In dem vorliegenden Bericht wird sowohl eine neuartige Merkmalsextraktion als auch ein HMM-Klassifizierer vorgestellt, die es ermöglichen, Piktogramme rotationsinvariant zu erkennen, ohne mittels komplizierter Vorverarbeitungsschritte die Piktogramme auszurichten.

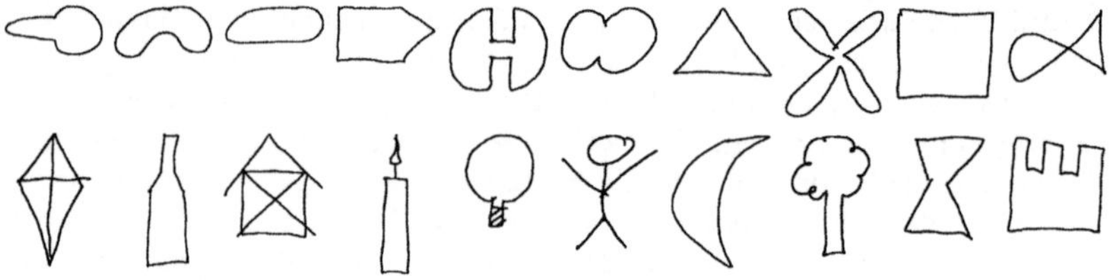

Abb. 1. Beispiele aus den verwendeten 20-Klassen-Piktogramm-Datenbasen. Die ersten acht Klassen werden auch in [2] und [3] verwendet.

1.2 Hidden-Markov-Modelle

Hidden-Markov-Modelle sind endliche stochastische Automaten, die erfolgreich angewendet werden z.B. in der Erkennung fließend gesprochener Sprache [4] und Handschrift [5] sowie neuerdings auch für inhaltsbasierte Bilddatenbankabfragen [6]. Ein Hidden-Markov-Modell $\lambda(\vec{\pi}, \overleftrightarrow{a}, \vec{b})$ besteht aus einer endlichen Menge von Zuständen und Übergangswahrscheinlichkeiten und ist vollständig beschrieben durch den Ausgabeverteilungsvektor $\vec{b}$, der Matrix der Übergangswahrscheinlichkeiten $\overleftrightarrow{a}$ und dem Vektor der Wahrscheinlichkeiten für die Einnahme eines Anfangszustandes $\vec{\pi}$. Die Ausgabeverteilungen sind üblicherweise Gaußsche Mischverteilungsdichten der Form

$$b_j(\vec{o}) = \sum_{m=1}^{M} c_{jm} \mathcal{N}\left(\vec{o}, \vec{\mu}_{jm}, \overleftrightarrow{\Sigma}_{jm}\right) \tag{1}$$

wobei $b_j(\vec{o})$ die Ausgabewahrscheinlichkeit für den Zustand j bei gegebenem Merkmalsvektor $\vec{o}$ ist, c_{jm} der Koeffizient der m-ten Gaußverteilung im Zustand

j und $\mathcal{N}(\vec{o}, \vec{\mu}_{jm}, \overleftrightarrow{\Sigma}_{jm})$ eine multivariate Gaußverteilung mit dem Mittelwertvektor $\vec{\mu}_{jm}$ und der Kovarianzmatrix $\overleftrightarrow{\Sigma}_{jm}$ ist. Nach erfolgtem Training der Markov-Modelle mit dem Baum-Welch-Algorithmus können unbekannte Muster durch die folgende Maximum-Likelihood (ML) Entscheidung klassifiziert werden:

$$\mathrm{p}^{\star} = \underset{\mathrm{p}}{\operatorname{argmax}} \left(Pr\left(\vec{O} \,|\, \lambda_{\mathrm{p}}\right)\right) \tag{2}$$

Entsprechend Gl. 2 wird eine Merkmalssequenz $\vec{O}$ der Klasse $\mathrm{p}^{\star}$ zugeordnet. $Pr(\vec{O}\,|\lambda)$ wird mit dem sog. Viterbi-Algorithmus angenähert, der diese Wahrscheinlichkeit, basierend auf der wahrscheinlichsten Zustandsfolge, bestimmt. Die HMM-Methoden werden in [4] detailliert erläutert. Für die in diesem Bericht vorgestellte rotationsinvariante Erkennung ist es notwendig anzumerken, daß die HMMs miteinander verbunden werden können, um hierarchische Strukturen zu modellieren. Beispielsweise können in der Handschrift-Erkennung Worte mittels vorgegebener Graphenfolgen und zugeordneter Graphen-HMMs modelliert werden.

2 Neuartige Merkmalsextraktion

Die verwendete Merkmalsextraktion ist schematisch in Abb. 2 dargestellt und kann auf beliebige Piktogramme oder Skizzen angewendet werden, ist also nicht auf Konturen beschränkt. Ein vorgegebenes Muster wird in eine Vektorsequenz $(\vec{o}_1, \ldots, \vec{o}_T)$ überführt, indem ein das Muster umgebender Kreis in T rechteckige Streifen und jeder Streifen in N kleinere Blöcke unterteilt wird. Für die Blöcke wird dann der Anteil an schwarzen Pixeln ermittelt und somit ergibt sich eine Merkmalssequenz der Länge T mit jeweils N-dimensionalen Merkmalsvektoren. Der erste Vektor dieser Sequenz beschreibt dabei stets den Streifen, der einer zwölf-Uhr-Position entspricht und die folgenden Vektoren werden im mathematisch negativen Drehsinn entnommen. Der Mittelpunkt des Kreises wird in den Flächenschwerpunkt des Piktogramms gelegt. Die Länge der einzelnen Streifen wird auf das jeweilige Piktogramm adaptiert und entspricht der Strecke zwischen dem Flächenschwerpunkt und dem relativ dazu am weitesten entfernten Piktogramm-Punkt. Durch diesen Adaptionsvorgang sowie den Bezug zum Flächenschwerpunkt wird eine Größen- und Translationsinvarianz erreicht, wohingegen die Rotationsinvarianz erst durch Maßnahmen bei der Modellierung erreicht wird.

3 Rotationsinvariante Modellierung mit Hidden Markov Modellen

Die grundlegende Idee bei der rotationsinvarianten Modellierung mit HMMs ist es, die entsprechend Kap. 2 erzeugte Merkmalssequenz zu duplizieren und diese Sequenz $(\vec{o}_1, \ldots, \vec{o}_T, \vec{o}_{T+1}, \ldots, \vec{o}_{2T})$ mittels der kombinierten Segmentierungs- und Klassifizierungsfähigkeiten des HMM-Erkenners Modellen zuzuordnen, die einen ersten unvollständigen Teil einer Klasse, gefolgt von der Klasse selbst

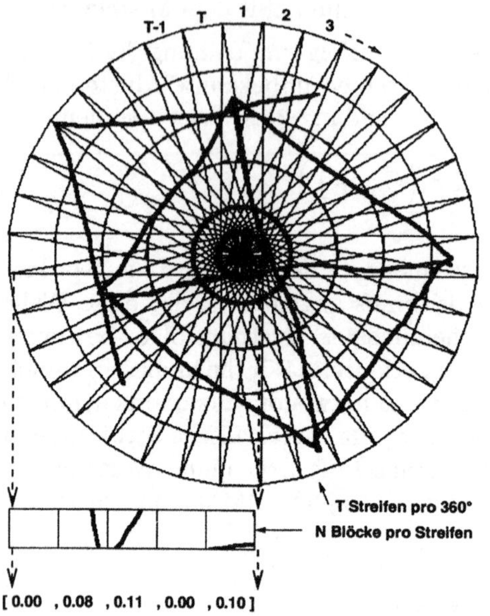

Abb. 2. Schematische Darstellung der Merkmalsextraktion

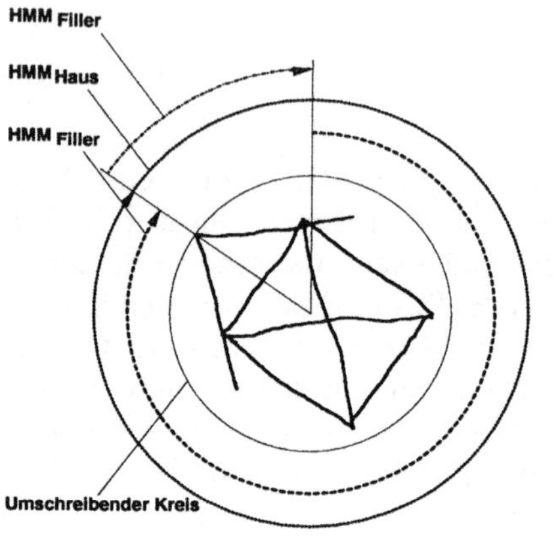

Abb. 3. Zuordnung der Elemente der zweimalig präsentierten Merkmalssequenz zu den HMMs *Filler* und *Haus*

und schließlich den verbleibenden Anteil der Klasse repräsentieren. Die beiden unvollständigen Anteile des Piktogramms können dabei entweder von sogenannten *Filler*-Modellen oder mittels kopierter und modifizierter Piktogramm-HMMs beschrieben werden. Dies ist schematisch in Abb. 3 dargestellt, wo ein Piktogramm der Klasse *Haus* vorgegeben ist und dessen Segmentierung in Anteile, die den Markov-Modellen *Filler* und *Haus* zugeordnet werden, gezeigt wird. Der Abb. 3 kann entnommen werden, daß wenn die Zuordnung zwischen den Vektoren der Merkmalssequenz und den HMMs bekannt ist, der Rotationswinkel über die Anzahl an Merkmalsvektoren, die dem ersten oder zweiten *Filler*-Modell zugeordnet wurden, bestimmt werden kann. Diese Kombination aus Winkelbestimmung und Klassifikation während der Erkennungsphase ist wesentlich eleganter als die komplizierten Vorverarbeitungsschritte, die in [2, 3] verwendet werden. Der Aufbau der zur rotationsinvarianten Erkennung verwendeten HMMs wird im folgenden detailliert beschrieben.

3.1 Modellierung mit *Filler*-Modellen

Der erste vorgestellte Ansatz für die Modellierung rotierter Piktogramme besteht darin, das unrotierte Piktogramm-HMM mit *Filler*-HMMs zu umgeben, die durch ein erstes Training mit den halbierten Merkmalssequenzen $(\vec{o}_1, \ldots, \vec{o}_{T/2})$ aller Klassen initialisiert werden. Anschließend kann mit dem Baum-Welch-Algorithmus ein gemeinsames Training der aneinandergehängten HMMs, wie in Abb. 4 dargestellt, erfolgen. In Abb. 4 wird ein zusätzliches Modell gezeigt, das aus der Verkettung zweier Piktogramm-HMMs besteht und speziell für unrotierte Piktogramme verwendet wird. Wenn also eines dieser beiden HMMs für

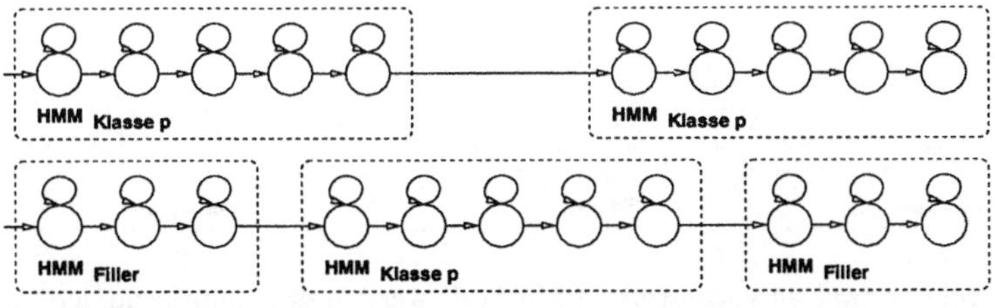

Abb. 4. Zwei Markov-Modelle repräsentieren eine einzige Piktogramm-Klasse. Das obere HMM modelliert unrotierte, das untere rotierte Piktogramme.

eine gegebene Merkmalssequenz die maximale Wahrscheinlichkeit $Pr(\vec{O}\,|\lambda)$ produziert, wird das Muster als zur *Klasse p* zugehörig klassifiziert. Die in unseren Experimenten verwendeten *Filler*-HMMs sind aus einer geringen Anzahl von Zuständen (3–5) aufgebaut, was zu einer schnellen Erkennung führt. Bedingt durch den Trainingsprozeß mit einer festen Anzahl von Beispielen werden von den *Filler*-HMMs die während des Trainings präsentierten Rotationswinkel bevorzugt. Dies kann bei einigen Anwendungen erwünscht sein, beispielsweise wenn in geringem Maße rotierte Ziffern wie "6" und "9" unterschieden werden sollen. Um eine vollständig rotationsinvariante Erkennung zu ermöglichen, kann das im nächsten Abschnitt beschriebene Verfahren eingesetzt werden.

3.2 Modellierung mit modifizierten Wahrscheinlichkeiten für die Anfangs- und Endzustände

Eine weitere Möglichkeit für eine rotationsinvariante Modellierung besteht darin, das Modell für die unrotierten Piktogramme zweimal zu duplizieren, und das original HMM mit diesen Modell-Kopien zu umgeben. Zusätzlich werden die Wahrscheinlichkeiten für die Anfangszustände des ersten Modells ($\vec{\pi}$) sowie die Wahrscheinlichkeiten für die Endzustände des dritten HMM entsprechend Abb. 5 verändert. Die Komponenten von $\vec{\pi}$ des ersten Modells werden alle auf den glei chen Wert 1/n gesetzt, wobei n die Dimension des Vektors $\vec{\pi}$ bzw. die Anzahl von HMM-Zuständen bezeichnet. Diese Maßnahme ermöglicht die rotationsinvariante Erkennung von Mustern ohne eine Bevorzugung eines bestimmten Winkels. Die Endzustände des dritten Modells in Abb. 5 werden entsprechend verändert und somit ist es möglich, daß sich das Markov-Modell in jedem seiner Zustände befinden kann, wenn der letzte Merkmalsvektor einer Sequenz präsentiert wurde. Dies wurde bei den bisher beschriebenen Modellen (z.B. Abb. 4) nicht gefordert.

3.3 Zyklische Vertauschung der HMM-Zustände

Die dritte Modellierungstechnik, die in diesem Bericht vorgestellt wird, verwendet eine erheblich höhere Anzahl von HMMs pro Klasse als die bisher beschriebenen Methoden, erfordert hingegen nur die Präsentation der nicht duplizierten Merkmalssequenz $(\vec{o}_1, \ldots, \vec{o}_T)$. Für jede Klasse wird das HMM für unrotierte Piktogramme n-mal kopiert und die Zustände, wie in Abb. 6 angedeutet, sukzessive zyklisch permutiert. Der Rotationswinkel kann nun nicht mehr wie in

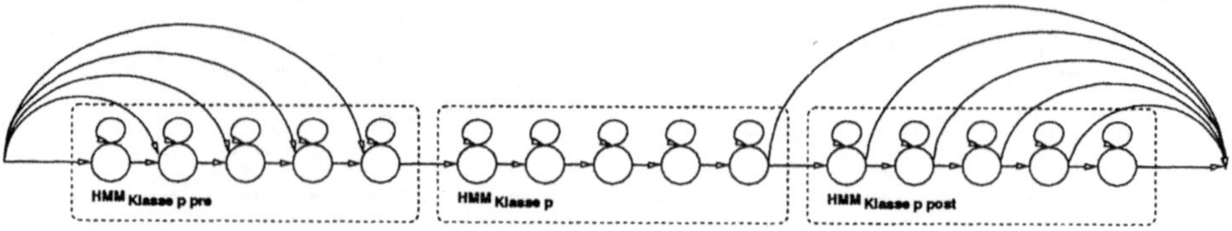

Abb. 5. Bei der zweiten vorgestellten Methode werden drei identische, jedoch bezüglich der Wahrscheinlichkeiten für die Anfangs- und Endzustände veränderte, HMMs verwendet.

Abschnitt 3 eingangs angedeutet, ermittelt werden, ist jedoch über die Konfiguration desjenigen HMMs mit der maximalen Wahrscheinlichkeit für eine gegebene Sequenz bestimmbar. Beispielsweise repräsentiert das Modell in Abb. 6 oben die Piktogramme der *Klasse p* mit einem Rotationswinkel von 0°, das Modell darunter einen Winkel von $360°/n$, usw. Die Winkel können bei diesem Ansatz also nur in Vielfachen von $360°/n$ angegeben werden und sind damit weniger genau bestimmbar als bei den in den Abschnitten 3.1 und 3.2 beschriebenen Methoden, wo die Quantisierung auf der Zuordnung der größeren Anzahl von Merkmalsvektoren zu den einzelnen Modellen beruht.

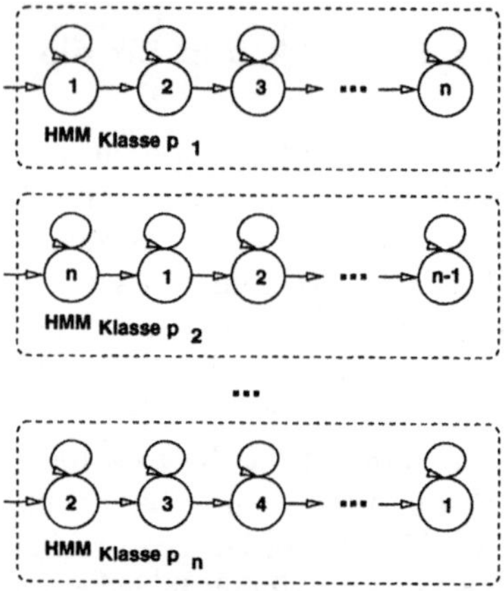

Abb. 6. Zyklisches Vertauschen der HMM-Zustände

Abb. 7. Fünf unrotierte und fünf rotierte Piktogramme der Klasse *Baum* aus der *stm*-Datenbank.

4 Experimente und Ergebnisse

Die Evaluierung der vorgestellten Methoden erfolgt mit Hilfe zweier großer Piktogramm-Datenbasen. Diese wurden von zwei Personen (im folgenden mit *stm* und *dib* bezeichnet) als Binärbild mittels eines Grafiktableaus aufgenommen. Beide bestehen jeweils aus 10 unrotierten und 20 rotierten handskizzierten Piktogrammen für jede der 20 in Abb. 1 vorgestellten Klassen. Die rotierten

Piktogramme sind in einen Test- und einen Trainingssatz aufgeteilt worden. Um die großen Formvariationen innerhalb der einzelnen Klassen zu demonstrieren, sind 10 Beispiele der Klasse *Baum* aus der *stm*-Datenbasis in Abb. 7 dargestellt. Die verwendeten Modelle bestehen aus 30 Zuständen für die Piktogramm-HMMs und fünf Zuständen für die *Filler*-HMMs, dabei ist jede Ausgabeverteilung aus vier multivariaten Gaußverteilungen zusammengesetzt. Die Merkmalsextraktion wird mit 36 Streifen und 5 Blöcken pro Streifen durchgeführt. Die mit diesen Parametern erzielten Erkennungsraten sind in Tabelle 1, getrennt für die *stm* und *dib*-Datenbasen, aufgeführt. Um eine detailliertere Bewertung der vorgestellten Methoden zu ermöglichen, wurden Experimente auf denselben Datenbasen mit invarianten Momenten und einem neuronalen Klassifizierer durchgeführt. Diese Experimente sind in [7] beschrieben und ergaben Erkennungsraten bis zu 99,0%.

Verwendete Methode	Siehe auch	*stm*	*dib*	Mittelwert
Modellierung mit *Filler* HMMs	Kap. 3.1	99,5%	98,5%	99,0%
Modellierung mit HMMs mit modifiziertem $\vec{\pi}$	Kap. 3.2	99,5%	98,5%	99,0%
Zyklisch permutierte HMMs	**Kap. 3.3**	**99,5%**	**99,5%**	**99,5%**

Tabelle 1. Erzielte Erkennungsraten

5 Zusammenfassung und Ausblick

In diesem Bericht wurde eine neuartige Merkmalsextraktionsmethode sowie Hidden-Markov-Modell-basierte Klassifizierer vorgestellt, mit denen translations-, skalierungs- und rotationsunabhängig handskizzierte Piktogramme erkannt werden können. Die Merkmalsvektoren werden in einem festen Winkelabstand entlang der Verbindungslinie zwischen dem Schwerpunkt und einem das Piktogramm einschließenden Kreis entnommen. Drei Variationen des HMM-Klassifizierers werden beschrieben, die neben der vollständig rotationsunabhängigen Erkennung auch eine Erkennung mit eingeschränkten Winkelbereichen ermöglichen. In den Experimenten wurden Erkennungsraten von bis zu 99.5% mit Piktogramm-Datenbasen erreicht, die aus 20 Klassen bestehen. Die vorgestellten Methoden können mit geringem Aufwand auf die Erkennung natürlicher Bilder erweitert werden. Abb. 8 zeigt beispielsweise erste Ergebnisse, die mit einem experimentellen Bilddatenbanksystem, das intuitiv über Skizzen des Benutzers abgefragt werden kann und das die in diesem Bericht vorgestellten Methoden verwendet, erzielt wurden. In der Abbildung wird in jeder Zeile zunächst auf der linken Seite die Anfrage-Skizze des Benutzers gezeigt, gefolgt von den Bildern der Datenbank, deren Hidden-Markov-Modell die höchste Wahrscheinlichkeit $Pr(\vec{O}\,|\lambda)$ ausgegeben hat. Vorher wurde während der Trainingsphase jeweils ein HMM mit der Topologie entsprechend Abb. 5 für jedes der 120 Grauwertbilder in der Datenbasis erstellt. Die Grauwertbilder der Werkzeuge können aufgrund des homogenen Hintergrundes sehr gut histogrammbasiert segmentiert werden, wodurch mögliche Schwankungen des Schwerpunktes weitestgehend vermieden werden.

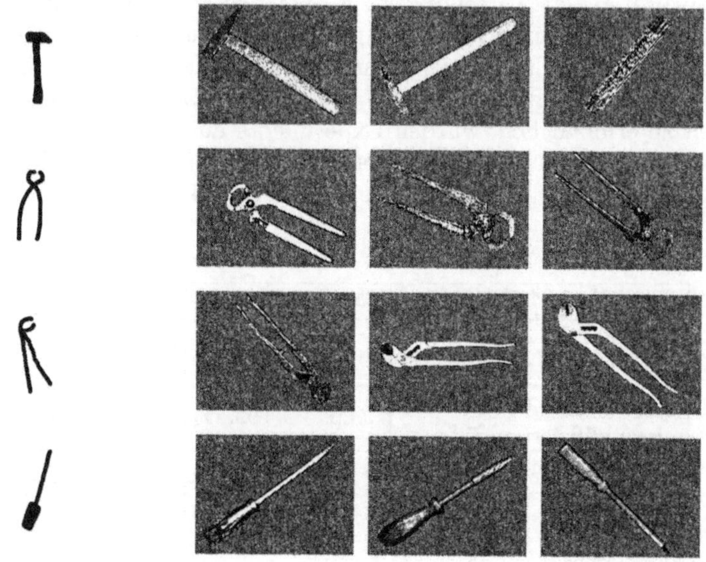

Abb. 8. Anfrage-Skizzen und die vom Datenbanksystem ermittelten drei ähnlichsten Grauwertbilder

Literatur

1. J. Wood, "Invariant Pattern Recognition: A Review", *Pattern Recognition*, Vol. 29, No. 1, 1996, pp. 1–17.
2. Y. He and A. Kundu, "2-D Shape Classification Using Hidden Markov Model", *IEEE Transactions on Pattern Analysis and Machine Intelligence*, Vol. 13, No. 11, 1991, pp. 1172–1184.
3. S. Lee and B. Lovell, "Modelling and Classification of Shapes in Two-Dimensions Using Vector Quantisation",*Proc. IEEE-ICASSP*, Vol. 5, 1994, pp. 141–144.
4. L.R. Rabiner, "A Tutorial on Hidden Markov Models and Selected Applications in Speech Recognition", *Proc. IEEE*, Vol. 77, No. 2, Feb. 1989, pp. 257–286.
5. K. S. Nathan, J. R. Bellegarda, D. Nahamoo and E. J. Bellegarda, "On-line Handwriting Recognition Using Continuous Parameter Hidden Markov Models",*Proc. IEEE-ICASSP*, Vol. 5, 1993, pp. 121–124.
6. H.-C. Lin, L.-L. Wang and S.-N. Yang, "Color Image Retrieval Based on Hidden Markov Models", *IEEE Transactions on Image Processing*, Vol. 6, No. 2, Feb. 1997, pp. 332–339.
7. S. Müller, G. Rigoll, A. Kosmala and D. Mazurenok, "Recognition of Hand-Drawn Pictograms Using HMMs with a Rotating Feature Extraction", *Intern. Workshop on Frontiers in Handwriting Recognition*, Taejon, Korea, 1998.

Attentive Face Detection and Recognition *

Volker Krüger, Udo Mahlmeister, and Gerald Sommer

University of Kiel, Germany,
Preußerstr. 1-9, 24105 Kiel, Germany
Tel: ++49-431-560496
FAX: ++49-431-560481
vok@informatik.uni-kiel.de,
WWW home page: http://www.informatik.uni-kiel.de/~vok

Abstract. In this paper we will present an approach for the attentive detection and recognition of faces in gray-value images. The approach is biologically motivated. The attentive face system, as we call it, shows great robustness with respect to scale, rotation, viewing orientation, changes in illumination, facial expressions, partial occlusions and other distortions caused, e.g., by glasses or a beard. The system has knowledge of several templates of different persons as well as of their exact relative positions. In a first low-level step the system detects relevant image features by evaluating a similarity measurement between local image features and known facial templates. In a second high-level step the system verified the consistency of these features by using the knowledge of the exact relative positions of the templates and reports whether a face was recognized, detected or whether no face was present.

Keywords: face detection, face recognition

1 Introduction

A face detection and recognition system that is supposed to work under the variety of different real world situations has to be robust and discriminative at the same time. It has to be robust in order to compensate changes in illumination and contrast in the image, changes in different expressions and orientations of the faces that is to be detected and changes in scale. Furthermore it has to be able to detect all different kinds of faces, including bearded faces, faces with glasses, faces with long hair, etc. Still, the system has to be discriminative for recognition purposes. In [9] it is argued that, in contrast to [1], neither a template based approach nor a feature based approach per se is able to accomplish such a task. According to them, rather a "hybrid" approach is able to meet the real world requirements. In their approach, they represent faces with graphs labeled with the responses of locally applied gabor wavelets, called jets. The graphs were matched using a graph similarity function [9]. In our paper we want to promote an attentive approach for a hybrid type of recognition and detection system.

* This work is partially supported by the DFG grand So 320/1-2

Given a gray-level image, our attentive face system (AFS) is able to detect or even recognize faces robustly in various aspects.

How the attention is directed and how the evaluation is done we will describe in detail in the next sections 2, 3 and 4. In section 5 we will show our experimental results while we will finish our contribution in section 6 with concluding remarks.

2 Methodology

In this section we will delve into details about the methodology and the mechanisms employed by our AFS.

The AFS uses as model knowledge facial templates and their exact relative positions. Facial templates can be derived from different persons and may contain various facial features. We assume for the sake of simplicity, that always the same number J of templates T_{ij} with the same relative positions are taken from each sample face so that T_{ij} represents the jth template at position $\mathbf{p}_j$ in the ith face.

Given a gray-level image the AFS uses attentional mechanisms in order to be attracted by certain for the recognition and detection process relevant image features while, at the same time, arbitrary image features do not have an attractive effect. This is done by defining the relevance by a similarity measurement between an image region and a template: For each template a similarity measurement, the *attention image* is calculated. Similar to the HVS the attentional mechanism of the AFS is very robust with respect to several situations and image formation parameters.

As the success of the AFS is highly depending on a good trade off between robustness and discriminability, the AFS verifies if the image regions that were detected in the previous step are facial features. This is done by using the knowledge about the exact relative positions of the templates. If the overall similarity of templates derived from one single person and positioned at exact relative positions is larger than a given threshold a_{rec}, then a face has been recognized. If, on the other hand, the overall similarity for templates that are derived possibly from different persons and that are positioned at exact relative positions is larger that a certain threshold a_{det}, then a face has been detected. An exact definition of the thresholds will be given in section 5.

The AFS contains therefore two stages that can be summarized as follows:

1. The *attention stage*: The attention stage is a low level stage that classifies the visual information within an image. Using known facial templates, local image features are rated with respect to their importance and relevance for the recognition and detection of a face. The attention stage assures the robustness of the AFS with respect
   - different viewing angles,
   - different colors and intensities of lightning,
   - facial expressions,
   - partial occlusion,
   - size and
   - pan, tilt and curl.

2. The *evaluation stage*: This stage is the high level stage that is responsible for the evaluation of the local image features that were classified in the preceding stage as "important". It discriminates between object features and arbitrary image features. Depending on the local features that were identified as facial features, this stage decides whether a face has been identified, detected or whether no face is present.

In the following two sections we will delve into details about the two stages. In section 3 the details on how images features and facial templates are represented is explained. In section 4 we will discuss the evaluation stage in detail.

3 Attention Stage

In this section we will explain, how local image features and facial templates are represented so that a high robustness of the AFS is guaranteed. The attention stage consists of two steps: In the first step the representations of local image features and facial templates are calculated. Facial templates consist of local facial image features such as the eyes, the cheeks, the mouth or nose of possibly different persons. Their exact relative positions are known to the system. Also it is known from which person they are derived. Facial templates and local image features are represented in the same manner. Then, in the second step, the representations of the facial templates are each compared with the representations of the local image features. A similarity measurement of the facial templates with local image features is used as the measurement for their relevance in the image. The result of the attention stage is an *attention image*. In figure 1 an example can be seen. The left image is given as a test image. In this image, the right eye of that person (figure 1, center) has to be detected. The application of the attention system results then in an attention image (figure 1, right) that shows a maximum of attention (black) at the position of the right eye. The grade of blackness in the attention image represents the grade of relevance or similarity for the test image.

It has to be pointed out that the representations of the local facial features have to be calculated just once. Therefore only for the given image, the corresponding representation needs to be calculated.

3.1 Representation of Local Image Features

As the basic image representation we use *local orientations* of gray-value patterns within an image. As inspired by Granlund [2] we assign to each point within the image a local amplitude a and a local orientation θ. With this, a correlates with local contrast while θ correlates with the angle of the predominant orientation. Each value is given as the sum of four orientation selective filters h_i, $i = 0 \dots 3$:

$$a = \left| \sum_{i=0}^{3} |h_i| \, e^{-j2\Theta_i} \right| \qquad \theta = \arg \sum_{i=0}^{3} h_i e^{-j\Theta_i} \tag{1}$$

where Θ_i denotes the orientation of the ith filter. It has to be pointed out that, in opposition to Granlund, equation (1) defines orientations ranging form $0° \dots 360°$, whereas Granlund defined orientations ranging form $0° \dots 180°$. In

Fig. 1. The left image shows a test image while the center image shows the right eye of the person in the image. For both images the representations of local image features are computed and compared. The attention image to the right shows a maximum of relevance (black) at the position of the right eye. White regions in the attention image represent a minimum of relevance.

Fig. 2. From left to right: points (black) without orientation, points with one predominant orientation, point with multiple orientations.

figure 2 the left image of figure 1 has been used as an example: The left image shows points without orientation, the center image shows points with one predominant orientation while the right image shows image points with multiple orientations.

Local orientations have been embedded into the steerable pyramids [6, 7] as introduced by [3]. This concept we call *local orientation pyramid* (LOP) which is the means we use for representation of facial templates and local image features. The LOP provides joint steerability in position as well as in orientation. Fig. 3 shows the block diagram of the steerable pyramid architecture. By replacing the filled dot with the dashed box a new level of recursion is entered. Four pyramid levels were used for our experiments. The orientations h_i (see eqn. (1)) are calculated with four band passes $\mathcal{B}_i,\ i = 0 \ldots 3$. Their responses supply a 4D-vector $(h_0\ h_1\ h_2\ h_3)^T$ for each point within the pyramid. The band passes $\mathcal{B}_i,\ i = 0 \ldots 3$ are scale dependent and are designed as rotated copies at orientations $\theta_i = \frac{1}{4} i\pi$.

Steering position enables the interpolation of missing samples which allows a conversion of the image pyramid into an image heap that contains images of even size. Steerability in orientation is used for calculating local amplitude and orientation as denoted in equation (1) while using the responses $h_i,\ i = 0 \ldots 3$ as basis functions.

3.2 Similarity Measurement of Local Image Features

In this subsection we will discuss how the similarity of facial templates with local image features is calculated. It has been shown in [5] that the use of color

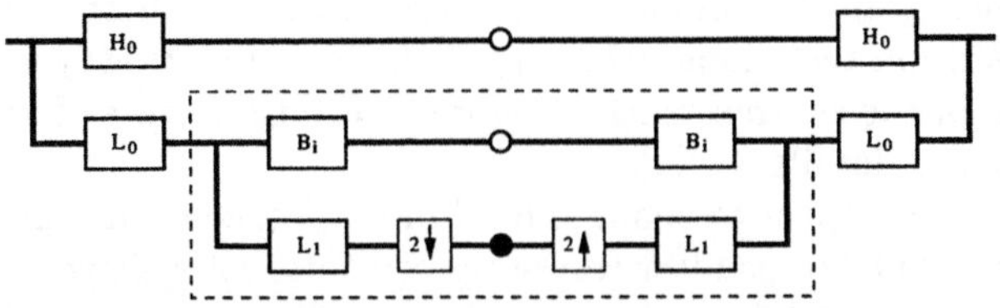

Fig. 3. Steerable pyramid architecture[7]: initial high-pass $\mathcal{H}_0$, low-pass $\mathcal{L}_0$, recursive subsystem (dashed) with low-pass $\mathcal{L}_1$, sub-sampling, and orientation selective band-passes $\mathcal{B}_i$.

histogram matching provides a fast and robust technique to determine the position of a specified object within a scene. As described in [3], we adapted the *intersection*-technique [5, 8] for multiple-scale local orientation. To be precise, the upper three pyramid layers of the LOP are transformed into a heap of even sized layers by enlarging the first and second layer to the size of the third pyramid layer. This is achieved by steering in position which allows the interpolation of the missing samples. With this, a $N \times N$ region in each heap layer can be associated with each pixel $\mathbf{x}$ of the top LOP layer ($N \times N$ is the size of the region of support of pixel $\mathbf{x}$).

The histograms are then calculated by first extracting 3D vectors $z^T = (z_0\ z_1\ z_2)^T$ from the upper three heap layers where the ith vector component z_i is derived from the ith heap layer. Then, the vector components are quantized to five different values: the "null-orientation" if the contrast of the corresponding pattern is smaller than a certain threshold $|z_i| < a_{\mathrm{thr}_i}$ and the four values 0°, 90°, 180°, 270°. The resulting histograms have therefore each a size of $L = 5^3 = 125$ bins. In the left image of fig. 1 regions can be seen that show patterns with very low contrast. To these regions the "null-orientation" has been assigned as shown in the left image of fig. 2. The threshold a_{thr_i} is assigned to a fixed fraction of the maximum amplitude occurring in each pyramid layer i. For each top layer pixel, the histogram is calculated over the three associate heap regions.

For example, the support of one pixel in the top layer of the LOP, in case of the four-layer pyramid of our AFS, is $N = 16$ image pixels. We calculate a histogram for each LOP top layer pixel. Each histogram then contains $3 \cdot 16^2 = 768$ values. For an image I of size 640×480 we get $640/2^4 \cdot 480/2^4 = 40 \cdot 30$ histograms $I^H(\mathbf{x})$ where $\mathbf{x} = (0 \ldots 39, 0 \ldots 29)$. A 32×32 template T is therefore described by $2 \cdot 2$ histograms $T^H(\mathbf{x})$ with $\mathbf{x} = (0 \ldots 1, 0 \ldots 1)$, respectively.

Given a template histogram T and an image histogram H the intersection is defined as [8]:

$$\cap(T, H) = \sum_i \min(T_i, H_i) \tag{2}$$

The attention map $A(I, T)$ for an image I and a template T is then given by evaluating at each position of the corresponding histogram image I^H the intersection with the histogram image T^H of the template T:

$$A(I,T)(\mathbf{x}) = \sum_{\mathbf{k}} \cap \left(T^H(\mathbf{k}), I^H(\mathbf{x}+\mathbf{k})\right) \ . \tag{3}$$

In [5] several different histogram techniques have been tested. Furthermore it is reported that the intersection technique needs sparse histograms. However, we have found that for our purposes this method shows the best results. For extensive experiments and more details see [4].

Insensitivity of the AFS with respect to changes in illumination color and intensity is achieved because the LOP basis filter responses are quite insensitive to illumination changes. Further insensitivity is achieved by considering primarily local orientation information during histogram evaluation while local amplitude, which carries contrast information, is discarded. In this respect our method differs from those primarily ignoring phase[10] or those using the filter responses directly[5]. Robustness to geometric distortions implies a mechanism which considers a large local context. This is realized by the sliding window for histogram calculation. The size of the sliding window determines to what extend neighboring local orientations are considered for the similarity measurement between the image and the template. Robustness to changes in orientation, changes in scale of the face, partial occlusion of changes of the viewing position are gracefully inherited from the similarity measurement[8].

4 Evaluation Stage

In this section we will delve into details about the evaluation stage. This stage is responsible for the evaluation of the local image features that were classified in the preceding stage as "relevant". It discriminates between object features and arbitrary image features. Depending on the local features that were identified as facial features, this stage decides whether a face has been identified, detected or whether no face is present.

As explained above J different facial templates of each face as well as their exact relative positions are known to the AFS. For simplicity, let T_{ij} be the jth template at position $\mathbf{p}_j$ of the ith face. Given an image I, attention maps $A(I, T_{ij})$ for each template T_{ij} were calculated. The AFS reports a face as recognized if the overall similarity of the templates T_{ij} that are derived from the ith person and that are positioned at the positions $\mathbf{p}_j$ is larger than a given threshold a_{rec}:

$$\sum_j A(I, T_{ij})\,(\mathbf{x} - \mathbf{p}_j) > a_{\mathrm{rec}} \ . \tag{4}$$

The AFS reports a face as detected if there exists a set of templates $S = \{T_{i_j j} |\text{ for each } j\}$ containing a template for each $j = 0 \ldots J-1$ of the i_jth face so that the overall similarity of the templates of set S with the image I is larger than a give threshold a_{det}:

$$\sum_S A(I, T_{i_j j})\,(\mathbf{x} - \mathbf{p}_j) > a_{\mathrm{det}} \ . \tag{5}$$

In practice the similarities are normalized with the maximal similarities of the templates with their original faces. With this, the threshold were empirically chosen.

5 Experiments

For our experiments presented here we used nine different templates of just one single person. The templates are positioned on a regular 3×3 grid within

Fig. 4. The above images show sample results of the AFS: Top row, from left to right: rotation, distortion, glasses. Bottom row: large, small, background without face.

the eye-mouth region of a sample face. As sample face the left image in fig. 1 was used. The templates were taken of constant size 32×32. The eye in the center image of fig. 1 is one of these templates. The size of the sliding window is adapted to the size of the template. Fig. 4 shows some sample results of the AFS. The top row shows (from left to right) a rotated face, a distorted face, and a face with glasses. The bottom row shows faces of different sizes. The table to the left shows the overall similarities of the example images. The similarities have been normalized to the maximal overall similarity of the templates with the original image. Thresholds were chosen empirically and were set in our experiments to $a_{\text{rec}} = 0.85$ and $a_{\text{det}} = 0.7$. As it can be seen the first two images were reported as recognized. By choosing templates of just one single example face demonstrate the great robustness/discriminability tradeoff of the AFS while using templates of more than one face improves the face detection rate.

image	rel. ov. sim.
rotation	0.9
distortion	0.89
glasses	0.81
large	0.71
small	0.72
background	0.65

It can be seen in fig. 4 that the box that indicates the facial position never encloses the inner face precisely. The reason for this is the large support for each histogram (32×32) and therefore for each pixel within the attention image. This does not allow a precise positioning. It needs to be mentioned that the large neighborhoods considered here for the histogram calculations are, in fact, the major drawback of the system: The size of an image region considered for a single histogram is 16×16 image pixels. Therefore, the size of a template is always a multiple of 16×16.

6 Conclusions

In this paper we have presented an approach for the attentive detection and recognition of faces in gray-value images. The approach is biologically motivated and showed in our experiments a very good robustness including robustness to scale, rotation, facial expressions and illumination changes. Robustness to illumination was realized by discarding local amplitude information while using local orientation pyramids instead. Robustness to geometric distortions, on the other hand, was achieved by using histogram techniques and by considering large local neighborhoods. It is well known that the local amplitude/phase representation decouples detection and classification of oriented patterns. Whereas a significant amplitude value indicates the presence (visibility) of a pattern the phase at that point yields a classification of its symmetry. With this local amplitude and phase have been accepted as two real feature dimensions. We believe that the role of local phase has to be re-estimated in the context of natural image statistics. In fact, we have kept the evaluation stage simple in order to not covering up the potency of the orientation approach.

The approach of our AFS has proven to be very potential for recognition and detection and will be adapted for a general object recognition system, which is future work.

References

1. Roberto Brunelli and Tomaso Poggio. Face recognition: Features versus templates. *IEEE Trans. Pattern Analysis and Machine Intelligence*, 15(10):1042–1052, Oct. 1993.
2. G. H. Granlund and G. D. Knutsson. *Signal Processing For Computer Vision.* Kluwer Academic Publisher, 1995.
3. U. Mahlmeister, M. Sandgaard, and G. Sommer. Sample-guided progressive image coding. http://www.informatik.uni-kiel.de/~uhm/research/icpr1.ps.gz, accepted in ICPR'98, Brisbane Australia, 1997.
4. M. Sandgaard. Attentionsgesteuerte progressive Übertragung von Bildern. Master's thesis, Institute of Computer Science and Applied Mathematics, University of Kiel, Germany, 1997. in German.
5. B. Schiele and J. L. Crowley. Object recognition using multidimensional receptive field histograms. In *Proc. Fourth European Conference on Computer Vision*, Cambridge, UK, April 15-18, 1996.
6. E. P. Simoncelli and W. T. Freeman. The steerable pyramid: a flexible architecture for multi-scale derivative computation. Technical report, GRAPS Laboritory, Philadelphia, 1995.
7. E. P. Simoncelli, W. T. Freeman, E. A. Adelson, and D. J. Heger. Shiftable multiscale transforms. *IEEE Trans. Information Theory*, 38(2):587–607, 1992.
8. M. J. Swain and D. H. Ballard. Color indexing. *International Journal of Computer Vision*, 7(1):11–32, 1991.
9. L. Wiskott, J. M. Fellous, N. Krüger, and C. v. d. Malsburg. Face recognition and gender determination. In *International Workshop on Automatic Face- and Gesture-Recognition*, Zurich, Switzerland, June 26-28, 1995.
10. L. Wiskott, J. M. Fellous, N. Krüger, and C. v. d. Malsburg. Face recognition by elastic bunch graph matching. *IEEE Trans. Pattern Analysis and Machine Intelligence*, 19(7):775–779, July 1997.

Schnelle Bestimmung von Kamerabewegungen in blockbasiert codierten Videosequenzen

J. Heuer+, A. Kaup+, H. Burkhardt++

+ Siemens AG, Fachzentrum Netze und Videokommunikation, 81730 München

++ Lehrstuhl für Mustererkennung, Institut für Informatik, Am Flugplatz Geb. 052, 79110 Freiburg i. Br.

Zusammenfassung Die vorliegende Arbeit beschreibt ein schnelles Verfahren zur automatischen Beschreibung der Kamerabewegung in einer blockbasiert codierten Videosequenz. Hierzu wird ein diesbezüglich aussagekräftiges Bewegungsmodell gewählt. Dessen Parameter werden aus den Bewegungsvektoren berechnet, die im Datenstrom der codierten Videosequenz enthalten sind. Hierbei werden die Vektoren entsprechend ihrer Zuverlässigkeit ausgewählt bzw. gewichtet. Trotz des für eine schnelle Berechnung nötigen einfachen Ansatzes zur Bewegungsbeschreibung erzielt das Verfahren eine hohe Genauigkeit.

1. Einleitung

In heutigen Videocodierungsstandards wie MPEG-4 [6] ist die Bewegungsschätzung mit Blockmatching ein Verfahren, das zur Redundanzreduktion genutzt wird. Die im Bitstream enthaltene Bewegungsinformation ist jedoch als Vektorfeld nicht zur aussagekräftigen Charakterisierung der Videosequenz zu nutzen. Eine Sequenzbeschreibung, wie sie Inhalt der MPEG-7 [1] Standardisierung sein wird, die aus diesen schon extrahierten Merkmalen berechnet werden kann, ist die globale Kamerabewegung. Globale Bewegungsbestimmung wird heute zum Beispiel in der Bildstabilisierung von Handheld-Kameras und in der Bildsegmentierung [4][5] eingesetzt. Hierbei wird entweder ein einfaches Bewegungsmodell wie beispielsweise nur Kameraverkippung angenommen. Um hohe Präzision zu erreichen, wird jedoch ein komplexeres Bewegungsmodell zugrundegelegt, das auf Pixelebene mit Hilfe der Gradienten im Bild bestimmt wird [2][3]. Ist jedoch die Kamerabewegung aus Sicht der Szenenbeschreibung zu analysieren, so sollten die gewählten Bewegungsparameter bezüglich der elementaren Kamerabewegungen möglichst gut interpretierbar sein und zudem geringen Rechenaufwand mit guter Präzision verbinden.

In dem hier untersuchten Verfahren werden die bei der Codierung in MPEG-4 durch unterschiedliche Verfahren des Blockmatching bestimmten Bewegungsvektoren als Berechnungsgrundlage herangezogen. Aufbauend hierauf werden die Parameter eines vierparametrigen Bewegungsmodells approximiert.

Dieser Beitrag gliedert sich in die folgenden vier Teile: zunächst wird das Bewegungsmodell erläutert und die Näherungen zur realen Bewegung aufgezeigt. Anschließend wird das im Verification Model (VM) des ACTS-Projekts „Mobile Multimedia System“ MoMuSys[6] verwendete Blockmatchingverfahren beschrieben. Danach wird auf den verwendeten Approximationsalgorithmus eingegangen und abschließend die erzielten Ergebnisse analysiert.

2. Vier-Parameter-Modell für die globale Kamerabewegung

Ziel dieses Kapitels ist es, die Auswirkungen der Kamerabewegung in der Bildebene so zu modellieren, daß die Bildveränderung mit guter Genauigkeit von Parametern beschrieben werden, die sich aus den Bildern in der Bildebene berechnen lassen und einen einfachen Rückschluß auf die vorliegende Kamerabewegung geben.

Bildet man eine natürliche, dreidimensionale Szene mit einer Kamera auf eine zweidimensionale Projektionsebene ab, so kann man für die Abbildung eines Punktes $\vec{p}_0 = (x_0, y_0, z_0)^T$ folgende Gesetzmäßigkeit angeben:

$$\begin{pmatrix} X \\ Y \end{pmatrix} = \frac{F}{z_0}\begin{pmatrix} x_0 \\ y_0 \end{pmatrix} \quad \wedge z_0 >> F, \tag{1}$$

wobei F die Brennweite darstellt und X, Y die Koordinaten des abgebildeten Punktes auf der Bildebene sind.

Wird nun die Kamera bewegt, so bleibt die Abbildungsvorschrift im synchron zur Kamera mitbewegten Koordinatensystem erhalten, die Koordinaten der Objektpunkte müssen jedoch in dieses Koordinatensystem transformiert werden. Da man alle Kamerabewegungen als Akkumulation von Rotation und Translation auffassen kann, läßt sich die Transformation vom ortsfesten Koordinatensystem (x, y, z) zum mitbewegten Koordinatensystem $(\tilde{x}, \tilde{y}, \tilde{z})$ wie folgt formulieren:

$$\begin{pmatrix} \tilde{x}_0 \\ \tilde{y}_0 \\ \tilde{z}_0 \end{pmatrix} = \begin{pmatrix} r_{11} & r_{12} & r_{13} \\ r_{21} & r_{22} & r_{23} \\ r_{31} & r_{32} & r_{33} \end{pmatrix} \cdot \begin{pmatrix} x_0 \\ y_0 \\ z_0 \end{pmatrix} + \begin{pmatrix} t_1 \\ t_2 \\ t_3 \end{pmatrix} \tag{2}$$

Diese affine Transformation zur Beschreibung der Kamera- bzw. Koordinatensystembewegung wird demnach durch zwölf Parameter beschrieben. Daraus ergibt sich für die Abbildungsvorschrift:

$$\begin{aligned} \tilde{X}_0 &= F \cdot \frac{r_{11}x_0 + r_{12}y_0 + r_{13}z_0 + t_1}{r_{31}x_0 + r_{32}y_0 + r_{33}z_0 + t_3} \\ \tilde{Y}_0 &= F \cdot \frac{r_{21}x_0 + r_{22}y_0 + r_{23}z_0 + t_2}{r_{31}x_0 + r_{32}y_0 + r_{33}z_0 + t_3} \end{aligned} \tag{3}$$

Aus der Abbildungsvorschrift geht nun hervor, wie sich x_0 und y_0 aus X_0 und Y_0 berechnen lassen, die Tiefeninformation z_0 liegt jedoch bei einem monocularen System nicht vor. Approximiert man die abgebildeten Objekte wie in [7] mit einer Fläche, so läßt sich diese Tiefeninformation aus den Größen x_0 und y_0 herleiten. Jedoch ist die daraus resultierende Abbildungsvorschrift nur für diese Fläche gültig. Zudem sind die resultierenden 8 Parameter abhängig von der Lage der Approximationsebene und somit nur schwer bezüglich der realen Kamerabewegung zu interpretieren. In der Bildstabilisierung liefert dieses Modell gute Ergebnisse. Jedoch sind die berechneten Parameter bezüglich der Kamerabewegung nicht aussagekräftig, da sie auch die Ebenenbewegung für die Objektapproximation beschreiben.

Um für die Kamerabewegung aussagekräftige Parameter zu erhalten, werden zunächst die elementaren Kamerabewegungen betrachtet. Dies sind translatorische Bewegungen entlang den Koordinatenachsen, Rotationen um diese und das Zoomen. Formuliert man dies in den aufgeführten Transformationen, erkennt man, daß die Translation entlang der z-Achse und das Zoomen bzw. die Translation entlang der x-, y-Achse und die Rotation um die y-, und x-Achse aufgrund der fehlenden Tiefeninformation in der Bildebene nur schwer voneinander unterschieden werden können. Daher wird vereinfachend angenommen, daß die betrachteten Szenen bezüglich der Raumtiefe „flach" sind. So kann, ausgehend von Gl.(3), die von der Kamerabewegung verursachte Bildveränderung wie folgt modelliert werden:

$$\begin{pmatrix}\Delta X\\ \Delta Y\end{pmatrix}=\begin{pmatrix}C_F\cos(\varphi_z)-1 & -C_F\sin(\varphi_z)\\ C_F\sin(\varphi_z) & C_F\cos(\varphi_z)-1\end{pmatrix}\cdot\begin{pmatrix}X\\ Y\end{pmatrix}+\begin{pmatrix}t_X\\ t_Y\end{pmatrix}, \tag{4}$$

wobei ΔX, ΔY die Koordinatenveränderung der Bildpunkte hervorgerufen in einem Zeitintervall Δt bei beschriebener Kamerabewegung ist und φ_z der Winkel um den die Kamera um die z-Achse in diesem Zeitintervall rotiert wurde. Der Faktor C_F modelliert die Brennweitenänderung bzw. die Translation entlang der z-Achse.

3. Bewegungsvektorfelder in MPEG-4 Sequenzen

Wie anfangs erwähnt, soll ein möglichst schnelles Verfahren zur Berechnung der Bewegungsparameter realisiert werden. Daher werden als Berechnungsgrundlage die Bewegungsvektoren, die zur Codierung durch Blockmatching gefunden wurden, aus dem MPEG-4 Datenstrom ausgelesen, wobei nur in I- und P-Frames codiert wurde. Da das verwendete Blockmatching nicht in MPEG-4 standardisiert wird, ist hier das verwendete Verfahren, das im VM 8.0 von MoMuSys implementiert ist, kurz dargestellt.

Für jeden 16x16 Macroblock wird eine volle Suche im codierbaren Bereich (±16 Pixel) durchgeführt. Der nach der L_1-Norm bezüglich der Summe aller Pixel des Macroblocks (SAD) beste Match wird gewählt. Nun wird für die 8x8 Unterblöcke in der Umgebung von ±2 Pixeln das Verfahren nochmals angewendet. Führt dies zu einer Verbesserung in der SAD, werden diese vier Bewegungsvektoren für den Makroblock codiert, andernfalls nur der zuerst Berechnete.

4. Bestimmung der Bewegungsparameter

Da das sich ergebende Gleichungssystem aus der Gl.(4) nichtlinear ist, ist die direkte Bestimmung der Parameter nicht möglich. Deswegen wurden zwei Methoden untersucht: die schnellere Berechnung über ein sechsparametriges Bewegungsmodell und die Berechnung durch Linearisierung der sin- und cos- Terme in Gl.(4). Anschließend wird ein morphologischer Filter genutzt, um Vektoren, die durch Mismatch oder durch bewegte Objekte hervorgerufen wurden, zu eliminieren. Somit können die Parameterwerte iterativ bestimmt und der zum Bewegungsmodell passende Hintergrund vom Vordergrund segmentiert werden. Wird die Bewegungsschätzung beim Codieren vorgenommen, können zusätzlich Informationen über die Zuverlässigkeit der geschätzten Bewegungsvektoren genutzt werden.

4.1. Bestimmung über das Sechs-Parameter-Modell

Das in Gl. (4) beschriebene Modell für die Kamerabewegung in der Abbildungsebene wird nun durch ein Bewegungsmodell mit sechs Parametern ersetzt:

$$\begin{pmatrix}\tilde{X}_0\\ \tilde{Y}_0\end{pmatrix}=\begin{pmatrix}r_{11}' & r_{12}'\\ r_{21}' & r_{22}'\end{pmatrix}\cdot\begin{pmatrix}X_0\\ Y_0\end{pmatrix}+\begin{pmatrix}t_x'\\ t_y'\end{pmatrix} \tag{5}$$

Das hieraus mit den Daten des Bewegungsvektorfeldes entstehende Gleichungssystem kann nun durch lineare Regression gelöst werden, wobei die Komplexität einer Invertierung einer symmetrischen 3x3 Matrix entspricht. Hat man auf diese Weise die Parameter r_{11}', r_{12}', r_{21}', r_{22}', t_x' und t_y' bestimmt, so lassen sich hieraus die Parameter der Gl.(4) folgendermaßen approximieren:

$$\bar{T} = \bar{T}', \quad C_F^{\,2} = \det\begin{pmatrix} r_{11}' & r_{12}' \\ r_{21}' & r_{22}' \end{pmatrix}, \quad \varphi_z = \arcsin\frac{1}{2}\left(r_{21}' - r_{12}'\right) \qquad (6)$$

Für die Berechnung von φ_z werden nur die Werte r_{21}' und r_{12}' genutzt, da diese bei i.a. vorkommenden kleinen Winkeln ω_z aufgrund des sinualen Zusammenhangs eine genauere Berechnung zulassen.

Bei der Berechnung der vier Parameter über das Sechs-Parameter-Modell ist jedoch zu beachten, daß die lineare Regression nur einen den quadratischen Fehler minimierenden Parametersatz bezüglich des sechs Parameter Modells berechnet. Durch die nachfolgende Transformation in das Vier-Parameter-Modell wird zusätzlich ein Fehler hizugefügt. Da das sechs-Parameter-Modell auch Bildveränderungen modellieren kann, die nicht durch die Kamerabewegung sondern z. B. durch Objektbewegung hervorgerufen wird, kann dies zu einem schlecht approximierten Vier-Parameter-Satz und zudem zu Fehlern bei der Segmentierung (s. u.) führen.

4.2. Parameterberechnung durch Linearisierung

Um die Nachteile der Optimierung bezüglich eines anderen Modells bei dem oben beschriebenen Verfahren zu vermeiden, werden die sin- und cos- Funktionen mit den ersten Termen der entsprechenden Tayler-Reihe linearisiert. So ergibt sich für kleine Winkel φ_z:

$$\begin{pmatrix} \Delta X \\ \Delta Y \end{pmatrix} \approx \begin{pmatrix} C_F - 1 & -C_F\varphi_z \\ C_F\varphi_z & C_F - 1 \end{pmatrix} \cdot \begin{pmatrix} X \\ Y \end{pmatrix} + \begin{pmatrix} t_X \\ t_Y \end{pmatrix} = \begin{pmatrix} R_1 & -R_2 \\ R_2 & R_1 \end{pmatrix} \begin{pmatrix} X \\ Y \end{pmatrix} + \begin{pmatrix} t_X \\ t_Y \end{pmatrix} \quad (7)$$

Da die Optimierung der Gleichungen für ΔX und ΔY nicht unabhängig voneinander ist, wird bezüglich der Summe der Fehlerquadrate minimiert, also

$$\sum_{\vec{V}} \left[\left(\Delta X_\eta - R_1 X_\eta + R_2 Y_\eta - t_X \right)^2 + \left(\Delta Y_\eta - R_2 X_\eta - R_1 Y_\eta - t_Y \right)^2 \right] \rightarrow \min. \qquad (8)$$

Die sich hieraus für die Berechnung ergebende Komplexität entspricht der Invertierung einer symmetrischen 4x4 Matrix zuzüglich acht Additionen und sechs Multiplikationen je ausgewertetem Bewegungsvektor.

4.3. Segmentierung der Bewegungsvektoren

Die äußeren Bewegungsvektoren entsprechen durch das in MPEG-4 durchgeführte Padding [6] häufig nicht den realen Bewegungen. Die inneren Bewegungsvektoren tragen zum einen nur wenig zur Bestimmung der Größen C_F und φ_z bei und werden oft durch Objektbewegungen beeinflußt. Daher wurde im ersten Berechnungsschritt eine Default-Maske für die Bewegungsvektoren gewählt, so daß die am Rand und in der Mitte liegenden Bewegungsvektoren nicht betrachtet werden (Abb. 1).

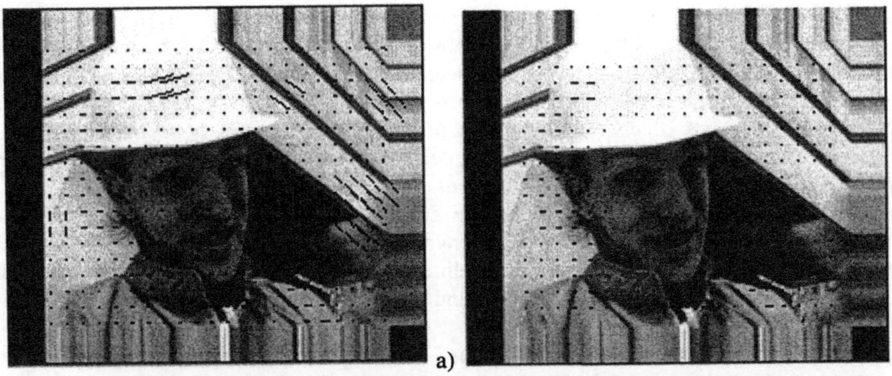

a) b)

Abb. 1:a)Bild mit allen Bewegungsvektoren der 8x8 Blöcke (Bilder durch Padding für das Blockmatching erweitert), b) mit Bewegunsvektoren nach einer Iteration mit Schwellwert 6 und Default-Maske als Start-Maske (s. Text).

Nachdem die Bewegungsparameter unter Beachtung dieser Maske berechnet wurden, werden in einem Iterationsschritt zusammenhängende Vektorfelder, die in der Abweichung vom Bewegungsmodell einen Schwellwert überschreiten, aus der Maske mit einem morphologischen Filter eliminiert. Dieser Filter eliminiert nur dann den Bewegungsvektor, wenn ebenfalls ein Nachbarvektor in der Abweichung diesen Schwellwert überschreitet. Auf diese Weise können Bewegungsvektoren von sich unabhängig bewegenden Objekten und Blockmatchingfehler bei unstrukturierten Blöcken eliminiert werden. Ein Beispiel einer resultierenden Maske ist in Abb.1b) gezeigt.

Aufgrund des iterativen Vorgehens (s. Abb.2) besteht die Gefahr, in ein lokales Minimum zu laufen, das nicht der realen Kamerabewegung entspricht. Zudem reduziert sich mit der Anzahl der verwendeten Bewegungsvektoren auch die Genauigkeit der Parameterbestimmung bei einer gegebenen Streuung der Werte. Aus diesen Gründen und um den Berechnungsaufwand gering zu halten, wurde in den im nächsten Kapitel aufgezeigten Ergebnissen nur ein Iterationsschritt zugelassen.

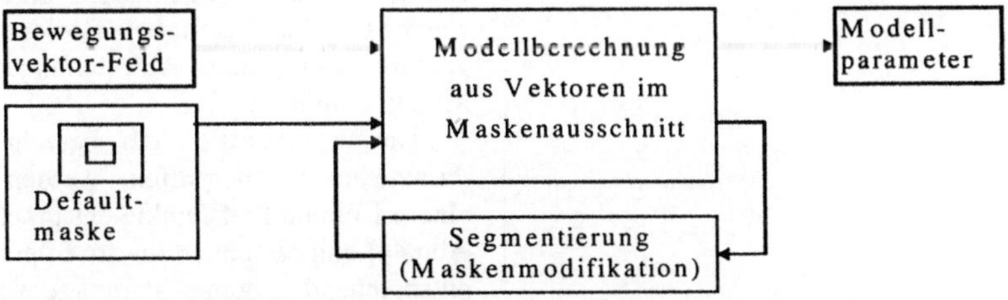

Abb. 2: Blockschaltbild der globalen Bewegungsschätzung

4.4. Gewichtung der Bewegungsvektoren

Das Blockmatchingverfahren zur Bestimmung der Bewegungsvektoren arbeitet in Abhängigkeit vom Blockinhalt in Bezug auf die reale Bewegung unterschiedlich zuverlässig. So zeigt sich in Abb. 1a), daß z.B. entlang Kanten erhebliche Fehler auftreten können. In [8] wird deshalb die Zuverlässigkeit der Bewegungsvektoren beurteilt. Dies kann während des Blockmatchings anhand der SAD der korrespondierenden Blöcke oder über die Varianz der SAD-Werte bezogen auf das Suchgebiet beim Blockmatching geschehen. Letzteres Kriterium wurde hier aufgrund der niedrigeren Empfindlichkeit gegenüber Belichtungsschwankungen gewählt, wobei zur schnelleren Berechnung die Standardabweichung gegenüber dem besten Match als Kriterium gewählt wurde. In Abb. 4e) ist die Standardabweichung der Bewegungsvektoren als Grauwertbild für den ersten P-Frame der in Abb. 4 gezeigten Sequenz über die Koordinaten der korrespondierenden 8x8 Blöcke aufgetragen. Gut zu erkennen ist die hohe Standardabweichung bei Helligkeitsschwankungen, wie sie an der Wand und dem Schreibtisch auftreten. Ein eventuell schlechter Match entlang der Kante wird durch den Iterationsschritt eliminiert.

Am einfachsten ist es, das Kriterium für die Zuverlässigkeit der Vektoren bei der Berechnung zu berücksichtigen, indem ein Schwellwertentscheider eingesetzt wird, der entsprechend unzuverlässige Vektoren in der Defaultmaske eliminiert. Jedoch ergibt sich aus einer weiteren Reduktion der Bewegungsvektoren bei Verwendung des QCIF-Formats eine ungenaue Bestimmung der Parameter.

Daher ist es besser, bei der Minimierung des Fehlers des Bewegungsmodells durch lineare Regression die Zuverlässigkeit als Gewichtungsfaktor *gm* einzubeziehen. Statt einer binären (Default)Maske wird also für die Vektorkomponenten getrennt eine Gewichtungsmaske eingesetzt. Als Fehlerminimierungsansatz ergibt sich aus Gl.(8) also:

$$\sum_{\vec{V}} gm^2 \cdot \begin{bmatrix} \left(\Delta X_\eta - R_1 X_\eta + R_2 Y_\eta - t_X\right)^2 + \\ + \left(\Delta Y_\eta - R_2 X_\eta - R_1 Y_\eta - t_Y\right)^2 \end{bmatrix} \rightarrow \min. \tag{9}$$

5. Diskussion der Ergebnisse

Das Verfahren wurde anhand von aufgenommenen Testsequenzen mit entsprechend elementaren Kamerabewegungen wie auch mit von MPEG-4 verwendeten Testsequenzen untersucht. Für das morphologische Filter hat sich als Fehlermaß die L1-Norm als praktikabel erwiesen. Als Schwellwertentscheider im morphologischen Filter (s. Kap.4.3) wurde der Wert 6 festgelegt. Wie Abb.1b) und Abb.3) zeigen, werden bei diesem Schwellwert vor allem die durch Mismatches und bewegte Objekte erzeugten Bewegungsvektoren mit großer Abweichung vom Modell eliminiert.

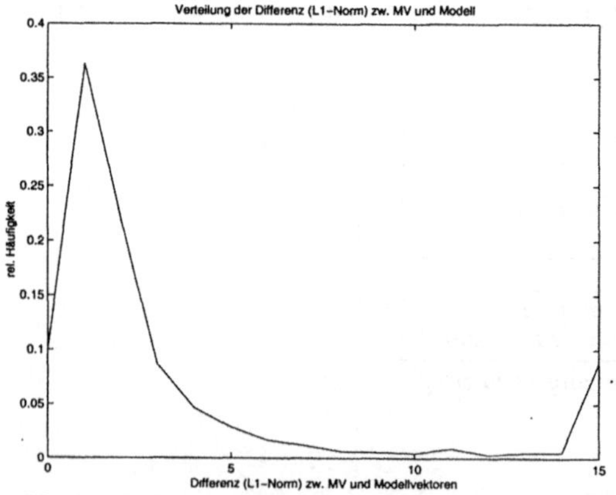

Abb. 3: Abweichung der Bewegungsvektoren eines P-Frames vom im ersten Schritt berechneten Bewegungsmodell bezüglich der L1-Norm

Um die Präzision der berechneten Parameter zu überprüfen, werden bei einem I-Frame fünf Punkte entsprechend Abb.4a) eingezeichnet, die im folgenden entsprechend dem Parametermodell bewegt werden. Bei guter Approximation durch das Bewegungsmodell sollten diese Punkte sich entsprechend dem Hintergrund bewegen. Da diese Punkte nur entsprechend dem Bewegungsmodell verschoben werden, akkumuliert sich der Fehler also von Frame zu Frame auf.

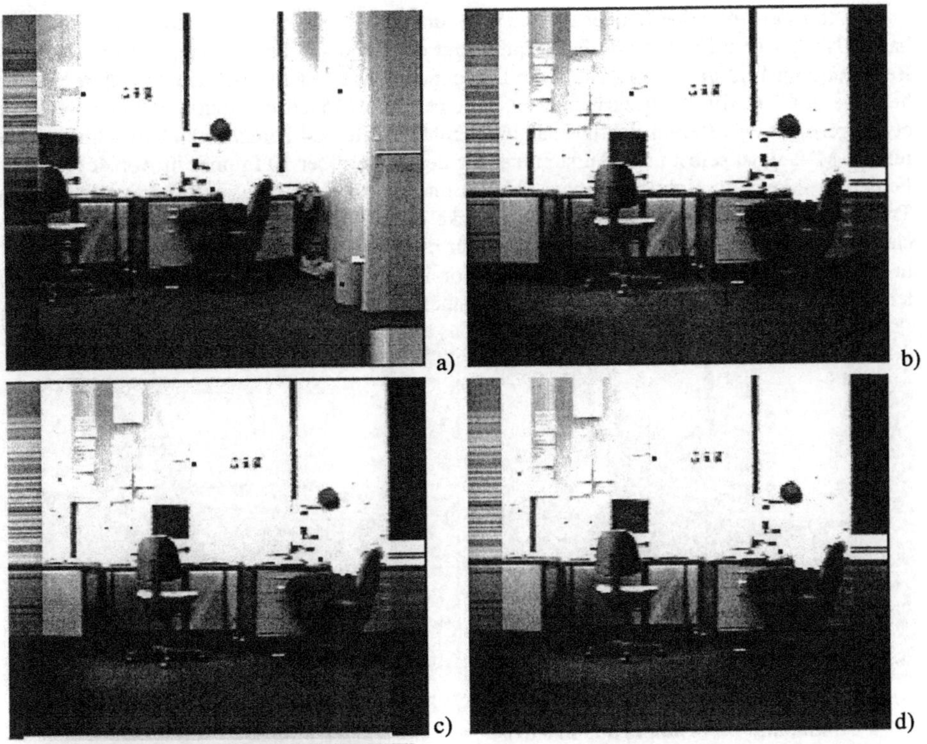

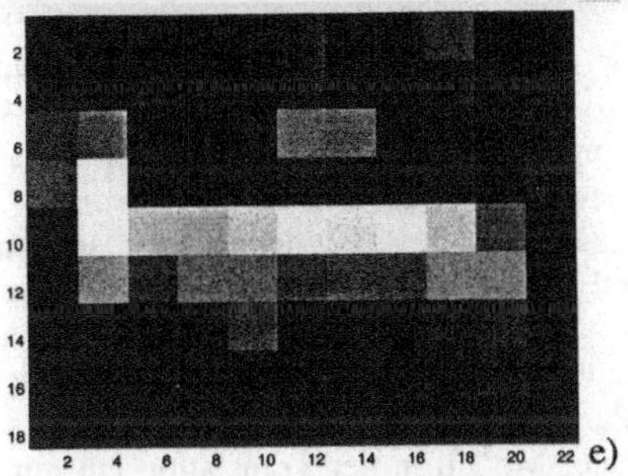

Abb. 4: Linksschwenk der Kamera, a) I-Frame, b) 10P-Frames, später mit Auswertlücke in der Bildmitte, c) mit Gewichtung der MV und d) mit Gewichtung der MV aber ohne Auswertlücke (Berechnung durch Linearisierung des 4-Parameter-Modells);
e) Gewichtungsfaktoren gm des ersten I-Frames als Grauwertbild über den 8x8 Blockindex aufgetragen.

	1.P/G	1.P	2.P/G	2.P	3.P/G	3.P	4.P/G	4.P
t_x	-6.06	-5.27	-5.68	-5.07	-5.65	-4.75	-5.94	-5.07
t_y	-0.10	-0.03	0.00	0.13	-0.02	-0.09	-0.23	-0.14
C_F^2	1.0011	0.9947	1.0023	0.9984	1.0061	1.0021	1.0047	0.9948
φ_z	.0015	.0093	.0013	.0077	.0008	.0061	-.0025	.0003

Tab. 1: Bewegungsparameter der ersten 4 P-Frames aus Abb.4 c) und b) mit und ohne Gewichtung gm der MV

In Abb.4) wurde eine Sequenz untersucht, die einen Schwenk von rechts nach links beinhaltet. Nach vier P-Frames wurde entsprechend der Sequenz die überwiegend translatorische Bewegung im Bild modelliert. Da in die Berechnung des Modells noch einige Vektoren eingehen, die auf ein Mismatch zurückgehen, fällt bei dem Modell die translatorische Bewegung etwas geringer aus. Berücksichtigt man die gemäß Kapitel 4.4 berechneten Gewichtungsfaktoren der MV, so ist selbst nach Aufsummierung der Fehler über 10 Frames in Abb.4c) kaum ein Fehler zu erkennen. Vergleicht man Abb.4c) und d), so sieht man, daß die Berechnung des Translationsvektors kaum von der Auswertlücke in der Mitte beeinträchtigt. Dies gilt entsprechend der Annahme aus Kap. 4.3 erst recht für die Kamera-Rotation um die Z-Achse, wie sie in Abb.5) dargestellt ist. Wie bei der Translation ist auch hier nach acht Bildern kaum ein Fehler in der Modellierung der Bewegung zu erkennen.

a)

b)

Abb. 5: Rotation der Kamera, a) I-Frame, b) 8P-Frames später mit Auswertlücke in der Bildmitte und Berechnung das Linearisierte 4-Parameter-Modell

Zur Analyse der Approximationsunterschiede zwischen der Berechnung der 4 Bewegungsparameter über das 6-Parameter-Modell und der Berechnung durch Linearisierung des 4-Parameter-Modells wurde die kritische Zoom-In Sequenz gewählt. Denn bei reiner Translation würden sich nur Unterschiede bei nicht in 4 Parameter modellierbaren Bildverzerrungen zeigen. Abb.6) zeigt das Resultat beider Verfahren nach 10 P-Frames. Betrachtet man nun in Abb.6), wie die entsprechend dem Bewegungsmodell bewegten fünf Punkte nach 10 P-Frames liegen, so stellt man fest, daß bei der Berechnung über das Vier-Parameter-Modell sie sich besser mit den optischen Merkmalen im Bild mitbewegt haben als es beim Sechs-Parameter-Modell der Fall ist. Betrachtet man in Tab.2 die Parameter für die ersten fünf P-Frames, so stimmt die Berechnungsmethode über das 6-Parameter-Modell in der Translation gut mit der Linearisierungsmethode überein, hingegen die Werte für C_F und φ_z zeigen Abweichungen voneinander. Die Berechnung über die Linearisierung erweist sich also, wie bereits in Kap.4 erläutert, als zuverlässiger.

	1.P/lin	1.P/6p	2.P/lin	2.P/6p	3.P/lin	3.P/6p	4.P/lin	4.P/6p
t_x	0.08	0.06	0.04	0.04	0.19	0.17	0.06	0.06
t_y	0.04	0.02	-0.20	-0.21	-0.04	-0.03	-0.29	-0.32
C_F^2	0.9654	0.9666	0.9688	0.9693	0.9623	0.9623	0.9702	0.9724
φ_z	-.0005	-.0003	-.0003	-.0002	-.0003	0.0001	0.0006	0.0007

Tab. 2: Bewegungsparameter der ersten 5 P-Frames aus Abb.7) für die Berechnung durch Linearisierung und über das Sechs-Parameter-Modell

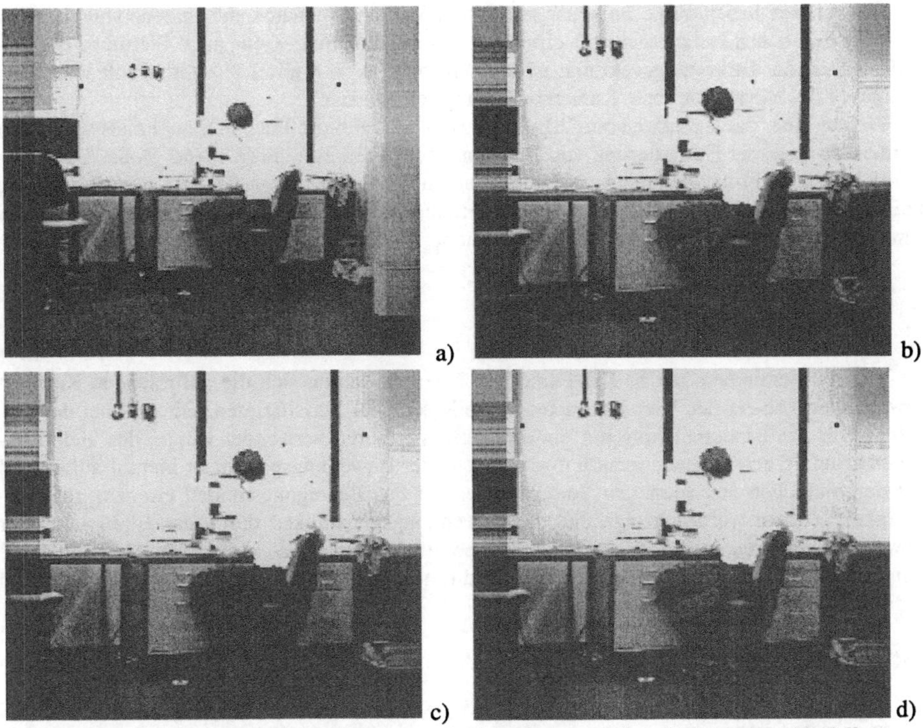

Abb. 6: Zoom-In Sequenz, a) I-Frame, b) nach 10 P-Frames, über 6 Parameter Modell, c) über Linearisierung, d) mit Gewichtungsfaktoren gm

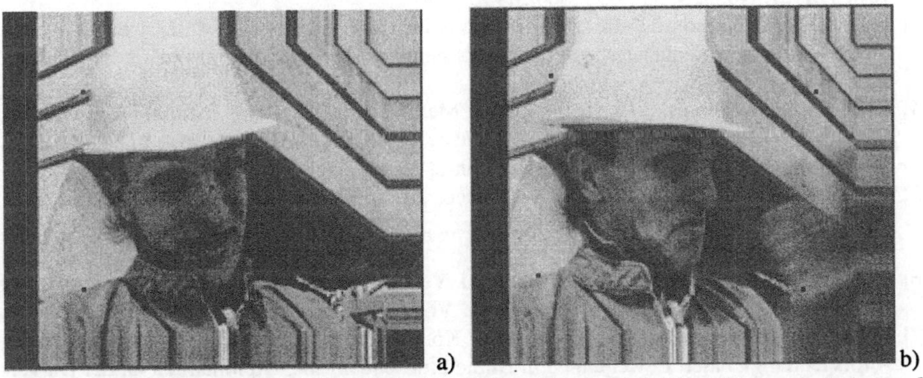

Abb. 7: Foreman-Sequenz, a) I-Frame, b) 7 P-Frames später mit leichter Kamerarotation und großer Objektbewegung

Neben den Sequenzen mit elementarer Kamerabewegung wurde auch die Foreman-Sequenz analysiert. Da in dieser Sequenz sehr viel Objektbewegung bei hoher Hintergrundabdeckung auftritt, stellt es eine gute Testsequenz für die Robustheit des Algorithmus dar. In Abb.7) ist eine Stelle in der Sequenz herausgegriffen, in der die Hand des Sprechers durch das Bild fährt. Trotzdem wurde die auftretende Kamerarotation korrekt erkannt und modelliert. Jedoch zeigt sich bei genauerer Betrachtung, daß nicht alle zur Hand gehörenden Bewegungsvektoren als nicht der Hintergrundbewegung zugehörig erkannt wurden.

Desweiteren haben obige Beispiele gezeigt, daß durch Mismatch entstandene Nullvektoren oft nicht durch den Iterationsschritt eliminiert werden konnten. Zieht man allerdings die Zuverlässigkeit der Bewegungsvektoren in Betracht, wie es in Kapitel 4.4 vorgestellt wurde, so beschreibt das Modell die reale Kamerabewegung sehr präzise.

Wendet man das Verfahren auf CIF-Sequenzen an, kann die Präzision nochmals verbessert werden, da trotz der Eliminierung von Bewegungsvektoren eine hinreichend große Anzahl im Vergleich zur Zahl der Parameter verbleibt. Zudem wird aufgrund der erhöhten Auflösung die Textur von dem nun auch oft großflächigeren Hintergrund erfaßt und so die Zuverlässigkeit der durch Blockmatching gefundenen Vektoren erhöht.

6. Zusammenfassung

Wie die Parametersätze in Tab.1 und Tab.2 zeigen, lassen sich die auftretenden Kamerabewegungen anhand des Vier-Parameter-Modells sehr gut klassifizieren. Obwohl bei der Codierung für das Blockmatching nur ein einfaches Verfahren verwendet wurde, das viele Mismatches liefert, erreichte die quantitative Aussage der Bewegungsparameter hierauf aufsetzend mit den einfachen und schnellen Berechnungen für das Bewegungsmodell ein sehr gute Genauigkeit. Zudem wurde gezeigt, daß die Berechnungsgenauigkeit durch das Eliminieren der Bewegungsvektoren in der Bildmitte nicht beeinflußt wird. Dies senkt zum einen aber die Empfindlichkeit gegenüber häufig in der Bildmitte existierender Objektbewegung und zum anderen den Berechnungsaufwand.

Literatur

[1] „MPEG-7: Context and Objectives", Public Document No. MPEG98/N2082, ISO/EC JTC1/SC29/WG11, Feb. 1998.

[2] S.S. Beauchemin, J.L. Barron, „The Computation of Optical Flow", ACM Computing Surveys Vol. 27, No. 3, S. 366-433, Sept. 1995.

[3] J. Dugelay, H. Sanson, „Differential Methods for the Identification of 2D and 3D Motion Models in Image Sequences", Signal Processing: Image Processing, Vol.7, S. 105-127, Sept. 1995.

[4] R. Mech, M. Wollborn, „ A Noise Robust Methode for 2D Shape Estimation of Moving Objects in Video Sequences Considering a Moving Camera", Workshop on Image Analysis for Multimedia Interactive Services, Louvain-la-Neuve, Belgien, Jun. 1997.

[5] S. Colonnese, A. Neri, G. Russo, C. Tabacco, „Adaptive Segmentation of Moving Object versus Background for Video Coding", Proc. of SPIE Annual Symposium, Vol. 3164, S. Diego, Aug. 1997.

[6] T. Sikora, „The MPEG-4 Video Standard Verification Model", IEEE Transactions on Circuits and Systems for Video Technology, Vol. 7, No. 1, Feb. 1997.

[7] M. Reder, M. Wollborn, „Schätzung und Kompensation der durch Kamerabewegungen verursachten globalen Bewegungen in Bildtelefonsequenzen", Diplomarbeit an der Universität Hannover, Dez. 1994.

[8] R. Mester, M. Hötter, „Zuverlässigkeit und Effizienz von Verfahren zur Verschiebungsvektorschätzung", Mustererkennung, 17. DAGM-Symposium, S. 285-294, Springer-Verlag, Berlin, 1995.

Untersuchungen zur Robustheit von Farbraummodellen bezüglich Szenengeometrie und Schwankungen in der Beleuchtungsintensität

Wilfried Kubinger, Markus Vincze, and Gerfried Zeichen

Institut für Flexible Automation,
Technische Universität Wien,
Gußhausstr. 27–29/361, A–1040 Wien
wk@flexaut.tuwien.ac.at
http://www.infa.tuwien.ac.at/

Zusammenfassung In dieser Arbeit wird ein allgemeines Verfahren vorgestellt, mit dem man Farbraummodelle auf ihre Robustheit bezüglich der Szenengeometrie sowie von Schwankungen in der Beleuchtungsintensität untersuchen kann. Es zeigt sich, daß lineare Farbraummodelle prinzipiell variant sind und daher Methoden der Bildverarbeitung stark beeinträchtigen. Die meisten der hier untersuchten nichtlinearen Modelle beinhalten unter der Voraussetzung, daß der Zusammenhang zwischen der einfallenden Lichtintensität und den Sensorsignalen *linear* ist, invariante Komponenten. Bei deren Verwendung ist eine robuste, von Szenengeometrie und Intensitätsschwankungen unabhängige Auswertung der Bilddaten möglich[1].

1 Einführung

Zur Erhöhung von Robustheit und Verläßlichkeit von Algorithmen und Verfahren in der Bildverarbeitung wird häufig die Farbinformation eines Objektes herangezogen. Da sich die von einer marktüblichen Farbkamera gelieferten Sensorwerte R (Rot), G (Grün) und B (Blau) nicht immer optimal zur Weiterverarbeitung eignen, wurde eine große Anzahl verschiedene Farbraummodelle entworfen, die eine bessere Darstellung der Farbinformation ermöglichen. Bis jetzt wurde jedoch, wie schon in [1] erkannt wird, nicht darauf geachtet, ob es zu einer Veränderung der Farbverteilung bei einer Änderung in der Szenengeometrie oder in der Intensität der Beleuchtung in Abhängigkeit des verwendeten Farbraummodells kommt. Vielmehr wurden meist verschiedene Modelle, basierend auf einem bestimmten Segmentierungsverfahren für eine bestimmte Anzahl von Testbildern verglichen, um das für diese Anwendung beste Modell auszuwählen [8, 10] oder ein neues Modell zu entwerfen [12].

Nur bei wenigen Anwendungen wurde für das verwendete Modell nachgewiesen, daß es unabhängig von der Szenengeometrie ist und dies als Auswahlkriterium angegeben [1, 2, 4] oder ein Modell gezielt so gewählt, daß es unabhängig

[1] Teile dieser Arbeit werden durch den österreichischen Fonds zur Förderung der wissenschaftlichen Forschung, Projekt P11420–MAT gefördert.

von der Beleuchtungsintensität ist [13]. Ein „optimales" Farbraummodell sollte gegenüber den folgenden Effekten invariant sein:

1. Änderungen in der Szenengeometrie, also der Konfiguration Lichtquelle–Objekt–Kamera,
2. Schwankungen in der Beleuchtungsintensität und
3. Variationen in der Farbe der Beleuchtung.

Die Untersuchungen hier beschränken sich auf die ersten beiden Punkt, der dritte Teil ist dem Forschungsgebiet „color constancy" zuzuordnen [1, 5] und wird hier nicht behandelt.

2 Robustheit von Farbraummodellen

Wir konzentrieren uns in dieser Arbeit auf zwei Effekte, die bei der Betrachtung natürlicher Szenen anzutreffen sind. Die Geometrie der Szene kann in der Regel nicht als feststehend betrachtet werden, da sich für jeden Objektpunkt zusammen mit dem korrespondierenden Bildpunkt eine andere Geometrie, üblicherweise beschrieben durch die photometrischen Winkel, ergibt. Des weiteren sind Schwankungen in der Intensität der Lichtquelle nicht auszuschließen, da einerseits die Kontrolle über die Quelle im Normalfall nicht gegeben ist und auch Abschattungen nicht verhindert werden können. Unter der Robustheit eines Farbraummodells verstehen wir in diesem Sinne die Invarianz der Komponenten des Modells gegenüber den beiden hier angeführten Effekten. Dies ist insofern sehr wichtig, da für die gute Funktion von Algorithmen ein stabiles Standbein, das in der Farbbildverarbeitung fast immer ein Farbraummodell ist, unbedingt notwendig ist. Ist das Modell jedoch nicht in der Lage, eine von den beiden Effekten unabhängige Beschreibung der Farbinformation zu liefern, so kann dies beispielsweise bei Segmentierungsverfahren dazu führen, daß einem farblich homogenen Zylinder mehrere Segmente zugewiesen werden. Dies ist meist nicht erwünscht und läßt sich durch genaue Analyse der zur Auswahl stehenden Farbraummodelle vermeiden.

2.1 Modellierung von Szenengeometrie und Schwankungen in der Beleuchtungsintensität

Die von einem Körper reflektierte Strahlung $R(g, \lambda)$ läßt sich, unter der Voraussetzung, daß man nicht direkt in eine Reflexion der Lichtquelle blickt, nach [4] mit

$$R(g, \lambda) = M(g)C(\lambda) \tag{1}$$

sehr gut annähern. In dieser Gleichung bezeichnet λ die Wellenlänge und $C(\lambda)$ ist der spektrale Reflexionsfaktor des Materials [6]. $M(g)$ gibt die Abhängigkeit von den photometrischen Winkeln g, also den Einfluß der Szenengeometrie, an. Der gemessene Wert eines Senorkanals einer Farbkamera, die als Ergebnis je

Pixel mit den Bildkoordinaten (x, y) die Primärfarben R, G und B liefert, läßt sich ganz allgemein mit

$$s(x,y) = \int_{\lambda} f_s(\lambda) L(\lambda) R(g,\lambda) d\lambda \qquad s = R, G, B \tag{2}$$

angeben. Dabei steht $f_s(\lambda)$ für die spektrale Empfindlichkeit des Sensorkanals s und $L(\lambda)$ beschreibt die spektrale Leistungsverteilung der Beleuchtungsquelle.

Eine Beleuchtungsquelle, deren Lichtintensität schwankt, also beispielsweise heller oder dunkler geschalten wird, behält näherungsweise ihre spektrale Verteilung bei [13], nur ist die dann vorhandene spektrale Leistungsverteilung

$$L_a(\lambda) = aL(\lambda), \qquad a \ldots skalar \tag{3}$$

nun proportional zu der vorherigen Leistungsverteilung. Aus den Beziehungen (1), (2) und (3) läßt sich nun ein Modell für die drei Kanäle angeben, in dem sowohl die Szenengeometrie als auch Beleuchtungsschwankungen wie folgt modelliert sind.

$$\begin{aligned} M_a(g,a) &= aM(g) \\ s(x,y) &= M_a(g,a) \int_{\lambda} f_s(\lambda) L(\lambda) C(\lambda) d\lambda \end{aligned} \tag{4}$$

Der Beziehung (4) ist zu entnehmen, daß die beiden hier modellierten Effekte sich durch einen skalaren Faktor $M_a(g, a)$ ausdrücken lassen, der nur von der Intensitätsschwankung sowie von der Szenengeometrie abhängt, aber unabhängig vom Sensorkanal ist. Diese Eigenschaft läßt sich nun dazu benützen, um bei einem Farbraummodell, bei dem pixelweise vom RGB– in einen beliebigen anderen Farbraum transformiert wird, analytisch festzustellen, ob das Modell invariant gegenüber den beiden angeführten Effekten ist.

2.2 Der Skalierungstest

Mit Hilfe des Modells in (4) läßt sich nun eine sehr einfache Überprüfung durchführen, ob die durch eine Farbraumtransformation berechnete Größe oder Größen invariant gegenüber Beleuchtungsschwankungen und Änderungen der Szenengeometrie sind. Hiebei wird (4) in der Form

$$s(x,y) = M_a(g,a) \cdot s_0(x,y) \qquad s_0 = R_0, G_0, B_0 \tag{5}$$

angeschrieben, wobei $s_0(x, y)$ den spektralen Integrationsprozeß repräsentiert, also alle wellenlängenabhängigen Komponenten beinhaltet. Der Wert $M_a(g, a)$ bewirkt in Abhängigkeit der Szenengeometrie sowie der Beleuchtung eine Skalierung von $s_0(x, y)$ und genau diese Eigenschaft von (5) wird für einen Skalierungstest verwendet, da $M_a(g, a)$ für eine beliebige Bildkoordinate (x, y) für alle drei Kanäle den selben Wert hat. Wenn ein Farbraummodell invariant gegenüber den obigen beiden Effekten sein soll, darf der Faktor $M_a(g, a)$ aus (5) keinen

Einfluß auf das neue Farbraummodell haben. D.h. es muß bei einer pixelweisen Transformation vom RGB– in einen anderen Farbraum für eine invariante Größe $P1(x, y)$ immer gelten:

$$P1(x,y) = f(R,G,B) = f(R_0,G_0,B_0) \tag{6}$$

Mit $f(\ldots)$ wird hier die Transformationsvorschrift vom RGB– in den Zielfarbraum bezeichnet. Ist die Gleichung (6) erfüllt, dann ist diese Größe $P1(x, y)$ invariant gegenüber Beleuchtungsschwankungen und Änderungen in der Szenengeometrie. Da man dies einfach durch beliebiges skalares Skalieren der drei Kanäle R, G und B mit derselben Größe und einfaches analytisches Auswerten der Transformationsvorschrift erreicht, bezeichnen wir dieses hier vorgestellte Verfahren als Skalierungstest.

2.3 Gegenüberstellung gängiger Farbraummodelle

Bei den heutzutage in der Bildverarbeitung hauptsächlich eingesetzten Farbraummodellen lassen sich prinzipiell zwei Hauptklassen unterscheiden, die *linearen* und *nichtlinearen* Farbraummodelle.

Bei der ersten Art handelt es sich um Modelle, die durch eine *lineare* Matrixtransformation der Form

$$\mathbf{p}(x,y) = \begin{pmatrix} P1(x,y) \\ P2(x,y) \\ P3(x,y) \end{pmatrix} = \mathbf{C} \cdot \mathbf{s}(x,y) = \mathbf{C} \cdot \begin{pmatrix} R(x,y) \\ G(x,y) \\ B(x,y) \end{pmatrix} \tag{7}$$

gewonnen werden. Die am häufigstem eingesetzten Modelle sind YUV, YIQ, YCbCr, XYZ, I1I2I3 und I1I2'I3'. Der Hauptgrund für die Verwendung dieser Modelle ist darin zu sehen, daß sich die 3x3 Matrix $\mathbf{C}$ sehr effektiv implementieren läßt und sich damit eine Farbraumkonversion in Echtzeit (also "on the fly") realisieren läßt. Wendet man nun den Skalierungstest (5), (6) auf diese Modelle, repräsentiert durch ihre Transformationsvorschrift (7), an, so folgt aus

$$\mathbf{p}(x,y) = \mathbf{C} \cdot \begin{pmatrix} R(x,y) \\ G(x,y) \\ B(x,y) \end{pmatrix} = \mathbf{C} \cdot \begin{pmatrix} M_a(g,a)R_0(x,y) \\ M_a(g,a)G_0(x,y) \\ M_a(g,a)B_0(x,y) \end{pmatrix} =$$

$$= M_a(g,a) \cdot \mathbf{C} \cdot \begin{pmatrix} R_0(x,y) \\ G_0(x,y) \\ B_0(x,y) \end{pmatrix} \neq \mathbf{C} \cdot \begin{pmatrix} R_0(x,y) \\ G_0(x,y) \\ B_0(x,y) \end{pmatrix}$$

sofort, daß diese Modelle vom Prinzip her nicht in der Lage sind, invariante und damit robuste Farbinformationen zu liefern. Damit sind diese Modelle aber prinzipiell gar nicht geeignet, beispielsweise zur Farbsegmentierung realer Szenen eingesetzt zu werden, da die Farbinformation, die etwa beim YUV–Modell in den Komponenten U und V steckt [14], ebenfalls eine Abhängigkeit von der Szenengeometrie sowie von der Intensitätsstärke der Beleuchtung aufweist. Auch das Modell I1I2I3, dessen einzelne Größen sehr gut dekorreliert sind, ist diesen

Effekten ausgesetzt und damit nicht brauchbar für Anwendungen in einer realen Umgebung. Diese Abhängigkeit von den beiden Effekten zeigt sich in [12] auch darin, daß die Autoren mit diesem Modell nicht in der Lage waren, der Oberfläche eines einfärbigen Zylinderteils ein einzelnes Segment zuzuordnen.

In die zweite Klasse fallen all jene Modelle, deren Transformationsgleichungen *nichtlinearer* Natur sind. Oft eingesetzte Modelle aus dieser Kategorie sind HSI, HSV, YT1T2 (L1 normalisierte Farben), YT3T4 (L2 normalisierte Farben), Lab, $U^*V^*W^*$ oder $r\Theta\varphi$ (RGB–Modell in Kugelkoordinaten). Am Beispiel des YT1T2–Modells [11] wird exemplarisch gezeigt, wie man den Skalierungstest auf ein Farbraummodell dieser Art anwendet. Bei diesem Modell enthält Y die Intensitätsinformation und T1 und T2 beinhalten die Farbinformation. Der Skalierungstest für dieses Modell ergibt, wobei $k = M_a(g, a)$ gesetzt wurde,

$$Y(x,y) = \frac{k(R_0(x,y) + G_0(x,y) + B_0(x,y))}{3} \neq \frac{R_0(x,y) + G_0(x,y) + B_0(x,y)}{3}$$

$$T1(x,y) = \frac{kR_0(x,y)}{kR_0(x,y) + kG_0(x,y) + kB_0(x,y)} = \frac{R_0(x,y)}{R_0(x,y) + G_0(x,y) + B_0(x,y)}$$

$$T2(x,y) = \frac{kG_0(x,y)}{kR_0(x,y) + kG_0(x,y) + kB_0(x,y)} = \frac{G_0(x,y)}{R_0(x,y) + G_0(x,y) + B_0(x,y)}$$

klarerweise eine Abhängigkeit der Intensitätskomponente von der Szenengeometrie und der Beleuchtung. Die beiden Farbkomponenten T1 und T2 sind jedoch invariant gegenüber diesen beiden Effekten. Dieser Test wurde für die gängigsten Farbraummodelle durchgeführt. Die Ergebnisse dieser Untersuchung sind in der Tabelle 1 zusammengefaßt. Es geht daraus hervor, daß viele Modelle keine invarianten Komponenten beinhalten und kein Modell ausschließlich invariante Anteile aufweist. Dies bedeutet aber beispielsweise für Segmentierungsverfahren, daß Invarianz gegenüber Szenengeometrie und Beleuchtungschwankungen und damit eine höhere Robustheit und Verläßlichkeit von Algorithmen erreicht werden kann, wenn nur die invarianten Komponenten (z.B. T1 und T2 vom YT1T2–Modell) herangezogen werden.

linear	Farbraummodell	RGB	YUV	YIQ	YCbCr	XYZ	I1I2I3	I1I2'I3'
	Referenz	[3]	[14]	[3]	[14]	[12]	[12]	[12]
	invariante Komp.	–	–	–	–	–	–	–
nicht-linear	Farbraummodell	HSI	HSV	YT1T2	YT3T4	$r\Theta\varphi$	Lab	$U^*V^*W^*$
	Referenz	[3]	[9]	[11]	[4]	[13]	[12]	[12]
	invariante Komp.	H,S	H,S	T1,T2	T3,T4	Θ,φ	–	–

Tabelle1. Gegenüberstellung von häufig benutzten Farbraummodellen und Angabe ihrer gegenüber Szenengeometrie und Schwankungen in der Beleuchtungsintensität invarianten Komponenten. Ein "–" bedeutet, daß keine invariante Komponente vorhanden ist.

2.4 Linearität der Kamerasignale, Gammakorrektur

Bei der Beziehung (4) handelt es sich um ein *lineares* Modell, womit der Zusammenhang zwischen der einfallenden Lichtintensität und dem vom Sensorkanal gelieferten Signal ebenfalls *linear* ist. Für ein Farbraummodell, von dem durch den Skalierungstest bekannt ist, daß es invariante Komponenten beinhaltet, bedeutet dies aber, daß die von einer Farbkamera gelieferten Signale R, G und B linear sein müssen. Dies ist allerdings in der Regel nicht der Fall, da alle heutzutage erhältlichen Kameras über eine eingebaute Gammakorrektur verfügen, die sich nur bei den wenigsten, sehr teuren Industriekameras abschalten läßt. Diese Korrektur dient zur Kompensation der nichtlinearen Kennlinie eines TV– oder Computerbildschirms, um für den Menschen höchstmögliche Bildqualität zu erreichen [14]. Diese nichtlineare Korrektur in der Kamera zerstört allerdings die lineare Beziehung in (4) und führt außerdem zu intensitätsabhängigen Farbverschiebungen, die sich besonders stark in dunklen Bereichen auswirken [7]. Auch sind die invarianten Größen aus Tabelle 1 nun variant bezüglich Szenengeometrie und Beleuchtungsschwankungen und verlieren damit durch die Gammakorrektur ihre Insensitivität gegenüber den hier genannten Effekten.

Aus den hier genannten Gründen ist es unbedingt notwendig, die Gammakorrektur an der Kamera zu deaktivieren oder die durch sie verursachte Nichtlinearität rechnerisch nachträglich zu eliminieren (siehe [7, 14]), um, basierend auf invarianten Farbraummodellen, verläßliche Verfahren realisieren zu können.

3 Experimentelle Verifikation

Um den hier gezeigten Einfluß der Szenengeometrie sowie von Schwankungen in der Beleuchtungsintensität auf die einzelnen Farbraummodelle deutlicher herauszustreichen, wird in diesem Abschnitt das Ergebnis einer Farbsegmentierung mit einem varianten und einem invarianten Farbraummodell präsentiert. Als Segmentierungsalgorithmus wurde ein Verfahren von [15], daß auf einer Hauptkomponentenanalyse einer vorgegebenen Initialisierungsregion des zu segmentierenden Körpers basiert, so erweitert, daß zur Segmentierung nur der chromatische Anteil des Modells (z.B. T1 und T2) verwendet wird. Der Vergleich dieses Verfahrens unter der Verwendung der Farbraummodelle YT1T2 (invariant) und YUV (variant) ist der Abbildung 1 zu entnehmen, wobei als Testkörper eine gelbe Kugel verwendet wurde, die zwar eine einheitliche Körperfarbe aber keine konstante Oberfächennormale und somit für jeden Körperpunkt eine andere Szenengeometrie besitzt. Bei der Verwendung des varianten YUV–Modells erkennt man ganz deutlich die Abhängigkeit von den beiden Effekten, da einerseits das Ergebnis der Segmentierung ringförmig ist und eher einer Isoirradianzlinie gleicht und andererseits das Ergebnis auch noch stark vom Initialisierungsort abhängig ist. Mit dem invarianten Modell YT1T2 wird hingegen fast die gesamte Kugel unabhängig von der Initialisierungsregion segmentiert, nur die Reflexion des Leuchtkörpers sowie die sehr dunklen Bereiche am Kugelrand wurden nicht erkannt. Letzteres ist damit zu erklären, daß bei der verwendeten Farbkamera die Gammakorrektur nicht abschaltbar war und erst nachträglich rechnerisch

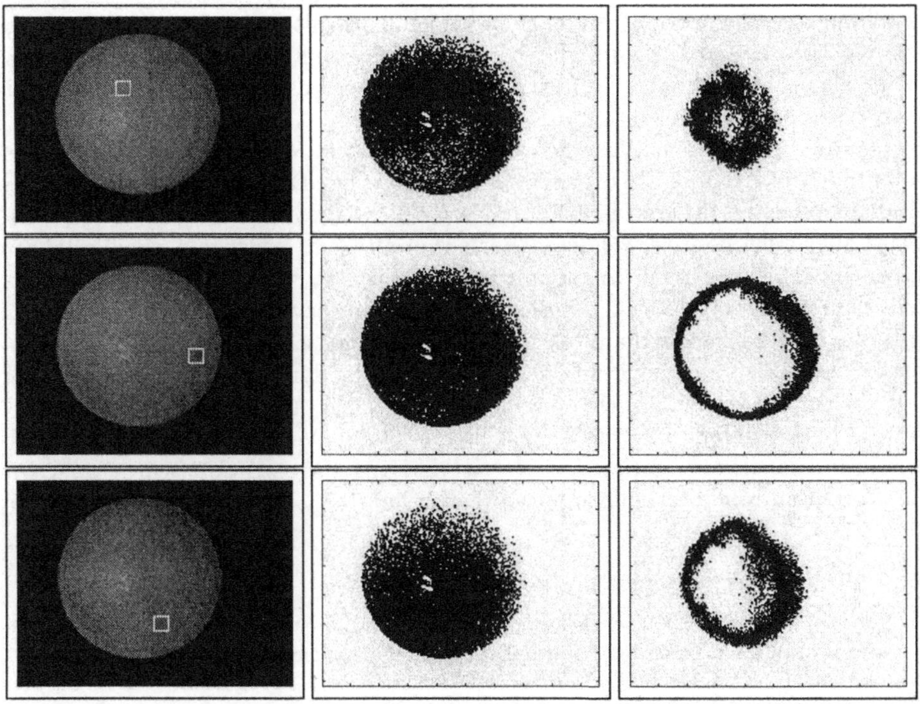

Abbildung 1. Vergleich eines Segmentierungsverfahrens unter Verwendung zweier verschiedener Modelle YT1T2 und YUV an einer realen, mit einer 3–Chip CCD–Farbkamera aufgenommenen, gelben Kugel. In der linken Spalte ist der Intensitätsanteil $I = \frac{R+G+B}{3}$ dargestellt (gammakorrigiert, $\gamma = 2.8$), der Initialisierungsbereich ist durch ein kleines Viereck markiert. Die mittere Spalte enthält das Ergebnis der Segmentierung für die drei verschiedenen Initialisierungspunkte basierend auf dem YT1T2–Modell und in der rechten Spalte ist das auf dem YUV–Modell basierende Ergebnis dargestellt. Ein segmentierter Punkt des Körpers wird dabei durch ein schwarzes Pixel angezeigt.

eliminiert werden konnte. Der dafür theoretisch korrekte Wert von $\gamma = 2.8$ für eine PAL–Kamera ist jedoch nach [14] für viele Farbkameras zu hoch und konnte dadurch nicht exakt korrigiert werden.

4 Zusammenfassung

Basierend auf dem hier vorgestellten Skalierungstest wurden die in der Bildverarbeitung am häufigsten eingesetzten Farbraummodelle analysiert. Es stellte sich heraus, daß alle linearen Modelle variant und damit abhängig von Szenengeometrie und Intensitätsschwankungen sind, womit deren Einsatz unter realen Umgebungsbedingungen als nicht zielführend erscheint. Es wurde gezeigt, daß nur einige nichtlineare Modelle über invariante Komponenten verfügen. Somit eignen sich nur jene Modelle, und nur deren invariante Komponenten, für eine robuste, von Szenengeometrie und Schwankungen in der Beleuchtungsintensität unabhängigen Auswertung von Farbbilddaten. Dies gilt unter der Voraussetzung, daß der Zusammenhang zwischen Lichtintensität und Sensorsignalen linear ist. Es ist also sehr wichtig, eine durch eine Gammakorrektur verursachte Nichtlinearität zu eliminieren.

Die hier getroffenen Aussagen helfen zwar, eine Vorauswahl bezüglich der potentiell einsetzbaren Farbraummodelle zu treffen. Die Entscheidung, welches Modell am besten für eine robuste und verläßliche Bildverarbeitung geeignet ist, hängt aber auch vom verwendeten Verfahren ab. Zum Austesten von Verfahren haben wir sehr gute Erfahrungen mit Kugeln gemacht (wie in Abschnitt 3 bereits kurz demonstriert), da sich bei diesem Testkörper sowohl lokale Eigenschaften (z.B. die Oberflächennormale) als auch globale Eigenschaften (z.B. Interreflexionen) einer realen Szene direkt in **einem** Testbild darstellen lassen. Basierend auf solchen Bildern läßt sich für einen entworfenen Algorithmus eruieren, wie gut die Leistungsfähigkeit der einzelnen Modelle in Verbindung mit diesem Algorithmus ist, also beispielsweise wie viel der Kugel segmentiert wurde oder wie das Verfahren mit Interreflexionen umgehen kann.

Literatur

1. Berwick, D., Lee, S.W.: A Chromaticity Space for Specularity, Illumination Color- and Illumination Pose Invariant 3-D Object Recognition. Proc. ICCV (1998)
2. Gevers, T., Smeulders, A.W.M.: Color Based Object Recognition. Lecture Notes in Computer Science, Springer **1310** (1997) 319–326
3. Gonzalez, R.C., Woods, R.E.: Digital Image Processing. Addison-Wesley Publishing Company, Reading, Massachusetts, 2. Edition (1993)
4. Healey, G.: Segmenting Images Using Normalized Color. IEEE Transactions on Systems, Man, and Cybernetics **27** (1) (1992) 64–73
5. Healey, G.: Recognizing Objects in Color Images. Proc. of the 10th Scandinavian Conference on Image Analysis **2** (1997) 823–829
6. Klette, R., Koschan, A., Schlüns, K.: Computer Vision. Vieweg Technik (1996)
7. Kubinger, W., Vincze, M., Ayromlou, M.: The role of Gamma correction in colour image processing. Proc. of the 9th European Signal Processing Conference (1998)
8. Lee, J.-H., Chang, B.-H., Kim, S.-D.: Comparison of colour transformations for image segmentation. Electronics Letters **30** (20) (1994) 1660–1661
9. Levkowitz, H., Herman, G.T.: GLHS: A Generalized Lightness, Hue, and Saturation Color Model. CVGIP: Graphical Models **55** (4) (1993) 271–285
10. Liu, J., Yang, Y.-H.: Multiresolution Color Image Segmentation. IEEE Trans. on PAMI **16** (7) (1994) 689–700
11. Nevatia, R.: A Color Edge Detector and Its Use in Scene Segmentation. IEEE Trans. on Systems, Man, and Cybernetics **7** (11) (1977) 820–826
12. Ohta, Y.-I., Kanade, T., Sakai, T.: Color Information for Region Segmentation. Computer Graphics and Image Processing **13** (1980) 222–241
13. Pla, F., Juste, F., Ferri, F., Vicens, M.: Colour segmentation based on a light reflection model to locate citrus fruits for robotic harvesting. Computers and Electronics in Agriculture **9** (1993) 53–70
14. Poynton, C.A.: A Technical Introduction to Digital Video. John Wiley & Sons (1996)
15. Rasmussen, C., Toyama, K., Hager, G.D.: Tracking Objects By Color Alone. Yale University, Tech. Report TR 1114, (1996)

Inkrementelle Verarbeitung von diskontinuierlichen Konstruktionen in gesprochener Sprache*

Susanne Kronenberg, Franz Kummert, Gerhard Sagerer

Technische Fakultät, AG Angewandte Informatik
Universität Bielefeld, Postfach 100131, 33501 Bielefeld
email: susanne@techfak.uni-bielefeld.de

Zusammenfassung Der vorgestellte Ansatz zur Verarbeitung von gesprochener Sprache ermöglicht, diskontinuierliche Konstruktionen inkrementell zu parsen. Charakteristisch für aufgabenorientierte Dialoge sind Konstruktionen, die von Produzent und Rezipient koordiniert produziert werden. Exemplarisch für solche Konstruktionen werden Extraversionen nach rechts untersucht. Bei diesen werden Konstituenten über Konstituentengrenzen oder über die Satzklammer nach rechts extravertiert. Der vorgestellte Ansatz, basierend auf einem LR(1)-Parser und einer Unifikationsgrammatik, bietet eine Möglichkeit, solche Konstruktionen inkrementell zu verarbeiten, indem der referentielle Bezug zwischen extravertierter Konstituente und Bezugskonstituenten nachträglich erstellt wird.

1 Einlcitung

Aufgabenorientierte Dialoge zeichnen sich durch syntaktische Konstruktionen aus, die von den beteiligten Agenten systematisch und situiert koordiniert werden. Ein integraler Bestandteil solcher Koordinationskonstruktionen sind Extraversionen nach rechts. Bei diesen werden Konstituenten, wie z.B. VP-Komplemente, VP-Adjunkte, Genitiv-NPs in der Rolle von NP-Modifikatoren und bestimmte NP-Argumente, über eine 'natürliche' Konstituentengrenze oder die Satzklammer nach rechts extravertiert. Abhängigkeiten, die zwischen extravertierter und nicht extravertierter Konstituente auftreten, beeinflussen entscheidend die Arbeitsweise des sprachverarbeitenden Systems. Besonderes Interesse gilt in diesem Zusammenhang den *crossed dependencies*, die auftreten, wenn die Bestandteile zweier diskontiniuierlicher Konstituenten kreuzweise ineinander verschachtelt auftreten. Im folgenden wird gezeigt, wie solche Abhängigkeiten inkrementell verarbeitet werden können.

Inkrementalität ist für die effiziente Verarbeitung gesprochener Sprache von zentraler Bedeutung. Teilergebnisse der vorhergehenden Verarbeitungseinheit können analysiert werden, bevor diese Verarbeitung abgeschlossen ist. Ebenso

* Die vorliegende Arbeit wurde im Rahmen des SFB 360 "Situierte Künstliche Kommunikatoren" von der Deutschen Forschungsgemeinschaft gefördert.

werden bereits Teilergebnisse an nachfolgende Komponenten weitergereicht. Eine inkrementelle Verarbeitungsweise von gesprochener Sprache reflektiert darüber hinaus, daß Koordiniationskonstruktionen konstituentenweise erstellt werden. Der Rezipient kann mit der syntaktischen Analyse einer Produzentenäußerung nicht warten, bis diese vollständig produziert ist. Der Produzent seinerseits formuliert stückweise Äußerungen, die dem Rezipienten die Möglichkeit bieten, die jeweils erreichte Folge von Teiläußerugen fortzusetzen. Für die Arbeitsweise des Parsers bedeutet dieses darüber hinaus, daß Turngrenzen nicht implizit als Äußerungsgrenzen betrachtet werden dürfen.

Der hier vorgestellte Parser für gesprochene Sprache basiert auf der Combinatorial Categorial Grammar (CCG) [AS82] und einem modifizierten LR(1)-Parser. Auf das Regelwerk der CCG wird nur soweit eingegangen, wie es für die Ableitung von Extraversionskonstruktionen notwendig ist, ansonsten werden paraphrasierte Phrasenstrukturregeln verwendet.

2 Linguistischer Hintergrund

Die Beispiele für Koordinationskonstruktionen und Extraversionen nach rechts, die im folgenden behandelt werden, stammen aus einem Korpus von Konstruktionsdialogen [SER94], der im Rahmen des Sonderforschungsbereiches *Situierte Künstliche Kommunikatoren* erstellt wurde. Generell treten in diesen Dialogen Extraversionen nach rechts weitaus häufiger auf, als solche nach links.

2.1 Koordinationskonstruktionen und Extraversionen nach Rechts

Koordinationskonstruktionen werden von Produzenten und Rezipienten gemeinsam produziert. In Beispiel 1 vervollständig der Rezipient *I* die vom Produzenten *K* angefangene Konstruktion.

Beispiel 1:
K: Wo kommt denn dann dieses Ganze
I: Gedöns mit dem Rad?
K: Ja.

Symbolisiert die Turngrenze die Äußerungsgrenze, so läßt sich die Konstituente *dieses Ganze* nur als abgeschlossene Nominalphrase interpretieren. Eine turnübergreifende Verarbeitungsweise gewährleistet, daß die Konstituente *das Ganze* als angefangene Nominalphrase erkannt wird, die durch die Konstituente *Gedöns* vervollständig wird. Die Möglichkeit, *Ganze* als nominalisiertes Adjektiv aufzufassen, wird damit verworfen.

Extraversionen nach rechts treten sowohl in Koordinationskonstruktionen (Beispiel 2), wie auch in separaten Äußerungen auf (Beispiel 3).[1]

[1] Die extravertierte Konstituente wird kursiv gedruckt.

Beispiel 2:
K: Also wenn die Rückseite jetzt zu mir zeigt, das Hinter, Hinterteil
I: Das Hinterteil zu dir.
K: *des Flugzeuges.*

Beispiel 3:
I: Also du hast doch eben einen Würfel angeschraubt *mit der roten Schraube.*

Beispiel 4 ist eine Koordinationskonstruktion, bei der das von *K* produzierte Nomen *Flügelschraube* kongruenzmäßig nicht zu der von *I* begonnenen Präpositionalphrase *mit einem orangenen* paßt. Da die von *K* gewählte Expansion keinen Ausnahmefall darstellt, muß ein Parser für gesprochene Sprache so robust sein, daß solche Äußerungen akzeptiert werden.

Beispiel 4:
I: Befestigst das unten mit einem orangenen
K: Flügelschraube.

2.2 "Crossed"- und "Nested"-Dependencies

Extraversionen nach rechts lassen sich in zwei Hauptkategorien unterteilen:

1. In *kontinuicrlichcn Extravcrsioncn* folgt dic cxtravertierte Konstituente direkt auf die Bezugskonstituente. In Beispiel 5 wird die Nominalphrase *eine von diesen Dreierstangen* um die Nominalphrase *diese scheibenförmigen Viecher* expandiert.

   Beispiel 5:
   I: Und nimmst noch eine von diesen Dreierstangen mit den drei Löchern, *also diese scheibenförmigen Viecher.*

2. In *diskontinuierlichen Extraversionen* treten zwischen extravertierter Konstituente und Bezugskonstituente noch weitere Konstituenten auf, wie in Beispiel 3 das infinite Verb.

Kontinuierliche Extraversionen stellen einen Spezialfall der diskontinuierlichen dar. Die Ableitung von kontinuierlichen Extraversionen ist somit in dem Ableitungsschema für diskontinuierliche enthalten. Da diskontinuierliche Extraversionen durch die auftretenden *long distance dependencies* syntaktisch und parsertechnisch die interessanteren Konstruktionen sind, werden im folgenden hauptsächlich diese behandelt.

Beispiel 6: (Modifiziertes Beispiel 3)
I: Du schraubst den Würfel an *mit der Schraube.*

Bedingt durch die extravertierte Stellung der Präpositionalphrase *mit der Schraube* ergeben sich in Beispiel 6 die folgenden Abhängigkeitsschemata:

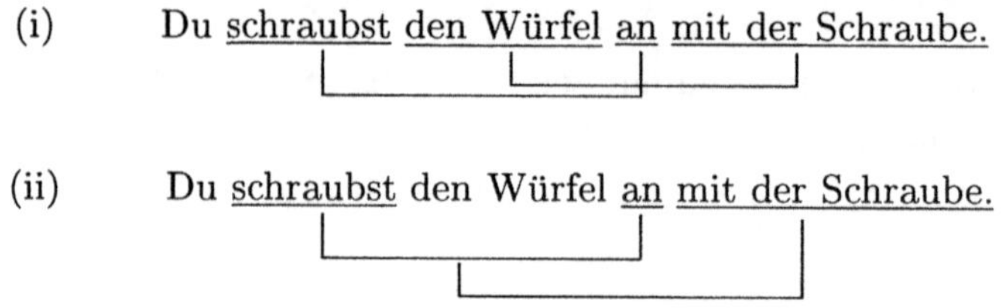

Abbildung1. Abhängiggkeitsschemata in extravertierten Konstruktionen

Beispiel 7:
I: Du schraubst den Würfel mit der Schraube an.

Betrachtet man eine nicht extravertierte Version des Satzes (Beispiel 7), so treten die folgenden Abhängigkeiten auf:

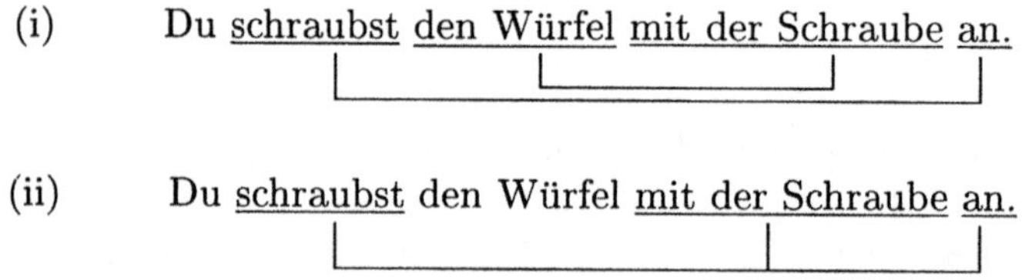

Abbildung2. Abhängigkeitsschemata in nicht extravertierten Konstruktionen

Der Parser muß somit in der Lage sein, *crossed dependencies* wie in Abbildung 1.i und *nested dependencies* wie in Abblidung 2.i zu verarbeiten. In Hinblick auf die inkrementelle Verarbeitungsstrategie bedeutet dies, daß bei Konstruktionen, bei denen *crossed dependencies* auftreten, konstituentenweise geparst und der referentielle Bezug nachträglich erstellt wird.

3 Der Parser

Der Parsingmechanismus basiert auf einem LR(1)-Parser. Syntaktische Mehrdeutigkeiten wie in Beispiel 1 werden konkurrierend weiterverarbeitet [Tom86]. Abhängigkeiten, wie sie in Beispiel 7 auftreten, äußern sich in einem shift-reduce-Konflikt [Per85]. Beide Parsingmöglichkeiten werden ebenfalls konkurrierend weiterverfolgt, um alle Lesarten zu erhalten. Für ein robustes Parsen von gesprochener Sprache ist es notwendig, daß nicht kongruente Konstruktionen wie in Beispiel 4 akzepiert werden. Die Grammatik, die Combinatorial Categorial Grammar, die dem Parser zugrunde liegt, ist lexikalisiert. Die meiste Information, die standardmäßig durch Phrasenstrukturregeln kodiert ist, wird somit im Lexikon durch Merkmalstrukturen repräsentiert. Eine reduce-Aktion des Parsers impliziert die Unifikation der zugehörigen Merkmalstrukturen. Nicht kongruente Merkmalstrukturen lassen sich unifizieren, indem deren Unifikation bewertet stattfindet. Um Extraversionen nach rechts, bei denen *crossed dependencies* auftreten, inkrementell parsen zu können, wird der LR(1)-Parser um zwei zusätzliche Regeln ((3.2) und (3.3)) erweitert. Diese zusätzlichen Regeln ermöglichen es, nachträglich auf Konstituenten zuzugreifen, obwohl diese bereits durch eine reduce-Aktion verarbeitet worden sind. Bei LR(1)-Parsern, die nur auf Grundlage von shift- und reduce-Aktionen operieren, besteht keine Möglichkeit, daß Konstituenten, wie die Nominalphrase *der Würfel* in Abbildung 1.i, die durch die Verbklammer eingeschlossen sind, durch nachfolgende Konstituenten modifiziert werden.

3.1 Ableitungsgrammatik

Die in Abbildung 3 angegebenen Grammatikregeln dienen nur zur Veranschaulichung der Arbeitsweise des Parsers bei Extraversionen nach rechts und erheben bei weitem nicht den Anspruch auf Vollständigkeit. [2] [3]

1	S_T	$\rightarrow$	NP S_I
2	$S_{T,I}$	$\rightarrow$	$S_{T,I}$ NP
3	$S_{T,I}$	$\rightarrow$	$S_{T,I}$ TPräfix
4	NP	$\rightarrow$	NP PP
5	$S_{T,I}$	$\rightarrow$	$S_{T,I}$ PP

Abbildung3. Grammatik zur Ableitung von Extraversionen nach rechts

Die Symbole repräsentieren die zugehörigen Merkmalstrukturen. Terminale Symbole werden durch die zugehörigen Merkmalstrukturen im Lexikon instantan zu nicht terminalen Symbolen reduziert. Die Nomenklatur spiegelt die Schreibweise der CCG wider. S_T und S_I symbolisieren Verben. S_I steht für ein Verb in erster Position des Satzes. Ist das Vorfeld besetzt, ändert sich während der Ableitung der Index von I zu T, wodurch keine weitere topikalisierte Konstituente akzeptiert wird. Dies ermöglicht eine inkrementelle Ableitung von links nach rechts. Haupt-, Imperativ- und Interrogativsätze können gleichermaßen geparst werden. Akzeptiert wird bei leerem Inputstring und leerer Subkategorisierungsliste.

3.2 Dekomposition

Bei der inkrementellen Verarbeitung von Konstruktionen, bei denen *crossed Dependencies* auftreten wie in Abbildung 1.i, ist die Bezugskonstituente *den Würfel*

---

[2] Aus Platzgründen wird davon ausgegangen, daß die Inputsymbole schon zu Konstituenten reduziert sind.

[3] Durch die Merkmalstrukturen wird gewährleistet, daß sich der trennbare Präfix TPräfix nur auf das Verb bezieht und nur ein trennbarer Präfix gebunden werden darf.

bereits geparst und mit dem Verb *schraubst an* zu einer Verbalphrase reduziert, bevor die extravertierte Präpositionalphrase *mit der Schraube* auf den Stack "geshiftet" wird. Somit ist es nicht mehr möglich, einen referentiellen Bezug zwischen der Nominalphrase und der Präpositionalphrase herzustellen. Die *Parametric Neutrality* der CCG ermöglicht die Dekomposition bereits verarbeiteter Konstituenten. Analog zur *Decompose*-Regel der CCG [Ste90] wird der Parser um eine Dekompositionsregel für Extraverionskonstruktionen erweitert:

Dekompositionsregel des Parsers:
$$X \quad Y \quad \Longrightarrow_{Dekomposition} \quad X_1 \quad X_2 \quad Y \quad , \text{mit } X = S.$$

Diese besagt, daß eine bereits geparste Sequenz von Inputsymbolen wieder in die beteiligten Konstituenten zerlegt werden darf, falls diese schon zu einem Satz reduziert wurde, der Inputstring aber noch nicht vollständig abgearbeitet ist.

An dem Beispielsatz *"Du schraubst den Würfel an mit der Schraube."* wird die Wirkungsweise dieser Regel veranschaulicht. Bis zum trennbaren Präfix *an* wird der Satz über die shift- und reduce-Regeln des LR(1)-Parsers abgeleitet (Abb. 4).

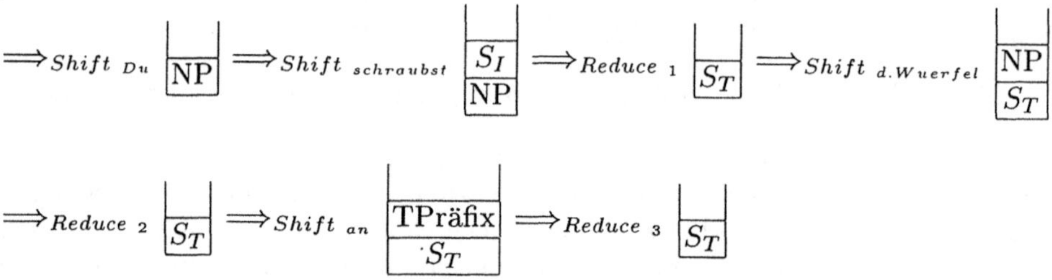

Abbildung4. Ableitung des ersten Teils des Satzes

Über die Grammatikregel 5 kann direkt weiter geparst werden, um die Lesart zu erhalten, bei der sich die Präpositionalphrase auf das Verb bezieht (Abb. 5).

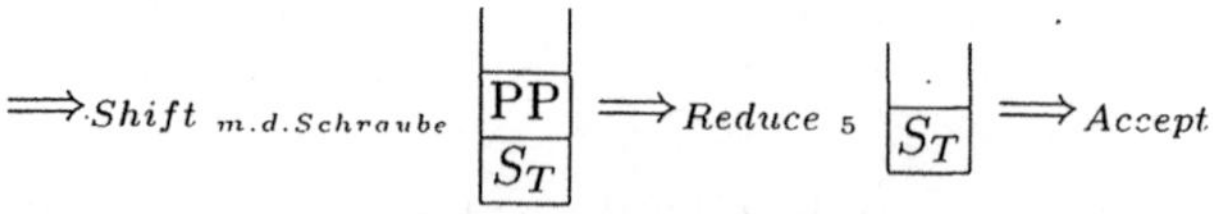

Abbildung5. Ableitung der Instrumentallesart

In Abbildung 4 sind die bereits geparsten Inputsymbole zu einem Satz S_T reduziert, der Inputstring ist aber noch nicht vollständig abgearbeitet. Dies bedeutet für den Parsingvorgang, daß die Dekompositionsregel angewandt wird, um alle Lesarten ableiten zu können. Die Präpositionalphrase *mit der Schraube* wird auf den Stack "geshiftet" und der bereits geparste Input dekomponiert (Abb. 6).

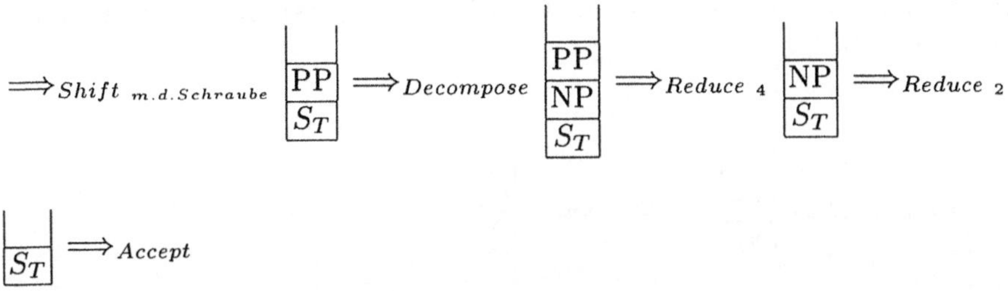

Abbildung6. Ableitung der Objektlesart

3.3 Koordination

In Sätzen wie Beispiel 5 sind Konstituenten extravertiert, die vorher bereits syntaktisch realisiert sind. Um solche Konstruktionen ableiten zu können, wird analog zu der *Coordination*-Regel der CCG [Ste90], eine Koordinationsregel für extravertierte Konstituenten benötigt:

Koordinationsregel des Parsers:
$X_1 \quad X_2 \quad \Longrightarrow_{Koordination} \quad X$, mit X_1, X_2= NP.

Diese Regel besagt, daß benachbarte Konstituenten desselben Typs, hier eingeschränkt auf Nominalphrasen, zu einer Konstituente zusammengefaßt werden dürfen. Diese Regel muß als eigenständige Regel realisiert werden, die nur im Zusammenhang mit der Dekompositionsregel angewandt werden darf, um eine Übergeneralisierung zu vermeiden. Würde eine entsprechende Regel, wie z.B. NP $\Longrightarrow$ NP NP, Bestandteil der Grammatik sein, so wäre die Ableitung von inkorrekten Sätzen wie " *Die Frau der Mann sieht das Kind*" möglich.

An dem Satz "*Ein Flugzeug kommt heraus, ein Propellerflugzeug.*" wird diese Regel erläutert. Bis zur rechten Satzklammer wird der Satz über die shift- und recuce-Regeln abgeleitet (Abb. 7).

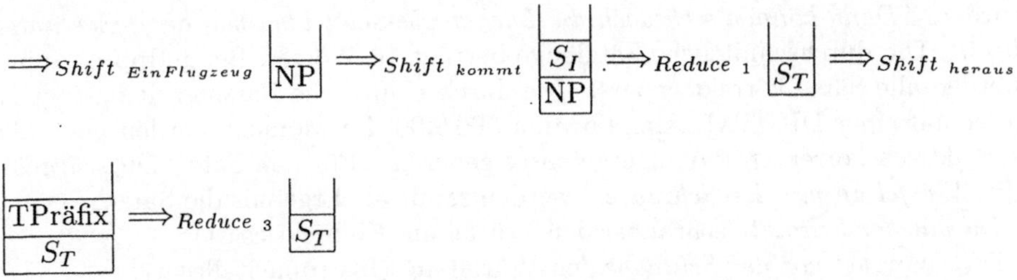

Abbildung7. Ableitung des ersten Teil des Satzes

Damit die Nominalphrase *ein Propellerflugzeug* mit der Nominalphrase *ein*

Flugzeug koordiniert werden kann, muß der bereits geparste Input dekomponiert werden (Abb. 8).

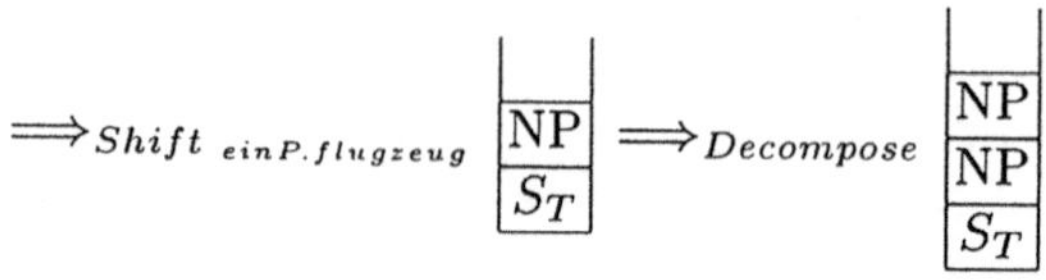

Abbildung8. Anwendung der Dekompositionsregel

Durch die Dekomposition sind die beiden Nominalphrasen benachbart und können koordiniert werden (Abb. 9).

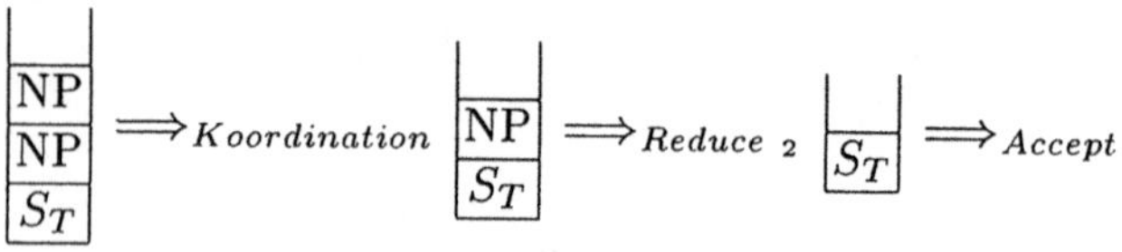

Abbildung9. Anwendung der Koordiantionsregel

4 Ergebnisse

Der vorgestellte Parser wurde an 25 Beispielsätzen [SK98] getestet. Innerhalb dieser Sätze tritt jeweils eine der Rechtsextraversionsarten, wie sie in Abschnitt (3.2) und (3.3) vorgestellt wurden, auf. Diese Sätze stammen aus einem Korpus von Konstruktionsdialogen [SER94]. Teilweise wurden diese Sätze um Reparaturen und Einschübe vereinfacht, da solche Konstruktionen nicht Gegenstand des hier vorgestellten Ansatzes sind. Das bedeutet, daß der Satz *"Dann können sich doch äh die Löcher überschneiden von der von dem, was unten liegt."* vereinfacht wird zu *"Dann können sich doch die Löcher überschneiden von dem, was unten liegt."*. Die durchschnittliche Sätzlänge beträgt 10 Wörter. Bei getippten Input werden alle Sätze korrekt geparst. Die durchschnittliche Parsingzeit beträgt 23 msec auf einer DIGITAL AlphaStation 500/400. Im Moment werden noch alle syntaktisch korrekten Parsingergebnisse generiert. Für den Satz *"Du schraubst den Würfel an mit der Schraube."* werden somit als Ergebnis die Sätze:
- *Du mit der Schraube* schraubst den Würfel an. (Subjektlesart)
- Du *schraubst mit der Schraube* den Würfel an. (Instrumentallesart)
- Du schraubst *den Würfel mit der Schraube* an. (Objektlesart)

generiert. Durch weitere Korpusanalysen sollen Kriterien gefunden werden, um die verschiedenen Lesarten so zu bewerten, daß nur noch die kontextuell sinnvollste erzeugt wird. Als Kontext sollen nicht nur Sprachdaten, sondern ebenfalls entsprechende Bilddaten verwendet werden. Sätze, bei denen keine Rechtsextraversionen auftreten, werden weiterhin korrekt geparst, so daß Übergeneralisierungen nicht auftreten.

5 Zusammenfassung

Extraversionen nach rechts sind typische Konstruktionen gesprochener Sprache. Der vorgestellte Parsingmechanismus bietet eine Möglichkeit, die dadurch induzierten *long-distance-dependencies* inkrementell zu verarbeiten. Für eine effiziente Verarbeitung von gesprochener Sprache ist eine inkrementelle Vorgehensweise von entscheidener Bedeutung. Inkrementalität reflektiert zum einen die gemeinsame Vorgehensweise von Produzent und Rezipient bei der Erstellung von Koordinationskonstruktionen, zum anderen kann die Interpretation der Äußerung beginnen, sobald die ersten Teilergebnisse von der spracherkennenden Komponente geliefert werden. Durch die Verwendung einer Unifikationsgrammatik, bei der die Unifikation Bewertungskriterien unterworfen ist, lassen sich nichtkongruente Strukturen parsen. Konstruktionen, die gemeinsam vom Produzenten und Rezipienten koordiniert werden, können dadurch geparst werden, daß Turngrenzen keine Äußerungsgrenzen implizieren. Solange die Inputsymbole nicht vollständig verarbeitet sind, werden diese soweit wie möglich mit Hilfe der Grammatik, der Dekompositions- und Koordinationsregel interpretiert. Weitere Analysen des Korpus sollen zeigen, wie viele Extraversionen nach rechts innerhalb einer Äußerung zulässig oder akzeptabel sind, um somit die Äußerungsgrenze bestimmen zu können.

References

[AS82] Anthony E. Ades and Mark J. Steedman. On the order of words. *Linguistics and Philosophy*, 4:517–558, 1982.

[Per85] Fernando Pereira. A new characterisation of attachment preferences. In D. Dowty, L. Kartunnen, and A. Zwicky, editors, *Natural Language Parsing: Psychological, Computational and Theoretical Perspectives*, ACL Studies in Natural Language Processing, pages 307–319. Cambridge University Press, 1985.

[SER94] G. Sagerer, H.-J. Eikmeyer, and G. Rickheit. "Wir bauen jetzt also ein Flugzeug ...": Konstruieren im Dialog - Arbeitsmaterialien. Technical report, Universität Bielefeld, 1994.

[SK98] K. Skuplik and W. Kindt. Ausklammerungskonstruktionen: Definition und empirische Untersuchung. Technical report, Universität Bielefeld, 1998.

[Ste90] Mark J. Steedman. Gapping as constituent coordination. *Linguistics and Philosophy*, 13:403–439, 1990.

[Tom86] M. Tomita. *Efficient Parsing for Natural Language*. Kluwer Academic Publishers, Hingham, Ma., 1986.

Bildbereichsbasierte Verfolgung von Straßenfahrzeugen durch adaptive Schätzung und Segmentierung von Optischen-Fluß-Feldern

H.-H. Nagel[1,2] und A. Gehrke[1]

[1] Institut für Algorithmen und Kognitive Systeme
Fakultät für Informatik der Universität Karlsruhe (TH)
Am Fasanengarten 5, D–76128 Karlsruhe, Germany
[2] Fraunhofer-Institut für Informations- und Datenverarbeitung (IITB)
Fraunhoferstr. 1, D–76131 Karlsruhe, Germany
hhn@iitb.fhg.de, angela@ira.uka.de

Zusammenfassung. Die Weiterentwicklung der Segmentierung eines adaptiv geschätzten Optischen-Fluß-Vektorfeldes unterstützt die Zuordnung von Segmenten zwischen aufeinander folgenden Aufnahmen. So lassen sich selbst kleine Abbilder von Straßenfahrzeugen in sogar leicht gestörten Bildfolgen normaler Konsum-Videokameras auch bei partieller Verdeckung durch Masten und Laub verfolgen. Die systematische Analyse der über mehrere hundert Aufnahmen hinweg verfolgten Objektmasken vertiefte das Verständnis lokaler Grauwertstrukturen, wodurch sich die Segmentierungs- und Schätzansätze bei Realwelt-Bildfolgen weiter verbessern ließen.

1 Einführung

Bei einer sichtsystemgestützten Erfassung von Vorgängen im Freien müssen oft mehrere Objekte, die sich an verschiedenen Stellen einer abgebildeten Szene in verschiedenen Richtungen bewegen, gleichzeitig erfaßt und verfolgt werden. Der Einsatz einer stationären Videokamera mit einem großen Gesichtsfeld wird in solchen Fällen sinnvoll, wenn es gelingt, die dann *zwangsläufig kleiner abgebildeten* bewegten *Einzelobjekte* dennoch zuverlässig zu detektieren und zu verfolgen. Man denke beispielsweise an die Aufgabe, Wechselwirkungen zwischen dem Verhalten verschiedener Straßenfahrzeuge an einer komplexen Kreuzung zu untersuchen. Die mit der Kleinheit der Abbildung zu verfolgender Objekte verbundenen Schwierigkeiten verstärken sich, wenn die Objekte zeitweilig partiell oder vollständig verdeckt werden oder wenn sie ihre Fahrtrichtung ändern und dadurch die Abbildung der Objekte sowie gegebenenfalls die ihres Schattens auf der Fahrbahn nicht mehr als zeitlich konstant angenommen werden darf.

Durch Rückgriff auf a-priori Wissen beispielsweise über die Struktur der abgebildeten Szene, der abgebildeten Objekte oder ihrer Bewegungen versucht man, solche Probleme in einer *modellgestützten* Bildfolgenauswertung in den Griff zu bekommen. Eine neuere Untersuchung [1] zeigte, daß noch verbleibende

Probleme bei einer solchen Vorgehensweise vorwiegend auf der *Initialisierung* durch Segmentierung eines Optischen-Fluß-Feldes beruhen.

Der Übergang auf einen *adaptiven* Schätzprozeß verbesserte die Ermittlung und Segmentierung von Optischen-Fluß-Feldern [6]. Darauf aufbauend berichten wir hier über erste Ergebnisse einer Weiterentwicklung dieses Ansatzes zu einer *rein bildbereichsbasierten* Verfolgung bewegter Objekte in Video-Bildfolgen.

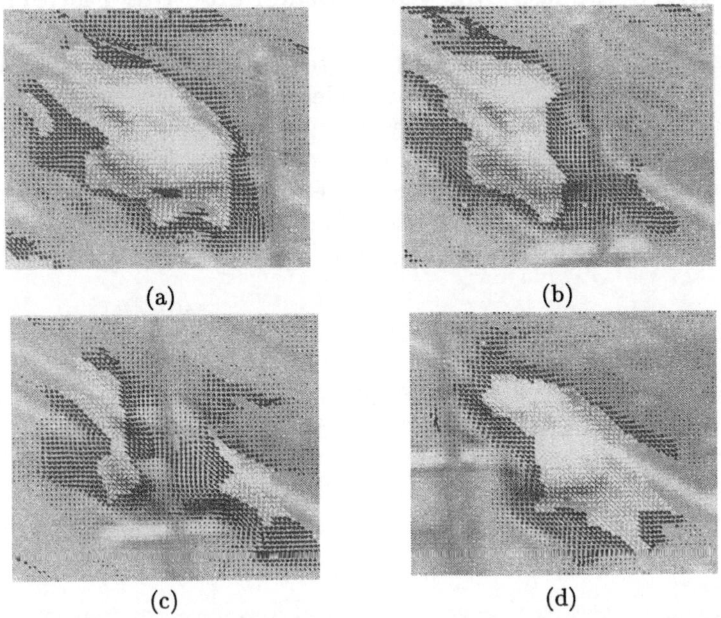

Abb. 1. Ausschnitte aus der Bildfolge 'Durlacher-Tor-neu', denen adaptiv geschätzte OF-Vektoren überlagert wurden. Die Pixelpositionen $\mathbf{x}$ werden aufgrund von $GLST(\mathbf{x})$ in verschiedene Kategorien eingeteilt: Nimmt der kleinste Eigenwert von GLST einen Wert an, der zu groß ist, um an dieser Stelle von einem homogenen OF-Feld auszugehen, so wird dies als Hinweis auf eine *OF_Diskontinuität (OF_D)* angesehen. Der an dieser Stelle geschätzte OF-Vektor wird schwarz gezeichnet. In Bereichen mit einer lokalen Grauwertstruktur, die zur zuverlässigen OF-Schätzung zu schwach ist, werden keine Vektoren angezeigt. Die in den übrigen Bereichen geschätzten OF-Vektoren werden weiß gezeichnet (*Regulärer_OF (R_OF)*). Der Ausschnitt aus Halbbild 398 (a, oben links) zeigt die Charakterisierung der OF-Vektoren kurz vor der Verdeckung des Fahrzeugs durch den Mast. Die folgenden Ausschnitte aus den Halbbildern 410 (b), 422 (c) und 440 (d) lassen deutlich werden, wie das Abbild des stationären Mastes als schwarze Diskontinuitätsfront durch das Abbild des Fahrzeugs wandert. Eine detailliertere Beschreibung der einzelnen Kategorien findet sich in [6]. Insbesondere die Stellen, an denen die lokale Grauwertstruktur zuwenig Information zur Schätzung des OF enthält, lassen sich weiter unterteilen.

Für eine ausführlichere Diskussion der Literatur verweisen wir aus Platzgründen auf [1,6]. Jähne und Mitarbeiter [3] ermitteln den Optischen-Fluß ähnlich wie wir als Eigenvektor eines von der lokalen Grauwertstruktur abhängigen Tensors, allerdings ohne einen – nach unseren Erfahrungen wichtigen – Adaptionsschritt. Die neueste Spezialliteratur (siehe z. B. [2]) über adaptive Verfahren im Bildbereich gibt keine Hinweise auf deren Nutzung zur Schätzung und Segmentierung von Optischen-Fluß-Feldern bei Realwelt-Bildfolgen.

2 Adaptive Schätzung von Optischen-Fluß-Feldern

Wir behandeln Optischen-Fluß (OF) – die scheinbare Verschiebungsgeschwindigkeit von Grauwertstrukturen in der Bildebene – als einen *dreidimensionalen* Vektor $\mathbf{u} = (u_1, u_2, 1)^T$ in einem (x,y,t)-Raum, der durch die Bildebenenkoordinaten (x, y) und die Zeitkoordinate t aufgespannt wird. Ein Ortsvektor in diesem Raum wird durch $\mathbf{x} = (x, y, t)^T$ gegeben. Grundlage eines Schätzverfahrens für den OF bildet die sogenannte 'Optische-Fluß-Beschränkungs-Gleichung' ('Optic Flow Constraint Equation (OFCE)', siehe [4]):

$$(g_x, g_y, g_t)\,\mathbf{u} = (\nabla g)^T \mathbf{u} = 0 \qquad \text{mit} \qquad g_x = \frac{\partial g(x,y,t)}{\partial x}, \quad \text{etc.} \quad . \tag{1}$$

Da diese Gleichung nur *eine* Beschränkung für die *zwei* unbekannten Komponenten u_1 und u_2 des OF darstellt, wird zusätzlich gefordert, daß das über eine bestimmte Umgebung des Punktes $\mathbf{x}$ gemittelte Quadrat der OFCE durch geeignete Wahl von $\mathbf{u}$ zu minimieren ist:

$$\overline{|(\nabla g)^T \mathbf{u}|^2} \stackrel{!}{=} \min_{\mathbf{u}} \; . \tag{2}$$

Um die formal immer noch zugelassene triviale Lösung $\mathbf{u} = \mathbf{0}$ auszuschließen, formen wir das Minimierungsproblem unter Einführung eines dreikomponentigen Vektors $\mathbf{e} = (e_1,\ e_2,\ e_3)^T$ mit der Randbedingung $||\mathbf{e}|| = 1$ folgendermaßen um:

$$\overline{|(\nabla g)^T \mathbf{e}|^2} + \lambda(1 - \mathbf{e}^T\mathbf{e}) = \mathbf{e}^T\ \overline{\nabla g(\mathbf{x})\,(\nabla g(\mathbf{x}))^T}\ \mathbf{e} + \lambda(1 - \mathbf{e}^T\mathbf{e}) \stackrel{!}{=} \min_{\mathbf{e}} . \tag{3}$$

Als Lösung erhalten wir den Eigenvektor $\mathbf{e}_{min} = (e_{min,1},\ e_{min,2},\ e_{min,3})^T$ zum kleinsten Eigenwert λ_{min} der Matrix $\overline{\nabla g(\mathbf{x})\,(\nabla g(\mathbf{x}))^T}$. Die Berechnung von $\nabla g(\mathbf{x})$ realisieren wir durch Faltung der gemessenen digitalen Grauwertverteilung g($\mathbf{x}$) mit partiellen Ableitungen der trivariaten Gauß-Funktion

$$G(\mathbf{x}) = \frac{1}{(2\pi)^{3/2}\sqrt{|\Sigma|}}\, e^{-\frac{1}{2}\mathbf{x}^T \Sigma^{-1} \mathbf{x}}, \quad \text{wobei zunächst } \Sigma_{init} = \begin{pmatrix} \sigma_x^2 & 0 & 0 \\ 0 & \sigma_y^2 & 0 \\ 0 & 0 & \sigma_t^2 \end{pmatrix} \text{ ist} . \tag{4}$$

Den 'Grayvalue-Local-Structure-Tensor (GLST)' (vgl. [6]) berechnen wir als

$$GLST(\mathbf{x}) = \int_{-\infty}^{+\infty} d\boldsymbol{\xi} \frac{\nabla g(\boldsymbol{\xi})\,(\nabla g(\boldsymbol{\xi}))^T}{(2\pi)^{3/2}\sqrt{|2 \cdot \Sigma|}} \cdot e^{-\frac{1}{2}(\boldsymbol{\xi} - \boldsymbol{x})^T (2\Sigma)^{-1} (\boldsymbol{\xi} - \boldsymbol{x})} . \tag{5}$$

Dabei sind wir [5] gefolgt und haben die Kovarianzmatrix 2Σ, die die *Ausdehnung des Mittelungsbereiches* durch eine dreidimensionale Gauß-Glocke festlegt, *doppelt so groß* gewählt wie diejenige, die zur *Glättung* im Zusammenhang mit der Bestimmung der *partiellen Ableitungen* verwendet wird. Aus Platzgründen verweisen wir auf [6] für eine ausführlichere Begründung, inwiefern der GLST eine gute Charakterisierung der lokalen Grauwertstruktur darstellt.

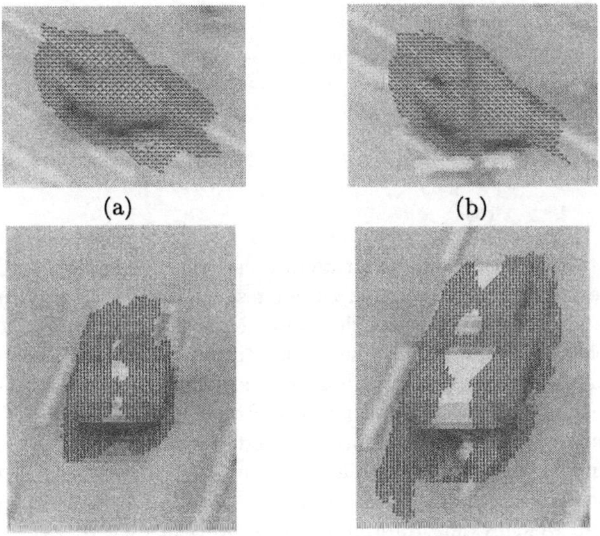

(a) (b) (c) (d)

Abb. 2. OIC-Masken für das in Abbildung 1 gezeigte Fahrzeug, und zwar für die Halbbilder 385 (a), 422 (b), 680 (c) und 790 (d). Man beachte, daß die OIC-Maske während der Verdeckung – siehe Teilbild (b) rechts oben – nicht wesentlich gestört ist. In den unteren beiden Teilbildern weist die OIC-Maske Löcher auf. Aufgrund der homogenen Grauwertverteilung im Inneren der Fahrzeugabbildung konnten in diesen Bereichen keine zuverlässigen OF-Vektoren geschätzt werden. Während der gezeigten Halbbilder ändert die Fahrzeugabbildung sowohl ihre Form als auch ihre Größe.

Wir können nun sowohl die Schätzung des Grauwertgradienten als auch die zur Mittelung herangezogene Umgebung des Punktes $\mathbf{x}$ an die lokal vorherrschenden Grauwertverhältnisse *adaptieren*, indem wir GLST($\mathbf{x}$) an Stelle einer konstanten, d. h. von $\mathbf{x}$ *unabhängigen*, inversen Kovarianzmatrix Σ^{-1} verwenden. Im Vergleich zu [6] vereinfachen wir die Begrenzung der Ausdehnung des GLST, indem wir dessen Eigenwerte $\lambda_i(\mathbf{x})$ durch eine vom normierten Eigenwert $\lambda_i^w(\mathbf{x})$ bestimmte konvexe Kombination aus kleinstem ($\sigma^2_{minsize}$) und größtem ($\sigma^2_{maxsize}$) zugelassenem Wert beschränken:

$$\alpha_i = \lambda_i^w(\mathbf{x}) \cdot \sigma^2_{minsize} + \left(1 - \lambda_i^w(\mathbf{x})\right) \cdot \sigma^2_{maxsize} \quad mit \quad \lambda_i^w(\mathbf{x}) = \frac{\lambda_i(\mathbf{x})}{Spur(GLST(\mathbf{x}))} \ , \qquad (6)$$

$i \in \{1,2,3\}$, vorausgesetzt $Spur(GLST(\mathbf{x})) \neq 0$. Bei den hier dokumentierten Experimenten verwendeten wir die Werte $\sigma^2_{minsize} = 0.5$ and $\sigma^2_{maxsize} = 4.0$ oder 6.0.

Der gesuchte OF-Vektor ergibt sich dann im Prinzip – bis auf eine gelegentlich erforderlich werdende differenziertere Vorgehensweise (siehe [6]) – durch Projektion des Eigenvektors $\mathbf{e}_{min}$ zum kleinsten Eigenwert λ_{min} des GLST in die Bildebene.

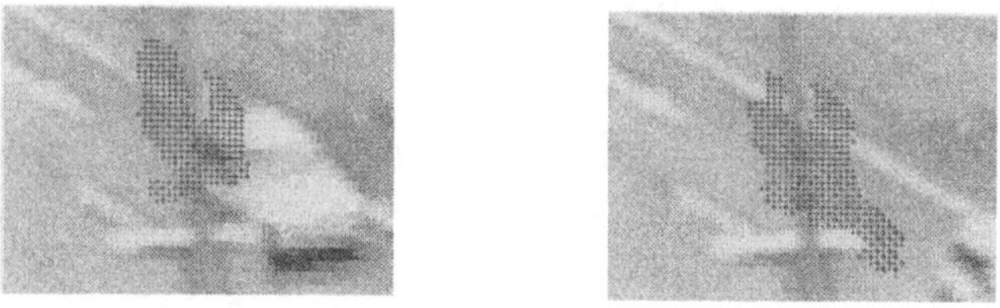

(a) (b)

Abb. 3. Die stationären Diskontinuitätshinweise, die innerhalb der OIC-Maske eines bewegten Objektes auftreten, werden zu einer sogenannten *Verdeckungsmaske* zusammengefaßt, die als Abbild einer das sich bewegende Objekt verdeckenden Szenenkomponente angesehen werden kann. Die Diskontinuitätshinweise erstrecken sich wegen der endlichen Ausdehnung der zur Charakterisierung einer lokalen Grauwertstruktur erforderlichen Umgebung über eine größere Breite als das Abbild der verdeckenden Szenenkomponente. Das linke Teilbild (a) zeigt die für das 435. Halbbild, das rechte Teilbild (b) die für das 460. Halbbild bestimmte Verdeckungsmaske. Die Diskontinuitätshinweise im rechten unteren Bereich von Teilbild (b) entstehen durch ein breites Wegweiserschild, welches den unteren Teil des Fahrzeuges verdeckt.

3 Bildbereichsbasierte Verfolgung

Abbildung 1 läßt deutlich werden, daß die Adaption der OF-Bestimmung an die lokale Grauwertstruktur deren Charakterisierung in intuitiv sehr einleuchtender Weise ermöglicht. Der in [6] behandelte Algorithmus zur Charakterisierung von OF-Vektoren, der zur Erstellung dieser Abbildung – mit zwischenzeitlich erarbeiteten Verbesserungen – verwendet wurde, liefert nicht nur eine brauchbare Grundlage zur rein datengetriebenen Segmentierung von OF-Feldern, sondern kann auch als Ausgangspunkt für einen rein im Bildbereich durchführbaren *Verfolgungsansatz* verwendet werden.

Die *Initialisierung* einer Verfolgung wird *zur Zeit noch indirekt interaktiv* durch Auswahl eines Bildausschnittes vorgenommen, in dem nur eine 4-zusammenhängende Komponente von R_OF-Vektoren auftritt, ein sog. *R_OF-Flecken.*

Die Lage eines R_OF-Fleckens wird entsprechend seinem mittleren Optischen-Fluß $\overline{OF}$ in das nächste Halbbild prädiziert, um einen Toleranzrand erweitert und als Bereich zur Berechnung von $GLST_{aktuelles-Bild}(\mathbf{x})$ herangezogen. Sofern ein $GLST_{vorangehendes-Bild}(\mathbf{x})$ an der Pixelposition $\mathbf{x}$ zur Verfügung steht, wird dieser für die Berechnung des neuen $GLST_{aktuelles-Bild}(\mathbf{x})$ verwendet, andernfalls wird dafür auf das Inverse der konstanten Kovarianzmatrix zurückgegriffen, die auch bei der Initialisierung eingesetzt worden ist.

Um alle aus $GLST_{aktuelles-Bild}(\mathbf{x})$ ermittelten R_OF-Flecken, die mit dem prädizierten R_OF-Flecken des vorherigen Halbbildes eine nichtleere Teilmenge bilden und deren mittlerer OF vom $\overline{OF}$ des prädizierten R_OF-Fleckens um nicht mehr als einen Schwellenwert differiert, wird eine konvexe Hülle gelegt. Die Größe $\overline{OF}_{aktuelles-Bild}$ berechnet sich nun aus den OF-Vektoren aller in dieser konvexen Hülle liegenden R_OF-Flecken.

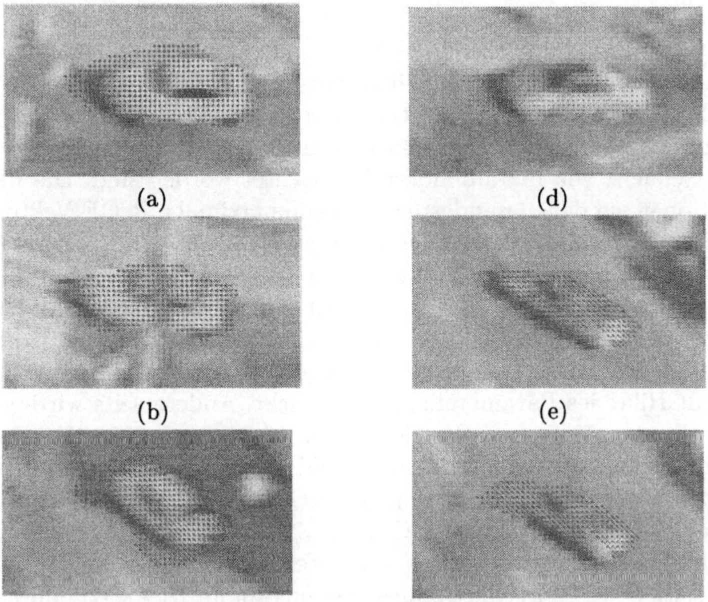

Abb. 4. Ausschnitte um zwei verschiedene Fahrzeuge aus einer am Nibelungenplatz in Frankfurt aufgezeichneten Bildfolge. Den Ausschnitten überlagert sind jeweils die durch rein bildbereichsbasierte Verfolgung bestimmten OIC-Masken, und zwar in der linken Spalte für ein erstes Fahrzeug zu den Halbbild-Zeitpunkten 125 (a), 174 (b) und 410 (c). Man beachte die OIC-Maske während der Verdeckung durch einen Mast in Teilbild (b) sowie die Tatsache, daß sich nicht nur die *Orientierung* des Fahrzeugs während dieser Verfolgung über fast 300 Halbbildzeitpunkte (ca. 6 sec) geändert hat, sondern das Fahrzeug gegen Ende dieser Periode auch in einen starken Schlagschattenbereich hineinfährt. Die rechte Spalte stellt Ergebnisse für ein zweites Fahrzeug dar, das auf einer – zu der vom ersten Fahrzeug befahrenen Fahrspur – parallelen Fahrspur fährt.

Zur weiteren Berechnung werden nur die innerhalb der konvexen Hülle liegenden OF-Vektoren beibehalten, deren Mahalanobis-Abstand vom $\overline{OF}_{aktuelles-Bild}$ eine Schwelle nicht überschreitet.

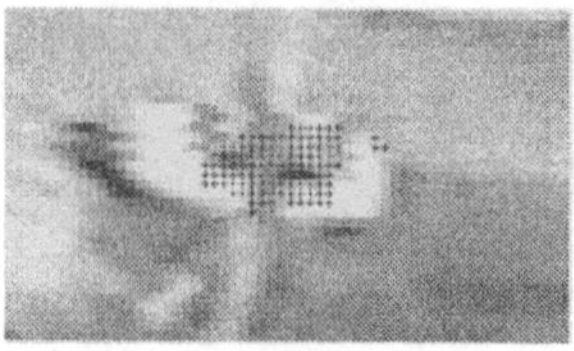

(a)

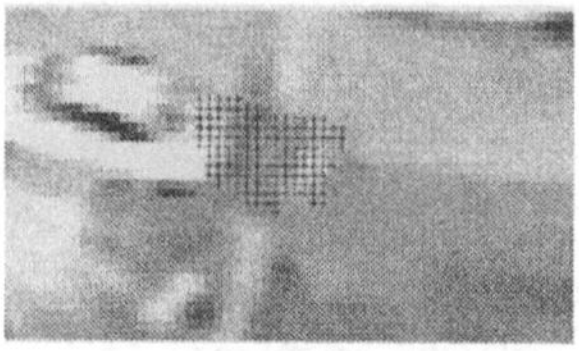

(b)

Abb. 5. Gezeigt werden Verdeckungsmasken, die ermittlet werden, wenn das erste Fahrzeug der Bildfolge Nibelungenplatz – siehe Abbildung 4(b) – vom Mast verdeckt wird, und zwar zum Halbbildzeitpunkt 175 (a) und 200 (b).

Nach n (= 25) Halbbildern werden für alle $i \in \{1, 2, \ldots, n-1\}$ die R_OF-Flecken$_i$ gemäß $\sum_{k=n-i}^{n-1} \overline{OF}_k$ auf einer diesem n. Halbbild entsprechenden Position überlagert. An jeder Pixelposition dieser Vereinigung aller n R_OF-Flecken wird bestimmt, wieviele R_OF-Vektoren an der betreffenden Stelle innerhalb dieser Teilsequenz von n Halbbildern beobachtet worden sind. Die Menge aller Pixelpositionen, an denen mindestens p % von maximal n R_OF-Vektoren akzeptiert worden sind, bildet eine *Object Image Candidate (OIC)*-Zwischenmaske. Mit einem analog ablaufenden Verfahren wird diese OIC-Zwischenmaske nun mit $\mathrm{OIC}_{vorangehende-Teilsequenz}$-Maske, d.h. der OIC-Maske, die n Halbbilder zuvor berechnet wurde, zur $\mathrm{OIC}_{aktuelle-Teilsequenz}$-Maske vereinigt. Auf diese Weise werden einerseits kurzfristige Schwankungen in Form und Größe der R_OF-Flecken mit Hilfe des Parameters p unterdrückt. Andererseits wird einem langsameren, aber beständigeren Wandel Rechnung getragen, wie ihn beispielsweise die bewegungsbedingte Änderung der Abbildung eines Fahrzeugs mit sich bringt.

Man beachte, daß die Berechnung der $\mathrm{OIC}_{aktuelle-Teilsequenz}$-Maske nur dann gute Ergebnisse liefert, wenn die $\mathrm{OIC}_{vorangehende-Teilsequenz}$-Maske, nachdem sie n Halbbilder gemäß des optischen Flusses vorgeschoben wurde, immer noch mit der Position des Fahrzeugabbildes übereinstimmt. Dies setzt eine sehr genaue Schätzung von $\overline{OF}$ voraus, was zum einen durch die Verwendung von adaptiven Masken erzielt wird und zum anderen dadurch zustandekommt, daß nur OF-Vektoren in R_OF-Bereichen zur Berechnung von $\overline{OF}$ verwendet werden. Abbildung 4 (f) zeigt eine OIC-Maske, die sich von der OIC-Maske aus Abbildung 4 (e) nur dadurch unterscheidet, daß sie gemäß des optischen Flusses vom 550. in das 572. Halbbild verschoben wurde. Die Position der OIC-Maske stimmt in beiden Abbildungen gleich gut mit der Fahrzeugabbildung überein.

Die auf diesem Wege erhaltenen Masken für das Abbild eines sich in der Szene bewegenden Objektes (einschließlich des Schattens) werden für das in Abbildung 1 gezeigte Fahrzeug in Abbildung 2 dargestellt. Abbildung 4 zeigt die OIC-Masken für zwei weitere Fahrzeuge einer anderen Bildfolge.

Eine aus einer Teilbildfolge erschlossene OIC-Maske wiederum erlaubt es, in Verbindung mit einem klar detektierbaren Teilbereich von Diskontinuitäts-Hinweisen im Inneren einer sich von Halbbild zu Halbbild verschiebenden OIC-Maske, auch eine Maske für das Abbild verdeckender Strukturen zu bestimmen – siehe Abbildungen 3 und 5.

4 Diskussion und Ausblick

Die auf einer Ermittlung lokaler Grauwertcharakteristika aufbauende *adaptive* Schätzung von OF-Vektoren führt zu deutlich klarer strukturierten OF-Feldern, was in einer weniger störanfälligen Segmentierung solcher OF-Felder resultiert. Dies unterstützt die Verfolgung relevanter Segmente von Halbbild zu Halbbild, wodurch sich eine brauchbare *OIC*-Maske für das Abbild eines sich bewegenden Objektes sowie deren zeitliche Positionsänderung in der Bildebene ermitteln läßt – und zwar selbst bei einer Verfolgung über mehrere hundert Halbbilder hinweg, während derer sich Form und Orientierung der Abbildung des Objektes sowie des von ihm auf die Straßenebene geworfenen Schattens erheblich ändern kann.

In Verbindung mit einem klar detektierbaren Teilbereich von Diskontinuitäts-Hinweisen im Inneren einer sich von Halbbild zu Halbbild verschiebenden OIC-Maske läßt sich auch eine Maske für das Abbild verdeckender Strukturen bestimmen. Damit werden wichtige qualitative Aussagen über die Relativlage von Szenenkomponenten unmittelbar aus bildbereichsbezogenen Auswertungsschritten erschließbar – *ohne* Rückgriff auf eine explizite dreidimensionale Repräsentation der Szenenstruktur. Darüber hinaus ergibt sich ein Kontext, innerhalb dessen einzelne OF-Vektoren genauer untersucht werden können. Phänomene, die früher leicht als Effekte von Störungen oder als Artefakte des Schätzprozesses angesehen wurden, können so systematisch analysiert werden.

Danksagung

Wir danken der Deutschen Forschungsgemeinschaft für eine Förderung von Teilen der hier vorgelegten Untersuchungen.

Literatur

1. Haag, M., Nagel H.-H.: Beginning a Transition from a Local to a More Global Point of View in Model-Based Vehicle Tracking. Proc. 5th European Conference on Computer Vision, 2–6 June 1998, Freiburg/Germany, Lecture Notes in Computer Science **1406**. Springer–Verlag: Berlin Heidelberg 1998, pp. 812–827.
2. ter Haar Romeny, B., Florack, L., Koenderink, J., Viergever, M. (Eds.): Scale-Space Theory in Computer Vision. Lecture Notes in Computer Science **1252**. Springer-Verlag: Berlin Heidelberg 1997.
3. Haußecker, H., Jähne, B.: A Tensor Approach for Precise Computation of Dense Displacement Vector Fields. In 'Informatik aktuell', Paulus, E., Wahl, F.M. (Hrsg.), Mustererkennung 1997, 19. DAGM-Symposium, Braunschweig. Springer-Verlag: Berlin Heidelberg 1997, pp. 199–208.
4. Horn, B.K.P.: Robot Vision. The MIT Press: Cambridge/MA and London/UK 1986.
5. Lindeberg, T., Gårding, J.: Shape-Adapted Smoothing in Estimation of 3-D Depth Cues from Affine Distortions of Local 2-D Brightness Structure. In Proc. 3rd European Conference on Computer Vision, Stockholm/S; J.-O. Eklundh (Ed.), Lecture Notes in Computer Science **800**. Springer–Verlag: Berlin Heidelberg 1994, pp. 389–400.
6. Nagel, H.-H., Gehrke, A.: Spatiotemporally Adaptive Estimation and Segmentation of OF-Fields. In Proc. 5th European Conference on Computer Vision, 2–6 June 1998, Freiburg/Germany, Lecture Notes in Computer Science **1407**. Springer–Verlag: Berlin Heidelberg 1998, pp. 86–102.

Erkennung von Körperbewegungsmustern durch Automaten

Kyungeun Cho[1,2] und Hyungje Cho[1]
[1]Dongguk University, Department of Computer Engineering,
Computer Vision Laboratory, Seoul, Korea
[2]Universität Regensburg, Institut für Medizinische Soziologie,
Universitätsstraße 31, D-93051 Regensburg

1 Einleitung

In den letzten Jahren ist das Interesse an der automatischen Erkennung von Körperbewegungen auf der Basis optischer Information ganz erheblich gestiegen. Die Vielfalt der Einsatzmöglichkeiten beweist, daß es sich nicht nur um das Interesse einer kleinen Schar von Experten handelt, z.B.: Ganganalyse, automatische Kommentierung von Sequenzen in Video-Datenbanken, die die Bewegungen von Menschen beinhalten [Cle95], drahtlose Mensch-Maschine-Schnittstellen für Virtual-Reality-Anwendungen [Bec97], Überwachungs-systeme, Interpretation von Zeichensprachen [Bra96][Sut96] und Choreographie von Tanz und Ballett [Cam95]. Für die automatische Erkennung der Körperbewegungen in Videobildern müssen mehrere Phasen durchgeführt werden: Initiale Detektion einer Person, Verfolgen der Person, Segmentierung des Körpers, Extrahierung der einzelnen Körperteile und Erkennung der Körperbewegungsmuster. Es sind schon mehrere Systeme in speziellen Gebieten entwickelt worden, die die Körperbewegungen erkennen können, z.B. Sensei [Bec97], ARGo [Bra96], First Sight [Leu95] usw. Zur Klassifikationsmethode von Bewegungsmustern wurden Phase Space Methode [Cam95], HMM (Hidden Markov Model) [Bec97][Bra96], Dynamic Time Warping [Gav95], Template Matching [Sut96] usw. angewendet.
Der Schwerpunkt dieses Beitrags liegt auf der Klassifikationsmethode von Körperbewegungsmustern mit dem Ziel, die nonverbalen Körpersignale zu erkennen. Eine Körperbewegung enthält gleichzeitig mehrere Körperteilbewegungen mit verschiedenen Bewegungsvektoren. Das hier angewandte "Berner System zur Zeitreihen-Notation nonverbaler Interaktion", im folgenden als Berner System referenziert, erfüllt dieses Prinzip in besonderer Weise [Fre81]. Es legt für alle Körperteile Kategorien fest, mit deren Hilfe jede einzelne Position dieser Körperteile sowohl räumlich als auch zeitlich mit hinreichender Genauigkeit beschrieben werden kann. Das Berner System eignet sich aufgrund seines hohen Formalisierungsgrades sehr gut als Grundlage der Klassifikationsmethode zur Erkennung von Körperbewegungsmustern. Die durch diese Methode erstellten Codes werden mit endlichen Automaten als eigene Körperteilbewegungen klassifiziert, und mittels einer Tabellenzuordnung und einem Körperbewegungsautomaten werden diese einzelnen Körperteilbewegungen zu einem Bewegungsprozeß kombiniert. Diese Klassifikationsmethode wurde mit Daten von einem Simulator, der für diesen Zweck entwickelt wurde, getestet. Diese Klassifikationsmethode kann von Systemen angewendet werden, die dreidimensionale Positionen einzelner Körpergelenke bestimmen können.

2 Anwendung des Berner Systems

Das Codierverfahren basiert auf dem folgenden Prinzip: "Da die verschiedenen Körperteile in sehr unterschiedlichem Ausmaß an der gesamten Bewegungsaktivität des Körpers beteiligt sind, ist es in der Regel ökonomischer, jeden Körperteil in einem eigenen Durchgang zu codieren" [Fre81]. Dies bedeutet zunächst, daß einzelne komplexe Körperbewegungen in ihre konstituierenden Elemente aufgeschlüsselt werden. Jedes einzelne Körperteil wird durch die Bewegungsdimensionen repräsentiert, innerhalb derer es zur Bewegungsvariation fähig ist. Im Berner System wurden für jedes Körperteil zwei bis neun Bewegungsdimensionen unterschieden. Da die verschiedenen Körperteile in unterschiedlichem Maße zu komplexer Bewegungsvariation fähig sind, wurden unterschiedliche Anzahlen von Dimensionen berücksichtigt. Zu den Bewegungsdimensionen gehören Sagital-, Rotational-, Vertikal-, Horizontaldimension usw. Für jede Dimension wird die kontinuierliche Körperbewegung als Folge diskreter Positionszustände codiert, wobei die räumliche und zeitliche Auflösung durch Regeln wählbar sind.
Im folgenden Beispiel wurden die Codierungsregeln für die Bewegungsanalyse des Oberkörpers vereinfacht. Hierzu wurden die aus der Originalarbeit bekannten Codierungsregeln angepaßt. Zur Bestimmung der räumlichen Komponente werden die Positionen der numerierten Körperteile »Kopf, Rumpf, Schultern, Ellbogen, Hände« in jeweils zwei bis drei Dimensionen bestimmt. Die hier benutzten Dimensionen sind Tiefen-, Vertikal- und Horizontaldimension (vgl. Abb. 1(b), 1(c) und 1(d)). Abb. 1(a) zeigt das Codierungsmodell. Hier ist ⑥ der Kopf, ⑧ der Rumpf, ① die rechte Schulter, ③ der rechte Ellbogen, ⑤ die rechte Hand, ⓪ die linke Schulter, ② der linke Ellbogen und ④ die linke Hand. Abb. 1(b) zeigt, daß die Hände vor dem Rumpf sind. Die Tiefendimension bedeutet in welcher Tiefe sich ein Körperteil befindet. Wenn ein Körperteil sich vor dem Rumpf befindet, bekommt er eine Codenummer 4 oder 5, wenn er sich in der Tiefe des Rumpfes befindet Codenummer 3 und wenn er sich hinter dem Rumpf befindet Codenummer 1 oder 2. In Abb. 1(c) werden die Körperteile entschieden, in welcher Höhe und in Abb. 1(d) auf welcher Breite sie sich befinden.

(a)Codierungsmodell (b)Tiefendimension (c)Vertikaldimension (d)Horizontaldimension

Abb. 1 Codierungsregeln

Scene	Head	Trunk	R_Shoulder	L_Shoulder	R_Ellbow	L_Ellbow	R_Hand	L_Hand
	u/d l/r	f/b l/r	u/d f/b	u/d f/b	f/b V H	f/b V H	f/b V H T	f/b V H T
0	2 2	3 2	2 3	2 3	4 6 2	4 6 8	5 6 2 0	5 6 8 0
1	0 0	0 0	0 0	0 0	0 5 0	0 0 0	0 5 0 0	0 0 0 0
2	0 0	0 0	0 0	0 0	3 4 0	0 0 0	4 4 0 0	0 0 0 0
3	0 0	0 0	0 0	0 0	0 3 0	0 0 0	3 3 0 0	0 0 0 0
4	0 0	0 0	0 0	0 0	0 2 0	0 0 0	0 2 0 0	0 0 0 0
5	0 0	0 0	0 0	0 0	0 0 0	0 0 0	0 1 0 0	0 0 0 0

Abb. 2 Codierungsbeispiel

Das Codierungsprinzip wird im Abb. 2 verdeutlicht. Die »Scene« repräsentiert die Zeitabfolge. Es sind die Körperteile »Head, Trunk, R_Shoulder, L_Shoulder, R_Ellbow, L_Ellbow, R_Hand und L_Hand« codiert worden. Dabei bedeutet »u/d (up/down)« Heben-Senken-, »l/r (left/right)« Links-Rechts-, »f/b (front/back)« Tiefen-, »V« Vertikal-, »H« Horizontal-, und »T (touch)« Berührungsdimension. Die Dimensionen »u/d, l/r, T« werden jedoch bisher noch nicht verwendet. In diesem Beispiel wurde das Heben des rechten Armes codiert. Der Code zur Bestimmung der rechten Handposition in der Vertikaldimension lautet beispielsweise: Die rechte Hand befindet sich in Scene 0 ›im Bereich des Rumpfes‹ (Codenummer 6), in Scene 1 ›im oberen Bereich des Rumpfes‹ (Codenummer 5) usw. (vgl. auch Abb. 1(c)).

3 Erkennung von Körperbewegungen

Die Methode zur Erkennung von Körperbewegungen läßt sich in mehrere Verarbeitungsphasen einteilen (vgl. Abb. 3). Im folgenden werden die einzelnen Phasen entsprechend der Bearbeitungsreihenfolge dargestellt.

3.1 Ruhephasen, Bewegungsrichtungen und Bewegungsdistanzen herausfinden

Jede einzelne Bewegungsdimension trägt mit Hilfe des allgemeinen endlichen Automaten zur Entscheidung der Ruhephase, der Bewegungsphase, der Richtungsveränderung von Bewegungen und der Bewegungsdistanzen bei. Der endliche Automat hat einen Anfangszustand, einen Endzustand und mehrere Zwischenzustände. Die Anzahl der Zwischenzustände sind für die Codierungsdimensionen unterschiedlich, beispielsweise, die Vertikaldimension hat neun, die Horizontaldimension sieben und die Tiefendimension fünf Zwischenzustände. Das allgemeine Zustandsdiagramm wird in Abb. 4 dargestellt. In Abb. 5 wird der endliche Automat für die Tiefendimension betrachtet. Hier bedeutet "-", daß es keinen gültigen Zustandsübergang gibt. Von den Eingabesymbolen bedeuten die Nummern die Codes der Bewegungsdimensionen und "E" bedeutet die Schlußkondition einer Bewegung. Wenn es keine Bewegung von einem Zwischenzustand gibt, ist das Eingabesymbol 0. Wenn alle Eingabesymbole in einer Scene 0 sind, ist es eine Ruhephase. Der Rest der Phasen, die keine Ruhephasen sind, sind Bewegungsphasen. Eine Richtungsveränderung von Bewegungen ist durch fallenden und steigenden Codenummern bestimmt. Wenn es sich auch nur in einem Körperteil die Bewegungsrichtung geändert hat, gilt es als Änderung der Bewegungsrichtung. Die Bewegungsdistanz ist durch die Differenz von zwei Codenummern eines Bewegungsdimensions gegeben. Die Schlußkondition ist eine Ruhephase oder eine Veränderung der Bewegungsrichtung. Wenn der Automat den Endzustand erreicht, wird er wieder vom Anfangszustand beginnen bis die letzte Scene erreicht wird. Die Phase 1 und 2 in Abb. 3 sind notwendig, weil man später die einzelnen "Körperbewegungsmuster" (Phase 5) in einem komplexen "Bewegungsprozeß" (Phase 7) herausfinden muß. Hier wird ein Bewegungsprozeß in mehreren "Körperbewegungsmuster" durch Ruhephasen und Änderungen der Bewegungsrichtung zerlegt.

Phase	Verarbeitungsphasen
0	Codierung nach der Anwendung des Berner Systems
1	Ruhephasen, Bewegungsrichtungen und Bewegungsdistanzen durch den allgemeinen endlichen Automat herausfinden (vgl. Abb. 4)
2	Bestimmung der Anzahl der Körperbewegungsmuster durch Ruhephasen und Änderungen der Bewegungsrichtung
3	Berechnung der einzelne "*Körperteil-Bewegungsvektoren*" durch die Kombination der Bewegungsdimensionen (vgl. Abb. 6)
4	Kombination der "*Körperbewegungsvektoren*" von einzelnen Körperteil-Bewegungsvektoren (vgl. Abb. 7 (a),(b))
5	Kombination der "*Körperbewegungsmuster*" von einzelnen Körperbewegungsvektoren (vgl. Abb. 7 (c))
6	Definition des "*Körperbewegungsautomat*" durch einzelne Körperbewegungsmuster (vgl. Abb. 8).
7	Erkennung des "*Bewegungsprozeß*" durch den Körperbewegungsautomat

Abb. 3 Verarbeitungsphasen

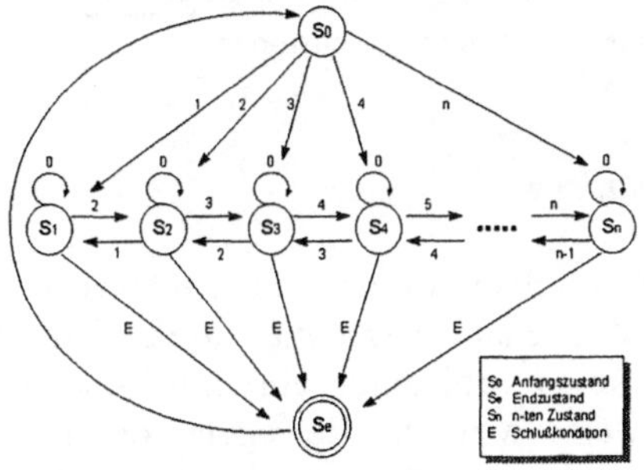

Abb. 4 Allgemeines Zustandsdiagramm

$$\text{Tiefendimension\_EA} = (\{S_0, S_1, S_2, S_3, S_4, S_5, S_e\}, \{1, 2, 3, 4, 5, E\}, \delta, S_0, \{S_e\})$$

δ	*1*	*2*	*3*	*4*	*5*	*E*
S_0	S_1	S_2	S_3	S_4	S_5	-
S_1	S_1	S_2	-	-	-	S_e
S_2	S_1	S_2	S_3	-	-	S_e
S_3	-	S_2	S_3	S_4	-	S_e
S_4	-	-	S_3	S_4	S_5	S_e
S_5	-	-	-	S_4	S_5	S_e

Abb. 5 Endlicher Automat für die Tiefendimension

3.2 Berechnung der einzelne "*Körperteil-Bewegungsvektoren*"

Durch die endlichen Automaten der Bewegungsdimensionen werden die Bewegungsrichtung und die -distanz herausgefunden. Dieses Ergebnis wird bei der Kombination der einzelnen Bewegungsdimensionen zu einem Körperteil-Bewegungsvektor benutzt. Bei der Kombination werden drei Faktoren von einzelnen Bewegungsdimensionen verwendet: Keine Veränderung der Codenummern (ø), fallende Codenummer (-), steigende Codenummer (+). Eine Kombination-Zahl wird herausgerechnet bei der Addierung der einzelnen Kombination-Codes (vgl. Abb. 6 links). Wenn die Bewegungsrichtung von Tiefenbewegungsdimensionen eines Körperteil-Bewegungsvektors steigt, ist die Kombination-Code 1, wenn sie fällt 5. Wenn sie eine Vertikalbewegungsdimension ist und steigt, ist die Kombination-Code 10 usw. Bei der Kombination werden die Kombination-Codes zu einer Kombination-Zahl summiert. Dadurch werden die einzelnen Bewegungsdimensionen zu 27 möglichen Körperteil-Bewegungsvektoren von einzelnen Körperteilen kombiniert (vgl. Abb. 6 rechts). Ein Körperteil-Bewegungsvektor beschreibt die 3D-Bewegung eines Körperteils. Die Richtung des Vektors gibt die Bewegungsrichtung an. Mit der Länge des Vektors kann die Bewegungsdistanz repräsentiert werden. Die Bewegungsdistanz wird hier noch nicht verwendet. Z.B.: Ob der Arm sich bis zum

Kopf bewegt oder bis über den Kopf bewegt, kann noch nicht bestimmt werden, dazu muß die Bewegungsdistanz angewendet werden.

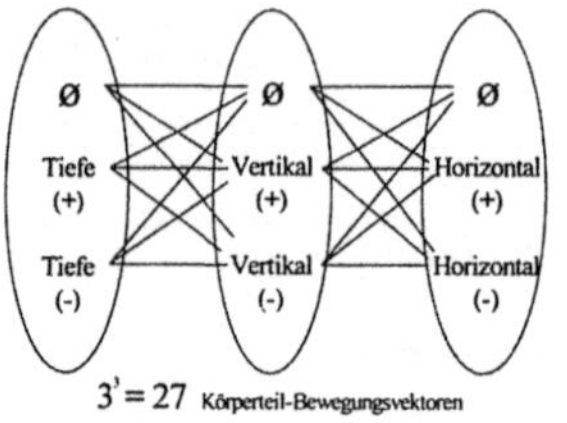

<Kombination-Codes>

Ø: 0 Tiefe (+): 1 Vertikal (+): 10 Horizontal (+): 100
Tiefe (-): 5 Vertikal (-): 50 Horizontal (-): 500

Kombination-Zahl				Kombination-Zahl				Kombination-Zahl			
0	ø	ø	ø	*1*	T(+)	ø	ø	*5*	T(-)	ø	ø
100	ø	ø	H(+)	*101*	T(+)	ø	H(+)	*105*	T(-)	ø	H(+)
500	ø	ø	H(-)	*501*	T(+)	ø	H(-)	*505*	T(-)	ø	H(-)
10	ø	V(+)	ø	*11*	T(+)	V(+)	ø	*15*	T(-)	V(+)	ø
110	ø	V(+)	H(+)	*111*	T(+)	V(+)	H(+)	*115*	T(-)	V(+)	H(+)
510	ø	V(+)	H(-)	*511*	T(+)	V(+)	H(-)	*515*	T(-)	V(+)	H(-)
50	ø	V(-)	ø	*51*	T(+)	V(-)	ø	*55*	T(-)	V(-)	ø
150	ø	V(-)	H(+)	*151*	T(+)	V(-)	H(+)	*155*	T(-)	V(-)	H(+)
550	ø	V(-)	H(-)	*551*	T(+)	V(-)	H(-)	*555*	T(-)	V(-)	H(-)

Abb. 6 Kombination von einzelnen Bewegungsdimensionen

3.3 Kombination der "*Körperbewegungsvektoren*"

Aufgrund der Eigenheit des menschlichen Körperbaus werden die Armbewegungen aus der Schultern-, der Ellbogen- und der Händebewegung kombiniert. Ein "Körperbewegungsvektor" (vgl. Abb. 3 Phase 4) wird durch einzelne "Körperteil-Bewegungsvektoren" kombiniert. Beispielsweise wird die Bewegung des rechten Arms durch die Kombination des rechten Ellbogen-Körperteil-Bewegungsvektors und des rechten Hand-Körperteil-Bewegungsvektors definiert. Alle Körperbewegungsvektoren werden mit dieser Kombination klassifiziert. Für die Anzahl der Körperbewegungsvektoren des rechten Arms ergeben sich in diesem Fall 27 x 27 Möglichkeiten. Wegen der großen Anzahl der Kombination wird zur Kombinationssuche eine Zuordnung über eine Tabelle verwendet (vgl. Abb. 7(a),(b)).

3.4 Kombination der "*Körperbewegungsmuster*"

Danach werden die einzelnen Körperbewegungsvektoren zu einem Körperbewegungsmuster kombiniert (vgl. Abb. 7(c)). Ein "Körperbewegungsmuster" (vgl. Abb. 3 Phase 5) ist die für die Bildung des Bewegungsprozesses einschlägige Körperbewegung, welche für sich allein keine bestimmte Bedeutung hat, z.B.: Arme gebeugt nach oben heben, Unterarme gebeugt nach innen bewegen usw. Sie werden durch einzelne "Körperbewegungsvektoren" kombiniert. Die schattierten Spalten sind die Kombinationscodes des rechten Arms, des linken Arms und der beiden Arme.

3.5 Erkennung des "*Bewegungsprozeß*" durch den Körperbewegungsautomat

Nachdem die "Körperbewegungsmuster" aus einzelnen "Körperbewegungsvektoren" kombiniert sind, werden sie in den "Körperbewegungsautomat" eingegeben (vgl. Abb. 8). Hier werden sie zu einen bestimmten "Bewegungsprozeß" erkannt. Hier wird ein "Bewegungsprozeß" (vgl. Abb. 3 Phase 7) als eine bestimmte Körperbewegung, die eine Bedeutung haben kann, definiert. Z.B.: Hände klatschen usw. Von den Eingabesymbolen bedeuten die Nummern die Kombinationcodes der Körperbewegungsmuster, und "E" bedeutet die Schlußkondition. Hier ist die Schlußkondition eine längere Ruhephase oder das Ende der Scene. Die Zwischenzustände sind die Körperhaltungen, die nach einer Bewegung, hier als Körperbewegungsmuster definiert, erreicht sind. Der Anfangzustand ist hier eine Körperhaltung mit einer Position Hände auf dem Tisch.

Körperteil- Bewegungs-vektoren		Körperbewegungsvektor	
R_Ellb	R_Hand	Kombinationscode	Bewegungsbedeutung
0	500	17	rechter Unterarm gebeugt nach außen
0	100	18	rechter Unterarm gebeugt nach innen
:	:	:	:

(a) Bewegung des rechten Arms

Körperteil- Bewegungs-vektoren		Körperbewegungsvektor	
L_Ellb	L_Hand	Kombinationscode	Bewegungsbedeutung
0	100	14	linker Unterarm gebeugt nach außen
0	500	15	linker Unterarm gebeugt nach innen
:	:	:	:

(b) Bewegung des linken Arms

Kombinationscode des rechten Arms	Kombinationscode des linken Arms	Kombinationscode der Arme	Körperbewegungsmuster der Arme
17	14	18	Unterarme gebeugt nach außen
18	15	19	Unterarme gebeugt nach innen
:	:	:	:

(c) Körperbewegungsmuster der Arme

Abb. 7 Beispiele für Zuordnungstabellen

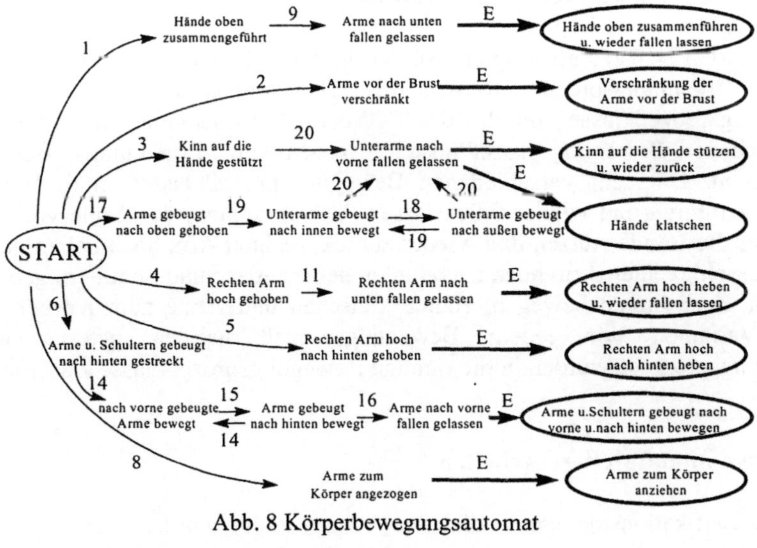

Abb. 8 Körperbewegungsautomat

4 Ergebnisse

Die Klassifikationsmethode, die in diesem Beitrag vorgestellt wird, wurde zuerst mit simulierten Eingabedaten getestet. Für diesen Zweck wurde ein Simulator entwickelt.

Hierbei wurden acht verschiedene Bewegungsprozesse mit jeweils drei modifizierten Ausprägungen simuliert. Im nächsten Schritt wurden dem System alle 24 Testbewegungen vorgegeben. Diese wurden der richtigen Bewegungsprozeßklasse zugeordnet.

4.1 Simulator

Die Idee, einen Simulator zu entwickeln, geht auf [Ben89] zurück. Dort wird ein auf dem Berner System basierender Ansatz zur Computersimulation des menschlichen Bewegungsverhaltens vorgestellt. Hier wird der Simulator zur Erstellung der verschiedenen Körperbewegungsbeispiele verwendet, mit denen die Klassifikationsmethode zu testen ist. In diesem Simulator werden die Bewegungscodes editiert, und das Simulationsergebnis der Codeeingabe wird am Bildschirm angezeigt. Abb. 9 ist ein Beispiel für das Simulationsergebnis der Codeeingabe vom Bewegungsprozeß "Hände oben zusammenführen und wieder fallen lassen".

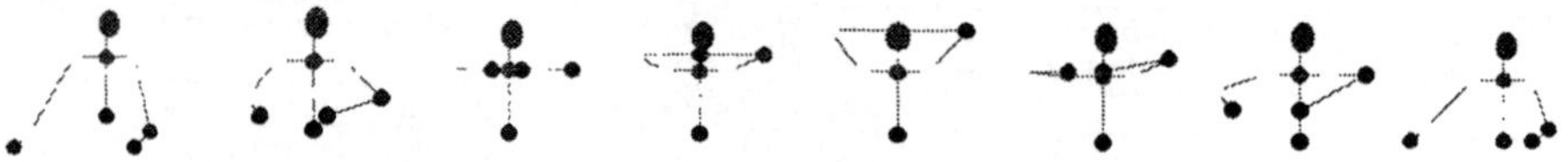

Abb. 9 Das Simulationsergebnis der Codeeingabe vom Bewegungsprozeß "Hände oben zusammenführen und wieder fallen lassen"

4.2 Klassifikation des Bewegungsprozesses

Die editierten Bewegungsbeispiele werden in den Erkennungsprozeß eingegeben. In diesem Erkennungsprozeß werden die Bewegungsprozesse in die richtige Bewegungsprozeßklasse durch die Klassifikationsmethode klassifiziert. Acht verschiedene Bewegungsprozeßmuster wurden im Erkennungsprozeß vorher eingegeben. Die acht verschiedenen Bewegungsprozeßklassen sind: Hände oben zusammenführen und wieder fallen lassen, Verschränkung der Arme vor der Brust, Kinn auf die Hände stützen und wieder zurück, rechten Arm hoch heben und wieder fallen lassen, rechten Arm hoch nach hinten heben, Arme und Schultern gebeugt nach vorne u. nach hinten bewegen, Hände klatschen und Arme zum Körper anziehen. Drei Varianten von jedem Bewegungsprozeß wurden getestet und alle Bewegungsprozesse wurden in die richtige Bewegungsprozeßklasse klassifiziert.

5 Diskussion und weitere Arbeiten

Die Klassifikationsmethode, die hier vorgestellt wurde, kann in mehrere Bewegungsprozeßklassen erweitert werden. Die Erstellung der Zuordnungstabellen und des "Körperbewegungsautomaten" muß wegen der großen Anzahl der Bewegungsvielfalt sehr sorgfältig bearbeitet werden. Schließlich müssen die Zuordnungstabellen und der "Körperbewegungsautomat" optimiert werden, so daß die ähnlichen Körperbewegungsmuster in dieselbe Klasse kategorisiert werden. Die weiteren Arbeiten werden sich mit einer verfeinerten Klassifikation von mehreren Bewegungsprozesse und der Anwendung auf realistische Videodaten beschäftigen. Die Anwendung auf realistische Videodaten wird mit der Ausgabe des Systems von [Mun98] gemacht (vgl. Abb. 10). Durch farbmarkierte Körpergelenke und Aufnahme von Stereokameras werden die Ausgabedaten in der Form sein, wie sie von dieser Klassifikationsmethode benötigt werden. Wenn die Anzahl der Bewegungsprozesse und der Kombinationen größer wird, sollte auch ein halbautomatischer Bewegungsklassifizierer vorhanden sein, der die Bewegungskombination automatisch generieren kann. Das Ziel dieser Arbeit ist, die nonverbalen Körpersignale in rohen Videodaten automatisch zu erkennen.

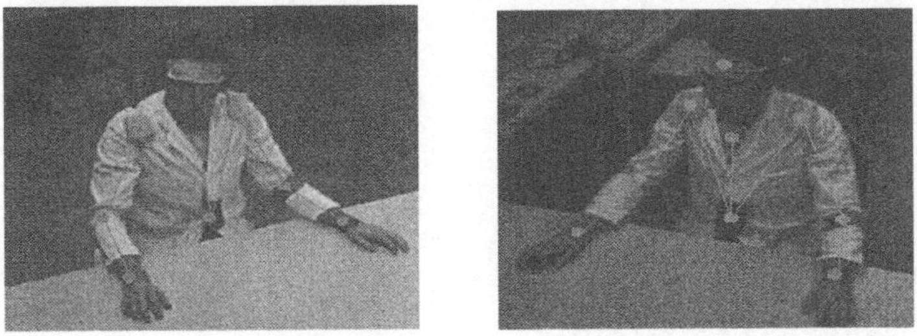

Abb. 10 Farbmarkierte Körpergelenke und die Aufnahme mit Stereokameras

6 Literatur

[Bec97] D.A. Becker, "Sensei: A Real-Time Recognition, Feedback and Training System for T'ai Chi Gestures", MIT Media Lab Perceptual Computing Group TR 426

[Ben89] G. Bente, "Facilities for the graphical computer animation of head and body movements", Behavior Research Methods, Instruments & Computers, S.455-462

[Bra96] A. Braffort, "ARGo: An Architecture for Sign Language Recognition and Interpretation", Proc. of Gesture Workshop 1996, pp. 17-30

[Cam95] L. Campbell, A. Bobick, "Recognition of Human Body Motion Using Phase Space Constraints", MIT Media Lab Perceptual Computing Group TR 309, 1995

[Cle95] E. Clergue, M. Goldberg, N. Madrane, B. Merialdo, "Automatic Face and Gestual Recognition for Video Indexing", Int'l Workshop on Automatic Face-and Gesture-Recognition, 1995, pp. 110-115

[Fre81] S. Frey, H.P. Hirsbrunner, J. Pool & W. Daw, "Das Berner System zur Untersuchung nonverbaler Interaktion: I. Die Erhebung des Rohdatenprotokolls", in P. Winkler (Hrsg.), Methoden der Analyse von Face-to-Face-Situationen, S. 203-236

[Gav95] D.M. Gavrila, L.S. Davis, "3-D Model-based recognition of human movement by dynamic time warping", Computer Vision Lab., Univ. of Maryland

[Leu95] M.K. Leung, Y. Yang, "First Sight: A Human Body Outline Labeling System", IEEE Pattern Analysis and Machine Intelligence 17(4), April, pp. 359-377

[Mun98] O. Munkelt, C. Ridder, D. Hansel, W. Hafner, "A model driven 3D image interpretation system applied to person detection in video images", 14'th Int'l Conf. on Pattern Recognition 1998, accepted

[Sut96] A. Sutherland, "Real-time Video-based Recognition of Sign Language Gestures using Guided Template Matching", Gesture Workshop 1996, pp. 31-38

Grundlagen und Algorithmen (Postersession)

Least Squares Fitting of Circle and Ellipse

SUNG JOON AHN, WOLFGANG RAUH and BEREND OBERDORFER

Fraunhofer Institute for Manufacturing Engineering and
Automation IPA, Nobelstr. 12, 70569 Stuttgart, FRG
Tel.: +49-711-970 1859, Fax: +49-711-970 1004
E-mail: {sja; wor; beo}@ipa.fhg.de

The least squares fitting of geometric features to given points minimizes the squares sum of error-of-fit in predefined measures. By the geometric fitting, the error distances are defined with the orthogonal, or shortest, distances from the given points to the geometric feature to be fitted. For the geometric fitting of circle and ellipse, robust algorithms are proposed which are based on the coordinate description of the corresponding point on the circle/ellipse for the given point, where the connecting line of the two points is the shortest path from the given point to the circle/ellipse.

1 Introduction

The fitting of geometric features to given points is desired in various fields of science and engineering. There is especially true of circle and ellipse, as these are practical geometric features for the application of image processing. In the past, for fitting problems the least squares method (LSM) has often been applied and has proven its usability in implementation and computing costs. In this paper we suggest robust algorithms for *least squares orthogonal distances fitting* of circle and ellipse to given points. The LS fitting minimizes the squares sum of error-of-fit in predefined measures. There are two main categories of LS fitting problems for geometric features, algebraic and geometric, differentiated by the definition of the error distances. By the *algebraic fitting* of a geometric feature, which is described by the implicit equation $F(\mathbf{x}, \mathbf{a}) = 0$ and the parameters vector $\mathbf{a} = (a_1, \cdots, a_q)$, the error distances are defined with the deviations of the implicit equation from the expected value, i.e. zero, at each given point. The nonequality of the equation indicates there is some error-of-fit. Most publications for the LS fitting of circle [1,2] and ellipse [3-8] are concerned with the squares sum of algebraic distances or their modifications:

$$\sum_{i=1}^{m} e_i^2 = \sum_{i=1}^{m} \left[(x_i - x_c)^2 + (y_i - y_c)^2 - r^2 \right]^2 \quad \text{for the circle fitting, and} \qquad (1)$$

$$\sum_{i=1}^{m} e_i^2 = \sum_{i=1}^{m} \left(a x_i^2 + 2 b x_i y_i + c y_i^2 + 2 d x_i + 2 e y_i + f \right)^2 \quad \text{for the ellipse fitting.} \qquad (2)$$

In spite of the advantages in implementation and computing costs, the algebraic fitting has drawbacks in accuracy, and in the physical interpretation of the fitting parameters and the error-of-fit [5,9]. The known disadvantages of the algebraic fitting are [4,6,10,11]:

- The definition of error distances does not coincide with the measurement guideline.
- The parameters are dependent on the coordinate transformation, and biased.
- The fitting errors are unwillingly weighted.
- The fitting procedure sometimes ends in an unintended geometric feature.

By the *geometric fitting*, named also as the best fitting, the error distances are defined with the orthogonal, or shortest, distances from the given points to the geometric feature to be fitted. The geometric fitting is possibly the only solution to all of the above problems of the algebraic fitting. The geometric fitting of circle and ellipse is a nonlinear problem and must be solved with iteration. For the geometric fitting of circle, there are some well established methods [12,13]. On the other hand, the geometric fitting of ellipse is a newly exploited field and is being continuously developed. Gander et. al. [13] have proposed a geometric ellipse fitting algorithm in parametric form, which has a large number of fitting parameters. Consequently, it has an unnecessarily deteriorative performance of convergence. The paper of Cui et. al. [14] describes a geometric ellipse fitting with a minimum variance estimator and parameter space decomposition technique. In this paper, we propose alternative algorithms for the geometric fitting of circle and ellipse. Our algorithms are based on the coordinate description of the corresponding point on the circle/ellipse for the given point, where the connecting line of the two points is the shortest path from the given point to the circle/ellipse.

2 Nonlinear least squares fitting

Suppose that q parameters $\mathbf{a}$ are assumed to be related to the p ($>q$) measurements $\mathbf{X}$ according to

$$\mathbf{X} = \mathbf{F}(\mathbf{a}) + \mathbf{e}, \tag{3}$$

where $\mathbf{F}$ represents some nonlinear continuously differentiable observation functions of $\mathbf{a}$, and $\mathbf{e}$ denotes errors with zero mean whose influence is to be eliminated. The nonlinear least squares estimate of $\mathbf{a}$ given $\mathbf{X}$ minimizes the performance index

$$\sigma_0^2 = [\mathbf{X} - \mathbf{F}(\hat{\mathbf{a}})]^{\mathrm{T}}[\mathbf{X} - \mathbf{F}(\hat{\mathbf{a}})]. \tag{4}$$

For convenience, the weighting matrix or noise covariance matrix is chosen as an identity matrix.

The existence of the unbiased optimal solution, which is generally nonlinear in the estimate $\hat{\mathbf{a}}$, can be referenced from elsewhere [15]. There are various methods for the numerical solution of above nonlinear estimation problem. We have chosen in this paper the Gauss-Newton iteration with initial parameters $\mathbf{a}_k$ and step-size parameter λ

$$\left.\frac{\partial \mathbf{F}}{\partial \mathbf{a}}\right|_{\mathbf{a}_k} \Delta\mathbf{a} = \mathbf{X} - \mathbf{F}(\mathbf{a}_k) \quad \text{and} \quad \mathbf{a}_{k+1} = \mathbf{a}_k + \lambda\Delta\mathbf{a} \tag{5}$$

with the terminating conditions [15]

$$\left\| \left.\left(\frac{\partial \mathbf{F}}{\partial \mathbf{a}}\right)^{\mathrm{T}}\right|_{\mathbf{a}_k} [\mathbf{X} - \mathbf{F}(\mathbf{a}_k)] \right\| \approx 0 \quad \text{or} \quad \|\Delta\mathbf{a}\| \approx 0 \quad \text{or} \quad \left.\sigma_0^2\right|_k - \left.\sigma_0^2\right|_{k+1} \approx 0. \tag{6}$$

The matrix of partial derivatives appearing in equation (5) is the Jacobian matrix **J**

$$J_{ij} \equiv \frac{\partial F_i}{\partial a_j}. \tag{7}$$

For this type of numerical solution the user must supply the function values vector **F** and the Jacobian matrix **J** at the nearest corresponding points on the geometric feature from each given point. This is a necessary requirement for the least squares orthogonal distances fitting with the performance index of equation (4). Unfortunately, in the case of ellipse fitting, it is not easy to locate the nearest point on the ellipse from a given point.

3 Geometric fitting of circle and sphere

The circle/sphere in n-dimensional space ($n \geq 2$) with center $\mathbf{X}_c$ and radius R can be described as below (fig. 1a)

$$\|\mathbf{X} - \mathbf{X}_c\|^2 = R^2. \tag{8}$$

For a given point $\mathbf{X}_i$, the nearest corresponding point $\mathbf{X}_i'$ on the circle/sphere is

$$\mathbf{X}_i' = \mathbf{X}_c + R\frac{\mathbf{X}_i - \mathbf{X}_c}{\|\mathbf{X}_i - \mathbf{X}_c\|}, \tag{9}$$

with the orthogonal error distances vector

$$\mathbf{X}_i'' = \mathbf{X}_i - \mathbf{X}_i' = \left[\|\mathbf{X}_i - \mathbf{X}_c\| - R\right]\frac{\mathbf{X}_i - \mathbf{X}_c}{\|\mathbf{X}_i - \mathbf{X}_c\|}. \tag{10}$$

From equation (9), the Jacobian matrix **J** for the Gauss-Newton iteration can be directly derived

$$\mathbf{J}_{\mathbf{X}_i',\mathbf{a}} = \frac{\partial \mathbf{X}_c}{\partial \mathbf{a}} + \frac{\mathbf{X}_i - \mathbf{X}_c}{\|\mathbf{X}_i - \mathbf{X}_c\|}\frac{\partial R}{\partial \mathbf{a}} - \frac{R}{\|\mathbf{X}_i - \mathbf{X}_c\|}\left[\mathbf{I} - \frac{(\mathbf{X}_i - \mathbf{X}_c)(\mathbf{X}_i - \mathbf{X}_c)^T}{\|\mathbf{X}_i - \mathbf{X}_c\|^2}\right]\frac{\partial \mathbf{X}_c}{\partial \mathbf{a}}. \tag{11}$$

We define the parameters vector **a**

$$\mathbf{a}^{\mathrm{T}} = \left(R, \ \mathbf{X}_c^{\mathrm{T}}\right), \tag{12}$$

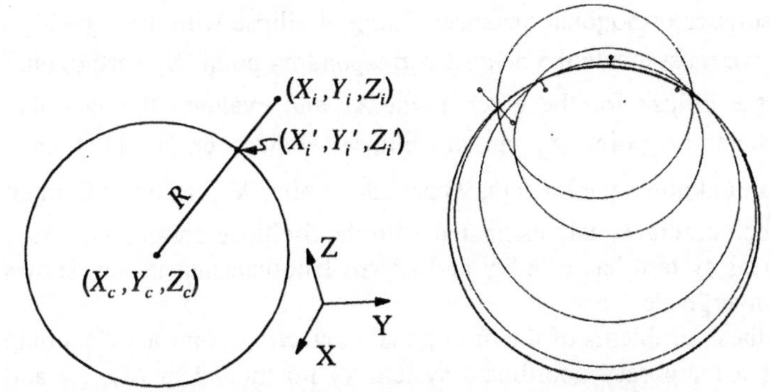

(a) (b)

Fig. 1: (a) Sphere in 3-dimensional space with center $\mathbf{X}_c$ and radius R;
(b) Geometric fitting of circle to the first 6 points in table 1.

Table 1:
8 coordinate pairs [13].

X	Y	X	Y
1	7	9	5
2	6	3	7
5	8	6	2
7	7	8	4

and, with the error distances vector $\mathbf{X}_i''$ of equation (10) and the Jacobian matrix $\mathbf{J}_{\mathbf{X}_i',\mathbf{a}}$ of equation (11) at each point $\mathbf{X}_i'$, we can construct $p = m \cdot n$ linear equations for the m given n-dimensional points. For the sphere fitting, the linear equations (5) look like

$$\begin{pmatrix} J_{X_1',R} & J_{X_1',X_c} & J_{X_1',Y_c} & J_{X_1',Z_c} \\ J_{Y_1',R} & J_{Y_1',X_c} & J_{Y_1',Y_c} & J_{Y_1',Z_c} \\ J_{Z_1',R} & J_{Z_1',X_c} & J_{Z_1',Y_c} & J_{Z_1',Z_c} \\ \vdots & \vdots & \vdots & \vdots \\ J_{X_m',R} & J_{X_m',X_c} & J_{X_m',Y_c} & J_{X_m',Z_c} \\ J_{Y_m',R} & J_{Y_m',X_c} & J_{Y_m',Y_c} & J_{Y_m',Z_c} \\ J_{Z_m',R} & J_{Z_m',X_c} & J_{Z_m',Y_c} & J_{Z_m',Z_c} \end{pmatrix} \begin{pmatrix} \Delta R \\ \Delta X_c \\ \Delta Y_c \\ \Delta Z_c \end{pmatrix} = \begin{pmatrix} X_1'' \\ Y_1'' \\ Z_1'' \\ \vdots \\ X_m'' \\ Y_m'' \\ Z_m'' \end{pmatrix} \tag{13}$$

The initial parameters vector starting the Gauss-Newton iteration may be supplied from an algebraic circle fitting. Since the above algorithm is very robust, we can also take the center of gravitation and the RMS central distance

$$\mathbf{X}_c = \overline{\mathbf{X}} = \frac{1}{m}\sum_{i=1}^{m}\mathbf{X}_i, \quad R = \sqrt{\frac{1}{m}\sum_{i=1}^{m}\left\|\mathbf{X}_i - \overline{\mathbf{X}}\right\|^2}\,. \tag{14}$$

For an experimental example of the circle fitting, we have taken $m = 6$ coordinate pairs ($n = 2, q = 3$) in table 1 [13], initial parameters $\mathbf{a}_k = (2.97, 4.50, 6.67)^{\mathrm{T}}$ from equation (14), and step size $\lambda = 1.3$. After 11 Gauss-Newton steps with norm of the terminal correction vector $\|\Delta\mathbf{a}\| = 5.32\times10^{-7}$, we have obtained (see fig. 1b)
$\hat{\mathbf{a}} = (4.71423, 4.73978, 2.98353)^{\mathrm{T}}$ and $\sigma_0^2 = 1.1080$.
A comparable algorithm, with initial parameters from an algebraic circle fitting, after 11 Gauss-Newton steps with $\|\Delta\mathbf{a}\| = 2.05\times10^{-6}$ delivers the same estimations [13].

4 Geometric fitting of ellipse

An ellipse in a plane can be uniquely described with $q = 5$ parameters, center coordinates X_c, Y_c, axis lengths a, b ($a \geq b$) and pose angle α ($-90° < \alpha \leq 90°$) (see fig. 2a). For the least squares orthogonal distances fitting of ellipse with the algorithm proposed in section 2, we must locate the nearest corresponding point $\mathbf{X}_i'$ (orthogonal contacting point) on the ellipse for the given point $\mathbf{X}_i$, and evaluate the Jacobian matrix at $\mathbf{X}_i'$. To locate the point $\mathbf{X}_i'$ on the ellipse, Gander et. al. [13] have introduced additional m unknown parameters, where each point $\mathbf{X}_i'$ has an individual angular parameter to be simultaneously estimated with the 5 ellipse parameters. As a consequence, their linear system has a bulky and sparse Jacobian matrix and shows only a deteriorative convergence.

We have solved these problems of the orthogonal contacting point and Jacobian matrix by introducing a temporary coordinate system xy positioned at (X_c, Y_c) and rotated with the angle α. The coordinate transformation between the two coordinate systems xy and XY is (see fig. 2a)

$$\mathbf{x} = \mathbf{R}(\mathbf{X} - \mathbf{X}_c) \quad \text{or} \quad \mathbf{X} = \mathbf{R}^{-1}\mathbf{x} + \mathbf{X}_c, \quad \text{where} \tag{15}$$

$$\mathbf{R} = \begin{pmatrix} C & S \\ -S & C \end{pmatrix} \quad \text{and} \quad \mathbf{R}^{-1} = \mathbf{R}^{\mathrm{T}}, \quad \text{with} \quad C = \cos\alpha, \quad S = \sin\alpha. \tag{16}$$

4.1 Orthogonal contacting point on ellipse

In the xy system, 3 of 5 ellipse parameters disappear (X_c, Y_c and α) and the ellipse will be described only with the axis lengths a, b as below (standard position)

$$\frac{x^2}{a^2} + \frac{y^2}{b^2} = 1. \tag{17}$$

For a given point (x_i, y_i), the tangent line at the orthogonal contacting point (x, y) on the ellipse, and the connecting line of the two points are perpendicular to each other

$$\frac{dy}{dx} \cdot \frac{y_i - y}{x_i - x} = \frac{-b^2 x}{a^2 y} \cdot \frac{y_i - y}{x_i - x} = -1. \tag{18}$$

Rewrite equations (17)-(18)

$$f_1(x, y) = \frac{1}{2}\left(a^2 y^2 + b^2 x^2 - a^2 b^2\right) = 0, \quad \text{and} \tag{19}$$

$$f_2(x, y) = b^2 x(y_i - y) - a^2 y(x_i - x) = 0. \tag{20}$$

The orthogonal contacting point (x, y) on the ellipse must simultaneously satisfy equations (19)-(20) (orthogonal contacting conditions). Safaee-Rad et. al. [4] have combined the two equations (19)-(20) into one quadric equation and chosen one solution which has the shortest connecting line length among the maximum 4 real solutions. We have solved this nonlinear problem using the generalized Newton method as below:

$$\mathbf{Q} = \begin{pmatrix} \frac{\partial f_1}{\partial x} & \frac{\partial f_1}{\partial y} \\ \frac{\partial f_2}{\partial x} & \frac{\partial f_2}{\partial y} \end{pmatrix} = \begin{pmatrix} b^2 x & a^2 y \\ (a^2 - b^2)y + b^2 y_i & (a^2 - b^2)x - a^2 x_i \end{pmatrix}, \tag{21}$$

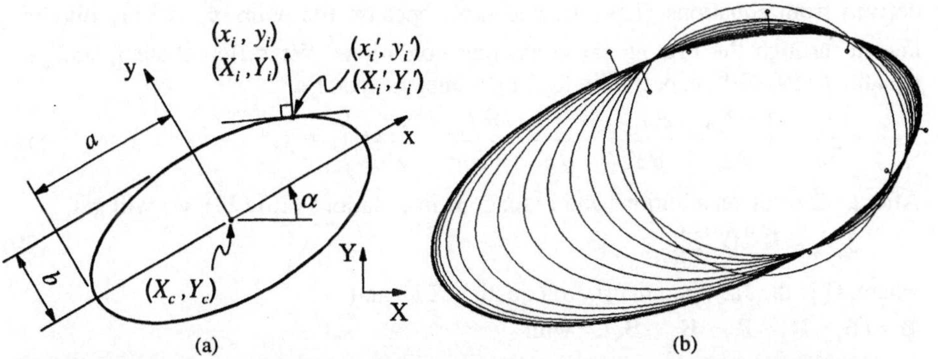

Fig. 2: (a) Ellipse in 2-dimensional plane; (b) Geometric fitting of ellipse to the 8 points in table 1.

$$\mathbf{Q}_k \Delta\mathbf{x} = \mathbf{f}(\mathbf{x}_k) \quad \text{and} \quad \mathbf{x}_{k+1} = \mathbf{x}_k - \Delta\mathbf{x}. \tag{22}$$

We supply the initial point on the ellipse starting the iteration of equation (22) as below, from the fact that the given point $\mathbf{x}_i$ and its nearest corresponding point $\mathbf{x}_i'$ on the ellipse lie in the same quadrant of the standard position.

$$\mathbf{x}_k = \begin{cases} \begin{pmatrix} x_i \\ sign(y_i) \cdot \frac{b}{a}\sqrt{a^2 - x_i^2} \end{pmatrix}, & \text{if } |x_i| \le a \\ \begin{pmatrix} sign(x_i) \cdot a \\ 0 \end{pmatrix}, & \text{if } |x_i| > a \end{cases}. \tag{23}$$

The iteration of equation (22) with initial values from equation (23) terminates in 3-4 cycles with adequately accurate coordinates for the orthogonal contacting point. After the given point $\mathbf{X}_i$ in XY system is transformed into $\mathbf{x}_i$ in xy system using equation (15), the orthogonal contacting point $\mathbf{x}_i'$ will be found by using the above generalized Newton method. Finally, we will have the point $\mathbf{X}_i'$ through the backward transformation of $\mathbf{x}_i'$ into XY system using equation (15). And the orthogonal error distances vector will be

$$\mathbf{X}_i'' = \mathbf{X}_i - \mathbf{X}_i'. \tag{24}$$

4.2 Jacobian matrix at the orthogonal contacting point on ellipse

If we define the parameters vector $\mathbf{a}$

$$\mathbf{a} = (X_c, Y_c, a, b, \alpha)^{\mathrm{T}}, \tag{25}$$

we get from the derivatives of equation (15) respectively to the parameters vector $\mathbf{a}$

$$\left(\frac{\partial \mathbf{x}_i}{\partial \mathbf{a}}\right) = \begin{pmatrix} -C & -S & 0 & 0 & y_i \\ S & -C & 0 & 0 & -x_i \end{pmatrix}, \quad \text{and} \tag{26}$$

$$\mathbf{J}_{\mathbf{X}_i',\mathbf{a}} = \left(\frac{\partial \mathbf{X}}{\partial \mathbf{a}}\right)\Big|_{\mathbf{X}=\mathbf{X}_i'} = \mathbf{R}^{-1}\left(\frac{\partial \mathbf{x}}{\partial \mathbf{a}}\right)\Big|_{\mathbf{x}=\mathbf{x}_i'} + \begin{pmatrix} 1 & 0 & 0 & 0 & -xS - yC \\ 0 & 1 & 0 & 0 & xC - yS \end{pmatrix}\Big|_{\mathbf{x}=\mathbf{x}_i'}. \tag{27}$$

The derivatives matrix appearing in the right side of equation (27) is to be derived from equations (19)-(20) and (26), because the point $\mathbf{x}_i'$ is only implicitly known through the orthogonal contacting conditions. We differentiate f_1 and f_2 of equations (19)-(20) respectively to the parameters vector $\mathbf{a}$

$$\frac{\partial f_{1,2}}{\partial \mathbf{a}} = \left(\frac{\partial f_{1,2}}{\partial X_c} \quad \frac{\partial f_{1,2}}{\partial Y_c} \quad \frac{\partial f_{1,2}}{\partial a} \quad \frac{\partial f_{1,2}}{\partial b} \quad \frac{\partial f_{1,2}}{\partial \alpha}\right) = \mathbf{0}. \tag{28}$$

After a series of substitutions and reductions in equations (26)-(28), we will get,

$$\mathbf{J}_{\mathbf{X}_i',\mathbf{a}} = \mathbf{R}^{-1}\mathbf{Q}^{-1}\mathbf{B}\big|_{\mathbf{x}=\mathbf{x}_i'} \tag{29}$$

where, $\mathbf{Q}$ is the Jacobian matrix of equation (21), and

$\mathbf{B} = (\mathbf{B}_1 \quad \mathbf{B}_2 \quad \mathbf{B}_3 \quad \mathbf{B}_4 \quad \mathbf{B}_5)$, with

$$\mathbf{B}_1 = \begin{pmatrix} b^2xC - a^2yS \\ b^2(y_i - y)C + a^2(x_i - x)S \end{pmatrix}, \quad \mathbf{B}_3 = \begin{pmatrix} a(b^2 - y^2) \\ 2ay(x_i - x) \end{pmatrix}, \quad \mathbf{B}_4 = \begin{pmatrix} b(a^2 - x^2) \\ -2bx(y_i - y) \end{pmatrix},$$

$$\mathbf{B}_2 = \begin{pmatrix} b^2xS + a^2yC \\ b^2(y_i - y)S - a^2(x_i - x)C \end{pmatrix}, \quad \mathbf{B}_5 = \begin{pmatrix} (a^2 - b^2)xy \\ (a^2 - b^2)(x^2 - y^2 - xx_i + yy_i) \end{pmatrix}. \qquad (30)$$

4.3 Orthogonal distances fitting of ellipse

With the error distances vector $\mathbf{X}_i''$ of equation (24) and the Jacobian matrix $\mathbf{J}_{\mathbf{X}_i',\mathbf{a}}$ of equation (29) at each point $\mathbf{X}_i'$, we construct $p\ (=2m)$ linear equations for the m given 2-dimensional points. For the ellipse fitting, the linear equations (5) look like

$$\begin{pmatrix} J_{X_1',X_c} & J_{X_1',Y_c} & J_{X_1',a} & J_{X_1',b} & J_{X_1',\alpha} \\ J_{Y_1',X_c} & J_{Y_1',Y_c} & J_{Y_1',a} & J_{Y_1',b} & J_{Y_1',\alpha} \\ \vdots & \vdots & \vdots & \vdots & \vdots \\ J_{X_m',X_c} & J_{X_m',Y_c} & J_{X_m',a} & J_{X_m',b} & J_{X_m',\alpha} \\ J_{Y_m',X_c} & J_{Y_m',Y_c} & J_{Y_m',a} & J_{Y_m',b} & J_{Y_m',\alpha} \end{pmatrix} \begin{pmatrix} \Delta X_c \\ \Delta Y_c \\ \Delta a \\ \Delta b \\ \Delta\alpha \end{pmatrix} = \begin{pmatrix} X_1'' \\ Y_1'' \\ \vdots \\ X_m'' \\ Y_m'' \end{pmatrix}. \qquad (31)$$

The initial parameters vector starting the Gauss-Newton iteration may be supplied from an algebraic ellipse fitting, or from a circle fitting. When the parameters from a circle fitting are to be used as the initial values for the ellipse fitting, we set

$$(X_c, Y_c)_{\text{ellipse}} = (X_c, Y_c)_{\text{circle}}, \quad a = b = R \quad \text{and} \quad \alpha = 0. \qquad (32)$$

If $a = b$, the elements of $\mathbf{B}_5$ in equation (30) are all zero and consequently the last column of the Jacobian matrix in equation (31) is filled with zero. In this case, $\Delta\alpha$ will have the solution $\Delta\alpha = 0$ through singular value decomposition (SVD) and succeeding backsubstitution [16]. During the iteration, if $a < b$ after an updating of the parameters, we simply exchange the two values and set $\alpha \leftarrow \alpha - sign(\alpha)\cdot\pi/2$.

For an experimental example of the ellipse fitting, we have taken 8 coordinate pairs in table 1 [13], initial parameters $\mathbf{a}_k = (4.84, 4.80, 3.39, 3.39, 0)^{\mathrm{T}}$ from the geometric circle fitting, and step size $\lambda = 1.2$. After 19 Gauss-Newton steps with norm of the terminal correction vector $\|\Delta\mathbf{a}\| = 4.17\times10^{-6}$, we have obtained (see fig. 2b)

$$\hat{\mathbf{a}} = (2.69961,\ 3.81596,\ 6.51872,\ 3.03189,\ 0.35962)^{\mathrm{T}} \quad \text{and} \quad \sigma_0^2 = 1.1719.$$

The comparable algorithm of Gander et. al. [13] gives the same estimations only after 71 Gauss-Newton steps. Besides, they could not directly use the parameters from a circle fitting, because of the singularity in their Jacobian matrix by $a = b$. To bypass the singularity problem, they have used $a = R$, $b = R/2$. But an arbitrary modification of one parameter without proper adjustments to the others can cause a divergence!

5 Discussion and Conclusions

We have proposed algorithms for the least squares orthogonal distances fitting of circle and ellipse. For the iterative circle fitting, the Jacobian matrix at the nearest point on the circle from a given point is directly available from the circle parameters. The proposed circle fitting is very robust and we can simply use the center of gravitation and the RMS central distance as the initial parameter values. However, for the ellipse fitting, it is not easy to locate the nearest point on the ellipse from a given point, and to evaluate the Jacobian matrix. We have overcome these difficulties by transforming the ellipse into the standard position and utilizing the orthogonal contacting conditions. The proposed ellipse fitting is robust and can take the parameters from a circle fitting without modification, and the comparison with other geometric ellipse fitting shows it has superior performances in convergence.

References

1. N. I. Chernov and G. A. Ososkov, Effective algorithms for circle fitting, *Computer Physics Communications* **33**, 329-333 (1984).
2. B. B. Chaudhuri and P. Kundu, Optimum circular fit to weighted data in multi-dimensional space, *Pattern Recognition Letters* **14**, 1-6 (1993).
3. F. L. Bookstein, Fitting Conic Sections to Scattered Data, *CGIP* **9**, 56-71 (1979).
4. R. Safaee-Rad, I. Tchoukanov, B. Benhabib and K. C. Smith, Accurate Parameter Estimation of Quadric Curves from Grey-Level Images, *CVGIP: Image Understanding* **54**(2), 259-274 (1991).
5. P. L. Rosin, Analyzing Error of Fit Functions for Ellipses, *Pattern Recognition Letters* **17**, 1461-1470 (1996).
6. K. Kanatani, Statistical Bias of Conic Fitting and Renormalization, *IEEE Trans. PAMI* **16**(3), 320-326 (1994).
7. J. Cabrera and P. Meer, Unbiased Estimation of Ellipses by Bootstrapping, *IEEE Trans. PAMI* **18**(7), 752-756 (1996).
8. A. W. Fitzgibbon, M. Pilu and R. B. Fisher, Direct Least Squares Fitting of Ellipses, *Proc. 13th Int. Conf. on Pattern Recognition*, Vienna, 253-257 (1996).
9. R. N. Goldman, Two Approaches to a Computer Model for Quadratic Surfaces, *IEEE Computer Graphics & Applications* **3**(9), 21-24 (1983).
10. DIN 32880-1, *Koordinatenmeßtechnik; Geometrische Grundlagen und Begriffe*, Deutsche Norm, Beuth Verlag, Berlin (1986).
11. M. Berman, Large Sample Bias in Least Squares Estimators of a Circular Arc Center and Its Radius, *CVGIP* **45**, 126-128 (1989).
12. A. vom Hemdt, *Standardauswertung in der Koordinatenmeßtechnik - Ein Beitrag zur Geometrieberechnung*, Dissertation, RWTH Aachen, FRG (1989).
13. W. Gander, G. H. Golub and R. Strebel, Least-squares fitting of circles and ellipses, *BIT* **34**, 558-578 (1994).
14. Y. Cui, J. Weng and H. Reynolds, Estimation of ellipse parameters using optimal minimum variance estimator, *Pattern Recognition Letters* **17**, 309-316 (1996).
15. H. W. Sorenson, *Parameter Estimation: Principles and Problems*, Marcel Dekker, New York (1980).
16. W. H. Press, B. P. Flannery, S. A. Teukolsky and W. T. Vetterling, *Numerical Recipes in C: The Art of Scientific Computing*, Cambridge University Press, Cambridge, UK (1988).

Real-Time Adaptive Smoothing with a 1-D Nonlinear Relaxation Network in Analogue VLSI Technology*

K. Wiehler[1], R.-R. Grigat[1]
J. Heers[2], C. Schnörr[2], and H.S. Stiehl[2]

[1] Technische Universität Hamburg-Harburg, Technische Informatik I
Harburger Schloßstraße 20, D-21071 Hamburg, Germany
phone: +49 (0)40 7718 -3536, -3125, fax: -2557
email: {k.wiehler, grigat}@tu-harburg.de
http://www.ti1.tu-harburg.de

[2] Universität Hamburg, FB Informatik, AB Kognitive Systeme
Vogt-Kölln-Str. 30, D-22527 Hamburg, Germany
phone: +49 (0)40 5494 -2576, -2581, -2453, fax: -2572
email: {heers, schnoerr, stiehl}@informatik.uni-hamburg.de
http://kogs-www.informatik.uni-hamburg.de

Abstract Reconstruction of given noisy data is an ill-posed problem and a computationally intensive task. Non-linear regularisation techniques are used to find a unique solution under certain constraints. In our contribution we present a parallel mixed-signal architecture which solves this non-linear problem within microseconds. By connecting all parallel cells in a circular manner it is possible to process noisy data vectors of infinite length. This is achieved by virtually shifting the non-linear adaptive filter kernel over the noisy data vector. A 1–D experimental chip has been fabricated using 0.8μm CMOS technology. On–chip measurements are shown to agree with results from numerical simulations. Results from applying the 1–D chip to nonlinear smoothing of image data will also be given correspondence.

1 Introduction

In [4] a variational approach to the unique reconstruction of images has been introduced. Minimisation of the underlying non–linear functional corresponds to nonlinear diffusion which adaptively smoothes images while preserving characteristic signal structures (e. g. contrast edges in images). The performance of the approach can be characterised by the following theoretical properties: unique solution, data adaption, continuous dependency of the result on both the input data and the regularisation parameters. However the theoretical achievement and

* Financial support by the Deutsche Forschungsgemeinschaft, research programme "System- und Schaltungstechnik für hochgradige Parallelverarbeitung" and Austria Microsystems GmbH (AMS) is gratefully acknowledged

the proven practicability contrast sharply with the high computational costs of sequential digital realisations.

As an alternative, analogue hardware concepts allow to map classes of early vision problems onto time-continuous high–speed analogue circuitry by the direct use of voltage, current and charge relationships of the physical devices [3].

In this contribution we present an analogue realisation of a 1–D relaxation network solving the non–linear smoothing problem. The advantages and limitations of analogue relaxation networks are presented. Implementing massive parallel analogue VLSI systems requires that deviations from the theoretical characteristics have to be taken into account in an early state of system design. Models of e. g. erroneous transfer functions as realized through system components have to be created and checked for consistency with theoretically proven transfer functions given by regularisation theory. With such models useful parameter ranges can be computed for different analogue designs. Sample mixed-signal VLSI circuits have been developed and fabricated. The circuits are fully functional in a laboratory set–up. Finally, applications to image processing are experimentally demonstrated.

2 Theory

In [4] a variational model for non-linear adaptive smoothing of multi-dimensional signals is described. The main properties are characterised by: (i) unique solution, (ii) data adaptation, (iii) continuous dependency of the result on both the input data and (iv) the regularisation parameters $(c_\rho, \lambda_h, \lambda_l)$. These are basic prerequisites for a successful design of an asymptotically stable relaxation network based on analogue hardware.

A suitable discretisation preserving the above characteristics is obtained by applying the Ritz-Galerkin method to the variational model. Let $g_i \in [0,1]$ denote input data and u_i the result after adaptive smoothing. Using piecewise linear basis functions we get for the 1–D case on an equidistant grid $\mathcal{D}$ $(i = 0..N-1)$ [4]

$$\dot{u}_0 = \frac{1}{6}\left(g_1 - u_1 + 2\left(g_0 - u_0\right)\right) + f_{nl}(u_1 - u_0) \tag{1}$$

$$\begin{aligned}\dot{u}_i =& \frac{1}{6}\left(g_{i-1} - u_{i-1} + 4\left(g_i - u_i\right) + g_{i+1} - u_{i+1}\right)\\ &+ f_{nl}(u_{i-1} - u_i) + f_{nl}(u_{i+1} - u_i) \quad : i = 1..(N-2)\end{aligned} \tag{2}$$

$$\dot{u}_{N-1} = \frac{1}{6}\left(g_{N-2} - u_{N-2} + 2\left(g_{N-1} - u_{N-1}\right)\right) + f_{nl}(u_{N-2} - u_{N-1}) \tag{3}$$

with the basic piecewise linear function

$$f_{nl}(t) = \begin{cases} \lambda_h^2 t & , 0 \le t \le c_\rho \\ \lambda_l^2 t + (\lambda_h^2 - \lambda_l^2) c_\rho \operatorname{sgn}(t) & , t > c_\rho \end{cases} \tag{4}$$

Reasonable parameter ranges for real data are: $3 < \lambda_h^2 < 100$, $\lambda_l^2 \ll 1$, $0.002 < c_\rho < 0.05$.

3 Analogue Implementation

Based on the 1–D discretisation of the convex functional (1)–(3) an architecture has been developed. The design has been driven by the required range of the parameter space (c_ρ and λ_h^2) and in addition relies on an perturbation analysis indicating the minimum accuracy necessary to achieve satisfying performance [7].

3.1 A dynamic circular architecture for processing infinite signals

One major problem in designing analogue circuits for signal processing is the interfacing to the signal source and the subsequent (digital) processing unit. When the signal is generated off-chip the signal must be fed into the parallel structure of the analogue unit. Usually this is done by simply writing the input to the isolated cells. During this time all cells act as memory elements, and afterwards the array *computes* the result within microseconds which is read out by the subsequent unit [2]. Such a design yields a rather poor computational performance since the effective computation time is low in comparison to the time used for data input/output. Another problem when computing signal vectors larger than the chip cell array is that the signal vector has to be broken apart into overlapping blocks which then have to be processed sequentially the other and merged by post-processing [1].

Our solution to these problems is based on a dynamic reconfigurable circular architecture which is shifted along the data vector at a very high rate [6]. For the 1–D case the neighbourhood $\mathcal{N}_r(i)$ of the cell C_i within a radius r on a grid $\mathcal{D}$ is defined by (Fig. 1 a)

$$\mathcal{N}_r(i) = \{C_j \mid \min(<i-j>_N, <j-i>_N) \leq r, j = 0, \ldots, N-1, j \neq i\}$$
$$i = 0, \ldots, N-1 \tag{5}$$

$< \cdot >_N$ is the modulo N operation. The shift mechanism of the network first replaces the static cell links with programmable switches and, second, allows each cell to performing different working modes (see (7)–(9) below).

Let u_i denote the state variable of the cell C_i. The dynamic behaviour of the cell is defined by

$$\frac{\partial u_i}{\partial t} = F\left(u_i, \mathcal{N}(i), \tilde{g}_i\right) \ . \tag{6}$$

$\mathcal{N}(i)$ is the local neighbourhood modulo the network boundaries as defined in (5). The terms constant during relaxation for node C_i are denoted by $\tilde{g}_i$ for centre cells and $\tilde{g}_i'$ for boundary cells. To operate the network in the dynamic manner as described above, each cell can be programmed to satisfy one of the following

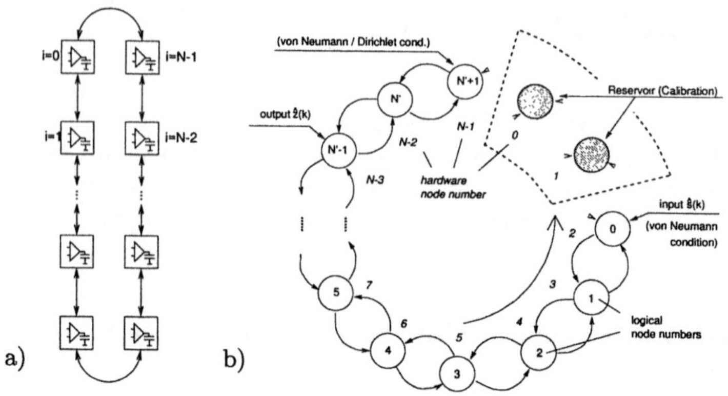

Figure 1. *a) 1–D circular network b) Dynamic circular loop: the network is reconfigured while the analogue state variables stay resident in their corresponding cells. Part of the network can be disconnected from the active network for (e.g. offset voltage) calibration purposes.*

conditions (all index calculations are modulo N):

$$\text{fully connected cell} \quad : \quad \frac{\partial u_i}{\partial t} = F\left(u_{i-1}, u_i, u_{i+1}, \tilde{g}_i\right) \tag{7}$$

$$\begin{array}{c}\text{left (right) boundary cell}\\ \text{v. Neumann (zero flux) cond.}\end{array} \quad : \quad \frac{\partial u_i}{\partial t} = F\left(u_i, 2\, u_{i+1(i-1)}, \tilde{g}_i'\right) \tag{8}$$

$$\begin{array}{c}\text{constant boundary cell}\\ \text{Dirichlet (fixed) condition}\end{array} \quad : \quad \frac{\partial u_i}{\partial t} = 0 \tag{9}$$

Furthermore, a cell can be disconnected from the active network for calibration purposes. In Fig. 1 the operation of a 1–D dynamic network is sketched. The analogue values reside in the corresponding cell while only the connections of input, output and boundary cells are reconfigured. The analogue network processes the data while it is dynamically reconfigured. Moreover, the reconfiguration process is simple because only a maximum number of six cells have to be accessed per cycle.

When no processing is performed, the network acts like an analogue shift register with a delay of N' multiplied by the sample time period. The effective length N' is the distance between the input and output node. It depends on the size of calibration pool and available boundary conditions (see Fig. 1). For optimal Dirichlet boundary condition ($\frac{\partial u_i}{\partial t} = 0$) at the end of the network, the effective length N' can be chosen as large as possible since there will be no reflection from that boundary node.

3.2 Signal propagation

Let u_i^k denotes an index operation modulo N in the kth step: $u_i^k = u_{<i-k>_N}$ with i the logical node number and $<i-k>_N$ the hardware node number. By applying equations (7)–(9) to the simplified 1–D discrete case of equation (1)–(3), the systems equations can be formulated as follows

$$\dot{u}_i^k = \frac{1}{2}(g_i^k - u_i^k) + f_{nl}(|u_{i+1}^k - u_i^k|) \qquad : i = 0 \qquad (10)$$

$$\dot{u}_i^k = (g_i^k - u_i^k) + f_{nl}(|u_{i-1}^k - u_i^k|) + f_{nl}(|u_{i+1}^k - u_i^k|) \qquad : i = 1..(N-2) \qquad (11)$$

$$\dot{u}_i^k = 0 \qquad : i = N-1 \qquad (12)$$

for $kT < t \leq (k+1)T$ while the cell inputs are $g_i^k = \hat{s}(k-i), \forall i = 0, \ldots, N-1$, and the initial values ($t = kT$) are given by the preexisting intermediate results

$$u_0^k(t = kT) = \hat{s}(kT) \qquad : i = 0 \qquad (13)$$

$$u_i^k(t = kT) = u_{i-1}^{k-1}(t = kT) \qquad : i = 1..(N-1)\ . \qquad (14)$$

$\hat{s}(k)$ is the input signal and T the sample time. The signal is virtually shifted over the network since the input signal g and the initial conditions (14) remain constant, respectively. The output can be extracted at each processing step by retrieving the state value of the Dirichlet boundary node $\hat{z}(k) = u_{N-1}^k$.

Furthermore an *area of influence* is defined in a neighbourhood of the cell C_j: $i - \Lambda < j < i + \Lambda$. The value of this crucial distance is determined by the characteristic space constant (e.g. diffusion constant) for the underlying signal processing task. As a matter of fact Λ is also limited by the size of the network and the required or achievable accuracy of the used analogue circuit.

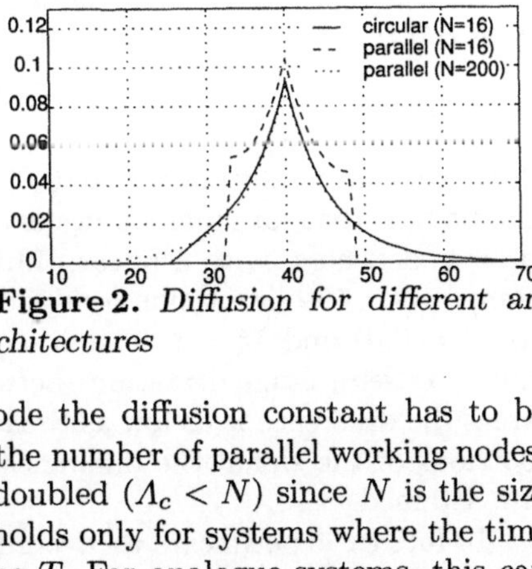

Figure 2. *Diffusion for different architectures*

In conventional parallel processing mode the diffusion constant has to be chosen to ensure $\Lambda_p < N/2$, with N being the number of parallel working nodes. For the dynamic circular mode Λ can be doubled ($\Lambda_c < N$) since N is the size of memory in the network. Of course this holds only for systems where the time constant τ is smaller than the sample time T. For analogue systems, this can be guaranteed by the extremely fast settling time of the network. In Fig. 2 the impulse response for different network architectures is shown. The dotted line is the correct propagation for a linear diffusion ($\Lambda \approx 30$) computed on a 1–D network of 200 nodes operating in parallel. The dashed line is valid for a parallel network of only 16 nodes, whereas the solid one holds for a network of the same size operating in the dynamic circular manner. It can be seen that the difference in impulse response between loop-mode and the correct propagation result is much smaller than between the parallel and ideal mode. Due to the fixed value boundary on the right, the memory length is extended to infinity resulting in a asymmetric impulse response.

3.3 1–D 0.8μm CMOS Realisation and 1–D/2–D Experiments

Based upon the algorithm and architecture described in the previous sections a 1–D prototype consisting of 32 identical analogue cells was fabricated (see Fig. 3). The major challenge was the design of the non-linear function which has to be very accurate to guarantee minimum impact of transistor mismatch. The piecewise linear function is achieved by cascading a standard OTA and a current limiter similar to [5]. Systematic errors are further reduced by applying minimal design cascode transistors in the output path. The design of this basic cell was done after careful consideration of optimal layout conditions. The dynamic routing resources are included by MOS switches which are controlled by a digital unit [6]. In Fig. 4 measured results for different values of c_ρ are

Figure 3. *Layout photograph of the chip: 32 cells are performing a non-linear filtering of analogue signals of infinite length (fabricated in 0.8μm CMOS, Austria Microsystems/AMS)*

depicted. c_ρ ranges from 0.01 to 0.05 for an approximate constant value of λ_h^2. It can be seen that for larger values of c_ρ the segmented areas are growing until the process degenerates to a linear diffusion. In Fig. 5 the measured signal can be compared to the simulation results computed by solving the non-linear equation for $c_\rho = 0.01$ and $\lambda_h^2 = 8$. The difference between these signal vectors is mainly due to offset voltage deviations between neighbouring cells. Whereas in smooth areas the difference is as low as the offset voltage itself (a few millivolt), in steep portions of the signal the difference can be large since different segmentations are attained.

In further experiments an overall random error of about $\sigma_v = 2.5$mV (standard deviation of the voltage differences between input and output of band limited signals) was measured. In [7] it was shown that this limits the minimum value for c_ρ to about 0.005. This first experimental design can successfully perform computationally intensive non–linear smoothing of 1–D signals.

In addition to the 1-D experiments, we carried out several experiments with 2-D gray-value images of size 256 x 256 pixels (see Fig. 6). The goal of these experiments was to investigate from a practical perspective to which extent even the 1-D chip with its reduced design complexity is capable of satisfactorily smoothing 2-D image data in a non-linear fashion. In this case, the image data was fed into the 1-D chip in a consecutive row-by-row and column-by-column-manner (note that by this means the result is not fully consistent with the

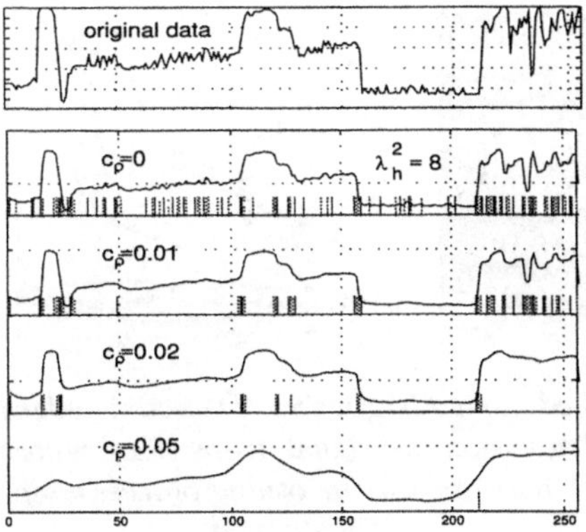

Figure 4. Measurement results for different values of c_ρ. The vertical lines is the segmentation which is inherent to the algorithm according to (4).

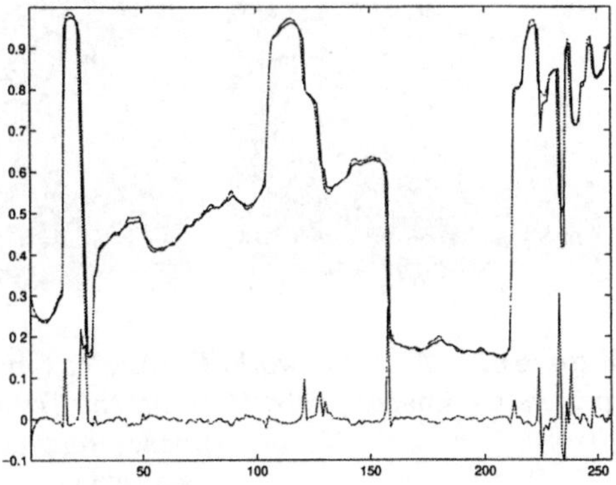

Figure 5. Measured and simulated results for $c_\rho = 0.01$ and λ_h^2=8. The largest differences between the simulation and measured data is located in steep signal portions due to different segmentation results.

2-D theory since only an orientation-constrained dual non-linear diffusion along orthogonal data vectors can be achieved). The promising result in Fig. 7 and Fig. 8 demonstrates the validity of this pragmatic approach since adaptive non-linear smoothing as well as preservation of contrast edges can be achieved. For this experiment, the partial derivatives u_x and u_y have been computed via simple finite differences whereas the threshold has been set to values slightly above the theoretical value of c_ρ such as to reduce the influence of the inevitable noise of the analogue components. Our current experience with the 1-D experimental chip indicates a peak performance of 500 kHz/pixel (equivalent to 130 msec/image). Since this performance is not limited by the relaxation time of the 1-D network of cells itself but solely by the rather weak performance of the currently used read-out circuitry, it can be even further improved in the future through appropriate design measures. Eventually the experiments showed that the 1-D chip has the potential of a powerful but cheap image preprocessing device.

4 Conclusion

An efficient analogue hardware implementation of a non–linear diffusion model has been proposed and fabricated. Substantial problems in the design of massive parallel analogue hardware have been overcome. The analogue network cells are connected in a circular structure. Due to the dynamic reconfiguration of the connections a non-linear adaptive filter kernel can be shifted virtually over a signal vector of *infinite* length. The boundary conditions have to be configured as von Neumann (zero flux) or Dirichlet (fixed) boundaries. Furthermore, our design has the potential for calibrating cells, that are not in the active part of the network, without disrupting the data stream. The storage duration is

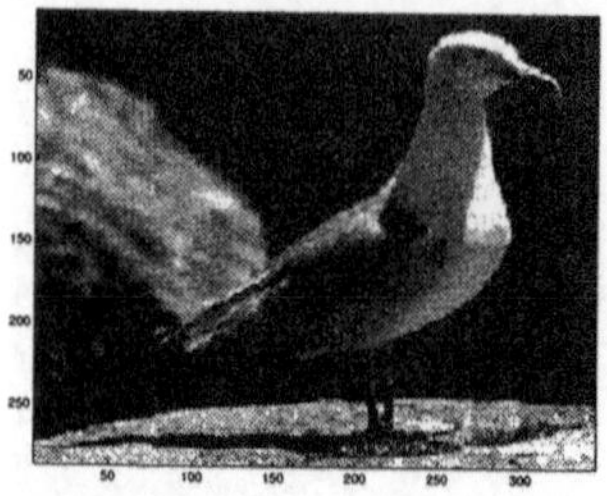

Figure 6. *A real-world gray-value image (256x256 Pixel).*

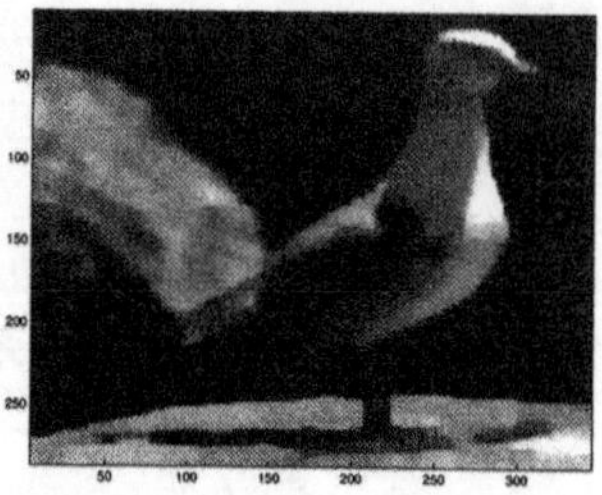

Figure 7. *Smoothed image using the 1D analogue non-linear relaxation network (see text).*

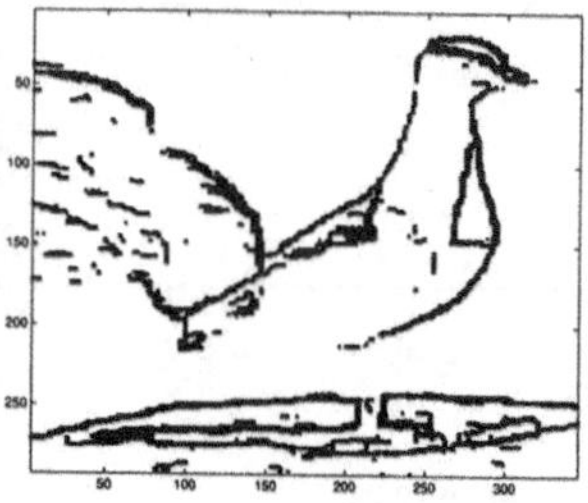

Figure 8. *Detected edges as signal transitions where the smoothing process adaptively stopped.*

constant for all analogue signal samples. Consequently linear systematic error effects resulting from leakage currents can be compensated as well. A prototype containing a 1–D non-linear network has been fabricated and is fully functional. The chip can also be used for 2–D image smoothing provided that the result of successive relaxation along the rows and columns is acceptable for a given task.

Further work will focus on a 2–D realisation fully consistent with the multi-dimensional theory. In addition, effective error reduction methods and efficient architectures for higher dimensions have to be investigated in order to exploit the potential of parallel analogue hardware for computer vision.

References

1. Jose Pineda de Gyvez, Lei Wang, and E. Sanchez-Sinencio. Large-image CNN hardware processing using a time multiplexing scheme. In *IEEE International Workshop on Cellular Neural Networks and their Applications*, pages 405–410, 1996.
2. Peter Kinget and Michiel Steyaert. *Analog VLSI integration of massive parallel signal processing systems.* Kluwer Academic Publishers, 1997.
3. Carver Mead. *Analog VLSI and Neural Systems.* Addison-Wesley Computational and Neural Systems Series. Addison-Wesley Publishing Company, 1989.
4. C. Schnörr. Unique reconstruction of piecewise-smooth images by minimizing strictly convex nonquadratic functionals. *Journal of Mathematical Imaging and Vision*, 4:189–198, 1994.
5. Bing J. Sheu and Joongho Choi. *Neural Information Processing and VLSI*, chapter 14, pages 397–466. Kluwer Academic Publishers, 1995.
6. Klaus Wiehler, Ralph Lembcke, Rolf-Rainer Grigat, Josef Heers, Christoph Schnörr, and H.-Siegfried Stiehl. Dynamic circular cellular networks for adaptive smoothing of multi-dimensional signals. In *IEEE International Workshop on Cellular Neural Networks and their Applications, London*, 1998.
7. Ralf Wöbke. Störungsanalyse eines eindimensionalen nichtlinearen Relaxationsnetzwerks zur Signalverarbeitung. Diplomarbeit, Universität Hamburg, Fachbereich Informatik, Arbeitsbereich Kognitive Systeme, 1997.

Ein verallgemeinerter Total Least Squares-Ansatz zur Schätzung der Epipolargeometrie

Matthias Mühlich und Rudolf Mester

J. W. Goethe-Universität Frankfurt, Germany
(Muehlich|Mester)@iap.uni-frankfurt.de

Zusammenfassung Kern dieses Beitrags ist (in Erweiterung von [6]) die statistische Analyse des $8+n$-Punkte Algorithmus' zur Bestimmung der Fundamentalmatrix. Dadurch gelingt es, eine Verbesserung dieses als sehr empfindlich geltenden Verfahrens zu erreichen, die über die Wirkung der von Hartley vorgeschlagenen Normalisierungstransformationen hinausgeht. An dem hier betrachteten „klassischen" Vision-Problem wird deutlich, daß die möglichst genaue statistische Analyse des Fehlerverhaltens eines Algorithmus' keine akademische Fingerübung, sondern eine zwingende Notwendigkeit auf dem Weg zu zuverlässigen und praxistauglichen Verfahren ist.

1 Ein altes Problem, neu betrachtet ...

Die Berechnung der Epipolargeometrie bzw. der relativen Orientierung gehört zu den zentralen Schritten der Bewegtbild-Analyse. Ausgehend von den klassischen Verfahren [5, 8] sind für diesen Zweck eine Reihe von nicht-iterativen[1] Algorithmen entwickelt worden, die sich durch geringen Rechenaufwand auszeichnen, jedoch den Nachteil haben, daß sie selbst beim Vorliegen vieler Punktkorrespondenzen äußerst empfindlich gegen Störungen in den Eingangsdaten sind.

Der Grund für die Empfindlichkeit liegt darin, daß das Standard-Verfahren bei verrauschten Daten statistische Eigenschaften dieser Daten voraussetzt, die überhaupt nicht gegeben sind. Die von Hartley [3] vorgeschlagenen Transformationen der Eingangsdaten beseitigen dieses Problem weitgehend. Sie werden im vorliegenden Beitrag statistisch begründet und weiter verbessert. Abschließend gehen wir noch auf einige flankierende Experimente mit simulierten Daten ein. Wir weisen aber schon an dieser Stelle darauf hin, daß in Anbetracht der vielen verschiedenen Parameter (Rotation, Translation, Struktur des Objekts, Ansicht, Größe der Fehler, Anzahl der Punktkorrespondenzen) dem Experiment bestenfalls die Rolle zukommt, Schwachstellen in der Theorie zu finden oder ihren Gültigkeitsbereich zu ermitteln.

[1] Diese Algorithmen sind nicht-iterativ mit Ausnahme der Lösung von Eigensystem-Gleichungen, die allerdings den Kern der Algorithmen bilden.

2 Konventionelles Verfahren und Total Least Squares

Zunächst fassen wir kurz den 8-Punkte-Algorithmus[2] zusammen, so wie er in [3] sehr kompakt beschrieben wird:

Es seien $\boldsymbol{u} = (u_1, u_2, 1)^T$ und $\boldsymbol{v} = (v_1, v_2, 1)^T$ die homogenen Koordinaten der Projektionen eines Objektpunktes in zwei verschiedenen Bildern. Wenn diese Koordinaten fehlerfrei sind, gilt $\boldsymbol{v}^T \mathbf{F} \boldsymbol{u} = 0$ mit einer 3×3-*Fundamentalmatrix* $\mathbf{F}$. Diese Gleichung gilt für alle Paare von korrespondierenden Punkten. Unsere Aufgabe ist nun, die Matrix $\mathbf{F} = (f_{ij})$ zu bestimmen, in der alle benötigten Informationen über die Epipolargeometrie enthalten sind. Die Matrix $\mathbf{F}$ hat stets Rang 2 [8]. Faßt man die Elemente von $\mathbf{F}$ zu einem Vektor $\boldsymbol{f}$ zusammen, gilt $\boldsymbol{a}^T \cdot \boldsymbol{f} = 0$ mit $\boldsymbol{a}^T = (v_1 u_1,\ v_1 u_2,\ v_1,\ v_2 u_1,\ v_2, u_2,\ v_2,\ u_1,\ u_2,\ 1)$. Jede der N Punktkorrespondenzen $(\boldsymbol{u}_i \leftrightarrow \boldsymbol{v}_i)$ liefert eine Gleichung dieser Form; zusammengefaßt erhalten wir:

$$\mathbf{A}_0 \cdot \boldsymbol{f} = \mathbf{0}\ , \tag{1}$$

wobei die Koeffizientenvektoren $\boldsymbol{a}_i^T$ $(i = 1, \ldots, N)$ die Zeilenvektoren der $N \times 9$-Matrix $\mathbf{A}_0$ sind (das Subskript 0 steht für eine ungestörte Matrix). Wenn eine Lösung für $\boldsymbol{f}$ (außer $\boldsymbol{f} = \mathbf{0}$) existieren soll, muß $\mathsf{Rg}\,(\mathbf{A}_0) \leq 8$ gelten.[3]

Da wir jedoch nur die Matrix $\mathbf{A} = \mathbf{A}_0 + \mathbf{D}$ messen können, die mit einer unbekannten Fehlermatrix $\mathbf{D}$ gestört ist $(\boldsymbol{a}_i = \boldsymbol{a}_{i0} + \boldsymbol{d}_i)$, gilt in aller Regel $\mathsf{Rg}\,(\mathbf{A}_0) = 9$ und Gleichung $\mathbf{A}\boldsymbol{f} = \mathbf{0}$ hat im Gegensatz zu (1) keine nicht-triviale Lösung mehr. Das konventionelle Verfahren besteht daher aus zwei Schritten:

Schritt 1: Suche den Vektor $\boldsymbol{f}$, der $|\mathbf{A}\boldsymbol{f}|$ minimiert und dabei die Norm 1 hat. Dies ist der Eigenvektor von $\mathbf{A}^T\mathbf{A}$ zum kleinsten Eigenwert von $\mathbf{A}^T\mathbf{A}$ oder, in anderen Worten, der Rechtssingularvektor von $\mathbf{A}$ zum kleinsten Singularwert.[4]

Schritt 2: Setze aus $\boldsymbol{f}$ eine 3×3-Matrix zusammen und ersetze diese durch eine Rang 2 Matrix, so daß die Differenzmatrix minimale Frobeniusnorm hat. Diese *Rangreduktion* kann mit der SVD unter Verwendung des Eckart-Young-Mirsky-Theorems [6, 1] erreicht werden.[5]

Der erste Schritt läßt sich als Total Least Squares Problem formulieren [4]:

Definition 1 (Total Least Squares Problem). *Gegeben ist ein überbestimmtes Gleichungssystem mit N Gleichungen $\boldsymbol{a}^T\boldsymbol{x} = b$ mit M Unbekannten $\boldsymbol{x}$, zusammengesetzt zu einer Matrixgleichung $\mathbf{A}\boldsymbol{x} = \boldsymbol{b}$. Sowohl $\mathbf{A}$ als auch $\boldsymbol{b}$ seien fehlerbehaftet. Die Aufgabe besteht darin, eine $N \times M$-Matrix $\hat{\mathbf{A}}$ und einen M-Vektor $\hat{\boldsymbol{b}}$ zu finden, für die die Gleichung*

$$\hat{\mathbf{A}}\boldsymbol{x} = \hat{\boldsymbol{b}} \tag{2}$$

[2] Gemeint sind eigentlich $8 + n$ Punkte. 8 korrespondierende Punkte sind mindestens notwendig und jede weitere Punktkorrespondenz kann zur Reduktion der Fehler in den Ausgangsdaten genutzt werden.

[3] Für $\mathsf{Rg}\,(\mathbf{A}_0) < 8$ erhält man keine eindeutige Lösung, sondern eine lineare Mannigfaltigkeit.

[4] Eine sehr schöne Darstellung der Singularwert-Zerlegung (SVD) findet man bei [7].

[5] Die Rang 2 Bedingung und der freie Skalierungsfaktor entsprechen der Tatsache, daß die Fundamentalmatrix nur 7 Freiheitsgrade hat.

eine exakte Lösung hat unter der Bedingungen, daß $||(\mathbf{A}|\boldsymbol{b}) - (\hat{\mathbf{A}}|\hat{\boldsymbol{b}})||_F \longrightarrow min$ *gilt. Alle Lösungen* $\boldsymbol{x}$*, die dann (2) lösen, werden als Total-Least-Squares (TLS) Lösungen bezeichnet.*

Van Huffel [4] gibt einen Überblick über die statistischen Eigenschaften von TLS-Lösungen an. Insbesondere ist festzuhalten, daß sich eine erwartungstreue Schätzung der ungestörten Matrix $\mathbf{A}_0$ ergibt, falls die Fehler in $\boldsymbol{a}_i$ und $\boldsymbol{b}$ mittelwertfrei und unkorreliert sind und alle Fehlervektoren die gleiche Kovarianzmatrix $\sigma^2\mathbf{I}$ haben. TLS-Verfahren sind auch unter der Bezeichnung *orthogonale Regression* bekannt.

Der 'klassische' 8+n Punkte Algorithmus setzt – ohne daß dies bisher explizit erwähnt wurde – das TLS-Verfahren ein. 1995 und 1997 stellte Hartley in zwei Artikeln ein Verfahren vor [3], das durch geeignete Normalisierungstransformationen drastisch verbesserte Ergebnisse bei der Schätzung der Fundamentalmatrix erbringt. Hartleys Verfahren beruht darauf, die Datenvektoren $\boldsymbol{u}_i$ und $\boldsymbol{v}_i$ mit zwei nicht-singulären Matrizen $\mathbf{R}$ und $\mathbf{S}$ zu transformieren. Die ursprünglichen Forderung $\boldsymbol{v}^T\mathbf{F}\boldsymbol{u} = \mathbf{0}$ lautet dann

$$\underbrace{\boldsymbol{v}_i^T\mathbf{S}^T}_{\boldsymbol{v}_i'^T}\underbrace{(\mathbf{S}^T)^{-1}\mathbf{F}\mathbf{R}^{-1}}_{\mathbf{F}'}\underbrace{\mathbf{R}\boldsymbol{u}_i}_{\boldsymbol{u}_i'} = 0\,. \tag{3}$$

Man bestimmt nun die Matrix $\mathbf{F}'$ mit TLS, setzt daraus eine 3×3-Matrix zusammen, deren Rang dann auf 2 reduziert wird (s.o.). Mit der Rücktransformation $\mathbf{S}^T\mathbf{F}'\mathbf{R}$ ergibt sich das gesuchte $\mathbf{F}$.

Wenn die dritten Komponenten der homogenen Koordinaten von $\boldsymbol{u}_i$ und $\boldsymbol{v}_i$ nicht geandert werden sollen, lassen sich die Transformationsmatrizen jeweils in eine Skalierung, eine Translation und eine Rotation zerlegen. Hartley benutzt (heuristisch begründet, aber dennoch fast richtig) eine Verschiebung in den Schwerpunkt, eine Drehung in das Hauptachsensystem und eine Skalierung, die dafür sorgt, daß der mittlere quadratische Abstand der Datenpunkte von Ursprung im neuen Koordinatensystem 1 ist. Daraus ergeben sich folgende Bestimmungsgleichungen für die Hartley-Transformationen $\mathbf{R}$ und $\mathbf{S}$:

$$\sum_{i=1}^{N}\mathbf{R}\boldsymbol{u}_i\boldsymbol{u}_i^T\mathbf{R}^T = \mathbf{I} \quad \text{und} \quad \sum_{i=1}^{N}\mathbf{S}\boldsymbol{v}_i\boldsymbol{v}_i^T\mathbf{S}^T = \mathbf{I}\,. \tag{4}$$

3 Mathematischer Hintergrund weiterer Verbesserungen

3.1 Equilibrierung von rechts

In [6] konnte gezeigt werden, daß die Rechtssingularvektoren (d.h. die Lösung unseres TLS-Problems) in linearer Näherung erwartungstreu geschätzt werden, falls für die Störmatrix $\mathbf{D}$ die Bedingungen $\mathsf{E}[\mathbf{D}] = \mathbf{0}$ und

$$\mathsf{E}\left[\mathbf{D}^T\mathbf{D}\right] = \sum_{i=1}^{N}\mathbf{C}_{d_i} = c\,\mathbf{I} \tag{5}$$

gelten.[6] Dazu ist die zuvor gemachte Voraussetzung $\mathbf{C}_{d_i} = \sigma^2\,\mathbf{I}$ nicht notwendig. Was ist aber zu tun, wenn auch (5) nicht erfüllt ist?

Sei $\mathsf{E}\left[\mathbf{D}^T\mathbf{D}\right]$ nun also regulär, aber ansonsten beliebig. Wir fügen in Gleichung (2) zwischen $\mathbf{A}$ und $\boldsymbol{f}$ den Term $\mathbf{B}\mathbf{B}^{-1}$ ein. $\mathbf{B}$ darf natürlich nicht singulär sein, da $\mathbf{B}^{-1}$ existieren muß. Damit ändert sich unsere Forderung aus Schritt 2 zu $|\mathbf{A}\mathbf{B}\mathbf{B}^{-1}\boldsymbol{f}| = |\mathbf{A}'\boldsymbol{f}'| \to \min$. Eine geeignet gewählte Matrix $\mathbf{B}$ „equilibriert" dann die Statistik von $\mathbf{A}$. Wir haben so die Transformation $\mathbf{A}' \stackrel{def}{=} \mathbf{A}\mathbf{B} = \mathbf{A}_0\mathbf{B} + \mathbf{D}\mathbf{B} \stackrel{def}{=} \mathbf{A}_0' + \mathbf{D}'$ durchgeführt.[7] Wenn es nun gelingt, $\boldsymbol{f}'$ erwartungstreu zu schätzen, so ergibt sich durch die Rücktransformation der gesuchte Vektor $\boldsymbol{f} = \mathbf{B}\boldsymbol{f}'$. Erwartungstreue läßt sich durch geeignete Wahl der Equilibrierungsmatrix $\mathbf{B}$ erreichen, indem $\mathbf{B}$ so gewählt wird, daß

$$\mathsf{E}\left[\mathbf{D}'^T\mathbf{D}'\right] = \mathbf{B}^T\mathsf{E}\left[\mathbf{D}^T\mathbf{D}\right]\mathbf{B} \stackrel{!}{=} c\,\mathbf{I} \tag{6}$$

gilt. Die Voraussetzung $\mathsf{E}\left[\mathbf{D}'\right] = \mathbf{0}$ ist erfüllt, weil $\mathsf{E}\left[\mathbf{D}\right] = \mathbf{0}$ gilt. Sei $\mathbf{U}\mathbf{W}\mathbf{U}^T$ die SVD von $\mathsf{E}\left[\mathbf{D}^T\mathbf{D}\right]$ (symmetrisch!) mit $\mathbf{W} = \mathsf{diag}\{w_1, w_2, \ldots, w_9\}$. Mit $\mathbf{W}^{-1/2} = \mathsf{diag}\left\{1/\sqrt{w_i}\right\}$ und $\mathbf{B} = \mathbf{U}\mathbf{W}^{-1/2}$ erhalten wir dann wie gefordert:

$$\mathsf{E}\left[\mathbf{D}'^T\mathbf{D}'\right] = \mathbf{B}^T\mathsf{E}\left[\mathbf{D}^T\mathbf{D}\right]\mathbf{B} = \mathbf{W}^{-1/2}\mathbf{U}^T\mathbf{U}\mathbf{W}\mathbf{U}^T\mathbf{U}\mathbf{W}^{-1/2} = \mathbf{I}\,. \tag{7}$$

3.2 TLS mit festen Spalten (TLS-FC)

Von Demmel [1] und von Golub, Hoffman und Stewart [2] existieren zwei Algorithmen für den Sonderfall von TLS, bei dem bestimmte Spalten der Matrix $\mathbf{A}$ als fehlerfrei bekannt sind. Diese beiden Verfahren zeichnen sich jedoch durch ihren großen Rechenaufwand aufgrund ihrer hohen Komplexität aus. Wir wollen diese Algorithmen TLS-FC (TLS with Fixed Columns) nennen.

Wir nehmen ohne Beschränkung der Allgemeinheit an, daß die letzten Spalten von $\mathbf{A}$ fehlerfrei seien, so daß also $\mathsf{E}\left[\mathbf{D}^T\mathbf{D}\right] = \mathsf{diag}\{c, c, \ldots, c, 0, \ldots, 0\}$ gilt. Mit der Notation des letzten Abschnittes ergibt sich $\mathbf{U} = \mathbf{I}$. Die Matrix $\mathbf{W}^{-1/2}$ ist aber nicht definiert; $1/\sqrt{0}$ existiert nicht. Das läßt sich leicht beheben, indem man stattdessen $\mathsf{E}\left[\mathbf{D}^T\mathbf{D}\right] = \mathsf{diag}\{c, c, \ldots, c, \varepsilon, \ldots, \varepsilon\}$ mit einem beliebig kleinen $\varepsilon > 0$ annimmt, so daß $\mathbf{W}^{-1/2} = \mathsf{diag}\{1/\sqrt{c}, \ldots, 1/\sqrt{c}, 1/\sqrt{\varepsilon}, \ldots, 1/\sqrt{\varepsilon}\}$ gilt.

Durch diesen Kunstgriff können wir das TLS-FC-Problem mit einem besonderes einfachen Spezialfall der Equilibrierung lösen: durch Equilibrierung mit einer Diagonalmatrix. In der neuen Formulierung haben wir sogar ein noch mächtigeres Hilfsmittel zur Hand, denn es lassen sich auch leicht Probleme lösen, bei denen die Varianzen der gestörten Spalten nicht alle c, sondern verschieden sind.[8] Wir erhalten eine beliebig gute Näherung für TLS-FC mit einem kleinen Teil des Rechenaufwandes und bezeichnen unser neues Verfahren im folgenden mit TLS-NFC (nearly fixed columns).

---

[6] $\mathbf{C}$ steht immer für Kovarianzmatrizen der als Subskript genannten Größen.

[7] Der Rang von $\mathbf{A}$ wird dabei nicht beeinflußt, weil $\mathbf{B}$ vollen Rang hat.

[8] Anders gesagt: Wir multiplizieren vor der Rangreduzierung jede Spalte mit einer Zahl, die für die 'Kosten' steht, die eine Änderung an den Einträgen verursacht. Da $\varepsilon^{-1/2}$ sehr groß ist, werden Änderungen an den festen Spalten beliebig 'teuer', so daß sich im Endeffekt daran nichts ändert und die Spalten, wie gefordert, fest bleiben.

4 Verbesserungen am 8-Punkte-Algorithmus

Mit den gerade erarbeiteten mathematischen Methoden wollen wir nun den ersten Schritt des 8-Punkte-Algorithmus' und seiner Derivate untersuchen und verbessern. Wir betrachten zwei Fälle: Zuerst den Fall, daß nur die Meßwerte aus *einem* Bild als gestört betrachtet werden (hier: die $\boldsymbol{u}_i$-Vektoren), und dann den Fall, daß Meßwerte aus *beiden* Bildern ($\boldsymbol{u}_i$- und $\boldsymbol{v}_i$-Vektoren) gestört sind.

Es gilt also zunächst $\boldsymbol{u}_i = \boldsymbol{u}_{i0}$ und $\boldsymbol{v}_i = \boldsymbol{v}_{i0} + \boldsymbol{c}_i$. Die Störungen der Meßwertevektoren $\boldsymbol{c}_i$ sind mittelwertfrei, untereinander unkorreliert und haben alle die Kovarianzmatrix $\mathbf{C}_c = \mathsf{diag}\left\{\sigma^2, \sigma^2, 0\right\}$. In [6] konnte gezeigt werden, daß die Forderung (5) nicht erfüllbar ist, sondern nur die Gleichung

$$\sum_i \mathbf{C}_{d'_i} = \mathbf{C}_{c'} \otimes \mathbf{K}_{u'} \stackrel{!}{=} c\ \mathsf{diag}\{1,1,1,1,1,1,0,0,0\} = c\ \tilde{\mathbf{I}} \tag{8}$$

mit $\mathbf{K}_{u'} = \sum_i \boldsymbol{u}'_{i0}\boldsymbol{u}'^{T}_{i0}$. Die Matrix $\mathbf{A}$ hat drei feste Spalten, d.h. das Problem läßt sich mit TLS-FC bzw. TLS-NFC lösen. Der Operator $\otimes$ steht für das Kronecker-Matrizenprodukt. Um Forderung (8) zu erfüllen, müssen die folgenden beiden Gleichungen gelten (k_1 und k_2 sind beliebige Konstanten):

$$\mathbf{C}_{c'} = \mathbf{S}\mathbf{C}_c\mathbf{S}^T \stackrel{!}{=} \mathsf{diag}\{k_1, k_1, 0\} \quad \text{und} \quad \mathbf{K}_{u'} = \sum_{i=1}^{N} \mathbf{R}\boldsymbol{u}_i\boldsymbol{u}_i^T\mathbf{R}^T \stackrel{!}{=} k_2\ \mathbf{I} \tag{9}$$

Das bedeutet, daß $\mathbf{R}$ der Hartley-Transformation nach Gl. (4) entspricht und $\mathbf{S}$ eine beliebige Rotation und eine beliebige Translation enthalten kann, aber nur eine *isotrope* Skalierung enthalten darf.

Nun wollen wir zu dem allgemeineren Fall von zwei verrauschten Bildern übergehen. Sei $\boldsymbol{u}_i = \boldsymbol{u}_{i0} + \boldsymbol{b}_i$ und $\boldsymbol{v}_i = \boldsymbol{v}_{i0} + \boldsymbol{c}_i$ mit den mittelwertfreien Störvektoren $\boldsymbol{b}_i$ und $\boldsymbol{c}_i$. Diese Vektoren sollen die gleiche Kovarianzmatrix haben und auch untereinander unkorreliert sein, d.h. $\mathbf{C}_b = \mathbf{C}_c = \mathsf{diag}\left\{\sigma^2, \sigma^2, 0\right\}$ und $\mathsf{E}\left[\boldsymbol{b}_i\boldsymbol{c}_i^T\right] = \mathsf{E}\left[\boldsymbol{b}_i\right]\mathsf{E}\left[\boldsymbol{c}_i\right]^T = \mathbf{0}$. Wir definieren nun die Transformationsmatrizen $\mathbf{R}$ und $\mathbf{S}$ wie von Hartley vorschlagen und erhalten im neuen Koordinatensystem (gestrichene Koordinaten) $\boldsymbol{d}'^{T}_i = (\theta_{i11}, \theta_{i12}, c'_{i1}, \theta_{i21}, \theta_{i22}, c'_{i2}, b'_{i1}, b'_{i2}, 0)$ mit $\theta_{ipq} = c'_{ip}u'_{i0q} + v'_{i0p}b'_{iq} + c'_{ip}b'_{iq}$.

Die ursprüngliche Forderung $\sum_{i=1}^N \mathbf{C}_{d'_i} = c\ \mathbf{I}$, die im Falle von nur einem verrauschten Bild noch durch Gl. (9) bis auf die festen Spalten zu realisieren war, läßt sich nun nicht mehr erreichen. Wir haben jedoch oben gezeigt, daß man mit Hilfe der Equilibrierung auch mit beliebigem $\sum_{i=1}^N \mathbf{C}_{d'_i}$ zurechtkommen kann, d.h. eigentlich könnte auf die Hartley-Transformationen in Schritt 1 verzichtet werden.

Dennoch ist es sinnvoll, solche Transformationen zu benutzen. Man kann dadurch erreichen, daß $\sum_{i=1}^N \mathbf{C}_{d'_i}$ zumindest diagonal wird; damit ist dann der größte Teil des Problems schon gelöst und die Equilibrierungsmatrix besteht nur noch aus einer Diagonalmatrix.

$$\sum_{i=1}^{N} \mathbf{C}_{d'_i} = \sum_{i=1}^{N} \mathsf{E}\left[\boldsymbol{d}_i\boldsymbol{d}_i^T\right] \stackrel{!}{=} \mathsf{diag}\{k_i\}\ . \tag{10}$$

Wir definieren: $\mathbf{C}_{b'} = \mathsf{E}\left[\boldsymbol{b}'_i\boldsymbol{b}'^{\,T}_i\right]$, $\mathbf{C}_{c'} = \mathsf{E}\left[\boldsymbol{c}'_i\boldsymbol{c}'^{\,T}_i\right]$, $\mathbf{K}_{u'_{i0}} = \boldsymbol{u}'_{i0}\boldsymbol{u}'^{\,T}_{i0}$, $\mathbf{K}_{v'_{i0}} = \boldsymbol{v}'_{i0}\boldsymbol{v}'^{\,T}_{i0}$, $\mathbf{K}_{u'_i} = \boldsymbol{u}'_i\boldsymbol{u}'^{\,T}_i$ und $\mathbf{K}_{v'_i} = \boldsymbol{v}'_i\boldsymbol{v}'^{\,T}_i$.

Es stehen nur $\mathbf{K}_{u'_i}$ und $\mathbf{K}_{v'_i}$ zur Verfügung, aber man braucht eigentlich $\mathbf{K}_{u'_{i0}}$ und $\mathbf{K}_{v'_{i0}}$. Man kann jedoch in guter Näherung $\mathbf{K}_{u'_i} = \mathbf{K}_{u'_{i0}}$ und $\mathbf{K}_{v'_i} = \mathbf{K}_{v'_{i0}}$ setzen und erhält damit

$$\mathbf{C}_{d'_i} = \mathbf{C}_{c'} \otimes \mathbf{K}_{u'_i} + \mathbf{K}_{v'_i} \otimes \mathbf{C}_{b'} + \mathbf{C}_{c'} \otimes \mathbf{C}_{b'} \,. \tag{11}$$

Den letzten Summanden kann man gegen die ersten beiden vernachlässigen. In der Summe ergibt sich:

$$\sum_{i=1}^{N} \mathbf{C}_{d'_i} = \mathbf{C}_{c'} \otimes \underbrace{\sum_{i=1}^{N} \mathbf{K}_{u'_i}}_{\mathbf{K}_{u'}} + \underbrace{\sum_{i=1}^{N} \mathbf{K}_{v'_i}}_{\mathbf{K}_{v'}} \otimes \mathbf{C}_{b'} = \mathbf{C}_{c'} \otimes \mathbf{K}_{u'} + \mathbf{K}_{v'} \otimes \mathbf{C}_{b'} \,. \tag{12}$$

Damit diese Summe diagonal wird, müssen alle vier beteiligten Matrizen $\mathbf{C}_{b'}$, $\mathbf{C}_{c'}$, $\mathbf{K}_{u'}$ und $\mathbf{K}_{v'}$ diagonal sein. Das führt zu den Forderungen

$$\mathbf{C}_{b'} = \mathbf{R}\mathbf{C}_b\mathbf{R}^T = \mathbf{R}\ \mathsf{diag}\left\{\sigma^2, \sigma^2, 0\right\} \mathbf{R}^T \stackrel{!}{=} \mathsf{diag}\left\{\alpha_1^2\sigma^2, \alpha_2^2\sigma^2, 0\right\} \tag{13}$$

$$\mathbf{C}_{c'} = \mathbf{S}\mathbf{C}_c\mathbf{S}^T = \mathbf{S}\ \mathsf{diag}\left\{\sigma^2, \sigma^2, 0\right\} \mathbf{S}^T \stackrel{!}{=} \mathsf{diag}\left\{\beta_1^2\sigma^2, \beta_2^2\sigma^2, 0\right\} \,, \tag{14}$$

wobei wir die Skalierungsmatrizen als $\mathsf{diag}\{\alpha_1, \alpha_2, 1\}$ für die Skalierung von $\mathbf{R}$ und $\mathsf{diag}\{\beta_1, \beta_2, 1\}$ für die Skalierung von $\mathbf{S}$ annehmen. Diese Gleichungen erlauben eine beliebige Translation, eine beliebige Rotation und eine beliebige Skalierung. Aus den Forderungen für $\mathbf{K}_{u'}$ und $\mathbf{K}_{v'}$ erhält man

$$\mathbf{K}_{u'} = \mathbf{R}\mathbf{K}_u\mathbf{R}^T \stackrel{!}{=} \mathsf{diag}\left\{\alpha_1^2 k_{u1}, \alpha_2^2 k_{u2}, 1\right\} \tag{15}$$

$$\mathbf{K}_{v'} = \mathbf{S}\mathbf{K}_v\mathbf{S}^T \stackrel{!}{=} \mathsf{diag}\left\{\beta_1^2 k_{v1}, \beta_2^2 k_{v2}, 1\right\} \,. \tag{16}$$

Dies schränkt nun die Freiheit von $\mathbf{R}$ und $\mathbf{S}$ deutlich ein, indem Rotation und Translation festgelegt werden. Im Fall unsymmetrisch verteilter Störungen (nur in einem Bild) haben wir oben gesehen [6], daß diese Forderung eine Verschiebung in den Schwerpunkt sowie eine Hauptachsentransformation bedeutet. Die k_{u1} und k_{u2} (bzw. k_{v1} und k_{v2}) sind dabei die Längen der Hauptachsen der Hauptachsentransformationen. Die Skalierung bleibt jedoch weiterhin frei.

Da die Elemente auf der Diagonalen von $\mathsf{E}\left[\mathbf{D}^T\mathbf{D}\right]$ in der Regel nicht auf gleiche Werte gebracht werden können und man andererseits auch mit beliebigen Diagonalmatrizen zurechtkommen kann, können die Skalierungsfaktoren in beliebiger Art und Weise festgelegt werden. Hier bietet sich die Hartley-Transformation ($\beta_1 = 1/\sqrt{k_{u1}}$, $\beta_2 = 1/\sqrt{k_{u2}}$, $\alpha_1 = 1/\sqrt{k_{v1}}$, $\alpha_2 = 1/\sqrt{k_{v2}}$) an, denn dadurch werden die Formeln deutlich einfacher. Wenn man jetzt $\mathbf{C}_{c'} \otimes \mathbf{K}_{u'}$ und $\mathbf{K}_{v'} \otimes \mathbf{C}_{b'}$ berechnet und über diese beiden Terme summiert, dann σ^2 ausklammert und anschließend ohne Beschränkung der Allgemeinheit auf 1 setzt, erhält man

$$\sum_{i=1}^{N} \mathbf{C}_{d'_i} = \mathsf{diag}\left\{\beta_1^2 + \alpha_1^2, \beta_1^2 + \alpha_2^2, \beta_1^2, \beta_2^2 + \alpha_1^2, \beta_2^2 + \alpha_2^2, \beta_2^2, \alpha_1^2, \alpha_2^2, 0\right\} \,. \tag{17}$$

5 Zur Ergänzung einige Simulationsergebnisse

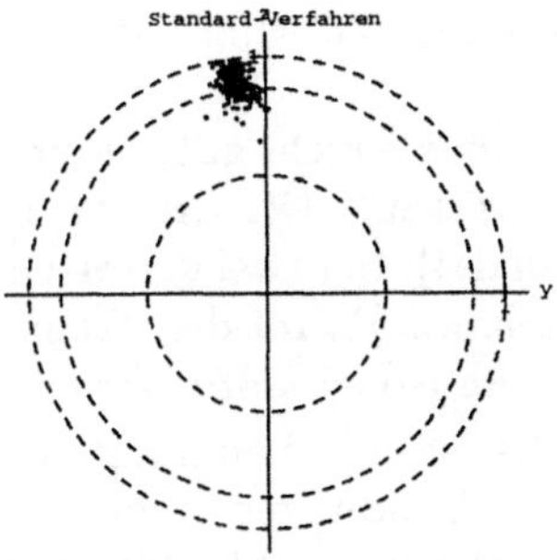

Abbildung1. x-Richtung, Standard-Verfahren

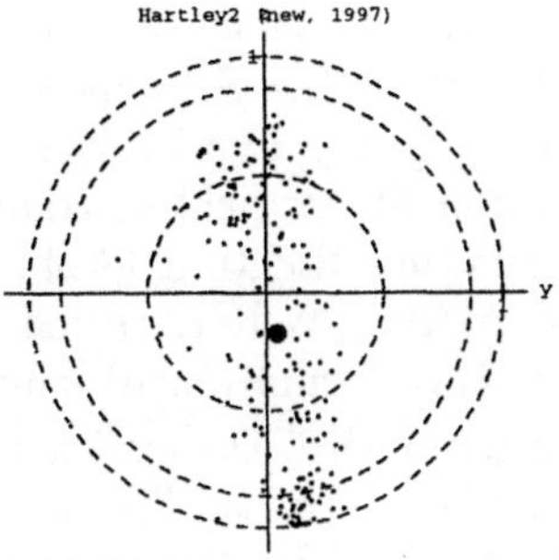

Abbildung2. x-Richtung, Hartley-Verfahren

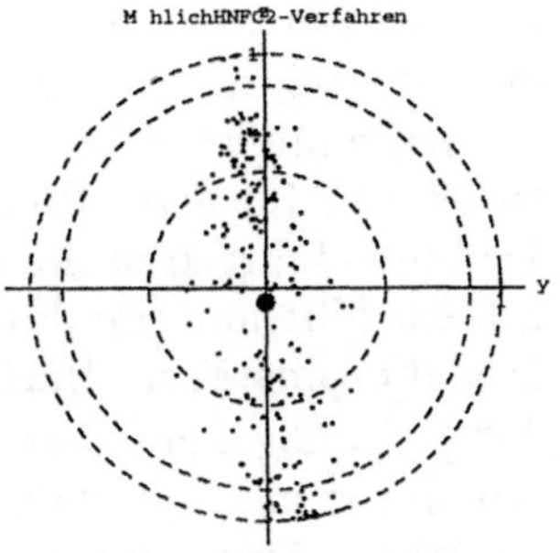

Abbildung3. x-Richtung, Mühlich-Verfahren

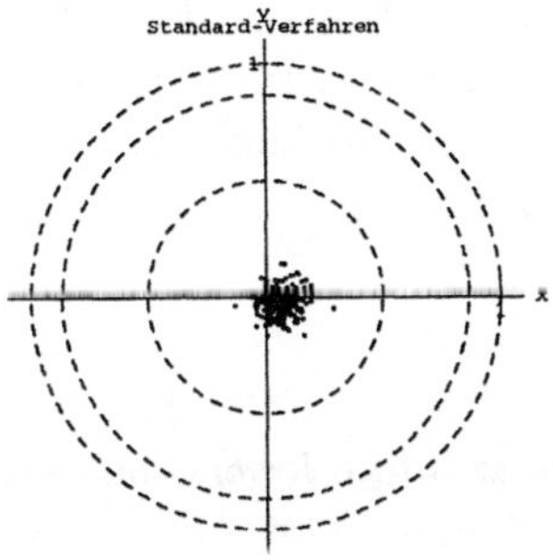

Abbildung4. z-Richtung, Standard-Verfahren

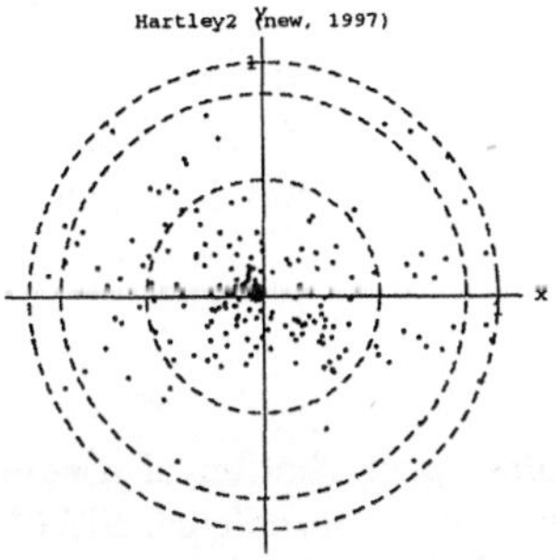

Abbildung5. z-Richtung, Hartley-Verfahren

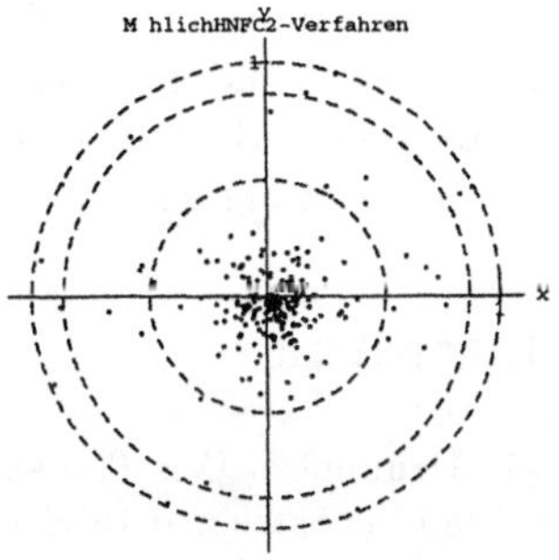

Abbildung6. z-Richtung, Mühlich-Verfahren

Zur Prüfung der obigen theoretischen Überlegungen haben wir eine Vielzahl von Experimenten mit synthetischen und somit hinsichtlich ihrer Fehlerstruktur vollständig kontrollierbaren Daten durchgeführt und dabei bisher keinen Widerspruch zu der dargestellten Theorie gefunden. Als Beispiele, die hier nur dazu dienen sollen, die Auswirkungen der Transformationen zu verdeutlichen, zeigen wir die Schätzung des Translationseinheitsvektors (wobei wir den Fall einer kalibrierten Kamera, also die Bestimmung der *Essential-Matrix* betrachten).

Für eine Bewegung in Blickrichtung (z-Richtung) und senkrecht dazu (x-Richtung) wurde jeweils ein Datensatz von 20 Punktkorrespondenzen 200 mal verrauscht (beide Bilder) und mit Standard-, Hartley- und unserem neuen Verfahren der Translationsvektor geschätzt; jeden geschätzten Vektor kann man sich als einen Punkt auf der Einheitskugel vorstellen. Zur Illustration sind die

Projektionen dieser Punkte auf die Ebene senkrecht zum wahren Translationsvektor gezeigt, d.h. wir sehen auf die Projektion einer Halbkugel, in deren Mitte der wahre Translationsvektor liegt. Ein Punkt in der Mitte entspricht einer sehr guten Schätzung, während ein Punkt auf dem äußeren Kreis maximalen Winkelfehler (90°) bedeutet. Die anderen beiden Kreise stehen für 30° und 60°. Der dicke dunkelgraue Punkt ist der mittlere Schätzvektor.

Herkömmliches TLS setzt $\sum_{i=1}^{N} \mathbf{C}_{d'_i} = c\,\mathbf{I}$ voraus. Da dieses nicht gilt, ist einfaches TLS für unser Problem nicht das richtige Handwerkszeug. Das Standard-Verfahren führt so zu einem Bias in z-Richtung (Abb. 1 und 4). Bei zwei gestörten Bildern hilft nur eine Equilibrierung weiter. Durch geeignete Wahl der Transformationsmatrizen (Hartley-Transformationen) kann man jedoch dafür sorgen, daß $\sum_{i=1}^{N} \mathbf{C}_{d'_i}$ zumindest diagonal wird. Die exakte statistische Behandlung mit Equilibrierung sorgt dafür, daß die Punktwolken in den Abbildungen dichter in die Mitte zusammenrücken. Das wird besonders am Vergleich der Abbildungen 5 und 6 deutlich. Die Varianz wird deutlich vermindert.

Schlußbemerkung

Nach unseren bisherigen Erfahrungen darf man davon ausgehen, daß die statistische Analyse von Schätzverfahren nicht nur beim Problem der Epipolargeometrie, sondern auch bei der gemeinsamen Verarbeitung von mehr als zwei Bildern eine Schlüsselrolle spielen wird. Die Forderung nach wirklich robusten und praxistauglichen Verfahren wird dies mit sich bringen.

Literatur

1. Demmel, J.W.: The smallest perturbation of a submatrix which lowers the rank and constrained total least squares problems. SIAM J. Numer. Anal. 24
2. Golub, G.H.; Hoffman, A.; Stewart, G.W.: A Generalization of the Eckart-Young-Mirsky Matrix Approximation Theorem, Linear Algebra and its Applications, 88/89: 317-327, 1987
3. Hartley, R.I.: In defence of the 8-point-algorithm. IEEE Trans. on Pattern Analysis and Machine Intelligence, Vol. 19, no.6, June 1997.
4. van Huffel, S., Vandewalle, J.: The Total Least Squares problem: Computational aspects and analysis. SIAM (Society for Industrial and Applied Mathematics), Philadelphia, 1991
5. Longuet-Higgins, H.C.: A computer algorithm for reconstructing a scene from two projections. *Nature*, vol. 293, September 1981, S.433-435
6. Mühlich, M.; Mester, R.: The Role of Total Least Squares in Motion Analysis. Proc. ECCV'98 (Vol.2), Springer, 1998
7. Strang, G: Linear algebra and its applications. 3rd edition, Harcourt Brace Jovanovich, Inc., 1988.
8. Tsai, R.Y.; Huang, T.S.: Uniqueness and estimation of three-dimensional motion parameters of rigid objects with curved surfaces. IEEE Trans. Pattern Analysis and Machine intelligence, vol.6, pp.13-27, 1984

Improving Phase-Based Disparity Estimation by Means of Filter Tuning Techniques*

Ingo Ahrns[1] and Heiko Neumann[2]

[1] Daimler Benz AG, Research and Technology, D-89081 Ulm, Germany,
ahrns@dbag.ulm.DaimlerBenz.COM

[2] University of Ulm, Dept. of Neural Inform. Proc., D-89069 Ulm, Germany,
hneumann@neuro.informatik.uni-ulm.de

Abstract. Phase differencing techniques have been proven to be fast and robust methods for estimating disparity between two views.
This disparity estimation depends on the quality of the local phase information which is a response of carefully designed frequency selective filter pairs for local phase estimation. Badly adjusted filter kernels yield responses with low amplitude and thus numerically instable phase information.
In this paper we investigate the role of filter tuning to avoid singular points. We present a new iterative algorithm to optimally adjust the local phase estimating filters and compare the results with other phase differencing techniques as well as an instantaneous frequency driven filter tuning. Various experiments demonstrate that the iterative filter tuning technique shows improved performance.

1 Introduction

Disparity estimation is necessary for depth extraction from stereoscopic views and, thus, plays an important role for many computer vision tasks.

During the last years, many different methods for disparity estimation have been proposed [DA89]. They can be divided into two main groups, area-based techniques and feature based techniques. The latter group includes methods which first extract features which are then matched by applying a similarity criterion. These methods are quite fast, but only produce sparse density information.

Area-based approaches solve the stereo correspondence problem by matching small areas of the original or band-pass filtered left and right images. This matching process can be performed by correlation based techniques or cepstral filtering. Both techniques suffer from missing subpixel accuracy and the need for algorithmic peak search.

Both drawbacks are avoided by the phase-based approach, which has been first introduced in [San88]. It has the following main advantages: First, it is a direct approach which does not need any explicit peak search, which in turn increases time complexity. Second, estimates are obtained from real-valued phase estimation and, thus yield subpixel accuracy. Third, phase information is more robust than amplitude information with respect to changes in contrast, scale and orientation [FJ93]. And at last, disparity estimation only requires convolution of the signal with frequency selective complex filters, e.g. Gabor filters. Fast implementations exist using special image processing hardware [HS96] or even for low-cost hardware utilizing spatial subsampling techniques [AN98].

* This study is part of the Advanced Servicing Robot (ASR) project at the Daimler Benz AG Research and Technology center and has been performed in collaboration with the SFB 527 at the University of Ulm.

Since Sanger proposed the first phase-based technique [San88], many improvements have been proposed in the literature. Fleet et. al. [FJJ91] proposed a Taylor expansion of the phase function and derived a modified differencing scheme to estimate the disparity. More recently, other authors proposed iterative techniques to estimate the disparity from phase information more precisely (e.g.[JJT91], [CM97]).

The main problem for all these algorithms is caused by phase singularities in scale-space [FJ90], i.e. configurations that produce low filter responses depending on the spatial location and the filter tuning, such that the phase information suffers from numerical instability. It has been reported [CM97] that error amplification is directly coupled to singularities in phase scale space. If singularities cannot be avoided, they must be detected in order to ignore the corrupted disparity information.

We propose two techniques, for tuning the filter kernels for local phase estimation, such that singularities are avoided and as a consequence disparity maps result at higher density and increased accuracy. The first method estimates the instantaneous signal frequency and uses this frequency to tune the filter kernels. This idea has been suggested in [HDS97], but never been discussed in detail. The second approach is new and applies a gradient ascent to tune the central frequency of the filter kernels to the direction which maximizes the amplitude of the filter response.

2 Phase Differencing Techniques

The simplest version of phase differencing techniques, proposed by Sanger [San88], is called the *constant frequency model* [TM94]. It is motivated by the shift theorem of the Fourier transform, which states that any spatial shift corresponds to a phase shift in frequency domain. Thus, disparity can be expressed in terms of phase differences. Since disparity normally varies over space the global Fourier transform must be replaced by local phase estimation techniques. The most commonly used approach is using complex Gabor filters $g_{\omega_0,\sigma}(x) = \frac{1}{\sqrt{2\pi\sigma}} e^{-\frac{x^2}{2\sigma^2}} e^{j\omega_0 x}$, which produce complex filter responses $R_i(x) = f_i * g_{\omega_0,\sigma}(x) = A_i(x)e^{j\varphi_i(x)}$ which are divided into amplitude $A_i(X)$ and phase $\varphi_i(x)$ for any signal f_i. Since the Gabor is a band pass filter tuned to the central frequency ω_0, we can approximate the disparity Δx by $\Delta x = \frac{[\varphi_1(x) - \varphi_0(x)]_{-\pi}^{\pi}}{\omega_0}$. Due to the frequency selectivity of $g_{\omega_0,\sigma}$ it is necessary to use a bank of Gabor filters tuned to different central frequencies and to combine the single disparity estimates weighted by the amplitudes of the filter responses [San88].

The next refinement of the phase-based approach was introduced in [FJJ91] and called the *local frequency model* by Theimer et. al. [TM94]. A Taylor series expansion of $\varphi_0(x+\Delta x)$ yields $\varphi_0(x+\Delta x) = \varphi_0(x) + \frac{\partial}{\partial x}\varphi_0(x)\Delta x + \mathcal{O}(\Delta x^2)$. If we assume fronto parallel shifting in a small neighborhood of x by Δx, we obtain $\varphi_0(x + \Delta x) = \varphi_1(x)$ and, hence, from the Taylor expansion follows $\Delta x = \frac{[\varphi_1(x) - \varphi_0(x)]_{-\pi}^{\pi}}{\frac{\partial}{\partial x}\varphi_0(x)}$. The derivative $\omega_{inst} = \frac{\partial}{\partial x}\varphi_0(x)$ is well known as the *instantaneous frequency* of the signal and leads to more accurate results in the disparity estimation [HDS97].

3 Definition of the scale space

The scale space in the context of local phase estimation has been first introduced in [FJ90] and is defined by $\Gamma_f(x, \lambda_0) = f * g_{\sigma(\frac{2\pi}{\lambda_0}), \frac{2\pi}{\lambda_0}}(x)$. Here, λ_0 denotes the wavelength belonging to the central frequency ω_0 of the local phase estimating filter. Figure 1 depicts the scale space for the local amplitude and phase. We observe that the phase information is generally stable in scale space, however some points exist where small changes in the wavelength result in dramatical changes of the phase behavior. The amplitude of the filter response becomes zero and the corresponding phase signal jumps discontinuously. An extensive discussion of this phase behavior in scale space can be found in [Wes95]. Those points are called singularities. They have to be avoided

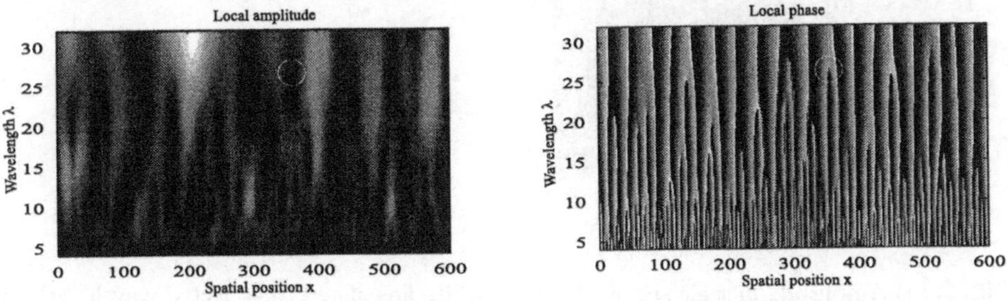

(a) (b)

Fig. 1. Scale space: (a) Local amplitude (b) Local phase (see text for details).

since no stable disparity estimation can be derived. Fortunately, singular points are coincident with low amplitudes and are therefore easily detectable. Figure 1 shows a singular point highlighted by a small circle in the amplitude and phase scale space for a random signal (figure 3 (b)).

4 Filter tuning

In [CM97] a further analysis of the numerical instability from phase singularities can be found. They assumed a small perturbation $\delta_i(x) = \delta_i^{Re}(x) + j\delta_i^{Im}(x)$ of the complex filter response $R_i(x)$ and derived the resultant perturbation $\delta A_i(x)$ in the amplitude and $\delta\varphi_i(x)$ in the phase. They have found $|\delta\varphi_i(x)| \leq \frac{|\delta_i(x)|}{A_i(x)}$. This inequality states that the phase is stable and errors are not amplified if $A_i(x) \geq 1$ holds. From this analysis follows the necessity for carefully designed filter kernels which maximize the amplitude of the filter response. In the following, we investigate two different approaches for this purpose. The first approach is based on the estimation of the instantaneous frequency which is then used to pick a Gabor filter out of a bank of filters with a central frequency tuned to the instantaneous frequency. However, one problem may occur if the instantaneous frequency is measured from a filter response whose filter was tuned to a singular wavelength (see figure 2 (b)). Then the instantaneous frequency may be corrupted and the method fails. For that reason, another method is investigated. Instead of using the instantaneous frequency, the central frequency of the filter kernels can be iteratively adjusted into a direction that maximizes the amplitude of the response.

4.1 Filter tuning to the instantaneous frequency

Assume a bank of $K+1$ Gabor filters tuned to different central frequencies $\omega_k = \frac{2\pi K}{k\lambda_{max}+(k-K)\lambda_{min}}$, where λ_{min} and λ_{max} denote the minimum and maximum wavelength of the filter kernels. In the first pass, we estimate the instantaneous frequency of the signal at position x by filtering with a Gabor tuned to fixed ω^*. The instantaneous frequency is then calculated as $\omega_{inst} = \frac{\partial}{\partial x}\varphi^*(x)$, where $\varphi^*(x)$ denotes the phase of the Gabor response. For the second pass we choose the Gabor filter with index $k_{inst} = K \cdot \frac{2\pi - \lambda_{min}\omega_{inst}}{\omega_{inst}(\lambda_{max} - \lambda_{min})}$ and estimate the phase using the filter $g_{\sigma(\omega_{k_{inst}}),\omega_{k_{inst}}}$.

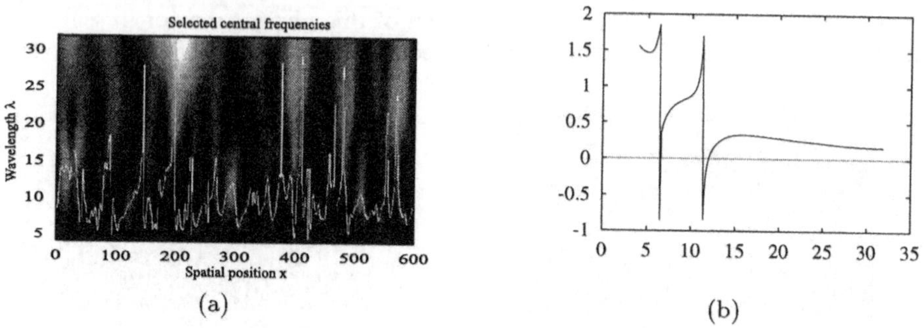

Fig. 2. (a) Amplitude in scale space where the white line shows the selected wavelength from a previous instantaneous wavelength analysis. (b) Instantaneous frequencies estimated with Gabor filters tuned to different central wavelengths. Two singularities are present which may lead to wrong measurements for the instantaneous wavelength.

4.2 Filter tuning by means of gradient search

Alternatively, an iterative optimization technique can be used to maximize the amplitude of the filter response. The objective function which has to be maximized reads $\mathcal{E}_f(\omega_0, x) = \left|f * g_{\sigma(\omega_0),\omega_0}(x)\right|^2$. If we assume constancy of $\sigma(\omega) = \sigma$ in a small neighborhood of ω_0, then the gradient of the objective function is approximated by $\frac{\partial}{\partial \omega_0}\mathcal{E}_f(\omega_0, x) \approx -2Im\left(\left((id_{\mathbb{R}} \cdot f) * g_{\sigma(\omega_0),\omega_0}\right) \times \left(f * g^*_{\sigma(\omega_0),\omega_0}\right)(x)\right)$, where $id_{\mathbb{R}}$ denotes the identity. Computing this gradient is quite time consuming because one convolution is necessary to determine the filter response $R(x)$ and another convolution with a modulated version of $g_{\sigma(\omega_0),\omega_0}$ is necessary. Therefore we decided to use a method that calculates the gradient above numerically. For that purpose we choose a constant stepwidth λ_{step} for the wavelength. For any current wavelength ω_0 the gradient $\frac{\partial}{\partial \omega_0}\mathcal{E}_f(\omega_0, x)$ is approximated by

$$\delta(\omega_0, x) = \frac{\lambda_{step}}{2\pi} \cdot \left(\left|f * g_{\sigma(\omega_o + \frac{2\pi}{\lambda_{step}}),\omega_0 + \frac{2\pi}{\lambda_{step}}}(x)\right|^2 - \left|f * g_{\sigma(\omega_o),\omega_0}(x)\right|^2\right). \quad (1)$$

In order to avoid two convolution operations for each iteration step, we only use the sign of the gradient information for the update of the central frequency $\omega_0' = \omega_0 + \text{sgn}\,(\delta(\omega_0, x))\,\frac{2\pi}{\lambda_{step}}$. Using a constant step width has the advantage, that we only need to apply a single convolution for each iteration step, if we use the amplitude information of the last iteration to compute the new gradient.

5 Experiments

In this section we now apply the proposed tuning techniques to different signals. For testing the approaches, we consider three different synthetic signal pairs with known disparity which are depicted in figure 3. Furthermore, we tested the algorithms for

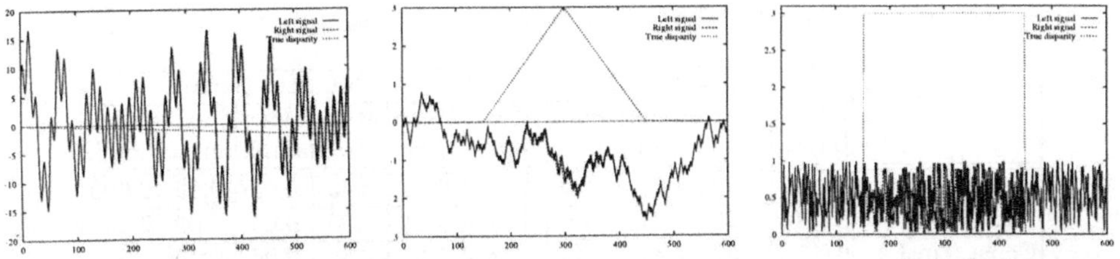

(a) (b) (c)

Fig. 3. Test signal pairs and true disparity (a) Harmonic signal pair and linear disparity function. (b) Random signal pair and its true disparity. (c) Random dot stereogram and a rectangular disparity function.

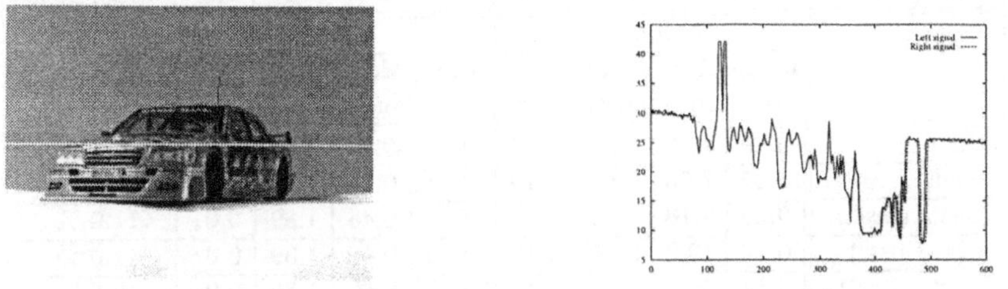

(a) (b)

Fig. 4. Intensity signal extracted from an (a) image of a toy car. (b) Extracted 1D signals of left and right view of the toy car scene.

real world signals by extracting the center row of an image of a toy car as shown in figure 4. Using these four signal pairs, we ran experiments by applying several disparity estimation methods. First, we applied Sanger's standard approach (the constant frequency model) for one Gabor filter tuned to $\omega_0 = \frac{1}{16}$. Second, we applied the local frequency model using the same central frequency. Then, both approaches have been refined by using the instantaneous frequency for filter tuning. For comparison we then applied the iterative filter tuning technique starting with the same ω_0. And finally we combined the iterative method and the instantaneous frequency based method. In fact, we used the instantaneous frequency as initial value for the iterative filter tuning process. Table 1 shows the results of these experiments. We protocolled the mean and maximum error in the disparity measurement and the mean, maximum and minimum amplitude and the amount of samples that shows an error reducing behavior, i. e. the amplitude is greater than 1. Looking at table 1 and all diagrams we observe that the local frequency model clearly outperforms the constant frequency model. However, both approaches suffer from severe outliers which are caused by a wrong choice of central frequencies at phase singularities. More interesting are the results concerning both filter tuning techniques. Both filter tuning techniques improve the constant and local frequency model. The mean disparity error can be reduced by a factor of 2 and more (see experiments for the random signal and the random dot signal). Both methods also improve the mean disparity error for the harmonic signal, however the error reduction is minimal due to the simplicity of the signal and the good result of the local frequency model without filter tuning. Comparing the two tuning techniques, the gradient-based tuning method yields slightly better results than the tuning to the instantaneous frequency. Only for the harmonic signal, due to the nearly perfect results of both techniques, the instantaneous frequency approach is slightly better. For the two random signals the gradient method outperforms the instantaneous fre-

Table 1. Table of error and amplitude statistics

	Harmonic signal					Random signal				
	Mean error	Max. error	Mean ampl.	Max. ampl.	Min. ampl.	Mean error	Max. error	Mean ampl.	Max. ampl.	Min. ampl.
Const.	0.240	0.92	1.71	1.89	1.07	0.18	4.43	0.05	0.09	0.003
Local	0.015	5.03	1.71	1.89	1.07	0.11	5.03	0.05	0.09	0.003
Const. (inst.)	0.018	0.37	2.41	2.52	1.44	0.11	2.55	0.06	0.13	0.002
Local (inst.)	0.009	0.47	2.41	2.52	1.44	0.09	5.03	0.06	0.13	0.002
Const. (iter.)	0.025	0.86	2.43	2.58	1.59	0.05	0.63	0.07	0.14	0.023
Local (iter.)	0.013	0.83	2.43	2.58	1.59	0.02	0.19	0.07	0.14	0.023
Const. (comb.)	0.026	0.88	2.43	2.58	1.59	0.07	2.55	0.07	0.15	0.003
Local (comb.)	0.013	0.85	2.43	2.58	1.59	0.06	5.03	0.07	0.15	0.003

	Random dot signal					Intensity signal			
	Mean error	Max. error	Mean ampl.	Max. ampl.	Min. ampl.	Mean ampl.	Max. ampl.	Min. ampl.	Ampl. ≥ 1
Const.	0.49	10.9	0.05	0.10	0.003	0.38	1.89	0.01	7.5 %
Local	0.27	7.95	0.05	0.10	0.003	0.38	1.89	0.01	7.5 %
Const. (inst.)	0.25	8.18	0.06	0.15	0.005	0.48	1.89	0.01	11.0 %
Local (inst.)	0.17	10.0	0.06	0.15	0.005	0.48	1.89	0.01	11.0 %
Const. (iter.)	0.14	3.78	0.07	0.18	0.010	0.60	1.89	0.01	18.0 %
Local (iter.)	0.11	3.73	0.07	0.18	0.010	0.60	1.89	0.01	18.0 %
Const. (comb.)	0.14	5.22	0.07	0.18	0.009	0.59	1.89	0.01	19.0 %
Local (comb.)	0.11	4.28	0.07	0.18	0.009	0.59	1.89	0.01	19.0 %

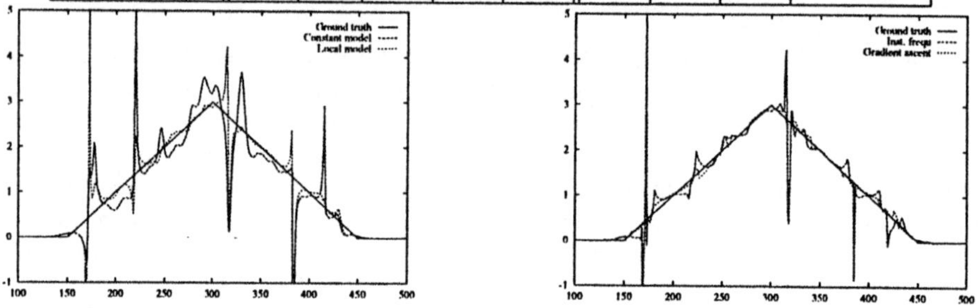

(a) (b)

Fig. 5. Test results for the random signal (a) Ground truth and results from the local and constant frequency model. (b) Ground truth and results from the local frequency model using iterative and instantaneous frequency based filter tuning.

quency method. The gradient method is not only able to reduce the mean disparity error but also reduces the number and magnitude of the outliers. The gradient ascent does not rely on a previously estimated instantaneous frequency, but iteratively improves the phase behavior. There is no danger that the finally tuned central frequency lies in the neighborhood of a singularity. Therefore severe outliers are dramatically reduced. For the intensity signal we can only observe the amplitude behavior since no ground truth is available. For all experiments both tuning methods were able to maximize the amplitude, and, thus reduce the error amplification. Again, the iterative method yields best results, whereas the instantaneous frequency approach yields less mean amplitudes. Especially this result is important for the intensity signal, where the number of pixels with amplitude ≥ 1, i. e. phase error reducing amplitudes could be increased from 7.5% to 18%. Finally we tried the combination of both techniques, however it could not improve the results.

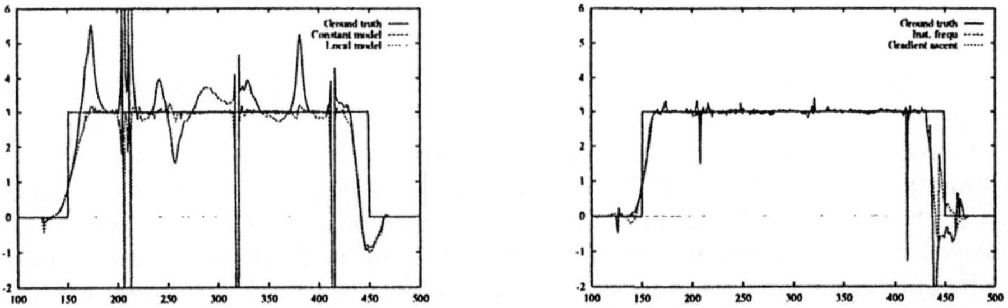

(a) (b)

Fig. 6. Test results for the random dot stereogram (a) Ground truth and results from the local and constant frequency model. (b) Ground truth and results from the local frequency model using iterative and instantaneous frequency based filter tuning.

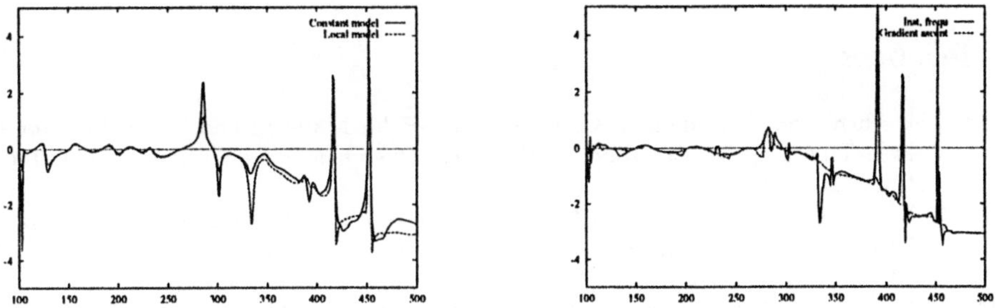

(a) (b)

Fig. 7. Test results for the intensity signal (toy car image) (a) Results from the local and constant frequency model. (b) Results from the local frequency model using iterative and instantaneous frequency based filter tuning.

Applying both tuning techniques to a stereo image pair is shown in figure 8. Figure 8 (a) depicts the disparity map derived from the local frequency model without any filter tuning. Figure (b) shows the result from the instantaneous frequency based method and (c) shows the result of the iterative method. Disparity estimates were suppressed if the corresponding amplitudes were below 0.15. No adaptation yields 25.7%, the first method 28% and the iterative method 29% of the amplitudes above 0.15. As shown in figure 7 (b) the number of outliers could be massively reduced.

(a) (b) (c)

Fig. 8. 2D disparity maps from the local frequency model (a) without filter tuning (25.7% of the amplitude of the image are ≥ 0.15) (b) with filter tuning to the instantaneous frequency (28.0% of amplitude ≥ 0.15) and (c) with filter tuning by means of gradient ascent (29.2% of amplitude ≥ 0.15).

6 Conclusion

In this paper we investigated the role of the phase behavior for phase-based disparity estimation. In contrast to other authors, we tried to minimize the area where phase singularities disturb disparity estimation, instead of suppressing disparity estimates at singular points. Since [CM97] have shown that the error amplification for the estimated phase is proportional to the inverse of the amplitude we looked for techniques that pick the central frequencies of the phase estimating filters where the amplitude has its maximum. For this purpose we investigated an approach that tunes the Gabor filter to the instantaneous frequency of the signal and compared the results to an iterative gradient ascent scheme. Both techniques were able to greatly improve the disparity estimates but the iterative method shows slightly better results than the instantaneous frequency driven method. We demonstrated in various experiments that the iterative approach yields less severe outliers with respect to disparity estimation and that the overall amplitude is increased by both techniques.

References

[AN98] I. Ahrns and H. Neumann. Combining Log-Polar Mapping and Gabor Filtering for Real-Time Depth Estimation. In *Workshop Dynamische Perzeption*, pages 119–126, Bielefeld, Germany, 1998.

[CM97] L.D. Cai and J. Mayhew. A Note on Some Phase Differencing Algorithms for Disparity Estimation. *International Journal of Computer Vision*, 22(2):111–124, 1997.

[DA89] U. R. Dhond and J. K. Aggarwal. Structure From Stereo - A Review. *IEEE Transactions on Systems, Man and Cybernetics*, 19(6):1489–1510, 1989.

[FJ90] D. J. Fleet and A. D. Jepson. Computation of Component Image Velocity from Local Phase Information. *International Journal of Computer Vision*, 5(1):77–104, 1990.

[FJ93] D. J. Fleet and A. D. Jepson. Stability of Phase Information. *IEEE Transactions on Pattern Analysis and Machine Intelligence*, 15(12):1253–1268, December 1993.

[FJJ91] D. J. Fleet, A. D. Jepson, and M. R. M. Jenkin. Phase based disparity measurement. *CVGIP*, 53(3):198–210, 1991.

[HDS97] M. Hansen, K. Daniilidis, and G. Sommer. Optimizing of Stereo Disparity Estimation Using the Instantaneous Frequency. In *Proceedings of the 7th. International Conference on Computer Analysis of Images and Patterns CAIP'97*, pages 321–328, 1997.

[HS96] M. Hansen and G. Sommer. Active Depth Estimation with Gaze and Vergence Control Using Gabor Filters. In *Proceedings of the 13th International Conference on Pattern Recognition ICPR'96*, pages 287–291, Wien, Österreich, 1996.

[JJT91] M. R. M. Jenkin, A. D. Jepson, and J. K. Tsotsos. Techniques for disparity measurement. *CVGIP: Image Understanding*, 53(1):14–30, 1991.

[San88] T. D. Sanger. Stereo Disparity Computation Using Gabor Filters. *Biological Cybernetics*, 59:405–418, 1988.

[TM94] W. M. Theimer and H. Mallot. Phase-Based Binocular Vergence Control and Depth Reconstruction Using Active Vision. *CVGIP: Image Understanding*, 60(3):343–358, 1994.

[Wes95] C.-J. Westelius. *Focus of Attention and Gaze Control for Robot Vision*. PhD thesis, Department of Electrical Engineering, Linköping University, Sweden, 1995.

Klassifikation und Neuronale Netze (Postersession)

Shape Decomposition and Shape Similarity Measure

Longin Jan Latecki and Rolf Lakämper

Institut für Angewandte Mathematik, Universität Hamburg
Bundesstr. 55, 20146 Hamburg
{latecki, lakaemper}@math.uni-hamburg.de

Abstract. We propose a simple and natural rule for decomposition of 2D objects into *parts of visual form*. The **hierarchical convexity rule** states that visual parts are enclosed by maximal convex boundary arcs (with respect to the object) at various levels of curve evolution. The proposed rule is based on a novel curve evolution method by digital linearization in which a significant visual part will become a convex part at some level of the evolution. The hierarchical convexity rule determines not only parts of boundary curves but directly the visual parts of objects, and the evolution hierarchy induces a hierarchical structure of the obtained visual parts.

Further, we derive a shape similarity measure based on the decomposition into visual parts and apply it to shape matching of object contours in an image database. The experimental results justify that our shape matching procedure is stable and robust with respect to noise deformations and gives an *intuitive* shape correspondence.

1 Introduction

We begin with decomposition of 2D objects into meaningful parts. According to Siddiqi et al. [13], "*Part-based representations allow for recognition that is robust in the presence of occlusion, movement, deletion, or growth of portions of an object. In the task of forming high-level object-centered models from low-level image-based features, parts serve as an intermediate representation.*"

There is also a strong evidence for part-based representations in human vision, see [13] for an overview. Hoffman and Richards [4] provide a strong evidence that contours are psychologically segmented at negative curvature minima. They state the following rule (cited from Hoffman and Singh [5]):

> "*Minima rule for silhouettes:* For any silhouette, all negative minima of curvature of its bounding curve are boundaries between parts."

Hoffman and Singh write in [5]: "*... the minima rule states in principle what these boundaries are, leaving open how in practice they may be computed from images despite noise and other resources.*"

One of the main motivations for our work was the question how to compute significant parts of boundaries from real digital images. Since objects we deal with in computer vision are mostly digital objects obtained by segmentation in digital images, our starting point is the discrete nature of digital objects.

The presented approach to multiscale shape decomposition is based on shape hierarchy obtained by a curve evolution. We present a novel approach to evolution of digital planar curves that is based on digital linearization. This method allows us first to eliminate noise influence without changing the shape of objects. If we continue to evolve the curve, we will linearize digital arcs that are relevant for the curve shape, which will result in a successive simplification of the curve shape, for example, see Figure 1 in [9] or on our www-side [10]. We use this shape simplification to derive a shape hierarchy for planar curves. The obtained shape hierarchy is a base for object decomposition into relevant visual parts. The parts obtained on the highest levels of the hierarchy determine the most relevant parts of the object (e.g., see Figure 2).

The cognitive motivation for the obtained shape hierarchy and shape decomposition is given by the similarity to well-known shape decompositions, which are justified by many psychological experiments (in particular, Hoffman and Richards [4], Hoffman and Singh [5], and Siddiqi et al. [13]). Although our shape decomposition rule yields for many objects similar parts as in [4], we do not use extremal points of contour curves to compute the parts. Since extremal points are obtained by local computation, thier computation is not robust in real digital images.

The usefulness of the obtained shape decomposition is demonstrated by numerous experiments with a similarity measure which is based on a correspondence of the obtained visual parts. The similarity measure is applied to shape matching of object contours in image databases. The experimental results justify that our shape matching procedure is stable and robust with respect to noise deformations and gives an *intuitive* shape correspondence.

For simplicity of the presentation, we analyze boundary curves of simple connected planar objects (i.e., objects with no holes). To decompose a digital curve into relevant parts on a given level of the shape hierarchy, we group the digital line segments of the curve into maximal convex arcs, which we will call **supported** arcs following Latecki and Rosenfeld [8]). Maximal supported arcs determine alternating and overlapping convex and concave parts (w.r.t. the object) of the boundary. For example, Figure 1 shows a decomposition of the boundary curve into maximal supported arcs: the arcs inside the object indicate the maximal concave arcs (w.r.t. the object) while the arcs outside the object indicate the maximal convex arcs (w.r.t. the object). In the following, the term **maximal convex arc** denotes a supported arc that is convex w.r.t. the object.

2 Shape Decomposition

The proposed approach to shape decomposition is based on a novel curve evolution method by digital linearization in which a significant visual part will become a convex part at some level of the evolution. For further details see [9] or our www-side [10].

Fig. 1. Decomposition of a polygonal arc into maximal supported arcs. Observe that the endpoints of the maximal convex arcs (w.r.t. the object) are close to points of minimal negative curvature.

It is a simple and natural observation that maximal convex parts of objects determine visual parts. However, the problem is that many significant visual parts are not convex, since a visual part may have concavities. This problems is solved by the hierarchical shape representation based on our curve evolution method, in which a significant visual part will become a convex part at some higher level of the evolution: A significant visual part may have concavities, whose boundary parts are supported concave arcs. Since these concave arcs contribute in a less relevant way to the shape of the object than the boundary arc enclosing the significant visual part, the concavities will disappear in an earlier stage of the boundary evolution. Thus, there exists an evolution stage at which a significant visual part is a convex part. We base our approach to shape decomposition into visual parts on the following rule:

- **Hierarchical convexity rule:**
  The maximal convex arcs (w.r.t. the object) at various levels of the evolution hierarchy determine parts of object boundary that enclose visual parts of the object.

Parts obtained by this rule are illustrated in Figure 2. The relevance of the parts depends on the level of the evolution hierarchy and on the relevance of the convex arcs. On the highest level of the shape hierarchy (a) in Figure 2, the evolved silhouette of the "fish" is decomposed into two maximal convex arcs, and the evolved silhouette of the "bear" is decomposed into three maximal convex arcs. The visual parts enclosed by these arcs represent the most relevant object parts, those are the body and the tail for the fish, and the upper body and the legs together with hips for the bear.

The shape hierarchy obtained by the evolution of a boundary curve induces a hierarchical structure of the visual parts:

- The lower is the level of the evolutional shape hierarchy, the less significant for the object shape are the parts determined.
- The parts obtained on a higher level of the shape hierarchy are inherited to the lower levels.

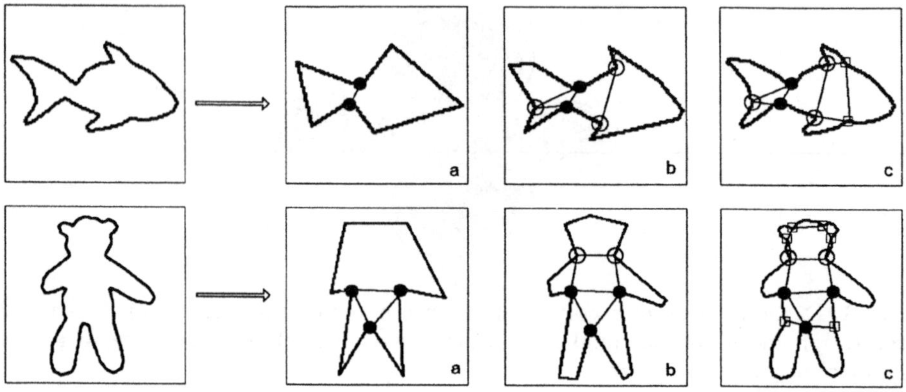

Fig. 2. Significant object parts obtained at various levels of the evolution of the boundary curve.

This means that on a lower level only the inherited parts are further divided. For example, on the level (b) in Figure 2, only the parts obtained on level (a) are further divided, e.g., the upper body of the bear is divided into the head and arms. The decompositions of the fish and the bear into significant object parts obtained by the hierarchical convexity rule are shown in Figure 2(c).

The parts of boundaries obtained by the hierarchical convexity rule correspond for many objects to the parts obtained using points of minimal negative curvature (Hoffman and Richards [4]), for example, see Figure 1. This is the case when the endpoints of convex arcs are located near the points of minimal negative curvature. Also for many objects, the obtained parts of objects correspond to limbs and necks in the theory of Siddiqi and Kimia [12]. For boundaries of continuous objects, the endpoints of maximal convex arcs correspond to inflection points (e.g., inflection points are used for shape description in Freeman [2]). The correspondence of the endpoints of maximal convex arcs to different kinds of critical points (in the sense of differential geometry) is possible, since we work in a digital space which does not exactly follow the rules of differential geometry. The correspondence of the decomposition into visual parts based on the hierarchical convexity rule to the well-known shape decompositions which are justified by many psychological experiments ([4], [13] [5]), gives cognitive motivation for the proposed rule.

In our approach, we do not need to decide which critical points have to be joint together in order to obtain object parts. The object parts obtained by joining two chosen negative curvature minima are called *part cats* (Beusmans et al. [1]). As argumented in Hoffman and Singh [5], a separate theory is necessary to determine the part cats knowing the boundary points of minimal negative curvature. In our approach, the part cats are automatically determined by maximal convex arcs. Our approach allows us additionally

- to give a relevance measure for an obtained part that is based on the level of the shape hierarchy and on relevance of the maximal convex arc,
- to obtain a shape decomposition stable with respect to noise, due to the levels of the shape hierarchy and due to the fact that the convex arcs represent global properties of object boundaries.

Therefore, convex arcs are significantly less influenced by noise than locally defined points of minimal negative curvature. Observe that part decomposition based on points of minimal negative curvature is binary, i.e., the obtained parts cannot be distinguished according to their salience. The relevance measure of visual parts we suggest here, is based on the levels of the shape hierarchy, as it is the case in Kimia et al. [6] (see also [12]). However, we additionally base the relevance measure of visual parts on the relevance of the contribution of maximal convex arcs to the object shape, which depends on their total turn (i.e., global curvature) and the length that is normalized with respect to the total length of the boundary curve.

Our shape decomposition corresponds to the hierarchy of parts based on the evolution of shocks in the entropy scale space in Kimia et al. [6]. As argumented in [6], the evolution of shocks requires not only boundary, but also region information. However, observe that our part decomposition requires only boundary information. This means that our approach to shape decomposition allows us to obtain the region information from the evolution hierarchy of the boundary curve. Thus, boundary information plus evolution hierarchy is sufficient for partitioning a two-dimensional shape.

3 Shape Similarity Measure

Our similarity measure is based on the decomposition into visual parts described above. The key idea is to find a right correspondence of the visual parts. We assume that a single visual part (i.e., convex arc) of one curve can correspond to a sequence of consecutive convex and concave arcs of the second curve. Thus, the correspondence can be a one-to-many or a many-to-one mapping, but never a many-to-many mapping. This assumption is justified by the fact that a single visual part should match to its noisy versions that can be composed of sequences of consecutive convex and concave arcs, or by the fact that a visual part obtained at a higher stage of evolution should match to the arc it originates from.

Let $arcs(C)$ denote the set of all subarcs of curve C, and similarly, $arcs(D)$ denote the set of all subarcs of curve D. We will call a **correspondence** any 1-1 relation R on $arcs(C) \times arcs(D)$ such that $\bigcup DomR = C$ and $\bigcup ImR = D$ and for any pair $(c, d) \in R$ either c is a maximal supported subarc of C or d is a maximal supported subarc of D. A **similarity measure** for curves C, D is defined as

$$S_c(C_1, C_2) = \min_R \sum \{S_a(c, d) : (c, d) \in R\}, \tag{1}$$

where the minimum is taken over all correspondences R and $S_a(c, d)$ is a similarity measure for arcs defined below.

Let c, d be digital arcs and $T(c), T(d)$ their tangent space representations, which we define in Section 4, below. The arc similarity measure is given by

$$S_a(c,d) = D_{L_1}(T(c), T(d)) \max(l(c), l(d)) \tag{2}$$

where D_{L_1} denotes the L_1-metric in the tangent space and l is the relative arclength of an arc with respect to the boundary length of the curve. Observe that we use measure (2) with the restriction that at least one of the arcs c, d is convex.

The similarity measure for curves C, D defined in (1) is computed using dynamic programming. Numerous experimental results show that the proposed similarity measure leads to intuitive arc correspondences, e.g., see Figure 3.

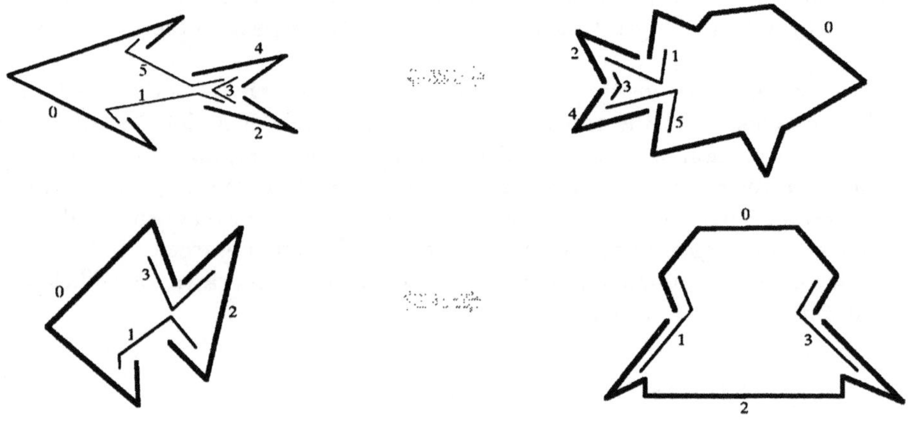

Fig. 3. The corresponding arcs are labeled by the same numbers.

We have applied this shape similarity measure to automatic object indexing and searching in image databases. First the shape of all objects in an image database is simplificated to obtain relevant supported arcs. This is done automatically by our contour evolution algorithm. The user query is given by a graphical sketch, which is also simplificated. An example is seen in Figure 4.

Since the newest advances in automatic object segmentation make it possible to extract objects of interest in some relevant classes of digital images, our technique can be applied to real image databases. This is demonstrated in Figure 5. First the original image (a) is segmented using color and texture segmentation in Pauwels et al. [11]. Then a watershed segmentation algorithm is applied to the segmented rabbit image (b). It produces the object contour (c) that is a simple closed 8-curve.[1] Our curve evolution method by digital linearization is then applied to the digital simple closed 8-curve. In (d), (e), (f), we see a few

[1] This watershed algorithm was successfully applied in image segmentation for object based data representation of video sequences in Lakämper and Seytter [7].

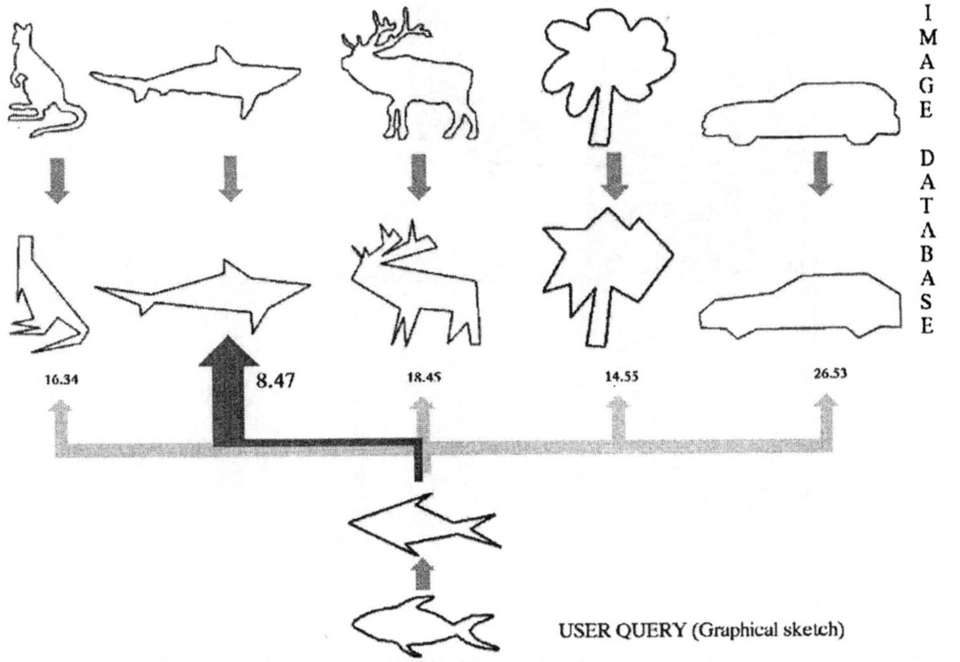

Fig. 1. An example of shape based search in image database.

stages of the curve evolution. The evolution hierarchy (f)>(e)>(d) yields the illustrated decomposition into visual parts.

A shape similarity measure based on correspondence of convex/concave arcs was used for classification of object contours in Ueda and Suzuki [14]. They use a standard scale-space curve deformation based on Gaussian smoothing to obtain convex/concave arcs at different evolution scales. Due to the properties of our curve evolution [9] our decomposition into convex/concave arcs seems to result in a more suitable shape decomposition. A comparison of our curve evolution approach to scale-space curve evolutions is given in Latecki and Lakämper [9]. We also use a different shape similarity measure than [14].

4 Tangent Space Representation

With every digital curve, a *tangent space representation* as a step function is associated. We use this representation as basis for the proposed similarity measure.

A polygonal curve is represented in the tangent space by the graph of a step function, where the x-axis represents the arclength coordinates of points in C

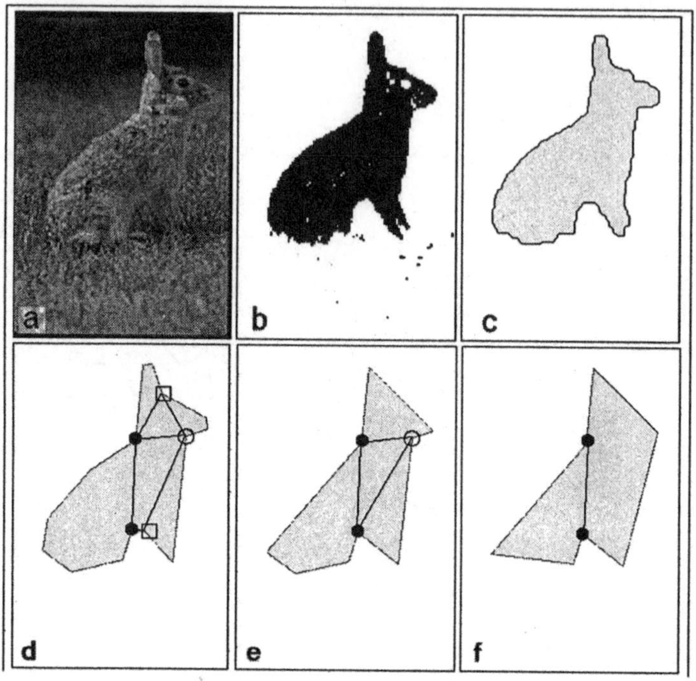

Fig. 5. From an image to a simplified shape representation and to visual parts. (a) → (b): color and texture segmentation (Pauwels et al.), (b) → (c): watershed segmentation, (c) → (f): boundary curve evolution.

and the y-axis represents the direction of the line segments in the decomposition of C. Formally, the **tangent space** is a torus $S_1 \times S_2$, where S_1 is a circle of length one that represent the length of a digital curve and S_2 is a circle that represents the angular direction of digital line segments. We will display the tangent space as a rectangle with the parallel sides identified in the standard way to obtain a topological torus. The y-difference between two adjacent steps in the tangent space represents the turn angle of the corresponding pair of line segments.

Any digital curve C can be interpreted as a polygonal curve with a possibly large number of vertices. For example, the vertices can be chosen to be the endpoints of maximal digital line segments contained in C. We obtain the vertices by the curve evolution method described in [9]. We assume that digital curve C is a polygonal curve composed of digital line segments $\mathcal{D}_m(C) = s_0, ..., s_{m-1}$. We denote $angle(s)$ to be the angular direction of line segment s in the standard coordinate system of the plane and we denote $l(s)$ to be the length of s normalized by the length of C.

We define a transformation T mapping a digital 8-curve to the tangent space. For illustration, see Figures 6(a) and (b). To each digital line segment $s \in \mathcal{D}_m(C)$,

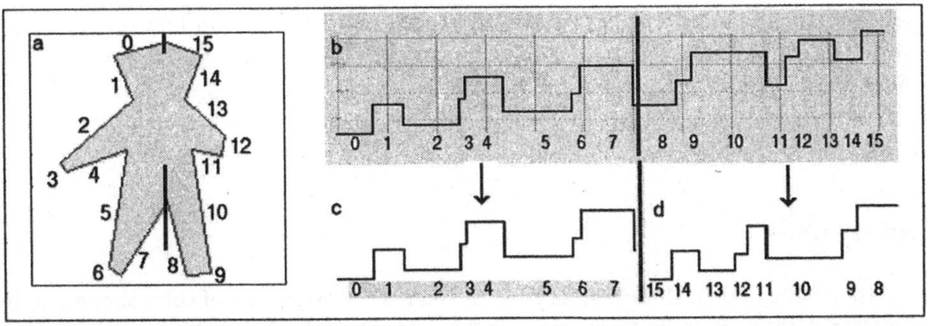

Fig. 6. A digital curve (a) and its step function representation in the tangent space (b). The similarity of (c) and (d) shows that the curve (a) is symmetric.

a line segment $T(s)$ in the tangent space is assigned such that

$$\pi_y(T(s)) = angle(s)$$

the length of $\pi_x(T(s))$ is equal to $l(s)$ and $\pi_x(T(s_i))$ is adjacent to $\pi_x(T(s_{i+1}))$, where π_x and π_y denote the projections on x- and y-axes in the tangent space, correspondingly. This means that y-value of all points in $T(s)$ is equal to $angle(s)$ and the projection of $T(s)$ on the x-axis has the same (normalized) arclength coordinates as s in C.

An analytical description of the transformation to the tangent space for continuous curves can be found in Zahn and Roskies [16]. They call the step function in the tangent space a *cumulative angular bend function.* Zahn and Roskies use this transformation only as an intermediate step to obtain Fourier descriptions of planar curves. They neither use the tangent space for analyzing the underlying planar curves nor for scale transformations, which we will describe below. Uesaka [15] uses the transformation of polygonal arcs to the tangent space to determine a circular approximation of polygonal arcs.

5 Conclusions

The presented approach allows us to describe a similarity measure for digital arcs. The hierarchical shape decomposition into visual parts makes it possible to apply our similarity measure for contours of planar objects in real images and to obtain a stable and robust shape matching procedure. It can be applied for automatic object grouping, indexing, and searching in image databases, since the newest advances in automatic object segmentation (e.g., [11]) make it possible to automatically extract objects of interest for some relevant classes of digital images.

The tangent space representation of planar curves can be also used to reduce the complexity of symmetry detection in presence of noise and occlusions. The symmetry of some boundary parts of a given object can be detected given the step function representations of the boundary curve: If boundary part A is a symmetric image of part B, then the step function of A is equal to the point reflection of the step function of B. For example, the point reflection of part 8–15 of the step function in Figure 6(b) shown in (d) is equal to part 0–7 shown in (c).

References

1. J. Beusmans, D. D. Hoffman, and B. M. Bennett. Description of solid shape and its inference from occluding contours. *Journal of the Optical Society of America* A. 4, 1155–1167, 1987.
2. H. Freeman. Shape description via the use of critical points. *Pattern Recognition* 10, 159-166, 1978.
3. I. Debled-Rennesson and J.-P. Reveilles. A Linear Algorithm for Segmentation of Digital Curves. *Intl. J. Pattern Recognition and AI* 9, 635–662, 1995.
4. D. D. Hoffman and W. A. Richards. Parts of Recognition. *Cognition* 18, 65–96, 1984.
5. D. D. Hoffman and M. Singh. Salience of visual parts. *Cognition* 63, 29–78, 1997.
6. B. B. Kimia, A. R. Tannenbaum, and S. W. Zucker. Shapes, shocks, and deformations, I: The components of shape and the reaction-diffusion space. *Int. J. of Computer Vision* 15(3), 189-224, 1995.
7. R. Lakämper and F. Seytter. Manipulation objektbasiert codierter Bilder als Anwendungsbeipspiel neuer Videostandards. *Proc. DAGM Mustererkennung*, Braunschweig, 427–434, Springer 1997.
8. L. J. Latecki and A. Rosenfeld. Supportedness and Supportedness: Differentialless Geometry of Plane Curves. *Pattern Recognition* 31, 607–622, 1998.
9. L. J. Latecki and R. Lakämper. Discrete Approach to Curve Evolution. *Proc. DAGM Mustererkennung*, Stuttgart 1998.
10. L. J. Latecki and R. Lakämper. http://www.math.uni-hamburg.de/home/lakaemper/shape
11. E. J. Pauwels, P. Fiddelaers, and F. Mindru. Fully Unsupervised clustering using center-surround receptive fields with applications to colour-segmentation. *Proc. of 7th Int. Conf. on Computer Analysis of Images and Patterns*, Kiel, 17–24, 1997.
12. K. Siddiqi and B. B. Kimia. Parts of Visual Form: Computational Aspects. *IEEE Trans. PAMI* 17, 239–251, 1995.
13. K. Siddiqi, K. Tresness, and B. B. Kimia. Parts of Visual Form: Ecological and Psychophysical Aspects. *Proc. IAPR's Int. Workshop on Visual Form*, Capri, 1994.
14. N. Ueda and S. Suzuki. Learning Visual Models from Shape Contours Using Multiscale Convex/Concave Structure Matching. *IEEE Trans. PAMI* 15, 337–352, 1993.
15. Y. Uesaka. A New Fourier Description applicable to open curves. *Trans. on IECE Japan A* J67-A, 166–173, 1984 (in Japanese).
16. C. T. Zahn and R. Z. Roskies. Fourier Descriptors for Plane Closed Curves. *IEEE Trans. on Computers* 21, 269–281, 1972.

Bildorientierte Videoindexierung mit Hidden Markov Modellen

Stefan Eickeler, Gerhard Rigoll

Gerhard-Mercator-Universität Duisburg
Fachgebiet Technische Informatik
Fachbereich Elektrotechnik
47057 Duisburg
e-mail: {eickeler,rigoll}@fb9-ti.uni-duisburg.de

Zusammenfassung In diesem Beitrag wird ein neuartiger Ansatz zur inhaltsorientierten Videoindexierung auf der Basis von Hidden Markov Modellen vorgestellt. Bei diesem Ansatz wird für jedes Bild der Videosequenz ein Merkmalsvektor berechnet, welcher mit Hidden Markov Modellen modelliert und klassifiziert wird. Dieser Ansatz hat viele Vorteile gegenüber anderen Videoindexierungsansätzen. Das System kann durch die Vorgabe von manuell indexierten Beispielen trainiert werden und ist dadurch sehr leicht an neue Erkennungsaufgaben adaptierbar. Das System benutzt ein Modell, das die Abfolge von Inhaltsklassen vorgibt, um die Erkennungsleistung zu verbessern.

1 Einleitung

Das wachsende Volumen an Videodaten in Multimediadatenbanken erfordert ein effizientes Verfahren, um auf die Daten automatisch inhaltsorientiert zugreifen zu können. In den letzten Jahren wurden viele unterschiedliche Ansätze zur Videoindexierung vorgestellt. Diese Ansätze lassen sich in hauptsächlich zwei Kategorien unterteilen:

Die erste Kategorie betrifft die Wiederverwendbarkeit von Videosequenzen in Fernsehstudios[1, 6]. Bei diesen Ansätzen wird die Videosequenz in einzelne Szenen segmentiert und ein key-frame pro Szene mit einer Referenz auf das Videoband in einer Datenbank gespeichert. Über ein Beispielbild kann in dieser Datenbank dann nach einer Szene mit gleicher Farbverteilung oder Textur gesucht werden.

Der Zweck der zweiten Kategorie von Videoindexierungssystemen ist die Suche nach Informationen in Videodatenbanken. Eine mögliche Anfrage wäre: "Suche die Hauptberichte aus den Nachrichten des letzten Monats." Der Benutzer würde auf diese Anfrage eine Liste von Berichten erhalten, die er sich ansehen könnte.

Ansätze der zweiten Kategorie sind in [5] und [8] beschrieben. Sie basieren auf einer zweistufigen Klassifikation: Die erste segmentiert die einzelnen Szenen und die zweite ordnet diese Szenen einer Reihe von Klassen zu. Das Problem bei dieser Trennung in zwei Stufen ist, daß ein Fehler in der ersten Stufe eine korrekte Klassifizierung der Szene in der zweiten Stufe unmöglich macht, da keine Rückkopplung zur ersten Stufe existiert. Dieses ist ein sehr wichtiger Punkt, da besonders in Fernsehnachrichten eine Vielzahl von Überblendeffekten auftritt, die nicht so einfach zu finden sind wie Schnitte.

Unser Videoindexierungsansatz segmentiert und klassifiziert die Videosequenz in einem Bearbeitungsschritt. Je ein HMM repräsentiert jede Inhaltsklasse. Für jedes Bild der Videosequenz wird ein Merkmalsvektor berechnet. Durch probabilistisches Dekodieren der HMMs wird die Sequenz von Inhaltsklassen bestimmt, für die die Wahrscheinlichkeit, daß sie die Vektorsequenz erzeugen, maximal ist.

2 Inhaltsklassen und das Videomodell der Fernsehnachrichten

Fernsehnachrichten können in mehrere Inhaltsklassen unterteilt werden. In diesem Ansatz haben wir sechs Hauptklassen definiert.

- Nachrichtensprecher (NEWSCASTER): Das Aussehen dieser Klasse ist bei den verwendeten Nachrichtensequenzen sehr ähnlich. Auf der rechten Seite sitzt der Nachrichtensprecher; auf der linken Seite ist eine Tafel mit der Überschrift und einem Bild der aktuellen Nachricht zu sehen.
- Reportage (REPORT): Eine Szene einer Nachrichtenreportage.
- Interview: Diese Klasse ist ein Interview zwischen dem Nachrichtensprecher und einer zweiten Person.
- Wettervorhersage (WEATHER FORECAST): Diese Klasse zeigt eine Karte mit animierten oder statischen Symbolen für das Wetter.
- Anfang (BEGIN): Vorspann der Nachrichtensendung.
- Ende (END): Abspann der Nachrichtensendung.

Zusätzlich wurden vier Klassen für Überblendeffekte definiert:

- Schnitt (CUT): Dieses ist ein abrupter Übergang von einer Aufnahme zur nächsten. Dieses ist der am meisten verwendete Übergang.
- Überblenden (DISSOLVE): Weiche Überblendung von einer Aufnahme zur nächsten. Dieser Effekt wird bei den verwendeten Sendern nur innerhalb einer Reportage eingesetzt.
- Wischeffekt (WIPE): Dieser Effekt trennt zwei aufeinanderfolgende Reportagen.
- Themenwechsel (WINDOW CHANGE): Der Themenwechsel des Nachrichtensprechers wird durch eine Änderung der Anzeigetafel neben dem Nachrichtensprecher angezeigt.

Um Apriori-Wissen über den Ablauf einer Nachrichtensendung in das Erkennungssystem einzufügen, wurde ein Videomodell definiert, das die Abfolge von Inhaltsklassen angibt. In Abb. 2 ist ein Videomodell für die Nachrichten dargestellt. Es enthält die definierten Klassen, mögliche Übergänge zwischen den Klassen sind durch Pfeile gekennzeichnet.

3 Extraktion der Merkmale

In diesem Abschnitt wird die Methode zur Berechnung der Merkmalsvektoren, die als Eingabe für die Hidden Markov Modelle verwendet werden, vorgestellt. Der Merkmalsvektor soll von möglichst kleiner Dimension sein, um das System zu beschleunigen, aber dennoch die wichtigsten Informationen zur Klassifizierung der Sequenz beinhalten.

Abbildung 1. Ein Bild der Klasse NEWSCASTER

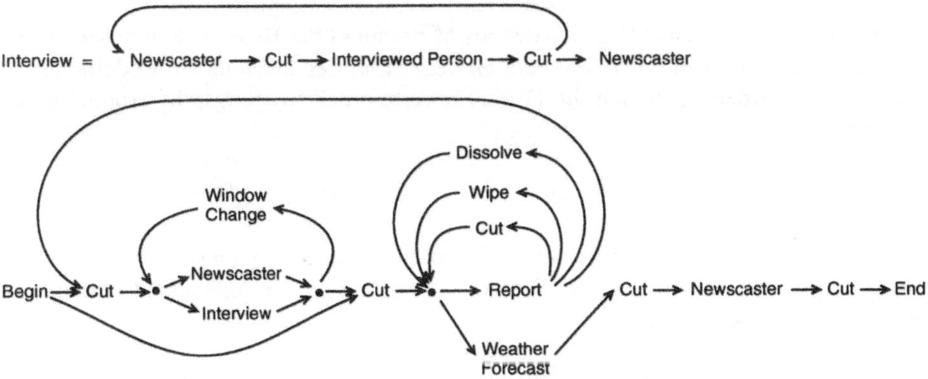

Abbildung 2. Videomodell der Fernsehnachrichten

3.1 Merkmale der Differenzbilder

Die wichtigsten Merkmale zur Klassifizierung werden aus dem Differenzbild berechnet. Sie spezifizieren die Bewegung des größten Objektes in der Bildsequenz. Das Differenzbild wird durch die Subtraktion der Grauwerte der Pixel aufeinanderfolgender Bilder der Nachrichtensendung berechnet.

$$d(x,y,t) = |i_Y(x,y,t) - i_Y(x,y,t+1)| \tag{1}$$

Ein Bild der Differenzbildsequenz enthält nur noch die sich bewegenden Teile des Originalbildes und gibt daher eine Annäherung für die Bewegung in der Bildsequenz.

Die Bewegung in der Differenzbildsequenz $d(x,y,t)$ kann durch die Momente der Verteilung beschrieben werden. Der Schwerpunkt der Bewegung $m(t)^T = [m_x(t), m_x(t)]$ des Differenzbildes $d(x,y)$ wird nach folgenden Formeln berechnet:

$$m_x(t) = \frac{\sum\limits_{x,y} x d(x,y)}{\sum\limits_{x,y} d(x,y)} \qquad m_y(t) = \frac{\sum\limits_{x,y} y d(x,y)}{\sum\limits_{x,y} d(x,y)} \tag{2}$$

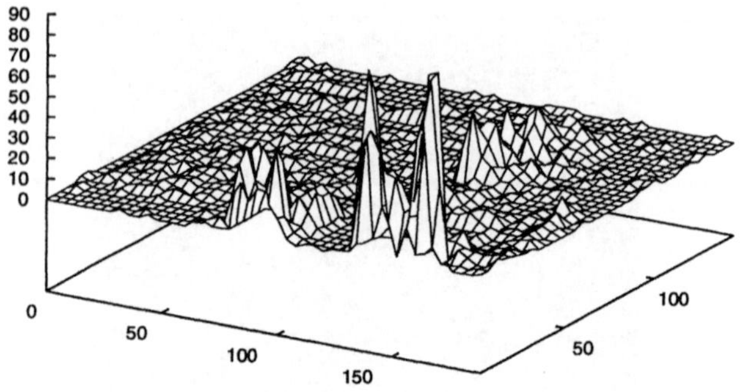

Abbildung 3. Verteilung eines Differenzbildes der Klasse NEWSCASTER

berechnet. Der Vektor $\boldsymbol{m}(t)$ kann auch als Mittelpunkt der Bewegung angesehen werden. Um die Fähigkeit der HMM, die Bewegung in der Sequenz zu modellieren, zu verbessern, werden zusätzlich die Differenzwerte des Bewegungsschwerpunktes verwendet:

$$\Delta m_x(t) = m_x(t) - m_x(t-1) \qquad \Delta m_y(t) = m_y(t) - m_y(t-1) \quad (3)$$

Ein anderes nützliches Merkmal ist die mittlere Abweichung $\boldsymbol{\sigma}(t)^T = [\sigma_x(t), \sigma_y(t)]$ der Bewegung aller Bildpunkte von diesem Bewegungsmittelpunkt. Sie ist folgendermaßen definiert:

$$\sigma_x(t) = \frac{\sum_{x,y} d(x,y,t)|(x-m_x(t))|}{\sum_{x,y} d(x,y,t)} \qquad \sigma_y(t) = \frac{\sum_{x,y} d(x,y,t)|(y-m_y(t))|}{\sum_{x,y} d(x,y,t)} \quad (4)$$

Dieses Merkmal kann auch als Ausdehnung der Bewegung angesehen werden. Es ist vergleichbar mit der Varianz der Bewegung, ist aber nicht so anfällig gegenüber Bildrauschen in der Bildsequenz.

Die Intensität der Bewegung, welche als Mittelwert des Differenzbildes berechnet wird, ist geeignet, um intensive Bewegungen von nahezu stationären Bildfolgen zu unterscheiden, und ist ein gutes Merkmal für die Detektion von Schnitten. Sie wird folgendermaßen berechnet:

$$i(t) = \frac{\sum_{x,y} d(x,y,t)}{\sum_{x,y} 1} \quad (5)$$

Da eine Videosequenz häufig Störungen in einzelnen Bildern enthält, teilweise durch die Übertragung, aber auch durch Schwierigkeiten bei der Aufnahme (z.B. Blitzlicht von Fotografen) verursacht, wird deren Auswirkung durch die Verwendung des kleineren Wertes für die Intensität der Bewegung zwischen den Bildern $(t, t+1)$ und $(t-1, t+2)$, reduziert. Durch dieses Verfahren wird verhindert, daß Bildstörungen, die in einem sehr hohem Wert für dieses Merkmal resultieren, als Schnitt erkannt werden.

$$i'(t) = \mathrm{MIN}\left(i(t), \frac{\sum\limits_{x,y} |i(x,y,t-1) - i(x,y,t+2)|}{\sum\limits_{x,y} 1} \right) \tag{6}$$

Die Merkmale $m(t)$,$\sigma(t)$ und $i(t)$ lassen sich zu einer grafischen Darstellung nutzen, die den großen Informationsgehalt dieser Merkmale zeigt. Eine Ellipse mit dem Mittelpunkt m und den Hauptachsen σ wird den Originalbildern überlagert. Die Intensität der Bewegung $i(t)$ wird als Helligkeit der Ellipse dargestellt. Abbildung 4 zeigt Beispielbilder verschiedener Inhaltsklassen mit überlagerter Ellipse.

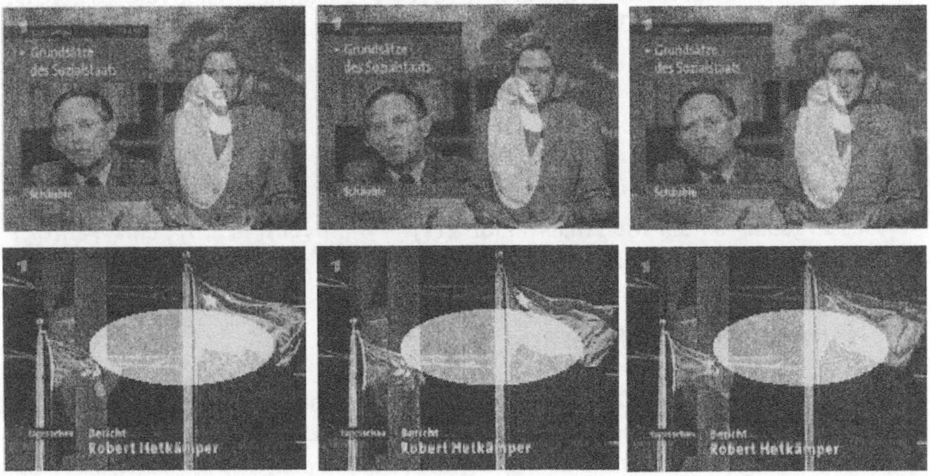

Abbildung 4. Beispiele aus den Klassen NEWSCASTER und REPORT mit überlagerten Ellipsen

3.2 Merkmale des Differenzhistogramms

In [4] wurde gezeigt, daß das Differenzhistogramm sehr gut für die Detektion von Schnitten geeignet ist. Die Intensität des Differenzhistogramms $h(t)$ ist definiert als:

$$hi(t) = \sum_g |h_g(t) - h_g(t+1)| \tag{7}$$

$h_g(t)$ ist die Klassenhäufigkeit des Grauwertes g in Bild t. Dieses Merkmal hat bei stationären Bildfolgen sehr kleine Werte, bei Schnitten hat es sehr hohe Werte. Um die Auswirkung von Bildstörungen zu reduzieren, wird hier das gleiche Verfahren wie bei der Bewegungsintensität verwendet:

$$hi'(t) = \mathrm{MIN}\left(hi(t), \sum_g |h_g(t-1) - h_g(t+2)| \right) \tag{8}$$

Um Detektion von Schnitten weiter zu verbessern, wird die Formel

$$hi''(t) = hi(t) - \text{MEDIAN}\,(hi(t-1), hi(t), hi(t+1)) \tag{9}$$

verwendet. Sie basiert auf der Median-Filterung, die impulsartige Störungen aus Bildern entfernen kann. In diesem Fall wird die Differenz zwischen Originalwert und gefiltertem Wert als Impulsdetektor verwendet.

3.3 Statische Merkmale

Die Merkmale der beiden vorhergehenden Abschnitte sind dynamische Merkmale. Die Verwendung von statischen Informationen kann die Erkennung weiter verbesseren. Die Mittelwerte der Y, U und V Farbkomponenten werden zum Merkmalsvektor hinzugefügt.

$$Y(t) = \frac{\sum\limits_{x,y} i_Y(x,y,t)}{\sum\limits_{x,y} 1} \qquad U(t) = \frac{\sum\limits_{x,y} i_U(x,y,t)}{\sum\limits_{x,y} 1} \qquad V(t) = \frac{\sum\limits_{x,y} i_V(x,y,t)}{\sum\limits_{x,y} 1} \tag{10}$$

Das Ergebnis ist ein Merkmalsvektor $\boldsymbol{x}(t)^T = [m_x, m_y, \sigma_x, \sigma_y, \Delta m_x, \Delta m_y, i',$ $hi', Y, U, V]$, der elf Merkmale enthält.

4 Statistische Klassifizierung

Hidden Markov Modelle haben sich als eine sehr effiziente Methode zur Modellierung von Mustersequenzen herausgestellt. Sie werden erfolgreich in der Sprach- [7], Handschrifterkennung und Gestikerkennung [2] eingesetzt. In dem hier präsentierten Ansatz werden Hidden Markov Modelle für die Modellierung der unterschiedlichen Inhaltsklassen verwendet. Für jede Klasse wird ein Modell mit dem Baum-Welch Algorithmus trainiert.

Es werden links-rechts Hidden Markov Modelle mit einer kontinuierlichen Ausgabeverteilung genutzt. Die Klassen REPORT, NEWSCASTER und WEATHER FORECAST werden mit einer relativ geringen Anzahl von sieben Zuständen trainiert. Die Klasse INTERVIEW wird aus den Klassen Sprecher1 und Sprecher2 zusammengesetzt, deren Modelle auch aus sieben Zustände bestehen. Die Klassen BEGIN und END haben 20 Zustände. Der Schnitt (CUT) besteht aus einem Zustand mit der Selbstübergangswahrscheinlichkeit von Null. Alle anderen Überblendeffekte werden durch Modelle mit 3 Zuständen dargestellt.

Das Videomodell der Fernsehnachrichten wird in ein übergeordnetes Hidden Markov Modell, mit den HMM für die unterschiedlichen Klassen als Ausgabe, umgewandelt. Übergänge in dem Videomodell, die mit Pfeilen gekennzeichnet sind, haben im Modell eine Übergangswahrscheinlichkeit von Eins, alle anderen Übergangswahrscheinlichkeiten sind Null. Um die Nachrichtensendung in die einzelnen Klassen aufzuteilen, wird der optimale Pfad, der das Ergebnis der Viterbi-Dekodierung [7] ist, ausgewertet. Das Videomodell ermöglicht nur sinnvolle Szenenabfolgen in der segmentierten Szenensequenz. So muß die Klasse BEGIN immer am Anfang der Nachrichtensendung sein, die Klasse END stets am Ende.

Sender	Länge	Segmentgrenzenerkennung			Erkennungs-rate	
		Insertions	Deletions	Accuracy		
ARD	15 min	9.90 %	3.19 %	0.22 frames	91.43 %	
ARD	5 min	6.45 %	0.00 %	0.03 frames	96.88 %	
ARD	15 min	4.95 %	4.95 %	0.35 frames	93.46 %	
ARD	15 min	1.61 %	6.15 %	0.20 frames	98.50 %	
ARD	15 min	1.87 %	0.94 %	0.03 frames	97.25 %	
ARD	15 min	3.03 %	11.11 %	0.13 frames	92.11 %	Senderabhängiger
ARD	15 min	6.09 %	12.90 %	0.36 frames	96.21 %	Test
ARD	15 min	1.71 %	1.71 %	0.17 frames	99.17 %	
ARD	5 min	6.25 %	11.02 %	0.22 frames	92.86 %	
ARD	10 min	0.00 %	0.00 %	0.04 frames	100.00 %	
ARD	insgesamt	3.99 %	5.38 %	0.19 frames	96.06 %	
ZDF	5 min	0.00 %	2.17 %	0.09 frames	97.87 %	
ZDF	5 min	0.00 %	6.82 %	0.80 frames	93.33 %	Senderunabhängiger
ZDF	5 min	0.00 %	15.09 %	0.96 frames	90.74 %	Test
ZDF	5 min	5.13 %	9.76 %	0.32 frames	90.91 %	
ZDF	insgesamt	1.18 %	8.70 %	0.55 frames	93.16 %	

Tabelle 1. Erkennungsergebnisse der einzelnen Fernsehnachrichten bewertet nach [3]

5 Experimente und Ergebnisse

Zum Test des Systems wurden Fernsehnachrichten von deutschen Sendern (ARD, ZDF) in einer Auflösung von 192×144 Bildpunkten im Farbraum YUV 4:2:2 bei einer Framerate von 12.5 fps aufgenommen. Die Länge der Nachrichten lag zwischen 5 und 15 min. Auf diesen Nachrichten wurde ein senderabhängiger und ein senderunabhängiger Test durchgeführt. Um die Datenmenge optimal für den senderabhängigen Test des Systems auszunutzen, wurde die hold-out Methode verwendet. Eine Nachrichtensendung wurde für den Test des mit neun Nachrichtensendungen trainierten Systems verwendet. Dieses wurde für jede Nachrichtensendung wiederholt. Für den senderunabhängigen Test wurden die Nachrichten vom ARD trainiert und die Nachrichten vom ZDF zum Test verwendet. Tabelle 1 zeigt die Erkennungsraten für die einzelnen Nachrichtensendungen. Zur Ermittlung der Ergebnisse wurde das Verfahren aus [3] verwendet. Tabelle 2 gibt die Erkennungsraten für die einzelnen Inhaltsklassen an. Die Rechenzeit für die Indexierung einer 15 min Nachrichtensendung beträgt insgesamt 5 min. Das bedeutet, daß das System dreimal schneller als Echtzeit arbeitet.

6 Zusammenfassung

In diesem Beitrag wurde ein neuartiger Ansatz zur inhaltsorientierten Video-Indexierung vorgestellt und am Beispiel von Nachrichtensendungen getestet. Es konnte ein System vorgestellt werden, das folgende Vorteile gegenüber anderen Systemen besitzt:

– Ausnutzung der Lernfähigkeit von HMM

Klasse	Correct Detect	False Detect
Newscaster	95.65 %	11.29 %
Report	97.82 %	2.30 %
Begin	83.33 %	0.00 %
End	100.00 %	0.00 %
Weather Forecast	69.23 %	0.00 %
Interview	86.11 %	4.44 %
Wipe	55.56 %	0.00 %
Window Change	100.00 %	6.45 %
Dissolve	72.73 %	42.86 %

Tabelle 2. Erkennungsraten der einzelnen Klassen für den senderabhängigen Test

- Klassifizierung und Segmentierung der Szenen in einem Verarbeitungsschritt
- Automatische Ausnutzung eines Videomodells
- Echtzeitfähigkeit

Die Erkennungsergebnisse zeigen, daß weitere Verbesserungen, besonders im Bereich der Wettervorhersage, notwendig sind. Zukünftige Arbeiten werden die Audiodaten in die Indexierung einbeziehen und auf komplizierteren Datenbasen (z.B. Sportnachrichten) arbeiten.

Literatur

1. E. Ardizzone, M. L. Cascia, and D. Molinelli. Motion and Color Based Video Indexing. In *Int. Conf. on Pattern Recognition (ICPR-96)*, Vienna, Austria, Aug. 1996.
2. S. Eickeler, A. Kosmala, and G. Rigoll. Echtzeitfähige Videosequenzerkennung mit statistischen Verfahren. In *19. DAGM-Symposium*, pages 105–112, Braunschweig, Germany, Sept. 1997.
3. S. Eickeler and G. Rigoll. Measurements for the Evaluation of Video Indexing Systems. Technical report, Faculty of Electrical Engineering - Computer Science, Gerhard-Mercator-University Duisburg, 1998. `http://www.fb9-ti.uni-duisburg.de/report.html`.
4. A. Hampapur, R. Jain, and T. Weymouth. Digital Video Indexing in Multimedia Systems. In *Proceedings of the Workshop on Indexing and Reuse in Multimedia Systems*. AAAI, Aug. 1994.
5. A. G. Hauptmann, M. J. Witbrock, and M. G. Christel. News-on-demand: An application of informedia technology, Sept. 1995.
6. R. W. Picard. A society of models for video and image libraries. Perceptual Computing Section Technical Report 360, MIT Media Lab, Cambridge, MA, 1996.
7. L. R. Rabiner. A Tutorial on Hidden Markov Models and Selected Applications in Speech Recognition. *Proc. of the IEEE*, 77(2):257–285, 1989.
8. S. W. Smoliar and H. Zhang. Content-Based Video Indexing and Retrieval. *IEEE Multimedia*, 1(2):62–72, 1994.

Konturbasierte Personenlokalisation mittels dreidimensionaler neuronaler Felder und steuerbarer Filter

Ulf-Dietrich Braumann, Andrea Corradini*,
Hans-Joachim Böhme und Horst-Michael Groß**

Fachgebiet Neuroinformatik, Technische Universität Ilmenau
PF 10 05 65, 98684 Ilmenau
{ulf,andreac,hans,homi}@informatik.tu-ilmenau.de
http://cortex.informatik.tu-ilmenau.de/

Zusammenfassung In dieser Arbeit wird ein Verfahren zur Lokalisation von Personen innerhalb unpräparierter visueller Szenen anhand typischer Konturen vorgestellt. Es wird sich dabei auf die äußere Kontur frontal ausgerichteter Personen im Bereich von Kopf und Schultern bezogen (Kopf-Schulter-Partie). Diese Kontur wird approximiert durch ein *räumlich verteiltes Arrangement steuerbarer (steerable) Filter*, das auf einer Anzahl von pyramidal abgestuften Auflösungen eines Bildes angewendet wird.

Unter den Bedingungen realer Umwelten führt der Einsatz dieses Filterarrangements allerdings häufig zu einem unscharfen und teilweise artefaktbehafteten Ergebnis, so daß ein Kriterium zur Selektion als Bedingung für eine gute Lokalisationsleistung benötigt wird. Daher werden zur Bildung einer optimalen Lokalisationsentscheidung *dreidimensionale neuronale Felder* verwendet, innerhalb derer eine dynamische Musterbildung mit dem Ziel der Selektion *einer* Person erfolgt. Mittels einfacher homogener feldinterner Wechselwirkungsmechanismen können robuste Lokalisationsentscheidungen (Richtung und Entfernung) gefunden werden. Die Aktivitäten der Feldneuronen können als interne Zustände angesehen werden, die eine permanente Lokalisation (und damit eine visuelle Verfolgung der Person) wesentlich vereinfachen.

1 Einleitung

Für ein visuell basiertes Gesten-Erkennungssystem, das auf unserer mobilen und autonom navigierenden Roboterplattform *MILVA* eingesetzt wird (siehe dazu auch [3]), hat die Lokalisation eines Nutzers essentielle Bedeutung. Insbesondere spielt die sichere visuelle Lokalisation des Kopfes dieses Nutzers eine wichtige

* gefördert durch ein TMR-Stipendium der Europäischen Union: Marie Curie Research Training Grant # ERB FMBI CT 97 2613

** gefördert durch das Thüringer Ministerium für Wissenschaft, Forschung und Kultur (TMWFK) im Rahmen des Projektes *GESTIK*

Rolle für alle weiteren gestenbezogenen Analysen. In diesem Kontext einer gestenbasierten Mensch-Maschine-Interaktion behandelt diese Arbeit die visuelle Lokalisation von Personen unter realen Umweltbedingungen. Dabei ist davon auszugehen, daß eine solche Lokalisationsaufgabe insbesondere dann relevant wird, wenn eher größere Entfernungen zwischen Roboter und Zielperson vorliegen. Daher rührt die Forderung her, daß die verwendeten Merkmale mit der zur Verfügung stehenden Ortsauflösung erfaßbar sein müssen, weswegen Details wie etwa Gesichtsstrukturen nur teilweise geeignet sind, sofern mit relativ großen Operationsentfernungen zu rechnen ist. Damit erweitert das hier vorgestellte Verfahren unsere bisherigen methodischen Ansätze (siehe [2] zur Lokalisation von Personen, besonders hinsichtlich des nutzbaren Entfernungsbereichs.

Wir sind der Ansicht, daß insbesondere unter Realwelt-Bedingungen die äußere Kontur einer Person die geeignetste Invariante darstellt. Unser einfaches Konturmodell besteht aus einem Arrangement orientierter Filter, das eine stückweise Approximation des Umrisses der Kopf-Schulter-Partie von frontal zur Kamera ausgerichteten Personen vornimmt. Die relevanten Positionen der steuerbaren Filter im Arrangement wurden basierend auf einem Satz von Portraits bestimmt. Durch die Anwendung dieses Filterarrangements auf mehrere Auflösungsstufen eines Bildes kann die Lokalisation eine Entfernungsbestimmung der Person einschließen.

Das zentrale Problem, die für die Personenlokalisation auffälligste Bildposition zu finden, wird mit einem *dreidimensionalen dynamischen neuronalen Feld* gelöst, das in einem iterativen Selektionsprozeß eine abgegrenzte Region aktiver Neuronen (Aktivitätsblob) herausbildet (dynamische Musterbildung, siehe [1]).

2 Arrangements steuerbarer Filter

2.1 Motivation und verwandte Arbeiten

Unser Ansatz beruht auf einer Beschreibung der äußeren Kontur von Kopf und Schultern. Demgegenüber stellen Oren et al. in [9] einen interessanten verwandten Ansatz vor, der sich unter Verwendung sog. *Haar Wavelet Templates* auf den kompletten Umriß von Personen (Fußgänger) bezieht. Der gemeinsame Aspekt beider Ansätze besteht darin, daß räumlich verteilte orientierte Filter genutzt werden, mit denen jeweils ein Orientierungsmaß innerhalb einer begrenzten Region (lokale Orientierung) bestimmt wird. Interessante Bezüge bestehen auch zu Arbeiten von Meller [8], der u. a. mittels verschiedener gitterartig angeordneter Gaborfilter bestimmte Teilmerkmale auf Personen extrahiert, deren lokale Relationen zueinander zur Personenlokalisation verwendet werden.

Die Idee des konturbasierten Ansatzes zur Bildung eines Auffälligkeitsmaßes fußt sowohl auf physiologischen Erkenntnissen als auch auf psychophysischen Effekten. Der visuelle Kortex besteht in verschiedenen Abschnitten aus Zellen, die orientierte rezeptive Felder aufweisen. Bezogen auf eine bestimmte retinale Position kann damit ein breiter Frequenzbereich abgedeckt werden. Eine Vielzahl von Untersuchungen hat gezeigt, daß das Profil der rezeptiven Felder von einfachen

Zellen im primären visuellen Kortex von Säugern durch bestimmte zweidimensionale mathematische Funktionen modelliert werden kann. *Gaborfunktionen*, mit denen Jones und Palmer [6] arbeiten, sowie *Gaußfunktionen* (einschl. Ableitungen niederer Ordnung), die Koenderink und van Doorn verwenden [7], erscheinen insbesondere zur Modellierung typischer Profile visueller rezeptiver Felder als geeignet. Derartige lokale Operatoren ermöglichen es, eine spektrale Zerlegung der visuellen Information vorzunehmen.

Psychophysische Aspekte des konturbasierten Lokalisationsansatzes, wie bspw. *gute Fortsetzung* und *Symmetrie*, die beide in den Gestaltgesetzen wiederzufinden sind, beschreiben offensichtlich Effekte, die aufgrund von Gruppierungsmechanismen zustande kommen können.

Da wir mittels des Filterarrangements eine Modellierung einer bestimmten Kontur vornehmen wollen, indem diese stückweise mit speziell orientierten Filtern approximiert wird, benötigen wir dafür unter Umständen eine Vielzahl verschieden orientierter Filter. Da das sehr bei diskreter Realisierung sehr berechnungsaufwendig wäre, verwenden wir dafür steuerbare Filter [5].

2.2 Bestimmung des Konturverlaufs

In unseren bisherigen Arbeiten [4] wurde ein heuristisch definiertes Konturmodell eingesetzt. Es war manuell entworfen und beschränkte sich auf vier Orientierungen. Es ist allerdings zu erwarten, daß unter Verwendung von mehr Orientierungen ein präziseres Modell bestimmt werden kann. Das macht es jedoch erforderlich, das Konturmodell eng an *reales Bildmaterial* anzulehnen.

Steuerbare Filter haben die angenehme Eigenschaft, daß eine initial feste Anzahl von Faltungen ausreicht, beliebige Orientierungsinformation aus einem Bild zu extrahieren. Sie erlauben, eine deutlich größere Anzahl von Orientierungen zu schätzen und dabei auf zusätzliche Faltungen zu verzichten.

Unser Datenmaterial besteht aus unter identischen Bedingungen vor einem homogenen Hintergrund aufgenommenen Portraits von insgesamt zehn Personen. Jede Person wurde direkt frontal sowie jeweils ($+10°$) und ($-10°$) davon abweichend aufgenommen. Um ein symmetrisches Konturmodell zu erhalten, wurde aus jedem dieser Bilder noch eine vertikal gespiegelte Kopie erzeugt, so daß insgesamt 60 Bilder zur Verfügung standen. Diese Grauwertbilder im Format 256×256 wurden entsprechend tiefpaßgefiltert und auf eine Größe von 16×16 unterabgetastet. Anschließend wurden mit einem Sobel-Operator die Kanten dieser Bilder hervorgehoben und schließlich über alle diese Kantenbilder gemittelt. Mit einer einfachen Schwellenoperation erfolgte durch eine Binarisierung die Bestimmung eines repräsentativen Konturverlaufs, der im Mittel einer Kontur der Kopf-Schulter-Partie der Personen aus dem Datensatz entspricht. Die optimale Schwellenhöhe wurde bestimmt, indem sie in einem automatisierten Verfahren solange angehoben wurde, bis durch die Binarisierung nur noch *ein einzelnes, lückenlos zusammenhängendes, binäres Kontursegment* entstand.

Damit liegt der interessierende Konturverlauf in Form einer 16×16-Binärmatrix $\mathbf{\Lambda}$ vor. Was noch offen ist, ist die lokale Orientierung der einzelnen Konturabschnitte. Wir bestimmen sie mittels steuerbarer Filter (s. u.), die als Faltungsoperatoren auf die Binärmatrix Λ angewendet werden. Somit wird korrespondierend zu jedem Element $\lambda_{i,j}$ ein dominanter Orientierungswinkel bestimmt.

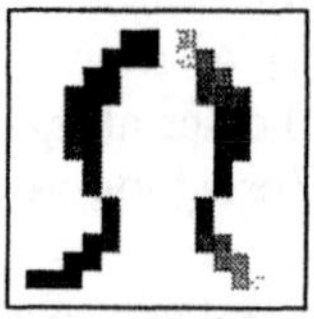

Abb. 1. Der ermittelte Konturverlauf Λ: die Orientierungswinkel sind in Grauwerten kodiert (0°: schwarz; 90°: mittelgrau; 180°: weiß). Man beachte, daß im Scheitelbereich Übergänge zwischen 180° und 0° auftreten. Die Kontur ist symmetrisch, da die Originalbilder jeweils zusätzlich gespiegelt vorlagen.

2.3 Die Anwendung steuerbarer Filter

Es wurde bereits darauf verwiesen, daß die Orientierungswinkel entlang der Kontur mittels steuerbarer Filter bestimmt werden. Diese steuerbaren Filter werden hierfür *ohne* Vorzugsrichtung eingesetzt und aus dem Ergebnis wird in einem Schritt die dominante lokale Orientierung bestimmt. Diese Maximierung könnte bspw. auch unter Nutzung von Gaborfiltern vorgenommen werden, jedoch wäre dies nur mit einer Vielzahl von diskreten Faltungen möglich, da für jede beliebige Richtung ein entsprechendes Quadraturpaar an Gaborfiltern benötigt würde. Je feiner man die Richtungsauflösung wählt, desto höher wird der Rechenaufwand. Daher ist es notwendig, einen akzeptablen Kompromiß zwischen Genauigkeit der Orientierungsbestimmung und der Anzahl der Filter und damit der Faltungen zu finden.

In dieser Arbeit gehen wir mit dem Einsatz von *steuerbaren Filtern* (siehe dazu z. B. die Arbeiten von Freeman und Adelson [5]) einen viel einfacheren Weg. Mit ihnen wird eine flexible und gleichzeitig effektive Orientierungsfilterung möglich, indem ein bestimmter Satz von wenigen, verschieden orientierten *Basisfiltern* eingesetzt wird, anhand derer eine Interpolation in beliebiger Richtung erfolgen kann. Daher sind steuerbare Filter insgesamt effizienter als nicht steuerbare, wie etwa Gaborfilter.

Im allgemeinen gilt eine Funktion $f(\cdot)$ als steuerbar, wenn die folgenden zwei Bedingungen erfüllt sind. Erstens muß die Menge der Basisfilter aus insgesamt M von $f(\cdot)$ gewonnenen, jedoch rotierten Kopien $f^{\alpha_1}(\cdot) \ldots f^{\alpha_M}(\cdot)$ mit festen Rotationswinkeln $\alpha_1 \ldots \alpha_M$ bestehen. Zweitens muß eine weitere, um einen beliebigen Winkel ϑ rotierte Kopie $f^{\vartheta}(\cdot)$ ebenfalls aus der Überlagerung der Funktionen der Basisfilter multipliziert mit Interpolationsfunktionen $k_j(\vartheta)$ zu bilden sein als

$$f^{\vartheta}(\cdot) = \sum_{j=1}^{M} k_j(\vartheta) f^{\alpha_j}(\cdot) \qquad (1)$$

Für unsere Arbeiten verwenden wir als Basisfilter ein Quadraturpaar aus der zweiten Richtungsableitung einer Gaußfunktion und einer Approximation ihrer Hilberttransformierten mittels eines Polynoms dritter Ordnung, das eine Gaußfunktion moduliert. Ausgehend vom Steuerbarkeitstheorem (siehe [5]) sind diese beiden Funktionen jeweils steuerbar. Man benötigt dafür $M = 7$ Basisfunktionen. Um nun die stückweise Orientierung entlang unserer betrachteten Kontur zu messen, verwenden wir die phasenlageunabhängige Summe der Quadrate beider Filterergebnisse. Dieses quadrierte Filterergebnis repräsentiert als eine Funktion der Filterorientierung am Punkt (x, y) eine *orientierte Energie* $E^{(x,y)}(\vartheta)$. Aus Gründen der jeweiligen Symmetrie der Funktionen des Filterpaars ist die Energie an jedem Punkt periodisch mit der Periode π.

Um eine genaue Schätzung der *dominanten* lokalen Orientierung zu erhalten, könnte man punktweise nun eine Maximierung der orientierten Energie vornehmen, indem man durch stetige Variation von ϑ für $\vartheta_{MAX}^{(x,y)} = \arg\max\{E^{(x,y)}(\vartheta) \mid \vartheta \in [0, \pi)\}$ eine optimale Lösung findet. Allerdings können wir auf eine iterative Suche des Maximums verzichten, da für dieses Problem von Freeman und Adelson in [5] bereits eine analytische Lösung zur Energiemaximierung zur Verfügung gestellt wurde. Im weiteren bezeichnen wir die Matrix, die alle zu einem Bildausschnitt korrespondierenden Orientierungswinkel $\vartheta_{MAX}^{(x,y)}$ enthält, als $\mathbf{\Theta}$.

Zusammenfassend kann gesagt werden, daß anders als beim Einsatz von Gaborfiltern bei Verwendung von steuerbaren Filtern keine zusätzlichen Faltungen im Anschluß an eine einmalige Faltung mit allen Basisfiltern erforderlich ist. Außerdem haben wir uns für diese steuerbaren Filter entschieden, da alle Basisfilter kartesisch separierbar sind, was den Rechenaufwand weiter erheblich reduziert.

3 Dynamische Neuronale Felder zur Lokalisation

3.1 Bestimmung der Eingabe des Neuronalen Feldes

Der vorangegangene Abschnitt beschreibt die Theorie und die Anwendung von steuerbaren Filtern. Mittels dieser Filter berechnen wir sowohl die Matrix $\mathbf{\Lambda}$, die den typischen Konturverlauf der Kopf-Schulter-Partie beschreibt, als auch die Matrix $\mathbf{\Theta}$, die zu dem jeweiligen Bildausschnitt korrespondiert, in dem eine Person zu detektieren ist und die die dominanten lokalen Orientierungswerte enthält.

Im folgenden wird ein Maß für das Vorhandensein der *visuellen Auffälligkeitskomponente Kopf-Schulter-Partie*, die durch den Faltungskern $\mathbf{\Lambda}$ repräsentiert wird, innerhalb der Matrix $\mathbf{\Theta}$ gebildet. Dafür nutzen wir ein Matchingverfahren, das auf einem *Ähnlichkeitsmaß* $m^{(x,y)}$ basiert. Wegen der π-Periodizität des Filterergebnisses der steuerbaren Filter einerseits, und um andererseits geeignet die „Ähnlichkeit" zweier Elemente von $\mathbf{\Lambda}$ und $\mathbf{\Theta}$ zu bestimmen, benötigt die Ähnlichkeitsfunktion dieselbe Periodizität.

$$m^{(x,y)} = \frac{\sum_{i=0}^{I-1} \sum_{\substack{j=0 \\ \lambda_{i,j} \neq 0}}^{J-1} \frac{1}{2} \left[\cos \left(2 \left| \lambda_{i,j} - \vartheta_{MAX}^{(x+i-\frac{I}{2}, y+j-\frac{J}{2})} \right| \right) + 1 \right]}{\operatorname{card}(\operatorname{supp}(\mathbf{\Lambda}))} \tag{2}$$

Hierin bezeichnet $\lambda_{i,j}$ ein Element aus $\mathbf{\Lambda}$ mit den Indizes (i, j) sowie $\vartheta_{MAX}^{(x+i-\frac{I}{2}, y+j-\frac{J}{2})}$ ein Element aus $\mathbf{\Theta}$ an der Stelle $(x+i-\frac{I}{2}, y+j-\frac{J}{2})$. $I = J = 16$ sind die Dimensionen der Matrix $\mathbf{\Lambda}$. Die Normierung auf die Kardinalität des Supports von $\mathbf{\Lambda}$ (der Support einer Matrix selektiert nur diejenigen Elemente daraus, die von Null verschieden sind) sichert $m^{(x,y)} \in [0,1]$ für die nachfolgenden Verarbeitungsschritte.

3.2 Das dreidimensionale nichtlineare dynamische Feld

Eine Lokalisation kann umso sicherer erfolgen, je stärker es abgegrenzte Häufungen von hohen Werten des Ähnlichkeitsmaßes gibt. Um eine akzeptable Lokalisation von Personen erzielen zu können, reicht aber die Bildung des Ähnlichkeitsmaßes allein nicht aus, sondern für eine definierte, sichere Lokalisationsentscheidung bedarf es eines *Selektionsmechanismus.* Dynamische neuronale Felder eignen sich gut für eine dynamische Selektion. Sie können mit einfachen homogenen feldinternen Wechselwirkungsmechanismen selektive einzelne, eng abgegrenzte Gruppen aktiver Neuronen herausbilden (sog. Blobs), deren Position im Feld als Lokalisationsergebnis genutzt werden kann.

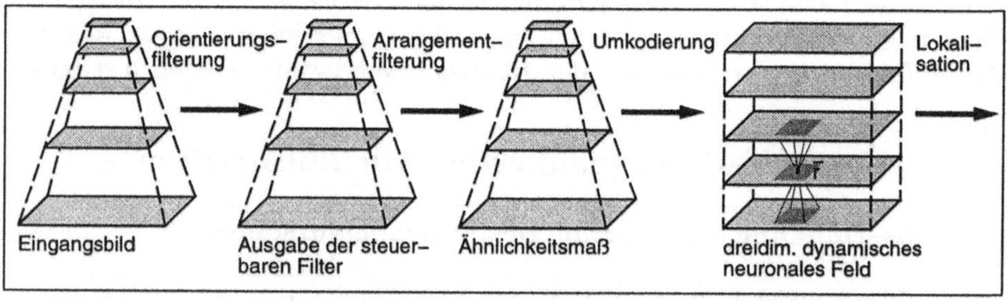

Abb. 2. Die Verarbeitungsstufen zur Personenlokalisation: Die Repräsentation eines Bildes in mehreren Auflösungsstufen wird ebenenweise mit einem Satz steuerbarer Filter verarbeitet. Mit dem anschließenden Einsatz des Filterarrangements wird ein Ähnlichkeitsmaß gebildet, was (nach Umkodierung) als Stimulus in ein dreidimensionales Feld dynamischer Neuronen geleitet wird. Der resultierende Blob (lokal abgegrenztes Muster aktiver Neuronen) wird zur Lokalisierung einer Person genutzt.

Für unsere Arbeiten haben wir ein derartiges Feld aus einer Anzahl von Schichten aufgebaut, da wir insgesamt fünf Auflösungsstufen verarbeiten (siehe Abb. 2). Somit wird ein dreidimensionales Feld aufgespannt, innerhalb dessen die Selektion abläuft. Daher ist es möglich, die Lokalisation auch hinsichtlich verschiedener Distanzen vorzunehmen. Das dynamische neuronale Feld F kann als rückgekoppeltes nichtlineares dynamisches System beschrieben werden. Ziel der Selektion ist die Ausbildung eines *einzelnen* lokalen Blobs aktiver Neuronen, die sich bei Annäherung an einen Gleichgewichtszustand gegenüber den übrigen behaupten können. Die folgende Gleichung beschreibt das System:

$$\tau \frac{d}{dt} z(\boldsymbol{r},t) = -z(\boldsymbol{r},t) - c_h h(t) + c_i x(\boldsymbol{r},t) + c_l \int_N w(\boldsymbol{r}-\boldsymbol{r}')y(\boldsymbol{r}',t)d^3\boldsymbol{r}' \qquad (3)$$

Hierbei steht $\boldsymbol{r}$ für eine dreidimensionale Koordinate einer Neuronenposition im Feld, $z(\boldsymbol{r},t)$ ist die Aktivierung des Neurons $\boldsymbol{r}$ zur Zeit t, $y(\boldsymbol{r},t)$ ist die Ausgabeaktivität dieses Neurons, die mittels einer Sigmoidfunktion bestimmt wird,

$x(\boldsymbol{r}, t)$ bezeichnet den externen Input (korrespondierend zum umkodierten Ähnlichkeitsmaß $m^{\boldsymbol{r}}$, siehe Gleichung 2), $h(t)$ ist eine feldglobale Inhibition zur Zeit t, die alle Neuronenaktivitäten im gesamten Feld $F \subseteq \mathbb{R}^3$ aufsummiert. $w(\boldsymbol{r} - \boldsymbol{r}')$ ist eine Funktion vom Mexikanerhut-Typ. Sie gewichtet die laterale Aktivierung eines Neurons $\boldsymbol{r}$ aus seiner Nachbarschaft $N \subseteq \mathbb{R}^3$. Für ein ausgewähltes $\boldsymbol{r}$ ist N in Abb. 2 (rechts) dunkel markiert. Die Konstanten c_h, c_l und c_i sind Parameter des Systems.

Wie auch in Abb. 2 dargestellt ist, ist zu beachten, daß innerhalb des Feldes die lokalen Korrespondenzen zwischen den Ebenen nicht verletzt werden. Deshalb nehmen wir eine Umkodierung in eine quaderförmige Struktur vor. Ein Nebeneffekt davon ist folgender: je gröber die Auflösungsebene in der Pyramide wird, desto räumlich unschärfer wird das Ähnlichkeitsmaß auf der Kubusschicht und desto ausgedehnter treten dementsprechend lokale Häufungen als Stimuli des dynamischen Feldes auf. Allerdings wurde dieser Effekt von uns nicht gesondert kompensiert. Es war lediglich zu beobachten, daß in den Schichten des Feldes, deren Input aus groben Auflösungsebenen entstammt, einige Schritte mehr zur Herausbildung eines Blobs bzw. Bestandteil eines Blobs benötigt werden, so daß die Anzahl der insgesamt benötigten Iterationsschritte bis zum Erreichen eines Gleichgewichts dadurch etwas erhöht werden kann.

3.3 Ergebnisse

Ein Ergebnis zur Personenlokalisation wir in Abb. 3 illustriert. Die rechte Spalte darin zeigt den Zustand aller fünf Schichten des dynamischen neuronalen Feldes in einer Momentaufnahme, zu der die Aktivitätsänderung des Neurons mit der höchsten Aktivität kleiner als 1% wurde. Im Mittel benötigt das System 11 Iterationszyklen bei Anwendung eines Eulerverfahrens zur numerischen Lösung des Differentialgleichungssystems. Die Ausdehnung des Blobs ist nicht nur auf eine Schicht des Feldes beschränkt. Um eine präzisere Entfernungsbestimmung vornehmen zu können, wäre es auch möglich, die genaue Lage des Blobzentrums zwischen zwei Schichten durch Interpolation zu präzisieren, was im Rahmen dieser Arbeit jedoch nicht realisiert wurde.

Die von uns vorgestellten Ergebnisse sind beispielhaft, wobei wir uns aus Platzgründen auf ein Beispiel beschränken mußten. Eine generelle klare quantitative Bewertung des Verfahrens erscheint uns nicht unproblematisch, da für Art und Komplexität von Person und jeweiligem Hintergrund ein geeignetes Bewertungsmaß benötigt wird. Der hier vorgestellte Ansatz erweist sich aus unserer Sicht als eine gute Lösung für das Problem der Lokalisation von Personen, insbesondere im Hinblick auf seine Toleranz gegenüber Änderungen von Umgebungsbedingungen. Seine Neuheit besteht im Einsatz eines dreidimensionalen dynamischen neuronalen Feldes, das einen wesentlichen Beitrag zur Robustheit des Verfahrens erbringt.

Danksagung Die Autoren danken Anja Brakensiek und Markus Krabbes vom FG Neuroinformatik der TU Ilmenau sowie Heiko Kempe vom Zentrum für Bild- und Signalverarbeitung e. V. Ilmenau für theoretische und praktische Anregungen.

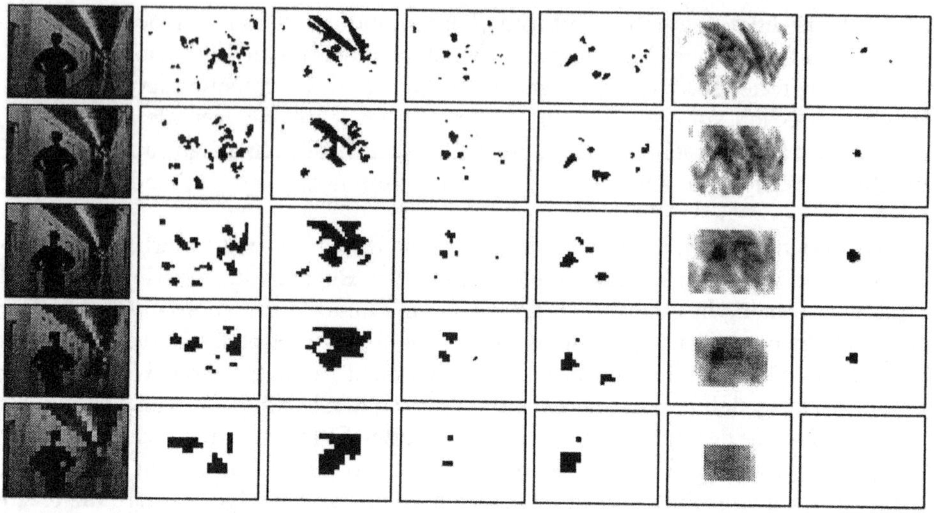

Abb. 3. Lokalisationsergebnis im Innenbereich: Die Lokalisation der Person erfolgt nicht scharf auf einer der Pyramidenebenen, der entstehende räumliche Blob (Spalte ganz rechts) ist am stärksten auf der mittleren der fünf Ebenen ausgebildet. In jeder Zeile sind alle Zwischenergebnisse und Ergebnisse für eine der fünf Auflösungsstufen (Abstand $1/\sqrt{2}$) dargestellt. Die sieben Spalten enthalten folgendes: Input, Ergebnis der Orientierungsfilterung mit steuerbaren Filtern (hier aus Darstellungsgründen beschränkt auf vier Richtungsintervalle zentriert um 0°, 45°, 90° und 135°), das Ergebnis der Anwendung des Filterarrangements (Ähnlichkeitsmaß) und schließlich das Ergebnis der Selektion innerhalb des dreidimensionalen Feldes dynamischer Neuronen.

Literatur

1. Amari, S.: Dynamics of Pattern Formation in Lateral-Inhibition Type Neural Fields. Biological Cybernetics **27** (1977) 77–87
2. Brakensiek, A., Braumann, U.-D., Böhme, H.-J. und Gross, H.-M.: Farb- und strukturbasierte neuronale Verfahren zur Lokalisierung von Gesichtern in Real-World-Szenen. Mustererk. 1997 – Tagungsband 19. DAGM-Symp. (1997) 113–120
3. Boehme, H.-J., Brakensiek, A., Braumann, U.-D., Krabbes, M. und Gross, H.-M.: Neural Networks for Gesture-based Remote Control of a Mobile Robot. Proc. IEEE Intern. Joint Conf. on Neural Networks (1998) Bd. 1, 372–377
4. Corradini, A., Braumann, U.-D., Brakensiek, A., Krabbes, M., Boehme, H.-J. und Gross, H.-M.: Visual Gesture Localization with Dynamic Neural Fields: Towards a Gesture Recognition System. Proc. X Italian Workshop on Neural Nets (1998)
5. Freeman, W. T. und Adelson, E. H.: The Design and Use of Steerable Filters. IEEE Transactions on Pattern Analysis and Machine Intelligence **13** (1991) 891–906
6. Jones, J. P. und Palmer, L. A.: An Evaluation of the Two-Dimensional Gabor Filter Model of Simple Receptive Fields in Cat Striate Cortex. Journal of Neurophysiology **58** (1987) 1233–1258
7. Koenderink, J. J. und van Doorn, A. J.: Receptive Field Families. Biological Cybernetics **63** (1990) 291–297
8. Meller, S.: Segmentierungsfreie Personendetektion. Diplomarbeit, Universität Erlangen-Nürnberg (1997)
9. Oren, M., Papageorgiou, C., Sinha, P., Osuna, E. und Poggio, T.: Pedestrian Detection Using Wavelet Templates. Proc. of the IEEE Computer Society Conference on Computer Vision and Pattern Recognition (1997) 193–199

Combining Contour and Region Information for Perceptual Grouping

Daniel Schlüter and Stefan Posch

Technische Fakultät, AG Angewandte Informatik
Universität Bielefeld, Postfach 100131, 33501 Bielefeld
{dschluet|posch}@techfak.uni-bielefeld.de

Abstract. Both contour and region segmentation have their own advantages and drawbacks. In this work we propose to combine contour approximation and contour-based groupings with region segmentation to enhance both contour- and region-based interpretation of the image data. Contour segments and regions are matched based on the distance between contours and region boundary. This information is used to hypothesise alternative regions and to generate additional closed contours within the contour-based grouping process. Initial results for real images are presented.

1 Introduction

The human visual system uses different types of visual cues for perception. Regarding computer vision, Sanocki, Bowyer, Heath and Sarkar [SBHS98] discuss the question, "Are edges sufficient for object recognition?". Their conclusion is, that contour information alone is not sufficient and they propose to use other sources of information like regions, surfaces or intermediate-level structures beside edge information to constrain each other.

Several algorithms have been developed to combine region techniques and edge based techniques to improve image segmentation. Pavlidis and Liow [PL90] apply criteria regarding contrast, gradient variation, and artefacts along region boundaries to verify and eliminate boundaries obtained from an oversegmenting split-and-merge method. In a similar way, [XAW95] uses the results of edge detection to verify and correct region boundaries generated by region-growing. The approach of [XYJ92] integrates gradient information into a region growing process. In contrast, [PFV96] proposes a method to create a fully connected line-drawing from an edge segmentation using a stochastical region growing. Other approaches using similar combinations of region and edge segmentation can be found in [BB92] and [Koh95]. While these approaches employ different region and edge detection techniques, all of them rely on pixelbased combinations of segmentation results.

Our proposal is to combine contour and region segmentation, where contours are represented by parametric models like lines, circular or elliptical arcs, because

* This work has been supported by the German Research Foundation (DFG) within SFB 360.

contours contain more abstract information about the global structure of the image. We describe, how the combination of region and contour segmentation can resolve incorrect segmentation results of both techniques and therefore support and constrain each other.

2 Contour Segmentation and Grouping

In this section we give an overview of the contour-based segmentation and grouping we use as a basis for the combination with a region segmentation. For the initial contour segmentation we use a standard Sobel operator and subsequent thinning with non maximum suppression and hysteresis threshold. The resulting edge elements are approximated with straight line segments and elliptical arcs with the method of Leonardis (see [Leo93]).

To overcome fragmentation and to obtain more abstract image primitives these contour segments are subsequently organised within a hierarchy of grouping hypotheses (see fig. 1) using various Gestalt laws. In the lowest level contour segments are combined to form collinear and curvilinear groups which are again approximated with straight line segments or elliptical arcs. Theses groups as well as the initial contour segments are denoted as *linear groups* subsequently. In addition, pairs of contour segments or linear groups may form a proximity grouping if their endpoints are spatially close. The next 2×1D level which is not considered further in the rest of this article deals with pairs of symmetric or parallel linear groups. To generate hypotheses within the 1D and 2×1D level of the hierarchy we employ the concept of *Areas of Perceptual Attentiveness* which are derived from a hand labelled training set (see [MP95]).

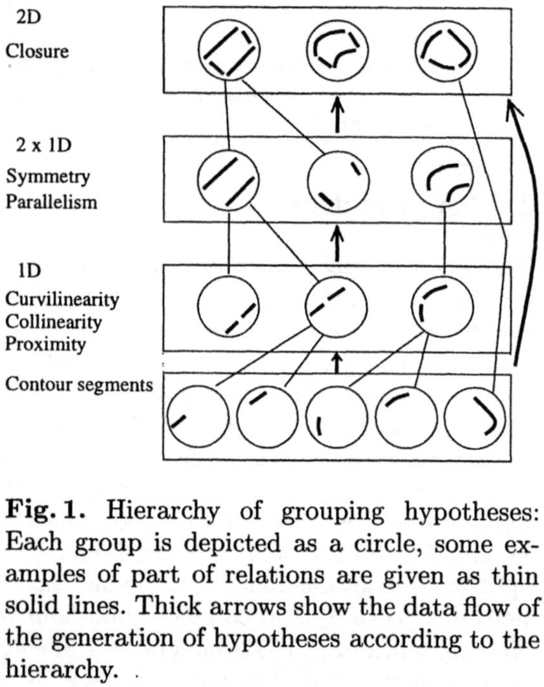

Fig. 1. Hierarchy of grouping hypotheses: Each group is depicted as a circle, some examples of part of relations are given as thin solid lines. Thick arrows show the data flow of the generation of hypotheses according to the hierarchy.

The last level organises linear groups into closed contours. To this end we construct an undirected marked graph from all linear groups hypothesised. This so-called *proximity graph* for a synthetic example is depicted in figure 2. Both endpoints of each linear group yield a node of the graph and both are connected by a *contour edge* representing the underlying contour. In addition, for each proximity grouping the two appropriate endpoints of both linear groups are connected by a *proximity edge.* In this graph a closed contour corresponds to a simple cycle with an alternating sequence of contour and proximity edges. In addition the underlying contour segments as well as the gaps between endpoints bridged by proximity edges may not intersect. To check for intersections, proximity edges are modelled as straight line segments between the endpoints and elliptical arcs are approximated with a polygon. In [HW91] and [UE96] contour segments are also represented in a graph to search for closed or longest paths. However, they model each contour segments as one node and therefore have to distinguish different types of edges for each node.

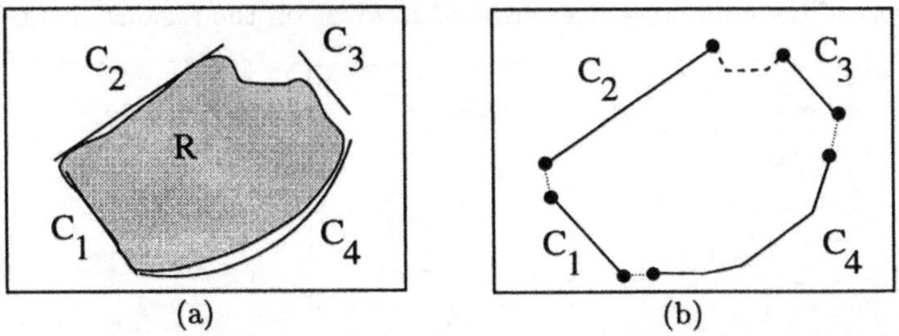

Fig. 2. (a) Synthetic region and contour segmentation; (b) Sample proximity graph: nodes are displayed as black circles, contour edges corresponding to contour segments with solid lines; for C_4 the polygonal approximation used to check against intersection is depicted; proximity edges from contour based grouping are shown with dotted lines, while the proximity edge connecting C_2 and C_3 derived from region information is shown as a dashed line (cf. section 3).

The generation of these grouping hypotheses within the hierarchy relies on local evidence of the image data only. Therefore, all hypotheses are subsequently judged in a more global context using a Markov Random Field. For details on this global judgement and the generation as well see [MPSS97].

The approach to combine contour and region information proposed in this work is based as well on grouping hypotheses and as on judged groups and will be described in the following section.

3 Approach

The aim of our approach is to resolve the drawbacks inherent in both segmentations by combining them. A typical drawback of region segmentation is depicted in figure 3. The region segmentation fails to segment the wooden bar as one region, because it is partially occluded by the cube: the segmentation produces two disconnected regions for the upper surface (fig. 3(c)). The collinear groupings connect the fragmented contour into linear groups (fig. 3(b)) and a closed path of linear groups for the surface of the bar is found within the proximity graph. Based on this closed contour and information, that the contours are nearby the underlying region boundaries, an alternative region can be hypothesised merging both regions of the surface. Another situation occurs if two wooden bars do overlap. Region segmentation techniques inherently tend to generate one region for both bars, because their projection in the image is of homogeneous colour

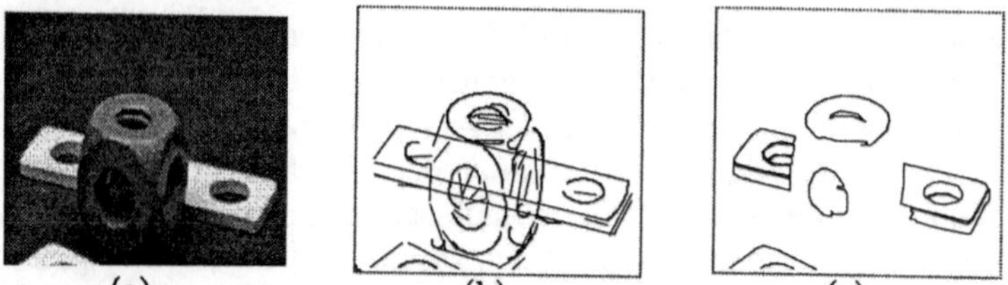

(a) (b) (c)

Fig. 3. (a) Image with parts of BAUFIX toys (cube and wooden bar), (b) contour segmentation and linear groups, (c) region segmentation depicted as closed polygons: some regions of the cube have been deleted to focus on the regions of the bar

(a) (b) (c)

Fig. 4. (a) Image of a wooden bar, (b) contour segmentation and linear groups, (c) region segmentation (depicted as closed polygons)

and the inhomogeneities at the boundary between both surfaces are usually insufficient for separating them. In this case the linear groups and detected closed contour paths for the surfaces of the two bars support a separation of the region into the correct surface projections of the bars. For contour segmentation, a characteristic disadvantage is shown in figure 4. The contour segmentation fails in situations where the gradient magnitude is not sufficient to locally support a discontinuity in the image (see lower right corner of the bar in fig. 4(a,b)). However, the region segmentation detects the boundary of the surface of the bar (fig. 4(c)). This is due to fact, that region segmentation tries to detect homogeneous areas in an image using more global information to segment the image, whereas the contour segmentation uses only local image data to detect discontinuities in an image. In this situation the region boundary can help to close a path of linear groups in the proximity graph traversing the boúndary to find the next linear group near the region boundary (see figure 2 for a similar example, where region information can be used to close a path between C_2 and C_3).

These two example show how we use a çombination of region and contour segmentation including linear groups to support and constrain each other. Our focus is on the latter case and will be described in detail in section 3.3. Some initial results concerning the former case are presented in section 4.

3.1 Notation

The combination of the two different segmentation techniques requires an appropriate representation. For both results, we use a polygonal representation, since this uniformity gives an efficient means to calculate various measures.

A linear contour groups is denoted as C and represented by open polygons $poly(C)$. To this end, elliptical arcs are approximated with appropriate polygons. The results of region segmentation are represented by closed polygons $poly(R)$ for each region R, approximating the region boundary.

3.2 Matching linear groups to regions

In the following we assign each linear group to one region if both probably belong to the projection of the same 3D-surface. Therefore each linear group is tested against all regions to find the region with minimum distance between its boundary and the linear group. To find the best matching region R for a given linear group C the sum of the minimum euclidean distance $d(v, poly(R))$ of each vertex v in the contour polygon $poly(C)$ to the region polygon $poly(R)$ is calculated[1].

$$d_{sum}(R,C) = \sum_{\forall v \in poly(C)} d(v, poly(R))$$

To enhance the efficiency of the calculation, the test is only applied to regions intersecting a neighbourhood around the linear group. The region R with minimum $d_{sum}(R,C)$ is established to match the linear group C. This selection only establishes a relation between a linear group and a region, where linear groups must not approximate the region boundary but can also be based on structure inside or nearby the region.

The correctness of each match is judged by three measures. The first and second measures are the euclidean distances $d_{start}(R,C)$ and $d_{end}(R,C)$ of the endpoints of the linear group to the region boundary. To compute an estimate of the average distance between the linear group and the region boundary, the polygon of the linear group $poly(C)$ is projected onto the polygon of the region $poly(R)$. The area A between the projection and the contour polygon is calculated. The third measure is then defined by: $d_{avg}(R,C) = \frac{A}{length(C)}$, where $length(C)$ is the length of the contour segment (see figure 5). This gives an estimate for the average distance between both polygons.

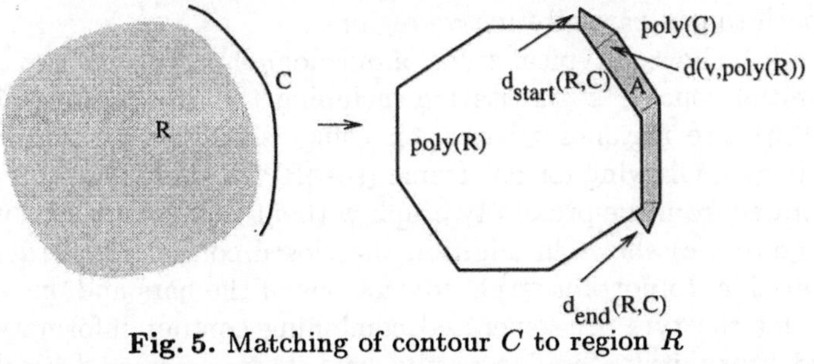

Fig. 5. Matching of contour C to region R

[1] We do not have to normalise the sum, since the number of vertices in $poly(C)$ is constant for the test against all regions R_i.

3.3 Using matches to detect closed paths of contours

We now describe how these matches can be used to generate cycles of linear groups even if the proximity information of the grouping process is not sufficient.

The search for closed paths of linear groups may fail in situations, where a proximity relation between two linear groups is missing to close a path. These situations can be resolved using the information which linear group match a region: In figure 2, the path $C_1 - C_2$ can not be closed to $C_3 - C_4 - C_1$, due to the missing contour based proximity information between C_2 and C_3. However, we can use the matching of C_2 and C_3 to the same region R to complete the path along the region boundary. We call this a *region based proximity* between the endpoints of C_2 and C_3. The following describes this idea in detail.

For this process matches between linear groups C and regions R are only considered, if the following condition is valid:

$$d_{avg}(R,C) < \theta_{err} \quad \wedge \quad d_{start}(R,C) < \theta_{dist} \quad \wedge \quad d_{end}(R,C) < \theta_{dist}$$

This insures a sufficiently small distance between the matching linear group and the region boundary. For each region all assigned linear groups are linearly ordered around the region. This order is defined by the projected start- and endpoints of the linear group onto the region polygon. Thus, for the endpoint of one matching linear group, the start of the next matching linear group $\tilde{C}$ can be found along the region boundary of the same matched region R. Adding region based proximities to the proximity graph for perceptual grouping of contour segments is controlled by the distance of the endpoints of C and $\tilde{C}$. Only if the distance between the endpoints of C and $\tilde{C}$ along the region boundary of R is smaller than the length of both segments and there is not already a proximity edge in the graph between them, a new region based proximity edge is inserted. These edges are used in the same way as proximity edges are used in the proximity graph as described in section 2.

4 Results

In this section we give results for using region information to enhance the quality of closed contours detected within the proximity graph and first results using this approach to generate alternative regions.

Figure 6(a) shows a typical scene of our domain, a wooden toy construction kit. The initial contour segmentation including the linear groups is displayed in figure 6(b), the region segmentation using Island-Growing and subsequent Sequence-Region-Growing for one frame (see [BP97]) in figure 6(c). Closed contours generated from the proximity graph without region information are shown in 6(d). Figure 6(e) shows in addition the closed contours generated using region information. Important structures for one of the bars and one of the cubes as well as for the tyre are generated combining contour information with region based proximities. Similar results have been detected for other images. Table 1 summarises runtimes for segmentation, grouping, and combining both segmentations. The main difference in the runtime of 2D-grouping is due to the

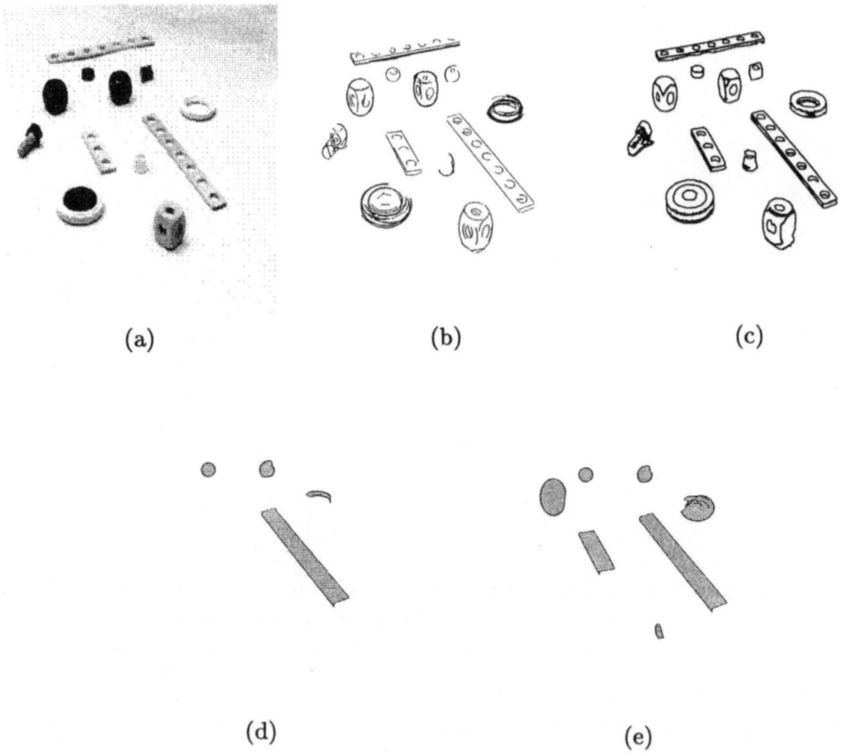

Fig. 6. (a) Image of a wooden toy construction kit; (b) initial segmentation and linear groups; (c) region segmentation; (d) hypotheses of closed contours without using region information and (e) with region information

	Contour		Grouping			Region
	Sobel + Thinning	Approxi-mation	1D contour	2D contour-based	contour+region-based	Segmentation
Time	748 ms	9164 ms	916 ms	50 ms	332 ms	1410 ms

Table 1. Runtime on Alpha 500/400MHz for image in figure 6(a)

matching of linear groups to regions ($\sim 80\%$). The contour-based approach uses 135 proximities to generate 4 closed path of linear groups, while the contour- and region-based approach uses 39 additionally region-based proximities to generate a total of 9 closed paths (3 overlapping closed paths for the tyre). As parameters we use $\theta_{dist} = 5$ and $\theta_{err} = 10$.

Figure 7 shows a typical situation where region segmentation fails to generate separate regions for the surfaces of the two overlapping wooden bars (see fig. 7(c)). Using significant, i.e. well judged, linear groups and closed contour paths (see [MPSS97]), and the information about matching regions with contours (fig. 7(d)) we get evidence to separate the underlying region. Therefore we propose alternative regions from closed contour paths, in which most of the constituting linear groups match to the same region. Figures 7(e), (f) show only a subset of alternative regions. There were also some for the projection of the

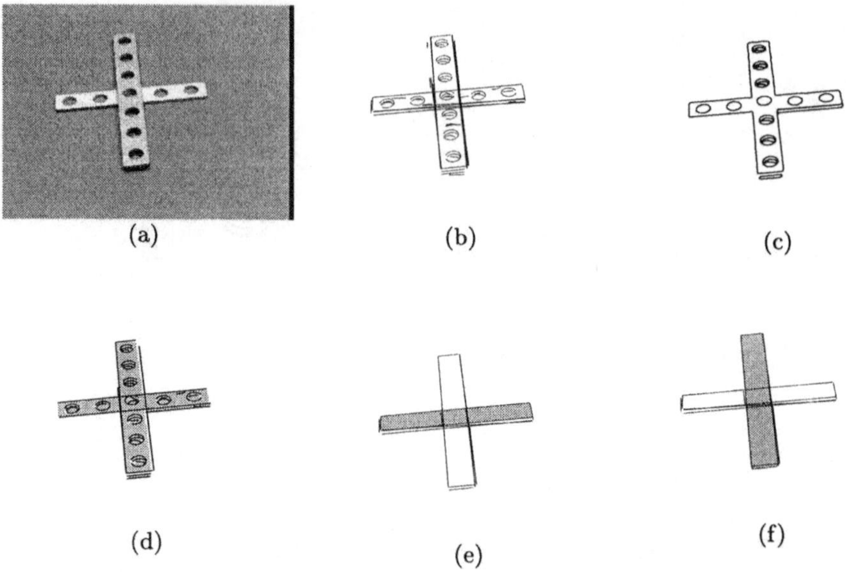

Fig. 7. (a) Image of a wooden toy construction kit; (b) initial segmentation and linear groups; (c) region segmentation; (d) region segmentation and matched linear groups; (e) , (f) subset of alternative contour-based regions generated in the proximity graph

3D-surfaces, e.g. two regions for the projection of the lower bar, and some incorrect ones, grouping the lower bar with the bottom part of the upper bar into a T-region. To generate the matches in this context we use $\theta_{dist} = 7$ and $\theta_{err} = 30$.

5 Conclusion

In this paper we propose to combine region- and contour based segmentation at a level, where contour segments have been approximated with straight lines and elliptical arcs and are already organised into more abstract linear groups and closed paths. Each linear group is matched to one region based on minimum distance of the underlying polygons. These matches are used to generate alternative region hypotheses in cases where the projection of one 3D surface decomposes into several regions or where projections of different 3D surface are merged because of homogeneous pixel values. Alternatively, the matches are employed to insert edges into the proximity graph of linear groups to add proximity information which is missing due to low local contrast during edge detection. Results for real images show significant improvements of the generated closed contours with only small computation overhead. Current work is concerned with realization of the former case and parameter selection.

References

[BB92] J. Benois and D. Barba. Image segmentation by region-contour cooperation for image coding. In *Proc. International Conference on Pattern Recognition*, volume C, pages 331–334, 1992.

[BP97] Thorsten Bomberg and Stefan Posch. Regionensegmentierung von farbbildfolgen. In E. Paulus and F.M.Wahl, editors, *Mustererkennung 1997. Proceedings 19. DAGM-Symposium*, Informatik Aktuell, pages 63–70. Springer, 1997.

[HW91] Daniel P. Huttenlocher and Peter C. Wayner. Finding convex edge groupings in an image. In *Proc. Computer Vision and Pattern Recognition*, pages 406–412, Lahaina, Maui, 1991. IEEE.

[Koh95] Peter Kohlhepp. Combining region and edge based segmentation for robust 3d localization in range images. In *Proc. 2nd Asian Conference on Computer Vision*, volume 2, pages 254–258, Singapore, 1995. IEEE.

[Leo93] Aleš Leonardis. *Image Analysis Using Parametric Models.* PhD thesis, University of Ljubljana, 1993.

[MP95] A. Maßmann and S. Posch. Mask-oriented grouping operations in a contour-based approach. In *Proc. 2nd Asian Conference on Computer Vision*, volume 3, pages 58–61, Singapore, 1995. IEEE.

[MPSS97] A. Maßmann, S. Posch, G. Sagerer, and D. Schlüter. Using markov random fields for contour-based grouping. In *Proc. International Conference on Image Processing*, volume II, pages 207–210. IEEE, 1997.

[PFV96] E.J. Pauwels, P. Fiddelaers, and L.J. Van Gool. Autonomous grouping of contour-segments using an adaptive region-growing algorithm. In *Proc. International Conference on Pattern Recognition*, volume B, pages 586–590, 1996.

[PL90] Theo Pavlidis and Yuh-Tay Liow. Integrating region growing and edge detection. *IEEE Transactions on Pattern Analysis and Machine Intelligence*, PAMI-12(3):225–233, March 1990.

[SBHS98] Thomas Sanocki, Kevin W. Bowyer, Michael D. Heath, and Sudeep Sarkar. Are edges sufficient for object recognition. *Journal of Experimental Psychology: Human Perception and Performance*, 24(1):1–10, 1998.

[UE96] A. Ude and T. Ekre. Stereo grouping for model-based recognition. In *Proc. International Conference on Pattern Recognition*, volume A, pages 223–227, 1996.

[XAW95] J. Xuan, T. Adali, and Y. Wang. Segmentation of magnetic resonance brain image: Integrating region growing and edge detection. In *Proc. International Conference on Image Processing*, volume C, pages 544–547, 1995.

[XYJ92] Yu Xiaohan and Juha Ylä-Jääski. Image segmentation combining region growing and edge detection. In *Proc. International Conference on Pattern Recognition*, volume III, pages 481–484, 1992.

Systematic Curvature Distortions in Stereo

Gregory Baratoff

Department of Neural Information Processing
University of Ulm, D-89069 Ulm, Germany
baratoff@neuro.informatik.uni-ulm.de

Abstract. Systematic distortions occur when incorrect parameters of a stereo configuration are used in the reconstruction of a three-dimensional scene from image correspondences. Consideration of the effect of errors on reconstruction is especially important for an active observer, since its stereo configuration changes continuously due to gaze shifts, and hence needs to be recomputed in real-time. For a fixating observer the largest distortions occur within epipolar planes; of particular interest among these is the fixation plane. We study the distortion of points in this plane on a global, qualitative, level, making use of properties of quadratic Cremona transformations, and on a local, quantitative level. We derive closed-form expressions for the distortion of curvature and determine the exact criteria governing changes in convexity of curves. Our simulations show that such distortions can be severe for very close distances; for the most part, however, curvature distortion depends roughly linearly on the fixation error, but only marginally on fixation angle or distance along the gaze line.

1 Introduction

One of the central tasks of a stereo vision system is to reconstruct the scene from the projections of environmental structures onto the two views. The reconstruction computation is based on two pieces of information : a set of corresponding points (or lines) in the two views, and the stereo configuration, i.e. the relative positions and orientations of the two cameras, and their internal parameters. Corresponding points are usually obtained by a matching procedure based on purely visual attributes, whereas the stereo configuration can be obtained from either visual or by non-visual sources of information.

Reconstruction and the Epipolar Constraint : The stereo configuration needs to be known exactly if the resulting reconstruction is to be veridical, i.e. if metric properties such as distances and angles are to be preserved. Unfortunately, obtaining the exact stereo configuration can be problematic. Estimates of the stereo configuration based on *non-visual* information are often not accurate enough. Furthermore, they are usually inconsistent with the epipolar constraint, which expresses the fact that corresponding rays issued by the two cameras should intersect in space. In order to exploit *visual* information, the cameras have to be calibrated individually, and their relative orientation has to be found, with both procedures in general requiring a non-linear minimization[11]. (The same applies for projective reconstruction from uncalibrated cameras.) Especially in the presence of noise these computations are plagued by instabilities, and convergence can not in general be guaranteed. The question we asked ourselves is whether this step could be avoided, and if so, what the nature of the resulting, necessarily *non-linear*, distortions would be.

Distorted Reconstruction in Space : In order to reconstruct points without calibrating the stereo rig beforehand, a triangulation method has to be employed that is defined even when the two corresponding rays do not intersect. We call such reconstructions *distorted reconstructions.* In [1, 3], we present an analysis of the *dominant camera* reconstruction model, in which the distorted reconstruction of a point is defined as the intersection of one of the rays and a plane containing the other ray. Alternative distorted reconstructions, such as the midpoint of the shortest segment connecting the two rays, are reviewed and evaluated in [6]. When we impose some restrictions on our analysis, then the choice of the reconstruction model becomes less important. In particular, we consider a constrained stereo configuration, namely that of the symmetrically fixating observer whose cameras are aligned in such a way that their Y-axes are parallel. (This is termed the "ground-plane motion" constraint in the motion analysis literature.) Furthermore, we focus on those (planar) sections of space where the distortion is highest. From the results in [1, 3], it follows that in the dominant camera model the planes of highest distortion are the epipolar planes. Similar reasoning can show that this also holds for the other models. In what follows, we therefore concentrate on distortion within such epipolar planes. Without loss of generality, these distortions can be studied in the fixation plane; the results carry over to other epipolar planes.

Overview . In section 2 we derive the planar shape distortion. It turns out to be a quadratic Cremona transformation. We discuss properties of such transformations in section 3, and apply them to the fixating observer configuration to explain some qualitative, global, properties of the shape distortion. Cremona transformations are birational transformations which are one-to-one almost everywhere. They do not in general preserve collinearity, except when they are of degree 1, i.e. when they are projective transformations. Concretely, this means that for example a line in general maps to a curve with non-vanishing curvature. In section 4, we derive a formula for the distortion of curvature, and present a quantitative analysis.

Related Work : The framework of distortions was introduced by [4] in the case of calibrated cameras, and under the assumption of infinitesimal motion. In [1, 2] we showed that the shape distortion transformation in the case of discrete displacements is a quadratic Cremona transformation for arbitrarily positioned and oriented, uncalibrated cameras, and used global properties of Cremona transformations to qualitatively explain the shape distortion. Quadratic distortions have been observed by photogrammetrists[10] in certain stereoscopes. In the Computer Vision literature, (plane) quadratic transformations have been used to describe the mapping between corresponding points in two image planes (e.g. [8]), but have not been applied in the analysis of transformations of reconstructed space.

2 Planar Shape Distortion

The distorted reconstruction from image correspondences involves what we call the *apparent stereo configuration*, different from the *true stereo configuration* used in the projection of the scene onto the cameras. Furthermore, we refer to an *apparent space*, in which objects live that were reconstructed using the apparent configuration, and a *true space*, in which objects live that were reconstructed using the true configuration. In order to distinguish objects from the different spaces, we will always use primed symbols (e.g. $\mathbf{X}', \mathbf{m}'$) to refer to objects in the apparent space.

Homogeneous Point and Line Coordinates : The use of homogeneous coordinates for points and lines in the plane simplifies expression considerably. We use bold-face letters for both, upper-case ones for points (e.g. $\mathbf{X}$), and lower-case ones for lines (e.g. $\mathbf{m}$). Homogeneous vectors for points and lines in the plane have three components, and they allow the representation of points and lines at infinity. Finite points $\mathbf{X} = (X\ Z\ W)^T$ satisfy $W \neq 0$, and points at infinity satisfy $W = 0$. The line at infinity $\mathbf{h}_\infty$ is given by the homogeneous vector $(0\ 0\ 1)$, so that $\mathbf{h}_\infty \cdot \mathbf{X} = 0 \Rightarrow W = 0$.

Two basic geometrical operations are the join $\mathbf{X} \vee \mathbf{Y}$ of two point $\mathbf{X}$ and $\mathbf{Y}$, and the meet $\mathbf{a} \wedge \mathbf{b}$ of two lines. The cross-product of homogeneous vectors implements both the join and the meet operators, and should therefore be substituted for either of them when formulas actually have to be calculated. The geometrical significance of formulas is, however, better expressed if different symbols are used to refer to join and meet operations. The scalar product $\mathbf{m} \cdot \mathbf{X}$ between a line and a point vector defines their scaled distance. Thus, $\mathbf{X}$ lies on $\mathbf{m}$ when $\mathbf{m} \cdot \mathbf{X} = 0$. It follows from this that the homogeneous coordinate vector of a point or of a line is only given up to scale, i.e. $\lambda\mathbf{X}$ represents the same point as $\mathbf{X}$, for any non-zero scalar λ.

Camera Geometry : We model a 1D-camera by a 2×3 matrix $A = (\mathbf{a}_1\ \mathbf{a}_2)^T$, where $\mathbf{a}_1$ and $\mathbf{a}_2$ represent the X-axis and the Z-axis, respectively, of the camera frame. From these, the camera center can be computed as $\mathbf{O}_A \doteq \mathbf{a}_1 \wedge \mathbf{a}_2$. The image projection of a point $\mathbf{X}$ is given by the homogeneous vector $\boldsymbol{\alpha} = (\mathbf{a}_1 \cdot \mathbf{X}, \mathbf{a}_2 \cdot \mathbf{X})^T = A\mathbf{X}$, and the camera ray passing through the image point $\boldsymbol{\alpha}$ is given by $\alpha_1\mathbf{a}_2 - \alpha_2\mathbf{a}_1$. Putting the last two operations together, one obtains for the camera ray passing through a point $\mathbf{X}$ the expression $(\mathbf{a}_1 \cdot \mathbf{X})\mathbf{a}_2 - (\mathbf{a}_2 \cdot \mathbf{X})\mathbf{a}_1$.

The Distorted Reconstruction : One arrives at a distorted reconstruction when the two camera matrices A and B used for the projection are different from camera matrices A' and B' used in the reconstruction. The transformation between the true and the distorted reconstruction is a quadratic plane Cremona transformation given by :

$$\mathbf{T} : \mathbf{X} \mapsto \mathbf{X}' = \mathbf{a}'(\mathbf{X}) \wedge \mathbf{b}'(\mathbf{X}) \tag{1}$$

where the *direct distorted rays* are

$$\mathbf{a}'(\mathbf{X}) \doteq (\mathbf{a}_1 \cdot \mathbf{X})\mathbf{a}'_2 - (\mathbf{a}_2 \cdot \mathbf{X})\mathbf{a}'_1 \ ; \ \mathbf{b}'(\mathbf{X}) \doteq (\mathbf{b}_1 \cdot \mathbf{X})\mathbf{b}'_2 - (\mathbf{b}_2 \cdot \mathbf{X})\mathbf{b}'_1 \tag{2}$$

Its reverse is simply obtained by exchanging the true and the apparent quantities :

$$\mathbf{T}' : \mathbf{X}' \mapsto \mathbf{X} = \mathbf{a}(\mathbf{X}') \wedge \mathbf{b}(\mathbf{X}') \tag{3}$$

where the *reverse distorted rays* are

$$\mathbf{a}(\mathbf{X}') \doteq (\mathbf{a}'_1 \cdot \mathbf{X}')\mathbf{a}_2 - (\mathbf{a}'_2 \cdot \mathbf{X}')\mathbf{a}_1 \;\; ; \; \mathbf{b}(\mathbf{X}') \doteq (\mathbf{b}'_1 \cdot \mathbf{X}')\mathbf{b}_2 - (\mathbf{b}'_2 \cdot \mathbf{X}')\mathbf{b}_1 \tag{4}$$

What happens qualitatively in the fixation plane under such a distortion can be seen in Fig. 1, where a rectangular grid is shown together with two distorted reconstructions corresponding to an under-, respectively an over-estimation, of the fixation angle. From this example, we see that straight line segments are not

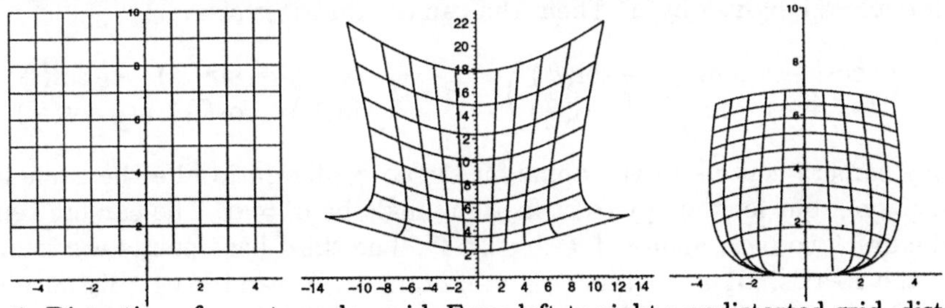

Fig. 1. Distortion of a rectangular grid. From left to right : undistorted grid, distorted grid due to underestimation of the fixation angle, distorted grid due to overestimation of fixation angle.

preserved, but map into conic sections. This is due to the quadratic nature of the shape distortion.

3 Planar Quadratic Cremona Transformations

Plane Cremona transformations are one-to-one mappings in general, with the exception of a finite number of *base points* at which they are one-to-many, and a finite number of *fundamental curves* where they are many-to-one, i.e. each point of a fundamental curve maps to the same point. A plane *quadratic* Cremona transformation $\mathbf{T}$ has three base points, which we denote by $\mathbf{B}_i$, $(i = 1, 2, 3)$. If they are all distinct, then $\mathbf{T}$ has three fundamental lines $\mathbf{f}_i$, each one passing through two of the base points. Similarly, there are three base points $\mathbf{B}'_i$ and three fundamental lines $\mathbf{f}'_i$ for the reverse transformation $\mathbf{T}'$. $\mathbf{T}$ and the reverse transformation $\mathbf{T}'$ establish a one-to-one correspondence between the base points of $\mathbf{T}$ and the fundamental lines of $\mathbf{T}'$, and between the base points of $\mathbf{T}'$ and the fundamental lines of $\mathbf{T}$.

Line Transformation : Since the transformation is of second order, a generic line maps to to a conic section. These conic sections all pass through the base points of the associated reverse transformation, and form a *net of conics* [9], i.e. a 2-parameter family of conics. Degenerate cases happen when the original line passes through one or two of the base points. In these cases the line maps to a line, respectively to a (base) point.

In [1] we show that the camera centers are base points of the planar shape distortion. The third base point is given by $\mathbf{B}_3 = \mathbf{a}(\mathbf{O}'_A) \wedge \mathbf{b}(\mathbf{O}'_B)$, the point erroneously reconstructed from the two distorted epipolar rays. The shape distortion maps camera rays to camera rays, whereas other lines map to conic sections passing through both camera centers. From this it follows that the distortion is particularly severe in the vicinity of the camera centers where the images of all lines converge. This clearly happens with the distorted grid in Figure 1, and suggests that the investigation of distortions should be of particular relevance for objects in the near range of the stereo rig.

The Fixating Observer : Let the two cameras be positioned on the X-axis separated by a base line of unit length at $x = -1/2$ and at $x = 1/2$, and let the fixation angle be given by α. Then, the camera matrices are :

$$A = \begin{pmatrix} \cos(\frac{\alpha}{2}) & \sin(\frac{\alpha}{2}) & \frac{1}{2}\sin(\frac{\alpha}{2}) \\ -\sin(\frac{\alpha}{2}) & \cos(\frac{\alpha}{2}) & \frac{1}{2}\cos(\frac{\alpha}{2}) \end{pmatrix} \quad B = \begin{pmatrix} \cos(\frac{\alpha}{2}) & -\sin(\frac{\alpha}{2}) & -\frac{1}{2}\sin(\frac{\alpha}{2}) \\ \sin(\frac{\alpha}{2}) & \cos(\frac{\alpha}{2}) & \frac{1}{2}\cos(\frac{\alpha}{2}) \end{pmatrix} \tag{5}$$

Let the camera centers in the apparent stereo configuration be the same as in the true one, but let the apparent fixation angle be $\alpha' \neq \alpha$. The camera centers are the first two base points of $\mathbf{T}$ and of $\mathbf{T}'$. The third base points are found to lie on the Y-axis, $\mathbf{B}'_3$ at distance $y_3 \doteq 1/2\tan(\Delta\alpha/2)$, and $\mathbf{B}_3$ at distance $-y_3$, where $\Delta\alpha = \alpha' - \alpha$. Thus, for underestimation of the fixation angle ($\Delta\alpha < 0$), $\mathbf{B}_3$ lies in front of the stereo rig and $\mathbf{B}'_3$ behind it, whereas for an overestimation the opposite holds.

4 Distortion of Curvature

Let $\mathcal{C}$ be a planar curve through the point $\mathbf{X}$, with its image $\mathcal{C}'$ under the planar shape distortion $\mathbf{T}$ passing through $\mathbf{X}' = \mathbf{T}(\mathbf{X})$. Let κ (κ') be the curvature of $\mathcal{C}$ ($\mathcal{C}'$) at $\mathbf{X}$ ($\mathbf{X}'$), and let κ_0 (κ'_0) be the curvature of the net conic through $\mathbf{X}$ ($\mathbf{X}'$) with the same tangent there as $\mathbf{C}$ ($\mathbf{C}'$). Then, the curvatures of $\mathcal{C}$ and $\mathcal{C}'$ are related as follows :

$$\kappa' = \kappa'_0 - \frac{\kappa'_0}{\kappa_0}\,\kappa \tag{6}$$

A proof of this can be found in [1]. From the curvature formula (6) we obtain a simple rule for the sign of the curvature of the distorted curve :

$$\operatorname{sgn}(\kappa') = \begin{cases} \operatorname{sgn}(\kappa'_0) & , \kappa < \kappa_0 \\ 0 & , \kappa = \kappa_0 \\ -\operatorname{sgn}(\kappa'_0) & , \kappa > \kappa_0 \end{cases} \tag{7}$$

Thus, the curvature κ' changes sign when the curvature κ of the original curve crosses the threshold κ_0 set by the curvature of the net conic. The relationship between the two curvatures is illustrated in Figure 2. The four curves through $\mathbf{X}$ on the left map under $\mathbf{T}$ into four curves through $\mathbf{X}'$ on the right. The existence of the middle intervals in Figure 2 is what distinguishes quadratic from projective (linear) transformations. It reflects the fact that a convex curve may transform

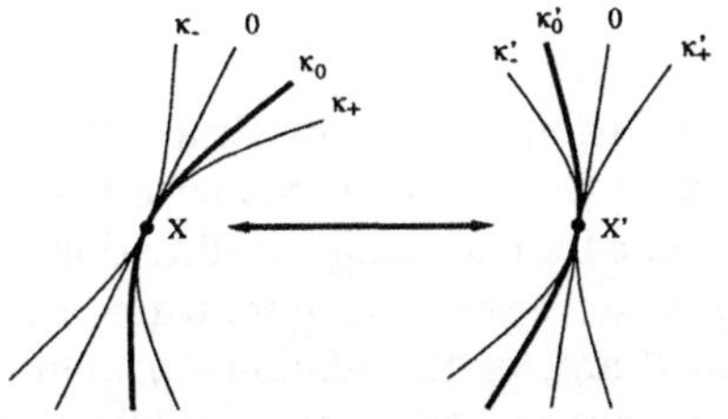

Fig. 2. Distortion of Curvature : The curves are labeled by their curvatures, and form the following pairs under $\mathbf{T}$: $\{(\kappa_-, \kappa'_-), (0, \kappa'_0), (\kappa_0, 0), (\kappa_+, \kappa'_+)\}$. The thick curves belong to the net of conics defined by the quadratic transformation.

into a curve with a concave segment, i.e. that changes of convexity can occur. This is illustrated in Figure 3, where a convex object (an ellipse) is transformed into a curve with a short concave segment. Figure 4 shows that the curvature of the ellipse and the graph of the convexity-violation threshold κ'_0 along the curve intersect at the two points where the curvature of the distorted curve has its roots.

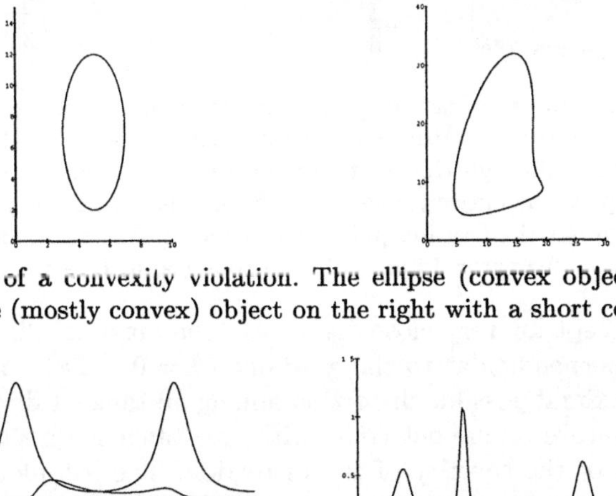

Fig. 3. Example of a convexity violation. The ellipse (convex object) on the left is distorted into the (mostly convex) object on the right with a short concave segment.

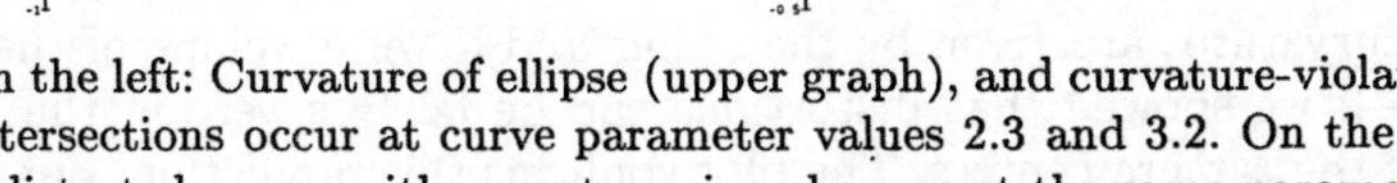

Fig. 4. On the left: Curvature of ellipse (upper graph), and curvature-violation threshold κ_0. Intersections occur at curve parameter values 2.3 and 3.2. On the right: Curvature of distorted curve, with curvature sign changes at the same parameter values

4.1 Curvature of Net Conics

Let a line be given in the apparent space with equation $\mathbf{m}' \cdot \mathbf{X}' = 0$. We define its image under the reverse distortion as : $\mathbf{c}(\mathbf{X}) \doteq \mathbf{m}' \cdot \mathbf{T}(\mathbf{X}) = 0$. This is a net conic. Its tangent line is $\nabla \mathbf{c}(\mathbf{X}) = \mathbf{T}_*(\mathbf{X})^T \mathbf{m}'$, where $\mathbf{T}_*(\mathbf{X}) \doteq \frac{\partial}{\partial \mathbf{X}}(\mathbf{T}(\mathbf{X}))$. As shown in [1], its curvature is given by $\kappa_0 \doteq \frac{\mathbf{P}_\infty^T \mathcal{H} \mathbf{P}_\infty}{\|\mathbf{P}_\infty\|^3}$, where $\mathcal{H} = \frac{\partial^2 \mathbf{c}(\mathbf{X})}{\partial \mathbf{X}^2}$ is the Hessian matrix of $\mathbf{c}(\mathbf{X})$, and $\mathbf{P}_\infty \doteq \nabla \mathbf{c}(\mathbf{X}) \wedge \mathbf{h}_\infty$ is the point at infinity on the tangent line to $\mathbf{c}(\mathbf{X})$ at $\mathbf{X}$. We use this formula for our simulations.

4.2 Simulations

We first consider the variation of the curvature distortion along the gaze line connecting the midpoint of the two cameras with the fixation point. At a given point, there is one net conic for each tangent direction, each with different curvature. On the left in Figure 5 we show the curvature of the net conic as a function of its tangent direction θ and of the distance y along the gaze line. The true and apparent fixation angles are fixed at $\alpha = 20°$ and $\alpha' = 15°$, respectively, corresponding to fixation distances of 2.84 and 3.80 baseline lengths. The plot

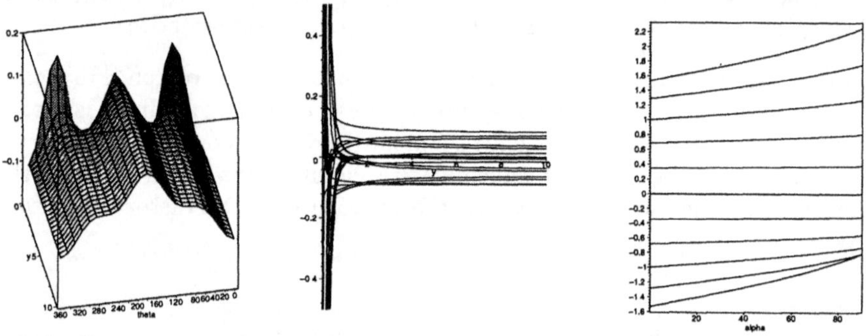

Fig. 5. *Left* : Curvature κ_0 of net conic as a function of line orientation θ and distance y along gaze line. *Middle* : Curvature κ_0 of net conic as a function of distance y along the ray $x = 1/2$ passing through the right camera center. Tangent direction θ is fixed for each individual plot; the entire range of $\theta = 0..180°$ is covered in steps of $10°$. *Right* : Maximal curvature at the fixation point as a function of fixation angle α and fixation error $\Delta\alpha = \alpha' = \alpha$. $\Delta\alpha$ varies from $-10°$ (bottom graph) to $+10°$ (top graph).

shows that - except for very close distances - the maximal distortion occurs for lines that are perpendicular to the gaze line ($\theta = 0, \pi, 2\pi$). Since we are interested in the maximal possible distortion among all tangent directions at a given point, the curvature of the net conic with this tangent direction is the appropriate measure of the severity of the distortion. The left plot also shows that the maximal curvature does not depend on the distance y from the observer. The middle plot shows the curvature distortion along a ray passing through the right camera center for different tangent directions. The maximal positive and negative curvatures are given by the upper and lower envelope of these graphs. In section 2 we showed that distortions can be fairly severe in the immediate vicinity of the camera centers. The plot confirms this prediction, but also shows that catastrophic distortions only happen at very close distances smaller than the baseline length.

So far we varied the distance to the observer for a constant fixation angle. In the last experiment, shown on the right of Figure 5, we vary the fixation angle and compute the maximal curvature at the fixation point only, and this along the gaze line. The plot shows the effect of the error in fixation angle $\Delta\alpha = \alpha' - \alpha$ on κ_0. One observes that the curvature distortion is roughly linear in $\Delta\alpha$, and that for fixation errors up to $8°$ it is only marginally dependent on the fixation angle.

5 Conclusions

For a fixating observer, systematic distortions arising from incorrect estimates of stereo parameters are largest within epipolar planes. Among these, the fixation plane is of primary importance. We had shown previously [1, 2] that such distortions are described by quadratic Cremona transformations. We analyzed the distortion of curves in the fixation plane on a qualitative, global level by making use of known properties of Cremona transformations. This yielded a prediction that the distortions should be particularly severe in the vicinity of the camera centers. Our simulations confirmed this fact, but also showed that the most severe restrictions were limited to very close distances of about one baseline length and less. The main effect the simulations showed was that beyond the near range the curvature distortion is largest for lines perpendicular to the gaze direction, and that is is approximately linear in the fixation error. On the other hand, the distance along the gaze line and the size of the fixation angle affect the maximal curvature distortion at a given point only marginally. Furthermore, we established the possibility of changes in convexity, and derived the exact criteria governing the change of convexity under the shape distortion.

References

1. G. Baratoff. Distortion of stereoscopic visual space. Technical Report CAR-TR-861, PhD Thesis, Center for Automation Research, University of Maryland, College Park, USA, May 1997.
2. G. Baratoff. Distortions of stereoscopic visual space and quadratic cremona transformations. In *Proc. Computer Analysis of Images and Patterns CAIP'97*, pages 239–246, 1997.
3. G. Baratoff and Y. Aloimonos. Changes in surface convexity and topology caused by distortions of stereoscopic visual space. In *Proc. European Conf. Comp. Vision ECCV'98*, pages 226–240, 1998.
4. L. Cheong and Y. Aloimonos. Iso-distortion contours and egomotion estimation. In *Proc. Int. Symp. on Computer Vision*, pages 55–60, 1995.
5. O. Faugeras. What can be seen in three dimensions with an uncalibrated stereo rig. In *Proc. Europ. Conf. on Computer Vision ECCV'92*, pages 563–578, 1992.
6. R. I. Hartley and P. Sturm. Triangulation. *Comp. Vision Image Understanding*, 68(2):146–157, 1997.
7. Q. T. Luong and O. Faugeras. The fundamental matrix : Theory, algorithms, and stability analysis. *Int. J. Computer Vision*, 17:43–75, 1996.
8. S. Maybank. *Theory of Reconstruction from Image Motion.* Springer-Verlag, Berlin, 1993.
9. J.G. Semple and L. Roth. *Introduction to Algebraic Geometry.* Oxford University Press, London, 1949.
10. C. C. Slama, C. Theurer, and S. W. Henriksen. *Manual of Photogrammetry.* American Society of Photogrammetry, 1980.
11. J. Weng, N. Ahuja, and T.S. Huang. Optimal motion and structure estimation. *IEEE Trans. Pattern Analysis and Machine Intelligence*, 15:864–884, 1993.

Ein hierarchischer neuronaler Klassifikator für die Erkennung von Einzelzeichen in mathematischen Formeltexten

Friedhelm Schwenker, Steffen Simon und Günther Palm

Universität Ulm, Abt. Neuroinformatik, D-89081 Ulm

Zusammenfassung In dieser Arbeit wird eine hierarchische neuronale Netzwerkarchitektur für die Musterklassifikation vorgestellt. Die verwendeten Basisarchitekturen sind kompetitive LVQ- (learning vector quantization) und RBF-Netzwerke (radiale Basisfunktionen). Die Gesamtarchitektur des Klassifikators ist zweistufig. Die erste Klassifikatorstufe ist ein LVQ-Netz zur Grobklassifkation. Klassen die auf dieser Stufe an Verwechslungen beteiligt sind, werden zu Oberklassen, den sogenannten Verwechslungsklassen, zusammengefasst. Für jede Verwechslungsklasse wird in der zweiten Stufe ein spezielles RBF-Netz aufgebaut, das zwischen den (wenigen) Klassen dieser Verwechslungsklasse diskriminiert. Die Verwechslungsklassen werden aus der Verwechslungsmatrix des LVQ-Klassifikators bestimmt. Sowohl die Gesamtarchitektur als auch die Parameter (= Gewichte der künstlichen Neuronen) der LVQ/RBF-Netzwerke werden während der zweistufigen Trainingsprozedur erzeugt. Der hierarchische Klassifikator ist für Klassifikationsprobleme mit sehr vielen Klassen einsetzbar. Die Initialisierung der künstlichen RBF-Neuronen durch die Prototypen des LVQ-Netzes liefert ein sehr effizientes Trainingsverfahren.

1 Einleitung

In dieser Arbeit wird ein hierarchischer neuronaler Klassifikator für die Einzelzeichenklassifikation vorgestellt. Die konkrete Aufgabe ist die Klassifikation von maschinengesetzten Zeichen, wie sie in mathematischen Formeltexten vorkommen. Dabei verstehen wir dieses Mustererkennungsproblem als ein Teilproblem der Texterkennungsaufgabe von mathematischen Formeltexten. Die Zeichen in mathematischen Formeltexten sind typischerweise: Lateinische Buchstaben, Griechische Buchstaben, Ziffern und Sonderzeichen. Diese Zeichen kommen sowohl im Fließtext als auch in den Formeln vor, stammen somit aus verschiedenen Fonts und sind häufig stark skaliert (siehe Abb. 2).

Betrachtet man nur den Zeitraum der letzten 100 Jahre, in dem in der heute üblichen Art in den Gebieten der Physik, der Mathematik, oder den Ingenieurwissenschaften veröffentlicht wird, so hat sich hier ein immenses Wissen angesammelt. Um dieses Wissen potentiellen Nutzern zugänglich zu machen wäre es hilfreich, es in elektronische Form zu überführen, beispielsweise in eine mathematische Formelsprache wie LaTeX oder eine Sprache zur symbolischen Verarbeitung mathematischer Formeln. Hierzu ist es zunächst einmal notwendig die mathematischen Texte einzuscannen, mit dem Ziel eine Klassifikation der Einzelzeichen durchzuführen.

Durch die heute verfügbaren OCR-Systeme (optical character recognition) kann maschinengesetzter Fließtext mit einer sehr geringen Fehlerrate klassifiziert werden. Bei der Verarbeitung mathematischer Texte sinkt die Einzelzeichenerkennungsleistung solcher Systeme allerdings drastisch ab, teilweise bis auf unter 10 Prozent. Die Gründe hierfür sind der zweidimensionalen Aufbau mathematischer Formeln, sowie die starke Variation der Zeichengröße und der Zeichenabstände [3]. Texterkennungssyteme für mathematische Formeln bzw. Formeltexte sind bereits seit Ende der 60er Jahren entwickelt worden. Hier sind vor allem die Arbeiten von Anderson zu nennen [1,2], der auf der Basis von Koordinatengrammatiken einen interaktiven 2D-Parser zur Verarbeitung mathematische Gleichungen entwickelte. Zu diesen frühen Entwicklungen zählen auch die Arbeiten von Martin [12] und Lewis [11]. In der letzten Zeit wird an verschiedenen Orten an Sytemen für Erkennung mathematischer Formeltexte entwickelt [5,10].

Für die Einzelzeichenklassifikation werden künstliche neuronale Netzwerke bereits seit mehreren Jahren eingesetzt [4,9]. Bei einem künstlichen neuronalen Netz werden in der Trainingsphase die Netzparameter mit Hilfe eines überwachten Lernverfahrens durch eine Menge von Lernbeispielen $S = \{(u^\mu, t^\mu) | u^\mu \in \mathbb{R}^n, t^\mu \in \{1, \dots, l\}, \mu = 1, \dots, m\}$ bestimmt. Hierbei ist u^μ die Eingabe des Netzes und t^μ die zugehörige Ausgabe bzw. Sollklassifikation. Eine Klasse wird i.a. durch genau ein Ausgabeneuron des Netzes repräsentiert, d.h es gibt genau l Ausgabeneuronen. In der Klassifikationsphase wird die Netzausgabe zur Eingabe $u \in \mathbb{R}^n$ berechnet. Diese ist bei Mehrschichtnetzwerken, wie Multilayerperzeptronen oder RBF-Netzen, ein reeller Vektor mit l Komponenten. Dieser wird als Wahrscheinlichkeitsvektor interpretiert, d.h. die Klasse zur Eingabe u ist durch die Position des größten Eintrags im Ausgabevektor gegeben. In einer kompetitiven Netzwerkarchitektur, ist die Netzausgabe durch die Klasse desjenigen Prototypen gegeben, der zur Eingabe u den kleinsten (euklidischen) Abstand hat (Nächster-Nachbar-Klassifikator).

Hierarchien neuronale oder probabilistischer Klassifikatoren werden seit vielen Jahren untersucht (siehe beispielsweise das Buch von Schürmann [15] oder die Artikel von Jordan und Jacobs [8,7]). Der hier verwendete hierarchische neuronale Klassifikator ist eine zweistufige prototypbasierte Architektur. Die erste Stufe besteht aus einem LVQ-Klassifikator zur Vorklassifkation. Klassen die auf dieser Stufe an Verwechslungen beteiligt sind, werden zu Verwechslungsklassen, zusammengefasst. Für jede Verwechslungsklasse wird in der zweiten Stufe ein spezielles RBF-Netz trainier, das zwischen den Klassen dieser Verwechslungsklasse diskriminiert.

Die Arbeit ist wie folgt aufgebaut: Zunächst werden die neuronalen Basisarchitekturen und die verwendeten Verfahren zur Adaptation der Netzparameter vorgestellt. Der Aufbau und die Funktion der zweistufige Netzwerkarchitektur wird im Kapitel 3 beschrieben. Schließlich werden Resultate für die konkrete Klassifikationsaufgabe vorgestellt.

2 Neuronale Klassifikatoren

Die neuronalen Basisarchitekturen sind RBF-Netzwerke und LVQ-Netzwerke.

2.1 Überwachte Vektorquantisierung (LVQ)

Das Ziel der ersten Klassifikationstufe ist es, möglichst viele Einzelzeichen korrekt zu klassifizieren. Andererseits, soll dieser Klassifikator mit geringem Aufwand trainiert werden können. Das zur Grobklassifikation verwendete Verfahren ist das von Kohonen vorgeschlagene Verfahren *Learning Vector Quantization* (LVQ) [9]. Es sind verschiedene LVQ-Varianten definiert worden, wobei hier das sogenannte OLVQ verwendet wird.

Zur Initialisierung der Klassenprototypen werden die Merkmalsvektoren einer einzelnen Klasse j einer k-means Clusteranalyse unterzogen. Die berechneten Clusterzentren $c_1, \ldots, c_{L_j}$ sind die initialen Positionen der Klassenprototypen. Anwendung der k-means Clusteranalyse auf jede der l Klassen ergibt eine Menge von Prototypen $\{c_1, \ldots, c_L\} \subset \mathbb{R}^n$, $L = \sum_j L_j$, wobei jeder Prototyp genau einer Klasse zugeordnet ist.

In der OLVQ Trainingsprozedur geschieht die Adaptation der Prototypen $c_1, \ldots, c_L$ bei Präsentation eines Musters $u^\mu \in \mathbb{R}^n$ zusammen mit dem zugehörigen Lehrersignal T^μ (gegeben durch einen Klassenindex k) durch:

1. Distanzberechnung und Gewinnerdetektion:

$$i^* = \mathrm{argmin_i} \|\mathrm{u}^\mu - \mathrm{c_i}\|$$

2. Anpassung der Gewichte

$$\Delta c_{i^*} = \begin{cases} \eta_{i^*}(u^\mu - c_{i^*}) & : \text{ Klasse von } i^* = \text{Klasse von } u^\mu \\ -\eta_{i^*}(u^\mu - c_{i^*}) & : \text{ sonst} \end{cases}$$

3. Anpassung der Lernrate des Gewinners:

$$\eta_{i^*}(t+1) = \begin{cases} \dfrac{\eta_{i^*}(t)}{1 + \eta_{i^*}(t)} & : \text{ Klasse von } i^* = \text{Klasse von } u^\mu \\ \min\left(\dfrac{\eta_{i^*}(t)}{1 - \eta_{i^*}(t)}, \eta\right) & : \text{ sonst} \end{cases}$$

Die individuelle Lernraten der Prototypen und die Lernratenanpassung in Abhängigkeit von der Klassifikation des Gewinnerprototypen führen dazu, daß korrekt klassifizierende Prototypen die Tendenz haben ihre Position weniger stark zu verändern, als falsch klassifizierende Prototypen.

2.2 Radiale Basisfunktionen

RBF-Netze spielen eine immer größere Rolle für Klassifikationsaufgaben [14, 6, 15]. Ein RBF-Netz besteht aus einer Eingabeschicht, einer Schicht mit RBF-Neuronen und einer Ausgabeschicht. Die Transferfunktion der RBF-Neuronen ist radialsymmetrisch, häufig verwendet wird die Gaußs'sche Dichtefunktion

$$h(u) = \exp(-\sum_{i=1}^{n} \frac{(u_i - c_i)^2}{\sigma_i^2}).$$

Diese beschreibt einen Hyperellipsoid um den Punkt $c \in \mathbb{R}^n$, dessen Form durch den positiven Vektor $\sigma \in \mathbb{R}^n$ bestimmt wird. Bei Eingabe eines Merkmalsvektors u erfolgt die Berechnung der Netzausgabe in zwei Schritten:

1. Berechnung der Aktivität des RBF-Neurons $j \in \{1, \ldots, L\}$ in der Zwischenschicht:

$$x_j = \sum_{i=1}^{n} \frac{(u_i - c_{ij})^2}{\sigma_{ij}^2}$$

$$y_j = e^{-x_j}$$

2. Berechnung der Aktivität des Neurons $k \in \{1, \ldots, l\}$ in der Ausgabeschicht:

$$z_k = \sum_{j=1}^{m} y_j w_{jk}$$

Die Position des maximal aktivierten Ausgabeneurons wird als Netzwerkklassifikation für die Eingabe u verwendet.

Während der Trainingsphase des RBF-Netzes werden die Netzparameter, d.h. die Positionen $c_1, \ldots, c_L \in \mathbb{R}^n$ und die Ausdehnungen $\sigma_1, \ldots, \sigma_L \in \mathbb{R}^n$ der RBF-Neuronen, sowie die Ausgabegewichte $w_1, \ldots, w_l \in \mathbb{R}^L$ durch Minimierung der quadratischen Fehlerfunktion

$$E(c_1, \ldots, c_L, \sigma_1, \ldots, \sigma_L, w_1, \ldots, w_l) = \sum_{\mu} \|T^{\mu} - z^{\mu}\|^2$$

bestimmt. Hierzu werdeen numerische Optimierungstechniken (z.B. (konjugiertes) Gradientenverfahren) eingesetzt. Wir setzen hierzu ein inkrementelles gradientenbasiertes Traingsverfahren für die Adaptation der Netzwerkparameter ein (siehe [6] Kapitel 8).

Die initialen Positionen der RBF-Neuronen sind durch die Prototypen $c_1, \ldots, c_L$ des LVQ-Netzes bestimmt [16]. Für das j-te RBF-Neuron werden die Ausdehnungen $\sigma_j \in \mathbb{R}^n$ durch die Streuung der Komponenten der Merkmalsvektoren aus dem Trainingsdatensatz initialisiert. Hierbei werden nur die Vektoren herangezogen, die dem Prototypen c_j durch Gewinnerdetektion des LVQ-Klassifikators zugeordnet werden.

3 Hierarchische LVQ-RBF-Netzarchitektur

Die Trainingsprozedur in der hier vorgestellten Architektur ist zweistufig. Neben der Anpassung der Netzwerkparameter ist es zusätzlich erforderlich, die hierarchische Netzstruktur zur Laufzeit aufzubauen. Hierzu werden nach dem OLVQ Netzwerktraining folgende Schritte durchgeführt:

1. Verwechslungsmatrix V des OLVQ-Klassifikators bestimmen
2. Verwechslungsklassen bestimmen
3. Merkmalsreduktion innerhalb der Verwechslungsklassen
4. RBF-Netz für jede Verwechslungsklasse initialisieren und trainieren

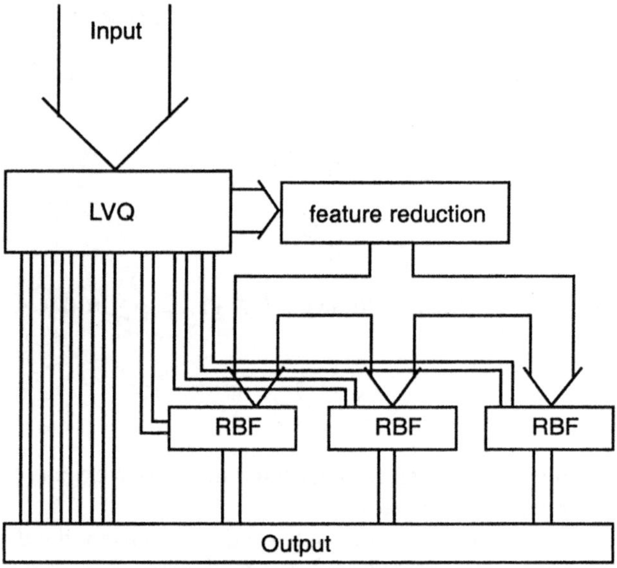

Abbildung1. Die hierarchische Netzwerkarchitektur mit dem LVQ-Klassifikator und den RBF-Klassifikatoren zur Diskrimination der Klassen innerhalb der einzelnen Verwechslungsklassen. Die Merkmalsreduktionen ist dem LVQ-Klassifikator nachgeschaltet und erfolgt für jede Verwechslungsklasse separat.

3.1 Bestimmung der Verwechslungsklassen

Durch Präsentation des Trainingsdatensatzes wird zunächst die Verwechslungsmatrix V des OLVQ-Klassifikators auf der Trainingsmenge berechnet.

Aus der Matrix V und einer Schwelle $\Theta > 0$ wird nun eine Boolsche $l \times l$-Matrix V' wie folgt definiert:

$$V'_{ij} = \begin{cases} 1 & : \quad V_{ij} \geq \Theta \\ 0 & : \quad \text{sonst} \end{cases}$$

Ein Eintrag $V'_{ij} = 1$ repräsentiert Verwechslungen von Klasse i nach Klasse j. Ausgehend von V' lässt sich jede der l Klassen zu genau eine der Verwechslungsklassen $K_1, \ldots, K_v$ zuordnen. Dabei ist v, die Anzahl der Verwechslungsklassen, durch die Schwelle Θ regulierbar. Die Verwechslungsklassen werden wie folgt gebildet:

1. $V'_{ij} = 1$, dann $\{i, j\} \subset K_p$ für ein p (Symmetrie)
2. $V'_{ij} = 1$ und $V'_{jk} = 1$, dann $\{i, j, k\} \subset K_p$ für ein p. (Transitivität)

Nach Festlegung der Verwechslungsklassen bildet der OLVQ-Klassifikator ein Eingabemuster u entweder auf eine Verwechslungsklassen K_p ab, oder auf eine Klasse k, für die gilt $V'_{kj} = 0$ und $V'_{jk} = 0$, für alle $j \neq k$. Bei Zuweisung zu einer Verwechslungsklasse K_p wird das Muster u dem nachgeschalteten RBF-Klassifikator präsentiert. Dieser diskriminiert nun zwischen den Klassen der Verwechslungsklasse K_p (siehe Abb. 1).

3.2 Merkmalsreduktion innerhalb der Verwechslungsklassen

Zu Fehlklassifikationen kommt es dann, wenn Muster verschiedener Klassen benachbart sind, beispielsweise, wenn sich diese Muster nur auf einer Teilmenge der Merkmale wesentlich unterscheiden. An dieser Stelle ist es möglich die Zahl der Merkmale zu reduzieren.

Zur Merkmalsreduktion können Methoden wie die *Hauptachsentransformation* (PCA = principal component analysis), oder die *lineare Diskriminanzanalyse* (LDA) eingesetzt werden. Das Resultat dieser Verfahren ist eine Transformationsmatrix, mit der die Muster vermöge einer Matrixmultipikation transformiert werden. Um eine solche Multiplikation zu verweiden, gehen wir nach einer einfacheren Methode vor. Es werden für jede Verwechslungsklasse die N Merkmale ausgewählt, welche die N größten Komponenten auf der Diagonalen der *between-class scatter matrix* aufweisen. Somit ist diese Merkmalsreduktion eine spezielle *Projektion* auf eine Untermenge der Merkmale.

Für jede Verwechslungsklasse K_p werden nach dieser Vorgehensweise N_p Merkmale ausgewählt. Dann wird für jede Verwechslungsklasse K_p ein RBF-Netz aufgebaut. Dabei werden die Positionen und die Ausdehnungen der RBF-Neuronen auf die zugehörigen Merkmalsräume projeziert.

4 Anwendung und Diskussion

Die Trainings- und Testdaten sind maschinengesetzte Einzelzeichen, gleichverteilt im Bereich von 6pt bis 12pt (ca. 2.2 mm bis 4.5 mm) mit einer Scannerauflösung von 300dpi. Diese Zeichen sind auf eine 16x16 Graustufen-Pixel-Matrix mit einer Auflösung von 256 Graustufen skaliert. In einem Vorverbeitungsschritt sind die 256 Merkmale durch eine Parallelprojektion mit anschließender Filterung (siehe [13] Kapitel 8) auf 98 Merkmale reduziert worden.

Der neuronale hierarchische Klassifikator wurde durch den beschriebenen zweistufigen Trainingsprozeß mit insgesamt 2700 Merkmalsvektoren zusammen

1234567890
ABCDEFGHIJKLMNOPQRSTUVWXYZ
ÄÖÜ
abcdefghijklmnopqrstuvwxyz
äöü
≠<>~+-.,:;*/∞øʃ()[]{}|
ΔΦΓΛΠΘΣΩΞΨ
αβχδεφγηιφκλμνπθρστυϖωξψζ

Abbildung2. Eine Beispielmenge von zu klassifizierenden Einzelzeichen.

mit zugehörigen Zeichenklasse als Lehrersignal trainiert. Die Klassifikationsleistung wurde anschließend auf einer unabhängigen Testmenge von 1000 Datensätzen ermittelt. Neben der hierarchischen LVQ/RBF-Architektur wurden 3 weitere Architekturen realisiert:

LVQ Klassifikation mit einem LVQ-Netz
RBF Klassifikation mit einem RBF-Netz
LVQ/LVQ LVQ als Grob- und LVQ Feinklassifikation
LVQ/RBF LVQ als Grob- und RBF als Feinklassifikation

Bei den LVQ-Architektur und RBF-Architektur wurden jeweils 198 Prototypen bzw. RBF-Neuronen (bei $l = 135$ Klassen) benutzt, ebenso die LVQ/LVQ-und LVQ/RBF-Architekturen. Alle Berechnungen wurden auf einem Power Macintosh 7500/100 durchgeführt mit jeweils 100 Lernepochen.

	LVQ	LVQ/LVQ	LVQ/RBF	RBF
Trainingszyklen	100	100/100	100/100	100
Klassifikationsfehler Trainingsmenge	7.9 %	5.1 %	1.1%	0.8 %
Klassifikationsfehler Testmenge	10.1 %	7.8 %	2.1 %	1.7 %
Trainingszeit	6 min	8 min	12 min	8h40min

Erwartungsgemäß unterscheiden sich die Klassifikationsleistungen, vorallem aber auch die Trainingszeiten der Architekturen beträchtlich. Es zeigt sich, daß die hierarchische LVQ/RBF-Architektur für Klassifikationsaufgaben mit sehr vielen Klassen deutliche Vorteile gegenüber einer RBF-Architektur hinsichtlich der Trainingszeit besitzt. Mit dieser Architektur lassen sich Klassifikationsresultate erzielen, die mit denen des RBF-Verfahrens in vergleichbar sind, obwohl die Klassifikationsleistung der ersten Klassifikatorstufe (LVQ) nicht zufriedenstellend ist.

Die Initialisierung von RBF-Neuronen durch vortrainierte LVQ-Prototypen bietet die Möglichkeit RBF-Netze effektiv zu trainieren. Ferner läßt sich die Dimension der LVQ-Prototypen innerhalb der Verwechslungsklassen durch Merkmalsauswahl reduzieren. Die merkmalsreduzierten LVQ-Prototypen können ihrerseits für die Initialisierung der nachgeschalteten RBF-Netze verwendet werden. Insgesamt zeigt sich, daß prototypbasierte Architekturen wie LVQ- und RBF-Netze effektiv zu hierarchischen neuronalen Klassifikatorhierarchien aufbauen lassen.

Literatur

1. R. H. Anderson. Syntax directed recognition of hand-printed two-dimensional mathematics. *Interactive Systems for Experimental Applied mathematics*, pages 436–459, 1968.
2. R. H. Anderson. *Two-Dimensional Mathematical Notation*, pages 147–177. Springer, 1977.
3. B. P. Berman and R. J. Fateman. Optical character regocnition for typeset mathematics. In *ISSAC94*, 1994.
4. C. M. Bishop. *Neural Networks for Pattern Recognition*. Clarendon Press, Oxford, 1995.
5. R. J. Fateman and T. Tokuyasu. Progress in recognizing typeset mathematics. In *Proc. of SPIE 96*, 1996.
6. S. Haykin. *Neural Networks: A Comprehensive Foundation*. Macmillan College Publishing Company, 1994.
7. R. A. Jacobs. Methods for combining experts probability assessments. *Neural Computation*, 7:867–888, 1995.
8. M. I. Jordan and R. A.Jacobs. Hierarchical mixtures of experts and the EM algorithm. *Neural Computing*, 6:181–214, 1994.
9. T. Kohonen. *Self-Organizing Maps*. Springer, 1995.
10. S. Lavirotte and L. Pottier. Graph grammar for mathematical formula recognition. Technical report, INRIA, 1997.
11. H. R. Lewis. Two applications of hand-printed two-dimensional computer input. Master's thesis, Harvad College, 1968.
12. W. A. Martin. *Computer Input/Output of mathematical Expressions*. PhD thesis, MIT, 1967.
13. A. Rosenfeld and A. Kak. *Digital Picture Processing*, volume 1. Academic Press, 1982.
14. R. J. Schalkoff. *Pattern Recognition – Statistical, Structural and Neural Approaches*. John Wiley & Sons, Inc., 1992.
15. J. Schürmann. *Pattern Classification – A Unifier View of Statistical and Neural Approaches*. John Wiley& Sons, 1996.
16. F. Schwenker, H. A. Kestler, G. Palm, and M. Höher. Similarities of LVQ and RBF learning - A survey of learning rules and the application to the classification of signals from high-resolution electrocardiography. In *Proc. IEEE International Conference on Systems, Man and Cybernetics*, pages 646–651, 1994.

Aktive Exploration (Postersession)

Navigation mit Schnappschüssen*

Matthias O. Franz[1], Bernhard Schölkopf[1], Hanspeter A. Mallot[1], Heinrich H. Bülthoff[1] und Andreas Zell[2]

[1] Max-Planck-Institut für biologische Kybernetik, Spemannstr. 38, D-72076 Tübingen
[2] Wilhelm-Schickard-Institut für Informatik, Lehrstuhl Rechnerarchitektur, Köstlinstr. 6, D-72074 Tübingen

Zusammenfassung Es wird ein biologisch inspirierter Algorithmus vorgestellt, mit dem sich ein Ort wiederfinden läßt, an dem vorher eine 360°-Ansicht der Umgebung aufgenommen wurde. Die Zielrichtung wird aus der Verschiebung der Bildposition der umgebenden Landmarken im Vergleich zum Schnappschuß berechnet. Die Konvergenzeigenschaften des Algorithmus werden mathematisch untersucht und auf mobilen Robotern getestet.

1 Einleitung

Für viele Navigationsaufgaben ist die Verwendung externer Informationen unumgänglich, da bestimmte Vorgänge - wie z.B. Drift beim Flug oder Schlupf der Antriebsräder beim Fahren - nicht immer intern erkannt werden können. Ein möglicher Ansatz ist die Identifizierung von Orten durch lokale Ansichten. Dabei sind zwei Fälle zu unterscheiden: Zum einen kann die lokale Ansicht den Ort selbst abbilden, zum anderen kann die Ansicht die umgebende Szene vom Ort aus gesehen darstellen. Die erste Möglichkeit bietet sich an, wenn der Ort durch auffällige visuelle Merkmale markiert ist, die man z.B. zur Zielanfahrt verwenden kann. Ist dies nicht der Fall, dann muß die Lage des Ortes aus seiner Relation zu den umgebenden Landmarken bestimmt werden (Abb. 1 links).

Eine Vielzahl von Experimenten aus der Verhaltensforschung hat gezeigt, daß bereits Wirbellose wie z.B. Bienen oder Ameisen diese Fähigkeit besitzen (s. Übersicht in Collett, 1992). Die experimentellen Befunde legen nahe, daß diese Tiere ihr Ziel an Orten suchen, an denen der momentane Anblick am besten mit einem dort vorher gespeicherten *Schnappschuß* der Umgebung übereinstimmt. Cartwright & Collett (1983) zeigten anhand eines einfachen Computermodells, daß die beobachteten Suchmuster sich erklären lassen, wenn man annimmt, daß ein Schnappschuß die einzig verfügbare Information ist. Die Richtung des Ziels wurde dabei aus der Veränderung des Sehwinkels und der Größe der umgebenden Landmarken berechnet. Die spätere Implementierung ähnlicher Mechanismen auf mobilen Robotern zeigte, daß sich Schnappschüsse auch für die Roboternavigation eignen (Hong et al., 1992; Röfer, 1997; Franz et al., 1997).

Im folgenden wird ein robuster Algorithmus beschrieben, mit dem sich ein durch einen Schnappschuß der Umgebung markierter Ort anfahren läßt. Dazu

* Gefördert durch das Human Frontier Science Program (HFSP).

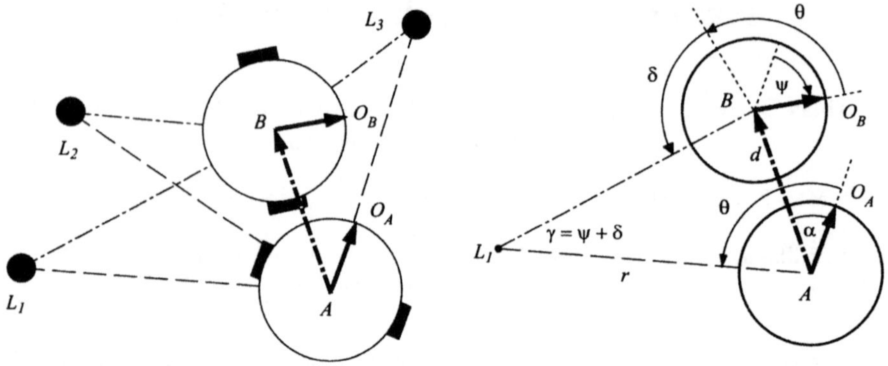

Abbildung 1. Links: Ist das Ziel nicht durch auffällige visuelle Merkmale definiert, muß die Fahrtrichtung aus der Veränderung der Sehwinkel der umgebenden Landmarken berechnet werden. Rechts: Bewegt sich der Sensorring vom Ausgangspunkt A nach B, so verändert sich der Sehwinkel der Landmarke L_1 um γ. Eine zusätzliche Orientierungsänderung um ψ führt zu einer Gesamtverschiebung des Bildes von L_1 um $\delta = \gamma - \psi$.

werden zunächst die mathematischen Grundlagen der Aufgabenstellung dargestellt, bevor auf die Details des Algorithmus eingegangen wird. Danach beschreiben wir die Implementierung auf mobilen Robotern und schließen mit einer Diskussion der Ergebnisse.

2 Mathematische Beschreibung des Problems

Als vereinfachtes Modell eines mobilen Agenten nehmen wir einen omnidirektionalen Sensorring an, dessen Bewegungen auf die Ebene beschränkt sind. Mit Hilfe des Sensors kann der Agent lokale 360°-Ansichten aufnehmen, die die Grauwerteverteilung entlang des Horizontes wiedergeben. Die Position eines Bildpunktes auf dem Sensorring sei durch den Winkel θ beschrieben. In diesem Zusammenhang bezeichnen wir alle im Bild identifizierbaren Umgebungspunkte als Landmarken. Dies stimmt nicht immer mit der üblichen Vorstellung einer Landmarke als einem physikalischen Objekt überein, d.h. ein Objekt kann durchaus mehrere Landmarken enthalten.

Der Sensorring bewege sich nun von seiner Ausgangsposition A in Richtung α zu dem in Entfernung d liegenden Punkt B und ändere seine Orientierung um den Winkel ψ (s. Abb. 1 rechts). Durch die Bewegung wandert das an Position θ sichtbare Bild der in Entfernung r liegenden Landmarke L_1 zu einer neuen Position $\theta + \delta$. Aus dem Dreieck AL_1B ergibt sich mit Hilfe des Sinussatzes die Beziehung

$$\frac{r}{d} = \frac{\sin(\pi - \theta + \alpha - \psi - \delta)}{\sin(\psi + \delta)} = \frac{\sin(\theta - \alpha + \psi + \delta)}{\sin(\psi + \delta)}. \tag{1}$$

Sind r/d und ψ bekannt, so läßt sich die Bewegungsrichtung α und damit die

Richtung $\beta = \alpha - \psi + \pi$ zurück zum Ausgangspunkt aus der Bildverschiebung δ berechnen.

Um die Verschiebung δ zum Wiederfinden eines durch einen Schnappschuß markierten Ortes verwenden zu können, müssen zwei grundsätzliche Probleme gelöst werden: 1. Dem von uns betrachteten visuell navigierenden Agenten stehen lokale Ansichten als einzige Information zur Verfügung. Der unbekannte Parameter r/d muß daher über geeignete Annahmen zu typischen Distanzverteilungen in der Umwelt des Agenten geschätzt werden. 2. Für die Berechnung von δ müssen Korrespondenzen zwischen dem Schnappschuß und der aktuellen Ansicht gefunden werden.

In dem hier vorgestellten Ansatz wird angenommen, daß alle Landmarken in ungefähr gleicher Entfernung vom Ausgangspunkt liegen. Auf den ersten Blick erscheint dies nicht sehr realistisch, dennoch beeinflussen die resultierenden Fehler die tatsächliche Navigationsfähigkeit des Agenten nur wenig (s. Abschn. 3.2). Darüberhinaus führt diese Annahme zu einem einfachen globalen Verfahren zum Auffinden von Korrespondenzen.

3 Berechnung der Zielrichtung

3.1 Parametrisierte Verschiebungsfelder

Unter der Annahme einer typischen Landmarkendistanz R läßt sich Gl. (1) in

$$\tan(\psi + \delta) = \frac{\frac{d}{R}\sin(\theta - \alpha)}{1 + \frac{r-R}{R} - \frac{d}{R}\cos(\theta - \alpha)} = \frac{\rho\sin(\theta - \alpha)}{1 + \epsilon - \rho\cos(\theta - \alpha)}. \tag{2}$$

überführen, wobei zur Erleichterung der Notation $\rho = d/R$ und $\epsilon = (r - R)/R$ gesetzt wird. Die Fälle $\rho = 0$, $\epsilon = -1$ und $\rho = 1 + \epsilon$, $\theta = \alpha$ müssen dabei ausgeschlossen werden. Praktisch bedeutet dies, daß der Agent nicht dieselbe Position wie eine Landmarke einnehmen darf.

Wie oben erwähnt, gehen wir von der vereinfachenden Annahme aus, daß alle Landmarken gleich weit vom Ausgangspunkt in Entfernung R liegen, d.h. wir setzen in Gl. (2) $\epsilon = 0$. Die durch

$$\tan(\psi + \delta) = \frac{\rho\sin(\theta - \alpha)}{1 - \rho\cos(\theta - \alpha)} \tag{3}$$

beschriebenen Verschiebungsfelder $\delta(\theta)|_{\epsilon=0}$ werden durch diese Annahme stark vereinfacht und hängen nur noch von den drei freien Parametern α, ψ und ρ ab. Die auf diese Weise parametrisierten Verschiebungsfelder können durch Variation von α, ψ und ρ an das real vorhandene Verschiebungsfeld angepaßt werden. Da die Bewegungsrichtung α einer der angepaßten Parameter ist, kann der Ausgangspunkt mit Hilfe des folgenden Algorithmus wiedergefunden werden:

1. Für alle Kombinationen von α, ψ und ρ wird ein Satz von Verschiebungsfeldern nach Gl. (3) berechnet. Mit Hilfe dieser Verschiebungsfelder werden die Bilder der Landmarken im momentanen Anblick verschoben und dadurch

ein Satz neuer Bilder generiert. Wäre die Annahme der gleichen Landmarkendistanz exakt erfüllt, dann würden diese Bilder genau dann entstehen, wenn sich der Agent tatsächlich nach den Parametern bewegt hätte.
2. Der Satz der generierten Bilder wird mit dem am Ausgangspunkt gespeicherten Schnappschuß verglichen, um den Anblick zu finden, der den Schnappschuß am besten rekonstruiert. Als Maß für die Qualität der Rekonstruktion wird das Skalarprodukt zwischen Schnappschuß und Ansicht verwendet.
3. Der Parameter α, der zur besten Rekonstruktion des Schnappschusses führt, wird zur Schätzung der Fahrtrichtung $\beta = \alpha + \pi$ zurück zum Ausgangspunkt herangezogen.
4. Der Agent bewegt sich einen Schritt weit in Richtung β und wiederholt den Vorgang, bis die Ausgangsposition wieder erreicht ist. Dies ist der Fall, wenn entweder das Skalarprodukt zwischen dem momentanen Anblick und Schnappschuß eine vordefinierte Schwelle überschreitet oder sich die berechnete Zielrichtung β nach Passage des Ausgangspunktes um 180° dreht.

Damit die drei Unbekannten in Gl. (3) bestimmt werden können, müssen mindestens drei Landmarken sichtbar sein. Ist dies nicht der Fall, so kann die Fahrtrichtung nur unter Hinzunahme weiterer Informationen aus Kompassen oder Odometrie berechnet werden.

3.2 Fehleranalyse und Konvergenz

Wie eingangs erwähnt basiert der vorgestellte Algorithmus auf der Annahme einer gleichen Landmarkendistanz. Obwohl diese Annahme im Normalfall kaum erfüllt sein wird, behindert der resultierende Fehler das Wiederfinden des Ausgangspunktes nur in geringem Maße. Dies liegt v.a. daran, daß der Fehler um so kleiner wird, je näher der Agent seinem Ziel kommt. Dies folgt direkt aus Gl. (2), wenn wir annehmen, daß der Agent bereits den offenen Bereich um das Ziel erreicht hat (d.h. $\rho < 1 + \epsilon$). In diesem Fall ist die Verschiebung δ durch

$$\delta = \arctan\left(\frac{\rho\sin(\theta-\alpha)}{1+\epsilon-\rho\cos(\theta-\alpha)}\right) - \psi \tag{4}$$

gegeben. Der Fehler durch die Vernachlässigung der individuellen Abweichungen der Landmarkendistanzen von R ist dann $E(\epsilon,\rho) := \delta(\epsilon,\rho) - \delta(0,\rho)$. Sowohl δ als auch E sind stetige Funktionen in ρ und ϵ für $\epsilon > -1$. Beide gehen asymptotisch gegen Null für $\rho \to 0$. Daraus folgt, daß für jedes $\epsilon > -1$, gegebenes θ und α und eine beliebige Fehlerschranke $E_0 > 0$ ein ρ_0 existiert, für das $|E(\epsilon,\rho)| < E_0$ aus $\rho < \rho_0$ folgt. In anderen Worten bedeutet dies, daß sich das Ziel beliebig genau anfahren läßt, auch wenn die umgebenden Landmarken unterschiedliche Entfernungen vom Ziel haben - vorausgesetzt, der Agent ist hinreichend nahe am Ziel. Jeder Schnappschuß ist also von einem Gebiet umgeben, in dem sich der Ort des Schnappschusses beliebig genau bestimmen läßt. Das gesamte Gebiet, von dem aus sich das Ziel erreichen läßt, wird *Fangbereich* des Schnappschusses genannt. In der praktischen Anwendung ist die theoretisch unbegrenzte Genauigkeit natürlich nicht erreichbar, da hier das Sensorrauschen die wahrnehmbaren

Abbildung 2. Links: Khepera mit omnidirektionalem Sensor. Rechts: RWI Pioneer-Plattform mit Industrie-PC zur Bildverarbeitung.

Unterschiede zwischen Schnappschuß und dem momentanen Anblick begrenzt (Franz et al., 1997).

4 Experimente mit mobilen Robotern

4.1 Experimenteller Aufbau

Die hier gezeigten Daten wurden mit Hilfe eines modifizierten Khepera-Miniaturroboters gewonnen, der über ein Kabel mit einer Workstation verbunden war (Abb. 2 links). Als Landmarken dienten acht Modellhäuser in einer 118×102 cm großen Arena. Um den Algorithmus in realitätsnäheren Umgebungen zu testen, wurde ein komplett autonomer Prototyp für Büroumgebungen entwickelt. Hierzu wurde ein Industrie-PC mit Framegrabber-Karte als Host auf eine käufliche RWI Pioneer-Basis montiert (Abb. 2 rechts). Das Kamerasystem und die Kontrollalgorithmen waren auf beiden Plattformen identisch.

Das bildgebende System beider Roboter besteht aus einer Miniatur-Videokamera, die auf einen polierten Metallkegel gerichtet ist (Abb. 3, vgl. Chahl & Srinivasan, 1996). Das ringförmige 360°-Sichtfeld dieser Anordnung erstreckt sich von 10° unter- bis 10° oberhalb des Horizontes. Die Videobilder werden mit 25 Hz in vier Ringen entlang des Horizontes mit einer Winkelauflösung von 4.6° ausgelesen und radial gemittelt, um eventuelle Verkippungen des Roboters auszugleichen. Die aus den so gewonnenen Bildern berechneten Bewegungskommandos werden anschließend seriell zu der Roboterplattform übertragen (12 Hz Khepera, 10 Hz Pioneer).

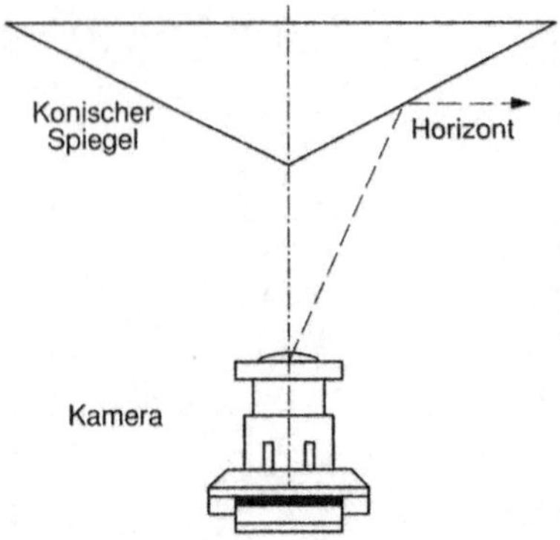

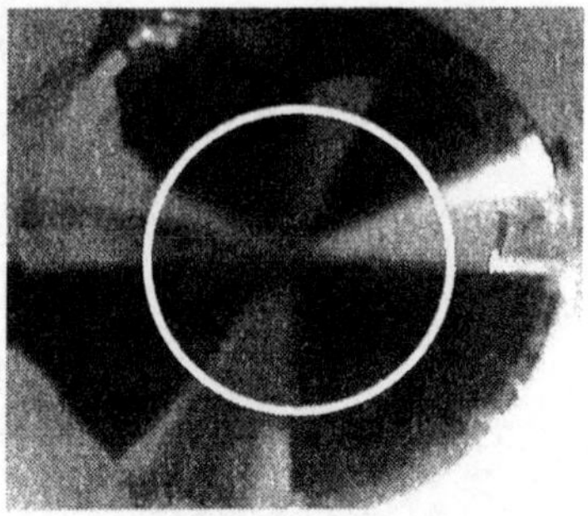

Abbildung 3. Die omnidirektionalen Ansichten werden mit einer Kamera aufgenommen, die auf einen polierten Kegel gerichtet ist. Rechts ist der Spiegel von der Kamera aus gesehen gezeigt. Der weiße Ring markiert den Auslesebereich entlang des Horizontes.

Die Position des Khepera in der Arena wurde mit einer kalibrierten Farbkamera verfolgt. Sowohl Bild- als auch Positionsdaten wurden mit Zeitmarkern aufgenommen und später synchronisiert. Während der Experimente standen beiden Robotern keine absoluten Positionsdaten zur Verfügung.

4.2 Ergebnisse

Um die Anwendbarkeit des Algorithmus zu testen, fuhr der Khepera alle zugänglichen Positionen innerhalb der Arena an und nahm dabei 10000 lokale Ansichten auf, die die gesamte Arena relativ gleichmäßig abdecken. Die Größe des Fangbereichs eines Schnappschusses kann mit Hilfe dieser Daten auf folgende Weise dargestellt werden: Für einen gegebenen Schnappschuß wird mit Hilfe des Algorithmus aus jeder lokalen Ansicht ein Vektor in Richtung des Ausgangspunktes berechnet (s. die Vektorkarte in Abb. 4). Ein Raumpunkt wird als Bestandteil des Fangbereiches betrachtet, wenn es einen Pfad entlang der Zielvektoren gibt, der bis zum Ausgangspunkt führt. Abbildung 4 zeigt, daß der Fangbereich in günstigen Fällen den gesamten offenen Raum um das Ziel abdecken kann. Der Fangbereich wird kleiner, wenn die Schnappschüsse näher an Objekten aufgenommen werden, da sich hier die momentane Ansicht bei Bewegungen schneller verändert. Dennoch funktionierte der Algorithmus in allen Gebieten zuverlässig, in denen der Roboter nicht mit Hindernissen kollidierte. Dies konnte anhand des folgenden Experiments gezeigt werden: Zunächst wurden an 20 zufällig gewählten Stellen Schnappschüsse aufgenommen und der Khepera anschließend in Entfernungen von 5 bis 25 cm passiv versetzt. Während 100 Testläufen wurde alle 83 ms ein neuer Zielvektor berechnet und die Fahrtrichtung dementsprechend korrigiert[3] Ein Versuch wurde als Erfolg gewertet, wenn der Roboter kollisions-

[3] Auf einem Standard-PC (133MHz Pentium) kann alle 40 ms ein neuer Vektor berechnet werden. Wegen der langsamen seriellen Kommunikation konnte dies allerdings

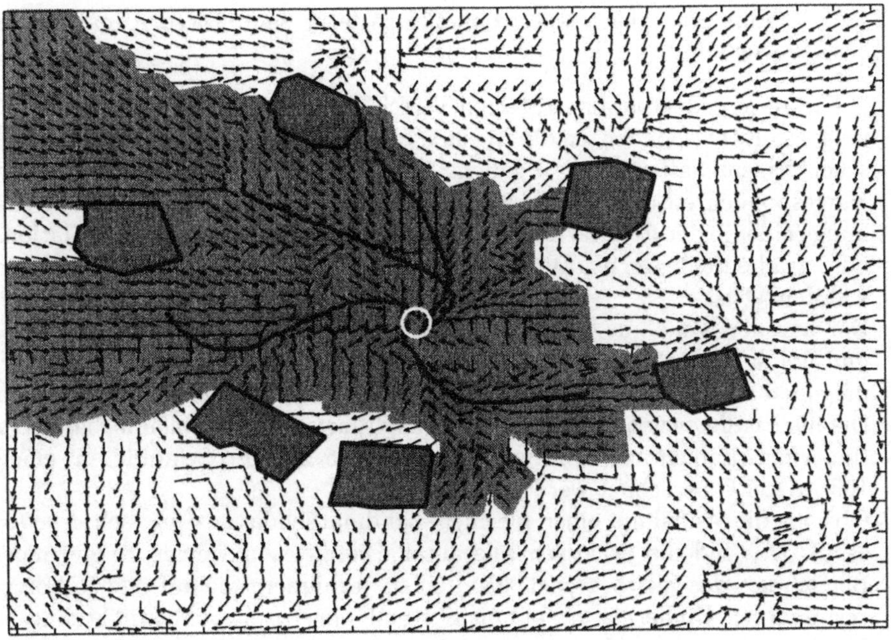

Abbildung 4. Durch den Algorithmus berechnete Zielvektoren für einen zentralen Schnappschuß in der Mitte der Arena (durch Kreis markiert) mit Trajektorien des Khepera bei der Zielanfahrt. Der Fangbereich ist grau dargestellt, die Landmarken sind schwarz umrandet.

frei einen Kreis mit 1 cm Radius um das Ziel innerhalb von 30 s erreichte. Es zeigt sich, daß der Algorithmus bis zu einer Entfernung von 15 cm mit einer Erfolgsrate von 95% funktioniert. Danach sinkt die Erfolgsrate bis auf 50% bei einer Entfernung von 25 cm. Hier befand sich der Roboter oft außerhalb des offenen Gebietes um das Ziel, so daß aufgrund von Verdeckungen in der momentanen Ansicht keine sinnvollen Korrespondenzen zu den Schnappschüssen gefunden werden konnten. Ähnliche Erfolgsraten wurden auch mit dem größeren Prototyp innerhalb der Büroumgebung unseres Institutes erzielt. Die maximale erreichte Distanz, von der aus das Ziel wiedergefunden wurde, lag hier bei ca. 2 m.

5 Diskussion

Der hier vorgestellte biologisch inspirierte Algorithmus ermöglicht das Anfahren eines durch einen 360°-Schnappschuß markierten Zielortes. Die Zielrichtung wird aus der Verschiebung der umgebenden Landmarken im Vergleich zum Schnappschuß berechnet. Da die für die Berechnung der Zielrichtung benötigten Entfernungen der Landmarken unbekannt sind, müssen diese über sinnvolle Annahmen geschätzt werden. Der Algorithmus basiert auf der vereinfachenden Annahme,

nicht direkt zur Korrektur der Fahrtrichtung benutzt werden.

daß alle Landmarken gleich weit vom Ziel entfernt sind. Diese Annahme hat mehrere Vorteile:

1. Die resultierenden parametrisierten Verschiebungsfelder erlauben die Erstellung von Korrespondenzen durch einen globalen Bildvergleich. Die geometrischen Eigenschaften der Aufgabenstellung gehen bei der Berechnung der Korrespondenzen direkt mit ein. Daher können alle berechneten Verschiebungsfelder in der Realität vorkommen, was bei den allgemeinen korrelations- oder gradientenbasierten Methoden zur Korrespondenzberechnung nicht gewährleistet ist.
2. Der aus der Annahme resultierende Fehler geht bei der Zielanfahrt gegen null. Es läßt sich beweisen, daß jeder Schnappschuß von einem Fangbereich umgeben ist, der den gesamten offenen Bereich um das Ziel einnehmen kann.
3. Die niedrigdimensionale Parametrisierung führt auf ein schnell lösbares Optimierungsproblem, so daß die Zielrichtung 25 mal pro Sekunde auf einem 133-MHz-PC berechnet werden kann.
4. Obwohl die Annahme einer gleichen Landmarkendistanz nur selten exakt erfüllt ist, sind die praktisch relevanten Distanzen auf einen ringförmigen Bereich um das Ziel begrenzt: Zum einen verhindern die auf den Robotern verwendeten Hindernisvermeidungssysteme eine zu starke Annäherung an Landmarken, zum anderen wirken sich weit entfernte Landmarken auf die Berechnung der Zielrichtung und auf den Gesamtfehler nur wenig aus.

Natürlich bleibt ein auf einem einzelnen Schnappschuß basierendes Navigationssystem auf die unmittelbare Umgebung des Zieles beschränkt. Durch Kombination mehrerer benachbarter Schnappschüsse zu einer graphenartigen Repräsentation der Umwelt lassen sich aber auch großräumige Navigationsaufgaben mit Schnappschüssen bewältigen (Mallot et al., 1997; Franz et al., 1998).

Literatur

1. Cartwright, B.A., Collett, T.S.: Landmark learning in bees. J. Comp. Physiol. A **151** (1983) 521–543
2. Chahl, J.S., Srinivasan, M.V.: Visual computation of egomotion using an image interpolation technique. Biol. Cybern. **57** (1996) 405–411
3. Collett, T.S.: Landmark learning and guidance in insects. Phil. Trans. R. Soc. Lond. B **337** (1992) 295–303
4. Franz, M.O., Schölkopf, B., Mallot, H.A., Bülthoff, H.H.: Learning view graphs for robot navigation. Autonomous Robots **5** (1998) 111–125
5. Franz, M.O., Schölkopf, B., Bülthoff, H.H.: Homing by parameterized scene matching. In Proc. 4th European Conf. on Artificial Life (1997) 236–245
6. Hong, J., Tan, X., Pinette, R., Weiss, R., Riseman, E.M.: Image-based homing. In Proc. IEEE Conf. Robotics and Automation (1991) 620–625
7. Mallot, H.A., Franz, M.O., Schölkopf, B., Bülthoff, H.H.: The View-Graph Approach to Visual Navigation and Spatial Memory. In Proc. 7th Intl. Conf. on Artificial Neural Networks (1997) 751–756
8. Röfer, T.: Controlling a wheelchair with image-based homing. Dept. Computer Sc., Manchester University, Techn. Report UMCS-97-4-1 (1997)

Entwicklung verteilter Musteranalysesysteme mit DACS

Gernot A. Fink, Nils Jungclaus, Helge Ritter, Gerhard Sagerer

Technische Fakultät,
Universität Bielefeld, Postfach 100131, 33501 Bielefeld
email: gernot,nils@techfak.uni-bielefeld.de

Zusammenfassung In traditionellen Ansätzen zu paralleler oder verteilter Verarbeitung werden meist wohlstrukturierte Probleme vollständig in einer Programmierumgebung implementiert. Im Gegensatz dazu ist es bei der Entwicklung großer Musteranalysesysteme meist erforderlich, viele heterogene Software-Module zu einem Gesamtsystem zu integrieren. Wir stellen ein Kommunikationssystem für verteilte Anwendungen vor, das speziell für diesen Zweck entwickelt wurde. Es erlaubt vollständig symmetrische Modulinteraktion, verwendet strukturierte und typisierte Daten und stellt verschiedene Aufrufsemantiken zur Verfügung. Die Leistungsfähigkeit des Systems wird anhand der Entwicklung des Prototyps eines "situierten künstlichen Kommunikators" demonstriert.

1 Einleitung

Im Rahmen des Sonderforschungsbereichs 360 "Situierte künstliche Kommunikatoren" werden kognitive Merkmale menschlicher Intelligenz in aufgabenorientierten Dialogen untersucht. Ein wesentliches Ziel ist dabei, die entwickelten Modelle auf künstliche informationsverarbeitende Systeme zu übertragen und diese so entweder zu validieren oder neue Grundlagen intelligenter Musteranalysesysteme zu schaffen.

Als Domäne für den sogenannten "situierten künstlichen Kommunikator" wurde die kooperative Konstruktion eines Spielzeugflugzeugs aus Teilen eines Holzbaukastens für Kinder gewählt. Die Maschine soll dabei in der Lage sein, akustische und visuelle Eingaben zu verarbeiten und durch sprachliche Ausgabe oder Manipulation von Objekten in der Szene angemessen reagieren zu können. Ihre Aufgabe ist es ausschließlich, die vom menschlichen Instrukteur gegebenen Konstruktionsanweisungen umzusetzen, so daß eine selbständige automatische Planung des Konstruktionsprozesses nicht erforderlich ist.

Um einen solchen "künstlichen Kommunikator" zu realisieren, müssen viele verschiedene Module für Signalverarbeitung, Mustererkennung, Bildanalyse, Sprachverstehen und Robotik zusammenarbeiten. Müßten Teilmodule vollständig neu implementiert werden, ließe sich die Entwicklung eines derartigen Systems

[1] Die vorliegende Arbeit wurde im Rahmen des Sonderforschungsbereichs 360 von der Deutschen Forschungsgemeinschaft (DFG) gefördert.

jedoch nicht in begrenzter Zeit bewältigen. Wir griffen daher auf bereits bestehende Teillösungen zurück (z.B. [2,5]) und setzten spezielle Entwicklungswerkzeuge für Mustererkennungs- und -analyseprobleme ein (z.B. [6]).

Das verwendete Kommunikationssystem muß daher in der Lage sein, heterogene Softwaremodule, die evtl. in unterschiedlichen Programmierumgebungen und nicht notwendiger Weise mit speziellem Augenmerk auf verteilte Verarbeitung erstellt wurden, zu einem verteilt arbeitenden Gesamtsystem zu integrieren.

Client-Server-Architekturen sind im Soft- und Hardwareengineering wegen der Möglichkeit zur gemeinsamen Nutzung von Betriebsmitteln von großer Bedeutung. Für eine komplexe Musteranalyseanwendung bringt eine solche Architektur jedoch unnötige Restriktionen für den Problemlösungsprozeß mit sich. Um die Beschränkung auf ein *einzelnes* Modell für verteilte Verarbeitung zu vermeiden muß daher vollständig symmetrische Kommunikation möglich sein.

In einem verteilten Musteranalysesystem erzeugen bzw. verarbeiten die einzelnen Module in der Regel komplex strukturierte Daten. Um diese korrekt zwischen heterogenen Hardwareplattformen austauschen zu können, muß Datentypinformation vorliegen. Wird diese zusammen mit den Binärdaten übertragen und steht ein Mechanismus zur generischen Bildung zusammengesetzter Datentypen zur Verfügung, so läßt sich Typ- und Strukturinformation unabhängig von der Anwendung in den kommunizierten Daten kodieren. Damit ist eine Typüberprüfung zur Laufzeit möglich und es können generische Entwicklungswerkzeuge eingesetzt werden, um Datenstrukturen zu inspizieren oder zu manipulieren.

Betriebsmittel in einem verteilten Musteranalysesystem werden am besten über symbolische Namen referiert. Im Gegensatz zu traditionellen Namensraumverwaltungen für z.B. Benutzer-Accounts oder Netzwerkdateisysteme muß die Aktualisierung der bereitgestellten Information möglichst schnell erfolgen. Dies ist wichtig für die Entwicklungsphase und unabdingbare Voraussetzung für die dynamische Rekonfiguration eines verteilten Gesamtsystems zur Laufzeit.

2 Entwicklungswerkzeuge für verteilte Systeme

Von den verschiedenen Werkzeugen zur Entwicklung verteilter Systeme kommen in der Praxis nur wenige für ein Integrationsprojekt wie es im SFB 360 durchgeführt wird in Frage. Rein programmiersprachliche Lösungen sind ungeeignet, da bereits bestehende Module zu integrieren sind, und auch die verbreiteten bibliotheks-basierten Systeme haben konzeptuelle Nachteile, die einen einfachen und effizienten Aufbau verteilter Systeme aus bestehenden Modulen erschweren:

DCE [1] bietet Kommunikationsmöglichkeiten über *Remote Procedure Calls (RPC)*. Die Argumentdatentypen sind statisch und müssen zur Übersetzungszeit bekannt sein. Ein dynamischer Austausch von Modulen zur Laufzeit läßt sich nur schwer erreichen, nicht zuletzt aufgrund der sich konzeptuell bedingt nur langsam verbreitenden Informationen im DCE-eigenen Namensraum.

ONC-RPC [11] erlaubt synchrone Prozeduraufrufe in einem typischen Client-Server Konzept. Die übertragenen Datentypen können nahezu beliebig strukturiert sein, müssen jedoch auch zur Übersetzungszeit bekannt sein und werden zur Laufzeit nicht überprüft. Die Entwicklung dynamischer konfigurierbarer Systeme wird durch die nötige explizite Angabe des Servers erschwert.

PVM [12] ist ein bekanntes Standard-System, das typischerweise zur verteilten Implementierung komplexer Algorithmen eingesetzt wird. Der Einsatz eines Dämon-Prozesses auf jedem in das System integrierten Rechner vereinfacht die Verwaltung und Überwachung des Gesamtsystems, da hier ein zentraler Ansatzpunkt für entsprechende Werkzeuge existiert. PVM besitzt keine zentrale Verwaltungsinstanz, d.h. daß z.B. Konfigurationsänderungen jeweils im Gesamtsystem propagiert werden müssen und damit einen zusätzlichen Kommunikationsaufwand erzeugen. Die Möglichkeiten zum Austausch strukturierter Daten sind sehr elementar gehalten, eine Typüberprüfung geschieht nicht. PVM stellt jedem Nutzen sein eigenes System zur Verfügung, das nicht mit Systemen anderer Benutzer interagieren kann. Das ist nützlich zur Implementierung verteilter Algorithmen, behindert jedoch die kooperative Entwicklung mit mehreren Entwicklern.

CORBA [9], eine von der *Object Management Group* spezifizierte Architektur zur Realisierung verteilter Objekte in heterogenen Umgebungen, bietet ähnlich wie DCE Client-Server-orientierte Kommunikationssemantiken in Form von Prozeduraufrufen, jedoch basierend auf einem objektorientierten Ansatz. Übertragene Datentypen müssen wie bei ONC-RPC in einer speziellen Definitions-Sprache beschrieben werden und zur Übersetzungszeit bekannt sein. Eine Erweiterung auf dynamische Interfaces ist zwar mittlerweile erfolgt, jedoch noch nicht in jeder Implementierung verfügbar. Obwohl die CORBA-Spezifikation Anbindungen an verschiedene Programmiersprachen beschreibt, bieten die meisten Implementierungen bislang lediglich eine C++-Anbindung, was die Integration bestehender Software erschwert. Weiterhin sind CORBA-Implementierungen bis auf eine Ausnahme nur als kommerzielle Produkte erhältlich.

Aufgrund der genannten Unzulänglichkeiten dieser Werkzeuge entwickelten wir ein auf unsere Anforderungen zugeschnittenes System, das eine einfache und effiziente Integration verteilter Module in heterogenen Umgebungen erlaubt.

3 Systemarchitektur

Der strukturelle Aufbau des ***Distributed Application's Communication System*** (DACS) [4] ähnelt dem von PVM (Abbildung 1). Auf jedem in das System integrierten Rechner sorgt ein Dämon-Prozeß für den Datenaustausch zwischen lokalen Applikationen. Kommunikation über Rechnergrenzen hinweg erfolgt ausschließlich über die jeweiligen Dämonen, die ihrerseits Verbindungen zu den beteiligten lokalen Applikationen unterhalten. Die dazu notwendigen Verbindungen werden dynamisch instantiiert. Die Systemkonfiguration wird in einem zentralen Directory-Service verwaltet, was den Kommunikationsaufwand bei Rekonfigurationen minimiert. In den einzelnen DACS-Dämonen sorgt eine Namens-Cache

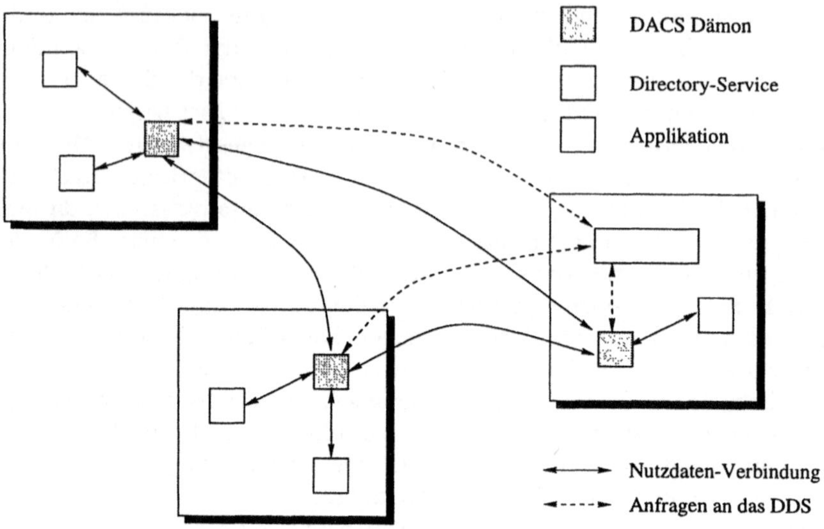

Abbildung1. Struktureller Aufbau von DACS

für möglichst wenige Zugriffe auf den Directory-Service. Ein zweiter Stand-By Directory-Service sorgt für Ausfallsicherheit.

Die Abwicklung aller Kommunikationsvorgänge durch den DACS-Dämon zieht zwar Leistungseinbußen nach sich, bietet jedoch pro Rechner einen zentralen Ansatzpunkt für Werkzeuge zur Wartung und Überwachung des Systems, die mit zunehmender Größe des Gesamtsystems unverzichtbar sind. Der Einsatz prozeßinterner Parallelität mit Hilfe von Threads gewährleistet zudem die optimale Ausnutzung der Kommunikationskanäle und erlaubt den Einsatz auch in zeitkritischen Umgebungen wie der Echtzeit-Roboteransteuerung.

Ein wesentliches Entwicklungsziel war die einfache Anwendbarkeit von DACS. Dieses wird zum einen durch eine schlanke Programmierschnittstelle und zum anderen durch die Verwendung von Threads innerhalb der Bibliothek erreicht, die den Applikationen die Kommunikationsfunktionalität bereitstellt. Die im Hintergrund ablaufende Kommunikation mit dem DACS-Dämon beeinflußt den Kontrollfluß der Applikation daher nur indirekt. Ein DACS-System kann von mehreren Benutzern gleichzeitig verwendet werden, wobei eine Sicherung gegenüber Unbefugten durch ein kryptografisches Schlüsselverfahren besteht. DACS ist für Digital-Unix, Linux, Solaris, Irix und HP-UX verfügbar.

4 Strukturierung und Typisierung von Daten

In vielen Entwicklungswerkzeugen für verteilte Systeme wird die Konsistenz von Daten zwischen Modulen sowie die Umwandlung zwischen Binärformaten unterschiedlicher Hardwareplattformen durch automatisch erzeugten Code für die

Interaktion sichergestellt. Dieser wird meist durch die Kompilation einer Schnittstellenbeschreibung gewonnen. Module müssen daher mit identischen Schnittstellendefinitionen erzeugt werden, was jedoch nicht garantiert werden kann.

Wie zuvor beschrieben erfolgt die gesamte Datenkommunikation in DACS vermittelt durch einen Dämonprozeß pro Maschine. Selbst wenn alle Schnittstellendefinitionen konsistent gehalten werden, wären diese im anwendungsunabhängigen Dämon nicht bekannt. Datenumwandlungen müßten daher bereits in der Applikation erfolgen, was entweder detailiertes Wissen über die Kommunikationswege erfordern oder unnötigen Aufwand verursachen würde.

Um dieses Problem zu umgehen, werden Daten in DACS in zwei Stufen kodiert. Programmiersprachenabhängige Datenstrukturen werden auf Applikationsebene in eine sogenannte ***Netzwerk-Daten-Repräsentation*** (NDR) mit expliziter Kodierung von Datentyp und -struktur überführt. Diese Transformation kann entweder durch vorhandene Routinen für primitive NDR-Datentypen erfolgen oder durch automatisch erzeugte Umwandlungsfunktionen. Letzteres Vorgehen entspricht in etwa dem anderer Schnittstellen-Compiler mit dem wesentlichen Unterschied, daß nicht nur eine Serialisierung der Daten erfolgt, sondern Typ- und Strukturinformation anwendungsunabhängig mitkodiert werden.

Sofern eine Binärkonversion der Daten erforderlich ist, kann diese in einem optionalen zweiten Kodierungsschritt ohne Kenntnis der Originaldatenstrukturen durch den DACS-Dämon ausgeführt werden. Aufgrund der in NDR-Daten eingebetteten Typinformation werden diese dann in ein plattformunabhängiges Binärformat transformiert, aus dem auf Empfängerseite wieder die lokale Binärrepräsentation erzeugt wird.

NDR stellt Basiskonvertierungsoperationen für skalare Datentypen (`char`, `byte`, `short`, `int`, und `float`) sowie für Zeichenketten (`string` und `symbol`) zur Verfügung. Da in Signalverarbeitungsanwendungen häufig ein- oder mehrdimensionale Vektoren verwendet werden, existieren für diese kompakte Kodierungen (`vektor` und `hyper`). Außerdem lassen sich Listen beliebiger Datentypen bilden (`list`). Als Konstruktor für zusammengesetzte Datentypen steht der aus der Programmiersprache LISP bekannte Paarbildungsoperator zur Verfügung.

Die Möglichkeit, Listen- und Baumrepräsentationen zu erzeugen zusammen mit einem Umsetzungsschema erlaubt es, beliebige programmiersprachliche Datentypen in NDR zu kodieren. Diese sind dann unabhängig von dem System, auf dem sie erzeugt wurden und bei geeignet gewählten Umsetzungsschemata auch unabhängig von einer speziellen Programmierumgebung.

5 Aufrufsemantiken

Wesentliche Eigenschaft von Werkzeugen zur Entwicklung verteilter Systeme sind die verfügbaren Aufrufsemantiken. Für viele Kommunikationsaufgaben sind wenige ausgewählte Standardmethoden ausreichend. Daher bietet DACS neben einfachem Message-Passing für funktionale Interaktionen RPCs an sowie eine weitere, besonders zur Übertragung von Datenströmen geeignete Kommunikationssemantik, sogenannte *Demand-Streams*. Wesentlich für eine einfache und

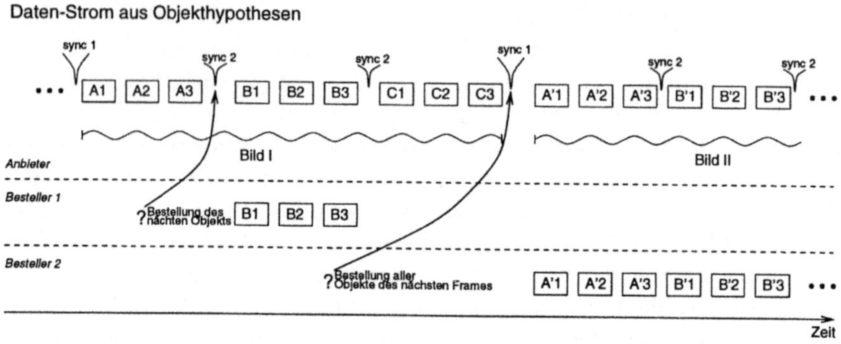

Abbildung2. Zeitliche Funktionsweise von Demand-Streams

flexible Integration von Modulen ist dabei, daß alle Aufrufsemantiken innerhalb einer Applikation *gleichzeitig* verfügbar sind.

Intern werden Aufrufsemantiken mit Hilfe von asynchronem Message-Passing implementiert. Dieses steht dem Benutzer direkt zur Verfügung, wobei zur Kanalisierung von Nachrichten benannte Kommunikationsendpunkte verwendet werden können. DACS-RPC bietet neben gewöhnlichen synchronen Prozeduraufrufen auch die Möglichkeit asynchroner Aufrufe, mit denen z.B. der parallele Aufruf mehrerer Prozeduren gleichzeitig möglich ist. RPC-Server können wahlweise eingehende Aufrufe sequentiell oder parallel bearbeiten und ohne die Applikation zu blockieren im Hintergrund laufen.

Speziell für den Einsatz in signalverarbeitenden Systemen sind die Demand-Streams konzipiert. Ein Datenstrom wird unter einem symbolischen Namen bereitgestellt und kann von anderen Modulen in Teilen oder vollständig bestellt werden. Die Auslieferung der Daten wird über die Message-Passing Funktionalität durchgeführt. Jedem Demand-Stream liegt ein frei definierbarer Basisdatentyp zugrunde, der mit Hilfe von Sychronisationsmarken in hierarchisch organisierte Gruppen zusammengefaßt werden kann.

Abbildung 2 zeigt schematisch einen Demand-Stream aus Objekthypothesen, bei dem je drei konkurrierende eine Gruppe bilden und alle Hypothesen zu einem Bild in einer weiteren Hierarchiestufe zusammengefaßt sind. Zu unterschiedlichen Zeitpunkten gehen von zwei Bestellern Anforderungen nach den nächsten drei konkurrierenden bzw. allen Objekthypothesen für das nächste Bild ein. Abhängig vom Bestellzeitpunkt wird die nächste erreichbare Gruppe der entsprechenden Hierarchiestufe ausgeliefert. Die entstehende asynchrone Interaktion erlaubt es, Module mit unterschiedlichen Verarbeitungsgeschwindigkeiten ohne Gefahr von Datenstaus in den Kommunikationswegen zu koppeln.

6 Die Entwicklung des SFB360-Prototypen

Der aktuelle Prototyp des "situierten künstlichen Kommunikators" ist die Weiterentwicklung eines Systems, das für die erste Begehung des Sonderforschungsbereichs realisiert wurde [3]. Es können drei logische Verarbeitungsblöcke für Bildanalyse, Sprachverstehen und Robotik unterschieden werden.

Auf der Basis von Stereo-Farbbildern einer statischen Szenenansicht wird durch eine Reihe kooperierender Bildverarbeitungmodule eine 3D-Repräsentation der auf der Arbeitsfläche vorhandenen Bauteile erzeugt. Alle diese Module arbeiten rein datengetrieben auf Eingabedatenströmen, die als Demand-Streams von anderen Modulen bereitgestellt werden. Die unterschiedlichen Verarbeitungsgeschwindigkeiten werden durch die mit Demand-Streams einfach zu realisierende asynchrone Koppelung berücksichtigt.

Die Verarbeitung beginnt mit der Berechnung von Farbregionen ausgehend von farbklassifizierten Eingabebildern. Parallel dazu werden holistische Objekthypothesen berechnet [5]. Sie dienen zusammen mit den Farbregionen als Eingabedaten für die schritthaltende Erzeugung strukturierter 2D-Objekthypothesen [7]. Parallel zur regionenbasierten Verarbeitungslinie werden auf Grauwertbildern Kantenhypothesen erzeugt und mit perzeptiven Gruppierungsverfahren zu größeren Konturhypothesen zusammengefaßt [8]. Die Kombination von regionenbasierten 2D-Objekthypothesen und Konturinformation aus zwei Stereoansichten dient als Grundlage für eine modellbasierte 3D-Rekonstruktion [10].

Die Interpretation spontansprachlicher Anweisungen geschieht ausgehend von den Ergebnissen eines inkrementellen Erkennungsmoduls, das bereits einfache Konstituentenhypothesen erzeugt [13]. Es schließt sich eine syntaktisch-semantische Analyse an. Ein Großteil der Interaktion des Systems mit dem Instrukteur übernimmt eine Dialogkomponente. Mit Hilfe von Bayes-Netzwerken werden Beziehungen zwischen sprachlichen und visuellen Analyseergebnissen hergestellt [10]. Die Interaktion auf diesen Ebenen erfolgt synchron über RPCs.

Die konzipierte Anbindung an die Robotikkomponente wird gegenwärtig realisiert wie auch die Einbindung eines Moduls zum aktiven Sehen.

Derzeit partizipieren bis zu 15 Module verteilt in einem lokalen Netzwerk am Gesamtsystem des Prototypen. Ein Großteil kommuniziert allein über Demand-Streams, was die besondere Bedeutung dieser Interaktionsform für Aufgaben der Musteranalyse unterstreicht. Durch die NDR-Kodierung der ausgetauschten Daten und die Flexibilität der Demand-Stream-Schnittstelle lassen sich Analyseergebnisse mit generischen Werkzeugen aufzeichnen und bei Bedarf replizieren. Dies ist von entscheidender Bedeutung für die Systementwicklungsphase, da so einfach Teilsysteme mit realen Ergebnisdaten getestet werden können. Für eine größere Gruppe von Forschern wäre sonst das gemeinsame Entwickeln an einem umfangreichen System kaum möglich.

7 Zusammenfassung

Wir stellten DACS vor, ein flexibles Entwicklungswerkzeug für verteilte Anwendungen mit speziellen Anpassungen an die Belange von Musteranalysesystemen. Es ist ausgelegt, die Integration existierender Module zu erleichtern. Daten werden zusammen mit Typ und Strukturinformation ausgetauscht und sind damit im System anwendungs- und hardwareunabhängig repräsentiert. DACS verwendet einen zentralen Kommunikationsmanagement-Dämon pro Maschine. Der Namensraum mit Informationen über beteiligte Applikationen und Schnittstellen wird von einem zentralen Modul effizient verwaltet.

Die Leistungsfähigkeit des Systems sowie die Tragfähigkeit der realisierten einfachen aber doch mächtigen Prinzipien wurde anhand der Entwicklung des umfangreichen Prototypen im Sonderforschungsbereich 360 "Situierte künstliche Kommunikatoren" demonstriert. Eine frühere Version dieses Systems wurde erfolgreich den Gutachtern der Deutschen Forschungsgemeinschaft (DFG) während der Begehung des Sonderforschungsbereichs im März 1996 präsentiert.

Literatur

[1] M. Bever, K. Geihs, L. Heuser, M. Mühlhäuser, A. Schill: *Distributed Systems, OSF DCE, and Beyond*, in A. Schill (Hrsg.): *Lecture Notes in Computer Science*, Bd. 731, Springer, Berlin Heidelberg, 1993, S. 1–20.

[2] G. A. Fink: *Integration von Spracherkennung und Sprachverstehen*, Bd. 103 von *Dissertationen zur Künstlichen Intelligenz*, infix, Sankt Augustin, 1995.

[3] G. A. Fink, N. Jungclaus, F. Kummert, H. Ritter, G. Sagerer: *A Distributed System for Integrated Speech and Image Understanding*, in R. Soto (Hrsg.): *International Symposium on Artifical Intelligence*, Cancun, Mexico, 1996, S. 117–126.

[4] G. A. Fink, N. Jungclaus, H. Ritter, G. Sagerer: *A Communication Framework for Heterogeneous Distributed Pattern Analysis*, in *International Conference on Algorithms And Architectures for Parallel Processing*, Brisbane, 1995, S. 881–890.

[5] G. Heidemann, H. Ritter: *A Neural 3-D Object Recognition Architecture Using Optimized Gabor Filters*, in *Proc. 13th Int. Conf. on Pattern Recognition*, Bd. IV, IEEE Computer Society Press, Vienna, 1996, S. 70–74.

[6] F. Kummert, H. Niemann, R. Prechtel, G. Sagerer: *Control and Explanation in a Signal Understanding Environment, Signal Processing, special issue on 'Intelligent Systems for Signal and Image Understanding'*, Bd. 32, 1993, S. 111–145.

[7] F. Kummert, G. A. Fink, G. Sagerer: *Schritthaltende hybride Objektdetektion*, in E. Paulus, F. M. Wahl (Hrsg.): *Mustererkennung 97, 19. DAGM-Symposium Braunschweig, Informatik aktuell*, Springer-Verlag, Berlin, 1997, S. 137–144.

[8] A. Maßmann, S. Posch, G. Sagerer, D. Schlüter: *Using Markov Random Fields for Contour-Based Grouping*, in *International Conference on Image Processing*, Bd. II, IEEE, 1997, S. 207–210.

[9] Object Management Group: *The Common Object Request Broker: Architecture and Specification*, OMG, Juli 1995, Version 2.0.

[10] G. Socher: *Qualitative Scene Descriptions from Images for Integrated Speech and Image Understanding*, Bd. 170 von *Dissertationen zur Künstlichen Intelligenz*, infix, Sankt Augustin, 1997.

[11] Sun Microsystems, Inc.: *Network Programming*, SUN, Mountain View, CA, 1990.

[12] V. S. Sunderam: *PVM: A Framework for Parallel Distributed Computing*, Department of Math and Computer Science, Emory University, Atlanta, GA 30322.

[13] S. Wachsmuth, G. A. Fink, G. Sagerer: *Integration of Parsing and Incremental Speech Recognition*, in *Proc. European Signal Processing Conference*, Rhodos, 1998, erscheint.

Anwendungen
(Postersession)

Bildmatching und Bewegungskompensation bei Fundus-Bildern

K.Voss, W.Ortmann, H.Süße
E-Mail: nbs@uni-jena.de
Friedrich-Schiller-Universität Jena
Fakultät für Mathematik und Informatik
Lehrstuhl "Digitale Bildverarbeitung"
Ernst-Abbe-Platz 1-4
07740 Jena

1 Einleitung

Bildgestützte Diagnosen des Augenhintergrundes werden oft durch eine instabile Fixation der Augen des Patienten erheblich erschwert. Deshalb ist es bei einer Aufnahmefolge des gleichen Augenhintergrundes unbedingt notwendig, die stabile Fixation durch eine Bewegungskompensation zu garantieren. Dies ist eine klassische Aufgabe der Bildregistrierung (siehe z.B. [1]). Da in den Bildern keinerlei Referenzpunkte oder Referenzobjekte sicher bestimmt werden können, scheiden alle merkmalsbasierten Matching-Verfahren aus, die auf Referenzen von Punkten bzw. Objekten beruhen. Daher müssen wir uns auf die signalbasierte Bewegungskompensation beschränken. Hier gibt es in der Literatur eine Fülle von Verfahren, die im wesentlichen in die differentiellen und die flächenbasierten Verfahren eingeteilt werden, siehe stellvertretend [12]. Auch alle Verfahren, die die "Mitnahme der Grauwerte" voraussetzen, sind für die vorliegende Aufgabenstellung ungeeignet, beispielsweise der "Optische Fluß" oder der Algorithmus in [11]. Ein großer zeitlicher Abstand zwischen zwei Aufnahmen des Augenhintergrundes bedingt erhebliche photometrische Schwankungen. Physiologisch bedingte Veränderungen der Retina führen zu starken Differenzen in den Grauwerten, so z.B. das Füllen/Leeren von Adern des Augenhintergrundes, siehe Abb.1. Die zwischen den Einzelbildern auftretenden geometrischen Transformationen bestehen im wesentlichen aus Translationen, aber auch geringe Rotationen, Skalierungen und sogar Deformationen sind vorhanden. Als mathematisches Entzerrungsmodell ist deshalb eine affine Transformation ausreichend, die durch gewichtete Referenzpunktepaare zu bestimmen ist. Diese gewichteten Referenzpunktepaare sind lokal durch Blockzuordnungen zu ermitteln. Dabei ist eine solche Methode zu finden, die möglichst robust gegenüber den oben genannten Grauwertstörungen ist und gleichzeitig so effizient, daß der Arzt sofort nach den Aufnahmen die entzerrten Bilder zur Verfügung hat. Zusätzlich soll der Algorithmus den Fall erkennen, wann eine Entzerrung völlig versagt. Eine Bildserie umfaßt dabei ca. 30 Aufnahmen mit Bildern der Größe 1580×1280. Zunächst beschränken wir uns auf die näherungsweise zu bestimmende Translation in einem Block eines Bildes. Damit die Translation detektiert werden kann, muß man lokal mit

---

Diese Arbeit entstand in Zusammenarbeit mit CZ Jena (Ophthalmologie)

"nicht zu großen Blöcken" arbeiten. Daher kommt als Standardmethode natürlich ein Block-Matching-Algorithmus in Frage, der entsprechend den obigen Forderungen ein robustes Matching-Maß besitzen muß. Mit dem normierten Korrelationskoeffizienten

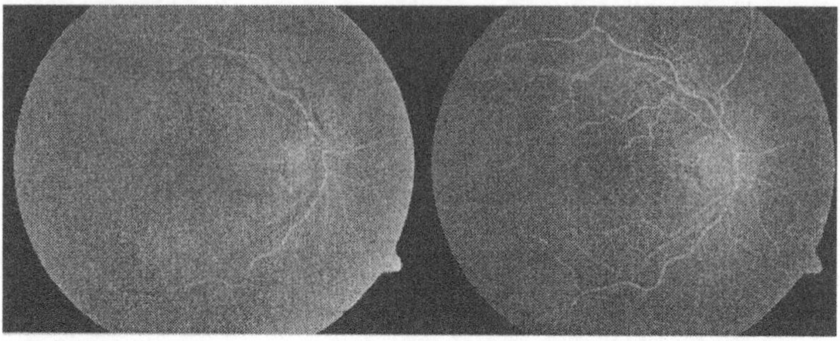

Abbildung 1: Zwei Aufnahmen aus einer Bildserie mit "Grauwert-störungen"

als robustes Gütemaß (siehe z.B.[5]) konnte die Entzerrung erfolgreich durchgeführt werden, allerdings mit solch einem extrem hohen Rechenaufwand, daß Block-Matching-Algorithmen für die obige Aufgabenstellung praktisch nicht relevant sind. Generell soll ein Suchraum, wie beim Block-Matching und den phasenbasierten Techniken vermieden werden. Die Referenz soll allein aus zwei fest vorgegebenen Blöcken bestimmt werden. Der Preis dafür, daß man den Suchraum einspart, sind die unangenehmen Randeffekte ("wrap around effects") bedingt durch die zyklische Verschiebung. Die verbreitesten Methoden ohne Suchraum sind die "Peak"-Bestimmungen mittels der Kreuzkorrelation (siehe z.B. [5]) und die mittels der Cepstrum-Methode (siehe [8] oder [9]). Diese Methoden setzen jedoch eine problemabhängige Vorverarbeitung der Blöcke voraus, wie z.B. die Wahl einer geeigneten Fensterfunktion und eine Bandpaßfilterung, um "wesentliche Strukturen" zu selektieren. Diese Techniken sollen im folgenden durch "Bildrestaurationstechniken"' ersetzt werden, bei denen Maßnahmen zur Dämpfung der Randeffekte und die Bandpaßfilterung in der Vorverarbeitung völlig entfallen können.

2 Aufgabenstellung

Der Einfachheit halber beziehen sich alle Formeln auf diskrete Funktionen einer Veränderlichen, die an n äquidistanten Stützstellen definiert sind (die Verallgemeinerung aufs Zweidimensionale ist unmittelbar möglich). Gegeben seien zwei "Frames", d.h. zwei Bildblöcke f und f'. Es sei δ der Einheitsimpuls und $\mathbf{S}_d$ der Verschiebungsoperator mit einer ganzzahligen Verschiebung d. Dann gilt

$$f' = \mathbf{S}_d f = \mathbf{S}_d f * \delta = f * \mathbf{S}_d \delta \tag{1}$$

bzw. bei additiver Überlagerung mehrerer verschobener Bilder

$$f' = \mathbf{S}_{d_1} f + \mathbf{S}_{d_2} f + \mathbf{S}_{d_3} f + \ldots = f * \{\mathbf{S}_{d_1}\delta + \mathbf{S}_{d_2}\delta + \mathbf{S}_{d_3}\delta + \ldots\} \; . \tag{2}$$

Damit ist die zyklische Verschiebung auf die Faltung von f mit dem verschobenen Einheitsimpulsen zurückgeführt. Die in der Literatur bekannten globalen phasenbasierten Methoden gehen im Grunde genommen von der Gleichung (1) aus, nutzen diese aber nur zum Teil. Wenden wir auf (1) das Faltungstheorem an, so erhalten wir das Verschiebungstheorem. Das Verschiebungstheorem bildet die Grundlage aller phasenbasierten Techniken, siehe [3], [4] und [14]. Da bei Verwendung des Verschiebungstheorems nur mit einer Frequenz, also nur mit einer Auflösungsstufe gearbeitet wird, wird die Wirkung einer integrierenden Funktion der Entfaltung verschenkt. Daran ändert sich auch nichts, wenn sukzessive und hierarchisch mit verschiedenen Auflösungsstufen gearbeitet wird. Die lokale Fouriertransformation mit verschiedenen Fensterfunktionen bringt nur die Ortsabhängigkeit ins Problem. Es wird dann wie beim Block-Matching zu einem gegebenen Block ein Referenzblock gesucht, der die gleiche Phase eines einzigen Fourierkoeffizienten besitzt. Weng zeigt in [14], daß die Nutzung von Gabor-Filtern sogar problematischer ist als die Nutzung der lokalen Fouriertransformation mit Rechteck-Fenstern, da Gabor-Filter noch nicht einmal die "Quasilinearität" der Phase einer Auflösungsstufe garantieren. Es muß aber wie beim Block-Matching eine Suchstrategie durchgeführt werden, die oft bei kleinen Disparitäten Nullstellenverfahren benutzt.

3 Methode

Betrachten wir das Modell (1), so fallen die Ähnlichkeiten zur Bildrestauration auf, wobei sich das Restaurationsproblem zunächst nur auf die reine Entfaltung beschränkt. Daher interpretieren wir f' als gemessenes oder aufgenommenes Bild, f interpretieren wir als Punktverbreiterungsfunktion (PSF), und der verschobene Einheitsimpuls $\mathbf{S}_d\delta$ ist das zu restaurierende Originalbild, welches wir noch nicht kennen.
Wie in der Bildrestauration tritt auch hier Rauschen auf, welches mit der reinen Entfaltung (z.B. durch inverse Filterung) noch verstärkt wird. Damit müssen wir wie in der Bildrestauration unbedingt Rauschen mit im Modell berücksichtigen, um stochastische Störungen der Grauwerte modellieren zu können. Dies ist aber nur ein Aspekt von Störungen. Wenn wir einen Block im ersten Bild f und einen zweiten an der gleichen Position im zweiten Bild f' positionieren und annehmen, daß die Modelle (1) und (2) richtig sind, so setzen wir die zyklische Verschiebung voraus. Die zyklische Verschiebung ist aber bei der vorliegenden Aufgabenstellung nicht annähernd realisiert. Vorhandene Strukturen verschwinden aus dem Block und neue Strukturen tauchen auf. Diese Randeffekte, die der Preis für die Einsparung des Suchraumes sind, versucht man oft mit Fensterfunktionen zu dämpfen. Wir wollen dagegen diese Randeffekte als additive Störungen mit ins Modell übernehmen. Folglich erweitern wir die Modelle (1) und (2) mit dem Störungsfeld N zu

$$f' = f * \mathbf{S}_d\delta + N, \tag{3}$$

wobei wir im folgenden der Einfachheit halber nur Modell (1) benutzen. Das Feld N enthalte somit Rauschen, physiologisch bedingte Grauwertänderungen und die Randeffekte. Damit ist N weder signalunabhängig noch dekorreliert und es ist infolge der Randeffekte stark ortsabhängig. Um zusätzlich photometrische Einflüße zu berücksichtigen, müssen wenigstens lineare Grauwerttransformationen $g'=c\cdot g+b$ mit ins Modell übernommen werden. Daher führt (3) sofort zu

$$f' = f * (c \cdot \mathbf{S}_d \delta) + b + N \tag{4}$$

Der Kontrast c bewirkt lediglich eine Stauchung oder Streckung des Einheitsimpulses. Infolge der Voraussetzungen an das Störungsfeld N können wir nicht bedenkenlos irgendeine Restaurationsmethode anwenden, z.B. das Wienerfilter. Daher entscheiden wir uns für die "Restauration unter Zwang". Vom Störungsfeld N nehmen wir im folgenden nur an, daß die "mittlere Störstärke", also $E(\|N\|^2)$, bekannt ist. Dann bilden wir folgendes Extremalproblem:

$$\begin{aligned} &\|\mathbf{L}g\| \rightarrow \mathit{Minimum} \\ &\text{bei } \|f'' - f * g\|^2 = E(\|N\|)^2 \end{aligned} \tag{5}$$

Dabei ist $f'' = f' - b$, und g sei die Lösung der Extremalaufgabe als Approximation für $c\mathbf{S}_d\delta$. Der Operator $\mathbf{L}$ sei ein vorzugebender LSI-Operator, der uns unter allen zulässigen Lösungen die für unser Problem günstigste selektiert. Auf Grund der LSI-Eigenschaft können wir $\mathbf{L}$ eine Funktion l zuordnen. Die allgemeine Lösung von (5) lautet (f^s bezeichne die gespiegelte Funktion von f):

$$\left[f^s * f + \beta \cdot l^s * l\right] * g = f^s * f' - f^s * b \tag{6}$$

Dabei ist β der Regularisierungsparameter, der in Abhängigkeit von der mittleren Störungsstärke zu bestimmen ist. Die Faltungsgleichung (6) lösen wir aus Effektivitätsgründen (FFT, zwei Blöcke mit einer Transformation) im Frequenzraum. Die Gleichung (6) beinhaltet noch vier unbekannte Parameter. Die Skalierung c brauchen wir nicht zu kennen, sie wird sozusagen mit restauriert. Den Regularisierungsfaktor β, der vom Rauschen abhängt, müssen wir an Hand der konkreten Aufgabe schätzen, dieser kann sich von Block zu Block ändern. Den Operator $\mathbf{L}$ müssen wir konkret vorgeben. So bleibt lediglich die Verschiebung d zu bestimmen. Wir bilden nun aus (6) das Spektrum mit der symmetrischen, diskreten eindimensionalen Fouriertransformation (n-Funktionswerte):

$$\alpha_k(g) = \frac{\overline{\alpha_k(f)} \cdot \alpha_k(f')}{|\alpha_k(f)|^2 \cdot \sqrt{n} + \beta \cdot |\alpha_k(l)|^2 \cdot \sqrt{n}}, \ k \neq 0 \ . \tag{7}$$

Es ist trivial, daß bei Normierung des 0-ten Fourierkoeffizienten die Helligkeit b eliminiert wird. Durch Rücktransformation dieses normierten Fourierkoeffizienten und des in (7) errechneten Spektrums erhalten wir den "optimal restaurierten" verschobenen (evtl. skalierten) Einheitsimpuls, den es mit geeigneten Methoden zu detektieren gilt. Zuvor müssen wir aber noch den Operator $\mathbf{L}$ und damit die Fourier-Koeffizienten $\alpha_k(l)$

festlegen. Wählen wir den identischen Operator, also $l = \delta$, so erhalten wir aus (7)

$$\alpha_k(g) = \frac{\overline{\alpha_k(f)} \cdot \alpha_k(f')}{|\alpha_k(f)|^2 \cdot \sqrt{n} + \frac{\beta}{\sqrt{n}}}, \quad k \neq 0 \; . \tag{8}$$

Dies ist das bekannte Pseudo-Invers-Filter. Es verbleibt der Regularisierungsparameter, der in Abhängigkeit von der mittleren Störungsstörke zu wählen ist. Um ein Gefühl für die Größe von β zu bekommen, betrachten wir das Wiener-Filter. Ins Wiener-Filter geht zusätzlich die Autokorrelation des zu restaurierenden Bildes ein. Vom verschobenen Einheitsimpuls ist die Autokorrelation aber der nichtverschobene Einheitsimpuls. Folglich stimmt formal das Wiener-Filter mit dem Pseudo-Invers-Filter (8) überein. Setzen wir nun formal dekorreliertes Rauschen mit der Varianz σ^2 an, so gilt $\beta = n \cdot \sigma^2 = E(\|N\|^2)$.

In Abbildung 2 sind die Restaurationsergebnisse für verschiedene Regularisationsparameter dargestellt. Je geringer die überlappende Fläche zweier Blöcke ist, desto stärker ist die mittlere Störungsstärke. Der Regularisierungsparameter hängt demnach von der Verschiebung selbst ab und drückt für unser Modell den "Abstand" von g zum verschobenen Einheitsimpuls aus. Für $\beta \to 0$ erhalten wir aus (8) die "reine Entfaltung" und für $\beta \to \infty$ erhalten wir die "reine Kreuzkorrelation" zwischen f' und f. Liegt eine nahezu ideale Faltung vor (Verschiebung gering, Störungen gering), dann wird mit "kleinem" β der Einheitsimpuls nahezu ideal restauriert. In diesem Fall ist für alle möglichen β die Kreuzkorrelation das schlechteste Maß, da sie am meisten vom Einheitsimpuls abweicht (das heißt nicht, daß man nicht dennoch das richtige Maximum findet). Sind die Verschiebungen groß, daher auch die Störungen, dann wird ein "mittleres" β geeignet sein und sowohl Entfaltung als auch die Kreuzkorrelation sind schlechtere Maße. Sind schließich die Störungen so drastisch, daß $\beta \to \infty$, so ist die Kreuzkorrelation das richtige Maß, d.h. in diesem Falle liefert die Kreuzkorrelation noch das dem Einheitsimpuls "ähnlichste" Bild. Für unsere β-abhängigen Modelle bedeutet dies, daß im allgemeinen die Kreuzkorrelation eigentlich das schlechteste aller Modelle ist. In Abbildung (2) ist dies deutlich zu sehen.

Eigentlich müßte man nun ein objektives Maß finden, welches das restaurierte Bild im Vergleich zum Einheitsimpuls bewertet und dann in einem Optimierungsprozeß das optimale β bestimmt. Da ein Entzerrungsalgorithmus oft hierarchisch abläuft, muß der Parameter β von Hierarchieebene zu Hierarchieebene monoton fallen. Wir können uns auch noch anders behelfen, wenn wir ausschließlich Modell (1) und nicht (2) betrachten, indem wir den LSI-Operator **L** mit unserem A-priori-Wissen über die Amplitude in Verbindung bringen: g soll dem verschobenen Einheitsimpuls möglichst nahe kommen, daher ist ideal $|\alpha_k(g)| = constant$. Folglich wählen wir **L** bzw. l derart, daß das Amplitudenspektrum von g zunächst konstant ist und dämpfen dies mit einem "flachen" Tiefpaß, wodurch wir nicht den idealen Einheitsimpuls, sondern eine "relativ steile" Gaußfunktion beschreiben. Folglich wird die folgende Formel (9), die aber nur für Modell (1) gültig ist, ein wichtiger Spezialfall:

$$\alpha_k(g) = const \cdot \frac{\overline{\alpha_k(f)} \cdot \alpha_k(f')}{|\alpha_k(f)| \cdot |\alpha_k(f')|} \cdot e^{-\sigma^2 k^2} \quad , k \neq 0 \ , \ \overline{\alpha_k(f)} \cdot \alpha_k(f) \neq 0 \qquad \mathbf{(9)}$$

Der Parameter σ ist sehr klein zu wählen. Die Formel (9) ohne den Exponential-Term wurde erstmals von Kuglin und Hines (1975) in [7] erwähnt. Die auf (9) basierende Methode heißt *Phasenkorrelation.* Die Robustheit dieser Methode wird in der Literatur nur verbal begründet, die vorliegende Arbeit leitet dagegen modellmäßig die ungewöhnliche Robustheit der Phasenkorrelation als Spezialfall ab. Für die Fundus-Bilder kann das Modell (1) statt (2) angenommen werden, daher können wir auch Formel (9) statt (8) benutzen. Natürlich sollen ähnlich dem Block-Matching Blöcke positioniert werden und mit der Formel (9) ist die lokale Disparität zu bestimmen. Dazu muß aber geklärt werden:

a) Die Wahl der Blockgröße in Abhängigkeit von der Größe der Verschiebung,
b) Die Zuverlässigkeit der Bestimmung des "Peaks", die Bewertung des "Peaks" als Wahrscheinlichkeitsmaß.

Abbildung 2 a) "Reine Entfaltung" b) "richtiger" Parameter c) Reine Kreuzkorrelation

Der eigentliche Entzerrungsalgorithmus wird nun wie folgt angewendet (eine Hierarchie-Ebene reicht aus):

a) Zu Beginn wird ein relativ großer Block (z.B. 256×256) in der Mitte beider Bilder plaziert. Damit kann eine grobe globale Verschiebung bestimmt werden.
b) Nun werden im ersten Frame kleinere Blöcke (64×64) plaziert. Im zweiten Frame werden die zugehörigen Blöcke unter Berücksichtigung der Anfangsverschiebung gesetzt. Dadurch haben die Blöcke eine große Flächenüberlappung. Dies ist auch der Grund, warum die Blöcke relativ klein dimensioniert werden können. Pro Blockpaar werden nun die Verschiebungen mittels (9) zusammen mit dem zugeordneten Wahrscheinlichkeitsmaß ermittelt.
c) Mittels gewichteter Ausgleichsrechnung der einzelnen Blockverschiebungen wird eine affine Transformation zwischen den beiden Bildern bestimmt.

Bei der experimentellen Überprüfung der Metode wurde in allen Beispielen bewußt auf eine Fensterfunktion und Bandpaßfilter verzichtet und es wurde direkt mit den originalen Fundus-Grauwerten gerechnet. Die folgende Aussage hängt natürlich vom Bildinhalt ab, aber im Mittel aller Testergebnisse konnte festgestellt werden, daß von 100 bis 50 Prozent überlappender Fläche zweier Blöcke der Einheitsimpuls immer noch sicher detektiert werden konnte und dies ohne jegliche Vorverarbeitung der Blöcke.

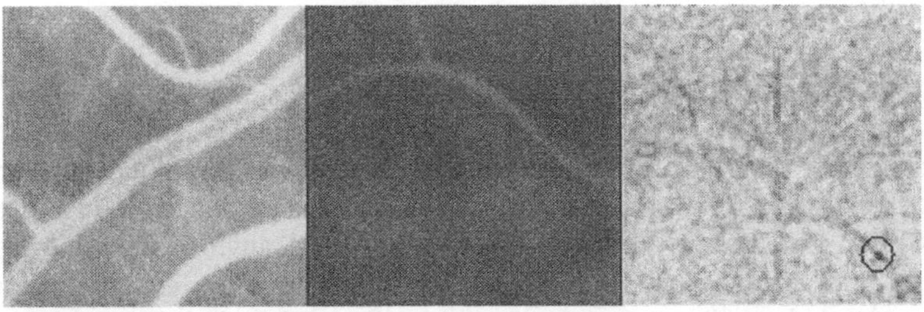

Abbildung 3 Zwei Frames mit massiven Grauwertstörungen, Restauration des Einheitsimpulses mit Formel (9)

4 Affine Transformationen

Sind die geometrischen Verzerrungen nicht mehr lokal durch Verschiebungen modellierbar, so ist es notwendig, mittels eines zentralen großen Blockes eine grobe affine Entzerrung durchzuführen und dann den dargelegten Matchingalgorithmus anzuwenden. Dazu wird die bekannte Eigenschaft genutzt, daß das Amplitudenspektrum translationsinvariant, aber homogen affin variant ist.

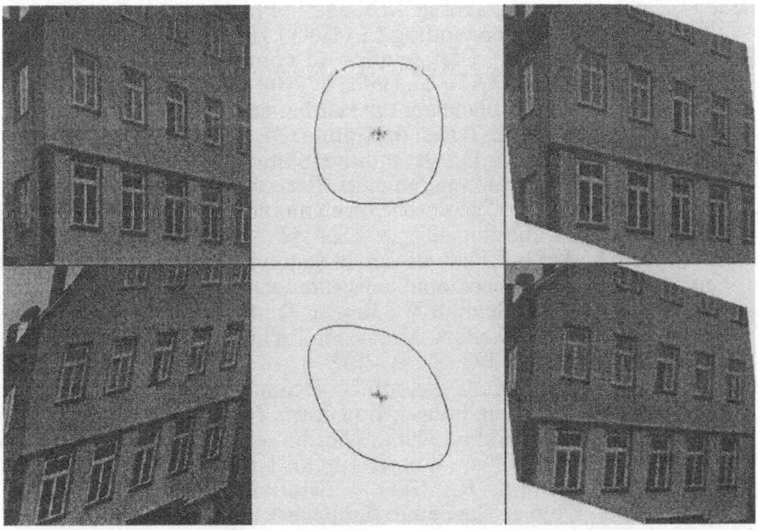

Abbildung 4 Affines Matching von Bild 1 und Bild 4 mittels Schwerpunktkontur. Bild 3 und Bild 6 zeigen das Ergebnis.

Wählen wir als spezielle affine Transformation eine Rotation und isotrope Skalierung, und überführen das Amplitudenspektrum in ein Log-Polar-Bild (Fourier-Mellin-Transformation), so haben wir Rotation und Skalierung auf eine zweidimensionale

Verschiebung im Log-Polar-Bild zurückgeführt, siehe auch De Castro in [2]. Auf Grund der vielen Störungen im Log-Polar-Bild benötigt man jedoch eine robuste Methode zur Translationsbestimmung, wofür sich die in dieser Arbeit dargelegte Methode ideal eignet.
Bezüglich affiner Transformationen beziehen sich die Arbeiten [6,10,15] ebenfalls auf das Amplitudenspektrum, sind aber nicht robust genug. Daher schlagen wir für ein affines Matching eine affin-equivariante Binarisierung der beiden zu vergleichenden Amplitudenspektren vor: Es werden für alle Halbräume jedes Amplitudenspektrums die (grauwertbezogenen) Schwerpunkte ermittelt. Diese Punkte bilden eine zusammenhängende Kontur. Da die Schwerpunkte derselben affinen Transformation wie das Amplitudenspektrum unterworfen sind, ist auch das durch die jeweilige Kontur umgrenzte Objekt B_i durch diese Transformation bestimmt. Bezüglich eines "Normalobjektes" B_0 gilt für die beiden zu vergleichenden Binärobjekte also $B_1=\mathbf{T}_1B_0$ und $B_2=\mathbf{T}_2B_0$. Durch die in [13] beschriebene Normalisierungsmethode können die Transformationen $\mathbf{T}_1$ und $\mathbf{T}_2$ ermittelt werden, so daß das affine Matching mittels $B_2=\mathbf{T}_2\ \mathbf{T}_1^{-1}B_0=\mathbf{A}B_0$ erreicht wird. Die Transformation der Originalbilder ist dann mit $G_2=\mathbf{A}^TG_1$ durchzuführen.

6 Literaturverzeichnis

[1] Brown L.G.:A Survey of image registration techniques: ACM Computing Surveys, Vol. 24, No. 4 (1992) 327-376

[2] De Castro E., Morandi C.: Registration of translated and rotated images using finite Fourier Transform. IEEE Trans. PAMI 9 (1987) 700-703

[3] Fleet D.J., Jepson A.D., Jenkin M.R.M.: Phase-based disparity measurement. CVGIP:Image Understanding 53 (1991) 198-210

[4] Fleet D.J., Jepson A.D.: Computation of component image velocity from local phase information. IJCV 5 (1990) 77-104

[5] Jain J.R., Jain A.K.: Displacement measurement and ist application in interframe image coding. IEEE Trans. on Comm. 29 (1981) 1799-1808

[6] Krüger S.A., Calway A.D.: A multiresolution frequence domain method for estimating affin motion parameters. Proc. ICIP 1996, Vol I, pp.113-116

[7] Kuglin C.D., Hines D.C.: The phase correlation image aligment method. Proc. IEEE on Cybernetics and Society, pp. 163-165, New York 1975

[8] Lee D.J., Mitra S., Krile T.F.: Analysis of sequential complex images, using feature extraction and 2-dimensional cepstrum techniques. JOSAA 6 (1989) 863-870

[9] Lehmann T., Goerke C., Schmitt W., Repges R.: Rotations- und Translationsbestimmung durch eine erweiterte Kepstrum-Technik. Proceedings 17. DAGM- Symposium, Bielefeld 1995, S. 395-402

[10] Licchese L., Cortelazzo G.M., Monti C.: Estimation of affine transformations between images pairs via Fourier transform. Proc. ICIP 1996, pp. 715-718

[11] Petrig B., Bigun J., Curchod M.G.: Motion estimation of ocular fundus images. Proceedings of ICIP'96 , Vol. III, pp. 691-694, 1996

[12] Sezan M.I., Lagendijk R.L.(Eds.) : Motion Analysis and Image Sequence Processing. Kluwer Academic Publishers, London 1996

[13] Voss.K, Süße H.: Invariant fitting of planar objects by primitives. IEEE Trans. PAMI 19 (1997) 80-84

[14] Weng J.J.: Image matching using the windowed Fourier phase. IJCV 11 (1993) 211-236

[15] Xiong Y., Shafer S.A.: Hypergeometric filters for optical flow and affine matching. IJCV 24 (1997) 163-177

Computergestützte Endokarderkennung echokardiographischer Bildsequenzen

A. Teuner [1], J. E. Santos Conde [1], O. Pichler [2] und B. J. Hosticka [1]
[1] Fraunhofer-Institut für Mikroelektronische Schaltungen und Systeme, Duisburg
[2] Bosch Telekom GmbH, Backnang, vormals [1]

Kurzfassung

In der Ultraschalldiagnostik müssen für den Nachweis koronarer Herzerkrankungen mit Hilfe der Streß-Echokardiographie die Bewegungen des Endokards während Systole und Diastole vollständig nachvollziehbar sein. Bedingt durch die zeitlich und örtlich stark verrauschten Aufnahmen in der Echokardiographie ist die Erkennung des Endokards durch den Diagnostiker oftmals nur bei laufender Bildsequenz möglich. Im einzelnen Standbild hingegen ist in Ermangelung einer zeitlichen Interpolationsmöglichkeit der Endokardverlauf nur unvollständig sichtbar. In diesem Beitrag wird ein Verfahren vorgestellt, das unter Verwendung eines an die besonderen Gegebenheiten von Ultraschallsequenzen angepaßten adaptiven Rangordnungsfilters, einer lokalen 3D-Grauwertäqualisation sowie eines Gebietswachstumsverfahrens die automatische Erkennung des Endokards realisiert. Es wird gezeigt, daß zur Rauschunterdrückung die Residuum-Filterung mit Hilfe eines mehrstufigen radialen Median-Filters sehr gute Ergebnisse liefert. Der ermittelte Endokardverlauf wird nach Anhalten der Sequenz in das aktuelle Frame eingeblendet, so daß er auch im Einzelbild für eine Diagnose zur Verfügung steht. Die vorgestellten Simulationsergebnisse zeigen, daß die algorithmisch ermittelten Endokardverläufe gut mit den subjektiv empfundenen Verläufen übereinstimmen.

1 Einleitung

Die Endokarderkennung stellt in der Herzmedizin ein wichtiges Verfahren zum Nachweis koronarer Herzerkrankungen dar. So kann beispielsweise mit Hilfe des erkannten Endokardverlaufs während der systolischen und der diastolischen Phase die Ermittlung der für die Diagnostik wichtigen Parameter Schlagvolumen und Ejektionsfraktion erfolgen [1,2]. In der Praxis erfolgt die Endokarderkennung meist manuell durch Betrachtung der Ultraschallsequenz. Nach Anhalten einer Ultraschallsequenz ist aber in Ermangelung der nunmehr fehlenden zeitlichen Interpolationsmöglichkeit eine ortsgenaue und sichere Erkennung des Endokardverlaufes nicht mehr möglich. Die manuelle Auswertung ist folglich subjektiv und sehr zeitraubend. Zudem hängt die Güte der Auswertung von der Ausbildung und Erfahrung des Diagnostikers ab.

Zur Unterstützung der Diagnostik wird in diesem Beitrag ein automatisches Verfahren zur Endokarderkennung vorgestellt, das nach Anhalten der Ultraschallsequenz den berechneten Endokardverlauf in das aktuelle Frame einblendet. Weil Ultraschallsequenzen verfahrensbedingt hohe Rauschanteile besitzen, werden in Kapitel 2 Methoden zur Rauschunterdrückung basierend auf adaptiven Rangordnungsfiltern vorgestellt, bei denen die Maske des Residuum-Filters den bedeutsamsten Rauschquellen angepaßt ist, so daß sie besonders wirksam zur Verbesserung des

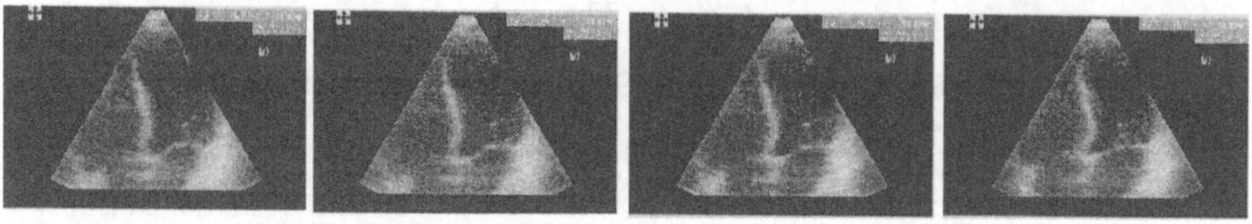

Abb. 1: Zeitlich aufeinanderfolgende Frames einer Ultraschallsequenz. Dargestellt ist der apikale Vierkammerblick des Herzens.

visuellen Eindrucks eingesetzt werden können. Dabei konzentrieren sich die Untersuchungen auf die mit Hilfe des Sector-Scan-Verfahrens gewonnen Ultraschallsequenzen, die insbesondere zur Diagnose akustisch schwer zugänglicher Organe wie z. B. dem Herzen eingesetzt werden. Anschließend werden die Ergebnisse für verschiedene Rausch-Filter vorgestellt und diskutiert, bevor in Kapitel 3 ein zur Kontrastanhebung geeignetes Verfahren, die 3D-Grauwertäqualisation, und ein zur Endokardverlaufserkennung geeignetes Gebietswachstumsverfahren eingeführt werden, die die algorithmische Endokardverlaufserkennung vervollständigen. In Kapitel 4 werden einige Simulationsergebnisse vorgestellt, die die Leistungsfähigkeit des gesamten Algorithmus demonstrieren. Der Beitrag schließt mit der Zusammenfassung der gewonnenen Erkenntnisse und mit einem Ausblick.

2 Rauschunterdrückung in echokardiographischen Sequenzen mit Rangordnungsfiltern

Ultraschallsequenzen besitzen neben technisch bedingten Artefakten, die von den Untersuchungsgeräten verursacht werden, Artefakte, die auf das Prinzip der Echoerfassung zurückgehen und daher nicht vermeidbar sind. Eine der wesentlichsten Störquellen, die sich im visuellen Erscheinungsbild von Ultraschallaufnahmen besonders bemerkbar macht, sind Ultraschallstrahlen, die auf Gewebestrukturen mit einem hohen Impedanzsprung treffen. Diese erfahren eine seitliche Ablenkung, die dazu führt, daß dem Sende-/Empfangskristall benachbarte Kristalle die Echosignale ebenfalls empfangen. Dieser Effekt macht sich in Ultraschallbildern, wie sie Abb. 1 zeigt, durch eine zum Sende-/Empfangskopf tangentiale Verschmierung bemerkbar.

Die zur Entstehung einer Ultraschallsequenz $g(x,y,t)$ beitragenden Prozesse besitzen in örtlicher als auch in zeitlicher Dimension in hohem Maße einen nichtstationären Charakter. Setzt man voraus, daß der Rauschprozeß $n(x,y,t)$ statistisch unabhängig von der Originalsequenz $f(x,y,t)$ sei [3], so erhält man für die Ultraschallsequenz folgendes Rauschmodell:

$$g(x,y,t) = f(x,y,t) + n(x,y,t)\,. \qquad (2.1)$$

Eine Bildverbesserung mit Hilfe linearer Bildrestaurationsverfahren, wie z. B. der Wiener-Filterung, führt bei Ultraschallsequenzen in der Regel nicht zum Erfolg, weil hierfür die *a priori* Kenntnisse der 3D-Autokorrelationsfunktion der Originalfunktion $f(x,y,t)$ notwendig sind. Ferner besitzen die Bildsequenzen unterschiedliche örtliche und zeitliche statistische Abhängigkeiten. Während diese im Ortsbereich in der Regel nichtstätionärer Natur sind, so existieren in statischen Bildregionen, in denen keine Bewegung auftritt, durchaus hohe zeitliche Korrelationen, in dynamischen Bereichen mit starken Bewegungen hingegen überwiegt wiederum das nichtstationäre

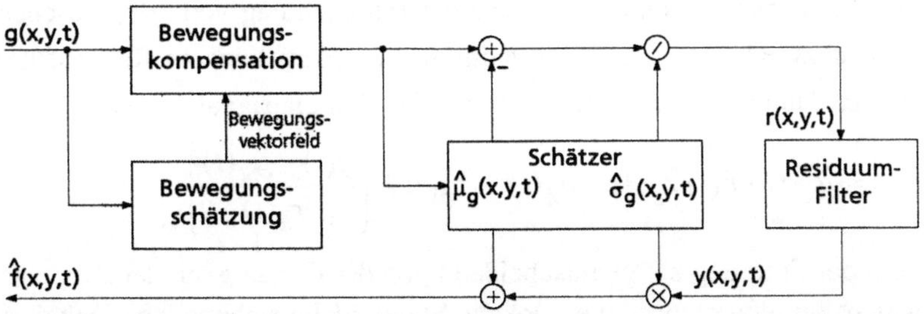

Abb. 2: Signalflußdiagramm des adaptiven Rangordnungsfilter. Die grau hinterlegte optionale Bewegungskompensation ist aufgrund ihrer Rauschsensitivität nicht implementiert.

Verhalten des dreidimensionalen Ultraschallsignals. Letzterem kann zwar durch bewegungskompensierende Verfahren zur Rauschunterdrückung entgegengewirkt werden. Da aber - wie auch Untersuchungen im Rahmen dieser Arbeit gezeigt haben - Algorithmen zur Bewegungsschätzung aufgrund des Blenden- und Korrespondenzproblems [4] gegenüber den in Ultraschallsequenzen auftretenden Rauscheinflüssen sehr sensitiv sind, konzentrieren sich die von uns durchgeführten Arbeiten auf die Betrachtung der Klasse dreidimensionaler adaptiver Rangordnungsfilter, die zum einen neben starken zeitlichen auch örtliche Korrelationen innerhalb einer Bildsequenz berücksichtigen und zum anderen über die für die Ultraschalldiagnostik eminent wichtige Eigenschaft verfügen, Bildkanten zu erhalten [5]. Gegenüber anderen Filtertypen besitzen Rangordnungsfilter zudem den Vorteil, zu einem Objekt gehörige Bildpunkte zu gruppieren. Dadurch bleiben Kanten und konstante Nachbarschaften erhalten, impulsförmige Rauscheinflüsse hingegen werden eliminiert.

Die Idee der adaptiven Rangordnungsfilterung beruht auf der Zerlegung der instationären Pixeltrajektorien $g(x,y,t) = g_{x,y}(t)$ in einen rauschfreien nichtstationären Signalanteil und eine im weitesten Sinne stationäre rauschbehaftete Signalkomponente, die Residuum genannt wird [6 - 9]. Dieses wird dann mit einem gewöhnlichen linearen oder nichtlinearen Filter weiterverarbeitet, siehe Abb. 2. Hierzu wird das instationäre Signal wie folgt zerlegt:

$$g(t) = \mu_g(t) + \sigma_g(t) y(t), \tag{2.2}$$

wobei der Einfachheit halber $g_{x,y}(t) = g(t)$ gesetzt wird. $\mu_g(t)$ kennzeichnet den lokalen Mittelwert des zeitlich instationären Signals $g(t)$ und $\sigma_g(t)$ die lokale Standardabweichung zum Zeitpunkt t. Das Residuum $y(t)$ besitzt einen zeitinvarianten Mittelwert von Null und eine zeitinvariante Standardabweichung von Eins. Setzt man voraus, daß das Rauschen den Mittelwert Null hat, so ist der lokale Mittelwert $\mu_g(t)$ rauschfrei und das Residuum $y(t)$ enthält vollständig den stätionären Rauschanteil. Dann gilt:

$$y(t) = \frac{f(t) - \mu_g(t)}{\sigma_g(t)} + \frac{n(t)}{\sigma_g(t)} = r(t) + v(t). \tag{2.3}$$

Da in der Praxis das Residuum $y(t)$ aus der Beobachtung von $g(t)$ geschätzt wird, und der Schätzwert $\hat{y}(t)$ für das Residuum mit einem linearen Rauschfilter $F\{\circ\}$ gefiltert wird, lautet die Schätzung $\hat{f}(t)$ für das Originalsignal $f(t)$:

$$\hat{f}(t) = \hat{\mu}_g(t) + \hat{r}(t)\hat{\sigma}_g(t) = \hat{\mu}_g(t) + \hat{\sigma}_g(t)\, F\left\{\frac{g(t) - \hat{\mu}_g(t)}{\hat{\sigma}_g(t)}\right\}. \tag{2.4}$$

Der Erfolg der Filterung hängt entscheidend von der Genauigkeit der Schätzungen für den lokalen Mittelwert und die lokale Standardabweichung ab. Dabei muß der Schätzalgorithmus so konzipiert sein, daß er insensitiv gegenüber Rauschen und sensitiv für Kanten ist. Den einfachsten Ansatz zur Schätzung der lokalen Statistiken stellt das Box-Averaging-Verfahren dar, bei dem innerhalb einer kubischen Nachbarschaft der lokale Mittelwert und die lokale Standardabweichung bestimmt werden. Innerhalb instationärer Bildbereiche, z. B. als Folge des Auftretens einer Bildkante, sind die Schätzungen dieser beiden Parameter jedoch zu ungenau. Außerdem wird eine kubische Schätzmaske den spezifischen Gegebenheiten von Ultraschallsequenzen, daß tangentiale Verschmierungen in Bildsektor vorliegen, nicht gerecht. Aus diesem Grund ist die von uns verwendete Schätzmaske an der Hypotenuse von $g(t)$ ausgerichtet und, wie Abb. 3 zeigt, in radialer Richtung größer gewählt als in tangentialer Richtung, um das in tangentialer Richtung korrelierte Hintergrundrauschen nicht zu hoch zu bewerten. Die lokalen Statistiken für die Ensemblewerte $g_{(i)}(t)$, die für die Residuum-Filterung notwendig sind, können durch lineare Regression ermittelt werden, wenn die Verteilung der Grauwerte auf der Pixeltrajektorie bekannt ist. Da die zur Bestimmung der Regressionsgeraden notwendigen Ensemblewerte in der Regel nicht bekannt sind, werden diese unter der Voraussetzung lokaler Ergodizität durch die lokalen Orts-Zeit-Werte ersetzt. Die Verteilung der Residuen für jede Pixeltrajektorie kann adaptiv ermittelt werden [10]. Innerhalb statischer Bildbereiche entspricht dann die Wahrscheinlichkeitsdichtefunktion des Rauschprozesses der Wahrscheinlichkeitsdichtefunktion der Pixeltrajektorie und der lokale Mittelwert dem konstanten Grauwert des Hintergrundes, da mittelwertfreies Rauschen angenommen wurde. Außerdem ist die lokale Standardabweichung identisch mit der Standardabweichung des Rauschprozesses. Innerhalb aktiver Pixelbereiche ist die Wahrscheinlichkeitsdichtefunktion schwieriger zu ermitteln. Um darüber hinaus den Einfluß von Ausreißern zu minimieren, wurde das Verfahren der linearen Regression [9] um die in [10] beschriebene Methode des *Iterated Weighted Least Square* (IWLS) erweitert. Hierbei werden die sortierten Ensemblewerte $g_{(i)}(t)$, die sog. Ranks, gewichtet, um bei der Schätzung der Originalsequenz auch zu berücksichtigen, daß es sich auch bei dem aktuellen Pixel $g(t)$ um einen Ausreißer handeln kann. Die Filterstruktur nach Gl. (2.4) wird dann wie folgt modifiziert:

$$\hat{f}(t) = \hat{\mu}_g(t) + \hat{\sigma}_g(t)\, F\left\{\frac{g(t) - \hat{\mu}_g(t)}{\hat{\sigma}_g(t)}\, w_g^2(t)\right\}, \tag{2.5}$$

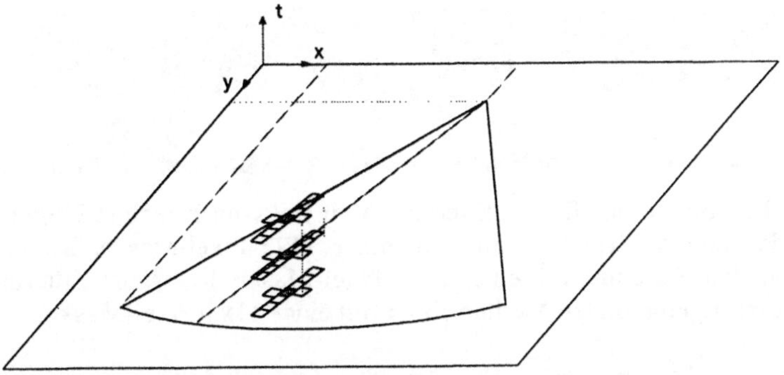

Abb. 3: 3D-Schätzmaske zur adapativen Rangordnungsfilterung.

wobei $w_g^2(t)$ den im IWLS-Algorithmus berechneten Gewichten von $g(t)$ entsprechen. Anders als in [10] beschrieben basiert die Schätzung des Initialvektors $\hat{\vec{\theta}}_0(t)$ hier auf der Minimierung der euklidschen Norm des Residuenvektors

$$\vec{\varepsilon}(t) = \vec{g}_{(\,)}(t) - [\,\vec{1} \quad E\{\vec{y}_{()}(t)\}\,]\cdot[\,\mu_g(t) \quad \sigma_g(t)\,]^T \equiv \vec{g}_{(\,)}(t) - \mathbf{A}\cdot\vec{\theta}(t)\,, \tag{2.6}$$

so daß der Initialvektor folgendermaßen bestimmt werden kann:

$$\vec{\theta}_0(t) = (\mathbf{A}^T\mathbf{C}_{yy}^{-1}\mathbf{A})^{-1}\,\mathbf{A}^T\,\mathbf{C}_{yy}^{-1}\,g_{()}(t)\,. \tag{2.7}$$

$\mathbf{C}_{yy}$ bezeichnet die Kovarianzmatrix des Vektors $\vec{y}_{()}(t)$. Diese Maßnahme führt zu einer schnelleren Konvergenz.

Für die Residuum-Filterung wurde ein mehrstufiges radiales Median-Filter gewählt, da dieses ungeachtet einer fehlenden Bewegungskompensation fein strukturierte Bildmerkmale bewahrt. In der ersten Stufe der Filterkaskade erfolgt eine zeitliche und radiale Median-Filterung. Anschließend wird der Median aus dem aktuellen Bildpunkt, dem zeitlichen und dem räumlichen Medianwert gebildet. Abb. 4 zeigt vergleichsweise die Filterergebnisse für ein Frame der Ultraschallsequenz. Eine einfache örtliche radiale Median-Filterung führt gegenüber einer mehrstufigen radialen Median-Filterung mit zeitlichen Fenstergröße von 3 Frames zu schlechteren Ergebnissen, weil zeitliche Korrelationen der Bilddaten nicht berücksichtigt werden. Eine deutliche Verbesserung wird durch eine Residuum-Filterung unter Verwendung des mehrstufigen radialen Median-Filters erzielt.

3 Endokarderkennung mit Gebietswachstumsverfahren

Zur Endokarderkennung müssen die Frames einer Ultraschallsequenz in zwei homogene Bildbereiche unterteilt werden, den linken Ventrikel und den Rest des Herzens. Das Endokard entspricht dann dem Verlauf der Grenzlinie der segmentierten Bildbereiche. Diese Forderung impliziert, daß keine isolierten Gebiete entstehen dürfen und die Segmentierung regionenorientiert erfolgen muß. Daher wurde ein Gebietswachstumsverfahren in Anlehnung an den in [11] vorgestellten Ansatz implementiert.

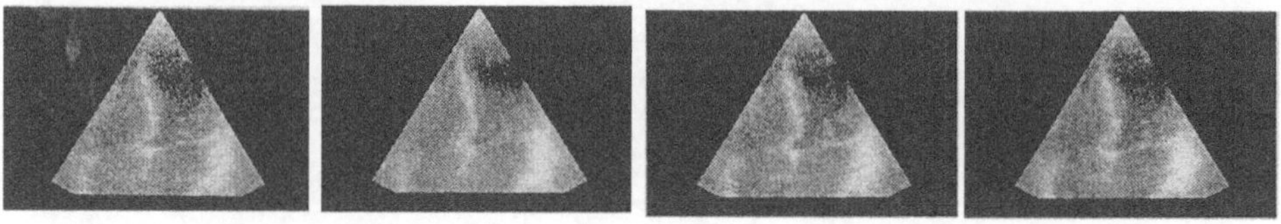

Abb. 4: Ergebnisse der Rauschfilterung. Von links nach rechts: Ungefilterter Frame. Radiale Median-Filterung mit einer 11x3-Pixel-Maske. Mehrstufige radiale Median-Filterung mit einer 11x3-Pixel-Maske. Residuum-Filterung mit dem mehrstufigen radialen Median-Filter mit einer 11x3-Pixel-Maske.

Zunächst werden die Repräsentanten R_i einer Region, die sog. Saatpunkte mit $1 \leq i \leq N$ definiert, die als Initialpunkte für das Gebietswachstum verwendet werden. N entspricht der Anzahl Regionen, in die das Bild unterteilt werden soll. Dann erfolgt die Segmentierung in Abhängigkeit von der Positionierung der Saatzellen wie folgt: ein Pixel wird einem der Bereiche R_i zugeordnet. Nun sei T die Menge der Bildpunkte ohne bisherige Zuordnung, die an mindestens eine Region angrenzen:

$$T = \left\{ \vec{p} \in \bigcup_{i=1}^{N} R_i \,|\, N(\vec{p}) \cap \bigcup_{i=1}^{N} R_i \neq \varnothing \right\}. \tag{3.1}$$

$N(\vec{p})$ entspricht der 8er-Nachbarschaft des Pixels $\vec{p}$. Grenzt ein Pixel der Menge T nur an eine Region an , so wird es dieser Region zugeordnet. Andernfalls wird es der angrenzenden Region zugeordnet, für die das Kriterium

$$\xi(\vec{p}) = \left| \hat{f}(\vec{p}) - \text{Mittelwert}(\vec{r} \in R_{i(\vec{p})}) \, [\hat{f}(\vec{r})] \right| \tag{3.2}$$

minimal ist, wobei $i(\vec{p}) \in \{1, 2, \ldots, N\}$ den Index bezeichnet, für den gilt:

$$N(\vec{p}) \cap R_{i(\vec{p})} \neq \varnothing \,. \tag{3.3}$$

Nach Zuordnung des Pixels wird der Mittelwert normalerweise neu berechnet. Da aber im vorliegenden Fall die Saatpunkte die Repräsentanten der Regionen darstellen, erfolgt keine Aktualisierung der Mittelwerte der veränderten Regionen. Ferner wurde das Zuordnungskriterium (Gl. 3.2) derart modifiziert, daß nicht der Grauwert eines einzelnen Pixels, sondern der Mittelwert seiner Umgebung verwendet wird. Dadurch wird die Robustheit der Methode erhöht. Das Gebietswachstum wird dann solange fortgesetzt, bis alle Bildpunkte einer Region zugeordnet sind. Zur weiteren Erhöhung der Robustheit wurden Saatzellen anstelle von Saatpunkten verwendet. Liegt beispielsweise ein Saatpunkt auf einem Ausreißer, wird das Segmentierungsergebnis negativ beeinflußt. Weil die Lage des linken Ventrikels im apikalen Vierkammerblick bekannt ist, kann die Endokarderkennung automatisch erfolgen. Die erste Saatzelle wird in den Bereich gelegt, der mit maximaler Wahrscheinlichkeit dem linken Ventrikel zugeordnet werden kann. Die restlichen Saatzellen werden außerhalb des linken Ventrikels gesetzt. Nach der Segmentierung werden die Regionen R_i mit $i > 1$ miteinander verschmolzen. Die Verwendung mehrerer Saatzellen führt zu einer genaue-

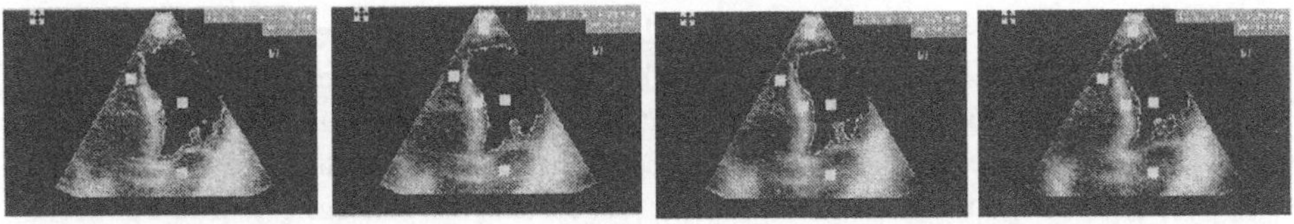

Abb. 5: Simulationsergebnisse der zeitlich aufeinanderfolgenden Frames einer Ultraschallsequenz.

ren und stabileren Endokarderkennung. Alternativ ließe sich die Lage der Saatzellen auch interaktiv durch eine graphische Eingabe vom Diagnostiker definieren.
Die Endokarderkennung kann qualitativ verbessert werden, wenn das Gebietswachstum um eine vorherige Grauwertäqualisation erweitert wird [12]. Hierzu wird die geschätzte Bildsequenz in eine Sequenz mit gleichverteilten Grauwerten transformiert. Die Transformationskennlinie wird hier aus der vollständigen Bildsequenz abgeleitet, wobei Bereiche außerhalb des Ultraschallsektors ausgeblendet werden.

4 Simulationsergebnisse

Die visuelle Erkennung des tatsächlichen Endokardverlaufes ist ein subjektiver Prozeß und hängt in hohem Maße von der Erfahrung des Diagnostikers ab. Ein potentielles objektives Kriterium zur Bewertung des Segmentierungsergebnisses wäre daher die Bestimmung der Flächendifferenz zwischen den visuell gefundenen und den algorithmisch bestimmten Regionen.
In Abb. 5 wird die Leistungsfähigkeit des Verfahrens illustriert. Der Endokardverlauf und die verwendeten Saatzellen wurden in die rauschreduzierte und äqualisierte Sequenz eingeblendet. Der algorithmisch ermittelte Endokardverlauf stimmt mit dem subjektiv empfundenen gut überein. Feine und für die Diagnose wichtige Strukturen wie die Mitralklappe bleiben nach der Filterung erhalten. Damit erhält man auch die Möglichkeit, durch Auswertung der end-systolischen und end-diastolischen Bilder das Schlagvolumen und die klinisch bedeutsame Ejektionsfraktion automatisch zu bestimmen. Die eingesetzten Algorithmen wurden unter Khoros/Unix implementiert. Die Dauer für die Rauschfilterung bei Verwendung des adaptiven Rangordnungsfilter einschließlich Grauwertäqualisation und Konturerkennung betrug auf einer SPARC-20-Workstationen für 28 Bilder mit einer Größe von 312 x 208 Pixeln ca. 3-5 min. Eine Echtzeitfähigkeit der Bildverarbeitung ist aber nicht erforderlich, da die Diagnose üblicherweise off-line erfolgt.

6 Zusammenfassung und Ausblick

In diesem Beitrag wurde eine neuartiges Verfahren zur computergestützten Endokarderkennung in digitalen echokardiographischen Bildfolgen vorgestellt, das nach Anhalten der Sequenz den algorithmisch ermittelten Endokardverlauf in die Einzelbilder einblendet. Zur Verbesserung des visuellen Eindrucks als auch der Resultate der Bildsegmentierung wurde ein auf die spezifischen Eigenschaften von Ultraschallbildern zugeschnittenes Rauschfilter verwendet. Auf die bewegungskompensierte Rauschfilterung wurde dabei bewußt verzichtet, weil eine Bewegungsschätzung

im Fall stark verrauschter Eingangssignale zu fehlerhaften Bewegungstrajektorien führen, wodurch sich die Schätzungen für die Originalsequenz verschlechtern.
Aufgrund zeitlicher und örtlicher Instationaritäten der Bildsequenzen wurde hier ein adaptives Rangordnungsfilter verwendet. Diese Filter besitzen den Vorteil, eine Segmentierung der Signale vorzunehmen. Für die Residuum-Filterung wurde ein mehrstufiges radiales Median-Filter eingesetzt, daß die bildgebenden Merkmale einer Ultraschallsequenz in besonderem Maße berücksichtigt. Gegenüber konventionellen nichtlinearen Rauschfiltern konnten die Bildsequenzen deutlich verbessert werden ohne fein strukturierte Objekte zu eliminieren.
Zur Endokarderkennung wurde ein Gebietswachstumsverfahren verwendet, daß automatisch oder interaktiv die Einblendung des Endokards ermöglicht. Die Qualität der Konturerkennung konnte durch eine vorherige 3D-Grauwertäqualisation der gefilterten Ultraschallsequenzen verbessert werden. In experimentellen Untersuchungen konnte gezeigt werden, daß die algorithmisch ermittelte Lage des Endokards gut mit dem subjektiv empfundenen Verlauf übereinstimmt.

Danksagung

Für die freundliche Bereitstellung des Bildmaterials sowie für zahlreiche Diskussionen und Anregungen, die zur Erstellung dieser Arbeit wesentlich beitrugen, möchten wir uns herzlich bei Herrn Dr. R. Ott, Städtische Krankenanstalten Krefeld, und Herrn J. Hackstein, Städtisches Krankenhaus Solingen, bedanken.

Literaturverzeichnis

[1] G. Sold: Klinische Doppler Echokardiographie. Stuttgart: Schattauer-Verlagsgesellschaft 1988.

[2] J. Hackstein, O. Pichler und A. Teuner: „Rechnergestützte Analyse von Ultraschallsequenzen des linken Ventrikels zur Bestimmung der Ejektionsfraktion", Biomedizinische Technik 41, S. 118 - 122, 1996.

[3] H. Peters, K.-H. Deeg und D. Weitzel: Die Ultraschalluntersuchung des Kindes. 3. Nachdruck. Berlin: Springer Verlag 1994.

[4] B. Jähne: Digitale Bildverarbeitung. 3. Auflage. Berlin: Springer Verlag 1993.

[5] J. C. Brailean, R. P. Kleihorst, S. N. Efstratiadis, A. K. Katsaggelos und R. L. Lagendijk: „Noise Reduction Filters for Dynamic Images Sequences: A Review", Proceedings of the IEEE, Vol. 83, No. 9, S. 1272 - 1292, 1995.

[6] R. P. Kleihorst, R. L. Langendijk und J. Biemond: „Noise Reduction of Image Sequences Using Motion Compensation and Signal Decomposition", IEEE Transactions on Image Processing, Vol. 4, No. 3, S. 274 - 284, 1995.

[7] A. K. Katsaggelos, R. P. Kleihorst, S. N. Efstratiadis und R. L. Lagendijk: „Adaptive Image Sequence Noise Filtering Methods", SPIE Proceedings Visual Communications and Image Processing, Vol. 1606, S. 716 - 727, 1991.

[8] R.-P. Kleihorst, J. Biemond und R. L. Lagendijk: „;Motion-Compensated Filtering of Noisy Image Sequences", Proceedings of the 4th International Workshop on Time-Varying Image Processing and Moving Object Recognition, Florence, S. 75 - 86, Juni 1994.

[9] E. H. Loyd: „Least-Square Estimation of Location and Scale Parameters Using Order Statistics", Biometrika, Vol. 39, S. 88 - 95, 1952.

[10] R. P. Kleihorst, R. L. Langendijk und J. Biemond: „Noise Reduction of Severely Corrupted Image Sequences", Proceedings of the IEEE International Conference on Acoustics, Speech and Signal Processing, Tampa (Florida), S. 375 - 378, 1985.

[11] R. Adams und L. Bischof: „Seeded Region Growing", IEEE Transactions on Pattern Analysis and Machine Intelligence, Vol. 16, No. 6, S. 641 - 647, 1994.

[12] W. K. Pratt. Digital Image Processing. Second Edition. New York: John Wiley & Sons, 1991.

Verfahren zur Erkennung der Herzkontraktion anhand von MRT-Aufnahmen

Christian Lenz[1], Rainer Brokordt[1] und Peter Barth[2]
[1] Universität GH Paderborn, Computergraphik, Visualisierung und Bildverarbeitung, Fürstenallee 11, D-33102 Paderborn
[2] Institut für Molekulare Biophysik und Radiopharmazie, Universitätsklinik der Ruhr-Universität Bochum, Herz- und Diabeteszentrum NRW, Georgstr. 11, D-32545 Bad Oeynhausen

Zusammenfassung: Die Untersuchung der Herzbewegung ist ein fundamentaler Schritt zum Verständnis von Herzmuskelerkrankungen. Die Erkennung und die anschließende visuelle, quantitative Darstellung der Bewegung erlauben es, frühzeitig Rückschlüsse über die Erkrankung des Herzmuskelgewebes zu ziehen. Wir präsentieren hier ein Verfahren zur Erkennung der Herzbewegung aus Tagged-MRT-Bildsequenzen und die anschließende Darstellung der Bewegung.

1. Einleitung

Die Untersuchung der Herzbewegung ist ein fundamentaler Schritt zum Verständnis der Physiologie des normalen Herzens und der Erkennung von Herzmuskelerkrankungen.

Invasive Methoden haben gezeigt, daß diese Bewegung aus komplizierten Drehungen und ungleichmäßigen Ausdehnungen des Herzmuskels besteht. So sind beispielsweise die Herzwanddicke und die Herzwandlänge Standardgrößen, durch deren Veränderungen die Herzfunktion analysiert wird. Mit der Entwicklung der Magnet-Resonanz-Tomographie wurde es möglich, beliebige Schichten des Herzens aufzunehmen. Durch die Technik des *Tagging*[1] ist es möglich geworden, bestimmte Bereiche des Herzens nichtinvasiv zu markieren und bei der Bewegung zu verfolgen. Die Markierung zeigt sich in Form eines Gitters im Bild, das mit der Herzbewegung verzerrt wird und somit eine Bewegungsverfolgung erlaubt.

Ziel dieser Arbeit ist es, aus vorliegenden Tagging-Aufnahmen die Bewegung des Herzens zu erkennen und entsprechend zu visualisieren. Datengrundlage sind lediglich die Bilder des Scanners. Da die Tagging-Struktur sehr schnell in einer Aufnahmesequenz an Qualität verliert, ist es notwendig, dem Benutzer die Möglichkeit zu einer manuellen Korrektur bei der Verfolgung der Gitterkreuzungspunkte zu geben.

In dieser Anwendungsarbeit wird auf bekannte und bewährte Methoden und Algorithmen zur Erkennung und Visualisierung der Bewegung zurückgegriffen.

---

[1] „Tagging": (engl.) identifizierend, markierend; bewirkt eine Gitterstruktur im Bild, siehe Kap. 2

2. Tagged-MRT

Um qualitative und quantitative Aussagen über die Bewegung eines Herzens machen zu können, ist es notwendig eine Sequenz von aufeinanderfolgenden Bildern zu akquirieren. Betrachtet man eine solche Sequenz der Herzbewegung in einer Animation, so sind aufgrund der recht komplexen Verschiebungen und Verdrehungen des Herzmuskels nur qualitative Aussagen möglich. Um konkrete Angaben über die Herzbewegung machen zu können, müssen Bereiche des Herzens markiert werden, um sie von einem Bild zum nächsten in einer Sequenz verfolgen zu können. Das Tagging macht es möglich, diese Markierungen unter *in vivo* Bedingungen und *nichtinvasiv* zu erzeugen.

Die MR-Markierungen (Tags) werden in einer sog. Präparationsphase vor der eigentlichen Bildaufnahme erzeugt. Durch Einstrahlen von frequenzselektiven RF-Impulsen in Verbindung mit linearen Gradientenfeldern wird dir longitudinale Magnetisierung in bestimmten Schichten (Tagging-Regionen) zerstört. In den orthogonal zu diesen Schichten aufgenommenen Bildern erscheinen diese Regionen als dunkle Streifen im Gewebe. Eine Bewegung des Gewebes hat somit auch eine Veränderung der Streifen zur Folge. Ort, Anzahl, Breite und Signalintensität der Tagging-Regionen können durch Modulation der RF-Impulse gesteuert werden. Die Signaldifferenz zwischen Tagging-Regionen und restlichem Gewebe nimmt allerdings mit der Zeit ab, da sich die Magnetisierung in den Tagging-Regionen langsam wieder aufbaut. Die Abbildung 1 zeigt die ersten 6 Aufnahmen einer entsprechend akquirierten Bildsequenz wie sie als Datengrundlage in dieser Arbeit dient.

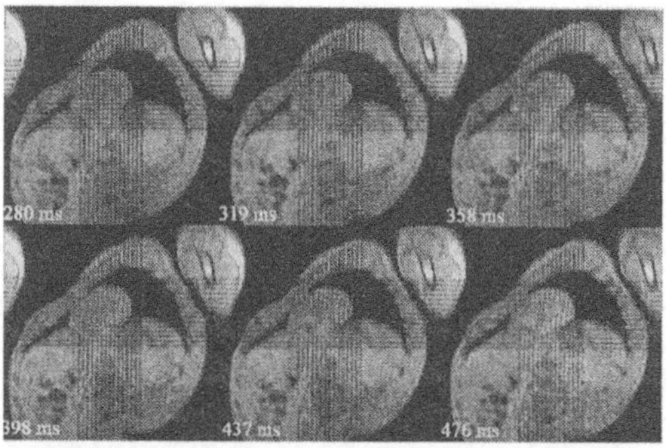

Abbildung 1: Tagging-Aufnahme des Herzen

3. Bisherige Auswertung

Die zur Zeit gebräuchlichen Verfahren zur Erkennung und Visualisierung der Herzbewegung aus Tagged-MRT-Aufnahmen basieren auf der Änderung der Herzwanddicke. Diese ergeben sich aus den Längenänderungen von Tagging-Linien, welche durch die Herzinnenwand (Endokard) und Herzaußenwand (Epikard) begrenzt werden.

In [BSD91] wird ein Tagging-Gitter zur Bestimmung der Herzwanddicke und Herzwandverkürzung benutzt. Dazu werden manuell Punkte in den Aufnahmen der

Herzwände identifiziert, welche zum Ende der Diastole und der Systole akquiriert wurden. Die Abbildung 2 zeigt ein schematische Darstellung einer Tagging-Aufnahme zur Bestimmung dieser Gitterpunkte. So wird z.B. die Wanddicke der linken Herzkammer durch den Abstand der Punkte x,X und y,Y voneinander bestimmt. Bei der Abstandsberechnung werden Krümmungen der Tagging-Linien zwischen diesen Punkten nicht berücksichtigt.

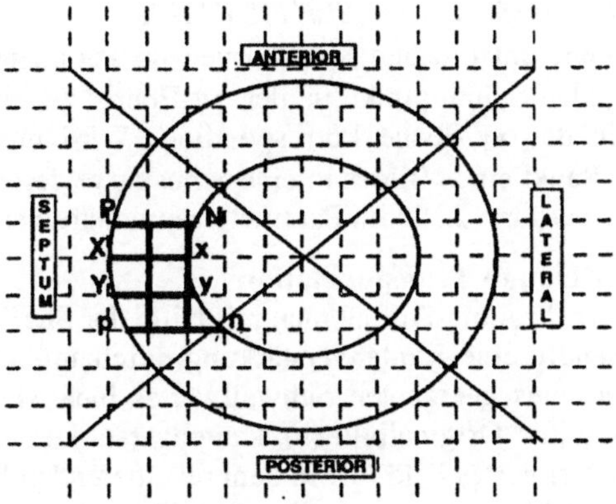

Abbildung 2: Schematische Darstellung der linken Herzkammer [BSD91]

Zur genauen Untersuchung müßten alle Kreuzungspunkte des Tagging-Gitters identifiziert, verfolgt und in ihrer Bewegung vermessen werden. Der Aufwand hierfür ist offensichtlich und motiviert das große Interesse an einem automatischen Verfahren. Es gibt noch eine Reihe weiterer Verfahren, die meistens parallel [AD89] oder sternförmig [ZPR88] verlaufende Tagging-Linien auswerten. Allerdings werden zur Erkennung und Verfolgung dieser Linien die Anfangspunkte und die zu erwartenden Signalintensitäten anhand der Parameter der verwendeten MR-Tomographen berechnet. Eine Erkennung lediglich auf Basis des Bildmaterials erfolgt nicht. Diese wissensbasierte Berechnung und Verfolgung der Tags verwendet dann Verfahren wie *„Snakes"* [ACC94] oder die Verfolgung der Tagging-Zwischenräume wie in [HFH94] beschrieben.

4. Das Verfahren

Im Gegensatz zu den im Kapitel 3 angesprochenen Verfahren bedient sich das im folgenden vorgestellte Programm nur der vom Tomographen gelieferten Bilder, die ein Größe von 256x256 Pixel, mit je 2 Byte/Pixel haben. Das verwendete Tagging-Muster besteht aus zwei senkrecht aufeinander stehenden Gruppen paralleler Linien. Die Kreuzungspunkte des Gitters werden benutzt, um die Bewegung zu erkennen und nachfolgend zu visualisieren. Das Herz wird in Kurzachsenschnitten dargestellt. Zur Erkennung der Herzbewegung sind drei Schritte notwendig.

1. Erkennen der Kreuzungspunkte des Gitters in der ersten Aufnahme
2. Verfolgen der erkannten Punkte in den nachfolgenden Aufnahmen
3. Ermitteln der Verschiebung der Gitterpunkte

4.1 Erkennung der Kreuzungspunkte

Bedingt durch die Qualität der Bilder ist ein direktes Auffinden der Gitterkreuzungspunkte, beispielsweise anhand der Grauwerte, nicht möglich. Um eine gute Startposition zu erhalten, werden die noch fast unverzerrten, gerade verlaufenden Linien im ersten Bild der Sequenz gesucht. Die dennoch auftretenden Abweichungen der ermittelten Punkten von der tatsächlichen Position läßt sich mit der später beschriebenen Verfolgung korrigieren. Zur Erkennung der geraden Linien dient die *„Straight Line Hough Transformation (SLHT)"* [Va94.

Punkten in der xy-Ebene entsprechen Sinuskurven im Parameterraum (r,φ) [Va94]. Dies bedeutet, daß jeder Schnittpunkt in der $r\varphi$-Ebene zu einer Linie durch die kollinearen Punkte in der xy-Ebene korrespondiert. Wird nun durch die Hough-Transformation jeder Punkt eines Bildes in eine Sinuskurve der $r\varphi$-Ebene konvertiert, wird die entsprechend berührte Zelle im Parameterraum inkrementiert.

4.1.1 Anwendung der Hough Transformation

Um in den vorliegenden Tagging-Aufnahmen mit Hilfe der SLHT die Tagging-Linien zu finden ist es notwendig eine Kantenverstärkung durchzuführen. Danach wird das erhaltene Bild mittels eines geeigneten Schwellwertes binarisiert. Die Abbildung 3 zeigt die einzelnen Schritte: Originalbild (a), kantenverstärktes Bild (b), binarisiertes Bild (c), Hough-Transformiertes Bild (d). In dem resultierenden Hough-Array werden dann die entsprechenden Maxima gesucht Die Abbildung 3 (e) zeigt das Resultat der Schnittpunktberechnungen der Tagging-Linien. Hier zeigt sich deutlich, daß nach diesem ersten Schritt nicht nur die Weiterverfolgung der Punkte notwendig ist, sondern auch die Überprüfung ihrer Gültigkeit. Dies ist in der ersten Aufnahme vor allem für den Bereich der Lungen und im weiteren Verlauf für den Bereich der Herzkammer notwendig[2].

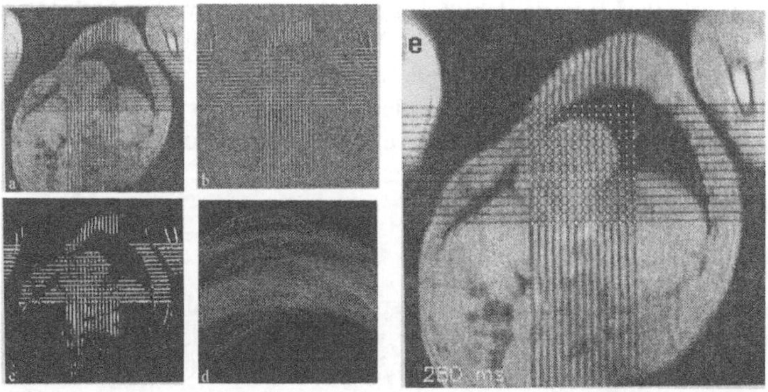

Abbildung 3: SLHT einer Tagging-Aufnahme (a-d) und erkannte Kreuzungspunkte(e)

[2] Durch Blutfluß sehr starke Bewegung, die aber nicht interessiert, bzw. nicht relevantes Gewebe.

4.2 Verfolgung der Kreuzungspunkte

Bei der Verfolgung der Kreuzungspunkte ergibt sich ein wesentliches Problem: Verschiebt sich ein Punkt von einem Bild zum nächsten so weit, daß er in die Nähe der ursprünglichen Position eines benachbarten Punktes gerät, so stellt sich die Frage, welchem der beiden Kreuzungspunkte die neue Position zuzuordnen ist. Um dieses Problem zu umgehen ist es nicht nur notwendig, die Position der Kreuzungspunkte, sondern auch den kompletten Verlauf der Tagging-Linien zu speichern. Ebenso spielt bei der Verfolgung der Kreuzungspunkte der Abstand der einzelnen Tagging-Linien eine Rolle. Ist die Verschiebung größer als die Hälfte des Gitterabstandes, ist keine Zuordnung mehr möglich.

Zur weiteren Erkennung der Kreuzungspunkte ist die SLHT nicht mehr verwendbar, da die Linien nicht mehr gerade sondern verzerrt verlaufen. Zur Verfolgung der Gitterlinien, deren Anfangsposition bekannt ist, werden deshalb *„Snakes"* benutzt.

4.2.1 Snakes

Snakes oder *„Aktive Kontur Modelle"* wurden als erstes von Kass, Witkin und Terzopoulos in [KWT88] vorgestellt. Sie stellen Spline-Kurven dar, deren Parameter von einer *internen* und *externen* Energie abhängen. Ist eine *„Spline"*-Kurve durch einen Vektor $v(s)=(x(s),y(s)),\ s \in [0,1]$ gegeben, so ist die Energie des entsprechenden Snakes definiert als:

$$E^*_{snake} = \int_0^1 E_{snake}(v(s))ds = \int_0^1 E_{intern}(v(s)) + E_{image}(v(s)) + E_{con}(v(s))ds$$

E_{intern} repräsentiert die interne Energie des Splines bezüglich der Krümmung, E_{image} die Bildmerkmale und E_{con} die externen Kräfte wie z.B. Benutzervorgaben. Über die interne Energie wird der Abstand der einzelnen *„Snaxels"*[3] und die Krümmung der Snakes minimiert. Die Bildenergie zwingt den Snake dazu, sich den Bildmerkmalen (Linien, Kanten oder bestimmte Grauwerte) anzupassen. Externe Kräfte können frei definiert werden. Ist die gesamte Energie eines Snakes minimiert [KWT88] [WS92], kann der Snake als optimal positioniert angesehen werden.

4.2.2 Anwendung der Snakes

Um eine bessere Anpassung der Snakes an die Tagging-Linien zu erreichen werden die Snakes mit einer Dicke von 2 Pixel berechnet. Somit besteht im vorliegenden Fall ein Snaxel aus 2 Pixeln. Die initiale Position der Snakes ergibt sich aus der Berechnung der Gitterlinien mittels der SLHT. Da die SLHT gerade Linien liefert, haben auch die Snakes einen geraden Verlauf. Die Snakes im Startbild liegen aus der SLHT bereits als Pixelkette vor.

Die Minimierung der Energie eines Snake bewirkt eine Anpassung des Snake an die Tagging-Linie und somit auch eine Verfolgung der Gitterbewegung.

Die Auswertung der Bildenergie, die maßgeblich an der Anpassung der Snakes an die Gitterlinien beteiligt ist, erfolgt über die Intensitäten der einzelnen Pixel. Da sich die Optimierung der Bildenergie für jedes Snaxel an zwei Pixeln orientiert, ist die Verfolgung wenig anfällig gegen Bildstörungen. Die Berechnung und Minimierung

---

[3] Snaxel: ein Punkt auf einem Snake

der internen Energie der Snakes kann nach der Approximationsmethode von Kass et. al. erfolgen [KWT88].
An den Gitterkreuzungen liegen die Snaxel nicht immer korrekt auf der unterliegenden Gitterlinie, da sie an diesen Stellen nach unten oder oben auswandern konnten. Um den genauen Punkt zu finden, an dem sich zwei Tagging-Linien kreuzen, wird eine Art Matching durchgeführt. Dabei wird von einem bestimmten, möglichst optimalen Punkt aus versucht, die vier abgehenden Liniensegmente zu finden. Diese bestimmen dann den eigentlichen Kreuzungspunkt. Über einen Linienfilter, der um einen Winkel von 360 Grad gedreht wird, wird eine Korrelation $c(\phi)$ berechnet:

$$c(\phi) = \sum_{"ueberT} \sum t(x\cos(\phi) - y\sin(\phi), x\sin(\phi) + y\cos(\phi))[i(x,y) - \bar{i}]$$

Ein guter Kreuzungspunkt erzeugt damit einen $c(\phi)$-Verlauf mit vier Maxima. Der Punkt mit den vier höchsten Maxima wird als neuer Kreuzungspunkt angesehen. Nach der Anpassung der Snakes an die korrigierten Kreuzungspunkte können diese zur weiteren Berechnung in den nachfolgenden Aufnahmen nach dem gleichen Verfahren verwendet werden. Damit sind die Positionen der Kreuzungspunkte in allen Bildern einer Sequenz bestimmt und somit die Verschiebung des Gewebes bekannt.

5. Visualisierung der Herzbewegung

Um die berechnete Verschiebung der Kreuzungspunkte darzustellen, bietet eine interaktive Benutzungsoberfläche die Möglichkeit, die ermittelte Bewegung als farbige Vektoren, Punkte oder Strecken anzuzeigen. Dabei muß in der Darstellung der Raumbezug zum Herzen gewahrt bleiben.

Bei der Visualisierung mit Pfeilen wird für jeden gefundenen Kreuzungspunkt in der Sequenz ein Pfeil von dessen Ausgangsposition in einem gewählten Anfangsbild bis zu einer Endposition im entsprechenden Endbild gezeichnet. Die Länge des Pfeiles entspricht dem Abstand zwischen den beiden Positionen des zugehörigen Gewebepunktes. Die Richtung des Pfeils entspricht der Richtung der Verschiebung. Die Visualisierung der Bewegung mittels Pfeilen zeigt allerdings nur die Bewegung des Gewebes in der „Luftlinie". Bei Verwendung von Strecken werden die vom Gewebe zurückgelegten Wege von Anfang bis zum Ende einer Sequenz dargestellt. Der Startpunkt wird durch ein kleines weißes Quadrat markiert. Ebenfalls erlaubt hier eine Farbkodierung das Erkennen von Entfernungsbereichen (vgl. Abbildung 4). Die Verwendung von Punkten dient weniger der Veranschaulichung der Gewebebewegung, sondern der interaktiven Korrektur durch den Benutzer. Hier sind im Gegensatz zu den anderen Darstellungsvarianten keine Tagging-Informationen durch die Visualisierung (Pfeile, Linien) verdeckt. Korrekturen können nun durch ein einfaches *Drag and Drop* vorgenommen werden. Um eine einfache Zuordnung eines gewählten Punktes zu gewährleisten, wird der Punkt beim Verschieben durch eine dünne Linie mit seinen unmittelbaren Nachbarn verbunden. Als Alternative ist die Einblendung des kompletten Gitters möglich.

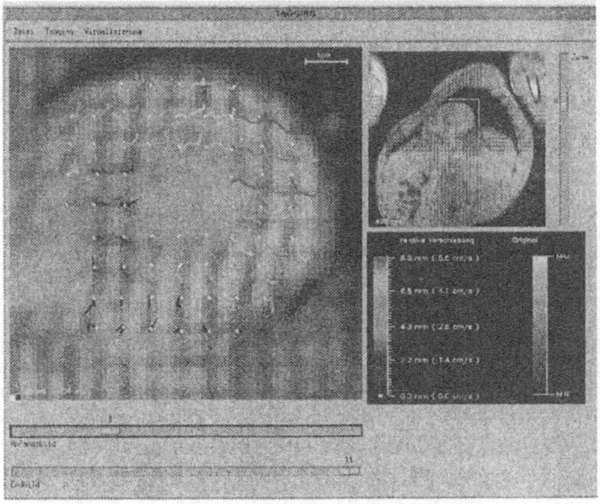

Abbildung 4: Visualisierung mit Strecken

6. Test und Ergebnisse

Zur Kontrolle des Verfahrens zur Erkennung der Gitterstrukturen wurden mehrere Berechnungen an Bildern eines Plexiglasphantoms durchgeführt. Dabei wurde das Verfahren auf Bilder unterschiedlicher Intensität angewandt. Die exakte Position des Gitters ist auch bei den Phantomaufnahmen nicht bekannt. Um die Ergebnisse des Verfahren analysieren zu können wurden zum Vergleich die Gitterkreuzungspunkte manuell markiert. Die mittlere Abweichung der vom Programm gefundenen Kreuzungspunkte zu den manuell gesetzten beträgt ab dem 6 Bild mehr als ein Pixel. Bei gedrehten und verschobenen Phantombildern liegt der Anteil der Abweichung von mehr als 2 Pixeln deutlich unter 1 Prozent.

Bei dem Test mit realen Patientendaten treten in den letzten Bilder große Abweichungen auf. Diese Fehler können vernachlässigt werden, da hier die Position der Fehlerstellen maßgebend ist. Diese Fehler treten in der rechten und linken Herzkammer auf. An diesen Stellen ist ein starker Blutfluß vorhanden. Aus diesem Grunde wird an diesen Stellen keine Bewegungsinformation erwartet.

Weitere Fehler treten an Punkten am Rand des Herzmuskels auf. Wegen des dunklen Lungenrandes kann das Programm nicht entscheiden, ob es sich um eine Tagging-Linie handelt oder nicht. Für diese Fälle ist eine manuelle Kontrolle durch den Benutzer nötig, wenn er Informationen über diesen Bereich wünscht.

7. Ausblick

Die Güte der Erkennung der Gitterkreuzungspunkte und somit der Visualisierung ist in erster Linie von der Qualität der akquirierten Datensätze abhängig. Eine weitere Verbesserung ist durch eine vorherige Segmentierung des Herzmuskelgewebes denkbar, die allerdings in allen Aufnahmen einer Sequenz mit gleicher Qualität vorliegen müßte. Ein wesentlicher Schritt in die Richtung der Erkennung von Herzfehlern wäre nach erfolgreicher Erkennung der Kreuzungspunkte eine dreidimensionale Analyse des gesamten Herzvolumens wie es schon als Ansatz in [GZM97] zu finden ist.

8. Anmerkung

Die Arbeiten wurden als Zusammenarbeit zwischen dem Herz- und Diabeteszentrum NRW in Bad Oeynhausen (Priv. Doz. Dr. G. Notohamiprodjo) und der Universität-GH Paderborn, FB 17 Informatik, Arbeitsgruppe für Computergraphik und Visualisierung (Prof. Dr. Gitta Domik) begonnen und durchgeführt. Die Mitarbeit von Herrn Lenz wurde durch Mittel des Forschungsministeriums des Landes NRW aus dem Verbundprojekt Bits, Bilder, Bedeutung finanziert.

Literatur:

[ACC94] A. Amini, R.W. Curwen, R.T. Constable and J.C. Gore; *Physics-Based Snake Tracking and Dense Deformations from Tagged MR Cardiac Images*; AAAI Symposium on Applications of Computer Vision to Medical Image Processing, Stanford University, Stanford, California, March 1994

[AD89] L. Axel and L. Dougherty; *MR Imaging of Motion with Spatial Modulation of Magnetization*, Radiology, 171(3):841, 1989

[BSD91] M.B. Buchhalter, C. Sims, A.K. Dixon , P. Weissberg, D.L. Stone, N.J. Shah, L.D. Hall, I.D. Wilkinson, D.J. Lomas and C.E.L. Freer; *Measurement of Regional Left Ventricular Function Using Labelled Magnetic Resonance Imaging*, The British Journal of Radiology, 64: 953 - 958, 1991

[GZM97] M.A. Guttmann, E.A. Zerhouni, A. Elias and E.R. McVeigh; *Tag and Contour Detection in Tagged MR Images of the Left Ventricle*, IEEE Computer Graphics and Applications,1: 30 - 38, 1997

[HFH94] J.C. Haselgrove, M. Fogel and J. Hofford; *Automatic Tracking of SPAMM Images*, Proc. Society of Magnetic Resonance. San Francisco, 1994

[KWT88] M.Kass, A. Witkin, D. Terzopoulos; *Snakes: Active Contour Models,* International Journal of Computer Vision, 321-331, 1988

[Va94] F. Van der Heijden; *Image Based Measurement Systems*, Verlag John Wiley & Sons, 1994

[WS92] D.J. Williams, M.A. Shah; *A Fast Algorithm for Active Contours and Curvature Estimation*, CVGIP: Image Understanding, 55(1): 14 - 26, 1992

[ZPR88] E.A. Zerhouni, M.D. Parish, W.J. Rogers, A. Yang and E.P. Sharpiro; *Human Heart: tagging with MR Imaging - A Method for Noninvasive Assessement of Myocardial Motion*, Radiology, 169(1): 59 - 63, 1988

Prüfsystem VINSPEC für die automatische Qualitätssortierung von Autospiegeln

Thomas Preußer
Leiter Oberflächeninspektion
VITRONIC Dr.-Ing. Stein
Bildverarbeitungssysteme GmbH
Hasengartenstr. 14a
65189 Wiesbaden
Tel.: 0611-7152-0, Fax.: 0611-7152-33

1. Einleitung

Die Anforderungen an Bildverarbeitungssysteme für Qualitätsprüfungen sind extrem gestiegen. Sowohl durch die kontinuierliche Weiterentwicklung der Analyseverfahren als auch durch die rasante Entwicklung der Hardwarekomponenten können komplexe Prüfaufgaben technisch und wirtschaftlich gelöst und realisiert werden. Ein wesentliches Problem bei der Realisierung komplexer Prüfanlagen ist die Handhabbarkeit für den Maschinenführer. Gerade bei Anlagen zur Prüfung ganzer Produktfamilien mit variierenden Qualitätsanforderungen, wie zum Beispiel Autorückspiegeln, muß der Maschinenführer in der Lage sein, seine Anforderungen schnell und gezielt umsetzen zu können.
Mit dem neuen Prüfsystem VINSPEC wurde ein allgemeiner Lösungsansatz gefunden, der diesen Anforderungen entspricht. Im weiteren werden zunächst am Beispiel der Qualitätssortierung von Autospiegeln die Anforderungen an das Bildverarbeitungssystem aufgezeigt, die Struktur der Bildverarbeitungssoftware erläutert und anhand der Kratzerprüfung die schwierigste Inspektionsaufgabe vorgestellt.

2. Beschreibung der Inspektionsaufgabe

Zur Herstellung von Autorückspiegeln sind viele Bearbeitungsschritte nötig. Ausgehend von verspiegelten, sphärischen und asphärischen Glaskalotten, werden die Spiegel geschnitten, die Schneidkanten gesäumt und versiegelt sowie teilweise durch einen Laser Linien und Symbole aufgebracht. Bedingt durch den schwierigen Werkstoff Glas kommt es bei jedem Bearbeitungsschritt und dem Handling im Produktionsdurchlauf zu einem Ausschuß in der Qualtiätssortierung von insgesamt 10% und mehr.
Als Qualitätskriterium gelten zum einen funktionelle Merkmale wie Maßhaltigkeit und Lebensdauer, die bei Beschädigung der Versiegelungsschichten stark reduziert ist, und zum anderen optische Merkmale. Fehlstellen, die entsprechend des Einsatzes

von Autorückspiegeln im Abstand von 800 mm vom Menschen erkennbar sind, müssen aussortiert werden. Dabei ist zu berücksichtigen, daß Spiegel bei unterschiedlichster Beleuchtung betrachtet werden: Tagsüber durch Sonnenstand und Wetterlage, Nachts durch rückwärtigen Verkehr. Sowohl für die manuell-visuelle Prüfung als auch bei der automatischen Inspektion durch ein Bildverarbeitungssystem müssen diese Forderungen durch den Einsatz mehrerer Beleuchtungsarten und Betrachtungswinkel erfüllt werden.
Die Fehlerklassen stellen sich im einzelnen so dar:

Konturfehler	Die Kontur wird dazu mit Kantenverfolgungsalgorithmen detektiert und mit vorgegebenen Toleranzen, basierend auf einem eingelernten Muster, verglichen.
Fehler der Säumung	Sie stellen ein breites Spektrum von Fehlerklassen dar. Neben der reinen Vermessungsaufgabe – die Saumbreite muß mit einer Prüftoleranz von 0,1 mm klassifiziert werden – können bei der Bearbeitung muschelförmige Ausbrüche auftreten, die sich durch eine lokale Schwankung der Saumbreite oder als Fehler der Reflexionsschicht äussern. Fehlstellen mit einem Durchmesser größer 0,15 mm müssen erkannt werden. Die Ausprägung der Fehler schwankt stark. Für die automatische Inspektion müssen möglichst viele Beleuchtungs- und Betrachtungswinkel realisiert werden, um diese Fehler sicher erkennen zu können.
Beschädigung der Schutzschichten	Rückseitig werden auf der eigentlichen Reflexionsschicht Schutzschichten und Lacke aufgebracht, die den Spiegel vor Umwelteinflüssen schützen sollen. Fehler äussern sich durch lokale Helligkeitsschwankungen.
Lokale Glasfehler – Blasen, Punkte	Die Spiegelklassifikation erfolgt über Fehlergröße und Anzahl. Punktförmige Fehler ab 0,2 mm müssen erkannt werden.
Kratzer	Sie werden nach Länge und Intensität klassifiziert. Wichtigste Maße für das Bildverarbeitungssystem sind die kleinste zu detektierende Kratzerbreite von 5µm und die Kratzerlänge von 1 mm.
Belagfehler	Durch diese Fehler variiert lokal das Reflexionsverhalten. Es resultieren bei geeigneter Beleuchtung wolkenartige Fehlerbilder.

Zerrungsfehler — In der Automobilindustrie werden heutzutage überwiegend sphärische und asphärische Spiegel eingesetzt. Durch die konkave Krümmung des Spiegels wird der Blickwinkel der Fahrers erweitert und optimal an die Erfordernisse im Verkehr angepaßt. Schwankungen in der Krümmung führen zu störenden optischen Verzerrungen. Die 3-D-Funktion der Spiegelkrümmung und damit die optische Übertragungsfunktion muß daher ermittelt und klassifiziert werden.

Beschriftungsfehler — Die auftretenden Fehler können durch eine Standard-Druckbildkontrolle mit Soll-Ist-Vergleich erkannt werden.

Die manuelle Sortierung ist extrem ermüdend, da die Fehler nur bei Fokussierung in unterschiedlichen Schärfeebenen zu sehen sind, zum einen auf die Spiegeloberfläche (z.B. Kratzer, Punkte), zum anderen bei Betrachtung einer strukturierten Fläche über den Spiegel (Zerrungen). Schwierig ist auch die Klassifikation nach Fehlergrößen, die personenabhängig schwankt. Da nahezu jeder Spiegel Fehlstellen aufweist, wird oft zu streng sortiert, der Ausschuß steigt. Anderseits werden durch Ermüdung grobe Fehler übersehen.

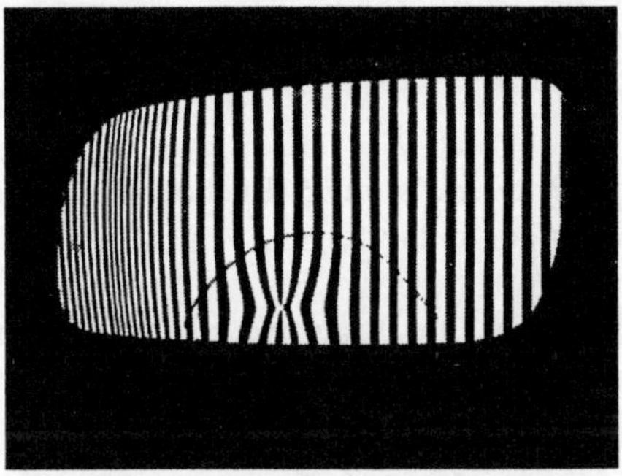

Bild 1: Streifenmuster auf asphärischem Spiegel mit Zerrungsfehler

Die von der Automobilindustrie geforderten Qualitäten von kleiner 500 ppm Fehlerteilen lassen sich manuell nur durch hohen, kostenintensiven Personaleinsatz erreichen. Abhilfe schafft die automatische Qualitätssortierung mit dem System VINSPEC. Durch den Einsatz von 22 Kamaras pro Linie und jeweils 6 verschiedenen Beleuchtungsarten können die oben beschriebenen Fehler kontrastiert, separiert und klassifiziert werden.

2.1 Weitere Randbedingungen

Die Lösung der Klassifikationsaufgabe deckt bei einem industriellen Einsatz der Bildverarbeitung nur einen Aufgabenbereich ab. Wirtschaftlichkeit und Akzeptanz hängen vor allem auch von Taktzeit, Verfügbarkeit und Handhabbarkeit ab. Diese Applikation charakterisiert sich folgerndermaßen:

Inspektionsbereich	maximal 160 mal 240 mm
Taktzeit	< 4s
Verfügbarkeit	> 99 %
Rüstzeit	< 1 min

Wichtigstes Kriterium für die Handhabbarkeit ist eine flexible, bedienerfreundliche Programmoberfläche, mit der vorhandene Prüfprogramme an neue Spiegeltypen angepaßt werden können. Zusätzlich muß durch Test- und Überwachungssoftware der fehlerfreie Betrieb sichergestellt werden. Vor allem die Tatsache, daß mit jedem neuen Kfz-Fahrzeugtyp ein neues Prüfprogramm erstellt werden muß, zwingt zu einer möglichst einfachen Bedienbarkeit des Systems.

3. Systemarchitektur

Durch die aufwendige Prüfaufgabe – es werden pro Spiegel mehr als 70 Kamerabilder ausgewertet – und die Taktzeit von nur 4s muß die Bildverarbeitung mit fünf Recheneinheiten parallelisiert werden. Die unterschiedlichen Beleuchtungskonzepte erfordern zusätzlich eine Verteilung auf sieben Inspektionsstationen. Die Qualitätssortierung setzt sich aus den Ergebnissen aller Prüfungen zusammen. Sie müssen deshalb zentral verwaltet werden, um eine abschließende Bewertung durchführen zu können.

Merkmale der Systemarchitektur

- Alle Teilsysteme arbeiten als autarke Bildverarbeitungs-Rechner, die ihre Ergebnisse an ein übergeordnetes Leitsystem übertragen.
- Die Basissoftware der Bildverarbeitungsrechner ist auf allen Teilsystemen identisch, die Prüfungen selbst werden durch eine einlernbare Prüfvorschrift vorgegeben.
- Die Prüfvorschrift setzt sich wiederum aus einzelnen Prüfmethoden zusammen, die in beliebiger Reihenfolge und Anzahl aufgabenspezifisch zusammengestellt werden können.
- Durch die objekt-orientierte Programmstruktur enthält die Prüfvorschrift neben der Parametrierung auch den Programmcode für die jeweilige Prüfung. Damit können neue Algorithmen integriert werden, ohne Basissoftware und vorhandene Prüfvorschriften zu verändern.
- Die Programmierung des Systems erfolgt zentral an der Leitstation.

Durch die große Anzahl von Fehlerklassen, die umfangreiche Hardware und die wechselnden Prüfanforderungen war es nicht sinnvoll, ein starres Programm für die Prüfung zu erstellen. In einer Programmstruktur nach dem klassischen monolithischen Konzept müssen alle möglichen Anwendung der Prüfanforderungen fest einprogrammiert werden. Eine feste Hardwarestruktur wird vorausgesetzt.

Erweiterungen oder Änderungen der bestehenden Anforderungen können nur durch Softwareänderungen realisiert werden.

3.1 Basisprogramm und Prüfvorschrift

Abweichend zu diesem Konzept wurde im neuen VINSPEC eine modulare, objektorientierte Programmstruktur gewählt. Ausgehend von einer zentralen Bedieneinheit können eine beliebige Anzahl von Bildverarbeitungsrechnern eingerichtet und gesteuert werden. Das System kennt alle zur Lösung der Bildverarbeitungsaufgabe benötigten Verfahren. Diese können frei programmierbar in einer produktspezifischen Prüfvorschrift zusammengestellt und parametriert werden.
Das komplexe System sieht nach außen wie eine Einheit aus. Sämtliche Methoden können für alle Teilsysteme und Kameras eingesetzt werden. Dies bietet zwei wesentliche Vorteile:
1. Die Mehrfachnutzung von Standardmethoden reduziert den Programmumfang.
2. Neben rudimentären Prüfmethoden muß der Bediener nur wenige applikationsspezifische Prüfmethoden konfigurieren, um das Prüfsystem auf seine Erfordernisse anzupassen.

Die oben beschriebene Aufgabe der Spiegelprüfung wird komplett mit elf Methoden durchgeführt. Davon sind wiederum sechs Methoden allgemeine Bildverarbeitungsmethoden und nur fünf applikationsspezifisch entwickelt worden.
Die zentrale Schnittstelle vereinfacht auch die Integration des Bildverarbeitungssystems an den Gesamtprozeß. Besonders hervorzuheben ist die installierte Fernwartungssoftware. Mit ihr kann das System in allen Funktionen über Modemanschluß geführt werden. Unterstützung des Bedienführers durch den Entwickler, Fehlerdiagnosen, Software-Updates usw. lassen sich schnell und kostengünstig durchführen.

4. Verfahren zur Kratzerdetektion

Die Kratzeranalyse ist für die Bildverarbeitung die größte Herausforderung. Wie oben erwähnt, müssen Kratzer ab einer Breite von 5µm erkannt werden. Selbst bei dem beschriebenen umfangreichen Hardwareaufwand liegt diese Breite weit unter der Auflösungsgrenze des Systems. Sie beträgt mit 100µm x 100µm das Zwanzigfache. Die Hauptbedeutung hat dabei die Kamera-Beleuchtungs-Anordnung. Durch geeignete Beleuchtungstechnik konnte erreicht werden, daß trotz dieses Mißverhältnisses Kratzer gegenüber dem ungestörten Hintergrund kontrastiert werden können.

4.1 Beleuchtungstechnik

Vergleichbar zum menschlichen Auge, das solche Strukturen auch nicht auflösen kann, führen starke lokale Überstrahlungen zu einer Aussteuerung der betroffenen Bildpunkte.

Genutzt wird für diese Anordnung der optische Effekt, den man Nachts beim Autofahren beobachten kann: Bei älteren Fahrzeugen weisen die Rückspiegel und die Windschutzscheibe eine Vielzahl von Haarkratzern auf. Fokussiert man nachts bei entgegenkommenden Verkehr auf eine solche Windschutzscheibe, kann man auf ihr Bereiche ausmachen, auf denen die Kratzer zu erkennen sind. Diese Kontrastverstärkung ist stark richtungsabhängig.

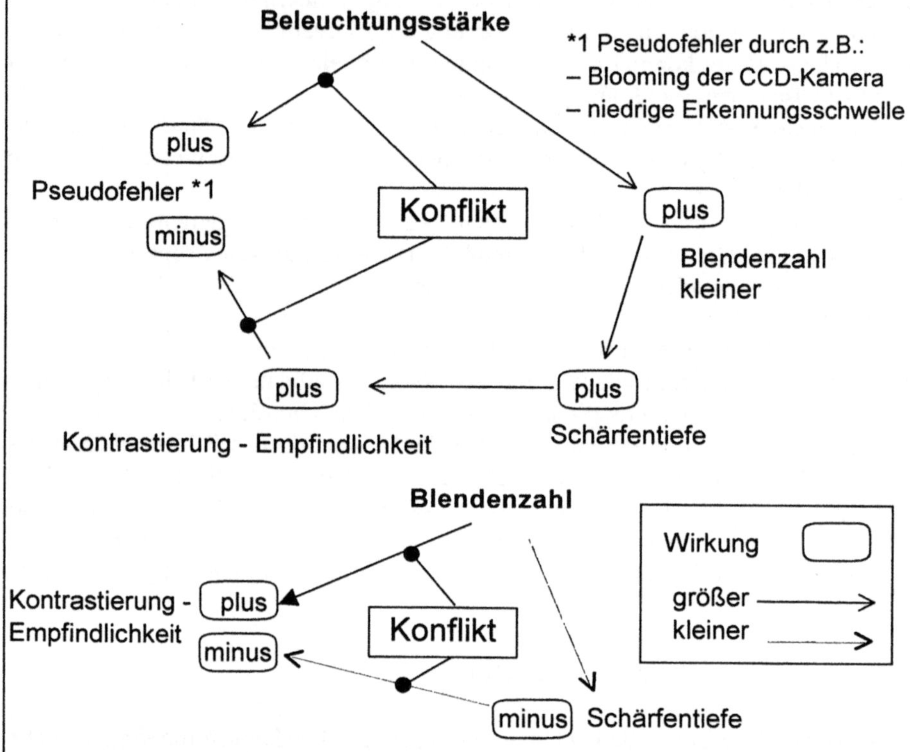

Bild 2: Zusammenwirkung von Beleuchtungs- und Kameraeigenschaften auf die Erkennungsleistung

Diesen Effekt nutzend, wurde eine Beleuchtungseinrichtung entwickelt, mit der der komplette Inspektionsbereich inspiziert werden kann.

Eine iterative Optimierung der gesamten Anordnung ist nötig, da wechselseitige Einflüsse zwischen den einzelnen Systemparametern bestehen. Hierbei ist vor allem noch zu berücksichtigen, daß aufgrund unterschiedlicher Spiegelstärken und Krümmungsradien eine Schärfentiefe von 6 mm erreicht werden muß. Dies setzt wiederum eine

minimale Blendenzahl und maximale Helligkeit des Inspektionsbereichs voraus, wobei eine zu helle Einstellung wiederum zu Blooming und damit zu Pseudofehlern führt (siehe Bild 2).
Unproblematisch ist das Anpassen an unterschiedliche Spiegeltypen mit variierender Reflexion. Nach der Optimierung kann die unterschiedliche Reflexion durch die Beleuchtungsanpassung ausgeglichen werden. Sie ist im Bereich von 0 bis 100 % einstellbar.

4.2 Fehlersegmentierung und Klassifikation

In den akquirierten Kamerabildern werden mittels angepaßter FIR-Filter kratzerförmige Strukturen verstärkt. Anschließend werden Hypothesen für mögliche Kratzer im gefilterten Bild gesucht und durch einen Kantenverfolgungsalgorithmus die Merkmale der gefundenen Hypothesen berechnet. Der Verfolgungsalgorithmus wurde so implementiert, daß feine, schwach kontrastierende Strukturen vollständig gefunden und kurze Unterbrechung überbrückt werden. Als Merkmale liegen der Konturverlauf, die Lage und der mittlere Kontrast vor. Abschließend erfolgt die Klassifikation entsprechend den eingestellten Fehlerschranken.
Natürlich ist mit dieser Fehlererkennung keine Vermessung der Kratzerbreite im herkömmlichen Sinn möglich. Entscheidend ist, daß dieses Vorgehen der menschlichen Wahrnehmung nahe kommt und durch Bewertung des mittleren Kontrastes ein Merkmal vorhanden ist, daß eine Aussage über den optischen Effekt liefert.

5. Ergebnisse

Wichtigste Kriterien zur Bewertung einer Bildverarbeitungsanlage im industriellen Einsatz sind die Erfüllung der Prüfanforderungen einerseits und die Akzeptanz der Anlage anderseits.
Für die Akzeptanz der Anlage ist neben der eigentlichen Inspektion die Handhabbarkeit des Systems wichtig. Für das Verständnis des komplexen Systems sind drei Grundlagen wichtig: Der Maschinenführer muß das oben beschriebene Konzept verstehen, die Aufteilung der Prüfaufgaben bzw. den Aufbau der Sensorik kennen und die bereitgestellten Bildverarbeitungsmethoden parametrieren können. Mit diesem Wissen kann der Maschineneführer flexibel die Anlage für unterschiedliche Spiegeltypen und Prüfanforerungen anpassen.
Die Prüfanforderungen werden durch die eingesetzte Sensorik und Bildverarbeitungsalgorithmen erfüllt. Wobei eine iterative Optimierung der Parametrierung nötig wurde, um die Empfindlichkeit der Anlage abweichend von den Vorgaben an das jeweilige Produkt anzupassen.
Die manuelle visuelle Gegenprüfung der maschinell sortierten Spiegel, bestätigte vor allem, daß das menschliche Auge hervorragend geeignet ist um Fehler zu detektieren. Die Klassifikation der Fehlstellen über Form und Ausdehnung ist im Gegensatz zum Inspektionssystem subjektiv und starken Schwankungen unterworfen.

Die beschriebene Sortieranlage läuft im 3-Schicht-Betrieb und erfüllt die aufgeführten Anforderungen vollständig. Nach nun mehr als eineinhalbjähriger Praxiserfahrung zeigen sich die großen Vorteile einer automatischen Inspektion. Die Ergebnisse der manuellen visuellen Prüfung werden um ein vielfaches übertroffen.

6. Übertragung auf andere Anwendungen

Auch wenn hier nur die Kratzeranalyse näher beschrieben wurde, zeigt die Liste der Fehlerklassen, daß allein für diese Applikation Algorithmen für Oberflächenprüfung, Konturvermessung, 3-D-Profilinspektion und Druckbildkontrolle zum Einsatz kommen. Darüber hinaus stehen allgemeine Verfahren zur Kantenvermessung und Vollständigkeitsprüfung zur Verfügung. Mit diesem Spektrum können viele Prüfaufgaben für die unterschiedlichsten Branchen direkt realisiert werden. Bisher wurden das neue VINSPEC erfolgreich in den Bereichen: Glas, Medizintechnik, Elektro, Automobil, Metall, Kunststoff und Werkzeugbau installiert. Durch den modularen Aufbau reicht das Spektrum von einfachen Einrechner-Lösungen, die gerade bei einfachen Vermessungsaufgaben und Vollständigkeitsprüfugen vom Anwender selbst konfiguriert werden können, bis hin zu hochkomplexen Prüfsystemen, wie hier beschrieben.
Der Software-Entwicklungsaufwand reduziert sich auf die applikationsspezifische Anforderung. Der Bildverarbeiter ist hauptsächlich für das Engineering gefragt. Bildaufnahme-, Beleuchtungstechnik, Systemintegration und Inbetriebnahme sind weiterhin Aufgabenbereiche, mit denen der fachfremde Anwender oft überfordert ist.

A Fast Image Processing Algorithm for Quality Control of Woven Textiles

Ch. Daul [1*], R. Rösch [1*], B. Claus [1†], J. Grotepaß [2] , U. Knaak [2], R. Föhr [3]

[1] Institut für Techno- und Wirtschaftsmathematik (ITWM) e.V., Erwin-Schrödinger-Straße, D-67663 Kaiserslautern

[2] Parsytec Computer GmbH, Auf der Hüls 183, D-52068 Aachen

[3] maVis GmbH, Auf der Hüls 183, D-52068 Aachen

1 Introduction

Textile manufacturers have to monitor constantly the quality of their products (i.e., woven fabrics) in order to maintain the rigorous quality standards required by their customers. Up to fifty different kinds of defects (depending on the type of the fabric) are usually considered when defining the textile quality. At present, the inspection of textiles is mainly realized offline (after the production process) by human controllers who accomplish a tedious and tiring work. Hence, the detection and classification of these defects becomes less accurate over time, and the defects are not classified according to systematic criteria since each controller perceives, analyzes and interprets the defects differently. In addition, the low control speed (about twenty to thirty meter/minute) when compared to the production speed (about sixty meters per minute) represents a problem in the work flow.
In this paper we describe an algorithm (developed in the framework of the European HIPERTEX project [1] and implemented on a multiprocessor system) which features the following improvements in the field of textile quality control:

- full inspection of the entire width of the fabric with a control speed of one meter per second,
- online detection and classification of defects,
- integration of the realtime system into the production process.

2 Problem description

The class of woven fabrics dealt with by our algorithm are textiles which exhibit a texture consisting of a deterministic part and a random part. The deterministic part of the texture is a result of the woven pattern in the fabric, like, e.g., parallel stripes in corduroy or denim fabric. The orientation of these stripes is typically either in vertical, horizontal or diagonal direction, and their

* E-Mail: daul@itwm.uni-kl.de, roesch@itwm.uni-kl.de
† *Current affiliation:* GE Corporate Research and Development, Niskayuna, NY 12309

width varies strongly from one textile type to another (it can even be so small that no texture is perceptible). For this kind of textiles defects correspond to perturbations (or inhomogeneities) occuring in a otherwise "mostly regular" texture which can not be attributed to the usual random part of the texture. This random part of the texture consists of inhomogeneities introduced, e.g., by the natural "roughness" of the fibers, measurement noise and inhomogeneities due to the illumination conditions. For our problem, three particular effects of the illumination, depending on the position of the camera, the light source and the textile, have to be considered. First, the local mean grey level of the image may not be constant, i.e., some regions of the image appear to be darker than others. Second, we may observe the same effect for the contrast in the image: the difference of the grey-level in neighboring stripes (of the woven pattern) may not be constant over the image. Finally, reflections on single threads complicate the defect detection.

A. Eldin [2] gives an overview on existing textile inspection systems based on computer vision methods. From this study a generally accepted sequential algorithm structure is obvious: the image is first corrected in a preprocessing step, a set of relevant features is then extracted from the image and defects are finally detected and classified using these feature vectors. The main differences between different algorithms lie in the choice and computation of the texture features ([3], [4], [5] and [6] give only few examples of features which can be used). Unfortunately, even though these algorithms achieve good results, they are often not suitable for online applications (see [7] and [8]). In the next section we present the system environment used for defect detection in textiles. In section 4 we introduce an algorithm that allows extracting the requested texture features in realtime (for our case of uniform and regular textures).

3 System environment

The algorithms (described in section 4) have been implemented on the inspection system platform of the HIPERTEX project. Figure 1 gives the general structure of the inspection system. It consists of the image acquisition unit (cameras, illumination, framegrabbers), the parallel computer, and the system software. The basis is a Parsytec parallel computer (CCi-5), consisting of 4 PentiumPro computing nodes, running WinNT. Each node is processing data input from 3 cameras. The 5th node serves as task manager for communication control and provides the GUI for the system operator. An IR darkfield illumination is used, where the infrared spectrum (880nm) provides nearly independence from colour aspects and delivers mostly pure object surface information. By using flashes in the range of 4-25 μm for lighting, textile vibrations due to preceding processing steps are covered, maintaining a sharp focus. The cameras used are standard CCD area cameras.

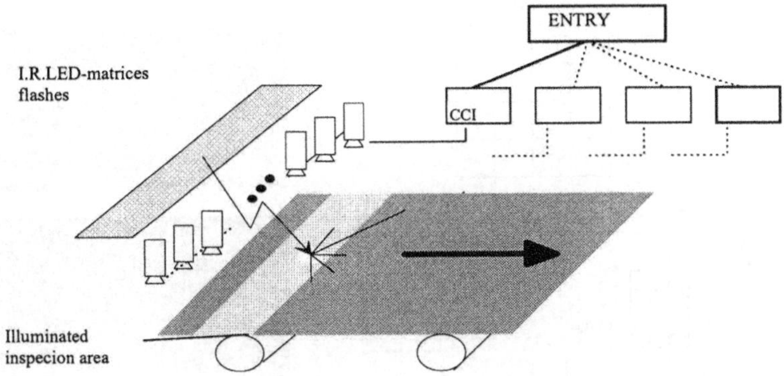

Fig. 1. Overview of the inpectionsystem.

4 The algorithm

We propose here a solution based on four steps, which basically corresponds to the general algorithm structure mentioned above. The first step consists of a preprocessing which is used to normalize the image in such a way that inhomogeneous lighting conditions are corrected. This allows us to use in subsequent steps procedures which do not have to be adapted to locally varying conditions in the image. In the second step we associate to each pixel a feature describing the local "regularity" of the texture. A simple thresholding operation is used to select the candidates for a possible location of a defect. This threshold is chosen according to the specific type of fabric considered. In a third step we cluster the candidate pixels in order to obtain rectangular regions of interest which describe (very crudely) the shape of the detected defect. As additional information we compute at this point further statistics of the features in the respective regions. Finally, in our last step, we classify the detected defects according to their shape and by also making use of additional criteria.

4.1 Preprocessing

The goal of the preprocessing step (sketched in Figure 2) is the correction of locally varying mean grey values and variances (the origins of these variations are given in Section 2). The idea here is to compute an image $M(x, y)$ which represents the local mean grey value (as it would be expected for that specific type of fabric under the given lighting conditions) and also an image $V(x, y)$ which represents the local variance of the underlying image of the fabric (see Figure 2). Once we have done this, we can subtract the "local-mean" image from the original $I(x, y)$, and then we divide the image by the square root of the "local-variance" image in order to obtain an image with approximately zero-mean (locally and globally) and unit-variance. This normalization reads in a

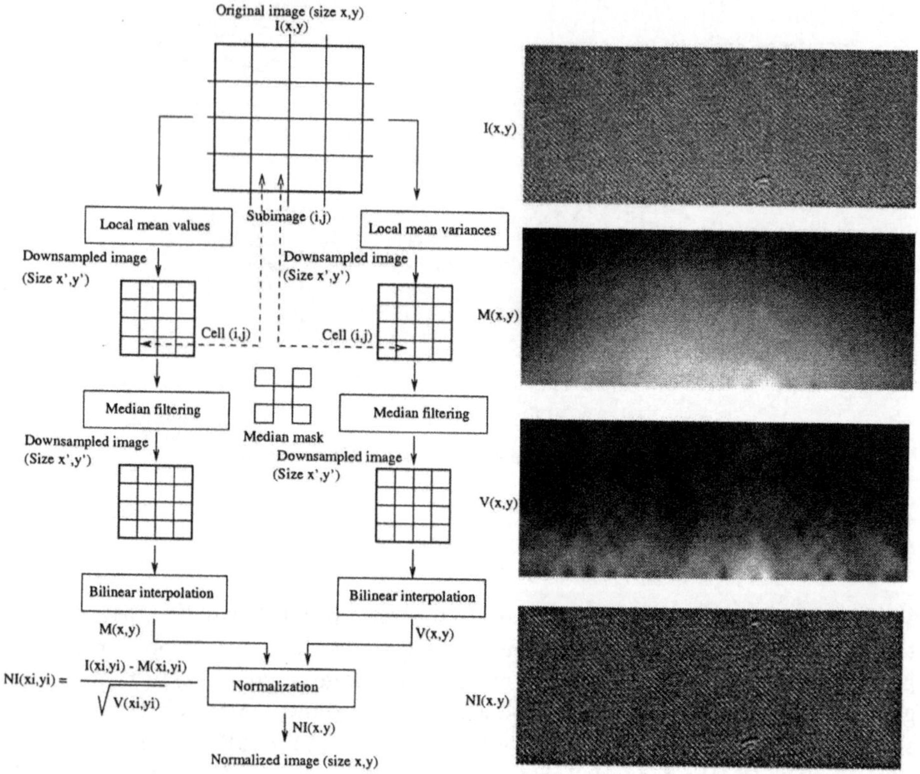

Fig. 2. Block diagram of the image normalization. The original image $I_{(x,y)}$ is divided into non-overlapping subimages $I(i,j)$. For each of these we compute the corresponding mean value and variance, which in turn represent the pixels/cells(i,j) of "downsampled" images. These are then median-filtered in order to eliminate the effects due to defects in the fabric. Using a bilinear interpolation we obtain images of the same size as the original image which are used to compute the normalized image ($NI(x,y)$).

mathematical notation:

$$NI(x,y) = \frac{I(x,y) - M(x,y)}{\sqrt{V(x,y)}}.$$

This approach is advantageous because it permits a subsequent evaluation of the image without having to adjust the parameters of the defect detection algorithm to locally varying conditions in the image.
Our method to compute these mean grey value or mean variance images, respectively, is motivated by the fact that we can assume (for a fabric without any defects or anomalies) that the local mean grey value and the local variance do vary slowly and smoothly across the image plane, i.e., that we do not have abrupt changes in the local characteristics of the image. We can therefore assume that

the mean and the variance computed in a subimage of appropriate size represent the "correct" local mean values if the following conditions are satisfied:

- no defects in the subimage are visible;
- the size of the subimage is small enough to capture the local variations of mean and variance;
- the size of the sub-image is large enough such that the (deterministic part of) the pattern in the fabric does not influence the result.

The last two conditions can be satisfied by simply choosing a window of appropriate size for computing the local mean and variance. The first condition, namely the fact that such a window shall not contain a defect in the fabric, has to be adjusted for in an appropriate way. We use for this elimination of the influence of defects a simple median filter (on the downsampled mean and variance images) with an appropriate mask (consisting of the considered pixel and its diagonal neighbors) which is chosen according to the characteristics of the defects we observe: they are either very small and well-localized, or they are long and narrow in direction of the fabric flow, or orthogonal to it. In all cases the median filter effectively eliminates the influence of subimages in which a defect is visible on the "local mean" and "local variance" images, respectively. The only "pathological" case which can occur in this setting consists of elongated defects, which are located such that they contribute (at some point) to at least three of the five values which are used to compute the median. This problem does not occur very frequently, and a solution to this problem can be provided by using a bigger mask for the computation of the median. We did not do this in our tests because the typical image size we used was to small to justify this step. The effect of the median filter on regions in the image without defects and where the local mean and local variance vary smoothly and slowly, is negligible.
During our tests we observed a relatively high number of isolated very bright and very dark pixels in the normalized image, which can be attributed to reflexions on single fibers or threads in the fabric. These "outliers" can have a significant and undesirable effect on the evaluation of the images. We perform a histogram clipping on the normalized image $NI(x, y)$ in order to eliminate these outliers.

4.2 Feature extraction

After the preprocessing step, we now discuss the evaluation of the image with the goal to detect defects in the fabric. Defects manifest themselves mostly by irregularities in the (deterministic) texture of the image, or by "black" regions in case of mechanical defects like "holes" in the fabric. Based on this observation we developed an operator delivering for each pixel a value describing the local "regularity" of the underlying texture. The idea of this operator is similar to the concept of directional edge detectors like the one proposed by Canny [9]. For this type of filters one smoothes in one direction of the image (y-direction) and calculates the derivative in the orthogonal direction (x-direction) in order to obtain the x-component of the image gradient (the y-component of the gradient is obtained correspondingly). Both components of the gradient are then used to determine edges in the image. In our case we replace the derivation operation by calculating a variance in the x and y directions, respectively (see Figure 3). The sum of these components provides a measure for the regularity of the texture in the neighborhood of the considered pixel.

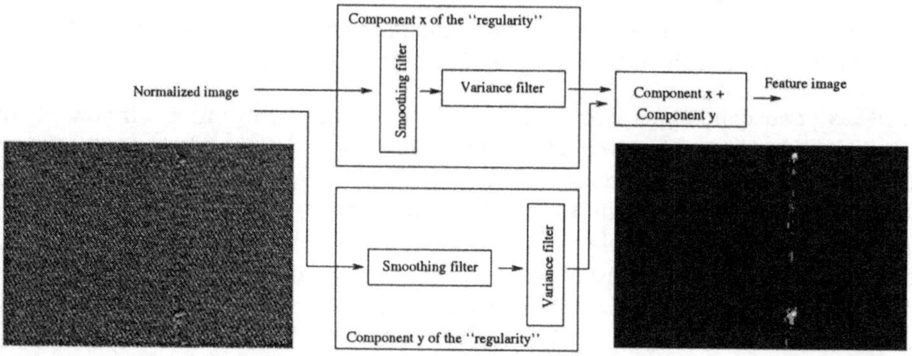

Fig. 3. The feature image gives an indication of the degree of "regularity" of the underlying image. All pixel where the threshold (which is constant for a given textile type) is exceeded represent candidate positions for a defect.

For each type of textile we defined characteristic window sizes for these operations (depending on the width of the corresponding stripe pattern). Choosing the appropriate window size for the smoothing filter leads to two different effects: the deterministic part of the pattern in "regular regions" of the fabric is cancelled out and the noise of the image vanishes to a large degree. The smoothed images should be "almost constant" in all "regular regions" of the images. In all regions where there are defects, the resulting image will not be constant, which is due to the fact that we have here either "dark regions" or irregularities in the woven pattern which prevent the "cancelling" of the pattern. Computing the local variance of this smoothed image provides a measure for the importance of structures which are not supressed by the smoothing step, and represents a measure of the probability of the presence of a defect.

4.3 Regions of Interest and feature vector

With the previously detected pixels we now construct regions of interest, i.e. rectangular regions which circumscribe the detected defects. The procedure we use here is the following:

1. We divide the image plane into square blocks by imposing a grid structure on the image - typically we choose here the same grid-size as for the normalization of the image.
2. Every pixel where the value of our "variance-feature" exceeds a given threshold S_1 (which is predefined for the specific fabric under consideration) represents a candidate pixel. All blocks where the number of candidate pixels

exceeds a given threshold S_2 are now considered to be part of some defect. In our tests we used a threshold value $S_2 = 0$, i.e., every candidate pixel "creates" a region which is considered to be a defect.
3. In a third step we perform a region labeling. This means that all blocks of the image which are associated with a defect are now analyzed with respect to their spatial relations, in particular we check for "neighborhood relations" between the image blocks. Adjacent blocks (horizontally, vertically, and diagonally - which is the standard second order neighborhood) are considered to be connected, i.e., it is assumed that neighboring "defect" blocks are caused by the same defect in the fabric, which "spreads out" over several blocks. We thereby obtain connected regions of blocks, and each region is assumed to correspond to exactly one defect in the textile.
4. Region of interest (RoI) - We now compute for each region the minimal circumscribing rectangle in order to obtain only rectangular regions which describe the localization and the spatial extent of the detected defects. For each of these regions of interest we compute in addition the number of candidate pixels and the mean value of the "regularity feature" with respect to the detected "defect blocks" in that region of interest.

4.4 Classification

Class description (and name)	Longitudinal (class L)	Transversal (class T)	Patch (class P)	Class L, T or P with low priority (class LP)	Small (class S)	Class S with low priority (class LS)
Critorion 1	$\frac{L(\text{RoI})}{W(\text{RoI})} > T_1$	$\frac{W(\text{RoI})}{L(\text{RoI})} > T_1$	$\frac{L(\text{RoI})}{W(\text{RoI})} < T_1$ or $\frac{W(\text{RoI})}{L(\text{RoI})} < T_1$	RoI $\neq$ single block	RoI = single block	RoI = single block
Criterion 2	$N_{\text{cand}} > T_2$	$N_{\text{cand}} > T_2$	$N_{\text{cand}} > T_2$	$N_{\text{cand}} \leq T_2$	$N_{\text{cand}} > T_2$	$N_{\text{cand}} \leq T_2$

Table 1. Classification criteria for the six macro-classes. $L(\text{RoI})$ and $W(\text{RoI})$ denote the lengh and the width of the region of interest. The threshold T_1 is used to decide if a region of interest is transversal or longitudinal (with respect to the "direction" of the fabric) or a patch. N_{cand} is the percentage of candidate pixels in the region of interest and T_2 is the corresponding threshold value.

In a final step we perform a classification of the detected regions of interest (RoI). In particular we introduce here six classes (called macro-classes in the textile industry), which are mostly concerned with the size and the geometric shape of the determined regions of interest. In addition to these purely geometrical considerations we also have an additional classification criterion which compares the percentage of candidate pixels in the region of interest with some given threshold. Table 1 provides the definition of the six classes. These classes are sufficient for a macro-classification of the defect. In order to realize a micro-classification (classification of the the defect types like holes, knot in warp, loom waste, etc.) we have to use additional features (more detailed geometric informations, statistical data, etc.). This will be realized in future developments.

5 Results and future developments

We provide here two examples (Figures 4 and 5) which demonstrate the ability of our system to detect and classify defects. The stripe width represents the major difference between our two examples. Furthermore we can see directly from the images that the lighting conditions in the image acquisition are far from being homogeneous, which implies that the preprocessing (normalization) of the image is absolutely necessary and crucial for a robust detection of defects in the fabric.

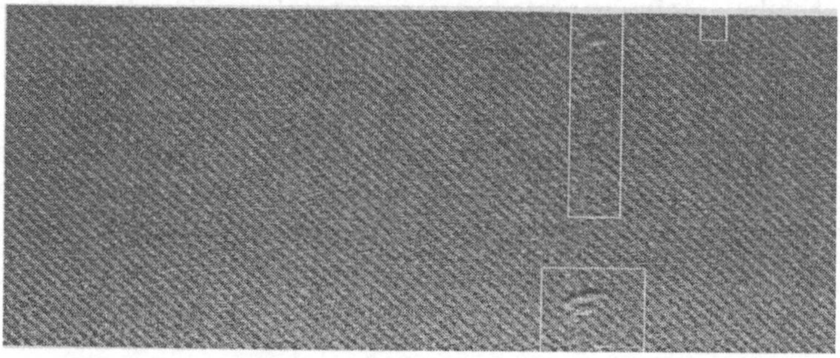

Fig. 4. A fabric with a very pronounced diagonal stripe pattern. We marked the detected regions of interest, which are assigned to three different classes. In particular the small region on the upper right corresponds to class LS, i.e., this region is of the same size as the blocks we used for computing the local mean and variance images.

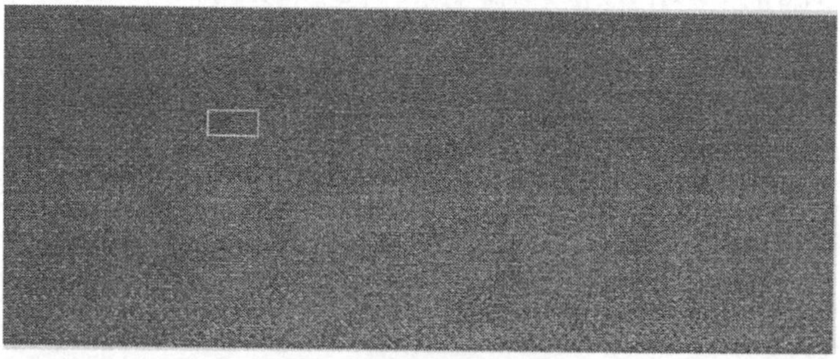

Fig. 5. A "flat" fabric, where we can see clearly the locally varying lighting conditions. Again the detected defect is marked by a rectangle.

With the described algorithm and depending on the textile type (denim, flat with light shade or flat with dark shade) seventy to ninety percent of the defects have been detected (in a factory environment) on a large number of samples consisting of a variety of textiles with different textures and defects. In order to increase the detection rate, additional features have to be extracted from the image. This will be realized in future developments. Currently a processing time of 500 ms/image is required, which *includes* the corresponding system overhead (operating system, image acquisition, etc.) when using a Pentium Pro PC (200 MHz). An optimization of the algorithm (this task is presently in progress) and faster PC's will result in a processing time which is sufficient for full realtime monitoring of the fabric

6 Acknowledgements

The research work presented in this paper has been performed in the framework of the ESPRIT project HIPERTEX. The goal of this project was research and development of a computer-based system for full automatic inspection of textiles, with the focus on denim and flat fabrics. The consortium was constituted by a textile manufacturer, Legler S.p.A. (I), the surface inspection companies Parsytec Computer GmbH (D) and maVis GmbH (D), which during the cooperation in the HIPERTEX project formed a joint venture, and the research institute IEI-CNR (I). The R&D results presented here were the results of the joint activities between the inspection companies and the subcontractor, ITWM.

References

1. CH. DAUL, J. GROTEPASS, U. KNAAK, R. FÖHR. Esprit project 21089, High Performance system for textile evaluation, midterm report, January 1998.
2. A. ELDIN. Computer vision for automated inspection of fabric products: methodology for feature extraction and classification, Phd Thesis of the university of Wuppertal, Germany, December 1988.
3. C. NEUBAUER. Segmentation of Defects in Textile Fabrics, *Int. Conf. on Pattern Recognition*, pp. 688-691, 1992.
4. CH. JACQUELIN, A. AURENGO, G. HEJBLUM. Evolving Descriptors for Texture Segmentation, *Pattern Recognition*, **30**(7):1069-1079, 1997.
5. J.G. CAMPBELL, C. FRALEY, F. MURTAGH, A.E. RAFTERY. Linear Flaw Detection in Woven Textiles using Model-Based Clustering, University of Ulster, Faculty of Informatics, Preprint INFM-97-001, 1997.
6. H.H. SOLIMAN, A.I. ABD ELFATTAH, N.A. HAMDY, A. ELDIN. Classification of flat product-defects using neural networks concepts, *Proc. of the Second IASTED Int. Conf. on Computer Applications in Industry*, vol.1, pp. 5-8, Zürich, Switzerland, May 1992.
7. F. DE NATALE. Rank-Order Functions for the Fast Detection of Texture Faults. *International Journal of Pattern Recognition and Artificial Intelligence*, **10**(8):971-984, 1996.
8. C. LU, P. CHUNG, C. CHEN. Unsupervised Texture Segmentation via Wavelet Transform, *Pattern Recognition*, **30**(5):729-742, 1997.
9. J. CANNY. A computational approach to edge detection. *IEEE Trans. Pattern Analysis and Machine Intelligence*, **8**(6):678-698, November 1986.

Tomographische 3D-Rekonstruktion von Gußfehlern

Carsten Lehr

Universität Hannover,
Institut für Theoretische Nachrichtentechnik und Informationsverarbeitung,
Prof. Dr.-Ing. C.-E. Liedtke, Appelstr. 9a, 30167 Hannover, Germany
lehr@tnt.uni-hannover.de

Zusammenfassung Voraussetzung für die tomographische Prüfung von Gußkörpern auf interne Defekte ist in der Regel eine große Anzahl um den Gußkörper verteilter Aufnahmen. Im Bereich der industriellen Qualitätsprüfung steht jedoch zum Teil aus Zeit-und Kostengründen nur eine sehr begrenzte Anzahl von Aufnahmen zur Verfügung. Zudem verhindert teilweise das Aspektverhältnis der Bauteile eine Prüfung von allen Seiten. Herkömliche Tomographieverfahren sind in solchen Fällen nicht einsetzbar. In diesem Beitrag wird ein neues Verfahren zur tomographischen 3D-Rekonstruktion von Gußfehlern aus wenigen Röntgenaufnahmen vorgestellt. Die Rekonstruktion wird in einem zweistufigen Verfahren durchgeführt. Zunächst wird das Rekonstruktionsgebiet auf Regions of Interest um die Defektbereiche eingeschränkt und anschließend die 3D-Rekonstruktion mit einem neu entwickelten iterativen Tomographieverfahren durchgeführt. Für die Regularisierung des tomographischen Rekonstruktionsproblems wird das Maximum Entropie Prinzip in Verbindung mit einem iterativen Verfahren zur Binärisierung der Rekonstruktion verwendet. Die Rekonstruktion wird unabhängig von a priori Annahmen über die Form der zu rekonstruierenden Gußfehler durchgeführt.

1 Einleitung

Die weitere Verbreitung tomographischer Prüfverfahren im Bereich der industriellen Qualitätskontrolle wird derzeit durch verschiedene Einflußfaktoren limitiert. Zum einen sind herkömliche Tomographieanlagen für viele Anwendungen zu aufwendig und teuer sowie die benötigen Prüfzeiten zu lange [1]. Zum anderen verhindert häufig das Aspektverhältnis der Bauteile eine für die Prüfung notwendige Durchstrahlung von allen Seiten. Es besteht daher großes Interesse an Tomographieverfahren, die auf der Basis einfacher Röntgendurchstrahlungsaufnahmen arbeiten (Reduzierung des Hardware-Aufwands) und eine Rekonstruktion aus wenigen Aufnahmen ermöglichen. Das tomographische Rekonstruktionsproblem ist in solchen Fällen stark unterbestimmt und auf Grund von verschiedenen Fehlereinflüssen (Kalibrierungsfehler, Rauschen) inkonsistent. Herkömmliche Tomographieverfahren (gefilterte Rückprojektion, Transformationsmethoden) sind daher für die Auswertung nicht geeignet. A priori Informationen über das zu rekonstruierende Objekt müssen zusammen mit einem Regularisierungsverfahren verwendet werden, um eine eindeutige Lösung zu erhalten.

Es werden in der Literatur verschiedene Lösungsansätze vorgeschlagen, die auf einer Beschreibung der zu rekonstruierenden Objekte durch parametrisierte Ansatzfunktionen beruhen (z.B. [2], [3]). Auf Grund der Vielfalt möglicher Gußfehler sind modellbasierter Ansätze für die Gußfehlerrekonstruktion jedoch ungeeignet, bzw. immer nur für eine stark eingeschränkte Klasse von Defekten zulässig. Dagegen stehen die Materialparameter des Gußkörpers als zuverlässige und wichtige a priori Information zu Verfügung.

Im folgenden wird ein zweistufiges Verfahren zur 3D-Rekonstruktion von Gußfehlern aus wenigen Röntgenaufnahmen vorgestellt, das auf einer systematischen Ausnutzung der a priori Kenntnis der Materialparameter beruht und ohne spezielle a priori Annahmen über die Form der zu rekonstruierenden Defekte auskommt. Idee dieses Verfahrens ist es, zunächst durch eine Beschränkung der Rekonstruktion auf Regions of Interest um die jeweiligen Defekte die Komplexität des Rekonstruktionsproblems zu reduzieren. Dazu werden die Defektbereiche segmentiert und die Ausdehnung der Defekte in Durchstrahlungsrichtung aus den gemessenen Daten berechnet. In einem zweiten Schritt werden die Defekte mit einem speziell für die Auswertung nur weniger Aufnahmen entwickelten iterativen Tomographieverfahren rekonstruiert. Die Regularisierung des Rekonstruktionsproblems erfolgt auf der Grundlage des Maximum Entropie Prinzips in Verbindung mit einem iterativen Verfahren zur Binärisierung des Rekonstruktionsergebnisses.

2 Vorverarbeitung

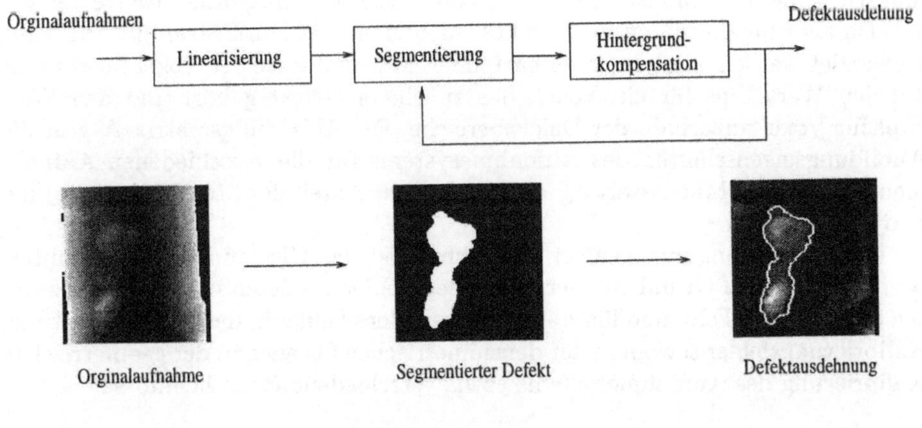

Abbildung1. Vorverarbeitung

Ziel der Vorverarbeitung ist die Einschränkung des Rekonstruktionsgebiets auf Regions of Interest um die Defektbereiche um die Komplexität der Rekonstruktionsproblems zu reduzieren. In Abbildung 1 ist das Prinzip der Vorverarbeitung dargestellt. Die Bilddaten werden zunächt unter Berücksichtung polychromatischer Strahlung und bekannter Materialparameter des Gußmaterials in durchstrahlte Materialausdehnungen umgerechnet (Linearisierung). Die Defektbereiche werden segmentiert (Segmentierung) und die Materialausdehnungen in den Defektbereichen aus den umgebenden Materialausdehnungen extrapoliert. Extrapolierte und gemessene Materialausdehnungen in Defektbereichen werden verglichen und daraus die Defektausdehnungen berechnet (Hintergrundkompensation) (vergleiche [4]). Die berechneten Defektausdehnungen werden iterativ für eine Verbesserung von Segmentierung und Hintergrundkompensation verwendet.

3 3D-Rekonstruktion

Ausgehend von der in der Vorverarbeitung wird die 3D-Rekonstruktion der Defekte berechnet. Dazu wird das Objektgebiet diskretisiert und die Abbildung der einzelnen Voxel in die verschiedenen Aufnahmen mit Hilfe einer Abbildungsmatrix beschrieben (Gleichung 1).

$$\boldsymbol{p} = \mathbf{A} * \boldsymbol{\mu} \tag{1}$$

$\mathbf{A}$: Abbildungsmatrix
$\boldsymbol{p}$: Vektor aller N Bildpunkte
$\boldsymbol{\mu}$: Vektor aller M Objektvoxel

Der Datenvektor $\boldsymbol{p}$ enthält die Bildwerte der verschiedenen Aufnahmen, d.h hier die in der Vorverarbeitung berechneten Defektausdehnungen in Durchstrahlungsrichtung. Der unbekannte Vektor $\boldsymbol{\mu}$ repräsentiert die Materialwerte der Voxel. Da als Eingangsdaten die Defektausdehnungen in Durchstrahlungsrichtung verwendet werden, sind die zu erwartenden Materialwerte der Voxel beschränkt auf den Wert Eins für ein Voxel, das zu einem Defekt gehört und den Wert Null für Voxel außerhalb der Defektbereiche. Die Abbildungsmatrix $\mathbf{A}$ gibt die Abbildungseigenschaften des Aufnahmesystems für die verschiedenen Aufnahmen wieder. Der Matrixwert a_{ij} entspricht dem Anteil des i.ten Voxels am j.ten Projektionsstrahl.

Bei Verwendung nur weniger Aufnahmen ist das Gleichungssystem 1 unterbestimmt und auf Grund von verschiedenen Fehlereinflüssen in den Eingangsdaten inkonsistent. Fehlereinflüsse sind insbesondere Fehler in der Vorverarbeitung, Kalibrierungsfehler sowohl in der densiometrischen als auch in der geometrischen Kalibrierung des Aufnahmesystems sowie verschiedene Rauscheinflüsse.

3.1 Maximum Entropie Prinzip

Das Maximum Entropie Prinzip ist ein in verschiedenen Anwendungsbereichen verwendetes Kriterium zur Bestimmung der Lösung eines unterbestimmten und inkonsistenten Gleichungssystems ([5], [6]). Die Entropie ist ein Maß für den Informationsgehalt einer Lösung. Im Sinne des Maximum Entropie Prinzips ist diejenige Lösung des Rekonstruktionsproblems optimal, die die maximale Entropie und damit den minimalen Informationsgehalt aller möglichen Lösungen aufweist (Gleichung 2). Diese Lösung ist maximal unbestimmt in Bezug auf nicht gemessene Projektionen, d.h. es werden nur solche Strukturen rekonstruiert, die eindeutig aus den gemessenen Daten hervorgehen.

Maximiere

$$Entropie = - \sum_{i=0}^{M-1} \mu_i * \ln(\mu_i) \tag{2}$$

mit

$$\boldsymbol{p} = \mathbf{A} * \boldsymbol{\mu};$$
$$0 <= \mu_i <= 1, \forall i \in \{0, M-1\}$$

Zur Berechnung der Lösung maximaler Entropie werden in der Literatur verschiedene Verfahren vorgeschlagen. Eigene Untersuchungen sowie die Ergebnisse von Subbarao ([7]) zeigen, daß das MART-Verfahren für die vorliegende Anwendung die besten Ergebnisse liefert. Der MART-Algorithmus beruht auf einer iterativen Anpassung der Rekonstruktionsdaten an die Bilddaten durch multiplikative Korrekturen entsprechend folgendem Algorithmus.

1. Berechne für jeden Bildpunkt p_j die geschätze Projektion $\tilde{p}_j = \sum_k a_{kj} * \mu_k$
2. Für jeden Voxel i im Projektionsstrahl j:

$$\mu_i^{neu} = \mu_i^{alt} * \left(\frac{\tilde{p}_j}{p_j}\right)^{\lambda * w_{ij}} \tag{3}$$

w_{ij}: Anteil des i.ten Voxels am j.ten Projektionsstrahl
λ: Relaxationsfaktor

Je nach Rauschanteil der Daten wird ein kleiner Relaxationsfaktor λ zwischen 0.01 und 0.1 gewählt, um eine stabile Konvergenz des Verfahren zu garantieren. Zusätzlich führt eine wiederholte Medianfilterung der Zwischenergebnisse zu einer deutlichen Verbesserung des Konvergenzverhalten (vergleiche [8]).

Die Materialwerte der nach Gleichung 3 bestimmten Lösung sind nicht auf Null oder Eins beschränkt. Die Lösung kann als eine Wahrscheinlichkeitsverteilung interpretiert werden, die angibt, wie wahrscheinlich ein Voxels zum zu rekonstruierenden Defekt gehört. Je gößer der Materialwert eines Voxels ist, desto wahrscheinlicher ist es, daß der Voxel zum Defekt gehört.

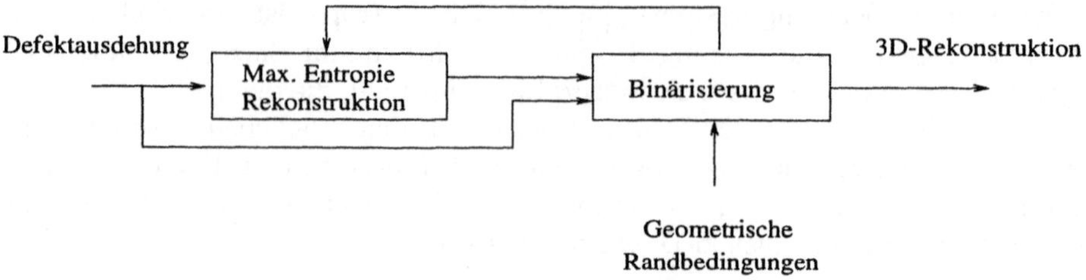

Abbildung2. Rekonstruktionsverfahren

3.2 Iterative Binärisierung

Für die binäre Rekonstruktion der Gußfehler wird das Maximum Entropie Prinzip mit der Nutzung der a priori Kenntnis der Materialparameter verknüpft. Der Ablauf des entwickelten Rekonstruktionsverfahrens ist in Abbildung 2 skizziert.

Zunächst wird die Löung maximaler Entropie berechnet. In einem Binärisierungsprozeß werden dann diejenigen Voxel mit den größten Materialwerten als zum zu rekonstruierenden Defekt gehörend identifiziert. Anschließend wird eine an die als Defekt identifizierten Voxel als Nebenbedingung angepaßte neue Lösung maximaler Entropie bestimmt. Dieses Verfahren wird iterativ bis zu einer vollständigen binären Rekonstruktion wiederholt. Für die Steuerung der Binärisierung werden einige natürliche geometrische Randbedingungen genutzt, die als sicheres a priori Wissen über die zu rekonstruierenden Gußfehler zur Verfügung stehen. So kann z.B. innerhalb des Defekts kein Material eingeschlossen sein. Liegen zusätzliche a priori Annahmen über die Form der Gußfehler vor, so werden diese in Form weiterer Nebenbedingungen in den Binärisierungsprozeß eingebracht. Entscheidend für den Erfolg des Verfahrens ist, daß in jedem Iterationsschritt nur relative wenige Voxel neu dem Defekt zugeordnet werden, um eine schrittweise Anpassung der Lösung an die Binärisierung zu gewährleisten.

4 Ergebnisse

Die Untersuchung des vorgestellten Verfahrens wird zunächst an simulierten Datensätzen durchgeführt. Als Testobjekte werden eine Kugel, ein Quader und ein in Abbildung 3 dargestelltes Objekt (Objekt 3) komplexer Geometrie verwendet. Die Rekonstruktionen werden jeweils aus 5 in einem Winkelbereich von 90° verteilten Aufnahmen berechnet. Die Lage des 90°-Winkelbereichs der Aufnahmen wird in verschiedene Versuchen variiert, um die Abhängigkeit der Ergebnisse von der Projektionsrichtung berücksichtigen zu können. Um den Einfluß fehlerhafter Meßdaten untersuchen zu können, wird den simulierten Daten gaussches Rauschen verschiedener Stärke überlagert. Als Fehlermaß zur Beurteilung der

	Rauschen 0%	Rauschen 5%	Rauschen 10%
Kugel	0%	1%	3%
Quader	0%	2%	3%
Objekt 3	2%	5%	7%

Tabelle1. Simulationergebnisse: Fehlerhaft rekonstruierte Voxel

Rekonstruktion eines Objekts wird die Anzahl der falsche rekonstruktierten Voxel bezogen auf das Volumen des Objekts verwendet.

In Tabelle 4 sind die Ergebnisse der Rekonstruktion aus simulierten Projektionen zusammengestellt. Die angegebenen Ergebnisse sind über mehrere Versuche gemittelt. Ohne Rauscheinfluß wird für Kugel und Quader eine exakte Rekonstruktion erreicht. Das komplexe Objekt 3 wird im Mittel ebenfalls sehr gut rekonstruiert. Je nach Wahl der Projektionsrichtungen treten jedoch leichte Abweichungen der rekonstruierten Form von der wahren Form auf. Diese von der Projektionsrichtung abhängigen Artefakte sind auf den stark eingeschränkten Winkelbereich der Aufnahmen zurückzuführen.

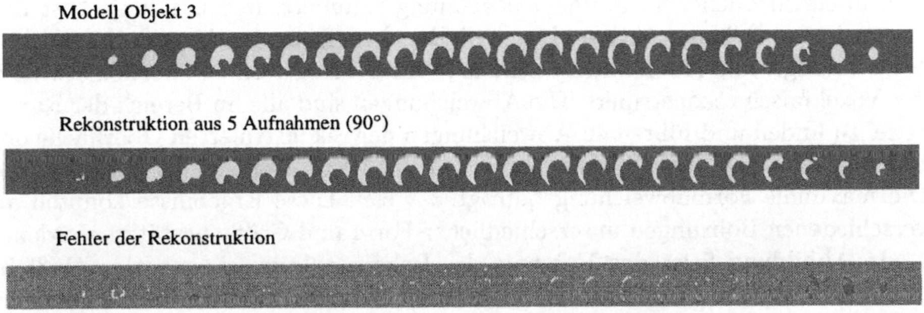

Abbildung3. Rekonstruktion Objekt 3

Mit zunehmenden Rauschen nimmt die Anzahl der falsch rekonstruierten Voxel zu, die Objektform wird jedoch in allen Fällen nur unwesentlich schlechter als im Fall idealer Daten wiedergegeben. In Abbildung 3 ist das Rekonstruktionsergebnis für Objekt 3 (10% Rauschen) schichtweise dargestellt. In der Darstellung der Rekonstruktionsfehler sind Voxel, die fehlerhaft als zum Defekt gehörende rekonstruiert wurden, weiß dargestellt. Voxel, die fehlerhaft als nicht zum Defekt gehörend bestimmt wurden, sind schwarz dargestellt.

Im Anschluß an die Simulationen werden Untersuchungen an Defekten bekannter Geometrie sowie an realen Gußfehlern durchgeführt. Die Aufnahmen

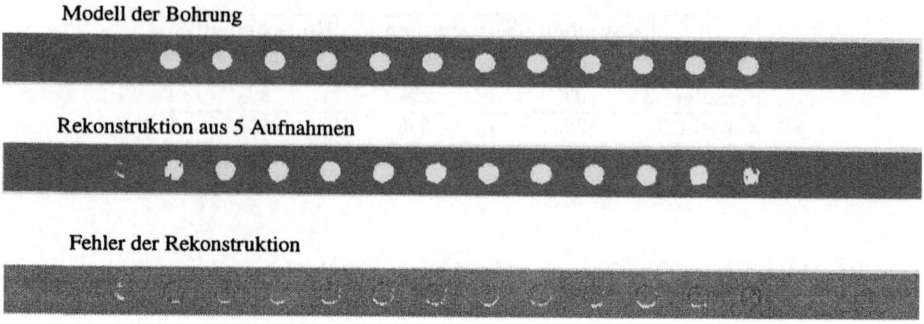

Abbildung4. Rekonstruktion einer zylindrischen Bohrung

wurden mit einem einfachen Radioskopiesystem bestehend aus Mikrofokus-Röntgenröhre, Bildwandler und CCD-Kamera aufgenommen. Die Kalibrierung der geometrischen Abbildungseigenschaften des Aufnahmesystems erfolgte über ein spezielles Kalibrierungsverfahren. Einzelheiten des Kalibrierungsverfahrens sind in [4] beschrieben.

Defekte bekannter Geometrie werden in einem Testkörper durch Bohrungen unterschiedlicher Tiefe und Ausdehnung simuliert. In Abbildung 4 ist das Ergebnis der Rekonstruktion einer zylindrischen Bohrung dargestellt. Bei der Verwendung von 5 Aufnahmen in einem Winkelbereich von 90° werden etwa 3% der Voxel falsch rekonstruiert. Die Abweichungen sind alle im Bereich der Randvoxel zu finden und führen zu Abweichungen der rekonstruierten Oberfläche der Bohrung von der tatsächlichen Form von im Mittel weniger als einem Voxel. Die maximale Formabweichung beträgt 2 Voxel. Diese Ergebnisse konnten an verschiedenen Bohrungen unterschiedlicher Form und Größe bestätigt werden.

In Abbildung 5 ist das Ergebnis der Rekonstruktion eines realen Gußfehlers wiederum aus 5 Aufnahmen in einem Winkelbereich von 90° dargestellt. Gegenübergestellt sind die Maximum Entropie Lösung und die binärisierte Rekonstruktion. Erst die Binärisierung läßt die tatsächliche Größe und Form des Defekts erkennen.

Die vorgestellten Versuchsergebnisse zeigen, daß die a priori Kenntnis der Materialparameter die 3D-Rekonstruktion von Gußfehlern aus wenigen Röntgenprojektionen ermöglicht ohne auf a priori Annahmen über die Form der zu rekonstruierenden Fehler zurückgreifen zu müssen.

5 Zusammenfassung

Es wurde ein Verfahren zur tomographischen Rekonstruktion von Gußfehlern aus wenigen Röntgenaufnahmen vorgestellt. Das zweistufige Verfahren beruht auf einer systematischen Nutzung der a priori Kenntnis der Materialparameter der Gußkörper in Verbindung mit der Beschränkung der Rekonstruktion auf Regions

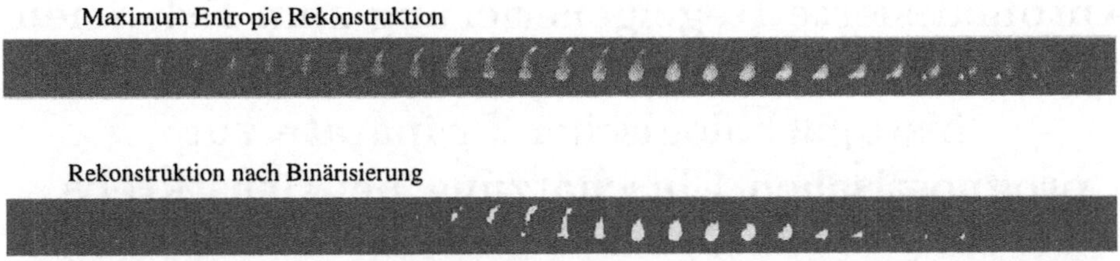

Abbildung5. Rekonstruktion eines Gußfehlers

of Interest um Defektbereiche. Für die Regularisierung des tomographischen Rekonstruktionsproblems wird das Maximum Entropie Prinzip in Verbindung mit einem iterativen Verfahren zur Binärisierung der Rekonstruktion verwendet.

Untersuchungen an simulierten Datensätzten sowie an Defekten bekannter Geometrie haben gezeigt, das das vorgestellte Verfahren eine Rekonstruktion aus sehr wenigen Aufnahmen ermöglicht. Bei einfachen Defektgeometrien ergibt sich eine hohe Rekonstruktionsgenauigkeit. Die Genauigkeit der Rekonstruktion komplexer Defekte wird derzeit in weiteren Versuchen untersucht.

Literatur

1. Hanke,R.: Ein Verfahren zu Bestimmung von Lage und Umfang von Volumendefekten in Echzeit am Beispiel der Gußteileprüfung. Dissertation Erlangen, 1996.
2. Milanfar, P.; Karl, W.C.; Willesky, A.S.: Reconstructing Binary Polygonal Objects from Projections: A Statistical View. Graphical Models and Image Processing, Vol 56, 5, 1994.
3. Wallingford, R.M.; Basart, J.P.: Reconstruction of Ellipsoidal Flaw Models from X-Ray Radiographs. Review of Progress in Quantitative Nondestructive Evaluation, Vol 10A, 1991.
4. Lehr, C.: 3D-Analyse von Gußfehlern. 19.DAGM Symposium, Braunschweig,1997.
5. Reis, M.L.; Roberty, N.C.: Maximum entropy algorithm for image reconstruction from projections. Inverse Problmes 8 (1992), S. 623-644.
6. Nailong Wu: The Maximum Entropy Method, Springer-Verlag Berlin, 1997
7. Subbarao, P.M.V.; Munshi, P., Muralidhar, K.: Performance of iterativ tomographic algorithms applied to non-destructive evaluation with limited data. NDT& E International, Vol. 30, 6,1997, S. 359-370
8. Eckert, G.: Untersuchung algebraisch iterativer Tomographieverfahren zur 3D-Rekonstruktion von Gußfehlern aus Röntgenprojektionen. Studienarbeit, Universität Hannover, 1998.

Automatisierte Regelgenerierung zum Erkennen von Mustern aus Merkmalen zyto- und histopathologischer Präparate zur prognostischen Einschätzung bei Brustkrebs

K. Gärtner[1], R. Schulze[1], S. Fuchs[1], F. Theissig[2], V. Dimmer[2], G. Haroske[2], K.-D. Kunze[2] und W. Meyer[2]

[1]Institut für Künstliche Intelligenz, Fakultät Informatik
[2]Institut für Pathologie, Universitätsklinikum
Technische Universität Dresden, Mommsenstr. 13, D-01062 Dresden

Zusammenfassung Infolge der Komplexität der Problemstellung ist die Formulierung von Entscheidungsregeln für ein Expertensystem zur Brustkrebsprognose aus histologischen und zytologischen Fallstudien sehr schwierig.
Zur Unterstützung dieses Prozesses wird hier ein neuronales Fuzzy-System untersucht. Um Muster in Form von unscharfen Regeln aus Datenmengen zu extrahieren, ist die Vorgabe der linguistischen Terme der ausgewählten Variablen durch einen Fachmann notwendig. In den Prozeß wird auch Expertenwissen eingebracht, jedoch nicht unter dem Zwang Entscheidungsregeln formulieren zu müssen, also in angemessener Form und mit Experimentiermöglichkeiten.
Zur Bewertung der erzeugten Regeln werden die Prognoseergebnisse verglichen, die einerseits mit von Experten formulierten Regeln und andererseits mit generierten Regeln des neuronalen Fuzzy-Systems erzielt werden. Die mit dem System erzeugten Regeln zeigen bezüglich der Fehlerrate ein deutlich besseres Entscheidungsverhalten als die Expertenregeln. Eine Optimierung für unsymmetrische Kosten steht noch aus.

1 Einleitung

In der medizinischen Versorgung gibt es einen deutlichen Bedarf nach einer automatisierten Entscheidungsunterstützung zur Bestimmung der Prognose von invasiven Karzinomen der Brustdrüse als Grundlage für eine individuelle Behandlung nach einer Operation. Durch die Definition von Risikogruppen soll die bisher unvermeidliche therapeutische Über- oder Unterbehandlung von Patientinnen eingeschränkt oder vermieden werden. Die konventionelle visuelle Beurteilung der in der Zyto- und Histopathologie vorliegenden komplexen Präparate ist im Hinblick auf eine prognostische Einschätzung in vielen Fällen nicht eindeutig und mit erheblichen Varianzen belastet.

Regelbasierte Expertensysteme ermöglichen die Befundkriterien zu präzisieren und zu vereinheitlichen, die Beobachtungen auf Konsistenz zu überprüfen und die Entscheidung nachvollziehbar und transportierbar zu machen. Die Gewinnung der Entscheidungsregeln für ein Expertensystem aus histologischen und zytologischen Fallstudien bleibt aber auf Grund der Komplexität der Problemstellung für den Pathologen ausgesprochen schwierig. Ein Grund dafür ist der ständige Fortschritt bezüglich der Bildanalysetechnik neuer Merkmale, mit denen noch wenig oder keine Expertise vorliegen. Ein anderer Grund ist die objektive und subjektive Unsicherheit. So werden maligne Tumore in der konventionellen morphologischen Diagnostik vorwiegend an histologischen Schnitten, aber auch an zytologischen Präparaten durch die Zuordnung zu einem bestimmten histologischen Typ und durch die Bestimmung des Grades der Malignität charakterisiert. Beide Charakteristika sind neben der Tumorgröße und dem Lymphknotenstatus für die Prognose eines Tumors und für die Festlegung der Therapieform von ausschlaggebender Bedeutung.

Die Bestimmung des Malignitätsgrades erfolgt nach qualitativen histologischen und zytologischen Kriterien, die die Differenzierung, das Wachstumsverhalten und die zelluläre Polymorphie eines Tumors widerspiegeln [7] [8]. Die meisten dieser Kriterien sind aber schlecht reproduzierbar. Ihre Anwendung für eine Festlegung der prospektiven Potenz und die Bestimmung des individuellen malignen Potentials eines Tumors ist mit Unsicherheit behaftet.

Weiterhin erfordert eine stärkere Orientierung auf individualisierte Therapiekonzepte die systematische Verknüpfung und Bewertung aller erreichbaren klinischen, paraklinischen und morphologischen Informationen, die zu einem gegebenen Tumorleiden gesammelt werden können, um mit höherer Konfidenz den weiteren Krankheitsverlauf und die Ansprechbarkeit auf bestimmte Therapieformen für den individuellen Patienten voraussagen zu können. Dies erhöht wieder die Komplexität und damit die Schwierigkeiten der Regelerstellung.

Zur Lösung dieses Problems werden auch für andere ähnliche Entscheidungsaufgaben, Methoden zur mehr oder weniger automatisierten Regelgenerierung untersucht [1] [2] [3] [4] [5] [9].

Einer der jüngeren Ansätze nutzt die Verarbeitungsstruktur und die Lernfähigkeit von neuronalen Netzen und die Modellierung der unscharfen Entscheidungen mit Fuzzy–Regeln. Dieser auch als Neuro–Fuzzy–Modell in der Literatur [2] [4] bekannte Ansatz ist hier wegen des einzubeziehenden, in verbaler Form vorliegenden medizinischen Fachwissens, der unbedingt gewünschten Interpretierbarkeit der Regeln durch Fachleute und auch wegen der Komplexität der Eingangsdaten zweckmäßig.

In dieser Arbeit zeigen wir, wie dieser Ansatz an die Aufgabe angepaßt wird und sehr ermutigende Ergebnisse für die Lösung des oben beschriebenen Problems liefert.

2 Datenmustererkennung mit Fuzzy–Mengen

Mittels der Datenmustererkennung soll aus großen, komplexen Datenmengen nützliches Wissen extrahiert werden. Der Schwerpunkt liegt hierbei darauf, Muster zu identifizieren, um daraus Regeln abzuleiten. Gewöhnlich nutzt man in solchen Fällen Standard–Clusterverfahren wie den K–Means–Algorithmus.

Die Spezifizierung linguistischer Terme in Form von Fuzzy–Mengen vor der Aufnahme von detallierten Analysen und dem Start rechnergestützter Algorithmen führt dabei zu einer fokussierten Entdeckung von Regelmäßigkeiten und Abhängigkeiten, die schon im Kontext durch den Nutzer aufgestellt werden.

Formal ist ein Entscheidungssystem durch eine Eingangsbeschreibung $\underline{x}$, eine Ausgangsbeschreibung $\underline{y}$ und ein Regelsystem $\underline{y} = f(\underline{x})$ definiert.

Im Anwendungsfall wurden kernbildanalytische Untersuchungen an 200 histologischen Präparaten durchgeführt. Aus den digitalisierten Zellkernbildern und weiteren klinikopathologischen Daten werden 80 Merkmale abgeleitet, die die Histo- und Zytopathologie des Tumors sowie das Tumorstadium der Patienten charakterisieren.

Nach einer Auswahl durch Korrelationsanalyse und Expertenbefragung werden in den Versuchsreihen 7 Merkmale als linguistische Eingangsvariable $x_1 \ldots x_7$ verwendet und die Meßwerte durch Fuzzifizierung in geeignete Fuzzy–Mengen umgewandelt.

Als linguistische Ausgangsvariable y wird die Prognoseabschätzung nach Brustkrebsoperationen anhand der 5–Jahres–Überlebensrate gebildet.

Im nachfolgenden Regelgenerierungsprozeß mit anschließender Termoptimierung werden in einer Entwicklungsumgebung für neuronale Fuzzy–Systeme, dem Fuzzy–Rule–Generator [6], unscharfe Regeln für das Expertensystem zur Brustkrebsprognose ermittelt (Abb. 1).

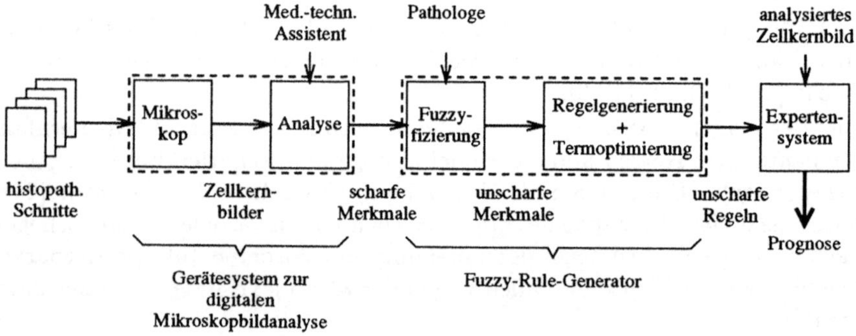

Abbildung1. Regelgenerierungsprozeß aus histopathologischen Schnitten für ein Expertensystem zur Prognose bei Brustkrebs

Der funktionale Zusammenhang zwischen den Merkmalen und der Prognose wird dabei mit einer Regelmenge der Form:

$$
\begin{array}{llllllll}
R_1\text{:} & \text{IF} & x_1 \text{ IS } A_{i_{1,1}} & \text{AND} \ldots \text{AND} & x_7 \text{ IS } A_{i_{7,1}} & \text{THEN} & y = B_{j1} \\
\vdots & & \vdots & \vdots & \vdots & & \vdots \\
R_r\text{:} & \text{IF} & x_1 \text{ IS } A_{i_{1,r}} & \text{AND} \ldots \text{AND} & x_7 \text{ IS } A_{i_{7,r}} & \text{THEN} & y = B_{jr} \\
\vdots & & \vdots & \vdots & \vdots & & \vdots \\
R_k\text{:} & \text{IF} & x_1 \text{ IS } A_{i_{1,k}} & \text{AND} \ldots \text{AND} & x_7 \text{ IS } A_{i_{7,k}} & \text{THEN} & y = B_{jk}
\end{array}
$$

widergespiegelt.

Dabei bezeichnet $A_{i_{p,r}}$ den in der Regel R_r ($r \in \{1 \ldots k\}$) verwendeten Term $A_i^{(p)}$ ($i \in \{1,2,3\}$) der linguistischen Variablen x_p und B_{jr} ($j \in \{1,2\}$) in analoger Weise den Term B_j der Variablen y. Mit $\mu_i^{(p)}$ bzw. ν_j werden die Zugehörigkeitsfunktionen der Terme der linguistischen Eingangs- und Ausgangsvariablen bezeichnet.

3 Fuzzy–Rule–Generator

Das Entwicklungswerkzeug Fuzzy–Rule–Generator verwendet zur Erkennung des in den Falldaten verborgenen Wissens das neuronale Fuzzy–Modell FALCON (Fuzzy Adaptive Learning COntrol Network) [2]. Die Grundlage für dieses Modell bildet die Transformation eines Fuzzy–Systems mit einer Regelbasis, wie sie im vorangegangenen Abschnitt angegeben ist, in eine neuronale Architektur (Abb. 2). Die Struktur eines FALCON–Systems entspricht einem vorwärtsgerichtetes, fünfschichtiges Netzwerk. Die Neuronen der Eingabeschicht repräsentieren die linguistischen Eingangsvariablen x_i und die Neuronen der Ausgabeschicht die linguistischen Ausgangsvariablen y_j. Die Neuronen der zweiten und vierten Schicht werden als Termneuronen bezeichnet und bilden die Zugehörigkeitsfunktion der Terme der linguistischen Variablen ab. Das hier betrachtete System basiert auf Zugehörigkeitsfunktionen der Form

$$\mu(x) = e^{-\frac{(x-m)^2}{\sigma^2}},$$

wobei m den Mittelwert und σ die Varianz der Glockenkurve bezeichnen. Die Neuronen der dritten Schicht repräsentieren die linguistischen Regeln.

Die Regeln werden durch einen Lernvorgang erzeugt, der auf eine initialisierte Startkonfiguration aufsetzt und sich aus einem Struktur- und Parameterlernagorithmus zusammensetzt (Abb. 3).

Vor dem Lernen ist die Anzahl der Terme für die linguistischen Ein- und Ausgangsvariablen vorzugeben und damit die Startkonfiguration des neuronalen Netzes zu erzeugen. Dieser Schritt ist jedoch nicht mit solcher Unsicherheit behaftet wie bei üblichen Neuronalen–Netz–Anwendungen. Er ist einerseits machbar durch das Expertenwissen und gibt andererseits die Chance solches Apriori–Wissen einzubeziehen.

In einer ersten Phase des Lernens werden zunächst die Fuzzy-Mengen der linguistischen Variablen durch ein Clusterverfahren, das die KOHONEN–Lernregeln

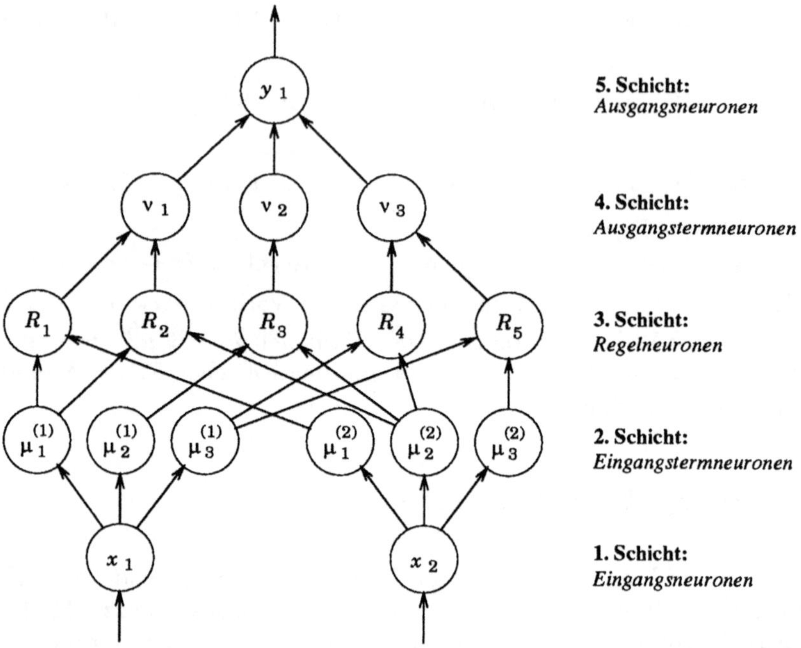

Abbildung2. Struktur eines FALCON-Systems mit zwei Eingangsvariablen und einer Ausgangsvariablen

und eine First–nearest–neighbor–Heuristik nutzt, initialisiert. Anschließend werden in einem Wettbewerbslernverfahren (GROSSBERG–Lernregel) die Entscheidungsregeln identifiziert und ihre Konklusionen festgelegt. Das nachfolgende Regelkombinationsverfahren entfernt überschüssige Regelneuronen aus dem System, wenn drei Bedingungen erfüllt sind:

- die Regeln haben die gleichen Konklusionen,
- die betrachteten Regeln weisen in der Prämisse mehrere Vorbedingungen auf, darunter mindestens eine gemeinsame und
- die Anzahl dieser Regeln ist gleich der Zahl der Kombinationen, die sich aus den Termen der linguistischen Variablen ergibt, die in den nicht gemeinsamen Vorbedingungen enthalten sind.

Die zweite Phase des Lernverfahrens dient der Einstellung der Parameter der mit den linguistischen Termen assoziierten Zugehörigkeitsfunktionen. Die Adaption der Parameter erfolgt unter Verwendung eines für das FALCON-Modell modifizierten Backpropagation-Algorithmus. Nach der Beendigung des Lernvorgangs wird vom Fuzzy–Rule–Generator die generierte Regelbasis automatisch aus der Netzwerkstruktur extrahiert und in eine deklarative Wissensrepräsentation überführt. Dabei benutzt das System eine reduzierte Version der Fuzzy Programmierung Language (FPL). Mit graphischen Darstellungen, die der

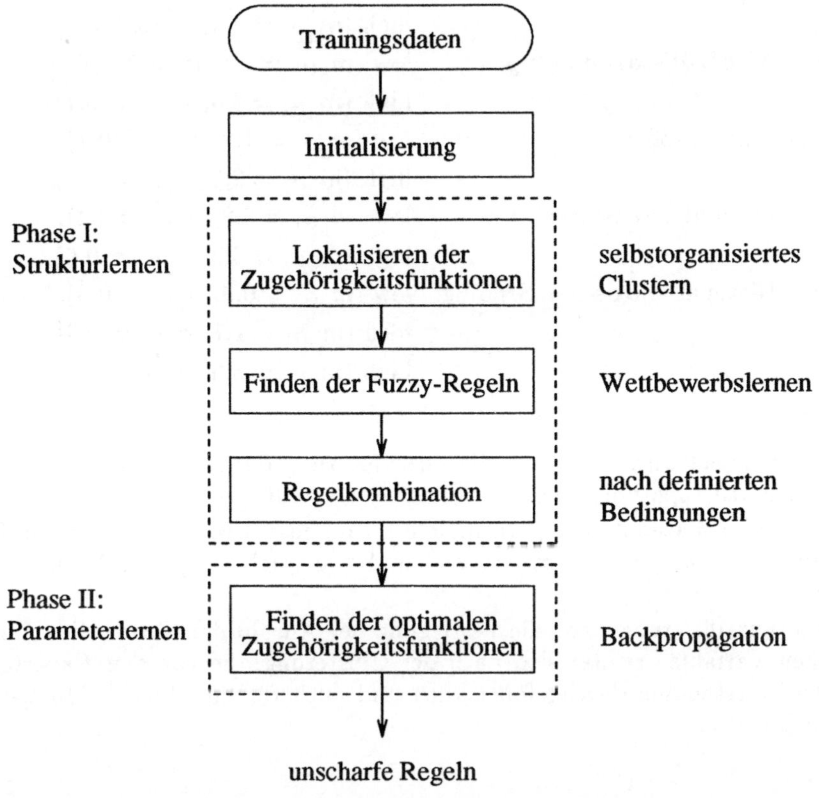

Abbildung3. Flußdiagramm des Lernalgorithmus

Fuzzy–Rule–Generator ausgibt (vgl. Abb. 4), wird dem Anwender ein visueller qualitativer Eindruck vom Charakter der linguistischen Variablen und der Zusammenhänge vermittelt.

4 Ergebnisse

4.1 Linguistische Variable und ihre Partitionierung

Ausgehend von der Korrelationsbewertung der Merkmale und nach Abstimmung mit den Pathologie–Experten werden folgende Eingangsvariablen in den Entscheidungsfindungsprozeß einbezogen und in folgender Weise mit linguistischen

Termen partitioniert:

x_1 - Kernfläche:	low ($m_{\mu_1^{(1)}} = 630$, $\sigma_{\mu_1^{(1)}} = 138$),
	high ($m_{\mu_2^{(1)}} = 907$, $\sigma_{\mu_2^{(1)}} = 138$)
x_2 - Zahl der Granula:	low ($m_{\mu_1^{(2)}} = 15$, $\sigma_{\mu_1^{(2)}} = 4.2$),
	high ($m_{\mu_2^{(2)}} = 24$, $\sigma_{\mu_2^{(2)}} = 4.4$)
x_3 - Rel. Maxima der Texturkurve:	low ($m_{\mu_1^{(3)}} = 3.9$, $\sigma_{\mu_1^{(3)}} = 0.4$),
	high ($m_{\mu_2^{(3)}} = 5.4$, $\sigma_{\mu_2^{(3)}} = 0.9$)
x_4 - AUER-Histogrammtyp:	low ($m_{\mu_1^{(4)}} = 1.5$, $\sigma_{\mu_1^{(4)}} = 0.6$),
	high ($m_{\mu_2^{(4)}} = 3.4$, $\sigma_{\mu_2^{(4)}} = 0.7$)
x_5 - Tumorgröße:	low ($m_{\mu_1^{(5)}} = 1.2$, $\sigma_{\mu_1^{(5)}} = 0.7$),
	high ($m_{\mu_2^{(5)}} = 2.3$, $\sigma_{\mu_2^{(5)}} = 0.6$)
x_6 - Lymphknotenstatus:	low ($m_{\mu_1^{(6)}} = 0.8$, $\sigma_{\mu_1^{(6)}} = 0.4$),
	high ($m_{\mu_2^{(6)}} = 3.2$, $\sigma_{\mu_2^{(6)}} = 0.6$)
x_7 - Histopathologisches Grading:	low ($m_{\mu_1^{(7)}} = 0.9$, $\sigma_{\mu_1^{(7)}} = 0.3$),
	med ($m_{\mu_2^{(7)}} = 1.7$, $\sigma_{\mu_2^{(7)}} = 0.6$),
	high ($m_{\mu_3^{(7)}} = 3.0$, $\sigma_{\mu_3^{(7)}} = 0.4$)

Als Beispiel wird in Abb. 4 die Fuzzyfizierung der linguistischen Eingangsvariablen "Histopathologisches Grading" dargestellt.

Die linguistische Ausgangsvariable y ("Prognose") wird in die beiden Terme "ungünstig" ($m_{\nu_1} = 1.0$, $\sigma_{\nu_1} = 2.2$) und "günstig" ($m_{\nu_2} = 5.1$, $\sigma_{\nu_2} = 1.8$) zerlegt.

Die Fuzzifizierung bzw. die Festlegung der Anzahl der Terme dieser linguistischen Variablen richtet sich nach der Clusterung, die von den Experten für ihr Fachverständnis durchgeführt wird und auch für ihre Entscheidungsregeln benutzt wurde.

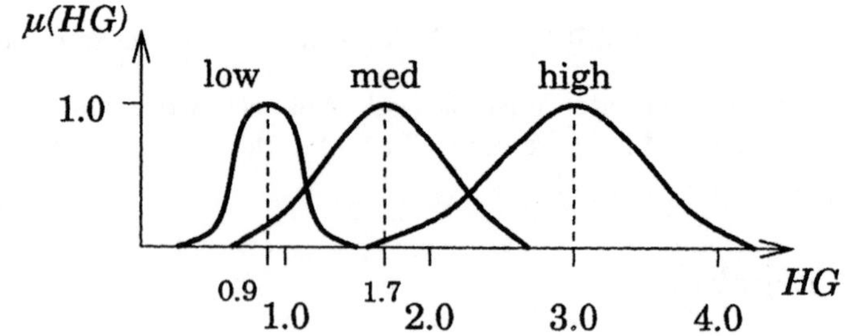

Abbildung4. Fuzzifizierung der linguistischen Eingangsvariable "Histopathologisches Grading" (HG)

4.2 Erzeugte regelbasierte Entscheidungssysteme

Mit zyklischem Austausch werden zwei Drittel der Fälle jeweils als Lernprobe und ein Drittel als Fremdprobe verwendet. Die in den nachfolgenden drei Experimenten mit dem FALCON-System generierten Wissensbasen (1.Versuchsreihe) beschreiben den funktionalen Zusammenhang zwischen den sieben gemessenen Merkmalen und der prognostischen Variablen mit 33/31/34 Regeln. Dabei sind in Versuch 1 und 3 alle Regeln des Experiments 2 enthalten.

Durch den in Abschnitt 3 beschriebenen Lernvorgang werden u.a. folgende Regeln R_1 und R_{12} erzeugt:

$$\begin{aligned} R_1 &: \text{IF } x_3 \text{ IS } \textit{high} \text{ AND } x_4 \text{ IS } \textit{high} \text{ AND } x_5 \text{ IS } \textit{high} \text{ AND } x_7 \text{ IS } \textit{med} \\ &\quad \text{THEN } y = \textit{low} \\ R_{12} &: \text{IF } x_1 \text{ IS } \textit{low} \text{ AND } x_2 \text{ IS } \textit{low} \text{ AND } x_3 \text{ IS } \textit{low} \text{ AND } x_7 \text{ IS } \textit{med} \\ &\quad \text{THEN } y = \textit{high} \end{aligned}$$

Nach Vorschlag der Pathologen wird in einer zweiten Versuchsreihe für die linguistischen Eingangsvariablen "Kernfläche", "Zahl der Granula" und "Relative Maxima der Texturkurve" jeweils ein zusätzlicher Term "MED" (Medium) aufgenommen und mit dem FALCON-System 48, 51 bzw. 46 Regeln generiert.

4.3 Bewertung der regelbasierten Entscheidungssysteme

Die beiden oben beschriebenen Entscheidungssysteme (1. und 2. Versuchsreihe) sollen untereinander und gegenüber dem Expertenregelsystem anhand der Erkennungsrate verglichen werden. Die Berechnung der Erkennungsrate setzt die Bestimmung der linguistischen Terme B_i und B_j der Ausgangsvariablen y voraus, für die die folgenden Bedingungen erfüllt sind:

$$\begin{aligned} \nu_i(\mathrm{y}) &= \arg\max_{\nu_1,\nu_2} \{\nu_1(\mathrm{y}), \nu_2(\mathrm{y})\} \quad \text{und} \\ \nu_j(\mathrm{d}) &= \arg\max_{\nu_1,\nu_2} \{\nu_1(\mathrm{d}), \nu_2(\mathrm{d})\}\,. \end{aligned}$$

Dabei bezeichnet y die Prognoseentscheidung des Systems und d den tatsächlichen Prognosewert. Die Entscheidung e wird durch Vergleich der Terme B_i und B_j ermittelt:

$$e = \begin{cases} 0 & \text{wenn } B_i = B_j \\ 1 & \text{wenn } B_i \neq B_j\,. \end{cases}$$

Dies wiederum impliziert eine symmetrische Kostenfunktion.

In Tab. 1 werden die Erkennungsergebnisse dargestellt, die einerseits mit den durch die Pathologen aufgestellten 42 Entscheidungsregeln und andererseits mit den FALCON-Regeln der beiden Versuchsreihen an der Teststichprobe (insgesamt 151 Fälle) erzielt werden. Die Ergebnisse belegen, daß Neuro-Fuzzy-Systeme (FALCON) in der Lage sind, relevante Regeln zu bestimmen. Darüber hinaus weisen die für die Aufgabe bestimmten Regelsysteme eine bessere Klassifkationsgüte für die Prognose auf als die Regeln der Experten. Es sind aber noch weitere Untersuchungen hinsichtlich einzubeziehender Merkmale, Terme sowie Regeln notwendig. Auch der Übergang zu unsymmetrischen Kostenfunktionen ist, wenn auch meist gescheut, für medizinische Aufgabenstellungen unabdingbar.

Erkennungsrate	Expertenregeln	FALCON-Regeln 1. Versuchsreihe	FALCON-Regeln 2. Versuchsreihe
Regeln	42	34	51
Fälle mit ungünstiger Prognose	94,8%	72,4%	72,4%
Fälle mit günstiger Prognose	49,5%	76,3%	88,2%
Gesamtteststichprobe	66,9%	74,8%	82,1%

Tabelle1. Vergleich der erzielten Erkennungsraten mit Expertenregeln und Regeln des Neuro-Fuzzy-Systems FALCON

Literatur

1. P. W. Hamilton, N. Anderson, P. H. Bartels und D. Thompson: Expert system support using Bayesian belief networks in the diagnosis of fine needle aspiration biospy specimens of the breast. Clin Pathol, 47 (1994) 329–336.
2. C. T. Lin und C. S. G. Lee: Neural Fuzzy Systems. A Neuro-Fuzzy Synergism to Intelligent Systems. Prentice Hall, London/New York, 1996.
3. B. Molnar, und Schäfer: Entscheidungsfindung durch neuronale Netze in der zytologischen Diagnostik. Forschungsbericht, Boehringer Mannheim GmbH, Forschungslabor Tutzing, 1994.
4. D. Nauck, F. Klawonn und R. Kruse: Neuronale Netze und Fuzzy-Systeme. Vieweg-Verlag, Braunschweig/Wiesbaden, 1996.
5. H. Petersohn: Vergleich von multivariaten statitischen Analyseverfahren und Künstlichen Neuronalen Netzen zur Klassifikation bei Entscheidungsproblemen in der Wirtschaft. Peter Lang GmbH, 1997.
6. R. Schulze: Fuzzy–Regelgenerierung in neuronaler Architektur am Beispiel der Ozonprognose. Diplomarbeit, TU Dresden, 1997.
7. F. Theissig: Konventionelle morphologische und kernbildanalytische Untersuchungen zur Metastasierung und Prognose des invasiven duktalen Mammakarzinoms. Habilschrift, Medizinische Akademie Dresden, 1990.
8. F. Theissig, V. Dimmer, G. Haroske, K.-D. Kunze und W. Meyer: Use of nuclear image cytometrie, histopathological grading and DNA cytometry to make breast cancer prognosis more objective. Analytical Celluar Pathology, 3 (1991) 351–360.
9. R. Weber: Entwicklung und Anwendung von Verfahren zur automatischen Akquisition unsicheren Wissens. VDI, 10 (211), 1992.

Visuelle Diagnose von Flammenfeldern in der Kohlenstaubfeuerung*

A. Heinze, K. Debes und H.-M. Groß

Technische Universität Ilmenau
Fachgebiet Neuroinformatik
D-98684 Ilmenau (Thüringen), Postfach 10 05 65
http://cortex.informatik.tu-ilmenau.de
Andrea.Heinze@informatik.tu-ilmenau.de

Zusammenfassung Gegenstand dieser Arbeit ist die Analyse der Farbwerte von Flammenbildern, um Rückschlüsse auf zugrundeliegende Prozeßsituationen zu treffen. Dazu erfolgt die Auswertung die Bildfarbwerte über einen am Datenmaterial belehrten neuronalen Clusterer in einem Histogramm, welches anschließend durch Verwendung eines Backpropagation Netzes bezüglich eingestellter Prozeßparameter klassifiziert wird.

1 Problemstellung

Ein wesentliches Ziel der Automatisierung der Prozeßführung von Feuerungsvorgängen in Kohlekraftwerken besteht in einer *Wirkungsgradmaximierung* bei gleichzeitiger *Emissionsminimierung*. Dazu wurden bislang nur prozeßfern gewonnene und daher totzeitbehaftete Punktinformationen über die Randprozesse der Feuerung (z.B. Dampftemperatur, Luftmenge, Emissionsmengen) genutzt. In Erweiterung dessen wäre aber vor allem die Analyse *primärseitiger Prozeßinformationen* (Farbbilder, spektroskopische und tomografische Aufnahmen), die direkt aus dem Flammenfeld und der Feuerungszone abzuleiten sind, vorteilhaft. Nur mit diesen Informationen kann frühestmöglich verhindert werden, die Anlage in energetisch ungünstige Prozeßsituationen, verbunden mit hohen Emissionen, zu führen. Probleme bei der Nutzung primärseitiger Prozeßinformationen bestehen sowohl in deren sensorischer Erfassung als auch in der Auswertung der räumlich-zeitlich sehr komplexen Daten. Daher werden gegenwärtig im Rahmen eines BMBF-Verbundprojektes unter Federführung der Dipl.-Ing. Kurt-Henry-Mindermann GmbH Ratingen Methoden zur Gewinnung und Auswertung *prozeßnaher „bildhafter“ 2D/3D-Feldinformationen* über die komplexen und hochgradig nichtlinearen dynamischen Vorgänge in der Feuerungszone untersucht.

2 Farbwertbasierte Klassifikation von Prozeßzuständen

Das Ziel bei der Untersuchung von Korrelationen zwischen Flammenbildern und Prozeßsituationen ist es, Rückschlüsse von Flammenbildern auf eingestellte Prozeßparameter zu treffen. Dies betrifft hierbei die Detektion des „Lastbereiches“ und somit indirekt auch des Anteils „nichtverbrannter Kohle“. Dabei können die Bilddaten aufgrund ihrer hohen Komplexität nicht unverarbeitet in die Prozeßsteuerung eingebunden werden. Daher ist eine Komplexitätsreduktion im Sinne einer Detektion relevanter Merkmale, die wesentliche globale Eigenschaften der Flamme, wie deren Geometrie oder Farbverteilung, beschreiben, notwendig.

* gefördert durch das BMBF, Projekt-Nr.: 032 6843 B

2.1 Wahl eines Farbraumes

Zur Analyse der Farbinformationen eines Bildes ist die Nutzung eines spezifischen Farbraumes, wie der RGB- oder physiologische Farbraum [Pomierski 96], sinnvoll. Die Vorteile des physiologischen Farbraumes bestehen in einer dekorrelierten Repräsentation von Helligkeit und Farbinformation ebenso wie in der Möglichkeit einer separaten Ausprägung der vorrangig im Flammenbild auftretenden Farbwerte rot und gelb in zwei getrennten Merkmalskarten.
Da die mit beiden Farbräumen durchgeführten Untersuchungen zu qualitativ ähnlichen Ergebnissen führten, wird im folgenden der Nachweis einer Klassifizierbarkeit von Prozeßzuständen nur anhand der auf dem physiologischen Farbraum basierenden Ergebnisse geführt.

2.2 Allgemeines Verarbeitungsschema

Durch die Transformation der RGB-Bilddaten in den physiologischen Farbraum werden bild- und farbwertspezifische Punktwolken gebildet. Die Repräsentation der Farbwertverteilungen innerhalb dieser Punktwolken erfolgt durch einen am Datenmaterial erlernten, neuronalen Clusterer, der durch eine eindimensionale Kohonenkette [Kohonen 95] realisiert ist (Abbildung 1).

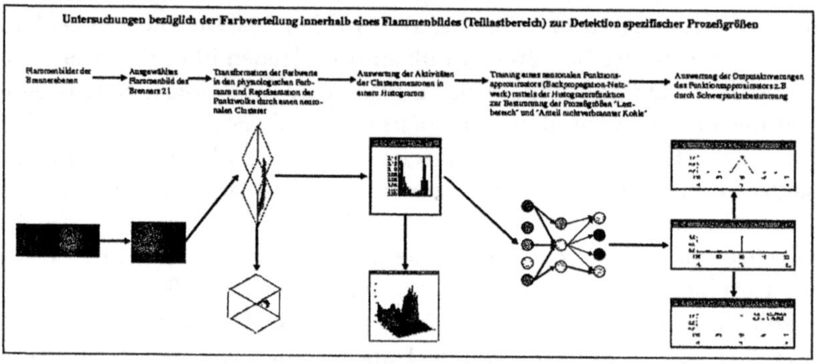

Abbildung1. *Schematische Darstellung der Klassifikation von Prozeßzuständen über eine niedrigdimensionale Repräsentation von Farbbildern und einer sich anschließenden Klassifikation mittels eines Backpropagation Netzwerkes.*

Aufgrund der durch die Kohonenkette erlernten niedrigdimensionalen Repräsentation können statistische Aussagen zur Farbwertverteilung innerhalb der Punktwolke getroffen werden. Zur Umsetzung dessen werden die Aktivitäten der Neuronen des Kohonenvektors bei Präsentation aller in den Farbraum transformierter Farbwerte eines Flammenbildes aufgrund des euklidischen Abstandes bestimmt und über jedes Neuron summarisch in einem Histogramm[1] ausgewertet. Dieses Histogramm ist somit ein Maß für die in einem Bild vorkommenden Farbwerte bezüglich der gelernten Referenzdatenmenge (siehe Abschnitt *Repräsentativität des Datenmaterials*). Durch die Auswertung der Häufigkeiten des Auftretens bestimmter, durch die Neuronen repräsentierter Farbwerte wie schwarz, rot

[1] Eine ebenfalls mögliche Histogrammbildung direkt über den Farbwerten würde unnötig viele Merkmale liefern, bzw. eine designerabhängige Reduktion bedeuten.

oder auch gelb können wiederum Rückschlüsse auf spezifische Prozeßsituationen getroffen werden. Entsprechend der auf visuellen Beobachtungen basierenden Hypothese, daß eine Erhöhung des „Anteils nichtverbrannter Kohle“ ebenso wie die Verschiebung des „Zündortes“ von der Brenneröffnung mit der Verringerung des „Lastbereiches“ korreliert, sollen diese Prozeßgrößen auch aus den Histogrammverteilungen abgeleitet werden. Dies erfolgt gegenwärtig durch einen neuronalen Funktionsapproximator, ein Backpropagation Netzwerk, das aufgrund gegebener Histogramme eine Abbildung auf die Prozeßgrößen *„Lastbereich“* und somit indirekt auch auf die Größe *„nichtverbrannte Kohlemenge“* lernt.

2.3 Neuronaler Clusterer als Grundlage für die Bestimmung von Histogrammen

Die in den Farbraum transformierten Flammenbilder bilden dort eine Punktwolke, die durch eine eindimensionale Kohonenkette repräsentiert wird. Im folgenden werden die wesentlichsten zu beachtenden Aspekte beim Lernen und bei der Auswertung dieser Repräsentation diskutiert.

Repräsentativität des Datenmaterials Eine wesentliche Voraussetzung für die Bestimmung geeigneter Histogrammverteilungen ist ein am Datenmaterial belehrter und somit repräsentativer neuronaler Clusterer. Daher muß bereits der Lernprozeß mit Datenmaterial erfolgen, das alle möglichen Situationen, hier im Sinne aller möglichen auftretenden Farbwerte, umfaßt. Da die Bilder aus dem Prozeßzustand „Vollast“ durch die vergleichsweise größte Vielfalt an Farbwerten gekennzeichnet sind, wurden diese auch zum Training der Kohonenkette genutzt. Dieser Aspekt ist bereits bei der visuellen Betrachtung der im physiologischen Farbraum gebildeten Punktwolken nachweisbar. So ist erkennbar, daß eine Punktwolke bei „Vollast“ sowohl entlang der Unbuntachse als auch bezüglich der Sättigung am stärksten aufgedehnt ist. Dagegen werden bei den Bildern, die andere „Lastbereiche“ repräsentieren, nur kleinere Punktwolken, die im allgemeinen in der Nähe des Schwarzbereiches der Unbuntachse liegen, gebildet.
Ebenso ist es zur Gewährleistung der Repräsentativität des Datenmaterials sinnvoll, mehrere zeitlich aufeinanderfolgende Bilder zum Lernprozeß zu nutzen, um so mögliche Abweichungen vom zu erwartenden Datenmaterial auszugleichen.
Weiterhin ist zu beachten, daß die Farbwerte benachbarter Bildpunkte stark korrelieren und somit in ihrem Informationsgehalt redundant sind. Daher müssen nicht alle Bildpunkte in den Lernprozeß einbezogen werden, und ein „Ausdünnen benachbarter Bildpunkte“ ist sinnvoll.

Parameterisierung der Kohonenkette In den bisherigen Untersuchungen wurden Kohonenketten mit einer fest vorgegebenen Größe von 20 Neuronen genutzt. Diese bildeten die Punktwolke sehr genau und ohne topologische Defekte ab. Dabei wurden die zwei die Kette begrenzenden Neuronen jeweils im Schwarz- bzw. Weißpunkt des Farbraumes „befestigt“ und im Lernprozeß nicht adaptiert. Dies ermöglicht ein Aufspannen der Kohonenkette über den gesamten Farbraum, wodurch auch deren Glattheit unterstützt wird. Sind dagegen die zwei die Kette

begrenzenden Neuronen adaptierbar, wird die Kette nach dem Lernprozeß gefaltet in der Punktwolke liegen.
Die Initialisierung der Neuronen erfolgte entlang der Unbuntachse, so daß im nachfolgenden Lernprozeß sowohl topologische Defekte weitestgehend verhindert als auch der Lernprozeß beschleunigt werden konnte.

Bildung von Histogrammen Da prozeßspezifische Flammenbilder jeweils typische Farbwerte besitzen, sind auch die entsprechenden Histogrammverteilungen durch charakteristische Merkmale gekennzeichnet, die wiederum Rückschlüsse auf den gegenwärtigen Prozeßzustand liefern können.
Die Farbwertauswertung von Bildern erfolgt so, daß jeder einzelne Bildpunkt der Kohonenkette[2] präsentiert, das Best-Matching Neuron ermittelt und seine Aktivität in einem Histogramm (Abbildung 2 links) ausgewertet wird. Alternativ dazu können auch die Aktivitäten aller Neuronen ($y_{rr'}$) berücksichtigt werden. Dies bewirkt, daß die Histogramme geglättet und „Ausreißer" verschliffen sind. Abbildung 2, 2.v.links (Gleichung 2), zeigt ein Histogramm, bei dem die Aktivität eines Neurons linear bzw. gaußförmig fallend (Abbildung 2, 3.v.links, Gleichung 1) in Abhängigkeit von seinem Abstand zum Best-Matching Neuron ($||r - r'||$) berechnet wird.

$$y_{rr'} = \exp\left(-\frac{||r - r'||}{2 \cdot \sigma^2}\right) \qquad \text{mit} \quad \sigma^2 = 3 \qquad (1)$$

$$y_{rr'} = \max\left(0, (1 - \frac{1}{\sigma^2} \cdot ||r - r'||)\right) \qquad \text{mit} \quad \sigma^2 = 3 \qquad (2)$$

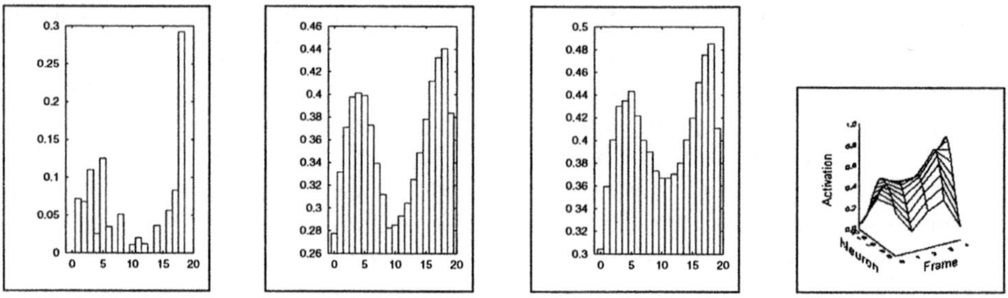

Abbildung2. *Histogramme mit relativen Häufigkeiten zur Repräsentation von Farbwerten durch einzelne Neurone, bezogen auf ein gleiches Bild. Eine Erläuterung der einzelnen Histogramme erfolgt im Text.* **Rechts:** *Durch die Auswertung von Histogrammen für zeitlich aufeinanderfolgende Flammenbilder können auch prinzipielle Änderungen im Verbrennungsprozesses erkannt werden.*

2.4 Prozeßidentifikation über neuronale Funktionsapproximatoren

Prinzipielles Ziel der Verwendung eines neuronalen Funktionsapproximators ist die Ermittlung einer Abbildungsvorschrift von den aus einem Verbrennungsprozeß gewonnenen Farbbildern auf spezifische Prozeßgrößen, wie den „Lastbereich". Als Funktionsapproximator wurde dabei ein zweischichtiges Backpropagation Netzwerk genutzt.

[2] Aufgrund der Vergleichbarkeit von Histogrammen unterschiedlicher Bilder dürfen die Gewichte der Kohonenkette nach dem Lernprozeß nicht mehr geändert werden.

Spezifikationen des Backpropagation Algorithmus Die hier genutzten Netzwerke bestanden aus zwei verarbeitenden Schichten mit 20 Input-, 3 bzw. 9 Output- und abhängig von der durchgeführten Untersuchung aus 2 bis 15 Hiddenneuronen. Die Adaptation der Gewichte der Netze erfolgte mittels des direct learnings.
Zur Bestimmung der Outputaktivitäten der Neuronen wurde jeweils eine Fermi-Funktion genutzt, durch deren Anstieg das Konvergenzverhalten der Neuronen in den Sättigungsbereich steuerbar ist. Dabei sind die in der Hiddenschicht genutzten Fermi-Funktionen durch einen wesentlich flacheren Verlauf als die für die Outputschicht festgelegten gekennzeichnet. Somit befinden sich die Aktivitäten der Hiddenneuronen auch bei hohen Zustandswerten, verursacht durch viele Eingangselemente mit hohen Inputwerten, im Arbeitsbereich und daher außerhalb ihrer Sättigung.
Zur Beschleunigung des Lernvorganges erfolgte weiterhin die Verwendung einer modifizierten Form der Berechnung der durch das Netz zurück zu propagierenden Fehler [Fahlman 88].

Abgeleitet aus der in Abbildung 3 dargelegten Möglichkeit zur visuellen Differenzierung der Histogramme wurden nachfolgend zwei prinzipielle Versuchsstrategien zur neuronalen Klassifikation der Bilder anhand ihres zugehörigen „Lastbereiches" durchgeführt.

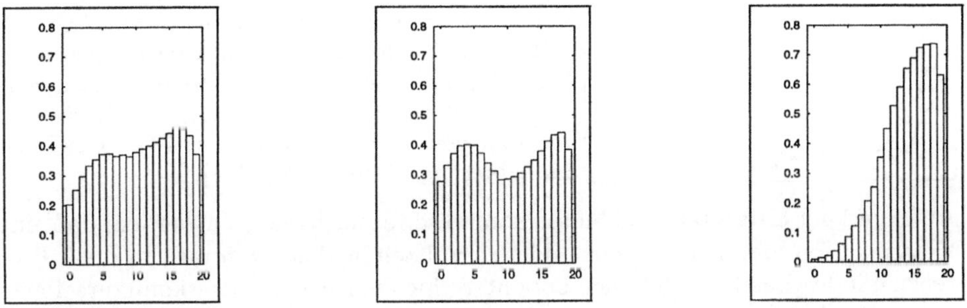

Abbildung3. *Basierend auf einem gelernten Kohonenvektor können Histogramme für beliebige Bilder, somit auch für solche aus unterschiedlichen Prozeßzuständen, ermittelt werden. Bereits die visuelle Betrachtung der Histogramme (hierbei erfolgte die Berücksichtigung der Aktivitäten aller Neuronen nach Gleichung 2) für Bilder aus unterschiedlichen „Lastbereichen" zeigte wesentliche, allerdings typische Unterschiede. Dabei sind die Histogramme von Bildern aus dem Bereich „Vollast"* (**links**) *durch eine fast gleichmäßig starke Aktivierung gekennzeichnet. Dieses Verhalten resultiert daraus, daß bereits der Kohonenvektor ausschließlich mit Bilddaten aus diesem Lastbereich trainiert wurde. Dagegen weisen die Histogramme von Bildern aus dem Prozeßzustand „Teillast"* (**mitte**) *eine typische bimodale Verteilung auf. Die Histogramme für Bilder aus dem Prozeßzustand „Schwachlast"* (**rechts**) *wiederum besitzen eine oder nur wenige starke Aktivierungen der Neuronen, die den Bereich in der Nähe des Schwarzpunktes des Farbraumes abdecken. Diese sehr ungleichmäßige Verteilung kann dabei durch die wesentlich schwächere Verbrennung als im Prozeßzustand „Vollast" erklärt werden.*

Grobklassifikation von Histogrammen Zunächst wurde eine Klassifikation der Histogramme in die 3 „Lastbereiche“ untersucht. Dazu wurde jedem Input (Histogramm), basierend auf der Angabe des „Lastbereiches“ für das zugrundeliegende Bild, ein Teachersignal zugewiesen (Gleichung 3).

$$\underline{t} = \begin{cases} 1.0 \quad 0.0 \quad 0.0 : \text{Histogramm für eine „Vollast“} \\ 0.0 \quad 1.0 \quad 0.0 : \text{Histogramm für eine „Teillast“} \\ 0.0 \quad 0.0 \quad 1.0 : \text{Histogramm für eine „Schwachlast“} \end{cases} \tag{3}$$

Das Ziel dieser Untersuchungen bestand dabei im prinzipiellen Nachweis der Möglichkeit der farbbildbasierten Klassifikation unterschiedlicher Prozeßzustände.

Im Ergebnis dieser Untersuchungen wurde festgestellt, daß die Klassifikationsaufgabe bereits durch ein Netzwerk mit 2 Hiddenneuronen in etwa 600 Epochen, unter Nutzung des direct-learnings, erlernt werden konnte (Abbildung 4). Eine Epoche umfaßte dabei die jeweils ersten 75% aller 519 Muster des Datensatzes. Die weiteren 25% der Muster dienten der nachfolgenden Validierung des Netzes.

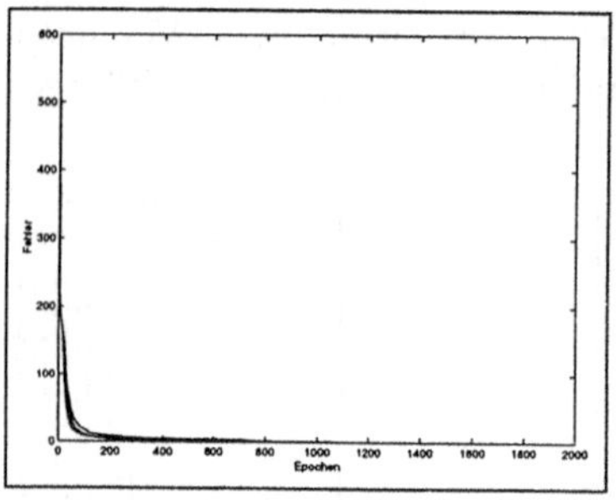

Abbildung4. *Typischer zeitlicher Verlauf des Epochenfehlers für ein Experiment mit drei Versuchswiederholungen. Der prozentuale Validierungsfehler, bezogen auf ein* 100% *falsch gelerntes Muster (maximaler absoluter Musterfehler von* 3*) betrug in diesen Versuchen nach* 2000 *Epochen durchschnittlich* 0.16% *pro Muster. Die diesem Versuch zugrundeliegenden Parameter sind* 2 *Hiddenneuronen, eine zeitlich fallende Lernrate mit* $\eta_{init} = 0.8$*, ein Momentum von* $\alpha = 0.2$ *und eine zufällige Gewichtsinitialisierung im Intervall* $[-1, 1]$.

Aufgrund der erreichten Ergebnisse konnte somit nachgewiesen werden, daß eine Klassifikation von den aus Farbbildern ermittelten Histogrammen in die 3 Prozeßzustände „Lastbereich“ auch bei unterschiedlichen Parameterkonfigurationen bereits nach wenigen Epochen möglich ist.

Feinklassifikation von Histogrammen Basierend auf diesem Ergebnis wurde nachfolgend auch eine Feinklassifikation der Histogramme bezüglich des „Lastbereiches“ untersucht. Das Ziel war es daher, die Bilder nicht nur anhand der definierten „Lastbereiche“, sondern auch bezüglich der Übergänge zwischen den vorgegebenen Zuständen zu klassifizieren, um so einen Trend im Verbrennungsprozeß erkennen oder auch Vergleiche zwischen mehreren Brennern realisieren zu können.

Aufbauend auf dem Nachweis, daß die Histogramme für die 3 vorgegebenen „Lastbereiche“ typische Verteilungen besitzen, wurde angenommen, daß auch die Histogramme für die Übergänge zwischen diesen Prozeßzuständen durch charakteristische Merkmale gekennzeichnet sind. Die dabei vermuteten Histogramme, ebenso wie die bereits real ermittelten, sind in Abbildung 5 schematisch dargestellt.

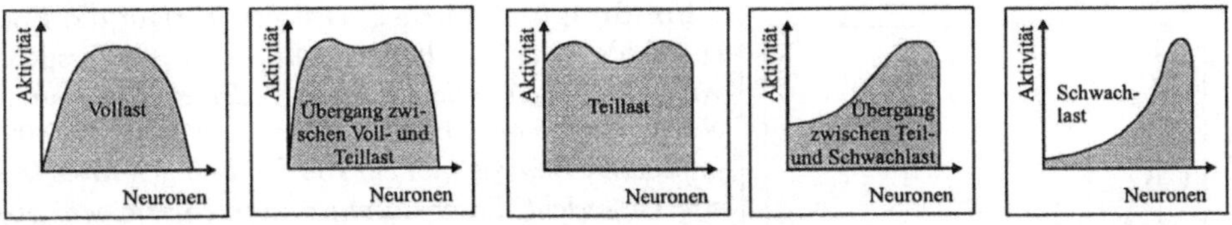

Abbildung5. *Schematische Darstellung möglicher Histogramme. Die bereits real ermittelten Histogramme für die 3 vorgegebenen „Lastbereiche" sind sowohl durch typische Verteilungen gekennzeichnet als auch durch signifikante Unterschiede voneinander separierbar. Weiterhin besteht die Hypothese, daß auch die Bilder von Prozeßzuständen, die jeweils die Übergänge zwischen den vorgegebenen „Lastbereichen" bilden, durch typische Histogramme gekennzeichnet sind. Es ist dabei zu vermuten, daß jede „Zwischenfunktion" die Charakteristik der beiden real ermittelten und sie begrenzenden Histogramme anteilig repräsentiert.*

Da die vorliegenden Bilddaten bezüglich des durch sie repräsentierten Prozeßzustandes nur grob in 3 Klassen unterteilbar waren, mußten die für eine Feinklassifikation notwendigen Histogramme aus den real ermittelten künstlich erzeugt werden. Diese Datengenerierung erfolgte, indem ein „künstliches" Histogramm aus einer prozentualen Überlagerung zweier realer Histogramme gebildet wurde. Ebenso wurde jedem Histogramm (Klasse) ein definiertes Teachersignal zugewiesen (Gleichung 4).

$$\underline{t} = \begin{cases} 1\,0\,0\,0\,0\,0\,0\,0\,0 & : \ 100\%\ \text{„Vollast"} \\ 0\,1\,0\,0\,0\,0\,0\,0\,0 & : \ 75\%\ \text{„Vollast"} \quad 25\%\ \text{„Teillast"} \\ 0\,0\,1\,0\,0\,0\,0\,0\,0 & : \ 50\%\ \text{„Vollast"} \quad 50\%\ \text{„Teillast"} \\ 0\,0\,0\,1\,0\,0\,0\,0\,0 & : \ 25\%\ \text{„Vollast"} \quad 75\%\ \text{„Teillast"} \\ 0\,0\,0\,0\,1\,0\,0\,0\,0 & : \ 100\%\ \text{„Teillast"} \\ 0\,0\,0\,0\,0\,1\,0\,0\,0 & : \ 75\%\ \text{„Teillast"} \quad 25\%\ \text{„Schwachlast"} \\ 0\,0\,0\,0\,0\,0\,1\,0\,0 & : \ 50\%\ \text{„Teillast"} \quad 50\%\ \text{„Schwachlast"} \\ 0\,0\,0\,0\,0\,0\,0\,1\,0 & : \ 25\%\ \text{„Teillast"} \quad 75\%\ \text{„Schwachlast"} \\ 0\,0\,0\,0\,0\,0\,0\,0\,1 & : \ 100\%\ \text{„Schwachlast"} \end{cases} \tag{4}$$

Mit der verfeinerten Auflösung der Prozeßzustände in 9 Klassen ist auch eine Anpassung der Netzwerkstruktur notwendig. Daher wurden die zu untersuchenden Netzwerke mit 20 Input- und 9 Outputneuronen strukturiert und der zu klassifizierende Input sowohl aus den bereits ermittelten als auch aus den künstlich generierten Histogrammen gebildet.
Es wurde festgestellt, daß diese Klassifikationsaufgabe Netzwerke mit einer Mindestanzahl von 8 Hiddenneuronen erfordert. Unter Nutzung des direct-learnings konvergierte der summierte Musterfehler je Epoche[3] nach etwa 3000 Epochen (1410 Muster je Epoche) gegen einen Fehlerwert von etwa 120, was einem durchschnittlichen Musterfehler von 0.08 entspricht. Bezogen auf ein 100% falsch gelerntes Muster (maximaler absoluter Fehler von 9) entspricht ein Musterfehler von 0.08 einem durchschnittlichen prozentualen Musterfehler von 0.94% (Abbildung 6).

[3] Eine Epoche umfaßte hierbei die jeweils ersten 75% aller Muster des Datensatzes. Die weiteren 25% der Muster dienten der nachfolgenden Validierung des Netzes.

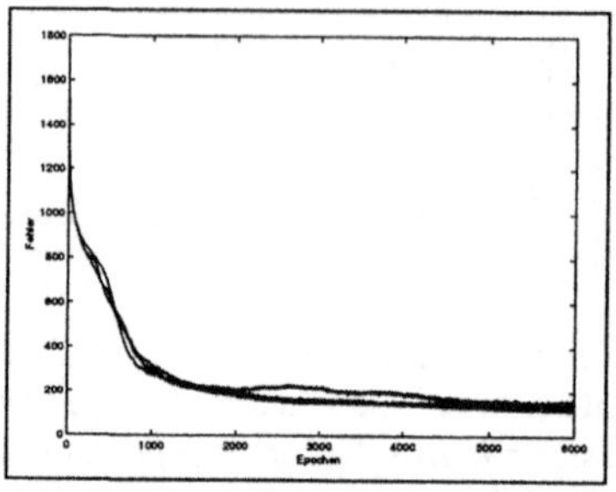

Abbildung6. *Typischer zeitlicher Verlauf des Epochenfehlers für ein Experiment mit drei Versuchswiederholungen. Hierbei betrug der durchschnittliche Epochenfehler nach* 6000 *Epochen* 140, *d.h.* 0.099 *je Muster. Bezogen auf ein* 100% *falsch gelerntes Muster entspricht dieser absolute Lernfehler einem prozentualen von* 1.1% *je Muster. Der prozentuale Validierungsfehler betrug dabei durchschnittlich* 4.9% *pro Muster. Die bei diesem Experiment genutzten Parameter sind* 15 *Hiddenneurone, eine zeitlich feste Lernrate von* $\eta = 0.8$, *ein Momentum von* $\alpha = 0.2$ *und eine zufällige Gewichtsinitialisierung im Intervall* $[1.0, -1.0]$.

Zusammenfassend wurde festgestellt, daß eine Klassifikation von den aus den realen Farbbildern ermittelten und den künstlich generierten Histogrammen in 9 Prozeßzustände möglich ist. Im Gegensatz zur Klassifikation von 3 Prozeßzuständen ist dieser Lernprozeß stärker parameterabhängig, und es wird ebenso eine größere Anzahl an Lernschritten benötigt.

3 Zusammenfassung der Ergebnisse und Ausblick

Das wesentliche Ergebnis dieser Untersuchungen ist somit der Nachweis, daß für die Bilder aus unterschiedlichen Prozeßzuständen „Lastbereich“ jeweils typische und klassifizierbare Histogramme existieren. Dies sollte in nachfolgenden Untersuchungen auch anhand realer Bild- und Prozeßdaten, die das gesamte Spektrum möglicher „Lastbereiche“ kennzeichnen, validiert werden.
Abgeleitet aus den bereits erreichten Ergebnissen sollten sich darauf aufbauende Untersuchungen im wesentlichen mit folgenden Aspekten befassen.

- Klassifikation der Prozeßzustände „Lastbereich“ durch ein einzelnes Neuron der Outputschicht, welches eine Klassenzugehörigkeit direkt über den analogen Wert (100% - 0% „Lastbereich“) seines Outputs kodiert.
- Extraktion der zur Unterscheidung von Prozeßsituationen signifikanten Bildbereiche. Dazu erfolgt die Ermittlung derjenigen Neuronen der Kohonenkette, die eine Unterscheidung der „Lastbereiche“ ermöglichen. Für diese Neuronen werden die durch sie kodierten Farbwerte und deren Position im Bild (z.B. in der Flamme oder auf der Brennerwand) bestimmt.
- Auswertung der Histogramme bezüglich weiterer korrelierter Prozeßgrößen, wie z.B. die „Luftmenge“.

Literatur

[Fahlman 88] **S.E. Fahlman** (1988). An empirical study of learning speed in back-propagation networks. In D.Touretzky, G.Hinton, T.Sejnowski (Eds.): Proc. of the 1988 Connectionist Models Summer School, June 17-26 1988, Carnegie Mellon University. Morgan Kaufmann.

[Kohonen 95] **T. Kohonen** (1995). Self-Organizing Maps. Springer-Verlag Berlin Heidelberg.

[Pomierski 96] **T. Pomierski and H.-M. Gross** (1996). Neural architecture for Chromatic Adaptation of Single Images and Picture Sequences. In *2. Nationaler Workshop „Farbbildverarbeitung“, TU-Ilmenau.* pp. 29-34.

3D-Geometrien
(Postersession)

Wissenserwerb für Produktionsnetze zur 3D-Erkennung von Fahrzeugen

E. Michaelsen, U. Wankmüller und U. Stilla

Forschungsinstitut für Informationsverarbeitung und Mustererkennung (FGAN-FIM)
Eisenstockstr. 12, 76275 Ettlingen
{mich,wan,usti}@gate.fim.fgan.de

Kurzfassung. Zur dreidimensionalen Detektion und Klassifikation von Fahrzeugen aus Bildfolgen wird ein modellbasiertes Erkennungsverfahren eingesetzt. Die verwendeten Modelle werden durch Produktionsnetze beschrieben. Der Wissenserwerb erfolgt aus CAD-Modellen. Zur Erstellung eines neuen Produktionsnetzes werden interaktiv Hierarchiegraphen erstellt und aus den CAD-Daten standardisierte Produktionen parametrisiert.

1 Einleitung

Für die 3D-Erkennung von *man-made objects* aus Luftbildern werden häufig modellbasierte Verfahren vorgeschlagen [6]. Zur Modellierung sind verschiedene Formen der Wissensrepräsentation bekannt [13], wie z.B. Produktionen, Frames oder semantische Netze. Beim FIM werden Produktionen zur Erkennung von Brücken [10, 2], Straßenkreuzungen [15] und Gebäuden [16, 17, 18] in Luftbildern verwendet. Wegen der großen Variationsmöglichkeiten dieser Objekte finden zur Klassifikation parametrische Modelle Verwendung.

Bei der Modellierung von Fahrzeugen können zur Beschreibung formfeste Modelle [17] verwendet werden, da die spezielle Geometrie eines Fahrzeugtyps als bekannt vorausgesetzt werden kann. Ein Beispiel für die Modellierung und 3D-Erkennung eines Fahrzeugtyps in einer Bildfolge ist in [11, 12] dargestellt.

Soll eine breite Palette verschiedener Fahrzeugtypen detektiert werden, so ist jeweils die spezielle Geometrie zu erfassen und durch entsprechende Produktionen zu modellieren. Da geometrische Beschreibungen von Fahrzeugen typischerweise in Form von CAD-Modellen vorliegen, liegt es nahe, dieses Wissen zur Automatisierung der Modellierung zu verwenden.

CAD-Modelle wurden häufig für Analysen auf der Basis von Aspekt-Graphen verwendet, wie z.B. in [3, 8, 9]. Im vorliegenden Beitrag wird die Wissenserwerbskomponente eines Analysesystems beschrieben, mit der Produktionen aus CAD-Modellen von Fahrzeugen parametrisiert werden.

2 Objekterkennung mit Produktionsnetzen

Symbolische Verfahren der Mustererkennung arbeiten häufig auf Strukturen, die durch Zeichenketten, Bäume, Felder oder Graphen beschrieben werden. Im Gegensatz dazu erfolgt die Objekterkennung mit Produktionsnetzen auf ungeordneten Mengen. Die Modellbeschreibung basiert auf der Theorie der *Koordinaten Grammatiken* [14, 11, 12].

Die Datenbasis, auf der die Produktionen arbeiten, besteht aus Objekten (Instanzen), die jeweils einen symbolischen Namen (Objekttyp) und einen Attributvektor beinhalten. Im Attributvektor einer Instanz sind z.B. Bild- oder Szenenkoordinaten, Orientierungen, Oberflächennormalen, Abmaße usw. abgespeichert. Ein Tupel solcher Instanzen bildet eine Konfiguration.

Produktion. Jede *Produktion* besteht aus einem *Bedingungsteil* und einem *Aktionsteil.* Der Bedingungsteil enthält Objekte und ein Prädikat, welches die zugehörenden Attribute erfüllen müssen. Der Aktionsteil enthält eine Attributberechnungsfunktion und erzeugt ein neues Objekt. Nehmen wir als Beispiel eine einfache Produktion der Form:

$$(Linie, Linie)\ \pi \xrightarrow{\phi} (Winkel)$$

Das Objekt *Linie* besitzt als Attribute Ortskoordinaten und eine Orientierung. Das Prädikat π beinhaltet die Relationen 'benachbart' und 'nicht parallel' und bildet zusammen mit dem Paar (*Linie*, *Linie*) den Bedingungsteil. Die Funktion ϕ bestimmt den Schnittpunkt der zu den Linien gehörenden Geraden. Dieser wird als Attribut für die entstehende Instanz des Objektes *Winkel* benötigt.

Produktionsnetz. Das Produktionsnetz zeigt den generellen Zusammenhang von Produktionen und Objekttypen (Konzepten). Ein Produktionsnetz ist ein gerichteter Graph, der ähnlich wie ein Petri-Netz [17] zwei verschiedene Arten von Knoten besitzt - Konzepte und Produktionen. Eine Kante führt von einem Konzept zu einer Produktion, wenn der Objekttyp im Bedingungsteil steht. Ist es mehrfach darin enthalten, wird die Kante entsprechend mehrfach aufgeführt. Eine Kante führt von einer Produktion zu einem Konzept, wenn der Objekttyp im ausführbaren Teil steht. Beispiele für Produktionsnetze finden sich in [15, 16, 17, 18].

Implementierung. Produktionsnetze werden in einer Blackboard-Architektur in der Systemumgebung BPI [10] implementiert. Jede Produktion ist durch ein Verarbeitungsmodul realisiert, das den Bedingungsteil auf einer Konfiguration prüft und bei Erfüllung eine neue Instanz generiert. Das System arbeitet akkumulierend, d.h. daß die *ersetzte* Eingangskonfiguration aus der Datenbasis nicht gelöscht wird. Einmal erzeugte Teilergebnisse stehen so zur Verfolgung verschiedener Hypothesen während der gesamten Analyse zur Verfügung. Das klassische *Backtracking* in Suchbäumen entfällt [12]. Die Datenbasis wird in einem Assoziativspeicher gehalten, der einen effizienten Zugriff auf Mengen von Instanzen erlaubt [16].

3 Vom Modellwissen zum Produktionsnetz

Die interaktive Konstruktion von Produktionsnetzen erfolgt unter Zuhilfenahme von CAD-Modellen der zu erkennenden Fahrzeuge. Abb. 1a zeigt als Beispiel ein CAD-Modell eines Lastwagens in vier verschiedenen Ansichten.

CAD-Modell. Durch Anbindung eines Moduls an das CAD-System AUTOCAD wurde eine Möglichkeit geschaffen, durch Markieren von Modellteilen interaktiv eine Untergliederung des Modells in eine Hierarchie von Teilen vorzunehmen. Neben der Teil_von-Beziehung muß hierbei vom Anwender auf jeder Hierarchieebene festgelegt werden, ob eine Unterteilung in notwendige oder hinreichende Teile erfolgt. Als terminale Teile werden Flächen im Raum (Viereck, Rechteck, Dreieck und Kreis) verwendet, die verschiedene Attribute besitzen. Bei einem Rechteck werden z.B. Länge, Breite, Schwerpunkt, Normalenrichtung und Richtung der Längsachse festgehalten. Die Untergliederung bzw. Gruppierung der Teile wird in einem sogenannten Hierarchiegraphen gespeichert.

Hierarchiegraph. Der Hierarchiegraph enthält somit zwei Typen von Kanten, nämlich notwendige Teil_von-Beziehungen und hinreichende Teil_von-Beziehungen. Notwendige Teilobjekte sind paarweise verknüpft. Ist für die Erkennung eines Objekts die Verknüpfung von mehr als zwei notwendigen Teilobjekten erforderlich, erfolgt dies durch die Einfügung von Zwischenstufen in der Hierarchie. Für manche Teile, die als Details im CAD-Modell vorkommen, ist eine Erkennung in Luftbildern unwahrscheinlich. Sind diese Teile jedoch zur Visualisierung nützlich, können sie durch eine spezielle Markierung in den Hierarchiegraphen aufgenommen werden.

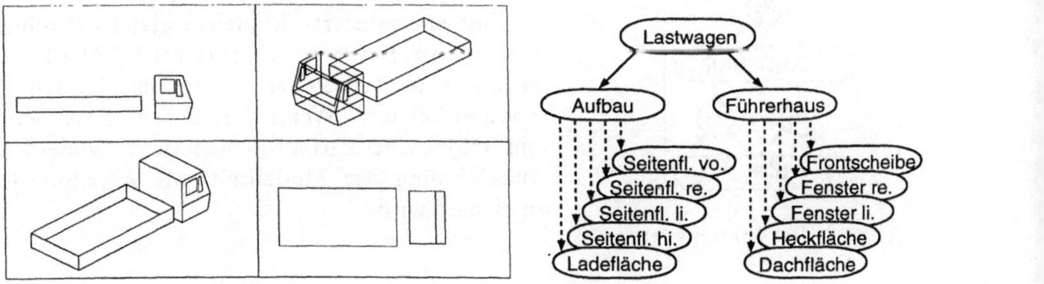

a b

Abb. 1: Beispiel Lastwagen. a) CAD-Modell, b) Hierarchiegraph

Den Hierarchiegraphen des modellierten Lastwagens zeigt Abb. 1b. Bei diesem Beispiel wurde das Objekt *Lastwagen* in die beiden Teilobjekte *Aufbau* und *Führerhaus* unterteilt. Diese wurden wiederum in solche Einzelteile zerlegt, die durch Flächen im Raum beschrieben werden können. Für die Erkennung des *Lastwagens* erscheint beispielsweise die Präsenz von *Aufbau* und *Führerhaus* erforderlich (durchgezogene Pfeile - notwendig). Für die Präsenz des

Aufbaus oder des *Führerhauses* genügt dagegen jeweils ein einziges der darunter liegenden Teile (gestrichelte Pfeile - hinreichend), etwa die *Ladefläche* bzw. die *Dachfläche.*

Ausgehend von einem Referenzpunkt und einer Referenzrichtung im Raum werden die Lage und die Orientierung der Teilobjekte relativ zu den jeweiligen Vorgängern in der Hierarchie aus dem CAD-Modell bestimmt. Die relativen Transformationen werden den Kanten des Hierarchiegraphen als Attribute zugewiesen.

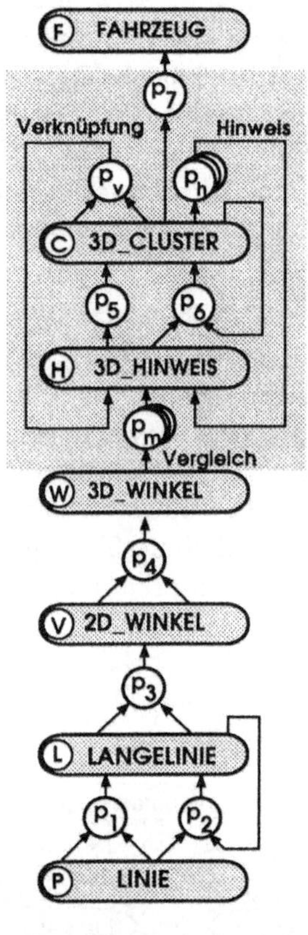

Abb. 2: Produktionsnetz

Produktionsnetz. Die in dem Produktionsnetz von Abb. 2 dargestellten Produktionen können im wesentlichen in zwei unterschiedliche Typen eingeteilt werden. Die Produktionen p_1 bis p_4 enthalten allgemeines Wissen für die 2D-Analyse im Einzelbild und für den Übergang in die 3D-Szene [17, 18]. Die verwendeten Parameter hängen von der Auflösung der Bilddaten ab. Die anderen Produktionen verarbeiten spezielles Wissen, wobei die parametrisierten Produktionen p_m, p_h, und p_v durch den Hierarchiegraphen spezialisiert werden.

Ausgehend von den Primitivobjekten LINIE werden in jedem Einzelbild durch sukzessive Linienverlängerung mit p_1 und p_2 die Objekte LANGELINIE erzeugt. Produktion p_3 liefert aus diesen Objekten die Objekte 2D_WINKEL. Korrespondieren zwei Objekte 2D_WINKEL in einem Bildpaar, so wird durch p_4 ein räumliches Objekt 3D_WINKEL generiert.

Eine spezialisierte Modellvergleichs-Produktion p_m prüft für ein Objekt 3D_WINKEL, ob es mit einem speziellen terminalen Modellteil kompatibel ist. Wenn dies der Fall ist, wird ein Objekt 3D_HINWEIS generiert, wobei die Bezeichnung des Modellteils als Attribut gespeichert wird.

Die räumliche Lage des Objektes wird durch Minimierung der Fehlerquadrate bestimmt. Läßt das primitive Modellteil durch seine Symmetrie und Orientierung zu den Freiheitsgraden mehrfache Lageinterpretationen zu, wie z.B. bei einer horizontalen Rechteckfläche, so werden entsprechend viele unterschiedliche Objekte 3D_HINWEIS generiert.

Beispielsweise prüft die Produktion für das Modellteil *Dachfläche*, ob das Objekt 3D_WINKEL (Abb. 3a) eine senkrechte Normale besitzt, die beiden Schenkel ungefähr einen rechten Winkel bilden und mit der Länge und Breite der Rechteckfläche ungefähr übereinstimmen. Ein erzeugtes Objekt

3D_HINWEIS (Abb. 3b), das die Hypothese Dachfläche repräsentiert, besitzt als Attribute die räumliche Position des Schwerpunktes und die Ausrichtung der Fläche. Da das Objekt 3D_WINKEL für die *Dachfläche* eine um 180 Grad gedrehte Lageinterpretation zuläßt, wird ein zweites Objekt 3D_HINWEIS (Abb. 3c) generiert.

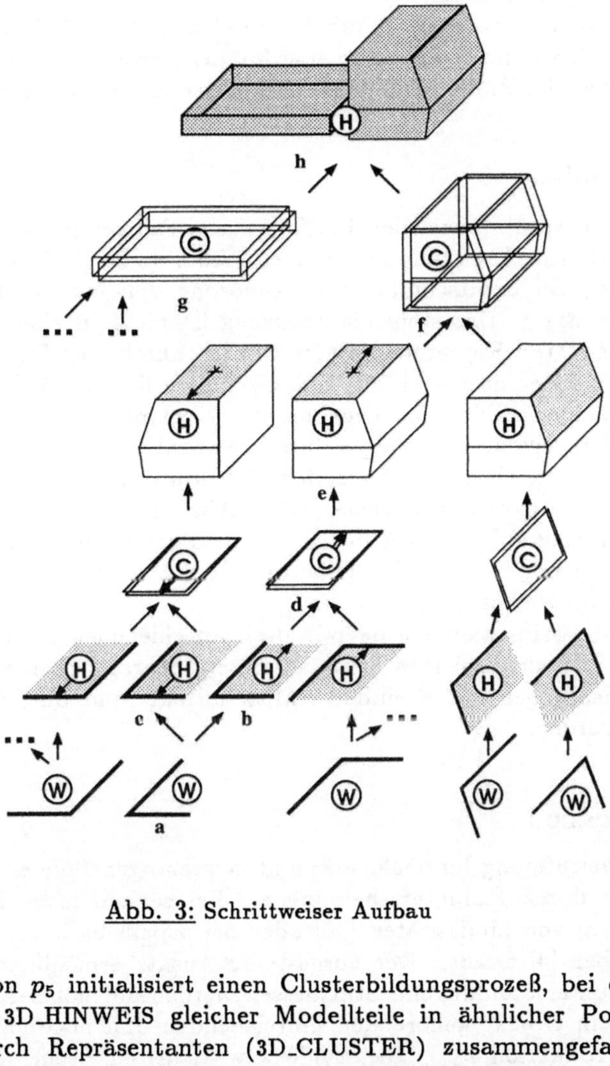

Abb. 3: Schrittweiser Aufbau

Die Produktion p_5 initialisiert einen Clusterbildungsprozeß, bei dem durch p_6 die Objekte 3D_HINWEIS gleicher Modellteile in ähnlicher Position und Ausrichtung durch Repräsentanten (3D_CLUSTER) zusammengefaßt werden (Abb. 3d). Die Attribute bestimmen sich durch Mittelwertbildung.

Während die Hinweis-Produktionen p_h Evidenz entlang der hinreichenden Kanten des Hierarchiegraphen transportieren, transportieren die Verknüpfungs-Produktionen p_v Evidenz entlang der notwendigen Kanten.

Beispielsweise wird durch eine Produktion p_h aus einem Objekt 3D_CLUSTER zu einer *Dachfläche* (Abb. 3d) ein Objekt 3D_HINWEIS zu einem *Führerhaus* (Abb. 3e) abgeleitet, ohne daß zusätzliche Bedingungen erfüllt sein müssen. Die Attribute zur Lage werden durch eine entsprechende Transformation berechnet, die aus dem Hierarchiegraphen stammt. Aus einem Objekt 3D_CLUSTER zum *Führerhaus* (Abb. 3f) und einem Objekt 3D_CLUSTER zum *Aufbau* (Abb. 3g) wird durch p_v ein Objekt 3D_HINWEIS zum *Lastwagen* (Abb. 3h) abgeleitet, wenn die räumlichen Positionen und Ausrichtungen der Objekte kompatibel sind.

Die Produktion p_7 erzeugt die Zielobjekte FAHRZEUG, wenn die Objekte 3D_CLUSTER zu einem *Fahrzeug* eine Mindestbewertung überschreiten. In die Bewertung geht z.B. die gefundene Anzahl der terminalen Modellteile ein.

4 Ergebnisse

Zur Verfahrensprüfung wurden Bildfolgen von Fahrzeugen in Schrägsichten mit einer Mittelformatkamera aufgenommen. Abb. 4a zeigt ein Beispiel, wobei die abgetasteten Bilderausschnitte eine Auflösung von jeweils 1000 x 1600 Bildpunkten besitzen. Das stehende Fahrzeug ist nicht in allen Bildern vollständig zu sehen. Die Kameraparameter wurden durch vier Paßpunkte bestimmt, die außerhalb der dargestellten Bildausschnitte liegen. Abb. 4b zeigt für die mittleren beiden Bilder das Ergebnis der Merkmalsextraktion. Die Generierung der Primitivobjekte erfolgte durch Segmentierung mit einem Mehrschwellenverfahren und einer nachfolgenden Konturapproximation mit einem dynamischen Split-Algorithmus (siehe [17]). Abb. 4c zeigt ein Resultat, das mit dem Hierarchiegraphen aus Abb. 1b und dem entsprechenden Produktionsnetz aus Abb. 2 erzielt wurde. Dafür wurde die baste resultierende Instanz in ein Bild der Folge rückprojiziert.

Die 3D-Klassifikation wurde mit drei verschiedenen Fahrzeugtypen untersucht, die in bodenbasierten Seitenansichten, Schrägsichten und aus senkrechten Blickrichtungen mit Kleinbild-, Mittelformat- und Luftbildkameras aufgenommen wurden.

5 Diskussion

Die Berücksichtigung hinreichender und notwendiger Teile erfolgt für die Analyse häufig durch Fallunterscheidungen. Bei semantischen Netzen wird dies z.B. in Form von Modalitäten [13] oder bei aspekt-basierten Systemen durch Untergraphen [4] erfaßt. Der vorgestellte Ansatz ermöglicht die Vermeidung dieser Fallunterscheidungen. Stattdessen werden auf höheren Hierarchiebenen die zu einem Objekt gehörenden hinreichenden und notwendigen Teile akkumuliert (Produktion p_6). Die Evidenz wird in der Szene, also im problemadäquaten Raum, zusammengefaßt, unabhängig davon welche Teilkonturen welcher Flächen in welchen Bildern gefunden werden. Durch diese Clusterung erhält der vorgestellte Ansatz auch seine Robustheit.

Im allgemeinen sind CAD-Modelle nicht für eine Erkennungsaufgabe erstellt worden. So sind bestimmte Bestandteile in Luftbildern nicht detektierbar (z.B. der Fahrzeugunterbau) oder aufgrund der Bildauflösung schlecht detektierbar

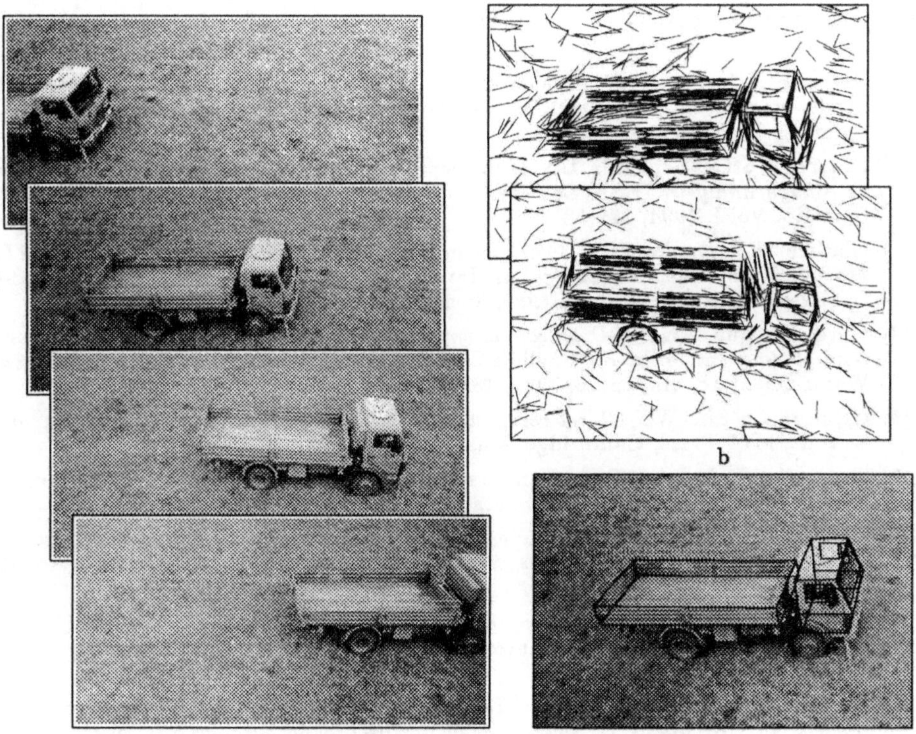

a b c

Abb. 4: Anwendungsbeispiel Lastwagen

(z.B. Fahrzeugdetails). Eine für die Erkennung nutzbare Untergliederung ist für komplexere Modelle in der Regel automatisch nicht ableitbar. Diese Aufgabe erfordert daher eine Interaktion mit dem Benutzer. Durch den teilautomatische Wissenserwerb können die Modelle jedoch rascher an neue Aufgaben angepaßt und erprobt werden.

6 Zusammenfassung

Der vorliegende Beitrag stellt einen Ansatz zum Wissenserwerb für die 3D-Objekterkennung mit Hilfe von Produktionsnetzen vor. Der Wissenserwerb erfolgt interaktiv mit Hilfe von CAD-Modellen. Um den Wissenserwerb hinreichend einfach und flexibel zu halten, wurde die Gestalt der Produktionsnetze gegenüber den früheren Untersuchungen eingeschränkt und standardisiert. Die vorhandenen Komponenten des Systems zur Beobachtung der entstehenden Statistiken, zur Fehlersuche, zur Erklärung und zur graphischen Ausgabe der Ergebnisse werden weiter verwendet.

Literaturverzeichnis

[1] Füger H.; Lütjen K.; Michaelsen E.; Schwan K.: *Strukturorientierte 3D-Szenenanalyse in Bildfolgen.* In: Großkopf R. (Hrsg): *Mustererkennung 1990*, Springer, Berlin, 1990, 659-666.

[2] Füger H.; Stein G.; Stilla U.: *Multi-populations evolution strategies for structural image analysis*, IEEE Conference on Evolutionary Computation (ICEC'94), Orlando, Vol I, 1994, 229-234.

[3] Glauser T.; Bunke H.: *Generierung von Entscheidungsbäumen aus CAD-Modellen für Erkennungsaufgaben.* In: Burkhardt H.; Höhne K. H.; Neumann B. (Hrsg): *Mustererkennung 1989*, Springer, Berlin, 1990, 334-340.

[4] Gmür E.; Bunke H.: *3-D object recognition based on subgraph matching in polynomial time.* In: Mohr R.; Pavlidis Th.; Sanfeliu A.: *Structural Pattern Analysis*, World Scientific, Singapore, 1989, 131-147.

[5] Grimson L.; Eric W.: *Object recognition by computer: The role of geometric constraints*, MIT Press, Cambridge, Mass., 1990.

[6] Gruen A., Baltsavias E.P., Henricsson O.: *Automatic extraction of man-made objects from aerial and space images (II)* Birkhäuser, Basel, 1997

[7] Guoqing Z.: *Recognizing primitives using aspect-interpretation model matching in a both CAD and LP-based measurement system*, ISPRS Congress, 1996, Vol. 31, B3, 1018-1027

[8] Ikeuchi K.: *Generating an interpretation tree from CAD model for 3D-object recognition in bin-picking tasks*, International Journal of Computer Vision, 1 1987, 145-165

[9] Lanser S., Munkelt O., Zierl C.: *Robust video-based object recognition using CAD models.* In: Rembold U, et al. (Hrsg.): *Intelligent autonomous systems*, IOS Press, Amsterdam, 1995, 529-536.

[10] Lütjen K.: *BPI: Ein Blackboard-basiertes Produktionssystem für die automatische Bildauswertung.* In: Hartmann G. (Hrsg.): *Mustererkennung 1986*, Springer, Berlin, 1986, 164-168.

[11] Michaelsen E.: *3D coordinate grammars.* In: Girod B.; Niemann H.; Seidel H.-P. (Eds.): *3D Image analysis and synthesis '96*, Infix, Sankt Augustin, 1996, 81-85.

[12] Michaelsen E.: *Über Koordinaten Grammatiken zur Bildverarbeitung und Szenenanalyse*, Diss., Techn. Fak., Univ. Erlangen, in Vorbereitung 1998.

[13] Niemann H.: *Pattern analysis and understanding*, Springer, Berlin, 1990.

[14] Milgram D. L.; Rosenfeld A.: *A note on 'grammars with coordinates'.* In: Nake F.; Rosenfeld A.: *Graphic languages*, North Holland, Amsterdam, 1972, 187-194.

[15] Stilla U.: *Map-aided structural analysis of aerial images.* ISPRS Journal of Photogrammetry and Remote Sensing, Vol 50, 1995, 3-10.

[16] Stilla U.; Michaelsen E.; Lütjen K.: *Structural 3D-analysis of aerial images with a blackboard-based production system.* In: Gruen A.; Kuebler O.; Agouris P.: *Automatic extraction of man-made objects from aerial and space images*, Birkhäuser, Basel, 1995, 53-62.

[17] Stilla U.; Michaelsen E.; Lütjen K.: *Automatic extraction of buildings from aerial images.* In: Leberl F.; Kalliany R.; Gruber M.: *Mapping buildings, roads and other man-made structures from images*, IAPR TC-7 Workshop, Oldenbourg, Wien, 1996, 229-244.

[18] Stilla U.; Michaelsen E.: *Semantic modeling of man-made objects by production nets.* In: Gruen A.; Baltsavias E. P.; Henricsson O.: *Automatic extraction of man-made objects from aerial and space images II*, Birkhäuser, Basel, 1997, 43-52.

Fehlererkennung der Oberflächenform durch 3D-Vermessung und Auswertung mit einem künstlichen neuronalen Netz

Tilo Lilienblum[1)], Georg Günther[2)], Michael Küchel[2) 3)] und Bernd Michaelis[1)]

[1)] Otto-von-Guericke-Universität Magdeburg, Institut für Prozeßmeßtechnik und Elektronik, PF 4120, D-39016 Magdeburg, e-mail: lilie@ipe.et.uni-magdeburg.de

[2)] Carl Zeiss Jena GmbH, D-07740 Jena

[3)] Aktuelle Adresse: Fachhochschule Aalen, Heinrich-Rieger-Str.22/1,D-73430 Aalen

Zusammenfassung: Der Beitrag beschreibt eine Methode für die Fehlererkennung der Oberflächenform durch eine 3-D-Vermessung und anschließende Auswertung der Daten mit Hilfe eines neuronalen Netzes.
Die Erkennung von Abweichungen der Oberfläche von der gewünschten Form eines Meisterobjektes erfordert normalerweise eine exakte Justierung der Objekte in einer Position. Insbesondere können Verformungen nicht ausgeglichen werden. Bei der hier vorgestellten Methode ist diese genaue Justierung nicht erforderlich.
Es werden zunächst mehrere Meisterstücke in unterschiedlichen Positionen und Variationen in ein neuronales Netz eingespeichert. Beim Abruf des Netzes mit den Daten der zu überprüfenden Teile, werden nicht gespeicherte Merkmale geglättet und dadurch quasi das Meisterstück in der aktuellen Lage rekonstruiert. Die Auswertung kann dann durch einfache Differenzbildung erfolgen.
Anhand eines Beispiels werden die Möglichkeiten, die diese Methode bietet, demonstriert.

1 Einleitung

Mit Hilfe optischer bzw. mechanischer 3D-Meßmethoden ist es möglich, eine Qualitätskontrolle der Oberflächenform von Werkstücken durchzuführen. Normalerweise wird dazu zunächst ein Meisterstück vermessen. Die Qualitätskontrolle kann dann durch die Berechnung der Differenz zwischen dem Meisterstück und dem zu überprüfenden Werkstück erfolgen.
Diese Verfahrensweise ist aber nur möglich, wenn für die zu vermessenden Werkstücke eine genaue Justierung realisiert werden kann. Die Genauigkeit der Justierung der Werkstücke ist jedoch oft nicht ausreichend oder kann nur mit erheblichem Aufwand erreicht werden. Sind die zu vermessenden Werkstücke gegenüber dem Meisterstück nur verschoben, kann diese Verschiebung in den Meßdaten entsprechend berücksichtigt werden. Meistens sind die Teile jedoch nicht nur verschoben, sondern auch in sich verbogen und verdreht. Ebenfalls ist es möglich, daß Fertigungstoleranzen erlaubt sind, die sich bei einer Differenzbildung viel stärker auswirken, als die eigentlichen Fehler. Solche Bedingungen erschweren eine automatische Klassifikation.
Deutlich wird diese Problematik bei dem im Punkt 0 beschriebenen Anwendungsfall. Mittels einer optischen 3D-Oberflächenvermessung soll das Höhenprofil eines

Prüfobjektes mit einer Fläche von ca. 1m x 1m vermessen werden. Auftretende Fehler zeichnen sich durch eine Abweichung der Höhenkoordinate von einer Sollform aus. Unter der Annahme, daß die Empfindlichkeit der Fehlererkennung bei 0,1 mm liegen soll ist bei der Größe des Prüfobjektes die notwendige Justierung nicht oder nur mit erheblichem Aufwand möglich. Ist die Justierung z. B. nur auf 5 mm genau möglich, so sind die durch fehlerhafte Justierung entstehenden Abweichungen 50 mal so groß wie der eigentlich zu detektierende Fehler.
Eine Möglichkeit, unter solchen Bedingungen Fehler zu erkennen, besteht in der Approximation der „glatten" Oberflächen im Bereich niedriger Ortsfrequenzen. Es erfolgt z. B. bei bisher realisierten Lösungen eine Polynomapproximation des Prüfobjektes. Anhand von Approximationsfehlern kann dann auf Fehler des Prüfobjektes geschlossen werden.
Auf Grund der Komplexität des zu vermessenden Prüfobjektes ist aber diese Verfahrensweise oft nicht ohne weiteres möglich. Kanten oder Feinstrukturen werden in diesem Fall auch als Fehler klassifiziert. Andererseits gibt es Fehler, die nicht als solche erkannt werden, z. B. großflächige Beulen, da sich die Polynome diesen anpassen können.
Im folgenden soll eine auf einem neuronalen Netz basierende Methode beschrieben werden, mit der auch bei ungenauer Justierung und erlaubten Fertigungstoleranzen Fehler automatisch erkannt werden können. Als Bezug wird in guter Näherung ein Meisterstück durch Recall eines künstlichen neuronalen Netzes in aktueller Lage rekonstruiert. Dieses Netzwerk, ein modifizierter Assoziativspeicher, wird schwerpunktmäßig für die modellbasierte optische 3D-Vermessung [1] sowie für andere technische [7] und biologische [3], [6] Anwendungen eingesetzt.

2 Fehlererkennung unter Nutzung eines neuronalen Netzes

Bild 1 zeigt den prinzipiellen Ablauf der Fehlererkennung. Zur Erkennung von lokalen Abweichungen der Oberflächenform eines Prüfobjektes von einer vorgegebenen Form wird folgendermaßen vorgegangen:
In der *Lernphase* (Bild 1, linke Seite) werden mehrere (fehlerfreie) Meisterstücke in unterschiedlichen Justierungen vermessen. Diese Meßdaten werden in einem künstlichen neuronalen Netz (Bild 2) gespeichert. Dadurch wird die gesamte Bandbreite der möglichen Variationen und Justierungen der Meisterstücke erfaßt. Zur Unterdrückung von Ausreißern ist eine entsprechende Vorverarbeitung der Meßdaten erforderlich. Bei zu geringem Umfang der Trainingsdaten kann durch Überlagern mit bandbegrenztem Rauschen die Form variiert und somit zusätzliche Trainingsdaten generiert werden (siehe Abschnitt 4).
In der *Recallphase* (Bild 1, rechte Seite) wird das Prüfobjekt vermessen. Die Meßwerte werden in das neuronale Netz eingegeben. In den Gewichten des neuronalen Netzes sind nach der Lernphase Meisterstücke in unterschiedlichen Positionen und globalen Verformungen gespeichert.
Am Ausgang des neuronalen Netzes bleiben beim Recall im wesentlichen nur diejenigen Anteile der Meßdaten übrig, die den gespeicherten Meisterstücken entsprechen. Fehler, also Abweichungen von diesen Meisterstücken werden geglättet.

Nach Differenzbildung zwischen den Daten des gemessenen Werkstücks und den durch das neuronale Netz verarbeiteten Daten bleiben die gesuchten Fehler übrig. Mit dieser Methode wird demnach zu jedem vermessenen Prüfobjekt ein zugehöriges Meisterstück generiert, welches in Position, globaler Verformung und Ausführung dem vermessenen Objekt weitgehend gleicht. Mit den in Punkt 1 genannten klassischen Methoden ist das nicht ohne weiteres möglich.
Nach der Differenzbildung zwischen Originalteil und Recallteil ist es oft sinnvoll, die überlagerten Störungen durch entsprechende Bildverarbeitungsroutinen zu reduzieren. Dadurch können die auftretenden Fehler dann besser klassifiziert werden.

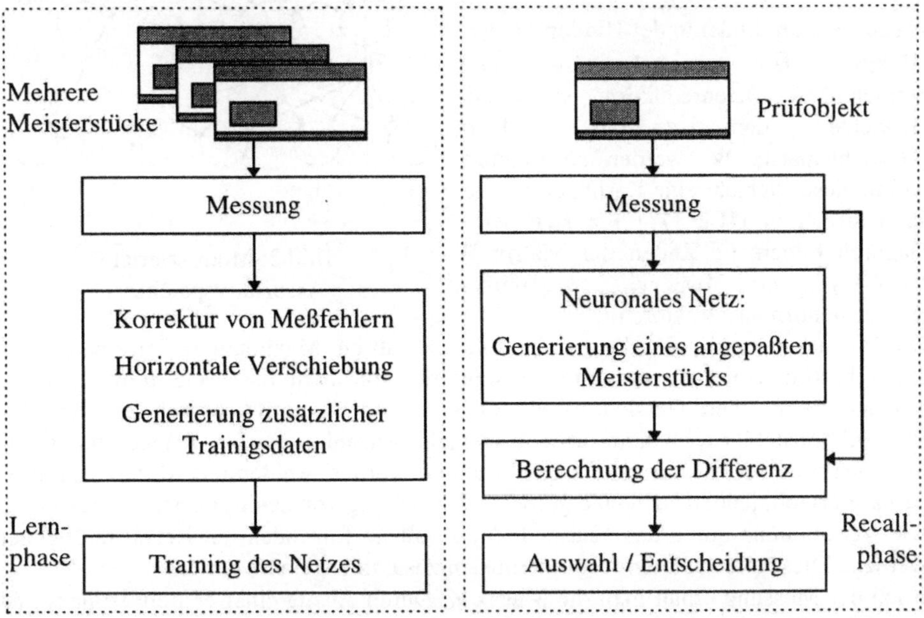

Bild 1: Lern- und Recallphase der intelligenten Fehlererkennung

3 Beschreibung des neuronalen Netzes

Im folgenden soll das genutzte neuronale Netz kurz beschrieben werden. Es wird ein linearer und modifizierter Assoziativspeicher verwendet [8], [9]. Im allgemeinen werden assoziative Netze genutzt, um spezielle Muster zu erkennen und zu rekonstruieren. Assoziativspeicher ohne Binarisierung und mit fehlender Rückkopplung interpolieren jedoch zwischen benachbarten und ähnlichen Mustern [4].
Dies entspricht sinngemäß einer Approximation mit angepaßten Funktionen, die unmittelbar aus der Vielfalt der Trainingsdaten generiert wurden. Daraus resultieren auch die wesentlich günstigeren Ergebnisse gegenüber einer einfachen Polynomapproximation.
In Bild 2 sind $z_1 \ldots z_l$ die Elemente des Eingangsvektors. $z_1^* \ldots z_l^*$ enthalten die Ausgangsdaten. Die Anzahl n der Ein- und Ausgänge sind gleich. Da das Netz auf

lineare Verbindungen festgelegt ist, resultieren einfache Beziehungen. Die Matrizen ***W*** und ***G*** enthalten die Gewichte der entsprechenden neuronalen Schichten.

Die korrigierten Werte $\vec{z}^*$ werden mit Hilfe der Matrizen ***W*** und ***G*** sowie den Ausgangswerten $\vec{a}$ der Neuronen der Hiddenschicht folgendermaßen berechnet:

$$\vec{a} = \boldsymbol{W}\vec{z}$$

da $\quad \vec{z}^* = \boldsymbol{G}\vec{a}$

folgt: $\quad \vec{z}^* = \boldsymbol{G}\boldsymbol{W}\vec{z}$

Wenn $m << n$, findet in der Hidden-Schicht (Vektor $\vec{a}$) normalerweise eine beträchtliche Datenreduktion statt. Die Gewichte der zweiten Schicht (Gewichtsmatrix ***W***) werden so trainiert, daß in dieser Schicht eine Karhunen-Loéve-Transformation [10], [11] realisiert wird. Dadurch bilden die Zeilen der Matrix ***W*** ($[w_{11}\ w_{12}\ ...\ w_{1l}]$ ... $[w_{m1}\ w_{m2}\ ...\ w_{ml}]$) einen Satz orthonormaler Vektoren.

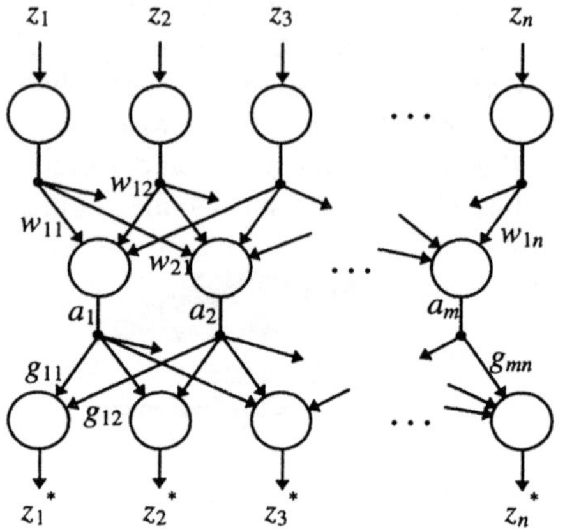

Bild 2: Modifizierter Assoziativspeicher

Die Datenkompression ist dann so organisiert, daß ein Maximum an Information für einen bestimmten Satz von Trainingsmustern, mit dem das Netz trainiert wird, erhalten bleibt. Die Gewichtsmatrix ***G*** wird mit $\boldsymbol{G} = \boldsymbol{W}^T$ berechnet. Der Rekonstruktionsfehler wird dann für jede spezielle Anzahl m von Hiddenneuronen minimal. Merkmale des Objektes, die nicht trainiert wurden, werden unterdrückt. Ein größeres m erlaubt jedoch auch eine größere Abweichung von den trainierten Objekten.

Die Verarbeitung durch das neuronale Netz soll im folgenden mit Recall bezeichnet werden. Dieser Recall kann folgendermaßen zusammengefaßt werden:

Nach der Messung erhält man die Matrix ***Z***. Durch Abzug einer Mittelwertmatrix ***M*** über alle P Trainingsdaten

$$\boldsymbol{M} = \frac{1}{P}\sum_{i=1}^{P}\boldsymbol{Z}_i$$

und anschließender Vektorisierung erhält man den Vektor $\vec{z}$. Der Recall des Assoziativspeichers

$$\vec{z}^* = \boldsymbol{W}^T\boldsymbol{W}\vec{z}$$

wird wieder in eine Matrix umgewandelt und mit der Mittelwertmatrix ***M*** addiert. Man erhält damit die Matrix $\boldsymbol{Z}^*$. Die Fehlermatrix ***F*** ist dann:

$$\boldsymbol{F} = \boldsymbol{Z} - \boldsymbol{Z}^*$$

4 Vorverarbeitung der Trainingsdaten

Bei idealen Trainigsdaten sind die in Bild 1 aufgeführten Schritte zur Aufbereitung der Meßdaten nicht notwendig. In diesem Fall stehen die Daten in ausreichender Menge zur Verfügung. Alle Meßwerte der Varianten eines fehlerfreien Werkstücks,

die auftreten können, liegen als Trainingsdaten vor, bzw. das Netz kann zwischen den vorhandenen interpolieren.
In der Praxis kann es jedoch sein, daß nicht genügend Meßwerte vorliegen. Außerdem können häufig zufällige Meßfehler auftreten, die bei einer zu geringen Menge an Trainingsdaten die statistischen Eigenschaften der Meßobjekte und des Meßverfahrens fehlerhaft wiedergeben. Aus diesem Grund sind in der Regel Schritte zur Vorverarbeitung notwendig.
Probleme ergeben sich vor allem durch kleine impulsförmige Störungen (Pfeffer- oder Salzrauschen), die bei der Messung auftreten können. Zur Beseitigung solcher Störungen eignet sich prinzipiell der Medianfilter. In dem untersuchten Anwendungsbeispiel (siehe Punkt 0) werden aber auch durch Anwendung des Medianfilters die Konturen der Objekte zu stark verändert, so daß diese Methode ungeeignet ist. Ursache sind die verhältnismäßig kleinen und dichten Störimpulse.
Günstiger ist eine Spline-Approximation. Die Objekte (480x480 Meßpunkte) wurden durch Splineflächen mit 40x40 Knoten approximiert. Das Verhältnis zwischen Glättung und Konturveränderung ist so besonders günstig. Nach der Approximation wird die erhaltene Funktion mit dem ursprünglichen Raster abgetastet. Eine solche Vorgehensweise ist nur bei Trainingsdaten möglich. Bei der eigentlichen Messung könnte es zur Unterdrückung von Fehlern kommen.
Problematisch kann es auch sein, wenn mögliche horizontale Verschiebungen durch die Trainingsdaten nicht erfaßt werden. In diesem Fall müssen durch Simulation der Verschiebung zusätzliche Trainingsdaten generiert werden. Zur Extrapolation der Randbereiche wird dabei die oben erwähnte Spline-Approximation genutzt.
Häufig reicht die Zahl der Trainingsdaten nicht aus und Meßfehler haben einen Einfluß auf das Lernergebnis des Assoziativspeichers. Durch Überlagerung der abgetasteten Meisterstücke mit bandbegrenztem Rauschen, das zusätzliche Variationen des Meisterstückes modelliert, lassen sich zusätzliche Lerndaten generieren [8].

5 Anwendungsbeispiel

Im folgenden soll ein industrielles Anwendungsbeispiel beschrieben werden, an dem die Leistungsfähigkeit dieses Algorithmus demonstriert werden kann. Dabei handelt es sich um tiefgezogene Autotüren aus Metall. Diese wurden bisher manuell auf ihre Qualität hin untersucht.
Es wurden dabei von vier Türen drei als fehlerfrei klassifiziert und eine als fehlerhaft. Zu dieser fehlerhaften Tür wurde eine Fehlerbeschreibung (Bild 3) angefertigt.

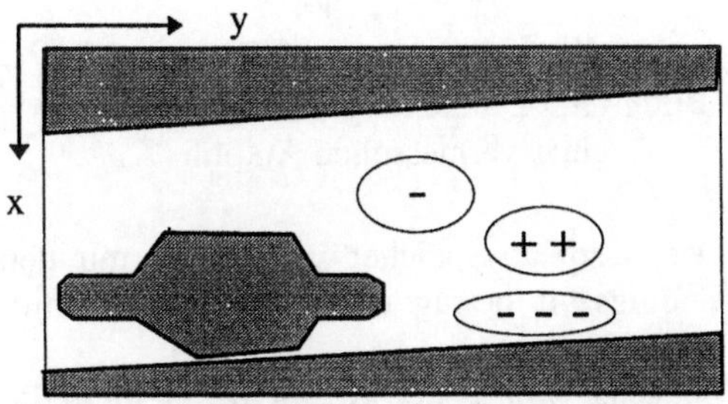

Bild 3: Prinzipielle Fehlerbeschreibung der als fehlerhaft klassifizierten Autotür („+“ und „-“ bezeichnen die Richtung der Abweichung. Regionen, die nicht vermessen werden konnten, sind grau markiert.)

Die einzelnen Türen wurden durch ein Streifenprojektionsverfahren [2], [5] vermessen. Die Abmessungen der Türen waren ungefähr 1x1 m. Bild 3 zeigt 3 fehlerhafte Regionen der als fehlerhaft deklarierten Tür. Regionen, die nicht vermessen werden konnten, sind grau markiert. Die maximale Abweichung von der Norm in Richtung der Oberflächennormale betrug ca. +- 200 μm.
Um die Messung durchzuführen, wurde die Tür an 3 Punkten fixiert, um eine Verdrehung weitgehend zu vermeiden. Die laterale Justiergenauigkeit betrug ca. 5 mm. Aus diesem Grund war die Erkennung der Fehlerregionen der fehlerhaften Tür mit herkömmlichen Methoden nicht mehr möglich.
Durch das Streifenprojektionsverfahren erhält man Datensätze von 480x480 Pixeln. Für das hier gezeigte Beispiel wurden aber nur ein ausgedünnter Datensatz von 96 x 96 Pixel (jeder 5.) verwendet. Von zwei Meisterstücken wurden 41 Datensätze der Teile in unterschiedlichen Positionen mit einer lateralen Toleranz bis zu 5 mm vermessen. Durch Verschieben der Meßpunkte wurden zusätzliche Datensätze generiert. Die z-Koordinaten der nicht vermessenen Bereiche wurden für die Darstellung auf Null gesetzt. Bild 4 zeigt die Daten einer fehlerhaften Autotür und Bild 5 die berechnete Differenz zwischen den z-Koordinaten der fehlerhaften und der fehlerfreien Autotür. Die Fehler sind im Bild 5 zum Teil erkennbar (Pfeil). Eine automatische Fehlererkennung ist jedoch nicht möglich, da die durch die ungenaue Justierung der Werkstücke entstehenden Fehler erheblich größer als die zu detektierenden Fehler sind.

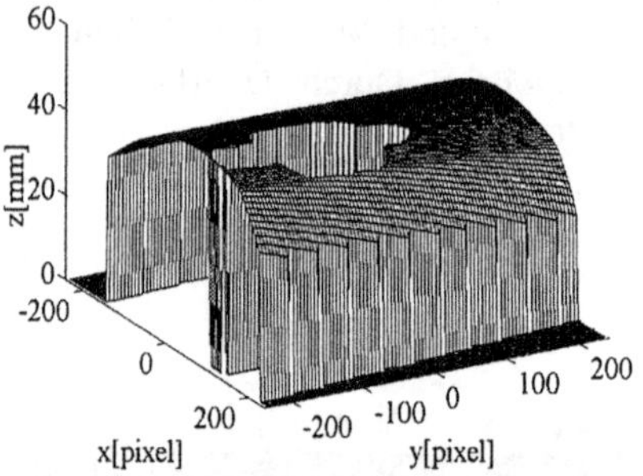

Bild 4: 3D-Darstellung des Datensatzes einer vermessenen Autotür

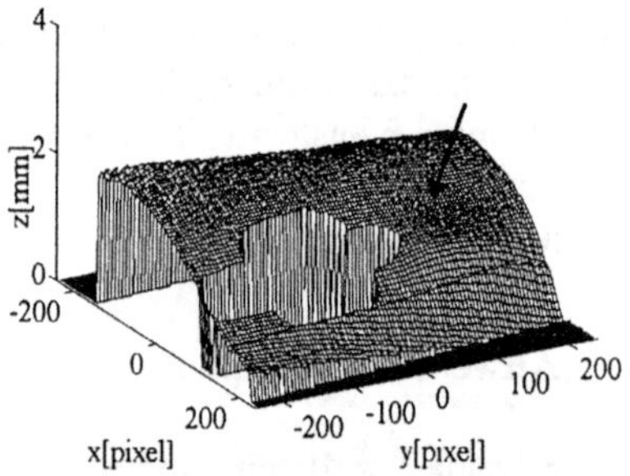

Bild 5: Berechnete Differenz der Daten einer fehlerhaften und eine fehlerfreien Autotür

Der Assoziativspeicher wurde nun mit den Daten der Meisterstücke trainiert. Die Trainigszeit betrug dabei ungefähr 1 Stunde (berechnet auf einer SPARC20 mit 80MHz).
Zunächst wurde der Recall des fehlerhaften Werkstücks berechnet und damit ein weitgehend fehlerfreier Datensatz mit gleicher Justierung erzeugt. Diese Berechnung dauerte ca. 1.5 s (auf einer SPARC20 mit 80MHz).
Bild 6 zeigt die berechnete Differenz zwischen dem Recall und dem originalen Datensatz. Die fehlerhaften Bereiche sind mit Pfeilen markiert. Die fehlerhaften Regionen können leicht erkannt werden. Die Daten der Bereiche, die nicht vermessen werden konnten (grau markiert in Bild 3), wurden in der Darstellung auf Null gesetzt.

Durch Überschwingen entstehen aber auch außerhalb der fehlerhaften Bereiche größere Abweichungen. Für eine gut/schlecht-Klassifikation stellt dieses Überschwingen aber kein Problem dar, da es nur bei fehlerhaften Teilen auftritt.

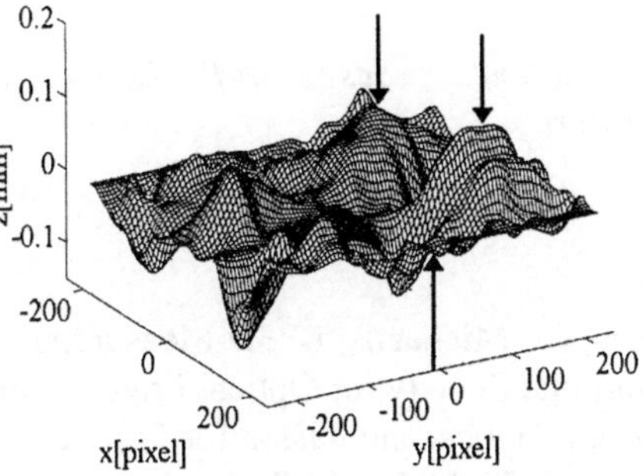

Bild 6: berechnete Differenz bei dem fehlerhaften Teil

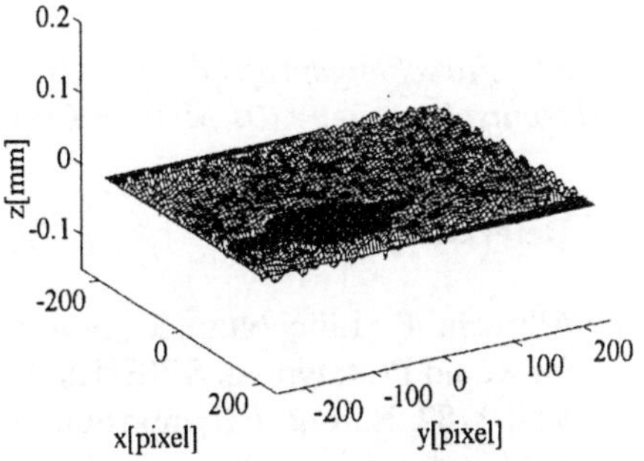

Bild 7: berechnete Differenz bei dem fehlerfreien Teil

Weiterhin wurde zur Kontrolle ein Recall mit den Daten einer fehlerlosen Tür durchgeführt. Um die Generalisierungseigenschaften des Verfahrens zu verifizieren, wurden diese Daten nicht zum Training des Assoziativspeichers verwendet. Bild 7 zeigt das Ergebnis der Differenzbildung zwischen Recall und Original für diesen Datensatz. Die Werte sind ungefähr um den Faktor 10 kleiner als bei feherbehafteten Türen. Es lassen sich bei den betreffenden Werkstücken Fehler von 0,1mm in *z*-Richtung nachweisen.

Bei einem Vergleich der Resultate des fehlerhaften Werkstücks mit dem fehlerfreien kann eine einfache Entscheidung zwischen „in Ordnung" und „nicht in Ordnung" erfolgen. Diese Entscheidung ist mit numerischen Standardmethoden innerhalb weniger Sekunden möglich. Auch eine quantitative Messung kann durchgeführt werden.

6 Zusammenfassung und Ausblick

Das vorgestellte Beispiel zeigt anschaulich, wie mit der vorgestellten Methode Fehler der Oberflächenform aus 3D-Meßdaten heraus erkannt werden können. Das neuronale Netz speichert die Daten unterschiedlicher Meisterstücke in verschiedenen Variationen und ist so in der Lage für ein vermessenes Werkstück einen weitgehend fehlerfreien Datensatz mit exakter Justierung zu erzeugen. Danach können fehlerhafte Werkstücke automatisch erkannt werden.

Weitere Arbeiten auf diesem Gebiet sind vor allem bei der Vorverarbeitung der Trainingsdaten erforderlich. Auf diese Weise kann z. B. das in Bild 6 auftretende Überschwingen sowie der Einfluß der durch fehlerhafte Meßwerte verursachten impulsförmigen Störungen vermieden werden.

In dem vorgestellten Beispiel wurden 9216 Pixel für die Berechnung verwendet. Werden mehr Pixel verwendet, würde der Recall zu lange dauern. In diesem Fall wäre eine Hardwareimplementierung des modifizierten Assoziativspeichers sinnvoll.

Bei Voruntersuchungen wurde festgestellt, daß mit Hilfe eines Neurocomputers CNAPS eine Fehlererkennung mit dieser Methode für größere Bilder (z. B. 512x512 Pixel) im Zeitraum unter einer Sekunde möglich ist.
Verbesserungen der Ergebnisse sind möglich durch Nutzung von noch mehr Trainingsdaten.

Dieses Forschungsprojekt wurde im Rahmen des BMBF-Leitprojektes „Elektronisches Auge“ (01 M 3018 E/3) gefördert.

7 References

[1] Albrecht, P.; Lilienblum, T.; Sommerkorn, G.; Michaelis, B.:3D-Measurement for Rapid Prototyping. SPIE-The International Society for Optical Engeneering, Vol. 2787, European Symposium on Lasers, Optics, and Vision for Productivity in Manufacturing I, Besancon, France, pp. 62-73, 10.06.-14.06.1996.

[2] Dörband B., Zilker A.„Fringe projection systems for real time testing of car bodies with 5 µm resolution“; VDI-Report Nr. 1118, pp 21-26 1994.

[3] Herzog, A.; Schütze, W.; Lilienblum, T.; Braun, K.; Michaelis, B.: 3D-Formrekonstruktion an der Auflösungsgrenze konfokaler Laserscan-Mikroskope. Mustererkennung 1997, Springer-Verlag Berlin Heidelberg New York, ISBN 3-540-63426-6, S. 219-226

[4] Hassoun, M. H.: Associative Neural Memories: Theory and Implementation. Oxford University Press, New York 1993

[5] Hof A., Hanßen A., Küchel M., Schütz J. „Streifenprojektionsmethoden in der 3-D-Meßtechnik“, VDI-Report Nr. 1006, pp 67-80, 1992

[6] Krell, G.; Tizhoosh, H. R.; Lilienblum, T.; Moore, C. J., Michaelis, B.: Fuzzy Image Enhancement and Associative Feature Matching in Radiotherapy. Proceedings of the International Conference on Neural Networks (ICNN ‘97), Houston, Texas, 09.06.-12.06.1997, pp. 1490-1495

[7] Lilienblum, T.; Zhang, W.; Michaelis, B.: Sensor Data Compression for Smart Mechanical Systems by Artificial Neural Networks. In: Gabbert, Ulrich (Hrsg.): Smart mechanical Systems - Adaptronics. Fortschr.-Ber. VDI Reihe 11 Nr. 244. Düsseldorf: VDI-Verlag 1997. pp. 235-244

[8] Lilienblum, T.; Albrecht, P.; Michaelis, B.: Improvement of 3-D Data with Neural Networks. `Journal of Microelectronic Systems Integration`, Volume 5 (1997), No. 1, pp. 53-64

[9] Lilienblum, T.; Albrecht, P.; Michaelis, B.: 3D-Measurement of Geometrical Shapes by Photogrammetry and Neural Networks. 13th International Conference on Pattern Recognition (ICPR '96), Wien, pp. 330-334, 25.08.-30.08.1996.

[10] Erkki Oja: A Simplified Neuron Model as a Principal Component Analyzer. J. Math. Bio., vol. 15, pp. 267-273, 1982

[11] Sanger, T.D.: Optimal Unsupervised Learning in a Single-Layer Linear Feedforward Neural Network. Neural Networks, Vol.2, pp. 459-473, 1989

Dynamische Analyse von rekonstruierten 3D Punkten zur Navigation eines Stereo-Vision basierten mobilen Systems

Dipl.-Ing. Oliver Schreer

Institut für Meß- und Automatisierungstechnik, TU Berlin
Sekr. EN11, Einsteinufer 17, D-10587 Berlin
e-mail: schreer@rtws18.ee.TU-Berlin.DE

1 Einleitung

Die verschiedenen Anwendungen im Bereich der Navigation von mobilen Systemen hängen u.A. von der Art des verfügbaren Wissens über die Umgebung ab. Anwendungen in Gebäuden verwenden häufig Landmarken oder -karten zur Orientierung oder man verfügt über eine genaue Kenntnis von Objekten, die sich in derartiger Umgebung befinden [2][6][9]. Ist nun kein *apriori*-Wissen über die Umgebung oder Objekte in dieser vorhanden, so sind allgemeine Ansätze zur Navigation gefordert. In diesem Beitrag wird ein Verfahren zur Navigation eines autonomen mobilen Systems in Innenräumen skizziert, das auf rekonstruierten 3D Punkten basiert. Es wird keine Kenntnis über die Umgebung vorausgesetzt. Grundlage dieses Verfahrens ist ein Liniensegment-Stereo-Algorithmus, sowie ein 3D Rekonstruktionsverfahren. Damit erhält man unsichere 3D Punkte einer Szene, die alle als mögliche Hindernisse aufgefaßt werden. Diese 3D Punkte werden mit einem kombinierten statistischen Verfahren analysiert, das schließlich zu einem Lenkwinkel für das mobile System führt. Um die Navigation robuster zu gestalten, wird der in [8] vorgestellte Ansatz um eine dynamische Analyse der rekonstruierten 3D Punkte erweitert. Das Verfahren wurde auf dem Experimentalsystem MOVILAR realisiert (MObile VIsion and LAser based Robot), welches über ein Stereokamerasystem verfügt.

2 Rekonstruktion von 3D Punkten

Das klassische Lochkameramodell beschreibt die Abbildung eines 3D Punktes auf ein 2D Bild und ist über die Projektionsmatrix definiert, welche die internen und externen Kameraparameter sowie die Rotation und Translation enthält.

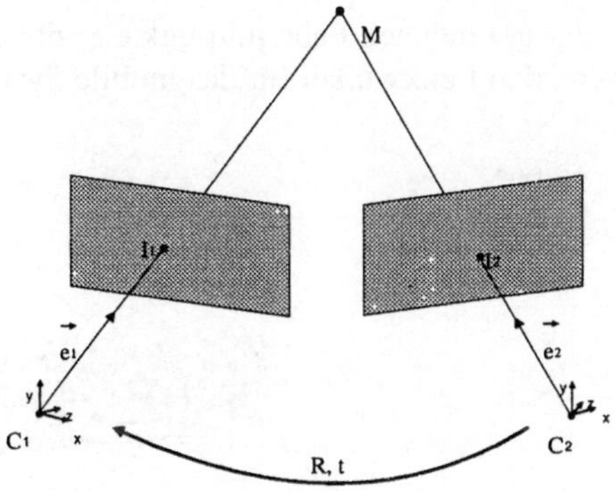

Abb. 1: Allgemeine Stereogeometrie

Die mittels Kalibrierung geschätzten Kameraparameter erlauben die Rekonstruktion eines 3D Punktes *M* für zwei korrespondierende Punkte $I_{1,2}$ im linken und rechten Steoreobild [4]. Derjenige 3D Punkt *M*, welcher am nächsten zu beiden optischen Strahlen liegt, kann dann mit Hilfe eines Least-Square-Ansatzes bestimmt werden [1][3][10]. Das Hauptproblem in der Stereobildverarbeitung ist die Lösung des Korrespondenzproblems zwischen verschiedenen Mustern im linken und rechten Stereobild. In dieser Untersuchung wird ein Liniensegment-basierter Stereoalgorithmus verwendet [1], wobei eine Erweiterung um zusätzliche Grauwerteigen-schaften vorgenommen wurde [7].

3 Fahrspurbestimmung

Die Navigationsaufgabe des mobilen Systems ist, geradeaus zu fahren oder den nächst möglichen freien Weg im Sichtbereich der Kameras zu suchen und Hindernissen auszuweichen. Alle weiteren Betrachtungen werden nur in der x/z-Ebene vorgenommen, da die Bewegung des Fahrzeuges sich auch nur in dieser Ebene vollzieht. Aufgrund der Unsicherheit der rekonstruierten 3D Punkte kann keine Fahrspurbestimmung mit einzelnen 3D Punkten durchgeführt werden. Daher wird die Gesamtheit aller 3D Punkte betrachtet, um über geeignete statistische Methoden den Einfluß von Rekonstruktionsfehlern zu minimieren.

3.1 Minimum Distanz Methode

Diese Methode geht davon aus, daß 3D Punkte von Hindernissen erkannt werden und ein möglicher Weg zwischen diesen möglich ist. Der Abstand beträgt mindestens d_w. Die Verbindungslinie zwischen zwei 3D Punkten wird *Fahrspursegment* genannt und der Punkt in der Mitte dieses Segmentes *Fahrspurpunkt.* 3D Punkte der gleichen Kante werden von der paarweisen Analyse ausgeschlossen.

$$
\begin{aligned}
&|P_i(x,z) - P_j(x,z)| > d_w \text{ und } |\beta| < 45^\circ \\
&\forall i, j \in N, i \neq j, \quad i \text{ und } j \text{ von versch. Kanten}, \quad N = \text{Anzahl der 3D Punkte}
\end{aligned}
\tag{1}
$$

Das gewichtete Mittel aller gefundenen Fahrspurpunkte ergibt dann die mittlere Fahrspur und dementsprechend den Lenkwinkel für das mobile System (Abb.2).

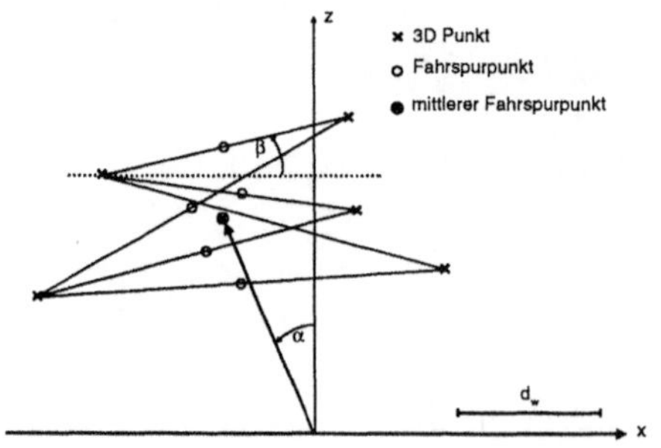

Abb. 2: Die Minimum Distanz Methode

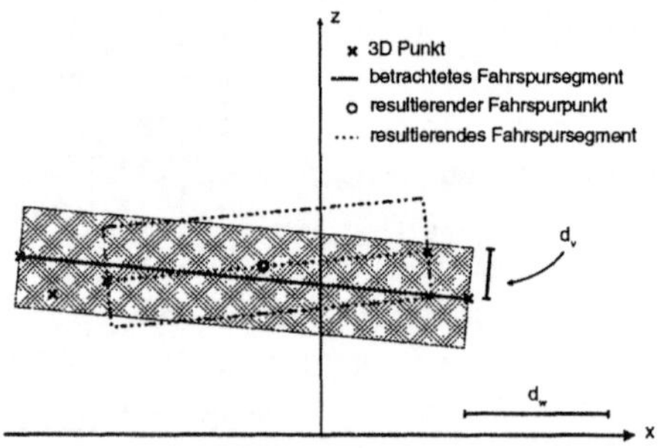

Abb. 3: Linear angeordnete 3D Punkte

Diese Methode liefert zuverlässige Ergebnisse, wenn die Hindernisse auf Bereiche beiderseits des mobilen Systems verteilt sind. Liegen die 3D Punkte jedoch auf einer Linie angeordnet, wie das z.B. bei Wänden der Fall wäre, so muß eine genauere Analyse der Lage der Punkte zueinander durchgeführt werden. Deshalb wird zusätzlich ein Bereich um jedes Fahrspursegment betrachtet, in dem kein anderer 3D Punkt liegen darf. Die Größe des Bereichs wird durch die Konstante d_v beschrieben, die einem minimalen Abstand der Fahrspurpunkte in der Tiefe des x/z-Raumes entspricht (Abb.3).
In den folgenden Abbildungen ist, anhand einer realen Szene, ein Beispiel für die Minimum Distanz Methode gegeben. Die Grundlage ist das Stereosegmentbild (links), wobei gestrichelte und durchgezogene Liniensegmente die Zugehörigkeit zum linken oder rechten Segmentbild zeigen. Die homologen Liniensegmentpaare (rechts) sind durch gleiche Nummern gekennzeichnet. Die Grafik unten stellt die 3D Daten in der x/z-Ebene dar, wobei die rekonstruierten 3D Punkte durch Kreuze, die berechneten Fahrspurpunkte durch Kreise und die geschätzte Fahrspur durch eine Linie dargestellt sind.

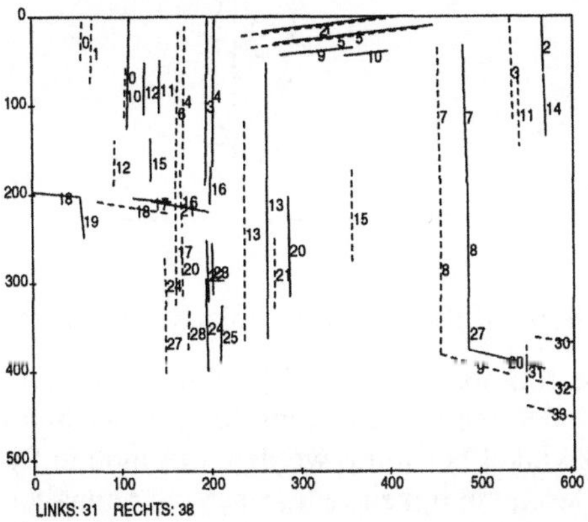

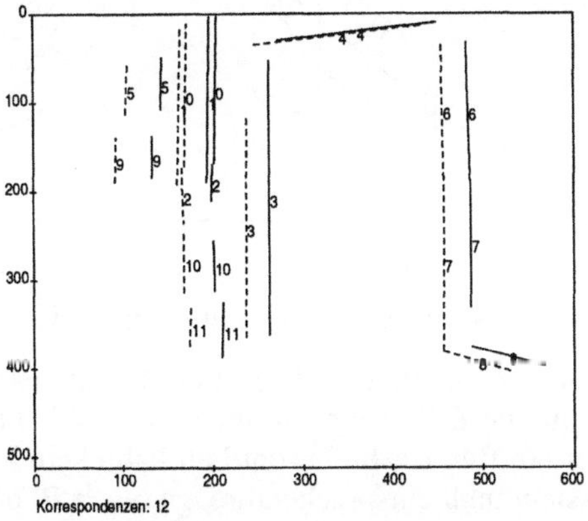

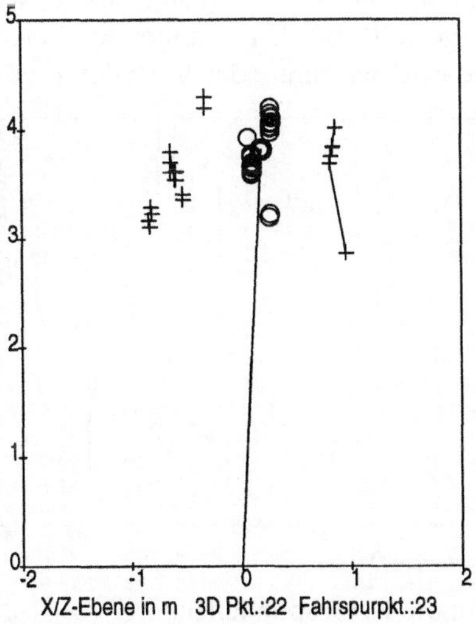

In nebenstehender Abbildung wurde die Minimum Distanz Methode angewendet, da auf der linken und rechten Seite 3D Punkte ermittelt wurden und ein Weg zwischen ihnen hindurch möglich ist. Liegen jedoch Hindernispunkte nur auf einer Seite des Fahrzeuges, so versagt diese Methode und eine weitere Analyse der 3D Punkte muß vorgenommen werden.

3.2 Punktverteilungsmethode

Diese Methode basiert auf der Kovarianzmatrix $\underline{C}$, die über alle rekonstruierten 3D Punkte berechnet wird.

$$\underline{C} = \frac{1}{N}\sum_{i=1}^{N}(\vec{p}_i - \vec{m}_0)\cdot(\vec{p}_i - \vec{m}_0)', \quad \vec{m}_0 = Mittelwert, \quad \vec{p}_i = P_i(x,z) \tag{2}$$

Die Eigenvektoren beschreiben die Orientierung der Verteilung. Die Quadratwurzel der Eigenwerte korrespondiert mit der Standardabweichung in Richtung der Hauptachsen der Verteilung, welche die Ausdehnung einer Ellipse angeben. Um aus der Punktverteilung eine Fahrspurinformation zu gewinnen, muß die Lage (Schwerpunkt m_0) und Orientierung (Winkel φ zwischen der größeren Hauptachse und der Z-Achse) der Ellipse klassifiziert werden. Der Lenkwinkel ergibt sich dann aus einem Endpunkt der Hauptachse der Ellipse und einem Sicherheitsabstand d_s.

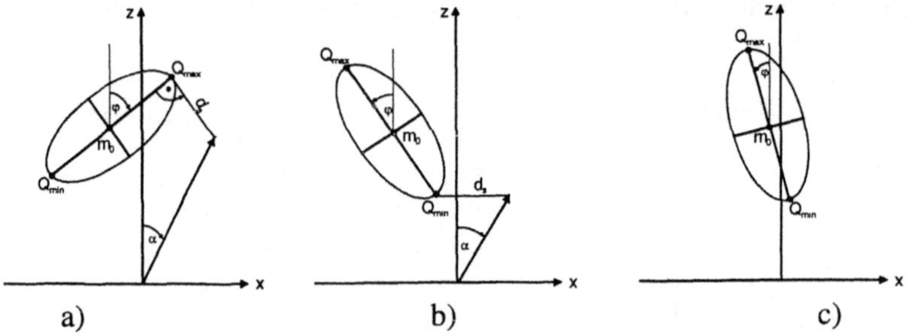

Abb. 4: Verschiedene Fälle von Punktverteilungs-Ellipsen für m_0 links der Z-Achse

In den Fällen a) und b) ist ein Lenkwinkel bestimmbar, im Fall c) liegt die Verteilung um die Z-Achse positioniert, so daß keine eindeutige Entscheidung für eine Richtung zu treffen ist. In diesem Fall kann kein Lenkwinkel bestimmt werden, das mobile System muß entsprechend reagieren, z.B. die Geschwindigkeit reduzieren oder anhalten. Die folgenden Abbildungen illustrieren die Methode anhand einer realen Szene, wobei die Punktverteilung durch eine Ellipse dargestellt ist. Die Länge der beiden Hauptachsen entspricht dem Dreifachen der Standardabweichung der Verteilung.

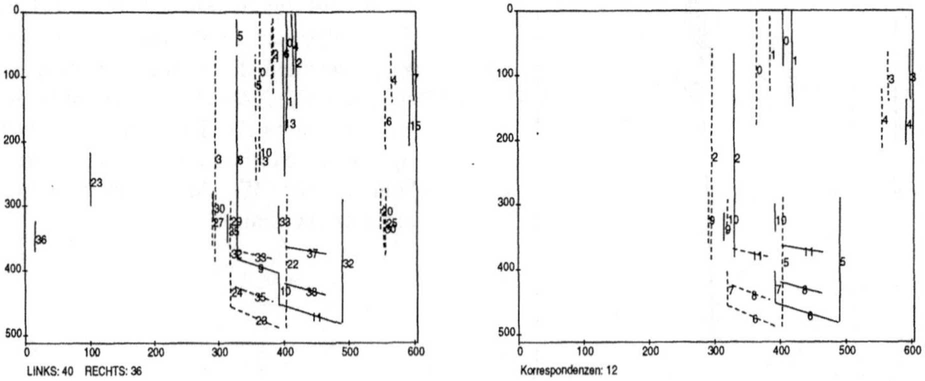

Abb. 5: Stereosegmente und berechnete Korrespondenzen

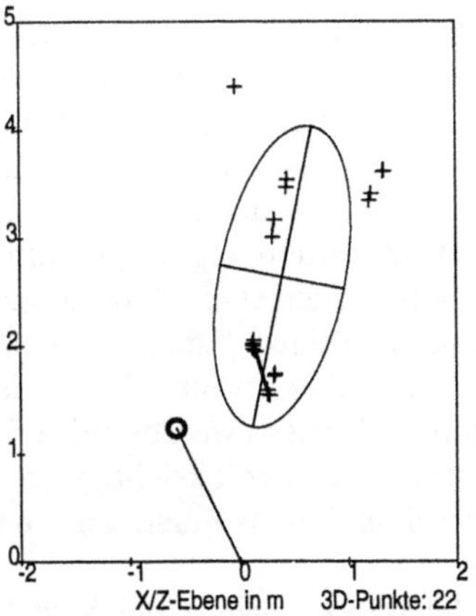

In nebenstehender Abbildung wurde, basierend auf den Ergebnissen aus Abb.5, die Punktverteilungsmethode angewendet, da nur Hindernisse rechts des mobilen Systems erkannt wurden. Die Punktverteilung wird durch eine Ellipse approximiert und ist in die Verteilung eingezeichnet.

4 Probleme bei einer statischen Analyse von rekonstruierten 3D Punkten

Das bisherige Verfahren, die sog. statische Analyse, basiert auf rekonstruierten 3D Punkten einer Szene. Das Verfahren führt zwar zu einer kollisionsfreien Navigation in Innenräumen, die geschätzten Lenkwinkel variieren jedoch relativ stark und es treten vor allem zwei Probleme bei der Fahrspurbestimmung auf.

1. Das Stereomatching-Verfahren liefert falsche Korrespondenzen, die zu fehlerhaft rekonstruierten 3D Punkten führen. Diese liegen unter Umständen in einer möglichen Fahrspur und werden bisher von dem Verfahren ignoriert, wenn die Anzahl der 3D Punkte gering ist. Liegt an dieser Stelle jedoch ein reales Hindernis mit dieser geringen Anzahl von 3D Punkten, so kann dieses nicht erkannt werden.
2. Es treten Szenen auf, die bedingt durch vorangegangene Verarbeitungsschritte wie Segmentierung oder Stereomatching, nur durch sehr wenig 3D Punkte beschrieben werden und damit für die Fahrspurbestimmung unbrauchbar sind.

5 Dynamische Analyse von rekonstruierten 3D Punkten

Bei der dynamischen Analyse werden die Ähnlichkeiten von 3D Punktverteilungen zwischen aufeinanderfolgenden Stereobildern untersucht, um die obengenannten Probleme zu lösen. Aufgrund der Ortsänderung ($\Delta x, \Delta z$) und Orientierungsänderung $\Delta\varphi$ des mobilen Systems zwischen aufeinanderfolgenden Stereobildpaaren ergibt sich im lokalen Koordinatensystem die entsprechende örtliche Verschiebung gleicher rekonstruierter 3D Punkte.

$$P_{t_{i+1}}(x,z) = P_{t_i}(x,z) \cdot \underline{R}_{\Delta\varphi} - \vec{t}, \text{ mit } \vec{t} = (\Delta x, \Delta z) \tag{3}$$

Ist die Orts- und Orientierungsänderung bekannt, so kann für jeden 3D Punkt bestimmt werden, ob dieser zu vorangegangenen Zeitpunkten schon existierte. Als Ähnlichkeitsmaß wird, unter Berücksichtigung der örtlichen Verschiebung, der euklidische Abstand verwendet, der sich innerhalb einer ε-Umgebung befinden muß. Dieser Ansatz ist aufgrund der Nächste-Nachbar-Betrachtung kein exaktes Tracking von 3D Punkten, genügt aber für entsprechend kleines ε zur Lösung des Navigationsproblems.

Das erste Problem fehlerhaft rekonstruierter 3D Punkte wird durch das vereinfachte Tracking aktueller 3D Punkte mit rekonstruierten 3D Punkten des vorangegangenen Zeitpunktes gelöst. Die wesentliche Annahme ist dabei, daß fehlerhafte 3D Punkte in aufeinanderfolgenden Szenen unter Berücksichtigung der Positionsänderung des mobilen Systems nicht an der gleichen Stelle auftreten. Diese Analyse führt jedoch andererseits dazu, daß korrekte 3D Punkte für eine Fahrspurbestimmung ausgeschlossen werden, die nicht in der vorangegangen Szene aufgetreten sind. Aus diesem Grund ist es notwendig, sichere 3D Punkte bei einer zukünftigen Fahrspurbestimmung zu berücksichtigen. Dabei gelten 3D Punkte als sicher, wenn sie in der aktuellen und zwei vorangegangenen Szenen aufgetreten sind. Diese werden dann für mehrere zukünftige Szenen bei der Fahrspurbestimmung berücksichtigt und ermöglichen damit die Navigation, wenn die Hindernisse bereits außerhalb des Sichtbereichs des Stereosystems liegen.
In Abb.6 ist anhand der Ergebnisse einer realen Szene die Wirkungsweise des Verfahrens zu erkennen. Im linken Bild liegen vorne rechts fehlerhafte 3D Punkte vor, die das statische Verfahren ignoriert und deshalb trotzdem eine korrekte Fahrspur bestimmt. Eine dynamische Analyse eliminiert diese fehlerhaften 3D Punkte und fügt weitere Punkte hinzu, die in den vorangegangenen Szenen als sichere 3D Punkte klassifiziert wurden. Damit erhöht sich auch die Anzahl der berechneten Fahrspurpunkte.

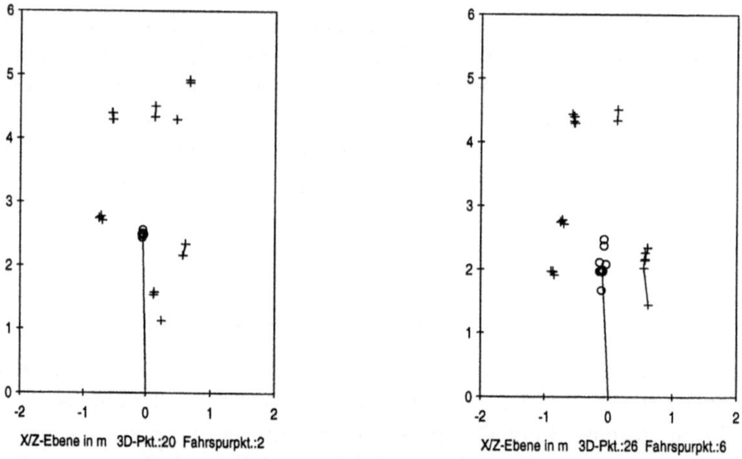

Abb. 6: Statische Analyse (links) und dynamische Analyse (rechts) von 3D Punkten

In Abb.7 und 8 sind die Ergebnisse der statischen und dynamischen Analyse einer 3D Punkt-Sequenz dargestellt. Sie resultieren aus 84 Stereobildpaaren einer Testfahrt durch die Räume des Instituts mit einer Dauer von ca. 2 min. Der Zeitabstand zwischen zwei aufeinanderfolgenden Datensätzen entspricht der gegenwärtigen Prozeßzyklusdauer von ca. 1.5s. Die Geschwindigkeit des mobilen Systems betrug ca. 10 cm/s. In Abb. 7 ist der Verlauf des ermittelten Lenkwinkels über der Zeit aufgetragen. Abb. 8 zeigt die Methode, die aufgrund der Lage der 3D Punkte angewandt wurde. Die Werte 0,1,2 für die Methode bedeuten: 0 = *keine Fahrspurbestimmung möglich*, 1 = *Punktverteilungsmethode*, 2 = *Minimale Distanz Methode*.

Die Unsicherheit bei den rekonstruierten 3D Punkten drückt sich im statischen Fall durch einen sehr schwankenden Lenkwinkelverlauf aus. Die dynamische Analyse führt hingegen zu einem wesentlich glatteren Verlauf, resultierend aus der Eliminierung unsicherer 3D Punkte und Verwendung sicherer 3D Punkte zu späteren Zeitpunkten.

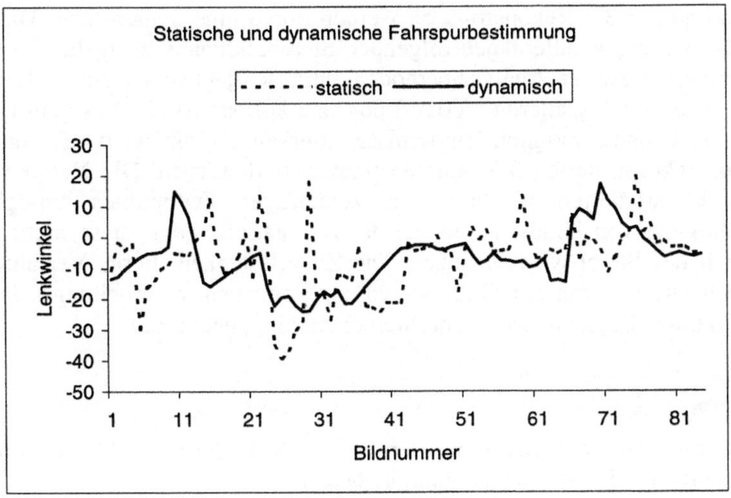

Abb. 7: Verlauf des Lenkwinkels bei statischer und dynamischer Analyse

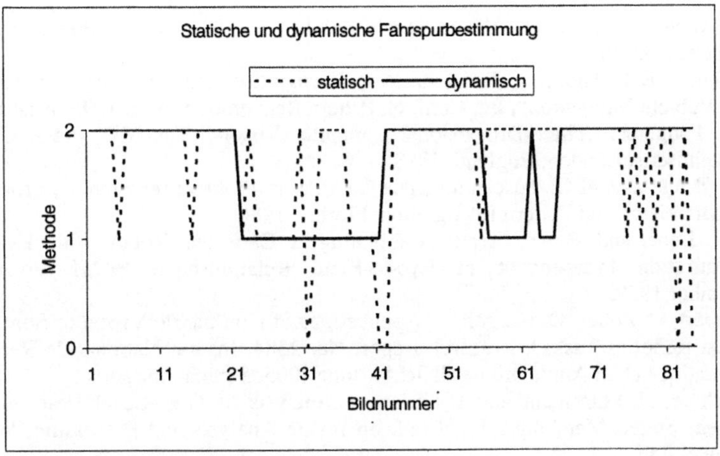

Abb. 8: Gewählte Methode bei statischer und dynamischer Analyse

Betrachtet man den Verlauf der gewählten Methode, so stellt man auch hier Schwankungen für den statischen Fall fest. Bei einigen Bildpaaren (Nr. 32,42,43,82,84) konnte keine Fahrspur bestimmt werden. Für die dynamische Analyse ändert sich die gewählte Methode wesentlich seltener und wird nur noch von der Szenerie bestimmt.

6 Zusammenfassung

Das vorgestellte Verfahren ermöglicht die Stereo-Vision-basierte Navigation eines mobilen Systems ohne *apriori*-Kenntnis der Umgebung. Ein Liniensegment-Stereo-Algorithmus mit anschließender 3D Rekonstruktion liefert unsichere 3D Punkte einer Szene, die alle als mögliche Hindernisse betrachtet werden. Die Fehler des Stereo-Matchings und der 3D Rekonstruktion werden durch eine dynamische Analyse von 3D Punktverteilungen aufeinanderfolgender Stereobildpaare weitgehend eliminiert. Die Kombination zweier Analysemethoden führt schließlich zu einem Lenkwinkel mit einen wesentlich glatteren Verlauf, der eine kollisionsfreie Navigation bewirkt. Weiterhin ist es damit möglich Hindernisse zu erkennen, dir nur durch eine geringe Anzahl von rekonstruierten 3D Punkten repräsentiert werden. Die Verwendung der als sicher klassifizierten 3D Punkte in zukünftigen Zeitpunkten ermöglicht die Berücksichtigung von Hindernissen bei der Navigation, wenn diese nicht mehr im Sichtbereich des Stereosystems liegen. Die Zuverlässigkeit dieses Verfahrens wird jedoch nach wie vor von der Güte der vorangegangenen Verarbeitungsschritte, wie Kantenextraktion, Segmentierung und Stereomatching, beeinflußt.

Danksagung

Ich möchte mich an dieser Stelle bei Herrn Prof. Dr. I. Hartmann für die volle Unterstützung bei meiner Forschungstätigkeit bedanken.

Literatur

[1] N. Ayache: „Artificial Vision for Mobile Robots", MIT Press, Cambridge, Massachusetts, 1991.

[2] M. Ekinci, B.T. Thomas: „Road Junction Recognition and Turn-Offs for Autonomous Road Vehicle Navigation", Int. Conf. on Pattern Recognition, Aug. 1996, Austria.

[3] O. D. Faugeras: „Three-Dimensional Computer Vision", The MIT Press, Cambridge, Massachusetts, London, England, 1993.

[4] O. D. Faugeras and G. Toscanini: „The Calibration Problem for Stereo", Proc. of Conf. Comput. Vision and Pattern Recognition, Florida, 1986.

[5] R. K. Lenz and R. Y. Tsai: „Calibrating a Cartesian Robot with Eye-on-Hand Configuration Independent of Eye-to-Hand Relationship", PAMI Vol.11, No.9, September 1989.

[6] L. Robert, C. Zeller, O. Faugeras: „Applications of Non-metric Vision to Some Visually Guided Robotics Tasks", Research Report No. 2584, Institut National de Recherche en Informatique et en Automatique (INRIA), June 1995, Sophia-Antipolis.

[7] O. Schreer, I. Hartmann and R. Adams: „Analysis of Grey-Level Features for Line Segment Stereo Matching", Int. Conf. on Image Analysis and Processing, Sept. 1997, Florence, Italy.

[8] O. Schreer: „Stereo Vision-Based Navigation in Unknown Indoor Environment", presented at Europ. Conf. on Computer Vision, June 1998, Freiburg, Germany.

[9] P. Weckesser, F. Wallner, R. Dillmann: „Position Correction of a Mobile Robot Using Predictive Vision", Int. Conf. on Intelligent Autonomous Systems, IAS-4, 1995.

[10] Z. Zhang, G. Xu: „Epipolar Geometry in Stereo, Motion and Object Recognition", Kluwer Academic Publisher, 1996, Netherlands.

Sprach- und Bildverstehen
(Postersession)

Orientation Constraints in Point-Based Elastic Image Registration

Mike Fornefett, Karl Rohr, Rainer Sprengel*, and H. Siegfried Stiehl

Universität Hamburg, Fachbereich Informatik,
Arbeitsbereich Kognitive Systeme
Vogt-Kölln-Str. 30, D-22527 Hamburg, Germany
fornefett@informatik.uni-hamburg.de

Abstract. We present a new approach to incorporate orientation attributes at landmarks in an elastic medical image registration scheme. The approach is based on a minimizing functional and can cope with images of arbitrary dimensions. Constraints due to orientations are incorporated through scalar products between the transformed orientation vectors in the source image and vectors which are perpendicular to the orientation vectors in the target image. For synthetic data sets we show that the structure of rigid objects in an otherwise elastically deformed image can be preserved. The application of our approach to tomographic data sets shows that improvements w.r.t. accuracy can be achieved provided a sufficient number of landmarks is well-distributed over the image.

1 Introduction

Landmark-based elastic registration relies on the information associated with landmark points. To improve the results of landmark-based approaches while maintaining their efficiency, work in the past has been done on including additional landmark attributes, like orientation. Especially in medical applications it is desirable to improve the registration result without adding further landmarks.

Bookstein and Green [1] incorporate orientations by approximating them through two landmarks. Mardia et al. [3] use a kriging method to include derivative information. Their interpolation scheme yields an exact solution, but orientation vectors are required to be unit vectors. In this paper, we describe a new approach which i) is based on a minimizing functional and ii) allows to incorporate orientation attributes. As in the approach of Mardia et al. [3], exact orientations at landmarks are incorporated. However, in our case the directional constraints are represented through scalar products and therefore the lengths of the orientation vectors need not to be normalized to unit vectors. Another advantage of our approach is that it covers the full range from interpolation to approximation using a regularization parameter λ. In Sect. 2, we give an overview of our approach for the general case of d-dimensional images and in Sect. 3, we demonstrate the applicability of our approach for 2D and 3D images.

* Now with debis Systemhaus GEI, Hamburg.

2 Approach

The inclusion of orientations at landmarks of two corresponding images means to constrain the transformation in the neighborhood of the landmarks in direction of the mapped orientations. Note, that it is possible to apply several constraining orientations at single landmarks. In our scheme the searched transformation function $\boldsymbol{u} : \mathbb{R}^d \to \mathbb{R}^d$, where d is the image dimension, minimizes the following functional:

$$\frac{1}{n}\sum_{i=1}^{n}(\mathbf{v}_i - L_i\boldsymbol{u})^2 + \lambda\, J_m^d(\boldsymbol{u}), \tag{1}$$

which can also be written as:

$$\frac{1}{dn+(d-1)n_\theta}\left(\sum_{i=1}^{n}(\mathbf{q}_i - \boldsymbol{u}(\mathbf{p}_i))^2 + \sum_{i=1}^{n_\theta}\sum_{j=1}^{d-1}\left((\mathbf{d}_i{}^T\nabla)\boldsymbol{u}^T(\mathbf{p}_{\theta_i})\,\mathbf{e}_{i,j}^{\perp}\right)^2\right) + \lambda\, J_m^d(\boldsymbol{u}), \tag{2}$$

where n is the number of pairs of point landmarks $(\mathbf{p}_i, \mathbf{q}_i)$. n_θ is the number of orientations $\mathbf{d}_i$ at locations $\mathbf{p}_{\theta_i}$ in the source image. The vectors $\mathbf{e}_{i,j}^{\perp}, j = 1 \ldots d-1$ are perpendicular to the corresponding orientations $\mathbf{e}_i$ at locations $\mathbf{q}_{\theta_i}$ in the target image. $J_m^d(\boldsymbol{u})$ is the smoothness-related energy term described in [4,5], where m is the order of derivatives used. $L_i, i = 1 \ldots n$ of (1) are bounded linear evaluation functionals. For the point constraints in the first sum of (2) we set $L_i\boldsymbol{u} = \boldsymbol{u}(\mathbf{p}_i)$, and for the orientation constraints in the double sum of (2) we set $L_{i,j}\boldsymbol{u} = (\mathbf{d}_i{}^T\nabla)\boldsymbol{u}^T(\mathbf{p}_{\theta_i})\mathbf{e}_{i,j}^{\perp}$. λ is the regularization parameter. In case of $\lambda = 0$, we obtain an interpolation, and in the case of $\lambda > 0$, we obtain an approximation of the point and orientation constraints and a larger smoothness of the transformation. In the latter case, the orientation vectors should be normalized.

The second term in (2) is a scalar product between $(\mathbf{d}_i{}^T\nabla)\boldsymbol{u}(\mathbf{p}_{\theta_i})$ and the vectors $\mathbf{e}_{i,j}^{\perp}$. $(\mathbf{d}_i{}^T\nabla)\boldsymbol{u}(\mathbf{p}_{\theta_i})$ is the transformed orientation vector of the source image and it should be perpendicular to $\mathbf{e}_{i,j}^{\perp}$, the j orthogonal vectors of the orientation vector of the target image, such that its contribution to the functional is zero.

Choosing vectors from the orthogonal space has the advantage that the corresponding scalar product is zero independent of the length of the vectors. Note however, that this holds true only in the case of interpolation but not for approximation. The $d-1$ perpendicular orientations $\mathbf{e}_{i,j}^{\perp}$ to the target orientation vector $\mathbf{e}_i$ constrain the orientation of the transformed source orientation vector $\mathbf{d}_i$ to lie on a line. Accordingly, the number of constraints is lower if an orientation is not constrained w.r.t. a line, but w.r.t. a plane, for example. The landmark pairs $(\mathbf{p}_{\theta_i}, \mathbf{q}_{\theta_i})$ with orientation information may, but need not be a subset of the landmark pairs $(\mathbf{p}_i, \mathbf{q}_i)$.

Application of the spline theorem in [5] yields the following solution to the minimizing functional in (2):

$$\begin{aligned} u(\mathbf{x}) = & \sum_{k=1}^{M}\sum_{j=1}^{d} a_{k,j}\phi_k(\mathbf{x})\,\varepsilon_j \\ & + \sum_{i=1}^{n}\sum_{j=1}^{d} w_{1,i,j}U(\mathbf{x}-\mathbf{p}_i)\,\varepsilon_j \\ & - \sum_{i=1}^{n_\theta}\sum_{j=1}^{d-1} w_{2,i,j}(\mathbf{d}_i{}^T\nabla)U(\mathbf{x}-\mathbf{p}_{\theta_i})\,\mathbf{e}_{i,j}^{\perp}, \end{aligned} \tag{3}$$

where $\phi_1, \ldots, \phi_M$ are the basis functions of the corresponding null-space of dimension M, and $U(\mathbf{x})$ are radial basis functions depending on the used function space. Note, that in order to obtain bounded functionals L_i the used function space has to be constrained. As a consequence, we obtain for both cases of 2D and 3D registration incorporating orientations the function $U(\mathbf{x}) = |\mathbf{x}|^3$ in (3) as the only possible solution for $m = 2$. The parameters $\mathbf{a} = (a_{1,1}, a_{1,2}, \ldots, a_{1,d}, a_{2,1}, \ldots, a_{M,d})$ and $\mathbf{w} = (w_{1,1,1}, \ldots, w_{1,n,d}, \ldots, w_{2,1,1}, \ldots, w_{2,n_\theta,d-1})$ can be computed as the solution of the following system of linear equations:

$$\begin{aligned} \mathbf{Kw} + \mathbf{Pa} &= \mathbf{v} \\ \mathbf{P}^T\mathbf{w} &= 0, \end{aligned} \tag{4}$$

where $\mathbf{K}$ and $\mathbf{P}$ are matrices consisting of elements of the functions $U(\mathbf{x})$, $(\mathbf{d}_i{}^T\nabla)U(\mathbf{x})$, $(\mathbf{d}_i{}^T\nabla)(\mathbf{d}_j{}^T\nabla)U(\mathbf{x})$, $\phi(\mathbf{x})$, and $(\mathbf{d}_i{}^T\nabla)\phi(\mathbf{x})$, respectively. $\mathbf{v}$ is a vector corresponding to the target landmarks as well as zero values due to the scalar product of the orientations.

3 Experimental Results

Experiments with 2D and 3D synthetic as well as 2D tomographic data sets have been carried out. In medical images it is often important to distinguish between rigid and elastic parts of the image, e.g. rigid parts may be bones while the elastic parts correspond to soft tissue. With our synthetic experiments we have studied whether rigid parts can be mapped onto each other, while the elastic parts deform smoothly inbetween.

In Fig. 1 a translation of 60 pixels in each direction (first and second image) as well as a rotation of 60° around a spatially diagonal axis (third and fourth image) of a cube are shown (image size 301 × 301 pixels), simulating both shape-preserving object shifts and rotations of objects in medical images, respectively. The edges of the cubes of the source images are drawn as outlines and each of the four images shows the transformed source image. Eight image border landmarks in the corners of the 3D images have been used to tie the whole image to its corners and another eight object landmarks have been used at the corners of the cube (see also the caption of Fig. 1).

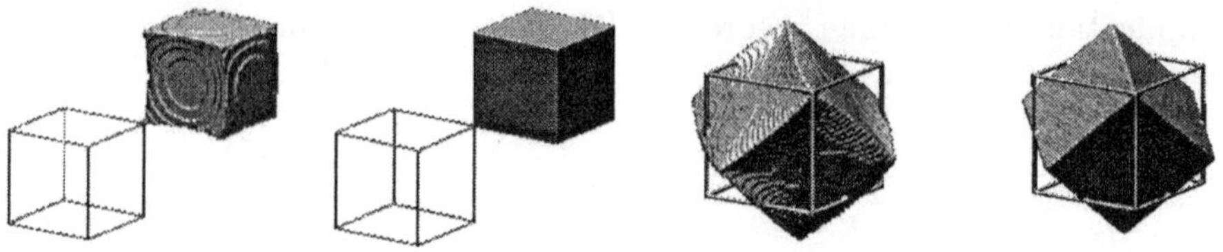

Fig. 1. Synthetic 3D data sets. Registration result of a translated (first and second image) and rotated (third and fourth image) cube. Thin-plate spline registration i) using only landmarks at each corner of the cube (object landmarks) and at each corner of the image (image border landmarks) (first and third image) as well as ii) with three orientations added to each object landmark (second and fourth image).

In case of translation (first and second image) we see that without orientation attributes (first image) the surfaces of the elastically transformed rigid cube become bended. In contrast, using orientation attributes at the cube's corners (second image), the surfaces of the rigid cube remain plane. Thus rigidity of the cube is preserved under translations. In case of rotation (third image) the surfaces and the edges of the cube become undulating. Using orientation attributes added at the corners of the cube, the cube's overall shape is much better preserved (fourth image).

In Fig. 2 an example of rotated rigid structures is shown (image size is 181×451 pixels). The five rigid structures simulate a human spine (according to Little et al. [2]). In the second row of the figure the registration results are represented. We see that registration with landmarks only does not preserve the rigidity of the squares. When using orientation attributes and setting $\lambda = 0$ (interpolation), we get a better registration result. The waviness of the outlines of the squares is reduced. In the last image of the figure we have used the approximation scheme with $\lambda = 100$ and the same orientations as before. In this case we yield the best registration result. The structure of the squares is nearly preserved and the mean distance between the transformed source landmarks and the target landmarks is 0.5 pixels (the maximum is 1.3 pixels).

Fig. 3 gives an example for real image data. The first row shows two sagittal slices of two MR images of the brain of two different patients (on the left side is the source and on the right side the target image). We manually specified a number of 13 landmarks. In the second row on the left the registration result using landmarks only is shown. The white outlines of the target image are overlayed on the transformed source image. Improvement of this result can be achieved if we add further landmarks. However, specifying additional landmarks is often not possible. Therefore, we investigate the incorporation of orientation attributes at landmarks. In the second row on the right we have added one orientation at each of the landmarks No. 1 and 13. As a result we can see that the region between landmarks No. 1 and 2 (at the corpus callosum) is better registered than using landmarks only. However, the incorporation of orientations has also strong global effects, e.g.,the region above the corpus callosum is deformed upwards.

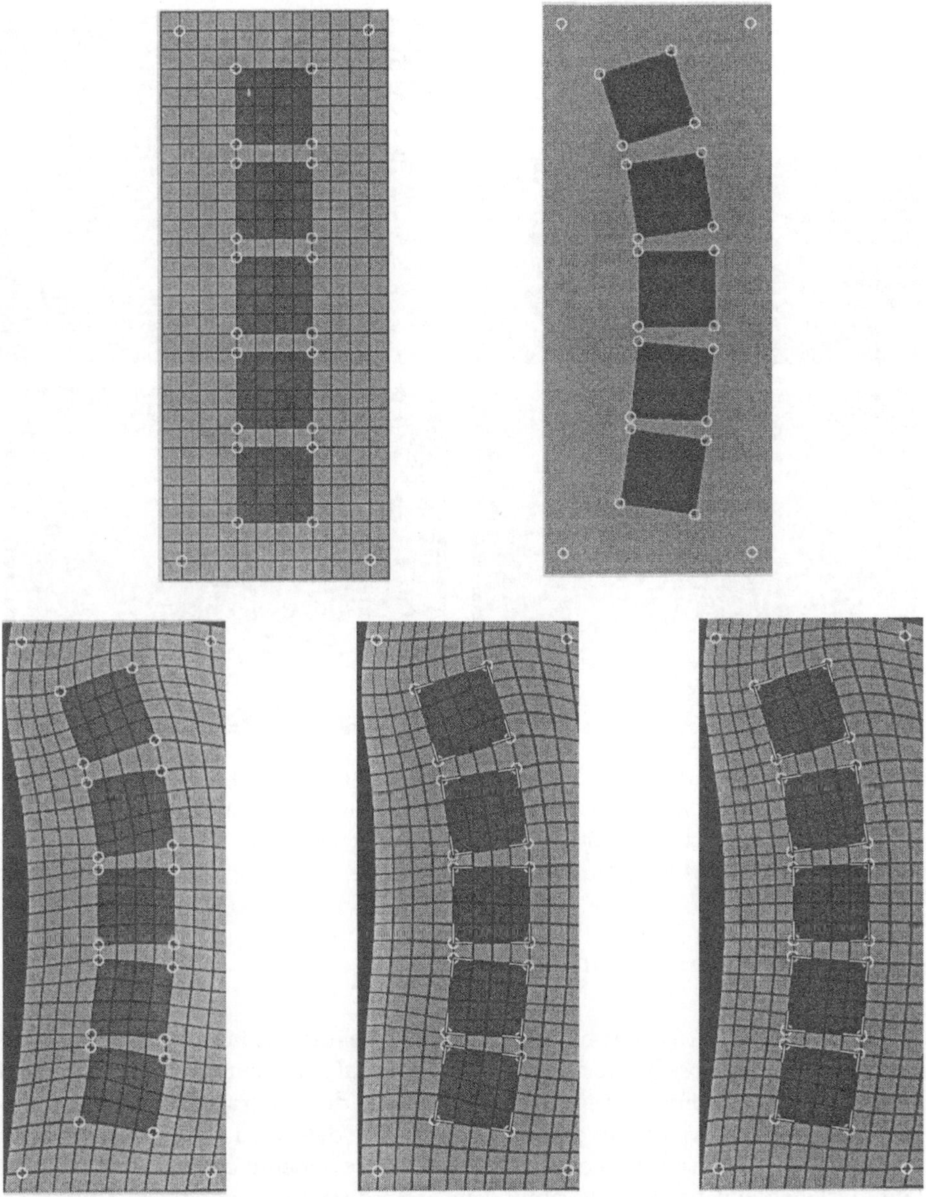

Fig. 2. Synthetic data set. Simulating a spine which is bended. First row: Source and target images. Second row: Registration using thin-plate splines with landmarks only (first), adding two orientations to each object landmark No. 5 to 24 (second), and using the approximation scheme with $\lambda = 100$ and the same orientations as before (third).

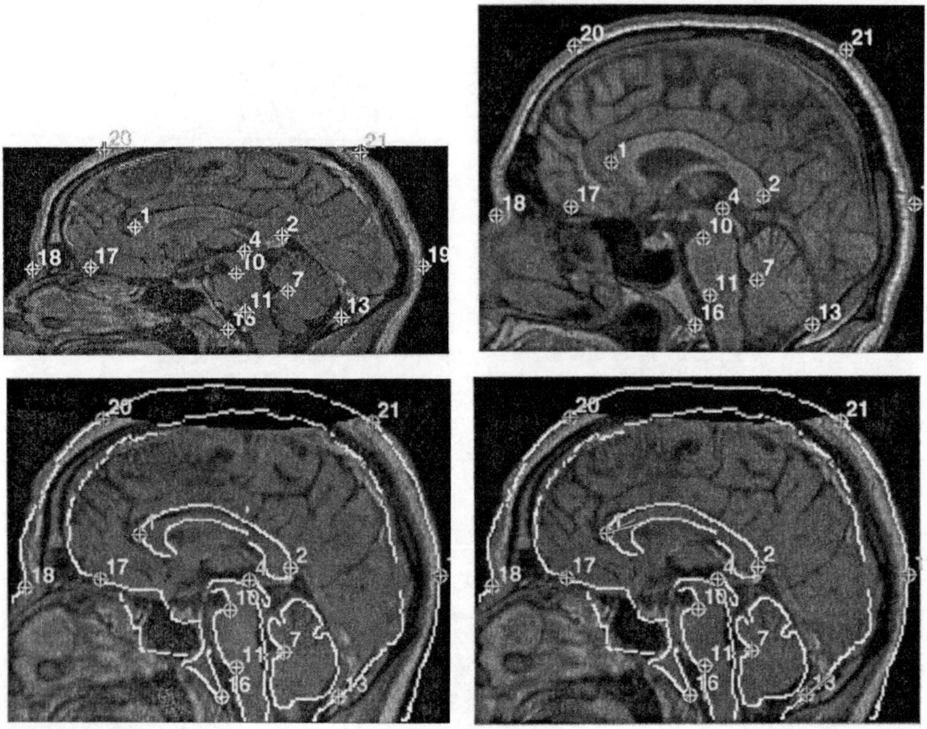

Fig. 3. Tomographic data sets. Two MR images of different brains with 13 landmarks, source and target images (first row). Second row: thin-plate spline registration using point landmarks only (left) and thin-plate spline registration using orientations at landmarks No. 1 and 13 (right).

4 Summary and Conclusion

We have demonstrated that by incorporating orientations in our elastic registration scheme it is possible to improve the registration accuracy without further adding landmarks. For synthetic images we have demonstrated that rigid structures can be preserved in an otherwise elastically deformed image. Experiments for tomographic data sets have shown that improvements can be obtained, but landmarks should be well-distributed in order to achieve good registration results. Future work will concentrate on how to limit the influence of the orientation constraints.

Acknowledgement

This work has been supported by Philips Research Hamburg, project IMAGINE (Image- and Atlas-Guided Interventions in Neurosurgery).

References

1. F. L. Bookstein and W. D. K. Green, "A Feature Space for Edgels in Images with Landmarks", *Journal of Mathematical Imaging and Vision*, **3**, pp. 231–261, 1993
2. J. A. Little, D. L. G. Hill, and D. J. Hawkes, "Deformations Incorporating Rigid Structures", *Computer Vision and Image Understanding*, **66**(2), pp. 223–232, 1997
3. K. V. Mardia, J. T. Kent, C. R. Goodall, and J. A. Little, "Kriging and splines with derivative information", *Biometrika*, **83**(1), pp. 207–221, 1996
4. K. Rohr, H. S. Stiehl, R. Sprengel, W. Beil, T. M. Buzug, J. Weese, and M. H. Kuhn, "Point-Based Elastic Registration of Medical Image Data Using Approximating Thin-Plate Splines", *Visualization in Biomedical Computing (VBC'96)*, Lecture Notes in Computer Science (1131), pp. 297–306, Springer, 1996
5. G. Wahba, "Spline Models for Observational Data", *Society for Industrial and Applied Mathematics*, Philadelphia, Pennsylvania, USA, 1990

Bilinguale Sprachverarbeitung mit Speedata

U. Uebler, H. Niemann

Bayerisches Forschungszentrum für Wissensbasierte Systeme (FORWISS)
Forschungsgruppe Wissensverarbeitung
Am Weichselgarten 7,
91058 Erlangen
E-mail: {uebler,niemann}@forwiss.de

Zusammenfassung In diesem Artikel wird das EU-Projekt SPEEDATA vorgestellt, in dem ein Prototyp für bilinguale Sprachverarbeitung entwickelt wurde. Neben deutschen und italienischen Muttersprachlern sind unter den Sprechern zahlreiche Nichtmuttersprachler und Dialektsprecher vertreten. Durch die Entwicklung und Anwendung von verschiedenen Algorithmen für die Adaption auf Sprecher- und Sprechergruppen sowie durch die Modellierung von Wortklassen für die Sprachmodellierung konnte bei 4704 deutschen und 3946 italienischen Wörtern eine Fehlerrate von 4.8 % für die italienische und 7.7 % für die deutsche Sprache erreicht werden.

1 Einführung

Mit SPEEDATA[1] wurde bilinguale Sprachverarbeitung in einer realen Anwendung in den Grundbuchämtern der Autonomen Regionen Trient und Südtirol in Italien eingeführt. In dieser Region wird sowohl italienisch als auch deutsch gesprochen. Die historischen Grundbücher müssen im Grundbuchamt in Südtirol in elektronischer Form gespeichert werden. Bei einer Eingabe der Daten mit Hilfe von Sprache werden laut einer Schätzung des Grundbuchamtes von den veranschlagten 285 Personenmonaten rund 40 % eingespart.

Die Eingabe muß teilweise in beiden Sprachen erfolgen, so daß ein hoher Anteil der Sprecher Nichtmuttersprachler sind. Darüberhinaus gibt es v. a. im Deutschen eine große Anzahl von Dialekten und eine große Abstufung zwischen den ursprünglichen Dialekten und der Standardsprache. Die Sprechertypen lassen sich grob in eine der vier Gruppen aus Bild 1 unterteilen.

Mit SPEEDATA können erstmals Äußerungen in zwei Sprachen gleichzeitig erkannt werden. Der entwickelte Erkenner behandelt die beiden Sprachen nicht mehr als getrennte Sprachen, sondern sieht sie als eine Sprache an. Dadurch können Wörter beider Sprachen in einer Äußerung erkannt werden und Wörter, die in beiden Sprachen auftreten, werden nur ein Mal modelliert. Dies stellt auch

[1] Telematics Application Programme der Europäischen Union, Projektreferenz-Nummer LE 1999. Weitere Partner: Informatica Trentina (Italien), Istituto per la Ricerca Scientifica e Tecnologica (Italien), Regione Autonoma di Trentino Alto-Adige/Südtirol (Italien) und Bundesministerium für Justiz (Österreich)

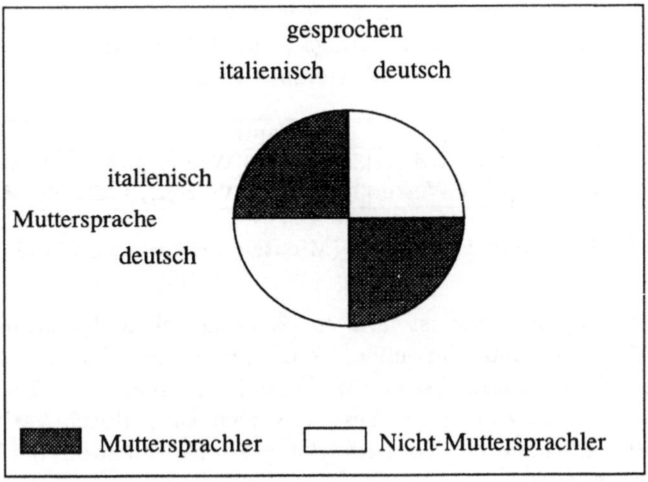

Bild1. Sprechergruppen in Abhängigkeit von der gesprochenen Sprache und der Muttersprache

einen neuen Ansatz für die Erkennung akzentbehafteter Sprache dar, da die Besonderheiten der Muttersprache eines Sprechers ebenso wie die Besonderheiten der anderen Sprache im Erkenner modelliert sind.

In diesem Artikel werden einige Schwerpunkte des Projekts vorgestellt, z. B. Aspekte der Aussprachemodellierung zwischen Standardsprechern, Nichtmuttersprachlern und Dialektsprechern vorgestellt. Schließlich werden zwei Aspekte zur Kombination von Sprachmodellen aufgeführt sowie die Bildung von Wortklassen für die Sprachmodellierung zur Repräsentation von selten auftretenden Wörtern.

2 Das Basis-System

Aus den bisher elektronisch gespeicherten Grundbüchern wurde der Wortschatz bestimmt sowie eine Anzahl von Texten gewählt, mit denen statistische Sprachmodelle trainiert werden können. In Zusammenarbeit mit den Benutzern wurde eine Schnittstelle entwickelt, die gleichzeitig Eingabe per Sprache, Maus und Tastatur ermöglicht, um für jeden Arbeitsschritt die effizienteste Methode wählen zu können.

Die Benutzeroberfläche enthält Eingabefelder für die einzelnen Daten, z. B. Felder für Namen, Datum oder Eingabe von ganzen Sätzen zur Beschreibung von Grundstücken. Für die Erkennung wurde für jedes dieser 21 Felder ein Sprachmodell entwickelt, 15 Sprachmodelle davon werden durch eine Grammatik erstellt (z. B. Datumsangaben), die restlichen werden statistisch trainiert.

Für die deutsche Sprache wurden aus bereits in elektronischer Form gespeicherten Grundbucheinträgen 4704 Wörter erfaßt, für das Italienische 3946; insgesamt besteht der Wortschatz aus 6509 bilingualen Wörtern, 2141 Wörter (v. a. Eigennamen) gibt es in beiden Sprachen. Für das Sprachmodelltraining stehen für die deutsche (italienische) Sprache Texte mit 62000 (71000) Wörtern zur

Verfügung. Die Menge der Trainings-, Adaptions- und Testdaten können Tabelle 1 entnommen werden. 40 Trainings- und 8 Adaptions/Testsprecher sind ausgeglichen nach Geschlecht und Muttersprache gewählt.

Sprache	Training		Adaption		Test	
deutsch	399 Min	1050 Wörter	39 Min	278 Wörter	77 Min	246 Wörter
italienisch	347 Min	2529 Wörter	36 Min	228 Wörter	73 Min	251 Wörter

Tabelle1. Akustische Daten (Minuten/verschiedene Wörter)

Der verwendete Erkenner ist u. a. in [2] beschrieben. Es werden semikontinuierliche Markovmodelle verwendet. Zur Sprachmodellierung, d. h. zur Bestimmung der Wahrscheinlichkeiten für Wortfolgen anhand von Texten, werden Bigramme verwendet. Von den n besten Ketten kann durch Verifikation mit Sprachmodellen mit höherem Kontext (Polygrammverifikation) die beste Äußerung bestimmt werden.

3 Bilinguale akustische Einheiten

Bei der Erweiterung eines einsprachigen Erkenners zu einer weiteren Sprache gibt es vornehmlich zwei naheliegende Alternativen:

1. Man betrachtet beide Sprachen getrennt bei Training und Erkennung. Die Entscheidung für eine Sprache fällt z. B. durch eine vorgeschaltete Sprachenidentifikation, danach wird mit dem entsprechenden einsprachigen System erkannt [4].
2. Man betrachtet die zu erkennenden Sprachen als eine einzige Sprache. Eine gewisse Unterscheidung zwischen den Sprachen findet durch Sprachmodelle statt. [1, 6].

Für das Training wurden zum einen monolinguale akustische Einheiten trainiert, eine Erkennung je nach Sprache mit einem der beiden Erkenner durchgeführt. Zum anderen wurden die in SAMPA nominell gleichen Phoneme beider Sprachen als identisch betrachtet und ein Erkenner mit akustischen Einheiten beider Sprachen trainiert. Die Anzahl der Phoneme pro Sprache und die Erkennungsrate ohne Sprachmodell für die Experimente ist Tabelle 2 zu entnehmen. Die Erkennung für italienisch ist besser als für die deutsche Sprache: zum einen liegt das an der geringeren Komplexität der italienischen Sprache [7]. Zum anderen mag dies vor allem an den verschiedenen Dialekten der Sprecher als auch am starken Akzent der Nichtmuttersprachler liegen (italienische Muttersprachler sprechen in der Regel mit stärkerem Akzent deutsch als umgekehrt). Des Weiteren wird ersichtlich, daß die Erkennung mit bilingualen akustischen Einheiten deutlich ansteigt mit 39 % statt 50 % Wortfehlerrate (Letzteres gemittelt aus den Fehlerraten der einsprachigen Erkenner mit 56% für die deutsche und 43 % für die italienische Sprache). Es wird ersichtlich, daß auch die Erkennung des Italienischen von den bilingualen Phonemen profitiert.

	italienisch	deutsch	zweisprachig
# Phoneme	50	63	89
Wortfehler (in %)	43	56	39

Tabelle2. Anzahl von Phonemen und Wortfehlerrate mit monolingualem bzw. bilingualem Training

Vor allem durch den starken Akzent der italienischen Muttersprachler ist es ratsam, die phonetische Information aus den italienischen Trainingsdaten auch für die deutsche Sprache weiterzuverwenden. Im folgenden werden in unserem System daher bilinguale akustische Einheiten verwendet.

4 Aussprachevarianten – Sprecheradaption

Ein anderes Charakteristikum bei SPEEDATA ist das Vorkommen von Dialekt und Akzent vor allem in der deutschen Sprache. Sprecher sind im allgemeinen bemüht, ihre Aussprache dem Hochdeutschen anzupassen, jedoch entsteht ein großer Unterschied im Grad der Anpassung [3]. Bei Dialekt und Akzent kann man Schemata erkennen, wie die Worte aus dem Hochdeutschen in den Dialekt oder Akzent verändert werden, s. a. Tabelle 3.

korrekt	gesprochen	Beispiel	korrekt	gesprochen	Beispiel
st	St	*Dienstag*	St	st	*Station*
h	-,x	*heute*	ts	dz	*zehn*
2	E	*Möglichkeit*	v	OU	*West*
y:	I	*über*	o:	O	*Rosenheim*

Tabelle3. Aussprachevariation im Deutschen (in SAMPA)

Ein üblicher Ansatz zur Anpassung des Systems an einen Sprecher ist die Adaption an die sprecherspezifischen Charakteristika. Im Gegensatz zum Nachtraining, wo die HMM-Parameter und das Codebuch neu geschätzt werden, schätzt unser Adaptionsverfahren nur die Mittelwertvektoren des Codebuches neu. Das verwendete Adaptionsverfahren [5] ist zweistufig: auf die gemeinsame affine Transformation der Mittelwertvektoren (Maximum-Likelihood-Schätzung) folgt eine Maximum-a-posteriori-Schätzung der Mittelwertvektoren.

Bei bilingualer Spracherkennung erscheint es auch sinnvoll, an eine Sprache oder Sprechergruppe zu adaptieren. In unseren Versuchen haben wir an einen Sprecher – mit/ohne Berücksichtigung der von ihm gesprochenen Sprache – adaptiert, sowie an die in Bild 1 dargestellten Sprechergruppen, d. h. nach gesprochener Sprache und bzw. oder Muttersprache. Von den 8 Sprechern und 2 Sprachen erhalten wir nun je nach Unterteilung 2 bis 16 Sprechergruppen. Bei einer spezifischen Sprechergruppe (z. B. einer Unterteilung nach Sprecher *und*

Sprache) ist weniger Adaptionsmaterial vorhanden als bei einer weniger spezifischen Sprechergruppe (z. B. einer Unterteilung nach der Muttersprache). Bei einer größeren Untergruppe (gesprochene Sprache und Muttersprache) wurde auch ein Nachtraining mit Sprechern aus der Lernstichprobe (andere Sprecher) mit den gleichen Charakteristika durchgeführt. Zum Vergleich wurde auch der monolinguale Erkenner adaptiert.

Generell kann man feststellen, daß Adaption italienischer Sprache eine größere Verbesserung erzielt als die deutsche. Ohne Sprecheradaption ist der bilinguale Erkenner dem monolingualen überlegen (11 bzw. 12 % Fehlerrate). Die geringste Fehlerrate wird hingegen mit dem monolingualen System und einer Adaption nach Sprecher (und Sprache) mit 8.61 % erreicht.

Die Adaption auf verschiedene Sprechergruppen führt zu ähnlichen Fehlerraten von unter 10 %. Hier kann man sehen, daß eine größere Menge von Adaptionsdaten die geringere Spezialisierung ausgleicht. Bei einer Adaption des monolingualen Erkenners auf 2 Gruppen hingegen kann man eine Verschlechterung der Erkennung feststellen. Hier scheint sich die Erkennungsleistung an die der monolingualen Erkenner anzugleichen.

Nachtraining führt ebenfalls zu einer Verbesserung der Erkennung gegenüber dem sprecherunabhängigen System. Die Erkennung erreicht mit 10.45 % allerdings nicht das Niveau der Adaption.

Ein weiterer Ansatz zur Erkennung der unterschiedlichen Aussprache der Sprecher liegt in der Modellierung von Aussprachevarianten. Die in Tabelle 3 ersichtlichen systematischen Abweichungen dienen zur Modellierung von alternativen Aussprachen der Wörter. Konkret heißt das, daß alle Wörter, die ein betreffendes Phonem enthalten, eine weitere Aussprachevariante erhalten, die aus dem Schema aus Tabelle 3 entsteht. So können alleine aus dem Wort Hof (*ho:f*) folgende Varianten gebildet werden (in SAMPA): *xo:f, o:f* bzw. *xOf, Of* für die Variation des deutschen */h/* und des Vokals */o:/*. Darüberhinaus fügen italienische Muttersprachler am Ende eines mit einem Konsonanten endenden Wortes häufig ein /@/ ein, so daß sich die theoretische Anzahl der Aussprachen erneut verdoppelt. Hiermit erhält man für dieses Wort 10 Varianten.

Drei Experimente mit Aussprachevarianten, die jeweils entweder *alle* möglichen Varianten oder nur die Varianten mit der maximalen Anzahl an Ausspracheveränderung (*xOf@, Of@*) oder die nur eine Veränderung pro Wort (*xo:f, o:f, hOf, ho:f@*) zulassen, haben eine große Steigerung des Wortschatzes um das bis zu Fünffache, aber keine Verbesserung in der Erkennung erzielt.

Weitere Experimente werden durchgeführt, um Aussprachevarianten bzw. realisierte Systematiken automatisch zu trainieren, zur Zeit liegen jedoch noch keine Ergebnisse dazu vor.

5 Verknüpfung von Sprachmodellen

Ein weiterer Schritt bei der Erkennung mehrerer Sprachen ist, inwieweit eine Verknüpfung in der Sprachmodellierung von Vorteil ist. Eine Möglichkeit, ein bilinguales Sprachmodell zu trainieren, ist, mit den Texten der beiden Sprachen ein gemeinsames Sprachmodell zu trainieren, das für die Erkennung beider Sprachen verwendet werden kann. Eine andere Möglichkeit ist, einen gemeinsamen Anfangszustand zu schaffen, von dem dann bei der Erkennung in einsprachige Sprachmodelle verzweigt werden kann [4].

Der Nachteil bei der ersten Möglichkeit ist die geringere Spezifität des neuen Sprachmodells gegenüber den einsprachigen. Der Nachteil der zweiten Möglichkeit besteht in der Größe des neuen Sprachmodells.

Die beiden Alternativen zur Verknüpfung von Sprachmodellen wurden untersucht. Des Weiteren wurde untersucht, bis zu welcher Anzahl von verknüpften Sprachmodellen die Erkennung akzeptabel bleibt. Für die Zusammenfassung zweier Sprachmodelle wurden die äquivalenten Sprachmodelle beider Sprachen verbunden. Aus den 6 Textarten pro Sprache wurden so 6 bilinguale textspezifische Sprachmodelle erzeugt. Für die Verbindung von mehr als zwei Sprachmodellen wurden einsprachige Sprachmodelle verknüpft, d. h. maximal 6 Sprachmodelle aus unterschiedlichen Textarten.

Die Experimente wurden mit und ohne Polygrammverifikation durchgeführt, da eine Verifikation der erkannten Hypothesen durch Polygramme [2] bessere Ergebnisse erzielt, eine explizite Polygrammverifikation aber noch nicht realisiert ist. Die Auswertung der Ergebnisse sind in Tabelle 4 dargestellt. Die Auswertung erfolgt für italienisch und deutsch als auch für das Mittel beider Sprachen.

Wortfehler	Basissystem			2 Sprachmodelle			6 Sprachmodelle		
	it. + dt.	ital.	dt.	it. + dt.	ital.	dt.	it. + dt.	ital.	dt.
explizit (ohne)	12.76	8.88	17.73	13.44	9.03	19.09	14.80	10.05	20.88
implizit (ohne)	12.76	8.88	17.73	19.83	15.38	25.53	23.90	18.85	30.38
implizit (mit)	9.83	7.82	12.40	15.71	13.86	18.07	21.86	18.32	26.41

Tabelle 4. Wortfehlerrate ohne Verknüpfung, bei Verknüpfung von 2 bzw. 6 Sprachmodellen; mit/ohne Polygrammverifikation

Bei der expliziten Verknüpfung von 2 Sprachmodellen kann man eine leichte Verschlechterung der Erkennung von 0.7 Prozentpunkten gegenüber dem Basissystem feststellen, bei der impliziten Verknüpfung ist eine größere Verschlechterung von 7 Prozentpunkten festzustellen. Mit Polygrammverifikation sinkt die Erkennung um 6 Prozentpunkte. Bei der Verknüpfung von 6 Sprachmodellen sinkt die Erkennung weiter, im expliziten Fall um 2 Prozentpunkte, im impliziten Fall um rund 11 Prozentpunkte.

Bei der Verbindung mehrerer Sprachmodelle erscheint die implizite Verbindung als nicht mehr tragfähig. Die Verschlechterung durch explizite Verbindung ist geringer, die Komplexität steigt jedoch an und verlangsamt u. U. den Erkennungsprozeß.

Bei einigen Äußerungen wurde die Erkennung mit einem bilingualen Sprachmodell verbessert, z. B. wenn italienische Ortsnamen vom italienischen Modell besser erkannt – da besser trainiert – wurden als beim deutschen Modell, in der

Gesamtauswertung jedoch ist die Erkennung bei Verwendung eines Sprachmodells pro Textart und Sprache am besten.

6 Bildung von Wortklassen für die Sprachmodellierung

Der Hauptgrund für die Untersuchungen in diesem Abschnitt ist die schlechte Abdeckung des Wortschatzes im Trainingstext: nur 4112 der 9392 Wörter treten im Trainingstext auf und können sinnvoll trainiert werden.

Eine Möglichkeit ist nun, für die Sprachmodellierung Wortklassen zu bilden, um auch die im Trainingstext nicht oder nur selten auftretende Wörter zu modellieren und Wahrscheinlichkeiten für bestimmte Kontexte zu schätzen.

Für die Modellierung von Wörtern, insbesondere von nicht im Trainingstext erscheinenden Wörtern, gibt es u. a. verschiedene Möglichkeiten:

1. Jedes häufiger als n Mal auftretende Wort stellt eine eigene Klasse dar, seltener (und nicht) auftretende Wörter bilden eine Restklasse.
2. Wörter mit einer ähnlichen semantischen Bedeutung oder Stellung werden in einer Klasse repräsentiert, z. B. Klassen mit Vornamen, Ortsnamen.
3. Wörter werden entsprechend ihrer morphologischen Zusammensetzung in einer Klasse repräsentiert, z. B. alle Wörter mit der Endung $-ung$.

Bei Wahl von Schranken für die Bildung von Prä- und Suffixen kann man auch zusammengesetzte Substantive klassifizieren, z. B. alle Wörter, die auf $-vertrag$ enden. Eine automatische morphologische Analyse des Wortschatzes nach Prä- und Suffixen ist allerdings oft fehlerhaft wegen falscher Abspaltungen oder weil Wortkerne nicht im Lexikon enthalten sind. Eine weitere Alternative ist, automatisch für verschiedene Längen die häufigsten Anfänge und Endungen zu bestimmen und diese gewissermaßen als Prä- und Suffixe zu verwenden. Die Ergebnisse sind in Tabelle 5 den obigen Alternativen entsprechend aufgeführt.

Fehler	Durchschnitt	italienisch	deutsch
gezählt (1)	9.83	7.82	12.40
Semantik (2)	9.25	8.05	11.88
Suffix (3)	6.60	4.04	9.88
Präfix	5.82	4.04	8.09
Prä- und Suffix	8.47	3.97	14.24
Prä- und Suffix und Abk.	5.77	3.74	8.38
Prä- und Suffix und Abk. und Großschreibung	11.42	5.82	18.60
Prä- und Suffix, automatisch	6.01	4.34	8.14

Tabelle5. Fehlerrate bei Verwendung von verschiedenen Wortklassen für das Training von Sprachmodellen

Bei der Auswertung der Ergebnisse wird die Fehlerrate mit rein gezählten Kategorien (1) in Tabelle 5 als Basis-System betrachtet und die weiteren Versuche daran verglichen. Die Wörter, die nach einer Einteilung in Wortklassen übrig bleiben, werden weiterhin wie (1) in Tabelle 5 je nach Häufigkeit in eine eigene Wortklasse oder in die Restklasse eingeteilt.

Die Zuhilfenahme von semantischen Kriterien wie Eigen- und Ortsnamen erzielt eine Verbesserung der Erkennung um einen halben Prozentpunkt. Bei einer Einteilung der Wortklassen nach Anfängen und Endungen wird eine drastische Verbesserung um 4 Prozentpunkte erzielt. Weiterhin kann man feststellen, daß die Verbesserung bei Verwendung von Anfängen oder Endungen ohne den jeweiligen Gegenpart eine deutliche größere Verbesserung erzielt als die Verwendung beider Kriterien zusammen. Wenn Abkürzungen zusätzlich eine Klasse bilden, ist eine weitere leichte Verbesserung zu bemerken. Wird bei den übrigen Wörtern je nach Groß- und Kleinschreibung unterschieden (statt Auszählung), so sinkt die Erkennung unter die des Basis-Systems.

7 Zusammenfassung

Durch den Einsatz der entwickelten und oben beschriebenen Techniken wird bei einem Wortschatz von 4704 deutschen und 3946 italienischen Wörtern eine Erkennungsrate für beide Sprachen von durchschnittlich 6.9 % für einen sprecherunabhängigen und von 6.0 % für einen sprecheradaptierten Erkenner erzielt.

References

1. U. Ackermann, F. Brugnara, M. Federico, and H. Niemann. Application of Speech Technology in the Multilingual SpeeData project. In *3rd Crim-Forwiss Workshop*, Montréal, 1996.
2. T. Kuhn, H. Niemann, and E. Schukat-Talamazzini. Ergodic Hidden Markov Models and Polygrams for Language Modeling. In *Proc. Int. Conf. on Acoustics, Speech and Signal Processing*, volume 1, pages 357–360, Adelaide, Australia, 1994.
3. H. Moser. Methodische Überlegungen zur Untersuchung des gesprochenen Deutsch in Südtirol. In H. Moser, editor, *Zur Situation des Deutschen in Südtirol*. Innsbrucker Beiträge zur Kulturwissenschaft, Innsbruck, Österreich, 1982.
4. E. Nöth, S. Harbeck, and H. Niemann. Language Identification in the Context of a Human Machine Dialog System. In R. Kuhn, editor, *Proc. of the 3rd CRIM FORWISS Workshop*, pages 85–95. Montreal, 1996.
5. M. Schüßler, F. Gallwitz, and S. Harbeck. A Fast Algorithm for Unsupervised Incremental Speaker Adaptation. In *Proc. Int. Conf. on Acoustics, Speech and Signal Processing*, volume 2, pages 1019–1022, München, 1997.
6. F. Weng, H. Bratt, L. Neumeyer, and A. Stolcke. A Study of Multilingual Speech Recognition. In *Proc. European Conf. on Speech Communication and Technology*, volume 1, pages 359–362, Greece, September 1997.
7. S. Young, M. Adda-Decker, X. Aubert, C. Dugast, J.-L. Gauvain, D. Kershaw, L. Lamel, D. Leeuwen, D. Pye, A. Robinson, H. Steeneken, and P. Woodland. Multilingual large vocabulary speech recognition: the European SQUALE project. *Computer Speech & Language*, 11:73–89, 1997.

Stereo Matching by RBF Networks with Occlusion Detection

Guo-Qing Wei, G. Hirzinger

Institute of Robotics and System Dynamics
German Aerospace Research Center (DLR)
82234 Oberpfaffenhofen
Germany

Abstract. In this paper, we propose a new solution to the stereo correspondence problem by using both features and intensity values. A cost function is formed by integrating intensity errors, gradient errors, and a smoothness error. Unlike previous use of a smoothness constraint, the smoothness term here is applied only to non-feature regions. To do function minimization, we propose to parameterize the disparity function by hierarchical Gaussians. To detect occlusions and disparity discontinuities, we propose to combine results from the left-to-right and right-to-left matchings without explicitly modeling discontinuities in the algorithm. Experimental results show efficiency of our method.

1 Introduction

Stereo algorithms used nowadays can be classified into area-based methods, feature-based methods, and pixel-based methods.

The main problem with area-based methods is the assumption of constant disparity within a window, despite the possibility to adapt the window size [8]. For feature-based methods [5, 1, 6, 10, 17], robust feature-localization is a precondition for matching, which poses a problem in smooth regions. Also disparity interpolations between features made no use of available intensity information. The pixel-based, or intensity-based methods, perform matching at each pixel, giving a dense disparity map. The earliest formulations were based on minimizing the left and right intensity differences, subject to a smoothness constraint (Horn [7] and Poggio et al. [12]). Refinements were made by detecting occlusions in the matching process [9],[3]. To minimize the error functions, different techniques have been employed. Horn's method [7] was found to be very slow in convergence and can easily get stuck into local minima. Simulated annealing [2] is computationally too expensive. Dynamic programming methods (e.g., [3]) can only be applied to each scan-lines independently.

In this paper, we propose a new solution to stereo matching by including features in the pixel-based method. To avoid the slow convergence and local minima occurring with variational or gradient approaches, we propose to parameterize the disparity function by a Radial-Basis-Function network (RBF). Because of the intrinsic smoothness of an RBF network, discontinuities in the disparity map cannot be modeled. We propose to model discontinuities and to find occlusions by combining results from a two-way matching.

2 Intensity and Gradient Based Matching

2.1 Intensity and Gradient Based Matching

Suppose $I_l(x,y)$ and $I_r(x,y)$ are the intensity images of the left and right cameras, respectively. For simplicity, we assume corresponding points in the two images to lie on the same horizontal scan-lines. (This can be achieved through camera calibration.) Denote the disparity at position (x,y) of the left image by $d(x,y)$; that is, (x,y) in the left image corresponds to $(x+d(x,y),y)$ in the right image. For the time being, we assume intensity values at corresponding points are the same. Then, an intensity constraint for the disparity function can be written as

$$I_l(x,y) = I_r(x + d(x,y), y) \tag{1}$$

By differentiating (1) with respect to x and y, we get

$$I_{lx}(x,y) = I_{rx'}(x',y)(1 + d_x(x,y)) \tag{2}$$

$$I_{ly}(x,y) = I_{rx'}(x',y)d_y(x,y) + I_{ry}(x',y) \tag{3}$$

where

$$x' = x + d(x,y), \tag{4}$$

which is the corresponding x-coordinate in the right image. Equations (2) and (3) have two physical meanings. First, they put emphasis on image features. Since image gradients contain information about neigbouring pixels, they have more disambiguity power than intensity values. But unlike conventional use of features for matching, the right hand sides of (2) and (3) represent *deformed* gradients of the right image, allowing *equality* instead of *similarity* to be used. The second meaning of (2) and (3) is hinted by the fact that the two equations also constrain the disparity gradients d_x and d_y. Constraints on disparity gradients are just what a smoothness term imposes. Therefore, (2) and (3) serve as an *intensity-adaptive* smoothness constraint, in contrary to a uniform smoothness. To see this, if we know that point P matches to point P', and that the (normalized) gradient I_x at P and P' are 0.5 and 0.2 respectively, (image gradients may change as the view point changes), then (2) implies that $d_x = 1.5$, constraining the local shape of the disparity surface at P.

Since (1), (2) and (3) do not uniquely constrain the disparity map in homogeneous regions, a smoothness constraint is needed in these regions to make proper interpolations. We first divide the image region into a feature region and a non-feature region. The feature region can be simply obtained by thresholding the gradient magnitude $g_l(x,y)$ of $I_l(x,y)$ as $F = \{(x,y)|g_l(x,y) > T_g\}$, where T_g is a threshold (we use 0.1 for normalized intensities on [0,1] in our experiments). The non-feature region is the complement of F, i.e., $\bar{F}$.

Now we compose our cost function as

$$E = \sum_{(x,y)\in D} (I_l(x,y) - I_r(x',y))^2 + \sum_{(x,y)\in F} \{[I_{lx}(x,y) - I_{rx'}(x',y)(1+d_x)]^2$$
$$+[I_{ly}(x,y) - I_{rx'}(x',y)d_y - I_{ry}(x',y)]^2\} + \lambda \sum_{(x,y)\in\bar{F}} (d_{xx}^2 + 2d_{xy}^2 + d_{yy}^2) \tag{5}$$

where D is the image region; x' is the same as in (4); λ is a smoothness weight; d_{xx}, d_{xy}, and d_{yy} are the 2nd order partial derivatives of $d(x,y)$ with respect to x and y, respectively. Because of the division of the image region into feature and non-feature areas, the variational principle becomes difficult to apply; whereas a gradient descent using finite differences has a low convergence rate and may easily get stuck in local minima. In the next section, we propose a parametric method which overcomes these drawbacks.

3 Disparity Parameterization and Function Minimization

3.1 Disparity Parameterization

We make use of a result from neural network research, that a Radial-Basis-Function (RBF) network is capable of universal approximations [11]. This motivates us to use an RBF network to represent the disparity. Here, we choose the Gaussian radial-basis-functions network, since a Gaussian function has finite support, which enables us to adopt an incremental estimation scheme to be described later. With the Gaussian basis function, the disparity $d(x,y)$ can be modeled as

$$d(x,y) = \sum_{k=1}^{N} w_k \phi_k + w_0, \tag{6}$$

$$\phi_k = e^{-\left(\frac{(x-t_{x,k})^2}{\sigma_{x,k}^2} + \frac{(y-t_{y,k})^2}{\sigma_{y,k}^2}\right)}, \tag{7}$$

where ϕ_k is the kth Gaussian, with its location and shape parameters as $\mathbf{t}_k = (t_{x,k}, t_{y,k})$ and $\mathbf{s}_k = (\sigma_{x,k}, \sigma_{y,k})$, respectively; w_k is the combination weight; w_0 is a constant shift (called the bias neuron in the Neural Networks literature); N is the number of Gaussians;

From computational considerations, here we let only w_k's (including w_0) as free parameters. The parameters $\mathbf{t}_k$'s and $\mathbf{s}_k$'s are set heuristically. We denote the weights $\{w_k\}$ by a vector $\mathbf{W}$. With this representation, (5) can be easily written as a function of the unknown weights $\mathbf{W}$. The purpose would be then to find $\mathbf{W}$ such that (5) is minimized.

3.2 Function Minimization

We use the stochastic gradient descent method for function minimization, which has been widely used for neural network training [14]. In our case, the Gaussian weights are updated, pixel for pixel, cyclically. To formalize, suppose E_i is the error function of the form (5) corresponding to the ith pixel, and $\mathbf{W}_i$ represents the active weights for that pixel. (Because of finite support, each neuron is responsible only for a finite area.) The update rule for pixel i is

$$\Delta \mathbf{W}_i^{(n)} = -\beta \frac{\partial E_i}{\partial \mathbf{W}_i} + \alpha \Delta \mathbf{W}_i^{(n-1)} \tag{8}$$

where n indexes the number of iterations; $\Delta\mathbf{W}_i^{(n-1)}$ represents the amount of update for the previous pixel; β is called the learning rate; while α is called the momentum term, which is used to damp possible oscillations [14]. Typical values for β and α in our experiment are $\beta = 0.1$ and $\alpha = 0.6$, respectively.

3.3 Incremental Schemes

We set the parameters $\mathbf{t}_k$'s and $\mathbf{s}_k$'s hierarchically in a scale-space fashion. The details can be found in our work on shape-from-shading using an RBF network [15].

In addition to the scale-space setting of the position and shape parameters, we estimate the weights $\mathbf{W}$ incrementally. That is, the weights at the coarsest hierarchy are first estimated. After convergence, the Gaussians at the next hierarchy are added, and the associated weights are estimated. This procedure repeats until the highest hierarchy is reached.

We also relaxed the equal intensity assumption by using a linear model to correct the left image intensity, and estimate the model parameters and the disparity map simultaneously.

4 Two-Way Matching

Since the function represented by an RBF network is intrinsically smooth, we cannot model discontinuities in a disparity map exactly with an RBF network. Here we propose to integrate matching results obtained from left-right and right-left matching. The scheme of left-right and right-left matching has been used previously. But it was only for verifying correct matches: if a point in a left image is matched to a point in the right image in the left-right searching, the same pair of points should be matched in the right-left search. Here we run our matching algorithm in the same two-way fashion; but results are interpreted quite differently. Suppose $d(x, y)$ is the true disparity map. To illustrate the idea of two-way integration, we show graphically, in Fig.1, the function $x_r = f_{lr}(x_l, y) = x_l + d(x_l, y)$ for certain y. This function maps a left scan-line to a right scan-line. Points on the x_r axis between a_r and b_r are not visible in the left camera and are called left-occluded points. Similarly, points on the x_l axis between b_l and c_l are right-occluded. The left occluded points cause a discontinuity in the function $f_{lr}(x_l, y)$. We notice, however, that if we inverse the function $f_{lr}(x_l, y)$, in the sense that we exchange the role of x_r and x_l as being the argument and the value of the mapping function, that is, if we write $x_l = f_{rl}(x_r, y)$, the discontinuity in $f_{lr}(x_l, y)$ is no longer a discontinuity in the function $f_{rl}(x_r, y)$. Therefore, we can do better modeling of the disparity curve around the point a_l by using right-to-left matching. Similarly, the discontinuity at position c_r on the x_r axis can be accurately modeled by using left-right matching.

Suppose $d_l(x_l, y_l)$ and $d_r(x_r, y_r)$ are the disparity maps obtained by running the matching algorithm in the left-to-right and right-to-left manner, respectively. The left-right mapping $f_{lr}(x_l, y)$ and the right-left mapping $f_{rl}(x_r, y)$ can be thus

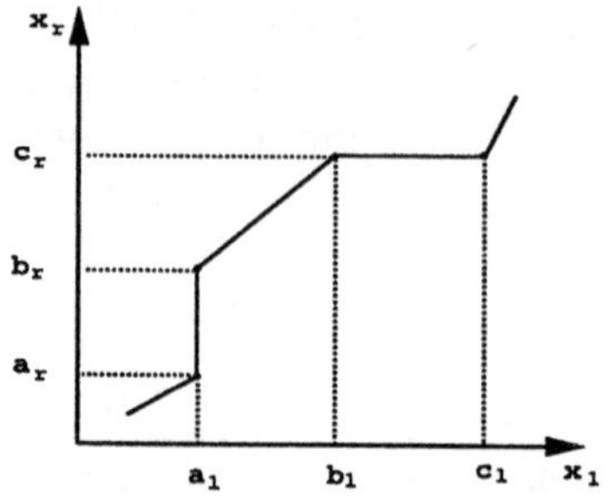

Fig. 1. Mapping between two scan-lines.

written as $x_l + d_l(x_l, y)$ and $x_r + d_r(x_r, y)$, respectively, where no distinction is made between y_l and y_r because of the assumption of horizontally aligned epipolar lines. Since a discontinuity in one mapping appears as a horizontal line in the other, we can detect such discontinuities by finding horizontal sections in the other mapping. Formally, define

$$f'_{lr}(x_l, y) = \frac{\partial f_{lr}(x_l, y)}{\partial x_l} = 1 + d_{l,x_l}(x_l, y), \tag{9}$$

and

$$f'_{rl}(x_r, y) = \frac{\partial f_{rl}(x_r, y)}{\partial x_r} = 1 + d_{r,x_r}(x_r, y). \tag{10}$$

We can segment out left occluded points as

$$L_o = \{(x_r, y_r) = (x_r, y) | f'_{rl}(x_r, y) < T_{o1}, f'_{lr}(x'_l, y) > T_{o2}\}, \tag{11}$$

where (x'_l, y) is the corresponding point of (x_r, y); T_{o1} and T_{o2} are two threshold values; we use $T_{o1} = 0.5$ and $T_{o2} = 1.2$ in our experiments. The first condition in (11) says that the local tangent of the right-left mapping should be nearly horizontal, while the second condition checks whether the left-right mapping curve has a steep orientation at the corresponding point in the left image. Because of the smoothness of an RBF function, we cannot expect this orientation to be nearly vertical; thus we can only use a very loose value for T_{o2}. Similarly, we can detect right occluded points as

$$R_o = \{(x_l, y_l) = (x_l, y) | f'_{lr}(x_l, y) < T_{o1}, f'_{rl}(x'_r, y) > T_{o2}\}, \tag{12}$$

where

$$x'_r = f_{lr}(x_l, y) \tag{13}$$

After occlusion points are detected, we can get a better modeling of the discontinuities by combining the left-right and right-left mappings. This is done as follows. For each point (x_l, y_l) at the left image, we find which point in the right image, that is not left-occluded, corresponds to the left point according to the right-to-left mapping $f_{rl}()$. Since the function $f_{rl}(x_r, y_r)$ cannot be inversed analytically, we find this point by trying each pixel in the corresponding scan-line in the right image and choosing the one whose mapped point is nearest to the left point. Formally, for each $(x_l, y_l) = (x_l, y)$, we find (x'_r, y'_r) as

$$(x'_r, y'_r) = (x'_r, y) : x_l - f_{rl}(x'_r, y) = min, \quad (x'_r, y) \notin R_o \tag{14}$$

If

$$f'_{lr}(x_l, y) < f'_{rl}(x'_r, y) \tag{15}$$

then the disparity value $f_{lr}(x_l, y_l)$ at (x_l, y_l) is retained; otherwise the disparity at this point is re-assigned as

$$d_l(x_l, y_l) = x'_r - x_l. \tag{16}$$

The reason to use (15) is that we expect the computation of a disparity map to be more accurate when the local orientation of the curve is less steep than the other at the corresponding point.

Now we want to expose the differences of our two-way combination from the conventional left-right and right-left checking. First, the conventional two-way checking cannot find out whether a mis-matching is due to left occlusion or right occlusion; that is, it cannot pick out the correct disparity in the case of an inconsistency. Second, our experience indicates that some points pass the two-way checking, but are occluded. Our use of *orientations* of the mapping cure, instead of only *values*, find more occlusion points than the simple left-right and right-left checking.

5 Experiments

We have tested our algorithm on different images of various complexities. Due to space limitations, we could only present one typical example.

IROCKS1-2 (Courtesy of INRIA): This stereo pair shows a collection of rocks resting on a platform, as shown in Fig.2 (a) and (b). The image size was reduced to 128×128. This is a very challenge stereo pair, because of the irregularity of the rocks. The disparity map obtained by the left-right matching is shown in Fig. 2(c), where brighter pixels represents points nearer to the viewer. It can be seen that discontinuities were smoothed out. By combing the left-right and right-left mappings, we detect the right- and left-occluded points, as shown in Fig. 2(c) and (d) as black points respectively. Fig. 2(d) shows the disparity map from the two-way combination. Discontinuities are now recovered. The results are consistent with human perception.

To assess the accuracy of matching results more objectively, we manually click on some left feature points via a computer mouse and check their mapped positions in the right image to see if they correspond to the correct features. We use both points and vertical lines for this mapping test. Corresponding points are marked with symbol '+'. To see the disparities, we also mark the positions of the left points/lines directly in the right image; points are marked by cross '×', and lines are re-drawn. Points are labeled by numbers. Fig. 3 shows the results

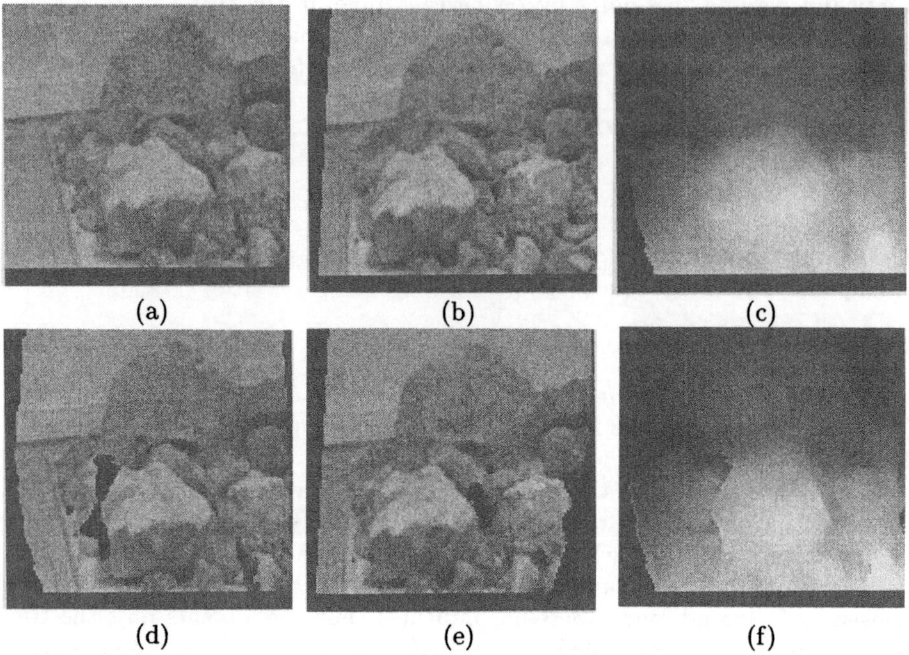

Fig. 2. IROCKS2 image. (a),(b) Left and right images. (c) left-to-right disparity map. (d) left occlusion. (e) right occlusion. (f) disparity map by the two-way combination; occlusion points are left extrapolated.

of the matching test by 15 points and 3 lines. It can be seen that the points have been matched correctly and that the deformations of the three lines conform very well with the human perception.

6 Conclusions

In this paper, we have presented a robust approach to stereo matching. Intensity and gradient are integrated in a unified manner. Another contribution of our work is the parameterization of the disparity function by hierarchical Gaussians. To model disparity discontinuities and to detect occlusions we propose to combine left-right and right-left matchings. Through a lot of experiments (not shown here) we found that our method can perform successfully on scenes with structure irregularity, with repetitive patterns, with homogeneous regions, with curved objects, and with fine structures.

It should be pointed out that not all occlusion points have been detected. Also, disparity discontinuities in the y-direction are not possible to recover. Future work should be toward the solution of these problems.

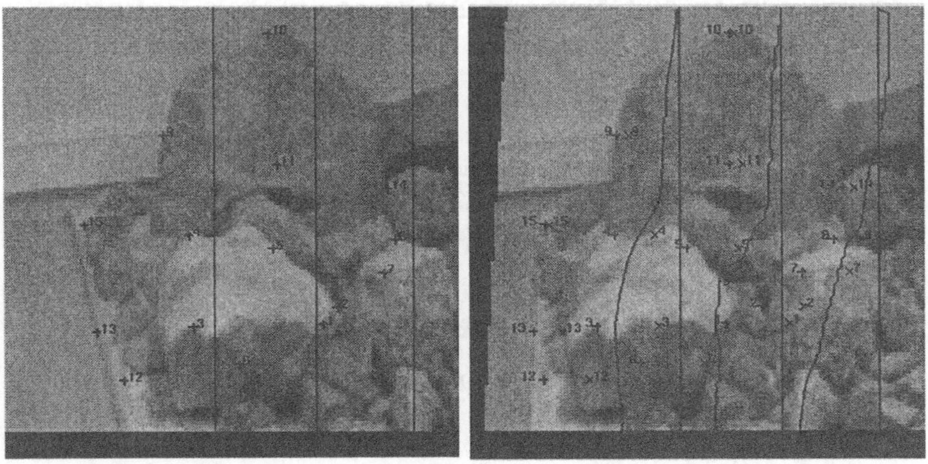

Fig. 3. Matching test. (a) Left image with vertical lines and feature points. (b) Mapping of the lines and feature points; check whether the corresponding lines and curves pass through the same intensity patterns, e.g., corners or edges.

Acknowledgment

Dr. Arbter at DLR is greatly acknowledged for his help in filling out the patent application.

References

1. S.T. Banard and W.B. Thompson, *IEEE Trans. PAMI*, Vol.2, pp.333-340, 1980.
2. S.T. Banard, *Int. J. Computer Vision*, Vol.3 pp.17-32, 1989.
3. P. N. Belhumeur, *Int. J. Computer Vision*, Vol.19, No.3, 1996, pp.237-260.
4. S.D. Cochran, G. Medioni, *IEEE Trans. PAMI*, Vol.14, No.10, pp.981-994, 1992.
5. W.E.L. Grimson, *IEEE Trans. PAMI*, Vol.7, No.1, pp.17-34, 1985.
6. W. Hoff and N. Ahuja, *IEEE Trans. PAMI*, Vol.11, pp.121-136, 1989.
7. B.K.P. Horn, *Robot Vision* , MIT Press, Cambridge MA, 1986.
8. T. Kanade, M. Okutomi, *IEEE Trans. PAMI*, Vol.16, No.9, pp.920-932, 1994.
9. A. Luo, H. Burkhardt, *Int. J. Computer Vision* , Vol.15, pp.171-188, 1995.
10. G. Medioni and R. Nevetia, *CVGIP* , Vol.31 pp.2-18, 1985.
11. J. Park, I.W. Sandberg, *Neural Computation* , Vol.3, pp.246-257, 1991.
12. T. Poggio, V. Torre, C. Koch, *Nature*, Vol.317, pp.314-319, 1985.
13. T. Poggio, F. Girosi, *Proc. IEEE* , Vol.78, No.9, pp.1481-1497, 1990.
14. D.E. Rumelhart, G.E. Hinton, and R.J. Williams, in D.E. Rumelhart, and J.L. McClelland (Eds), *Parallel Distributed Processing* Vol.I, pp.318-362, MIT Press, Cambridge MA, 1986
15. G.Q. Wei, G. Hirzinger, *IEEE Trans. PAMI*, Vol.19, pp.353-365, 1997.
16. Y. Yang, A.L.Yuille, *Artif. Intell.*, Vol. 78, pp.121-145, 1995.
17. Z.Zhang, R. Deriche, O.Faugeras, Q.-T. Luong, *Artif. Intell.*, Vol. 78, 1995.

Sonardaten-Integration für autonome mobile Roboter

Alexander Mojaev, Andreas Zell

Universität Tübingen, Wilhelm-Schickard-Institut für Informatik,
Lehrstuhl Rechnerarchitektur,
Köstlinstraße 6, D-72074 Tübingen
{mojaev,zell}@informatik.uni-tuebingen.de

Zusammenfassung. In dieser Arbeit wird ein Echtzeitverfahren zur Sonardaten-Integration und Aufbau von Gitterkarten anhand eines probabilistischen vierdimensionalen Sonar-Modells vorgestellt. Das Sonar-Modell berücksichtigt das Verhalten des Ultraschallsensors während der Roboterbewegung in der realen Umgebung. Experimente zeigen die hohe Genauigkeit aufgebauter Gitterkarten und die Robustheit des Verfahrens gegenüber den stark verrauschten Ultraschall-Daten.

1 Einleitung

Autonome mobile Roboter müssen die Einsatzumgebung auf eine interne Repräsentation abbilden können. Dafür ist die Fusion der Sensordaten notwendig. Die Gitterkarte, ein explizites Umgebungsmodell, wird in der mobilen Robotik oft verwendet und kann anhand von Ultraschallsensordaten aufgebaut werden. Trotz seiner vielen guten Eigenschaften hat der Ultraschallsensor den Nachteil, daß die Informationen über die gemessenen Entfernungen ziemlich unsicher sind. Verschiedene Faktoren können Meßfehler verursachen. Wie sich nach den Experimenten herausgestellt hat, wurden die Meßstörungen meist vom Meßobjekt und eher selten von den Sensoren selbst verursacht.

Da die Sonardaten immer redundant sind, kann das Rauschen durch die Fusion der Daten mehrerer Sensoren, sowie durch die Integration über die Zeit stark verringert werden. Das ist jedoch nur dann möglich, wenn ein der Realität gut entsprechendes Sensormodell zur Verfügung steht.

Es gibt viele verschiedene Ansätze, das Sonarmodell aufzubauen. Man verwendet sowohl ganz einfache Modellierungen der Ultraschalleigenschaften, wie die diskreten Sonarmodelle [Menezes95], [Howard96], als auch komplizierte, mehrdimensionale Modellierungen [Elfes87], die eine bessere Auflösung ergeben, dabei aber sehr zeitaufwendig sind. Ein anderer Ansatz wurde in [Thrun96] vorgestellt: als Sonarmodell wird ein probabilistisches neuronales Netz benutzt. Für die Integration der Daten wird normalerweise entweder die Evidenztheorie von Dempster-Shafer [Pagac96] oder die Wahrscheinlichkeittheorie von Bayes [Elfes87], [Soika96], [Thrun96] verwendet.

Die vorliegende Arbeit unterscheidet sich von bisherigen dadurch, daß für den Aufbau des Ultraschallmodells sowohl die theoretisch bestimmten Sonar-Eigenschaften wie Schallkeulenform und Luftdruckverteilung als auch die statistischen Daten wie Dichte der Wahrscheinlichkeit der während der Roboterbewegung entstehenden Entfernungsfehler benutzt werden.

:n der Fehlerwahrscheinlichkeit der Sonardaten wurde jeweils einer llsensoren des mobilen Roboters B-21 „COLIN“ verwendet. Diese von der Firma Polaroid hergestellt. Sie besitzen folgende technische ser: d = 38 mm, Sendefrequenz: f = 49.4 kHz, Wellenlänge: $\lambda = 6.95$ Ießbereich: $l_{max} = 6.5$ m, Halböffnungswinkel: $\theta_{max} = 4$ - 18°.
;winkel ist stark von den vorhandenen Meßobjekten abhängig den den durchschnittlichen Wert $\theta_{max} = 9.5°$ verwenden.
einen systematischen Fehler von Sonardaten zu bestimmen, wurde ;ssungen in einer statischen Umgebung durchgeführt. Das Meßobjekt :genstand mit einer Größe von 0.7x1 m^2 . Unter diesen Bedingungen te nahezu perfekt mit realen Entfernungen korreliert. Abweichungen t mit dem Erwartungswert

$$M = 0.125 l_s - 0.01 \qquad (1)$$

essene Entfernung ist, und mit der Standardabweichung $\sigma \approx 0.002$, ;ringfügigkeit nicht berücksichtigt wird. Dieser systematische Fehler :nsiert werden:

$$\hat{l}_s = 1.125 l_s - 0.01 \qquad (2)$$

gt die gemessenen und approximierten Entfernungen.
ıten sollen die Informationen über verschiedene Meßstörungen entrde eine spezielle Umgebung für den Roboter eingerichtet: ein :immer mit mehreren Papierrollen (1m hoch, 11cm ∅) und einem anlor. Der Roboter wurde mit der Hilfe eines Kollisionsvermeidungsal-1 von Laserscannerdaten gesteuert. Die genaue Entfernung in ırs wurde ebenfalls vom Laser-Entfernungsmesser gemessen.

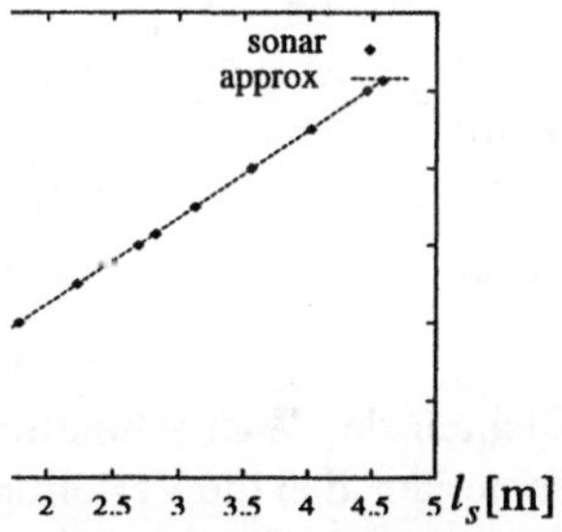

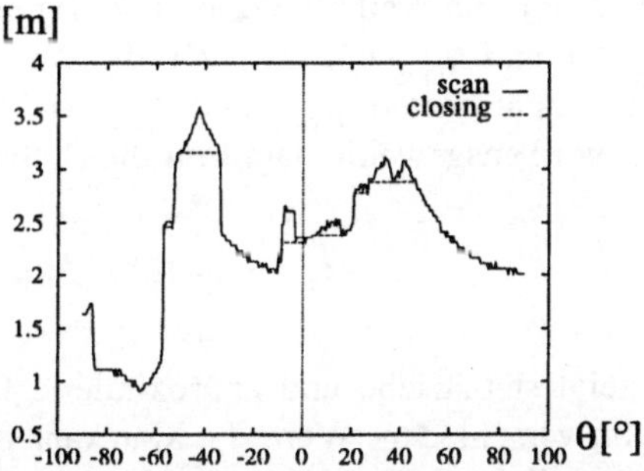

s: Kalibration von Sonardaten, rechts: Wirkung des Closing-Algorithmus

ɔnen zu berücksichtigen, in denen die Entfernung leicht mit dem La-;r mit den Ultraschallsensoren gemessen werden kann (z.B. enge ı Closing-Algorithmus verwendet, der die Entfernungen schließt, die ;sitzen, die kleiner als der Ultraschall-Öffnungswinkel ($\Delta\theta < 2\theta_{max}$) dieses Algorithmus ist am Beispiel eines Laserscans gezeigt (Abb. 1

rechts), bei dem Winkelentfernungen bis $2 \cdot 9.5 = 19°$ (z.B. zwischen -55° und -37°) geschlossen wurden.

Der Entfernungsmeßfehler wird durch E bezeichnet:

$$E = l - l_s \tag{3}$$

wobei l die reale Entfernung (vom Laserscanner) und l_s die vom Sonar gemessene Entfernung ist.

Anhand gesammelter Daten wurden Dichten der Zufallsgrößen E aufgebaut (Abb.2). Zu den empirischen Dichten passen Weibull-Verteilungen (4) [Bronstein87] am besten, wobei Änderungen der Zufallsgrößedichten mit l_s durch eine Anpassung der Verteilungsparameter berücksichtigt werden.

$$f(x) = \gamma p x^{p-1} \exp(-\gamma x^p) \tag{4}$$

Wir bezeichnen die Weibull-Verteilung als:

$$f(E, l_s) = \gamma(l_s) p(l_s) E^{p(l_s)-1} \exp(-\gamma(l_s) E^{p(l_s)}) \tag{5}$$

Da die Verteilung (4) nur für $x > 0$ bestimmt ist und bei $p < 1$ und $x \to 0$ gegen unendlich strebt, wurde die Wahrscheinlichkeitsdichte der Approximation der Entfernungsabweichungen als (6) definiert.

$$f(E, l_s) = \begin{cases} \min(f_{\max}(l_s), f_{\text{neg}}(-E, l_s)), & \text{falls } E < 0 \\ \min(f_{\max}(l_s), f_{\text{pos}}(E)), & \text{falls } E > 0 \\ f_{\max}(l_s), & \text{falls } E = 0 \end{cases} \tag{6}$$

wobei $f_{\text{neg}}(-E, l_s)$ die Weibull-Verteilung (5) negativer Abweichungen mit den Parametern $p_{\text{neg}}(l_s)$ und $\gamma_{\text{neg}}(l_s)$, $f_{\text{pos}}(E)$ die Weibull-Verteilung (5) positiver Abweichungen mit den Parametern p_{pos} und γ_{pos}, und $f_{\max}(l_s)$ der maximale Wert ist.

$f_{\max}(l_s)$ wird ausgewählt, damit für die Dichte gilt:

$$\forall l_s \qquad \int_{-\infty}^{\infty} f(E, l_s) dE = 1 \tag{7}$$

Abb. 2 zeigt statistische und approximierte Dichten der Wahrscheinlichkeit des Meßfehlers für verschiedene Werte l_s. Man kann bemerken, daß die Verteilung des negativen Fehlers (linke Seite von 0) von l_s abhängt, während die Abweichungsverteilung im positiven Bereich ungefähr gleich bleibt. Das kann offensichtlich damit erklärt werden, daß die negativen Abweichungen (reale Entfernungen sind kleiner) für verschiedene Sonarwerte durch Störungen verursacht werden, die sich in der Entstehung unterscheiden. Die so gemessenen Entfernungen, die im Bereich 3 - 6m liegen, sind oft (wenn nicht meistens) die Folge mehrerer Spiegelungen (Ecken, Wände) in der realen Umgebung. Die kleineren Werte (0.3 - 2m) vermitteln vorwiegend wahre Entfernungsinformationen. Mögliche Fehlerquelle hier sowie in dem positiven Abweichungsbereich (reale Entfernung größer) sind Crosstalk, Bewegungen und der Umstand, daß die Reflektionspunkte nicht unbedingt auf der Sonarachse liegen. Interessant dabei ist, daß die Meßfehler statistisch korreliert sind (lokale Maxima bei $l_s > 3\text{m}$). Die Korrelation entsteht vermutlich durch die Besonderheiten der heutigen Architektur, z.B. gleiche Abstände zwischen Türen, rechteckige Zimmer u.s.w.

Verwendete Approximationsparameter:

$p_{neg}(l_s) = 0.154 l_s$, $\gamma_{neg}(l_s) = \max(6 - 2l_s, 0.3)$, $p_{pos} = 0.28$, $\gamma_{pos} = 0.8$.

Der zweite Schritt war, die approximierte Verteilung auf ein vierdimensionales Modell zu erweitern. Es wurde wie bei [Elfes87] vorausgesetzt, daß die Wahrscheinlichkeit der richtigen Entfernungsmessung proportional zu Schalldruckverteilung ist und zwar entlang der Schallachse größer als an den Rändern. Obwohl es nicht ganz genau der Realität entspricht, weil im Sonar eine nichtlineare Signalverarbeitung durchgeführt wird, zeigt das Experiment deutlich unterschiedliche Empfindlichkeiten des Sonars abhängig vom Detektionswinkel. Dünne Gegenstände und konvexe Wandecken werden sogar bei kleinen Entfernungen nur im Winkel 3-7°detektiert, der viel kleiner als der Öffnungswinkel ist.

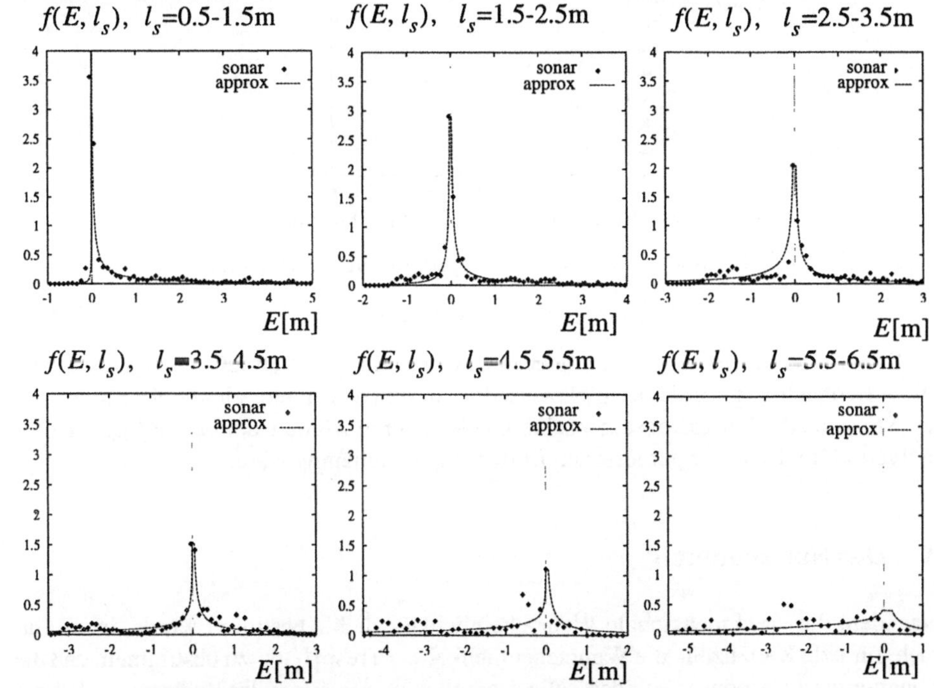

Abbildung 2. Sonarfehlerverteilungen für verschiedene Werte l_s

Die relative Schalldruckverteilung berechnet sich [Moita94] als:

$$V(\theta) = \frac{2J_1(k \cdot r \cdot \operatorname{asin}(\theta))}{k \cdot r \cdot \sin(\theta)} \tag{8}$$

wobei $k = 2 \cdot \pi / \lambda$ und $r = \frac{d}{2}$ der Radius des Ultraschallsensors und $J_1(x)$ die Besselsche Funktion erster Ordnung ist.

Die vierdimensionale Wahrscheinlichkeitsdichte ergibt sich zu:

$$f(E, \theta, l_s) = f(E, l_s) \cdot V(\theta) \tag{9}$$

Die Dichten werden durch die Anpassung $f_{max}(l_s)$ normiert, damit für jedes l_s gilt:

$$\int\int_{-\infty}^{\infty} f(E,\theta,l_s)dEd\theta = 1 \tag{10}$$

Die Abhängigkeit $f_{\max}(l_s)$ ist in Abb. 3 dargestellt. Interessant dabei ist, daß sich der Wert $f_{\max}(l_s)$ mit Vergrößerung von l_s ungefähr exponentiell verringert. Daraus folgt:

Die Glaubwürdigkeit der Sonarinformation sinkt exponentiell mit der gemessenen Entfernung.

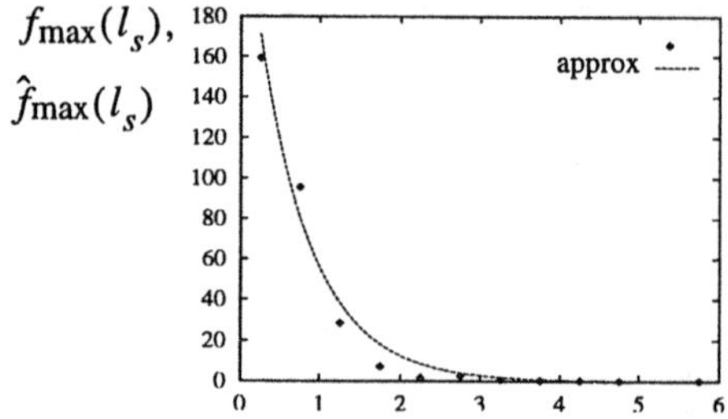

Abbildung 3. Normierung $f_{\max}(l_s)$ und ihre Approximation

Für die Modellierung wurde folgende Approximation verwendet:

$$\hat{f}_{\max}(l_s) = 250/\exp(1.5l_s) \tag{11}$$

Dabei müssen immer die Umgebungseigenschaften in Betracht gezogen werden. Diese Beobachtung würde beispielsweise für ein rundes Zimmer ohne Ecken nicht gelten, weil sich die Reflektionseigenschaften idealisieren würden, der Fehler folglich normalverteilt und von der gemessenen Entfernung unabhängig wäre.

3 Das Sonarmodell

Nachdem die vierdimensionale Wahrscheinlichkeitsdichte bestimmt wurde, ist es einfach, für jede Kartenzelle die Wahrscheinlichkeit $P(\text{resp}|l_s)_i$ zu bestimmen, daß die Sonarantwort (response) von der Zelle i erhalten wurde, unter der Bedingung, daß der Ultraschallsensor den Wert l_s geliefert hat:

$$P(\text{resp}|l_s)_i = \int\int_{S_i} f(E,\theta, l_s)(dE)d\theta \tag{12}$$

oder vereinfacht

$$P(\text{resp}|l_s)_i \approx f(E,\theta,l_s) \cdot S_i \tag{13}$$

Die Gleichung (13) wird im Algorithmus verwendet. Hier ist S_i die Fläche eines Kartenelementes und für $f(E,\theta,l_s)$ wird der maximale Wert der Verteilung in Grenzen der Kartenzelle ausgesucht.

Diese Wahrscheinlichkeit kann aber noch nicht für den Kartenaufbau verwendet werden, weil sie von anderen Kartenzellen, die sich im Sonarstrahl befinden, abhängt: die Sonaraussage kann nur von einer Kartenzelle erhalten werden und die Summe der Wahrscheinlichkeiten für alle Kartenzellen hinsichtlich des Sonarwertes l_s gleich 1 ist.

Das Sonarmodell ist eine Menge von bedingten Wahrscheinlichkeiten $P(\mathrm{resp}|\mathrm{occ})_i$, daß die Sonaraussage l_s von der Zelle erhalten wurde, unter der Bedingung, daß diese Zelle belegt ist, und unabhängig vom Zustand der anderen Kartenelemente. Der Übergang zu diesen Wahrscheinlichkeiten definieren wir als

$$P(\mathrm{resp}|\mathrm{occ})_i = \begin{cases} 0.5 + K(P(\mathrm{resp}|l_s)_i - P(0.5|l_s)_i), & l_i < l_s \\ 0.5, & l_i > l_s \end{cases} \tag{14}$$

Hier ist $P(0.5|l_s)_i$ die Wahrscheinlichkeit, die die Grenze zwischen $P(\mathrm{resp}|\mathrm{occ})_i$ und $P(\mathrm{resp}|\mathrm{occ})_i$ für die Kartenzelle i bestimmt. Die Werte $P(\mathrm{resp}|l_s)_i$, die gleich $P(0.5|l_s)_i$ sind, werden keine Änderung in dieser Zelle bewirken. In unserem Fall berechnet sich $P(0.5|l_s)_i$ über (15):

$$P(0.5|l_s)_i = T \cdot f_{\max}(l_s) \cdot P(\theta) \cdot S_i \tag{15}$$

Der Koeffizient T wird normalerweise auf 0.5 gesetzt und bestimmt die Wechselbeziehung zwischen Frei- und Belegtwahrscheinlichkeiten für die ganze Karte. Wenn z.B. die Aufgabe lautet, möglichst viel Freiraum zu finden, wobei Hindernisse weniger eine Rolle spielen, wird $T < 0.5$ ausgewählt und umgekehrt. K ist dabei der Koeffizient des Einflusses einer Sonaraussage auf die Modifikation der Belegtwahrscheinlichkeiten und wurde gleich 1 gesetzt.

Die Kartenzellen, die weiter als die vom Sonar gemessene Entfernung liegen, ($l_i > l_s$ in (14)) werden nicht geändert.

4 Fusionsverfahren

Um die Belegtwahrscheinlichkeit zu berechnen, wurde die Bayessche Regel in einer iterativen Variante verwendet:

$$P(\mathrm{occ})_i^{t+1} = \frac{P(\mathrm{resp}|\mathrm{occ})_i \cdot P(\mathrm{occ})_i^t}{P(\mathrm{resp}|\mathrm{occ})_i \cdot P(\mathrm{occ})_i^t + (1 - P(\mathrm{resp}|\mathrm{occ})_i) \cdot (1 - P(\mathrm{occ})_i^t)} \tag{16}$$

Die A-priori-Wahrscheinlichkeiten für alle Kartenzellen $P(\mathrm{occ})^{t=0}$ wurden auf 0.5 gesetzt.

5 Experimentergebnisse

Die Experimente wurden auf dem mobilen Roboter B-21 "COLIN" durchgeführt. Die Sonarinformationen wurden unregelmäßig mit einer Periode von 35 bis 300ms bei einer Robotergeschwindigkeit bis 0.3 m/s geliefert. Abb. 4 zeigt die aktuellen Ultraschalldaten eines Korridors. Zwei, meistens vier Sensoren zeigen die wahren Entfernungswerte während der Roboterbewegung. Mehrere Methoden zur Kartenaufbau wurden in dieser Umgebung ausprobiert und verglichen. In Abb. 5 (oben links) wird die mit „BeeSoft-Mapper" [BeeSoft97] aufgebaute Karte dargestellt Das Verfahren zeigt

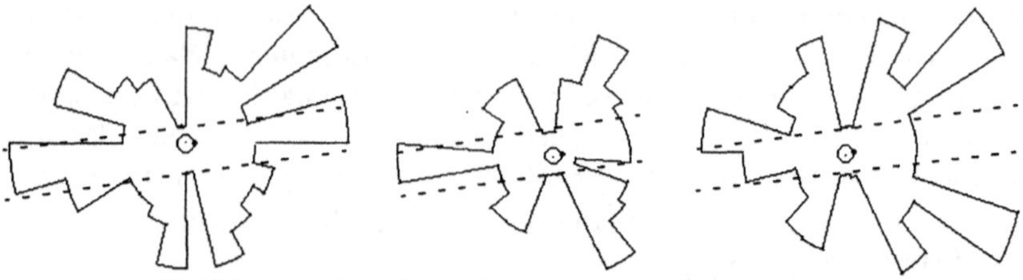

Abbildung 4. Beispiele primärer Sonardaten eines Korridors

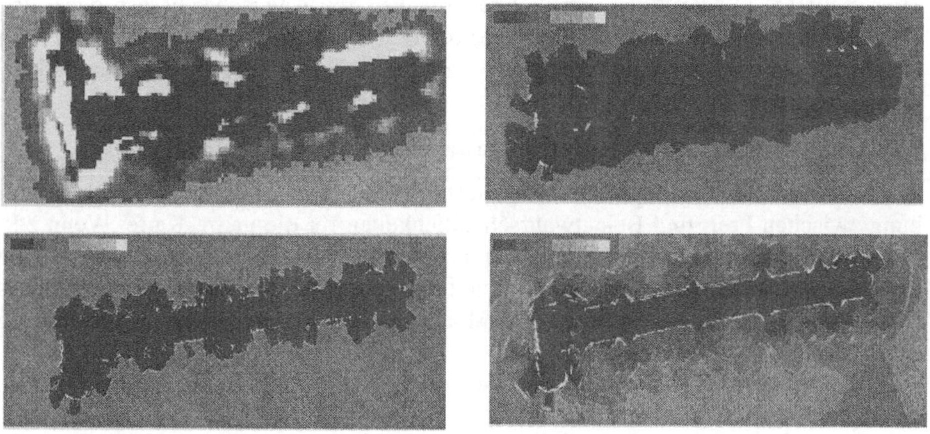

Abbildung 5. Sonarkarten des Korridors: oben links: mit „BeeSoft Mapper", oben rechts: nach [Menezes95] (S = 3m), unten links: nach [Menezes95] (S = 1.5m), unten rechts: mit dem beschriebenen Verfahren

eine niedrige Auflösung und kann die Wände nicht richtig abbilden. In der Arbeit von Menezes [Menezes95] wurde empfohlen, die Sonar-Entfernungswerte, die größer als 3m sind, einfach zu ignorieren. Abb. 5 (oben rechts) zeigt, daß auch unter dieser Bedingung die Sonardaten zu widersprüchlich sind, um korrekt integriert zu werden. Nur bei einem Schwellenabstand von 1.5 m (Abb. 5 unten links) (es wurden 48% Sonardaten im Fall des Korridors bearbeitet) enthält der größere Anteil der Sonarwerte wahre Entfernungsinformation und nahe Gegenstände können dabei wiedergegeben werden. Der Nachteil dieses Verfahrens ist offensichtlich: die Exploration ist nur innerhalb des kurzen Radius (<1.5m) möglich.

Die mit dem entwickelten Verfahren aufgebaute Karte wird in Abb. 5 (unten rechts) dargestellt. Der Algorithmus läuft online auf dem Roboter-PC (Pentium Pro 200MHz) mit einer relativ geringen Prozessor-Belastung. Er verwendet 100% der Sonardaten und zeigt eine hohe Robustheit und Genauigkeit bei der Wiedergabe der Umgebung. Die Wände des Korridors (weiße Zellen mit Belegtwahrscheinlichkeiten gegen 1) wurden in der Gitterkarte deutlich abgebildet, was als klarer Hindernisindikator für Navigation und Aufbau topologischer Karten verwendet werden kann. Ein weiterer Vorteil dieses Modells besteht darin, daß die bei der Exploration bereits aufgebauten Karten benachbarter Räume durch Fehlmessungen nicht zerstört werden. Die Exploration ist sowohl bei kleinen als auch großen Abständen in den Grenzen der Empfindlichkeit der Ultraschallsensoren möglich. Im Fall großer Entfernungen braucht das Verfahren mehr Zeit und Sonardaten für die Integration.

6 Zusammenfassung und Ausblick

Diese Arbeit stellt ein probabilistisches Verfahren zur Sonardaten-Integration vor. Mit dem entwickelten Sonarmodell ist es möglich, die fehlerhaften Entfernungsmeßwerte von Ultraschallsensoren mit hoher Genauigkeit zu integrieren und in der Gitterkarte wiederzugeben. Die Experimente mit dem mobilen Roboter „Colin" demonstrieren die Vorteile des Verfahrens bei den stark verrauschten Entfernungs-Informationen. In der Zukunft wird die Möglichkeit der Anwendung der beschriebenen Methode auch zur Integration den Daten von Infrarot-Sensoren untersucht.

Literatur

[Jorg94] Klaus-Werner Jörg: *Echtzeitfähige Multisensorintegration für autonome mobile Roboter*, Wissenschaftsverlag, 1994.

[BeeSoft97] *User's Guide and Software Reference*, Real World Interface, Inc,1997

[Bronstein87] Bronstein I.N., Semendjaew K.A.: *Taschenbuch der Mathematik*, Thun, Frankfurt/Main,1987.

[Elfes87] Elfes,A.: *Sonar-Based Real-World Mapping and Navigation*, IEEE, Journal of Robotics and Automation, Vol. RA-3, No. 3, June 1987, pp. 249-265.

[Howard96] Andrew Howard, Les Kitchen: *Sonar Mapping for Mobile Robots*, The University of Melbourne, Technical Report 96/34.

[Menezes95] Paulo Menezes, Jorge Dias, Helder Araujo,Anibal de Almeida: *Low Cost Sensor Based Obstacle Detection and Description Experiments with Mobile Robots using Grid Representation*, Proceedings of the Fouth international Symposium on Experimental Robotics, ISER'95, Stanford, USA, August 1995.

[Moita94] F.Moita,A.Feijao,U.Nunes,A.T.Almeida: *Modelling and Calibration of an Ultrasonic Ranging System of a Mobile Robot*, Associação Portuguesa de Controlo Automático Int. Conf. - Controlo 94, I.S.T., Lisbon, September 14-16, 1994.

[Pagac96] Pagac D., Nebot E., Durrant-Whyte H.: *A Evidential approach to probabilistic map-building*, IEEE International Conference on Robotics and Automation, April 1996, Minneapolis , Minnesota, pp. 745-750.

[Soika96] Martin Soika: *Gitterkartenbasierte Fehlererkennung und Kalibrierung für Umgebungssensoren autonomer mobilen Systeme (AMS)*, Autonome mobile Systeme, 1996, pp. 38-46.

[Thrun96] Thrun S., Bücken A.: *Integrating Grid-Based and Topological Maps for Mobile Robot Navigation*, Proc. of the Thirteenth National Conference on Artificial Intelligence AAAI, Portland, Oregon, August 1996.

Compressive Computation in Analog VLSI Motion Sensors

Rainer A. Deutschmann[1] and Oliver G. Wenisch[2]

[1] Walter Schottky Institute, 85748 Garching, Germany
[2] Technische Universität München, 85748 Garching, Germany
Rainer.Deutschmann@wsi.tu-muenchen.de

Abstract. We introduce several different focal plane analog VLSI motion sensors developed in the past. We show how their pixel-parallel architecture can be used to extract low-dimensional information from a higher dimensional data set. As an example we present an algorithm and corresponding experiments to compute the focus of expansion, focus of contraction and the axis of rotation from natural visual input. A fully integrated system for real-time computation of these quantities is proposed as well. In computer simulations it is shown that the direction of motion vector field is best suited to perform the algorithm even at high noise levels.

1 Analog VLSI Motion Sensors

In the past the computer vision community has invested much effort into developing motion detection algorithms; for a critical review see [BFB94]. Implementing these algorithms in real-time systems proved challenging for computational reasons. Additionally it has been realized that a motion vector field is useful mainly as starting point for further computation, such as segmentation from motion and determination of the focus of expansion.

We have developed and implemented several real-time motion sensors. Using different algorithms, our sensors share the following features: They are single-chip sensors, i.e. the photoreceptors and the motion computation circuitry sit in the focal plane. They are pixel-parallel implementations, i.e. motion computation is performed in synchrony by all pixels. No clock is required for motion computation and, in contrast to digital implementations, in our sensors transistors are used as analog computing elements (*analog* **V**ery **L**arge **S**cale **I**ntegrated). Previously we have proven that 1-D analog VLSI motion sensors can be used efficiently for measuring fast rotational velocity [BD97]. In this paper we experimentally show that our 2-D motion sensors can be used for estimating the focus of expansion (FOE), the focus of contraction (FOC), and the axis of rotation (AOR). We use an algorithm that ideally suits the pixel-parallel architecture of the sensors and effectively reduces the computational load from N^2 to $2\,N$. We also propose a fully integrated 2-D system for on-chip estimation of the FOE, FOC, and AOR, which ultimately compresses the computational load from N^2 to 2. Through computer

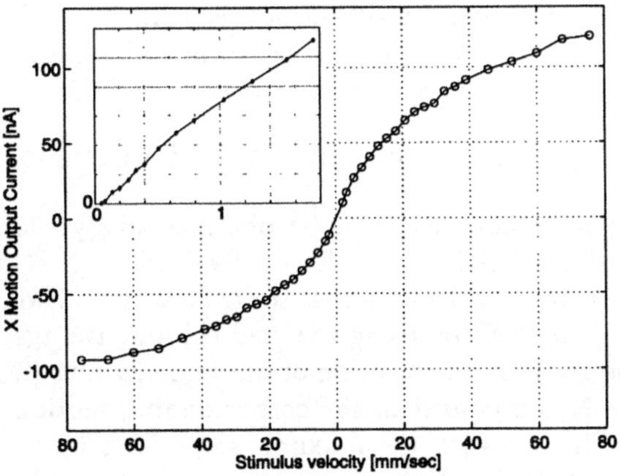

Fig. 1. Mean motion output u_x of one pixel of the Gradient2d sensor for different stimulus velocities. The inset shows the motion output for velocities up to 1.8 mm/sec.

simulations we have investigated the robustness of the proposed system to noise and low-contrast input.

Our motion sensors fall into two different categories:

1. **Feature-based sensors** generally look for features, such as edges, in the visual field and track them over time. Based on edge tracking for example the sensors reported in [DHK97] compute the direction of motion (DOM) vector field.
2. **Gradient-based sensors** instead use local temporal and spatial derivatives of the light intensity to compute motion. An implementation of the so-called gradient model, which yields velocity independent of spatial frequency and contrast, has been reported [DK98a]. The "Gradient2d" sensor implements an even simpler model for computing the motion vector field (u_x, u_y) by multiplying temporal and spatial derivatives [DK98b]:

$$u_x = u_o \frac{\partial I}{\partial t} \tanh\left(\frac{\partial I}{\partial x}\lambda\right) \tag{1}$$

$$u_y = u_o \frac{\partial I}{\partial t} \tanh\left(\frac{\partial I}{\partial y}\lambda\right) \tag{2}$$

where $I(x, y, t)$ is the light intensity distribution on the focal plane, $\partial I/\partial x$ and $\partial I/\partial y$ are both spatial derivatives, $\partial I/\partial t$ is the temporal derivative, u_o and λ are constants. The spatial derivatives are discretely approximated in the implementation. The sensor output u_x of one pixel for different stimulus velocities and a fixed orientation is shown in Figure 1. For small velocities the motion output is linear, for higher velocities it saturates due to low pass filtering in the photoreceptors and thus becomes direction of motion like.

The compressive computation algorithms of this work can be implemented with sensors of both categories. We present experimental results obtained with a Gradient2d 15×15 pixel array fabricated in a 1.2μm CMOS process.

2 Determining Axis of Rotation and Focus of Expansion

The FOE of a velocity flow field can be obtained analytically by solving an overdetermined system of $\mathcal{O}(\mathrm{N}^2)$ equations [HS93], where N×N is the array size. We extend the idea proposed in [Bor94] to general flow fields and reduce the computational load to $\mathcal{O}(\mathrm{N})$ by using row and column averages. The X component of the FOE is given by the position of the zero crossing (ZC) of the column average of u_x, the X component of the corresponding motion vectors. If there are multiple ZCs, the one with the maximal slope is to be taken. The FOE Y component and the AOR location are given accordingly:

$$x^{FOE} = ZC_x\left(\sum_y u_x(x,y)\right) \qquad y^{FOE} = ZC_y\left(\sum_x u_y(x,y)\right) \tag{3}$$

$$x^{AOR} = ZC_x\left(\sum_y u_y(x,y)\right) \qquad y^{AOR} = ZC_y\left(\sum_x u_x(x,y)\right) \tag{4}$$

Our motion sensors now can be used not only to compute the motion vector field $\boldsymbol{u}(x,y)$ in real time, but also the row and column averages: The pixel array is addressed by one row and one column scanner, and the motion vector components are represented as bidirectional electric currents. Thus by gating out one entire row or column, the vector sum is obtained automatically. All N row sums and N column sums are read into a computer, where the ZCs and thus the FOE or AOR is determined. Since only 2N instead of N^2 operations are required, the FOE and AOR can be determined very fast. We achieve frame rates beyond 400Hz.

2.1 Axis of Rotation

We start with describing experimental results on estimating the AOR. We rotate gray value images in the field of view of the sensor. With a digital computer we continously read out the motion vector field and the row and column averages as computed by the sensor. The vector field is simply read out and displayed for illustration. For AOR/FOE estimation only the averages are necessary. In Figure 2 we show a snapshot of the sensor data as a 'wagon wheel' stimulus was rotated in front of the sensor. In the lower right corner the stimulus as seen by the on-chip photoreceptors is shown. The solid curve in the horizontal graph represents the Y component of the column sum as computed by the sensor, the dotted line is obtained when the vector field is summed external of the sensor; correspondingly, the vertical graph represents the X component of the row averages. It can be seen that the motion vector field clearly reflects the motion of the stimulus. Most importantly, the zero crossing of the row and column average, as shown in Equation 4, obviously marks the location of the AOR.

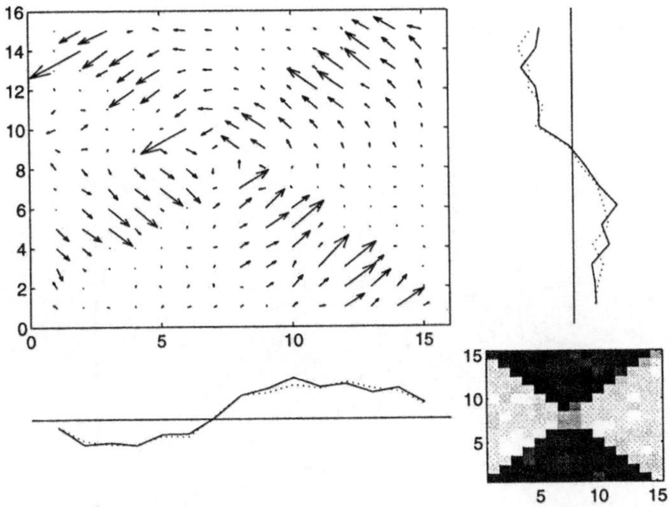

Fig. 2. Upper left: Motion vector field as computed by the sensor. Lower left: Column sum of the Y component, upper right: Row sum of the X component of the motion vectors. Lower right: Visual input as seen by the sensor.

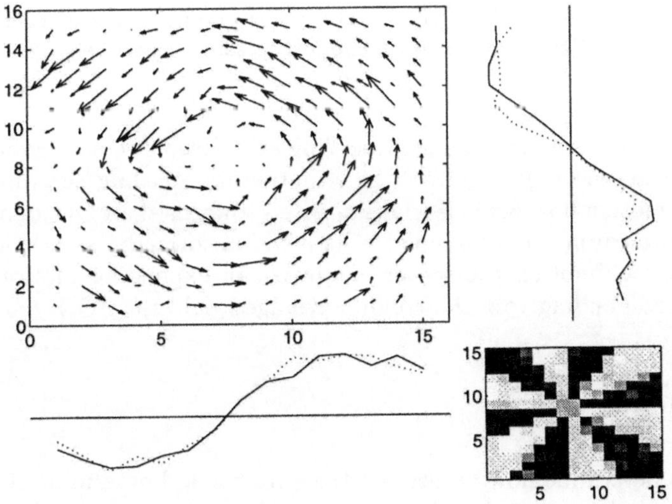

Fig. 3. Determining AOR from a dense flow field.

In Figure 3 a stimulus of higher spatial frequency was used. Due to the higher density of the flow field the slope at the ZCs of the row and column averages are steeper, thus reliably indicating the AOR.

2.2 Focus of Expansion

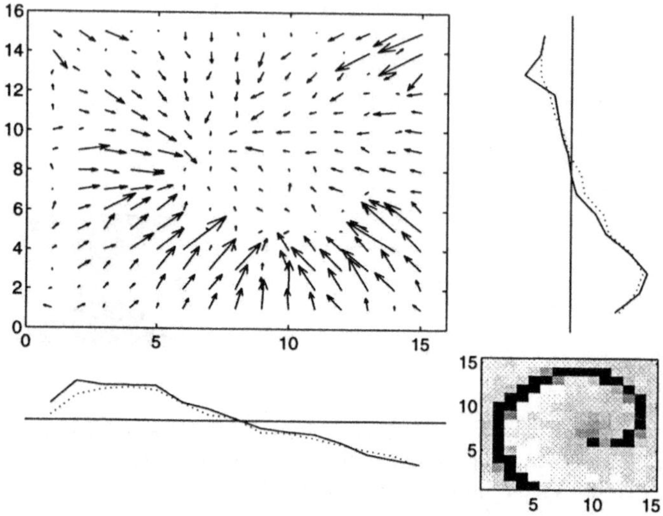

Fig. 4. Determining FOC. Horizontal and vertical graphs represent column and row averages of X and Y components of the motion vectors, respectively.

In order to simulate approaching or receding ego-motion we have rotated a spiral in front of the sensor. In Figure 4 the situation for receding ego-motion is captured: The motion flow field clearly is directed towards the FOC, and the ZCs of the row and column averages indicate its location correctly; see Equation 3. For approaching ego-motion the sensor computes an expanding motion flow field, and the corresponding row and column averages yield the FOE.

2.3 Accuracy

In order to determine how accurately the sensor could determine the AOR and FOE/FOC, we have recorded the row and column averages from the sensor chip several times while the same stimulus was rotating at a fixed location in the field of view of the sensor. As described above, the location of the AOR was determined from the zero crossing of the appropriate component of the row and column averages as computed by the sensor. We find a mean absolute error for the AOR X and Y components of 0.21 pixels and 0.29 pixels respectively, the relative error with respect to the array size being 1.4% and 1.9%. These low values demonstrate that our algorithm works reliably even with the inherently noisy motion vector fields computed by analog VLSI sensors.

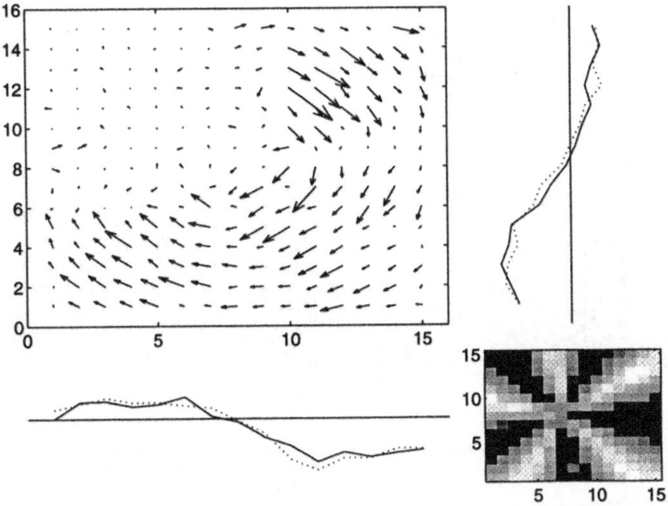

Fig. 5. Sensor output for occluded input. Lower right: Unoccluded input.

2.4 Occlusions

We have tested how the estimate of the AOR would be influenced by partial occlusions of the visual field of the sensor. For that purpose we were using a rotating sine wave stimulus and occluded about a quarter of it; cf. Figure 5. We have measured the AOR for several different occlusion scenarios, all of which only left three quarters of the full image information to the sensor. The surprising result is that the errors in estimating the AOR did not increase significantly as compared to the non-occluded situation: We find relative errors of 1.8% and 2.3% for the X and Y component, respectively. As can be seen from Figure 5, though, the averages obtained from the occluded areas are less strong, and the slope at the ZC becomes shallower. For that reason the estimate of the AOR becomes less reliable for stronger occlusions.

2.5 2-D FOE On-a-Chip

As we have shown that the compressive computation approach with our analog VLSI sensor is successful, we are now planning a single chip system that also takes over the task of finding the ZC from the row and column averages. The row and column averages are computed from the motion vector field and are then smoothed by a lateral resistive net. Similar to the 1-D design of Indiveri et. al. [IKK95] subsequently the zero crossing with the steepest slope is determined and its location is reported as a single 2-D vector. Thus AOR, FOE or FOC can directly be read off the sensor chip in true real time.

We have carried out computer simulations in order to determine the proposed systems' robustness against noise in the motion vector field, and to find out

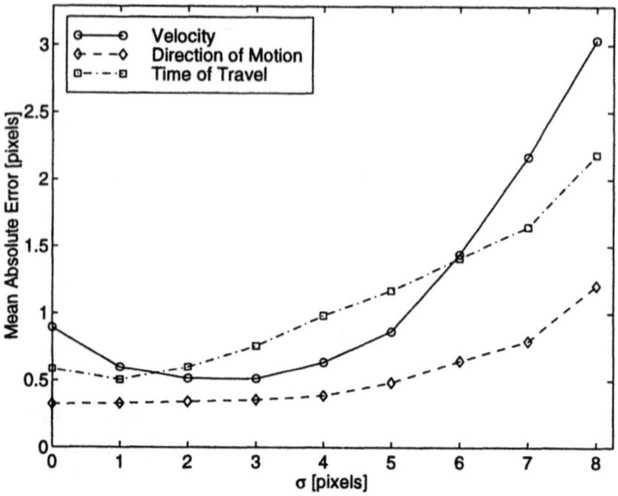

Fig. 6. Simulation result: Sensor performance for varying degree of smoothing.

which motion detection algorithm performs best. We find three main results. First, there is a minimum in the mean absolute error of the computed AOR from the true AOR depending on the smoothing length σ; cf. Figure 6. This is because of the tradeoff between the positive effect of noise rejection with short-scale smoothing and the erroneous shift of the ZC at large smoothing lengths. Second, it has turned out that the vector field of type direction of motion is best suited for determining the FOE, FOC, and AOR when using our proposed algorithm. The mean absolute error of about 0.3 pixels corresponds well with the measurement results presented in Section 2. Third, Figure 7 reveals that the direction of motion sensor is most insensitive against noise. The saturation of the error curves is due to the finite size of the pixel array.

3 Summary

After briefly introducing different types of analog VLSI motion sensors we have described how their pixel-parallel architecture can favourably be used for computational tasks where from a higher dimensional data set lower dimensional information is to be extracted. As an example we have described an algorithm for finding the FOE, FOC and AOR. Using a gradient based 2-D analog VLSI motion sensor we have experimentally demonstrated that these quantities can robustly be determined from natural visual input. Finally we have proposed a fully integrated 2-D system for real-time estimation of FOE, FOC and AOR and investigated in computer simulations its performance at the presence of noise. We find that direction of motion sensors perform most robust. Future work will concentrate on the implementation of the integrated system.

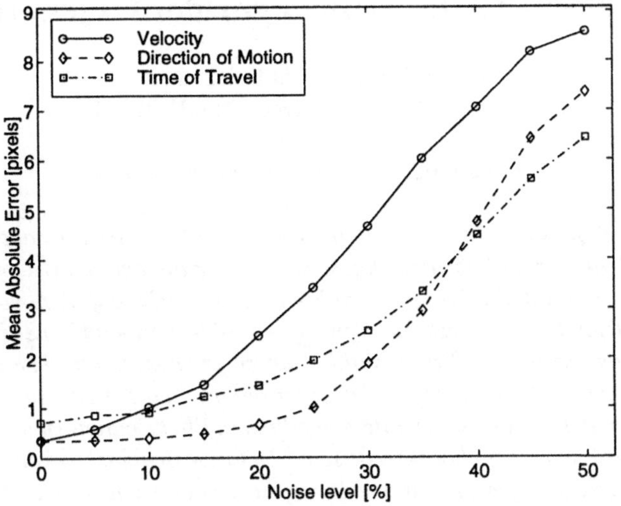

Fig. 7. Simulation result: Sensor performance for increasing noise level.

Acknowledgements: We would like to thank C. Koch for his support and J. L. van Hemmen for helpful discussions.

References

[BD97] C. Born and R. A. Deutschmann. Measurement of fast rotation by VLSI circuits. In *Proc. Deutsche Arbeitsgemeinschaft für Mustererkennung DAGM'97*, 1997.

[BFB94] J.L. Barron, D.J. Fleet, and S.S. Beauchemin. Systems and experiment: Performance of optical flow techniques. *Intern. J. Comp. Vis.*, 12:43–77, 1994.

[Bor94] C. Born. Determining the focus of expansion by means of flowfield projections. In *Proc. Deutsche Arbeitsgemeinschaft für Mustererkennung DAGM'94*, pages 711–719, 1994.

[DHK97] R. A. Deutschmann, C. Higgins, and C. Koch. Real-time analog VLSI sensors for 2-D direction of motion. In *Proc. Int. Conf. on Artificial Neural Networks ICANN'97*, volume 1327 of *Lecture Notes in Computer Science*, pages 1163–1168. Springer Verlag, 1997.

[DK98a] R. A. Deutschmann and C. Koch. An analog VLSI velocity sensor using the gradient method. In *Proc. IEEE International Symposium on Circuits and Systems ISCAS'98*, volume 6, pages 649–652, 1998.

[DK98b] R. A. Deutschmann and C. Koch. Compact real-time 2-D gradient based analog VLSI motion sensor. In *Proc. International Conference on Advanced Focal Plane Arrays and Electronic Cameras AFPAEC'98 Zurich*, 1998.

[HS93] R. M. Haralick and L. G. Shapiro. *Computer and Robot Vision*, volume II. Addison–Wesley, 1993.

[IKK95] G. Indiveri, J. Kramer, and C. Koch. Analog VLSI architecture for computing heading direction. *Proc. Intelligent Vehicles 1995*, 1995.

Automatische Generierung hybrider Objektmodelle

Ulrich Büker, Björn Kalkreuter
Heinz-Nixdorf-Institut, Universität-GH Paderborn,
33095 Paderborn
e-mail: bueker@get.uni-paderborn.de

Zusammenfassung. *In diesem Beitrag wird ein Verfahren zur automatischen Generierung hybrider Objektmodelle für die aktive 3D Objekterkennung vorgestellt. Unser Verfahren setzt direkt auf den von uns verwendeten Erkennungsstrategien auf. Diese basieren auf einer holistischen Auswertung der Kamerabilder auf einer subsymbolischen Ebene, die ergänzt wird durch eine Objektbeschreibung auf der Basis mehrerer Ansichten des Objektes aus unterschiedlichen Blickrichtungen und einer Reihe dazugehöriger charakteristischer Detailansichten. Die von uns gewählte Erkennungsarchitektur erlaubt die Generierung von Objektmodellen durch einmaliges Präsentieren der Objekte, so daß auf aufwendige Trainingssequenzen verzichtet werden kann. Dazu werden zunächst die Begriffe Objektansicht und Detailansicht geklärt, bevor ein Verfahren zur Auswahl charakteristischer Ansichten vorgestellt wird. Die Objektmodellierung erfolgt dann mit Hilfe dieser charakteristischen Ansichten in semantischen Netzwerken.*

1 Einleitung

Neben der zentralen Frage, wie das Wissen über Objekte in einem Bilderkennungssystem repräsentiert werden soll, ist auch die Frage, wie dieses Wissen in das System eingebracht werden kann, von entscheidender Bedeutung für die Akzeptanz und Einsetzbarkeit eines solchen Systems. Dies betrifft im besonderen die Erkennung komplexer dreidimensionaler Objekte. Das Erzeugen von Objektmodellen benötigt dabei häufig sehr umfangreiche Kenntnisse über die zu untersuchende Domäne, um das benötigte Wissen entsprechend strukturieren zu können. Verfahren zur automatischen Wissensakquisition sollen hierbei unterstützend wirken oder sogar diese Aufgabe vollständig übernehmen. Bei der Einteilung der existierenden Lernverfahren können zwei Klassen unterschieden werden: (i) symbolische Lernverfahren, die das in den Beispielen enthaltenen Wissen in eine explizite, symbolische Beschreibung der Objekte transformieren, z.B. in der Form von Regeln, Entscheidungsbäumen, semantischen Netzwerken, etc. [3], [7], [8] und (ii) subsymbolische Lernverfahren, die das Wissen aus den Beispielen in eine implizite Repräsentation überführen, wie dies z.B. bei den verschiedenen neuronalen Netzwerktypen der Fall ist [10], [11]. Beiden Lerntypen gemeinsam ist jedoch, daß üblicherweise eine große Zahl an Trainingsdaten zur Verfügung gestellt werden muß, was den Lernvorgang entsprechend aufwendig macht. Dagegen ist es unser Ziel, bereits durch einmaliges Präsentieren eines Objektes, ein Objektmodell aufzubauen, das für das Wiedererkennen unter veränderten Bedingungen (Position, Beleuchtung, etc.) geeignet ist.

Das eingesetzte Lernverfahren kann aber natürlich nicht losgelöst von der Objektrepräsentation betrachtet werden. Prinzipiell werden in der Objekterkennung zwei Klassen von Ansätzen zur Objektbeschreibung unterschieden:

- Objektzentrierte Ansätze setzen, wie der Name es bereits andeutet, das Objekt in den Mittelpunkt der Betrachtungsweise. Hierbei wird üblicherweise das Objekt in einfache 3D Komponenten zerlegt [12].
- Beobachterzentrierte Ansätze versuchen, das Objekt aus Sicht des Betrachters zu modellieren. Hierzu wird das Objekt mit Hilfe mehrerer 2D Ansichten repräsentiert [2], [4], [9].

Unser Bilderkennungssystem beruht auf einem solchen beobachter- oder ansichtenbasierten Ansatz. Dabei wird das Objekt zunächst einmal mit einer Reihe von 2D Ansichten aus unterschiedlichen Blickrichtungen beschrieben. Darüber hinaus werden aber auch einzelne Details durch gezielt „eingeengte" fokussierende Ansichten beschrieben. Dies erfolgt mit Hilfe einer subsymbolischen, neuronal motivierten Repräsentation. Auf einer übergeordneten symbolischen Beschreibungsebene werden die topologischen Relationen zwischen den einzelnen 2D Ansichten gespeichert. Bei der Erzeugung dieser Beschreibungen verfolgen wir auf beiden Ebenen das Ziel, ohne lange Trainingssequenzen ein Lernen quasi „auf einen Blick" zu realisieren.
Im diesem Beitrag soll nun speziell das Lernen auf der symbolischen Beschreibungsebene näher betrachtet werden. Zum besseren Verständnis werden dazu zunächst in Kapitel 2 die wesentlichen Eigenschaften unseres Erkennungsansatzes zusammengefaßt, bevor dann in Kapitel 3 die von uns durchgeführte Auswahl charakteristischer Ansichten erläutert wird. In Kapitel 4 wird dann die automatische Modellbildung auf der Basis dieser charakteristischen Ansichten beschrieben. Der Beitrag endet mit einer kurzen Zusammenfassung der Ergebnisse.

2 Aktive hybride Objekterkennung

In diesem Beitrag können nicht alle Details unseres aktiven hybriden Erkennungssystems beschrieben werden. Hierzu sei verwiesen auf ausführliche Darstellungen in [1] und [6]. Zum weiteren Verständnis unserer Strategien bzgl. der automatischen Generierung der Objektmodelle ist es jedoch notwendig, einige wesentliche Eigenschaften kurz vorzustellen.
Wir verfolgen für die Objekterkennung einen ansichtenbasierten Ansatz, bei dem ein komplexes 3D-Objekt durch mehrere 2D-Ansichten repräsentiert wird. Dabei setzen wir eine bewegliche Kamera ein, die mit Hilfe verschiedener Fovealisierungsroutinen und den daraus resultierenden Aufmerksamkeitspunkten gesteuert wird. Somit stehen uns für die Objekterkennung mehrere Bilder eines Objektes zur Verfügung. Die Verarbeitung eines einzelnen Bildes erfolgt dabei auf einer subsymbolischen Ebene, auf der die Aktivitätsmuster komplexer Zellen des visuellen Cortex nachgebildet werden. Diese Aktivitätsmuster zeichnen sich durch eine Toleranz gegenüber kleinen Verschiebungen des Fovealisierungspunktes und gegenüber leichten perspektivischen Verzerrungen aus. Durch den Einsatz normierender Abbildungen können daher relativ einfache Matchingroutinen für die Erkennung solcher 2D-Ansichten eingesetzt werden [6]. Abb. 1 zeigt, wie die tolerante Kantenrepräsentation eines im Raum verschobenen und leicht

rotierten Objektes nach einem Normierungsschritt ein großes Maß an Übereinstimmung (Match) mit nur einem geringen Mismatch liefert.

Bild a	Bild b	normalisierte Kantenrepräsentation der Bilder a und b	Match (Inneres Produkt)	Mismatch (Hamming Distanz)
(Distanzänderung, leichte Verdrehung um die Hauptachse)				

Abb. 1: Zwei Grauwertbilder einer Tasse, die normalisierten Aktivitätsmuster der komplexen Zellen sowie resultierender Match und Mismatch der Muster

Da jedoch für die robuste Erkennung komplexer 3D-Objekte in vielen Fällen die Auswertung von nur einer einzelnen 2D-Ansicht nicht ausreicht, erfolgt auf einer übergeordneten Ebene die Modellierung der Objekte mit Hilfe semantischer Netzwerke auf der Basis von 2D-Ansichten aus verschiedenen Blickrichtungen sowie auf der Basis von „eingeengten" Blicken, die sich gezielt auf markante Details des Objektes konzentrieren [1]. Das übergeordnete, steuernde KI-Modul regelt dabei auch den Zugriff auf den Roboter zum Bewegen der Kamera im Arbeitsraum und den Framegrabber zum Aufnehmen weiterer Bilder. Es wird somit nicht ein zuvor festgelegter Satz von Bildern ausgewertet. Die Bildaufnahme erfolgt vielmehr dynamisch und es wird erst während des Erkennungsprozesses entschieden, von welchen Positionen weitere Bilder benötigt werden.

Die Vorteile der Modellierung auf der Basis unserer subsymbolischen Erkennungsergebnisse können wie folgt zusammengefaßt werden:

- Die Modellierung kann sich auf einige wenige prägnante Teilansichten beschränken. Daraus ergibt sich ein geringer Modellierungsaufwand. In unseren Tests zur Erkennung von Autos reichen für die Modellierung eines Typs 10 Konzepte aus.
- Objektteilansichten tragen einen großen Informationsgehalt. Sie sind charakteristisch für die jeweilige Objektklasse und können daher nur Teilstruktur einiger weniger Objekte sein. Dies schränkt den Suchraum enorm ein.
- Den Teilansichten kann ebenso wie den Objektansichten eine Positions- und Orientierungsinformation zugewiesen werden. Dadurch kann zielgerichtet auf Position und Orientierung des Gesamtobjektes sowie weiterer relevanter Teilansichten geschlossen werden. Auch dies reduziert den Suchaufwand.

Auf der subsymbolischen Ebene ist aufgrund der Konstruktion unserer toleranten Merkmalsvektoren bereits ein Lernen auf einen Blick ohne lange Trainingssequenzen möglich. Die symbolischen Objektmodelle des KI-Moduls mußten jedoch bislang „per Hand" erzeugt werden. Die folgende Arbeit beschäftigt sich nun mit der automatischen Generierung dieser symbolischen Modelle. Diese konnte von uns soweit automatisiert werden, daß durch Vorgabe einiger weniger Parameter und durch einmaliges Präsentieren eines Objektes die für die Erkennung notwendigen semantischen Netzwerke erzeugt werden. Dabei wurde eine Zweiteilung des Generierungsprozesses durchgeführt. Im ersten Schritt wird eine Menge von Objekt- und Detailansichten erzeugt und analysiert. Hierbei werden die charakteristischen Ansichten eines Objektes extrahiert. Im zweiten Schritt werden dann mit Hilfe dieser Ansichten und ihrer topologischen Relationen die Objektmodelle erzeugt.

3 Charakteristische Ansichten

Aufgrund der Modellierung der zu erkennenden Objekte auf der Basis einiger charakteristischer Objekt- und Detailansichten, liegt die Hauptaufgabe der automatischen Modellgenerierung zunächst einmal in der Extraktion und Auswahl dieser charakteristischen Ansichten. Dabei geht es nicht darum, eine Auswahl von Ansichten zu finden, die die Kontur oder die Oberfläche eines Objektes vollständig abdeckt und beschreibt. Es geht auch nicht darum, eine Auswahl zu treffen, die keinerlei redundante Information über ein Objekt enthält. Es ist vielmehr unser Ziel, eine kleine Menge von Ansichten auszuwählen, die geeignet ist, ein Objekt von den anderen Objekten der Domäne zu unterscheiden. Hierzu ist eine vollständige Beschreibung nicht notwendig, während sich überschneidende Ansichten geeignet sind, im Falle von teilweisen Verdeckungen die Objekterkennung zu unterstützen. In der Verwendung von Detailansichten für die Objekterkennung unterscheidet sich unser Beitrag auch von dem in [5] beschriebenen Ansatz. Im folgenden seien zunächst einmal einige Begriffs-Definitionen gegeben.

3.1 Begriffe

Definition 1: Das Bild eines Objektes aus einer beliebigen Blickrichtung nennen wir **Objektansicht**, wenn das Objekt vollständig im Kamerabild enthalten ist.

Definition 2: Das Bild eines Objektes aus einer beliebigen Blickrichtung nennen wir **Detailansicht**, wenn lediglich ein Teil des Objektes im Kamerabild enthalten ist. Dabei ist die Aufnahmeentfernung typischerweise kleiner als bei einer Objektansicht. Dies ist aber nicht zwingende Voraussetzung.

Definition 3: Wir sprechen allgemein von einer **Ansicht** eines Objektes, wenn es sich um eine Objektansicht oder eine Detailansicht handelt.

Definition 4: Wir nennen eine Ansicht A **hart charakteristisch** für eine Ansichtenmenge $\mathcal{A}(K)$ einer Objektklasse K bzgl. eines Ähnlichkeitsmaßes d, wenn

1. zu jedem Objekt der Klasse K genau eine Ansicht $A' \in \mathcal{A}(K)$ mit großer Ähnlichkeit zu A existiert und wenn
2. für alle Objekte aus einer anderen Klasse $K' \neq K$ keine Ansicht $A' \in \mathcal{A}(K')$ mit hinreichend großem Ähnlichkeitsmaß zu A existiert.

Hinweis: Für die Festlegung eines großen Ähnlichkeitsmaßes wird verlangt: $d(A',A) > d_{min}$. Das Ähnlichkeitsmaß d und der Schwellwert d_{min} werden dabei üblicherweise durch den bei der Erkennung verwendeten Klassifikator bestimmt.

Definition 5: Wir nennen eine Ansicht A **weich-** oder **k-charakteristisch**, wenn in den Bedingungen 4.1 und 4.2 jeweils nur wenige, k Ansichten A_i' mit hinreichend großem Ähnlichkeitsmaß existieren.

Hinweis: Wenn wir im folgenden allgemein von charakteristischen Ansichten eines Objektes sprechen, so meinen wir damit weich- oder k-charakteristische Ansichten.

Es sei an dieser Stelle darauf hingewiesen, daß somit der Ansichten-Begriff vom Aspekt-Begriff unterschieden wird. Die Menge der Aspekte eines Objektes ist eine echte Teilmenge der Ansichtenmenge.

3.2 Auswahl charakteristischer Ansichten

Im folgenden wollen wir nun betrachten, wie wir näherungsweise aus der Ansichtenmenge eine geeignete Menge charakteristischer Ansichten für die Modellbildung extrahieren können. Da sich der Lernprozeß natürlich an den Erkennungsvorgang anlehnt, werden zur Erzeugung der Ansichten im ersten Schritt die Verfahren unserer subsymbolischen Erkennungsebene verwendet. Es werden also mit Hilfe der Fovealisierungsroutinen Aufmerksamkeitspunkte bestimmt, die von der Kamera angefahren werden. Die auf diese Weise aufgenommenen Detailansichten werden dann wie in Kapitel 2 beschrieben transformiert und gelernt. Mit Hilfe des Klassifizierers wird anschließend bestimmt, wie wahrscheinlich eine Fehlklassifikation eines einzelnen Merkmalsvektors innerhalb der betrachteten Domäne ist. Lediglich die Merkmalsvektoren, die nur sehr selten fehlklassifiziert werden, werden von uns als charakteristisch für ein Objekt betrachtet (weich oder k-charakteristisch nach Def. 5). Diese Vorgehensweise beschränkt sich zunächst einmal auf die Auswertung von Bildern aus einer Blickrichtung.

Für die Modellierung komplexer 3D Objekte sollen jedoch auch Objektansichten aus verschiedenen Blickrichtungen verwendet werden. Aus diesem Grunde wird zusätzlich mit einer vom Benutzer vorzugebenden Schrittweite die Ansichtensphäre eines Objektes abgetastet. Für jede dieser Objektansichten wird dann die oben beschriebene Vorgehensweise zur Erzeugung weiterer Detailansichten und deren Auswahl durchgeführt.

Auf diese Weise werden zunächst einmal verschiedene Objekt- und Detailansichten eines oder mehrerer Objekte der Domäne erzeugt. Dieses sind natürlich nicht alle Ansichten nach Def. 3 sondern nur eine kleine Teilmenge. Die Erzeugung aller Ansichten würde bei kleiner Diskretisierung der Ansichtensphäre zu einer nicht handhabbaren Menge von Ansichten führen. Wir müssen uns daher auf eine Teilmenge begrenzen. Die gewählte Einschränkung auf die Ansichten, die mit Hilfe von Fovealisierungsroutinen erzeugt werden können, ist sinnvoll, da auch bei der Erkennung nur solche „fovealisierbaren" Ansichten ausgewertet werden.

Im folgenden soll die Auswahl der Ansichten in formaler Form beschrieben werden. Wir betrachten zunächst einmal die Verarbeitung einer einzelnen Ansicht A eines Objektes O. Es seien nun $\mathcal{F}(A) = \{f_1, f_2, \ldots, f_n\}$ die Menge der Fovealisierungspunkte im Bild der Ansicht A, die für weitere detaillierte Aufnahmen betrachtet werden sollen. Es sei $\mathcal{N}(A)$ die Menge der normalisierten Detailansichten, die sich durch Anfahren der Fovealisierungspunkte ergeben. Hierbei werden Detailbilder mit konstanter Blickrichtung, aber veränderter Distanz aufgenommen. Die Normalisierung erfolgt bei den Detailbildern auf eine verringerte Referenzdistanz r', so daß eine bessere Auflösung der Details gewonnen wird. Die Merkmalsextraktion liefert uns Aktivitätsmuster der komplexen Neurone als Merkmalsvektoren $\mathcal{T}(A) = \{t_1, t_2, \ldots, t_n\}$. In Analogie zur subsymbolischen Klassifikation verwenden wir für die Auswertung der Merkmalsvek-

toren das Ähnlichkeitsmaß $d_{ij} = d(t_i, t_j)$, das sich aus dem inneren Produkt minus der Hamming-Distanz der beiden binären Merkmalsvektoren ergibt. Dieses Ähnlichkeitsmaß wird als Näherung für die Wahrscheinlichkeit betrachtet, daß zwei Merkmalsvektoren, respektive die zugrundeliegenden Ansichten verwechselt werden können. Mit $d_{ij} = d(t_i, t_j)$ ergibt sich die Wahrscheinlichkeitsmatrix $D = [d_{ij}]$.
Mit Hilfe der Schwellwertfunktion $u : \Re \rightarrow \{0,1\}$ mit $u(x) = 1$, falls $x > 0$ und $u(x) = 0$, falls $x \leq 0$ erhalten wir die Matrix $S = [s_{ij}] \in \{0,1\}^{n \times n}$ als $s_{ij} = u(d_{ij} - d_{min})$ durch Verwendung des Schwellwertes d_{min}. (Für unsere Experimente verwenden wir d_{min}=0.8).
Wir betrachten nun die Zeile $s_i = (s_{i1}, s_{i2},...,s_{in})$ mit der Norm $\|s_i\|_1 = \sum s_{ij}$. Da zumindest für die Diagonalelemente $s_{ii} = 1$ gilt, folgt unmittelbar $\|s_i\|_1 \geq 1$.
Mit Hilfe dieser Werte $\|s_i\|_1$ können wir nun drei verschiedene Klassen von Merkmalsvektoren t_i unterscheiden:

1. $\|s_i\|_1 = 1$: t_i wird nicht mit anderen Merkmalsvektoren verwechselt und trägt damit Information zur Orientierung und Lageschätzung des Objektes O.
2. $1 < \|s_i\|_1 < s_{max}$ mit einem Schwellwert s_{max} : Die Verwendung von t_i zur Orientierungs- und Lageschätzung von O kann zu falschen Hypothesen führen, da eine Verwechselungsgefahr mit anderen Merkmalsvektoren besteht. Der Merkmalsvektor t_i eignet sich jedoch zur Überprüfung mit Hilfe zuvor aufgestellter Hypothesen, da nur eine geringe Wahrscheinlichkeit zu Verwechslungen besteht. Der Einfluß dieser Ansichten auf die Erkennung wird durch eine Verringerung der Glaubwürdigkeit (Evidenz) verkleinert. (Wir verwenden als Schwellwert $s_{max} = n/10$.)
3. $\|s_i\|_1 \geq s_{max}$: Aufgrund der großen Verwechslungsgefahr sollte t_i nicht für die Erkennung von O verwendet werden.

Offensichtlich läßt sich dieses Verfahren direkt auf die Bearbeitung mehrerer Objekte O_k verallgemeinern. Dabei werden zunächst zu jedem Objekt eine Objektansicht und die hieraus resultierenden Detailansichten zu allen Objekten betrachtet. Für die 3D Verarbeitung werden darüber hinaus auch mehrere Objekt- und Detailansichten A_{kl} dieser Objekte in die Auswahl einbezogen.

4 Generierung von Objektmodellen

Die Zweiteilung des Generierungsprozeßes in eine erste Phase zur Auswahl der charakteristischen Ansichten und in eine zweite Phase der eigentlichen Modellbildung ermöglicht die einfache und schnelle Anpassung an Änderungen der Modellierungssyntax, an veränderte Modellierungsstrategien und auch die Verwendung für völlig andere Modellierungssprachen. So werden bei uns die ausgewählten Ansichten für die Objektmodellierung in einem semantischen Netzwerk verwendet (Abb. 2). Dabei werden die charakteristischen Objektansichten

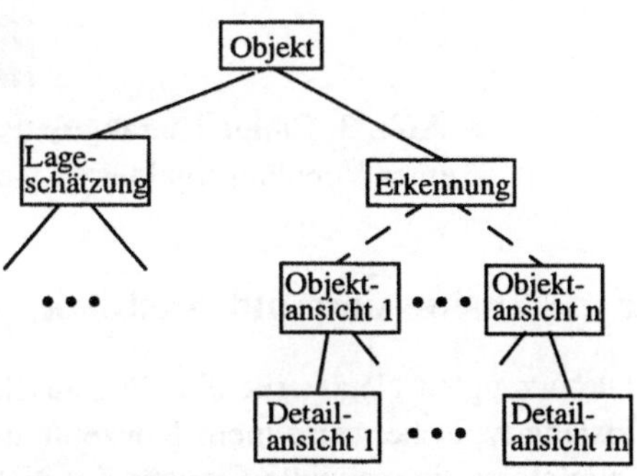

Abb. 2: Struktur der erzeugten semantischen Netzwerke

über eine Aspekt-Relation integriert, während die charakteristischen Teilansichten jeweils über eine Teil-von-Relation an die zugehörige Objektansicht gebunden werden. Jeder Ansicht wird dabei ein Konzept des Netzwerkes zugeordnet. In den Attributen der Konzepte wird vermerkt, mit welchen Operationen (Fovealisierung, Merkmalsextraktion, Klassifikator) die jeweilige Ansicht bearbeitet wurde. Die bei der Ansichtengenerierung verwendeten Blickrichtungen werden ebenfalls in den Attributen vermerkt, so daß auf diese Weise die topologischen Relationen der verschiedenen Ansichten gespeichert werden und bei der Erkennung ausgewertet werden können. Entsprechend unserer Erfahrungen mit dem Erkennungssystem haben wir die Modellierung eines Objektes so gewählt, daß sie sich aufteilt in ein Teilnetz zur Lageschätzung und eines für die darauffolgende eigentliche Erkennung über mehrere Ansichten des Objektes. Diese beiden Modellierungsaspekte werden dazu über eine Teil-von-Relation an das Objektkonzept gebunden (Abb. 2). Auf die Modellierung der Lageschätzung kann in diesem Beitrag jedoch nicht näher eingegangen werden. Abb. 3 zeigt einige der in das Netzwerk eingebundenen Objekt- und Detailansichten eines PKW. Die topologischen Relationen werden hierbei durch die geometrische Anordnung der Ansichten verdeutlicht.

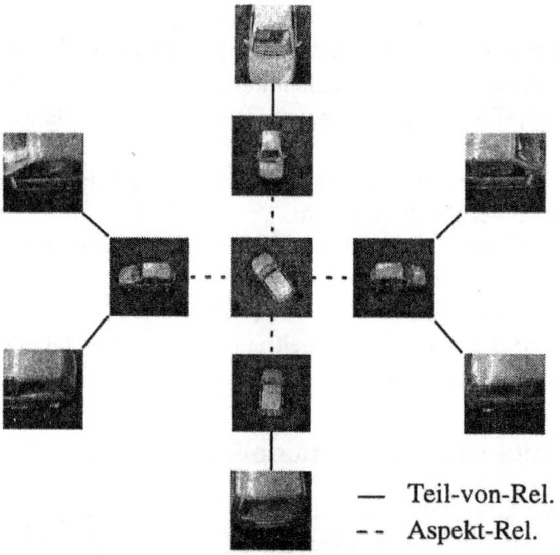

Abb. 3: Einige charakteristische Objekt- und Detailansichten eines Versuchsobjektes

5 Diskussion und Ausblick

Die erzeugten Netzwerke ähneln sehr stark den manuell erzeugten, enthalten jedoch bislang typischerweise mehr Konzepte und Ansichten. Dies macht offensichtlich, daß zum einen die manuelle Eingabe der Netzwerke zeitaufwendig ist und daher generell die Tendenz besteht, möglichst wenige Konzepte für die Modellierung zu verwenden. Zum anderen existieren jedoch möglicherweise beim Experten noch weitere Kriterien

zur Auswahl der Ansichten, die in unser Verfahren noch nicht eingeflossen sind. Interessanterweise konnten jedoch auch einige Ansichten ermittelt werden, die bislang vom Anwender als charakteristisch und nicht verwechselbar betrachtet wurden, die aber doch zu Verwechselungen mit anderen Ansichten neigen.
Eine nachträgliche Bearbeitung der Modelle war bei unseren Tests nicht notwendig. Es wurden quasi-zweidimensionale Werkstücke (Cranfield-Satz) und komplexe dreidimensionale Automodelle (Maßstab 1:24) untersucht. Sollte eine weitere Minimierung des Objektmodells gewünscht sein, so kann diese manuell problemlos durchgeführt werden. Ein Löschen einzelner Netzteile ist dabei natürlich deutlich einfacher als das manuelle Erstellen. Prinzipiell hat jedoch die Tatsache, daß die automatisch erzeugten Netze zusätzliche Konzepte enthalten, keinen Einfluß auf die Erkennungsleistung sondern lediglich auf die Laufzeit der Erkennung. So konnten die automatisch generierten Modelle von Autos bereits in Vorführungen unseres Erkennungssystems eingesetzt werden.
Eine automatische Minimierung der für die Modellierung verwendeten Ansichten soll durch ein Refinement des Netzwerkes nach Auswertung mehrerer Erkennungsdurchgänge erzielt werden, bei denen das Erkennungssystem Erfahrung über die betrachtete Domäne und das Instanziierungsverhalten der einzelnen Konzepte sammelt.

Literatur

[1] Büker, U.; Hartmann, G.: Knowledge based view control of a neural 3-D object recognition system. Proc. of the 13th ICPR, Vol.4, 1996, pp.24-29.

[2] Bülthoff, H.H.; Edelman, S.Y.; Tarr, M.J.: How are three-dimensional objects represented in the brain? Technical report A.I. Memo No. 1479, MIT, AI Lab, 1994.

[3] Carbonell, J.: Machine learning. Cambridge, MA (MIT-Press), 1990.

[4] Edelman, S.; Bülthoff, H.H.: Orientation dependence in the recognition of familiar and novel views of 3D objects. In: Vision Research, vol. 32, 1992, pp.2385-2400.

[5] Gremban, K.D.; Ikeuchi,K.: Planning multiple observations for object recognition. Int. Journal of Computer Vision, 12:2/3, 1994, pp.137-172.

[6] Hartmann, G.; et.al.: The SENROB Vision-System and its Philosophy. Proc. of the 12th ICPR, Vol. II, 1994, pp. 573-576.

[7] Michalski, R.; Carbonell, J.G.; Mitchell, T. (Hrsg.): Machine Learning: An Artificial Intelligence Approach. Vol. 1, Tioga, Palo Alto, 1983.

[8] Morik, K.: Knowledge Acquisition and machine learning. London, Acad. Press, 1993.

[9] Murase, H.; Nayar, S.K.: Visual Learning and Recognition of 3-D Objects from Appearance. In: International Journal of Computer Vision, 14, S.5-24, 1995.

[10] Ritter, H.; Martinetz, T.; Schulten, K.: Neuronale Netze. Bonn, Addison-Wesley, 1994.

[11] Rojas, R.: Theorie der neuronalen Netze. Berlin, Springer, 1993

[12] Sato, H.; Binford, T.O.: Finding and recovering SHGC objects in an edge image, Computer Vision Graphics and Image Processing, Vol. 57(3), S.346-356, 1993.

Erkennung und Lagebestimmung realer Objekte durch tolerante Repräsentation und modellbasiertes Gruppieren

N. Götze, S. Drüe, G. Hartmann
Universität-GH Paderborn, Fachbereich Elektrotechnik
Pohlweg 47-49, 33098 Paderborn
E-mail: {goetze | druee | hartmann}@get.uni-paderborn.de

Zusammenfassung. *Viele konturbasierte Erkennungsverfahren detektieren gut konturierte, starre Objekte in wenig strukturierten Szenen zuverlässig. Reale Objekte können aber ihre Erscheinungsform ändern und werden so weniger zuverlässig erkannt. Die Fehlerkennungsrate nimmt dabei mit steigender Szenenkomplexität zu. Zur exakten Lagebestimmung bereits detektierter Objekte entsteht ein über die Erkennung hinausgehender Aufwand. Wir stellen ein Erkennungssystem vor, das auf einer toleranten Konturrepräsentation beruht. Neben einer zuverlässigen Erkennung von Objekten mit leicht veränderlichen Konturen (z.B. aufgrund von Änderungen der Perspektive) auch in komplexen Szenen wird implizit deren genaue Lage bestimmt. Durch modellbasiertes Gruppieren wird eine teilobjektbasierte Erkennung von Gesamtobjekten gewährleistet, die einer direkten Erkennung des Gesamtobjektes in Bezug auf Effizienz und Robustheit überlegen ist.*[1]

1 Einleitung

Eine Vielzahl konturbasierter Erkennungsverfahren gehen von der Annahme starrer Objekte aus, die sich in einfacher Umgebung befinden (z.B. [Bal81], [Sto87], [Gri90]). Reale Objekte ändern aber z.B. aufgrund wechselnder Beleuchtung, Perspektive oder durch Okklusion ihre Erscheinungsform und sie kommen in mitunter stark strukturierter Umgebung vor. Eine Erkennung dieser Objekte wird dadurch schwieriger, und die Gefahr von Fehlerkennungen steigt gleichzeitig an ([Ols90]).

Viele Verfahren, die auf der General-Hough Transformation [Bal81] basieren, verwenden für das gelernte Objekt (=gelerntes Muster) und die zu untersuchende Szene (= zu untersuchendes Muster bzw. Eingangsmuster) die gleiche, meist aus einfachen Konturmerkmalen bestehende Repräsentation. Eine Hypothese über das Vorliegen des gelernten Objektes an einer bestimmten Position in der Szene wird über das Akkumulieren von Einzelhypothesen gebildet. Hierbei wird für jedes Merkmal im Eingangsmuster und jedes hierzu möglicherweise korrespondierende Merkmal des gelernten Musters die Erhöhung eines Akkumulatoreintrages an der Stelle vorgenommen, die zu der Objektposition relativ zu dem jeweiligen gelernten Mustermerkmal gehört. Nach der Abarbeitung des gesamten Eingangsmusters werden Einträge im Ergebnisspeicher, die über einer bestimmten Schwelle liegen, als Hypothesen für das Vorliegen des gelernten Objektes betrachtet. Werden zusätzlich zur Positionsinvarianz noch Orientierungs- und Skalierungsinvarianz gefordert, so werden Ergebnisse gewöhnlich in einem 4D-Akkumulator gespeichert.

Bei bereits geringen Unterschieden eines gelernten von dem zu untersuchenden Muster verteilen sich die Akkumulatoreinträge über einen umliegenden Bereich. Um dennoch ausgeprägte Maxima zu erhalten, werden unterschiedliche Verfahren verwendet.

1. Diese Arbeit wurde im Rahmen des BMBF Projektes 01 IN 506 B 2 DEMON gefördert.

Eine Mittelung via Faltung mit Gaußmasken ([Bal81]) ist sehr rechenintensiv, eine gröbere Quantisierung mit „Bins" (z.B. [Sto87], [Gri90]) oder eine Klassenbildung ([Sam97]) gehen einher mit ungenauer Lageschätzung und gesteigerter Fehlhypothesenbildung in komplexen Szenen.

In unserem System werden Prototypen in einem einzigen Schritt gelernt. Durch die in Abschnitt 2 vorgestellte besondere Repräsentation der Eingangsmuster werden geringe Unterschiede eines gelernten Musters zu einem Eingangsmuster, wie sie durch Perspektive, Verdeckungen, etc. entstehen können, toleriert, wobei implizit die im Mittel am Besten passende Lage ermittelt wird. Selbst bei stark strukturierten Szenen ist die Anzahl der Fehlhypothesen gering. Wie sich zeigen wird, kann durch Reduktion des Ergebnisraumes von vier auf zwei Dimensionen dieser auch bei lage- und skalierungsinvarianter Erkennung klein gehalten werden, wodurch die Bestimmung lokaler Maxima schnell durchführbar ist. Weitere Möglichkeiten der Effizienzsteigerung ergeben sich durch eine Grob-Fein Strategie, bei der Hypothesen in niedriger Auflösung generiert und in hoher Auflösung verifiziert werden. In Abschnitt 3 wird erläutert, wie durch modellbasiertes Gruppieren in Verbindung mit der toleranten Repräsentation eine teilobjektbasierte Erkennung effizient und robust durchgeführt werden kann.

Die Verfahren werden derzeit erfolgreich in einem Demontageprozeß zur Teiledetektion in Verbindung mit einem Stereokamerasystem eingesetzt [Ste98].

2 Erkennung mit toleranter Repräsentation

In unserem System verwenden wir zwei unterschiedliche, physiologisch motivierte Repräsentationen, um in Grauwertbildern vorkommende, orientierte Konturen darzustellen. In der „einfachen Repräsentation" wird das Verhalten simpler Neurone, in der „toleranten Repräsentation" das Verhalten komplexer Neurone des visuellen Cortex simuliert. Simple Modellneuronen stellen eine Kontur durch ein schmales Aktivitätsmuster dar. Demgegenüber fällt das gleiche Konturstück in mehrere seitlich versetzte Felder komplexer Neurone, so daß dieses durch ein breites, „wolkenförmiges" Aktivitätsmuster repräsentiert wird ([Hub88]). In diesem Abschnitt zeigen wir, wie die beiden Repräsentationen in einem Erkennungsvorgang Ortstoleranz gewährleisten können.

Ausgangspunkt für die Erzeugung der Repräsentationen ist das von der Szene aufgenommene Grauwertbild mit $N{=}M{*}M$ Elementen, das hier durch einen Vektor $\boldsymbol{x}$:= $(x_1, x_2, ..., x_M, x_{M+1}, ..., x_n, ..., x_N)$ beschrieben wird. Einem Bildpunkt x_n an einer beliebigen Position $n \in \{1, ..., N\}$ wird dabei entsprechend seiner Helligkeit ein Wert aus $Q := \{0, 1, ..., 255\}$ (255 entspricht „hell") zugewiesen. Durch die Verarbeitung des Bildes mit Gaborfilterbänken werden Grauwertkonturen in unterschiedlicher Orientierung und Auflösung detektiert ([Tra96]). Wir verwenden die Beträge der Filterantworten bei vierfacher, zweifacher und ohne Unterabtastung (grobe, mittlere und feine Auflösungsebene) sowie in $R = 12$ Orientierungen bei einer Schrittweite von 15 Grad. Bei Betrachtung einer bestimmten Auflösungsebene erhalten wir so an einer beliebigen Position n eine „Orientierungssäule" $\boldsymbol{g}_n := (g_{n1}, ..., g_{nr}, ... g_{nR})$, wobei g_{nr} also den Betrag der Gaborfilterantwort an der Position n für die Orientierung $r \in \{1, ..., R\}$ angibt (durch Normierung gelte $g_{nr} \in Q$). Die zu benachbarten Orientierungen gehörenden Elemente von $\boldsymbol{g}_n$ besitzen einander ähnliche Werte, was später für die Erzeugung der toleranten Repräsentation ausgenutzt wird. Durch das Aneinanderreihen dieser Vektoren für alle Positionen erhalten wir den gesamten Vektor $g := (g_1, ..., g_n, ..., g_N)$. Wir verwenden auch die alternative Indizierung, bei der für eine beliebige Orientierung r ein alle Positionen beinhaltender Vektor $g'_r := (g_{1r}, ..., g_{nr}, ..., g_{Nr})$ und schließlich $g' := (g'_1, ..., g'_r, ..., g'_R)$ gebildet wird. Ein schematisches Beispiel für ein Grauwertbild x mit Konturverlauf ist in Abb. 1a) gegeben. Die damit erzielte Filterantwort g'_r in passender Orientierung ist in Abb. 1b) zu sehen.

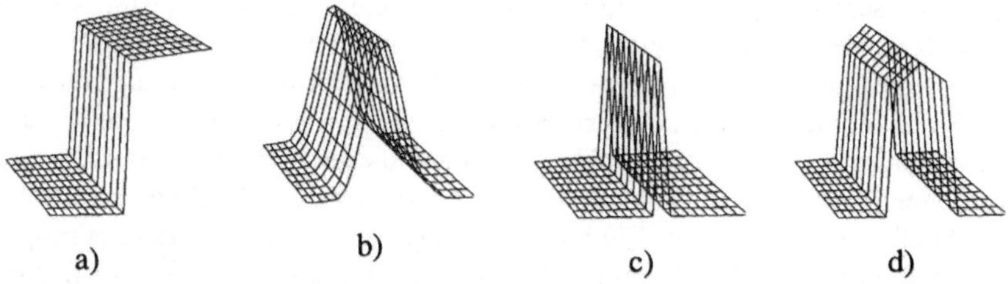

Abb. 1. Qualitative Darstellung von a) Grauwertkante x. b) Betrag der Filterantwort g'_r, c) einfache Repräsentation k'_r und d) tolerante Repräsentation c'_r, b)-d) nur passende Orientierung r gezeigt.

Mit den Grauwerten in x hängen auch die Filterantworten g bzw. g' von den Lichtverhältnissen ab. So erzeugen starke Kontraste hohe Filterantworten, deren Aktivitätswolken sich über einen weiten Bereich senkrecht zur jeweiligen Konturorientierung ausdehnen (s. Abb. 1c)). Niedrige Kontraste führen zu geringeren Antworten und schmaleren Aktivitätswolken ([Zie97], [Drü93]). Eine weitestgehend beleuchtungsinvariante, binäre Darstellung k'_r erreichen wir nun durch Skelettierung der g'_r senkrecht zu ihrer Orientierung r (s. Abb. 1c)). Analog zur obigen Nomenklatur stellt hier nun ein Element k_{nr} der sog. *einfachen Repräsentation k* dar, ob durch die Position n eine Kontur in der Orientierung r im Grauwertbild verläuft ($k_{nr} = 1$) oder nicht ($k_{nr} = 0$)[1]. Diese Repräsentation k entspricht einer Repräsentation durch simple corticale Zellen.

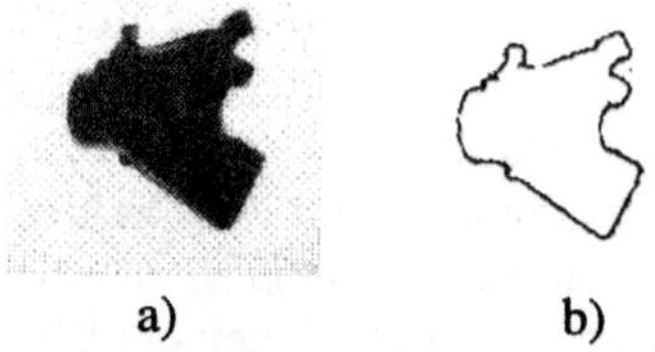

Abb. 2. Gelerntes Objekt in a) Grauwertdarstellung und b) einfacher Repräsentation (Maximum über die Orientierungssäule k_n an jeder Position n gezeigt)

Aus dieser erstellen wir die *tolerante Repräsentation c*, indem zunächst alle Komponenten c_{nr}=1 gesetzt werden, bei denen k_{nr}=1 ist. Zusätzlich werden senkrecht zum Konturverlauf weitere Komponenten c_{vr} mit geringerer Aktivität versehen. Der dabei entstehende rampenförmig nach beiden Seiten abfallende Aktivitätsverlauf ist in Abb. 1d) dargestellt. In der Repräsentation c sind jedoch nicht nur die mit der Konturorientierung r übereinstimmenden Elementen c_{nr} gesetzt, sondern auch Elemente $c_{nr\pm1}$ mit benachbarter Orientierung. Die Repräsentation c ist also orts- und orientierungstolerant. Das zur Bewertung verwendete Aktivitätsprofil kann dabei den Erfordernissen der

1. Dient die einfache Repräsentation k nicht zur Erzeugung der toleranten Repräsentation c (s. nächster Absatz), so wird in jeder Orientierungssäule k_n nur das Element der dominanten Orientierung auf 1 gesetzt.

Matching-Strategie (je nach Skalierungs- und Rotationsschrittweite und max. Objektausdehnung, s.u.) angepaßt werden. Bereits mit einem ad hoc verwendeten rampenförmigen Verlauf (s. Abb. 1d)) haben wir gute Ergebnisse erzielt.

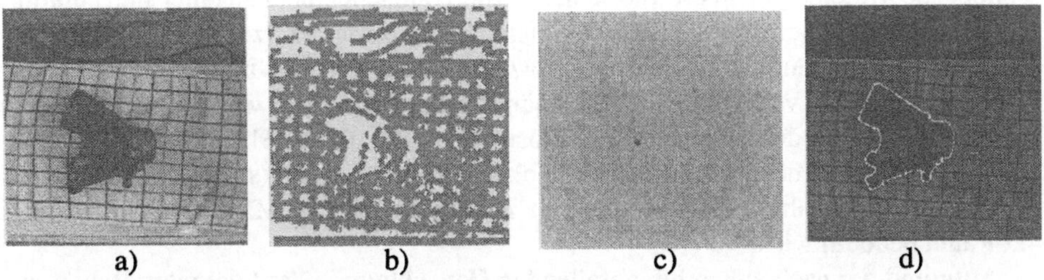

Abb. 3. Zu untersuchende Szene in a) Grauwertdarstellung und b) toleranter Repräsentation (2-fache Unterabtastung, Maximum über die Orientierungssäule c_n an jeder Position *n* gezeigt). c) 2D-Ergebnisspeicher mit sichtbarem Maximum (dunkle Werte sind hohe Matchwerte). d) Gelernte Kontur (weiß) in der ermittelten Lage, Grauwertbild dunkel eingeblendet.

Wesentlich ist für das weitere, daß das gelernte Muster in der einfachen Repräsentation (s. Abb. 2) dargestellt und mit dem in der toleranten Repräsentation befindlichen Eingangsmuster (s. Abb. 3) verglichen wird.

2.2 Matching

Bei Betrachten nur einer Orientierung gibt das Skalarprodukt der Vektoren $\boldsymbol{k}'_r$ und $\boldsymbol{c}'_r$ quantitativ die Übereinstimmung der repräsentierten Muster an. Aus den Aktivitätsprofilen in Abb. 1c) und d) wird zudem deutlich, daß bei geringfügigen Abweichungen des gelernten zum präsentierten Muster das Produkt $\langle \boldsymbol{k}'_r, \boldsymbol{c}'_r \rangle$ entsprechend der für die tolerante Repräsentation verwendeten Abstandsfunktion zwar kleiner wird, aber noch größer als Null ist. Allgemein verwenden wir als Maß für die Übereinstimmung (=Match) eines gelernten Musters *k* mit einem Eingangsmuster *c*:

$$\langle \boldsymbol{k}, \boldsymbol{c} \rangle = \frac{\langle \boldsymbol{k}_1, \boldsymbol{c}_1 \rangle + \ldots + \langle \boldsymbol{k}_r, \boldsymbol{c}_r \rangle + \ldots + \langle \boldsymbol{k}_R, \boldsymbol{c}_R \rangle}{\langle \boldsymbol{k}, \boldsymbol{k} \rangle},$$

wobei $\langle \boldsymbol{k}, \boldsymbol{k} \rangle = \langle \boldsymbol{k}_1, \boldsymbol{k}_1 \rangle + \ldots + \langle \boldsymbol{k}_r, \boldsymbol{k}_r \rangle + \ldots + \langle \boldsymbol{k}_R, \boldsymbol{k}_R \rangle$ als Normierungsfaktor dient. Somit haben das gelernte Muster und das Eingangsmuster auch bei leichter Veränderung des Eingangsmusters (z.B. durch Translation, Rotation, Skalierung, Perspektive, etc.) noch hohe Übereinstimmung.

Zur translationsinvarianten Erkennung muß lediglich der Match für verschiedene Translationen des gelernten Musters berechnet werden. Ebenso kann Orientierungs- und Skalierungsinvarianz erreicht werden, wobei das jeweils in Frage kommende Intervall aber nur in groben Schrittweiten durchgetestet werden muß, da ähnliche Orientierungs- bzw. Skalierungsparameter wegen der toleranten Repräsentation ähnliche Matchwerte ergeben (s.o.). Der zu testende Parameterraum (vertikale und horizontale Translation, Orientierung und Skalierung) ist dabei vierdimensional.

Da im gelernten Muster wie auch im Eingangsmuster in der jeweiligen Repräsentation nur wenige Elemente gesetzt sind, verwenden wir in der Implementierung die bzgl. der Ergebnisse äquivalente, aber effizientere General Hough Transformation [Bal81] ohne anschließende Mittelung. Durch die Verlagerung der Mittelung in die Toleranz der Repräsentation (s.o.) ist bei uns aber eine aufwendige 4D-Faltung ([Bal81]) oder die bei Clustering-Verfahren (z.B. [Sto87], [Gri90]) durchgeführte Quantisierung und anschließende Mittelung nicht erforderlich, und die Wahrscheinlichkeit einer sich zufällig ergebenden Fehlhypothese ist selbst bei reichlich strukturierten Szenen gering (s. Abb. 3d)). Der Ergebnisraum ist bei uns aber trotzdem klein, obwohl wir keine grobe Quantisierung in Buckets vornehmen. Er kann nämlich bei Bedarf von vier auf zwei Dimensionen reduziert werden, indem an einer Position nur die für alle Skalierungen und Orientierungen jeweils höchsten Matchwerte abgespeichert werden (s. Abb. 3c)). Dies ist bei Clustering-Verfahren nicht möglich, da zur abschließenden Mittelung der gesamte 4D-Raum benötigt wird.

Aufgrund der nach außen hin abfallenden Gewichtung in der toleranten Repräsentation wird dabei implizit bei denjenigen Lageparametern der höchste Match erzielt, bei denen das gesamte gelernte Muster mit dem Eingangsmuster am besten übereinstimmt. Eine explizite Kontur-Abstandsbestimmung wie z.B. bei Borgefors[Bor88] bzw. Olson und Huttenlocher[Ols97] ist daher nicht notwendig.

Eine weitere Effizienzsteigerung wird dadurch ermöglicht, daß die Repräsentationen auch in niedrigen Auflösungsebenen erzeugt werden können. So erlaubt z.B. eine 4-fache Unterabtastung in der Vorverarbeitung die Reduktion der anfallenden Datenmenge auf ca. 1/16. In einer Grob-Fein Strategie können so Hypothesen über das Vorliegen gelernter Muster schnell in niedriger Auflösung generiert und zielgerichtet in einer höheren Auflösung verifiziert werden.

3 Gruppierung von Merkmalen

Eine weitere Möglichkeit zur Steigerung der Erkennungssicherheit und der -geschwindigkeit, insbesondere bei stark strukturierten Szenen, besteht in der Gruppierung von Konturelementen zu höheren Merkmalen.

Hierzu werden einerseits datenbasierte Verfahren mit bottom-up Gruppierung angewendet (z.B. [Jac89], [Ols95]). Modellbasierte Verfahren nutzen andererseits explizites Wissen über die gelernten Objekte, um einfache Merkmale zu höheren zu Gruppieren (z.B. [Tur85], [Nis95]).

Wir verwenden einen zu den „composite shapes" von Ballard[Bal81] ähnlichen Ansatz des modellbasierten Gruppierens. Der Grundgedanke ist, ein Objekt aus mehreren Teilobjekten zu modellieren. Diese werden zunächst im Eingangsmuster detektiert (s. Abb. 4a)-d)) und als höhere Merkmal in einem zweiten Erkennungsschritt entsprechend der Anordnung im Gesamtobjekt gruppiert (s. Abb. 4e)-f)). Im Gegensatz zu Ballard verwenden wir anstelle aufwendiger Faltungsoperationen unsere tolerante Repräsentation des Eingangsmusters, lassen Skalierungen und Rotationen sowohl der Teil- als auch des Gesamtobjektes zu und wenden auf Zwischenergebnisse eine nichtlineare Abbildung an. Hierdurch wird die Erkennung robust gegenüber perspektivischen Verzerrungen, stark strukturierten Szenen, etc., so daß auch nicht-starre Teilobjektanordnungen toleriert werden.

Zu dem Verfahren sind einige Punkte anzumerken. Grundsätzlich können die Teilobjekte also durchaus unterschiedlich sein. Hierzu müssen die Ergebnisspeicher der Teilobjekterkennungen bei der Gruppierung zusammengeführt werden. Die implizit durch die Repräsentation gegebene Toleranz resultiert auch im Ergebnisspeicher der Teilobjekte in einer toleranten Repräsentation. Deswegen müssen auch zur Gruppierung nicht alle Skalierungen und Orientierungen durchlaufen werden. Durch das Zusammenführen der teilobjektbasierten Erkennungsergebnisse geht schließlich verloren, ob ein bestimmtes Teilobjekt vorhanden ist oder nicht. Das Vorhandensein dieses Teilobjektes kann aber zielgerichtet in dem jeweiligen Zwischenergebnisspeicher untersucht werden.

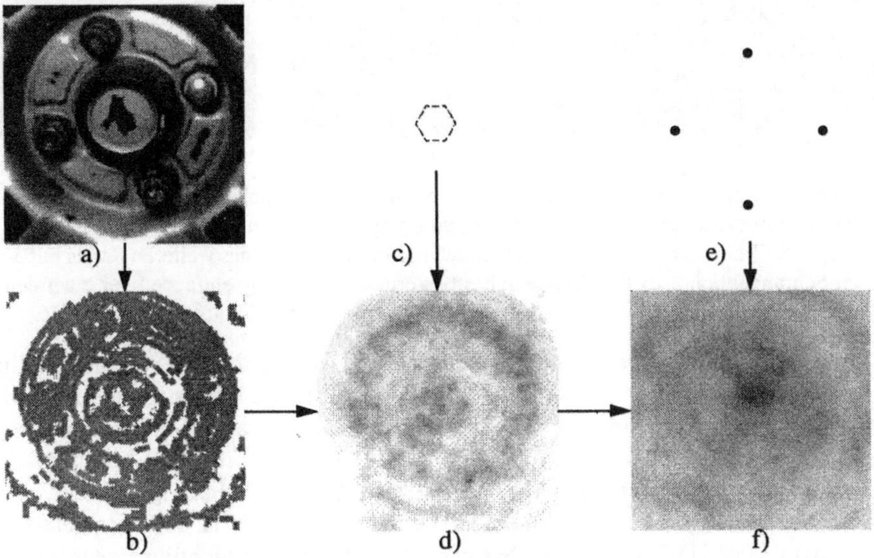

Abb. 4. Zu untersuchende Szene a) in Grauwertdarstellung bzw. b) toleranter Repräsentation. c) Gelerntes Teilobjekt in schmaler Repräsentation (überlagert), d) 2D-Ergebnisspeicher für Teilobjektdetektion nach nichtlinearer Abbildung. e) Gruppierung der Teilobjekte zum Gesamtobjekt, f) 2D-Ergebnisspeicher für Gesamtobjektdetektion.

4 Anwendungsbeispiel

Die betrachtete Aufgabenstellung besteht in der Erkennung und Lokalisierung von Radschrauben zur Raddemontage bei Altautos. Aufgrund der vielfältigen Ausprägungen der Schrauben selbst (Verschmutzung, Abnutzung etc.) sowie des geringen Kontrastes der Radschrauben zur Felge (z.B. Verschmutzung des Rades) ist eine direkte Detektion mit einer niedrigen Erkennungs- bzw. hohen Fehlerkennungsrate verbunden. Hier bietet es sich an, die symmetrische Schraubenanordnung zu berücksichtigen. Eine holistische Erkennung des Schraubenkranzes in einem Schritt ist allerdings aufgrund der anzutreffenden Freiheitsgrade der Schrauben und des Kranzes (Orientierung, Skalierung) und der damit verbundenen kombinatorischen Explosion von Prototypen ineffizient. Deswegen wird die Erkennung des Schraubenkranzes bei uns durch das in Abschnitt 3 beschriebene modellbasierte Gruppieren durchgeführt, wobei das Gesamtobjekt „Schraubenkranz" sich aus mehreren Teilobjekten „Schraube" zusammensetzt. Die

Gesamtheit der Schrauben in ihrer Anordnung ermöglicht damit die Detektion des Schraubenkranzes, der wiederum die Erkennung der Schrauben selbst unterstützt.

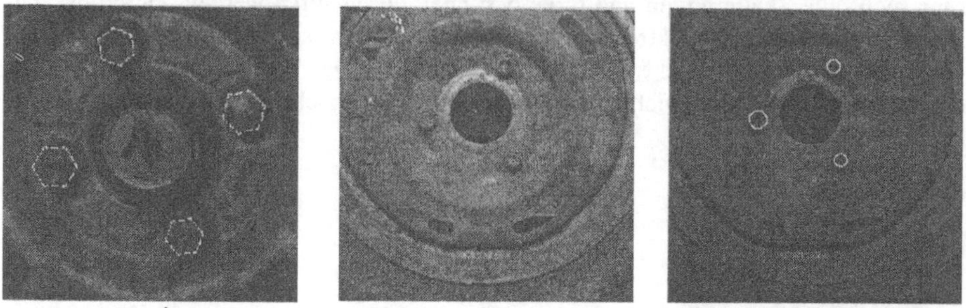

a) b) c)

Abb. 5. a) mit Konturen eingezeichnete Teilobjekte kennzeichnen ermittelte Lage des Gesamtobjektes sowie bestbewertete Orientierung und Skalierung der zugehörigen Teilobjekte; Grauwertbild dunkel eingeblendet. b) Grauwertbild eines weiteren Rades mit 3-er Schraubenkranz. c) Teilobjekte in bestbewerteter Skalierung, einfache Kreise wurden zur Bildung des Schraubenprototypen verwendet; Grauwertbild dunkel eingeblendet

Das in Abschnitt 3 vorgestellte Verfahren ist bezogen auf die Aufgabenstellung übersichtlich in Abb. 4 zusammengefaßt. Das Grauwert-Eingangsbild der Größe 256x256 (s. Abb. 4a)) wurde ohne Unterabtastung in die tolerante Konturrepräsentation überführt (Abb. 4b)), wobei das Aktivitätsprofil beidseitig 3 Elemente betrug (s. Abb. 1d)). Die Schraubenprototypen (Abb. 5c)) wurden in einer Skalierung von 0.85 bis 1.0 in 5 Schritten und symmetriebedingt von 0 Grad bis 60 Grad in 12 Schritten getestet. In Abb. 4d)) ist der 2D-Ergebnisspeicher nach Anwendung der nichtlinearen Abbildung vom Typ „$1 - \exp(-x^2)$" zu sehen, bei der alle Matchwerte über 40% verstärkt wurden. Hier ist zu entnehmen, daß aufgrund der Szenenkomplexität die Schrauben nicht direkt erkannt werden können. Werden diese höheren Merkmale nun gruppiert (Abb. 4e)), so wird das Gesamtobjekt „Schraubenkranz" bei einer Schwelle von 60% in der passenden Lage detektiert (Abb. 4f)), und es ergibt sich bei Berücksichtigung der Zwischenergebnisse (Abb. 4d)) die in Abb. 5a) gezeigte Lage der einzelnen Schrauben.

Ein weiteres Ergebnis, das mit 2-facher Unterabtastung des Grauwert-Eingangsbildes und mit einfachen Prototypen für die Teilobjekte erzielt wurde, ist in Abb. 5c) zu sehen.

5 Zusammenfassung

Wir haben ein Verfahren zur konturbasierten Erkennung vorgestellt, das aufgrund der verwendeten toleranten Repräsentation eine sichere Erkennung und Lagebestimmung von leicht veränderlichen Objekten selbst in schwierigen Szenen gewährleistet. Durch die Erweiterung mit einer modellbasierten Gruppierung können in komplexen Szenen auch Gesamtobjekte teilobjektbasiert detektiert werden.

Die Verfahren werden derzeit erfolgreich in einem Demontageprozeß zur Teiledetektion in Verbindung mit einem Stereokamerasystem eingesetzt [Ste98].

Das Modell zur Merkmalsgruppierung wird derzeit noch per Hand vorgegeben. Hier ist ein automatisches Zerlegen eines Objektes in signifikante, starre Teilobjekte wünschenswert. Geplant ist auch eine Erweiterung zur ansichtenbasierten Erkennung von 3D-Objekten.

6 Literatur

[Bal81] Ballard, D.H.: „Generalizing the Hough transform to detect arbitrary shapes." Pattern Recognition, vol. 13, no. 2, pp. 111-122, 1981.

[Bor88] Borgefors, G.: „Hierarchical Chamfer Matching: A Parametric Edge Matching Algorithm." IEEE Transactions on Pattern Analysis and Machine Intelligence, vol. 10, no. 6, pp. 849-865.

[Drü93] Drüe, S.; Hoischen, R.; Trapp, R.: „Tolerante Objekterkennung durch das Neuronale Active-Vision-System NAVIS". In: Kropatsch, W. G. (Ed.): Mustererkennung '94; Springer Verlag 1994.

[Gri87] Grimson, W.E.L., Lozano-Perez, T.: „Localizing Overlapping Parts by Searching the Interpretation Tree." IEEE Transactions on Pattern Analysis and Machine Intelligence, vol. 9, no. 4, pp. 469-482, 1987.

[Gri90] Grimson, W.E.L., Huttenlocher, D.P.: „On the sensitivity of the Hough transform for object recognition". IEEE Transactions on Pattern Analysis and Machine Intelligence, vol. 12, no. 3, pp. 255-274, 1990.

[Hub88] Hubel, D.: „Eye, Brain and Vision." Scientific American Library, 1988.

[Jac89] Jacobs, D.W.: „Grouping for Recognition." MIT AI Memo 1177, 1989.

[Ols95] Olson, C.F.: „On the Speed and Accuracy of Object Recognition when using imperfect grouping." International Symposium on Computer Vision, Florida, S. 449-454, 1995.

[Ols97] Olson, C.F., Huttenlocher, D.P.: „Automatic Target Recognition by Matching Oriented Edge Pixels." IEEE Transactions on Image Processing, vol. 6, no. 1, pp. 103-113, 1997.

[Sam97] Samal, A., Edwards, J.: „Generalized Hough transform for natural shapes." Pattern Recognition Letters, 18, pp. 473-480, 1997.

[Ste98] Stemmer, R., Trapp, R., Drüe, S., Hartmann, G.: „3D Lagebestimmung von zu demontierenden Objekten mit einem aktiven Stereokamerasystem." Angenommen zur DAGM 1998.

[Sto87] Stockman, G.: „Object recognition and localization via pose clustering." Computer Vision, Graphics, and Image Processing, vol. 40, pp.361-387, 1987.

[Tur85] Turney, J.L., Mudge, T.N., Volz, R.A.: „Recognizing partially occluded parts." IEEE Transactions on Pattern Analysis and Machine Intelligence, vol. 7, no. 4, pp. 410-421, 1985.

[Tra96] Trapp, R.: „Entwurf einer Filterbank auf der Basis neurophysiologischer Erkenntnisse zur orientierungs- und frequenzselektiven Dekomposition von Bilddaten." Interner Bericht NAVIS 01/96, Universität-Gesamthochschule Paderborn, 1996.

[Zie97] Ziegler, U.; Götze, N.; Drüe, S.; Hartmann, G.; Stemmer, R.; Bahro, J.: „Extrafoveale Erkennung von Industrieobjekten vor beliebig strukturiertem Hintergrund durch ein aktives Sehsystem." In: Groß, H.M.(ed): Selbstorganisation von Adaptivem Verhalten, VDI-Verlag, S. 186-193, 1997.

Autorenverzeichnis

GPSR Compliance

The European Union's (EU) General Product Safety Regulation (GPSR) is a set of rules that requires consumer products to be safe and our obligations to ensure this.

If you have any concerns about our products, you can contact us on ProductSafety@springernature.com

In case Publisher is established outside the EU, the EU authorized representative is:

Springer Nature Customer Service Center GmbH
Europaplatz 3
69115 Heidelberg, Germany

Batch number: 10844669

Printed by Printforce, the Netherlands